INTRODUCTION TO

Advanced Food Process Engineering

INTRODUCTION TO

Advanced Food Process Engineering

EDITED BY

Jatindra K. Sahu

CRC Press is an imprint of the
Taylor & Francis Group, an **informa** business

CRC Press
Taylor & Francis Group
6000 Broken Sound Parkway NW, Suite 300
Boca Raton, FL 33487-2742

First issued in paperback 2016

Version Date: 20131227

ISBN 13: 978-1-138-19967-5 (pbk)
ISBN 13: 978-1-4398-8071-5 (hbk)

Library of Congress Cataloging-in-Publication Data

Introduction to advanced food process engineering / [edited by] Jatindra Kumar Sahu.
pages cm
Includes bibliographical references and index.
ISBN 978-1-4398-8071-5 (hardback)
1. Food industry and trade--Technological innovations. 2. Food industry and trade--Quality control. I. Sahu, Jatindra Kumar.

TP370.I58 2014
338.4'7664--dc23 2013049553

Visit the Taylor & Francis Web site at
http://www.taylorandfrancis.com

and the CRC Press Web site at
http://www.crcpress.com

Contents

Preface

Food is a complex product of biological nature composed of various macronutrients and micronutrients. All food materials are available in many forms, such as solid, liquid, colloid, and viscous, which are processed prior to their consumption. Various processing technologies are employed for these purposes, and these technologies improve the shelf life and also maintain the safety and nutritional, sensory, physicochemical, and biological qualities of a food. However, the growing consumers demand for minimally processed, safe, microbiological-free, nutrient, shelf-stable, and convenient foods in the last two decades has created tremendous pressure to restructure the food processing industries by employing novel food processing technologies in a cost-effective manner. Therefore, food process engineering has a major role in processing, packaging, storing, distributing, and enhancing the quality attributes of food products. As food process engineering, in general, becomes more advanced and sophisticated, there is a huge need for specific knowledge of the raw materials and effects of various processing treatments on them rather than specific commodity technology.

In this context, the present book *Introduction to Advanced Food Processing Engineering* is intended as a general reference book for students and others who are interested in various aspects of food processing, packaging, storage, quality control, and assessment and distribution systems. The book describes the basic principles and major applications of emerging food processing technologies of modern research and development in the field of food process engineering. The book also provides a comprehensive overview of modern food processing technologies and their applications, covers advanced techniques for food safety and quality assessment, explores future directions and research needs, and provides solved problems and case studies for better understanding of the technologies.

The above processes are systematically described in three sections through 20 novel chapters from different areas of food process engineering. All the chapters have been prepared by high-profile, internationally renowned, and experienced professors and scientists throughout the world. Section I describes advanced food processing technologies such as osmo-concentration of fruits and vegetables; non-thermal processing technologies such as high-pressure processing, the pulse electric field, and ultrasonic system; novel drying technologies such as heat pump–assisted drying, intermittent batch drying, MW-vacuum drying, low-pressure super-heated steam drying, pulse combustion drying, spray-freeze drying, and atmospheric freeze drying of foods; controlled and modified atmosphere (CA and MA) storage of fruits and vegetables; membrane applications in food processing; nanotechnology in food processing; and computational fluid dynamics (CFD) in food process modeling. Section II describes quality control and food safety, and quality assessment using machine vision systems, vibrational spectroscopy, biosensors, and chemosensors. Section III describes waste management, byproduct utilization, and energy conservation in the food processing industry.

In brief, the focus of the book has been to emphasize novel food processing engineering. In each chapter, case studies and examples are included to illustrate state-of-the-art applications of the novel technologies.

In this endeavor, I would like to thank all the contributors who have rendered their well-defined chapters and responded in a timely manner. Finally, reader suggestions will be highly appreciated to improve the quality of the book in its further editions.

Jatindra K. Sahu

Editor

Dr. Jatindra K. Sahu is an associate professor in the Department of Agricultural Engineering in Assam University, Silchar, India. Dr. Sahu earned his postgraduate and doctoral degrees from the Indian Institute of Technology, Kharagpur, India, with specialization in dairy and food engineering and is a graduate in agricultural engineering from OUAT, Bhubaneswar, India. Dr. Sahu is an active, dynamic, enthusiastic, and committed academician and researcher in the field of food engineering. He has been honored with the prestigious Seligman APV Fellowship by the Society of Chemical Industries, London, for his significant contribution to the field of food engineering and for a research project he carried out at the University of Reading, Reading, United Kingdom. Dr. Sahu has also been conferred the BOYSCAST Fellowship by the Department of Science and Technology, Ministry of Science and Technology, Govt. of India, to undertake a research project in the Department of Biological Systems Engineering at Virginia Polytechnic and State University, Blacksburg, Virginia, USA. He is also a recipient of the Institute Research and GATE Fellowship by the Ministry of Human Resource Development, Govt. of India, to carry out doctoral and postgraduate programs at the Indian Institute of Technology Kharagpur.

The major focus of Dr. Sahu's research is to develop or improve technologies in the field of food process engineering. The majority of his research activities aim to upgrade the traditional methods of food processing to international standards through emerging food processing technologies. While working in various institutes and universities, all his research activities received national and international acknowledgment. His major research contributions to food industries include a continuous heat–acid coagulation unit, a continuous soft cheese–making unit, a *sandesh* (an Indian milk sweet) production line, technologies for utilization of traditional and nontraditional forest products, technology for production of vacuum-dried honey powder, among others. Dr. Sahu's current areas of research include high-pressure processing (HPP) of food materials, high-voltage pulsed electric field (PEF) processing of foods, application of light emitting diodes (LED) for food preservation, ultrasound food processing, nondestructive analysis of food quality, nanotechnology applications in food processing, and subcritical extraction of active food compounds.

Dr. Sahu has filed four Indian patents, completed many research projects, prepared one textbook in food engineering, published several book chapters and monographs on novel food processing technologies, published more than 50 national and international peer-reviewed research papers, participated in many national and international seminar and conference proceedings, and organized many national and international seminars and conferences. For his innovative ideas, Dr. Sahu has been bestowed with the prestigious Jawaharlal Nehru Award by the Indian Council of Agricultural Research, New Delhi. In addition, he is an active member of the National Academy of Agricultural Sciences, India; the Institutions of Engineers, India; the Association of Food Scientists and Technologists, India; Indian Dairy Association, New Delhi; Dairy Technology Society of India, Karnal; Society of Chemical Industry, United Kingdom; and the American Society of Agricultural and Biological Engineers, USA. Dr. Sahu is an editor of the *International Journal of Food Science, Nutrition and Dietetics*, the *Journal of Food Research and Technology*, the *Journal of Bioresource Engineering and Technology*, and the *Journal of Food Product Development and Packaging*.

Contributors

Faiyaz Ahmed
Department of Studies in Food Technology and Nutrition
University of Mysore
Mysore, India

D.S. Bunkar
Centre of Food Science and Technology
Institute of Agricultural Sciences
Banaras Hindu University
Varanasi, India

S. Chakraborty
Department of Agricultural and Food Engineering
Indian Institute of Technology
Kharagpur, India

H. Das
Department of Agricultural and Food Engineering
Indian Institute of Technology
Kharagpur, India

Wan Ramli Wang Daud
Department of Chemical and Process Engineering
Faculty of Engineering
Universiti Kebangsaan Malaysia
Bangi, Malaysia

Satyanarayan Dev
Department of Food Science and Nutrition
A'Sharqiyah University
Ibra, Sultanate of Oman

Catherine W. Donnelly
Department of Nutrition and Food Science
University of Vermont
Burlington, Vermont

Stephane Evoy
Department of Electrical and Computer Engineering
University of Alberta
Edmonton, Alberta, Canada

T.K. Goswami
Department of Agricultural and Food Engineering
Indian Institute of Technology
Kharagpur, India

M.K. Hazarika
Department of Food Engineering and Technology
Tezpur University
Tezpur, India

N.R. Swami Hulle
Department of Agricultural and Food Engineering
Indian Institute of Technology
Kharagpur, India

Abid Hussain
Department of Bioresource Engineering
McGill University
Ste-Anne-de-Bellevue, Quebec, Canada

D.S. Jayas
Department of Biosystems Engineering
University of Manitoba
Winnipeg, Manitoba, Canada

Sujata Jena
Department of Process and Food Engineering
College of Agricultural Engineering and Post-Harvest Engineering
Sikkim, India

Alok Jha
Centre for Food Science and Technology
Institute of Agricultural Sciences
Banaras Hindu University
Varanasi, India

C. Karunakaran
Department of Biosystems Engineering
University of Manitoba
Winnipeg, Manitoba, Canada

B. Pal Kaur
Department of Agricultural and Food Engineering
Indian Institute of Technology
Kharagpur, India

N. Kaushik
Department of Agricultural and Food Engineering
Indian Institute of Technology
Kharagpur, India

Riitta L. Keiski
Department of Process and Environmental Engineering
Mass and Heat Transfer Process Laboratory
University of Oulu
Oulu, Finland

Deepak Kumar
Department of Biological and Ecological Engineering
Oregon State University
Corvallis, Oregon

Chung Lim Law
Department of Chemical and Environmental Engineering
Faculty of Engineering
University of Nottingham
Kuala Lumpur, Malaysia

P. Kumar Mallikarjunan
Biological Systems Engineering
Virginia Polytechnic Institute and State University
Blacksburg, Virginia

S. Mangaraj
Central Agricultural Engineering Institute
Indian Council of Agricultural Research
Bhopal, India

Arun S. Mujumdar
Department of Chemical and Biomolecular Engineering
Hong Kong University of Science and Technology
Kowloon, Hong Kong

Ganti S. Murthy
Department of Biological and Ecological Engineering
Oregon State University
Corvallis, Oregon

Liisa Myllykoski
Department of Process and Environmental Engineering
Mass and Heat Transfer Process Laboratory
University of Oulu
Oulu, Finland

Balunkeswar Nayak
Department of Food Science and Human Nutrition
University of Maine
Orono, Maine

Tomás Norton
Department of Agricultural Engineering
Harper Adams University College
Newport, United Kingdom

J. Paliwal
Department of Biosystems Engineering
University of Manitoba
Winnipeg, Manitoba, Canada

Nóra Pap
MTT Agrifood Research Finland
Biotechnology and Food Research
Sustainable Bioeconomy
Jokioinen, Finland

Sanjib K. Paul
Department of Agricultural Engineering
Assam University
Assam, India

Eva Pongrácz
University of Oulu
and
Nortech Oulu
Oulu, Finland

Somayyeh Poshtiban
Department of Electrical and Computer Engineering
University of Alberta
Edmonton, Alberta, Canada

Vijaya Raghavan
Department of Bioresource Engineering
McGill University
Ste-Anne-de-Bellevue, Quebec, Canada

Mohammad Shafiur Rahman
Department of Food Science and Nutrition
College of Agricultural and Marine Sciences
Sultan Qaboos University
Al-Khod, Sultanate of Oman

P. Srinivasa Rao
Department of Agricultural and Food Engineering
Indian Institute of Technology
Kharagpur, India

Jatindra K. Sahu
Department of Agricultural Engineering, School of Technology
Assam University
Guwahati, India

Amit Singh
Northeastern University
Boston, Massachusetts

Brijesh Tiwari
Manchester Food Research Centre
Manchester Metropolitan University
Manchester, United Kingdom

Gopal Tiwari
Department of Agricultural and Biological Engineering
University of California
Davis, California

Ashutosh Upadhyay
Centre of Food Science and Technology
Institute of Agricultural Sciences
Banaras Hindu University
Varanasi, India

N.S. Visen
Department of Biosystems Engineering
University of Manitoba
Winnipeg, Manitoba, Canada

Lijun Wang
Biological Engineering Program
School of Agriculture and Environmental Sciences
North Carolina Agricultural and Technical State University
Greensboro, North Carolina

Wenbo Wang
Department of Biosystems Engineering
University of Manitoba
Winnipeg, Manitoba, Canada

N.D.G. White
Cereal Research Centre
Agriculture and Agri-Food Canada
Winnipeg, Manitoba, Canada

G. Zhang
Department of Biosystems Engineering
University of Manitoba
Winnepeg, Manitoba, Canada

ADVANCES IN FOOD PROCESS ENGINEERING

Chapter 1

Microwave and Radio Frequency Heating of Food Products: Fundamentals, Modeling, and Applications

Deepak Kumar, Ganti S. Murthy, and Gopal Tiwari

Contents

1.1 Introduction

The first patent for the use of microwaves in food applications was filed in 1945 by Percy Spencer (US Pat. No. 2495429). In the last few decades, the application of microwave (MW) and radio frequency (RF) energy for heating and drying of food materials has become increasingly popular. Although the use of MWs for domestic MW ovens is popular, application of MWs for drying and processing of foods on a commercial scale is still in a developing stage. The RF and MW heating processes involve interaction of electromagnetic (EM) fields with food materials. When food materials are placed in an alternating EM field, charges (free ions and electric dipoles) present in foods get displaced and attempt to follow changes in the external alternating EM field. Part of the EM energy absorbed by the material to carry out these displacements is dissipated as heat.

1.2 Fundamentals

MWs are the EM waves in the frequency range of 300 MHz to 300 GHz (Figure 1.1) with wavelengths of 1 m–1 mm (Drouzas and Schubert 1996). The two most commonly used MW frequencies for food applications are 915 MHz and 2.45 GHz.

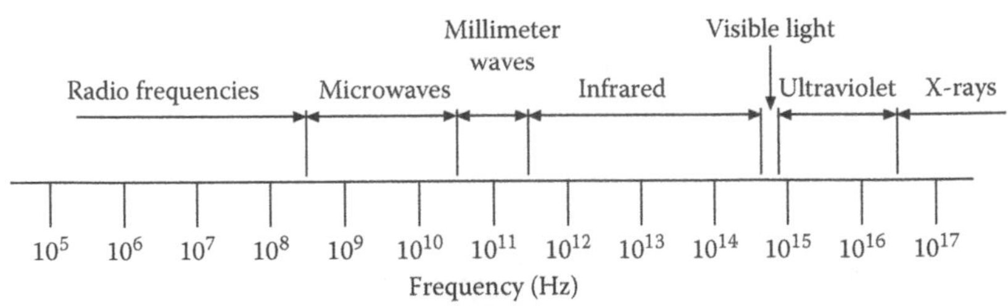

Figure 1.1 Electromagnetic spectrum. (From Datta, A.K. and Ramaswamy, C.A., *Microwave Technology for Food Applications*, Marcel Dekker, New York, 2001.)

Table 1.1 Frequency Bands for Industrial, Scientific, and Medical Applications

Frequency (± tolerance) MHz	*Country*
433 ± 0.2%	Austria, Netherlands, Portugal, Germany, Switzerland, Great Britain
896 ± 10	Great Britain
915 ± 13	North and South America
2375 ± 50	Albania, Bulgaria, Hungary, Romania, Czechoslovakia, Russia
2450 ± 50	Worldwide except in areas where 2375 MHz frequency is used
5800 ± 75	Worldwide

Source: Datta, A.K. and Ramaswamy, C.A., *Microwave Technology for Food Applications*, Marcel Dekker, New York, 2001; Mujumdar, A.S., *Drying Technology in Agriculture and Food Science*. Science Publishers Inc., Enfield, NH, 2000.

While 915 MHz is primarily used for industrial applications, 2.45 GHz is used in both domestic and commercial applications (Zante 1973; Galema 1997; Venkatesh and Raghavan 2004). Higher frequencies are used mostly for laboratory research purposes. MW furnaces that use variable frequencies from 0.9 to 18 GHz have also been developed (Murthy 2003). Narrow bands of frequencies within the MW spectrum are allotted for industrial, scientific, and medical applications (ISM) around the world, which differ by country (Table 1.1).

A MW heating system, in general, consists of three main components: the source, transmission line, and the applicator. MWs generated by the source are transmitted to the applicator through the transmission lines. Part of the MW energy is absorbed by the dielectric material in the applicator while the rest is reflected back to the source.

1.2.1 Microwave Sources

Generation of MW radiation is done by accelerating the electrical charges. To achieve the high power frequencies required for the MW heating application, most devices use vacuum tubes. The most common vacuum tubes are magnetrons, traveling wave tubes (TWTs), and klystrons. Magnetron tubes use resonant structures to generate the electromagnetic field and hence are capable of producing a fixed frequency EM field. Due to their high reliability and mass production, magnetron tubes are used in household MW ovens. They are mass produced, and they are the cheapest source of MWs available. In variable frequency MWs, TWTs are used to generate an EM field. The design of TWTs allows amplification of a broad band of MW frequencies in the same tube.

In a vacuum tube, the anode is set at a higher potential than the cathode. The potential difference causes a strong electric field between the electrodes. The cathode is heated to remove the loosely bound valence electrons. Once the electrons are removed from the cathode, they are accelerated toward the anode by the electric field. In a magnetron, an external magnet is used to create a magnetic field orthogonal to the electric field, and the applied magnetic field creates a circumferential force on the electrons as they are accelerated to the anode. The force causes electrons to travel in a spiral direction, and this creates a swirling cloud of electrons. As the electrons pass through the resonant cavities, the cavities set up oscillations in the electron cloud. The frequency of the oscillations of the electron cloud depends on the size of the cavities. EM energy is coupled from one of the resonant cavities to the transmission lines through a coaxial line or wave guide.

Two methods are commonly used to control the average power output of magnetron tubes. The output power of the magnetron can be controlled by adjusting the period of operation, the cathode current, or magnetic field strength. In household MW ovens, the magnetron is operated at full power. Current to magnetron is turned on/off for segments of cycle time to achieve different time averaged output MW power levels. This on/off cycle is referred to as a duty cycle, and this strategy helps in keeping the magnetron temperature within the limits. If continuous power output is required, the output power of the magnetron tube can be varied by changing the current amplitude of the cathode or by changing the intensity of the magnetic field. This allows variable control of the MW power within the range of the source.

For variable-frequency MWs, high-power TWTs are used as MW sources. Unlike magnetrons, in which the tube is used to create the frequency of the oscillations as well as to amplify the signal, the TWT serves only as an amplifier. A voltage-controlled oscillator generates the MW signal. The input voltage controls the frequency of the oscillator, and the signal is then sent to the TWT for amplification. These types of sources are able to switch the output frequencies rapidly as the oscillator determines the frequency.

1.2.2 Transmission Lines

The transmission lines couple the energy of the MW source to the applicator. In low-power systems, coaxial cables are used, as in television cables. However, energy losses in the coaxial lines become unacceptably high for high-frequency and high-power systems. Waveguides, which are hollow tubes of circular or rectangular cross sections, are commonly used as transmission lines. The most commonly used cross sections are rectangular.

There are two modes of MW propagation possible in the wave guides: transverse electric (TE) and transverse magnetic (TM). For TE mode, the electric field in the direction of propagation is zero, and for TM mode, the magnetic field is zero in the direction of propagation. The most common wave guide mode is TE10 mode. The mode indicates the number of maxima and minima of each field in the wave guide.

In addition to wave guides, there are several other transmission-line components that are used for equipment protection, sensing purposes, and coupling MWs with the material in the applicator. Circulators are the components that prevent the MWs from being reflected back to the magnetron and cause damage. The circulators can be considered as the MW equivalent of a diode. Directional couplers separate a small amount of forward and reflected waves for measurement of MW power. Tuners are used for matching the source impedance to that of the load impedance for maximum power absorption. Several tuners, such as irises, three stub tuners, and E-H plane tuners, are commonly used.

1.2.3 Microwave Applicators

The design of applicator is critical in MW heating because MW energy is transferred to the materials through the applicator. Material temperatures achieved using MW heating are inherently linked to the distribution of the electric fields within the applicator. Common MW applicators include wave guides, traveling wave applicators, single-mode cavities, and multimode cavities. For processing of materials, resonant applicators, such as single-mode and multimode cavities, are more popular because of their high field strengths.

1.2.3.1 Single-Mode Applicators

In a MW applicator, or cavity, theoretical analysis can be performed to describe the response of MW. Given the geometry of the applicator, it is often possible to solve the Maxwell's equations analytically or numerically with appropriate boundary conditions. The design of the single-mode cavity is based on the solution of the Maxwell's equations to support one resonant mode. Consequently, the size of the single-mode applicators is of the order of approximately one wavelength, and to maintain the resonant mode, these cavities require a source that has little variation in the frequency of output. Because the EM field can be determined analytically, the areas of high and low EM field are known, and single-mode applicators have nonuniform but predictable EM field distribution. In general, they have one hot spot where the field strength is high. This ability to design an applicator with known high and low field areas offers a distinct advantage in some cases. Through proper design, single-mode applicators can focus the high field to a given location. This technique has been exploited for joining of ceramic in which it is important to apply the heat at a given location without heating the bulk of the ceramic.

In addition, knowledge of the EM field distribution can allow materials to be placed in the area of highest field strength for optimum coupling. Therefore, these cavities are used for laboratory scale studies of interactions between materials and MWs. Although heating patterns of single-mode cavities are nonuniform, the different resonant modes have complementary heating patterns, thus mode switching can result in approximation of a uniform pattern in a single-mode cavity.

1.2.3.2 Multimode Applicators

Applicators that are able to sustain a number of high-order modes at the same time are known as multimode cavities. These types of applicators are used in home MW ovens. Unlike the designs of the single-mode applicators, which are based on the solutions of Maxwell's equations, the designs of multimode applicators are often based on trial and error, experience, and intuition. As the size of the MW cavity increases, the number of possible resonant modes also increases. Consequently, multimode applicators are larger than one wavelength. For a rectangular cavity, the mode equation for the resonant frequencies for different modes can be described as the solution to the following equation:

$$f_{nml} = c\left[\left(\frac{l}{2d}\right)^2 + \left(\frac{m}{2b}\right)^2 + \left(\frac{n}{2a}\right)^2\right]^{1/2} \quad (1.1)$$

The presence of multiple modes results in multiple hot spots within the MW cavity. To reduce the effect of multiple hot spots, several techniques are used to improve EM field uniformity. One such strategy is to increase the size of the cavity, which increases the number of possible resonant modes resulting in overlapping EM fields, thus leading to a more uniform EM field.

A second commonly adopted strategy is the use of mode stirrers and turntables (Tulasidas 1994). Mode stirrers resemble fans and rotate within the cavity, are reflectors of the EM waves, and hence tend to produce a more uniform EM field inside the MW applicator. The purpose of the turntable, on the other hand, is to pass the material through the regions of high and low fields and thus achieve time-averaged EM field uniformity.

For a specified size of cavity, there exists a particular dominant mode at the resonant frequency. There is also a frequency below which the waveguide does not allow the modes to propagate, and attenuation of waves takes place. This frequency is known as the cutoff frequency. There are also evanescent modes, in which the propagation is not sustained, and they decay rapidly to zero. Thus there is no wave propagation in evanescent mode. Some of the propagating modes in multimode cavity are evanescent modes, and some of them are dominant modes. The time-averaged intensity of the evanescent modes is zero whereas the dominant modes have strong positive magnitudes. Thus the heating occurs mainly due to the dominant modes. It becomes highly difficult and impractical to determine the dominant and evanescent modes by analytical methods for the partly loaded multimode cavities as this is dependent on a number of factors such as, but not limited to, the placement of load, mass of load, and dielectric properties of the load. MW heating in single- or multimode cavities occurs predominantly through dominant modes.

1.2.4 Maxwell's Equations

The Maxwell's equations are the theoretical basis for all electromagnetic phenomena. The importance of the Maxwell's laws can be understood from the fact that they summarize all the known laws of the EM field. The integral form of Maxwell's equations describes the underlying physical laws whereas the differential forms outlined below are commonly used for the solution of the problems. For a field to be qualified as an EM field, it must satisfy all four equations.

$$\nabla \times E = \frac{\partial B}{\partial t} \nabla \times H = \frac{\partial D}{\partial t} + J \nabla \mathrm{B} = 0; \quad \nabla \mathrm{D} = \rho \mathrm{D} = \varepsilon E; \quad B = \mu H;$$
$$J = \sigma EP = 2\pi f \varepsilon_0 \varepsilon'' E^2$$

$$\nabla \times \mathrm{E} = -\mu \frac{\partial H}{\partial t} \tag{1.2}$$

$$\nabla \times \mathrm{H} = \sigma \mathrm{E} + \varepsilon_0 (\varepsilon' - j\varepsilon'') \frac{\partial E}{\partial t} \tag{1.3}$$

$$\nabla \times \mathrm{D} = \rho \tag{1.4}$$

$$\nabla \times \mathrm{H} = 0 \tag{1.5}$$

$$\mathrm{D} = \varepsilon E; \quad B = \mu H; \quad J = \sigma \tag{1.6}$$

Equation 1.2 describes the generation of an electric field from a time-varying magnetic field (Faraday's law). Equation 1.3 describes the generation of magnetic fields from time-varying electric fields (Ampere's law). Equation 1.4 depicts the Gauss's law, which is a mathematical statement for the conservation of electrical charges. Equation 1.5 is the mathematical statement for the nonexistence of the isolated magnetic charge.

The analytical solution of a complex EM field problem is not always possible. Hence, numerical methods are generally used for the solution of complex EM field problems. Yee's algorithm (FDTD method) is most commonly used for the numerical solution of the equations.

1.3 Material Interactions with MW and RF

In any material, there exist either free or bound charges. The interaction of material with the EM field causes motion of the bound charge. This effect is called polarization and can be categorized as ionic, orientation, atomic, and electronic polarization. Movement of ions in the material under the influence of the EM fields is called ionic polarization. It is the predominant effect at frequencies below 1 GHz. MWs do not have sufficient energy to produce ionization (Zante 1973). Electric polarization occurs in atoms, in which electrons can be displaced with respect to the nucleus. This effect occurs in all substances to varying degrees. In atomic polarization, the atoms can be moved in crystals or molecules. When only electronic polarization and atomic polarization mechanisms are present, the materials remain almost lossless at the MW frequencies. Atomic, or vibration, polarization is related to the electronic polarization, but due to the higher masses moved in atomic polarization, the resonant frequencies occur in the lower range of the electromagnetic spectrum in the infrared range, whereas the electronic polarization has frequencies in the optical band (Nyfors and Vainikainen 1989).

Of all the possible forms of energy loss mechanisms, orientation polarization is most important at MW frequencies above 1 GHz. This type of polarization also influences the lower frequencies as well, and ionic loss mechanism, which is strongly temperature dependent typically predominates at frequencies below 1 GHz (Ryynänen 1995). The orientation polarization is found in many dielectric materials, such as water molecules that carry a permanent dipole moment. In a high-frequency EM field, the molecules try to align with the electric field causing them to lose the energy to the random thermal motion of the molecules. This lost energy appears as heat and the associated temperature rise. This effect is strongly temperature dependent whereas the electronic and atomic polarizations are not dependent on temperature. The relationship between the permittivity (ε) and the polarization (p) is described by Equation 1.7.

$$p = (\varepsilon - 1)\varepsilon_0 E \tag{1.7}$$

The relative permittivity is, therefore, a measure of the polarizing effect of the external electric field. Polarization of the electric charge in which the rotational motion is restricted results in a lag between the electric field and the polarization. This time lag is known as relaxation time and is proportional to the dissipation of energy as heat within the material. MW heating is a result of this dielectric relaxation.

Properties that contribute to the dielectric response of materials include electronic polarization, atomic polarization, and Maxwell-Wagner polarization mechanisms (Ryynänen 1995). At MW frequencies, dipole polarization is the most important mechanism for energy transfer at the molecular level.

Water is chiefly responsible for the MW heating effect due to dominant dipolar interactions with MWs compared to proteins and lipids, which interact weakly with MW energy. The magnetic field component of MW does not result in heating effects of food. However, some of the food packaging materials may interact with magnetic field (Datta and Ramaswamy 2001).

Both RF and MW heating involve interaction of EM fields with food materials. Part of the EM energy absorbed by the material is dissipated as heat due to the different mechanisms described above. In an EM spectrum, RF (3–300 MHz) and MW (300 MHz–300 GHz) bands occupy adjacent sections with RF waves having lower frequencies. The major advantage of RF heating over MW heating is its ability to penetrate deeper into food materials as the RF wavelength is much longer (e.g., ~11 m at 27.12 MHz) than the MW wavelength (e.g., ~12 cm at 2450 MHz). One of the challenges with the RF heating technology is its nonuniform heating (Tang et al. 2000). Different factors, such as sample dielectric and thermo-physical properties, size, shape, its position between the RF electrodes, and electrode configuration, may affect temperature uniformity in a RF-treated food product. Therefore, it is essential to understand the complex behavior of RF and MW heating mathematically as well as experimentally.

1.3.1 Dielectric Properties

For the heat to be generated within the material, the MW must penetrate and transmit energy into the material. The dielectric constant and the dielectric loss factor quantify the capacitive and conductive components of the material. The components are often expressed in terms of complex dielectric constant (Datta and Ramaswamy 2001; Wang et al. 2003). The dielectric constant is a measure of the capability of a material to couple with EM energy or to store that energy. The loss factor determines the ability of the material to dissipate electric energy and convert it to heat energy.

$$\varepsilon = \varepsilon' - j\varepsilon'' \tag{1.8}$$

Sometimes, the dielectric constant and loss factors are expressed in terms of their relative values in a vacuum.

$$\varepsilon'' = \varepsilon_0 \varepsilon_r'' \tag{1.9}$$

Another commonly used term for expressing the dielectric response is the loss tangent. It is defined as

$$\tan\delta = \frac{\varepsilon''}{\varepsilon'} \tag{1.10}$$

Materials with a high loss tangent absorb more EM power.

1.3.2 Factors Affecting Dielectric Properties

There are various factors, such as EM frequency, temperature, density, water content, and chemical composition of the material, which affect dielectric properties of the food materials.

The effect of the frequency of the electric field on material dielectric properties is due to its effect on polarization. The dielectric properties vary significantly with a change in frequency for most of the materials except low glossy materials (Nelson 1994). The dielectric properties of water for a wide range of frequencies can be calculated by using Debye equations for pure polar materials (Datta and Ramaswamy 2001).

The effect of temperature on the dielectric loss factor is dependent on the frequency. The dielectric loss factor may increase or decrease with an increase in temperature, depending on operating frequency. There is no analytical expression that can be used in Debye's equations to determine variation in properties with change in temperature (Buffler 1993).

In addition to the true density of the materials, the effect of bulk density is important as in the case of granular materials as they entrap air in the voids, which have a very low dielectric loss factor compared to water. In highly granular and low bulk density materials, a high amount of air, which has a low dielectric loss factor causes a decrease in the overall dielectric loss factor and ultimately results in a low heating rate in the MW field (Nelson 1994; Mujumdar 2000).

Water has a dielectric constant of approximately 78 at room temperature, which is very high as compared to other constituents, such as proteins, carbohydrates, and lipids of food materials that are not very active ($\varepsilon' < 3$ and $\varepsilon'' < 0.1$) (Tulasidas 1994). Thus, the water content of the materials largely determines the dielectric properties of the food materials (Wang et al. 2003). Some researchers have conducted studies to measure and to observe the effect of various factors on the dielectric properties of materials: grapes and sugar solutions (Tulasidas et al. 1995), garlic (Sharma and Prasad 2002), whey protein gel, liquid whey protein mixture, macaroni noodles, cheese sauce, and macaroni and cheese (Wang et al. 2003; Nelson 1994; Datta and Ramaswamy 2001).

The dielectric properties of many food products and their variation with moisture and temperature have been discussed in detail by several authors (Nelson 1994; Ryynänen 1995; Datta and Ramaswamy 2001; Venkatesh and Raghavan 2004). In general, the dielectric constant increases with the material's moisture content. Sharma and Prasad (2002) observed the increase in dielectric constant of garlic with the increase in moisture content at any temperature and found a quadratic relationship between moisture content and dielectric constant. Moisture had the largest effect on loss factor, and temperature effect was dependent on moisture content. Dielectric constant and loss factor both increase with an increase in temperature in low moisture-content food materials (Tulasidas 1994). Wang et al. (2003) observed a decrease in the dielectric constant in the MW range (915

Table 1.2 Dielectric Properties of Some Food Materials at 2.45 GHz

Food Product	*Moisture Content (%)*	ε′	ε″
Apple	88	54	10
Banana	78	60	18
Carrot	87	56	15
Grape	82	65	17
Mango	86	61	14
Potato	79	57	17
Milk fat (solid)	0	2.6	0.2
Distilled water	100	78	13.4

Source: Datta, A.K. and Ramaswamy, C.A. *Microwave Technology for Food Applications,* Marcel Dekker, New York, 2001.

and 1800 MHz) with the increase in temperature. However, for cooked macaroni noodles, the dielectric constant was found increasing in the MW range (915 and 1800 MHz) with an increase in temperature. The dielectric constant and loss factor of some food products are reported in Table 1.2.

1.3.3 Measurement of Dielectric Properties

The methods to determine the dielectric properties of materials can be categorized as resonant and nonresonant methods. The resonant methods have relatively higher accuracy compared to nonresonant methods, such as those using reflection techniques. Among the resonant techniques, the perturbation cavity technique is popular due to its accuracy, application to low lossy materials, experimental simplicity, direct evaluation, and high temperature capability (Buffler 1993; Liao 2002; Sharma and Prasad 2002). Additional techniques used to measure the dielectric properties of food materials are discussed in Nelson (1994), Ryynänen (1995), Nelson (1999), and Datta and Ramaswamy (2001).

1.3.4 Energy Conversion in MW Heating

The dielectric properties of the materials in combination with the applied EM field cause conversion of some part of the EM energy into heat. The power that is transmitted to a material can be determined by Poynting Vector Theorem. The transmitted power across the surface S of a volume V is given by the integration of the real portion of the time averaged Poynting vector:

$$P = \frac{1}{2}\int_s E \times H^* dS \tag{1.11}$$

The * in this case denotes the complex conjugate. If the material is homogeneous, the following equation can be obtained for transmitted power using the divergence theorem and Maxwell's equation.

$$P = \frac{1}{2}\int_v (\omega\mu HH^* + \omega\varepsilon'' EE^* + \sigma EE^*)dV \tag{1.12}$$

In dielectric materials, the magnetic permeability is usually very small and can be neglected. If the electric field is assumed to be constant throughout the volume, the following simplified equation is obtained:

$$P = 2\pi f\varepsilon_0\varepsilon'' E^2 \tag{1.13}$$

Equation 1.13 is valid only for very thin materials as the electric field decreases as a function of the distance from the surface of the material. The magnitude of electric field intensity decreases as the wave propagates into the material. The magnitude of the electric field intensity for any wave travelling in z direction through a slab can be given as Equation 1.14. Power dissipated in the material is proportional to the square of electric field intensity, so the power drop is very fast (Equation 1.15).

$$|E| = E_0 \exp(-\alpha Z) \tag{1.14}$$

$$P = P_0 e^{-2\alpha Z} \tag{1.15}$$

The term "penetration depth" is used to quantify the depth of material into which MWs can penetrate before significant attenuation takes place. Penetration depth (d_p) is defined as the distance at which the power density drops to $1/e$ (37%) of the original value at the surface (Datta and Ramaswamy 2001). Beyond this depth, the volumetric heating due to MWs is considered negligible. It is not a material property but a combination of various factors (Mujumdar 2000).

$$d_p = \frac{1}{2\alpha} = \frac{\lambda_0}{2\pi}\left\{\frac{2}{\varepsilon'\left(\sqrt{1+\left(\frac{\varepsilon''}{\varepsilon'}\right)^2}-1\right)}\right\}^{1/2} \tag{1.16}$$

1.3.5 Energy Conversion in RF Heating

An accurate calculation of RF power density is essential to determining temperature rise in food materials during RF heating. RF power density in the material is given by (Rowley 2001):

$$Q = 2\pi f \varepsilon_0 \varepsilon'' |E|^2 \tag{1.17}$$

Temperature rise in the food materials can be calculated by the unsteady heat transfer equation given by (Birla et al. 2008):

$$\frac{\partial T}{\partial t} = \nabla \alpha_1 \nabla T + \frac{Q}{\rho_m C_p} \tag{1.18}$$

It is clear from Equation 1.17 that determination of the electric field is necessary to calculate the RF power density inside the dielectric material. The electric field inside the RF applicator can be determined by solving Maxwell's equations (Equations 1.2 through 1.5). A general solution of Maxwell's equations leads to the wave nature of the RF electric field. Because RF wavelengths are substantially longer (e.g., ~11 m at 27.12 MHz) compared to practically available RF applicator sizes, Maxwell's equations can be simplified into a single equation, the Laplace Equation, by neglecting the effect of the magnetic field. The assumption results in the static RF electric field approximation inside the RF applicator (Birla et al. 2008):

$$\nabla(\sigma + j2\pi f \varepsilon_0(\varepsilon' - j\varepsilon'')\nabla V) = 0 \tag{1.19}$$

1.3.6 Mechanism of MW Drying

MW heating involves four simultaneous processes: heat generation, sensible heating, convective heat loss, and moisture transfer. The EM field setup in the cavity provides a heat source for the material. The single-mode cavity has one dominant mode, and the multimode cavity has many propagating modes as described earlier in Section 1.2.3.2. Heat and mass transfer processes inside the material mainly occur in three stages. During the first stage of heating, MW energy is converted into thermal energy. The heat generated by the MW causes the material to heat up, but the degree of temperature rise is governed by many factors including:

- Latent heat of evaporation of the evaporating liquid water inside the material
- Convective transport of the air, vapor, and liquid water across the cells
- The specific heat and thermal conductivity of the material
- The surrounding air temperature and humidity

The second stage of heating consists of rapid moisture loss due to multiple mass transfer processes. The main mass transfer mechanisms for liquid water, water vapor, and air are diffusion and convection. As the temperature rises, the vapor is continuously formed inside the material. Formation of vapor is governed by the moisture content and water activity of the material. This vapor, along with the air inside the pores of the material, causes an increase in the pressure within the material. Moisture migration is mainly due to the pressure gradient between the interior and outside of the food material (Tulasidas 1994). If the material is dry, the gases will escape through the pores. However, most of the food products heated in a MW oven are porous and contain high moisture. In such materials, pores may be saturated with water, and the pressure will build up inside the material in such cases. This building up of pressure would cause the pressure driven, Darcy-like flow in the material. This is true not only for the boundaries, but also for the interior of the material. If the material reaches saturation greater than one at any location, the liquid water is pushed out across the boundaries of the control volume in the direction of the decreasing pressure gradient. The food materials (or most biological materials) are hygroscopic in nature and therefore undergo a change of structure as the loss of moisture occurs. Thus, the porosity of the material is also dependent on the material's moisture content. Movement of liquid water, vapor, and air can also take place by the Fickian diffusion.

When liquid saturation at the surface is less than 100%, water is evaporated and conveyed away by the surrounding air, depending on its relative humidity. The same is true for the water vapor arriving at the surface. However, if the surface of the material is saturated with water, the water moves out across the boundaries without a change in the phase, and this represents a major fraction of water removed in a very short time (Ni et al. 1999; Ramaswamy and Holm 1999). Most of the drying during MW heating occurs during the second stage of heating.

During the third stage of heating, overall moisture transfer is reduced, and there is significant sensible heating of the sample, and the temperatures may rise above the boiling point of water (Zhang et al. 2006). The outward vapor flux results in higher drying rates and also helps in preventing the shrinkage of tissue structure, which is found in most conventional air-drying techniques (Zhang et al. 2006).

1.4 Mathematical Modeling of MW Heating

The process of MW heating in the MW oven cavity is governed by the EM field interactions, heat, and mass transfer occurring in the material. Understanding of heat and mass transfer inside the MW cavity is essential for optimization and good product quality. However, detailed analytical descriptions of MW heating are difficult to study because of the complex interaction between the MWs, the cavity, and the food (Datta and Ramaswamy 2001). Another challenge is that with change in local temperature and moisture content, the dielectric properties of the food material also change, which results in different EM field patterns leading to differential heating patterns. Therefore, a complete understanding of the MW heating problem requires a solution for the coupled EM, heat, and mass transfer problem.

Comprehensive modeling of MW heating and drying would require a completely coupled model integrating the EM model with the generalized heat and mass transfer models. A generalized mathematical model for MW heating or drying would consist of three coupled sets of differential equations as follows:

Part 1: Electromagnetic field distribution

Same as the Maxwell's equations (Equations 1.2 through 1.6) described earlier. The temperature and moisture dependence of various loss factors is incorporated.

Part 2: Heat transfer

Energy conservation equation

$$\frac{\partial}{\partial t}(c_a h_a + c_v h_v + c_w h_w + c_s h_s) + \nabla\ (n_a h_a + n_v h_v + n_w h_w) = \nabla\ (k_{eff} \nabla T) - \lambda i + \dot{q} \tag{1.20}$$

The first term on the left-hand side (LHS) of the equation describes the change in the enthalpy of the material. The second LHS term accounts for convective heat transfer due to movement of air, water vapor, and liquid water. The first term on the right-hand side (RHS) of the equation accounts for conductive heat transfer, and the second term describes the changes due to phase changes in liquid water. The third term is the heat source (MW power dissipation) and is directly related to the local EM field intensity ($\dot{q} = P = 2\pi f \varepsilon_0 \varepsilon'' E^2$ from Equation 1.13).

Part 3: Mass conservation

$$\frac{\partial c_a}{\partial t} + \nabla n_a = 0 \tag{1.21}$$

$$\frac{\partial c_v}{\partial t} + \nabla n_v = i \tag{1.22}$$

$$\frac{\partial c_w}{\partial t} + \nabla n_w = -i \tag{1.23}$$

The first terms on the LHS of the above equations describe the changes in the concentrations of air, water vapor, or liquid water. The second terms account for the convective and diffusive transfer of air, water vapor, or liquid water. Terms on the RHS describe the changes in concentration due to phase change.

The coupled nature of the sets of differential equations arises because the terms in the equations are affected by other values. For example, electromagnetic field distributions are used to obtain the local heat generation term ($\dot{q}$) within the material. This information is used as an input to the energy conservation equation that governs the heat transfer processes. Similarly, the fluxes of air, water vapor, and liquid water appear in both energy and mass conservation equations. Additionally, not only the fluxes, but also the parameters in the equations, such as dielectric properties, porosity, and thermal conductivity are strongly correlated to moisture content and temperature of the samples.

1.5 Research Efforts in Modeling MW Heating

It is almost impossible to obtain an analytical solution to this complex coupled problem for realistic cases. However, by employing numerical methods, solutions can be obtained after simplifications. However, in a majority of the applications, it is neither necessary nor feasible to solve the completely coupled model using numerical methods. Many of the research efforts that have been reported in the literature reflect this trend of simplification. However, with the increasing sophistication of numerical methods and challenges in industrial MW applications, a solution to coupled models can be obtained for the few essential cases. One of the numerical solution procedures (Figure 1.2), considering a completely coupled problem with a pseudo steady state assumption was suggested by Dincov et al. (2002) and Murthy (2003).

The efforts to model the MW heating in a comprehensive way started in the late 1970s (Whitaker 1977). The researchers primarily used Lambert's law, in which the exponential decay in the electric field intensity was assumed. This was considered to be valid for materials whose thickness was greater than the wavelength of the MW. However, Ayappa et al. (1991) showed that there could be significant deviations from Lambert's law for thin slabs due to the internal resonance that takes place in such slabs. Fu and Metaxas (1992) proposed a new definition of the power penetration depth based on the normalized values of the total power absorbed. Almost all the papers considering heat and mass transfer base their work on a pioneering paper by Whitaker (1977). The author derived the mechanistic coupled heat and mass transfer model in porous media based on a volume-averaging process at the microscopic level.

Hill and Marchant (1996) provided a comprehensive overview of the mathematical modeling efforts in MW heating. Thostenson and Chou (1999) reviewed the application of MW heating in the polymer industry. Ramaswamy and Tang (2008) reviewed the state of the art in MW and RF heating and identified the research needs. A summary of models reported in the literature and their major characteristics is provided in Table 1.3.

Jolly and Turner (1990) applied Maxwell's equations to a one-dimensional slab assuming zero electrical conductivity and constant magnetic permeability. The

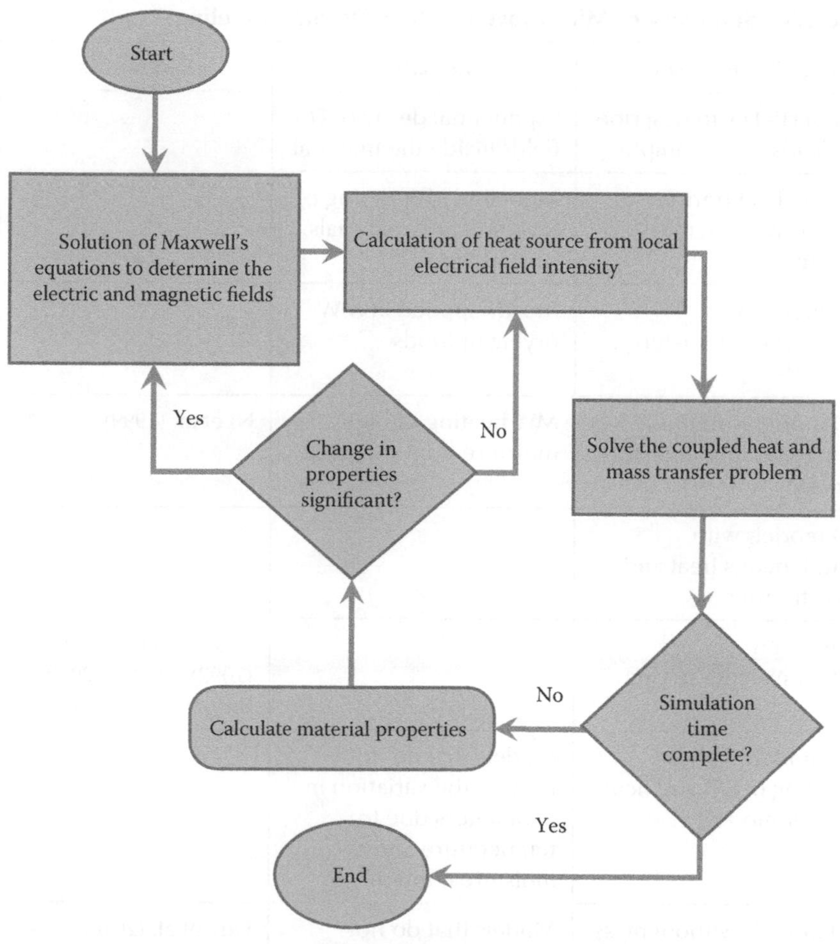

Figure 1.2 Algorithm to solve coupled MW heating problems.

equations were solved using the finite difference technique, and it showed multiple peaks and troughs in the power absorption that was a result of the standing wave patterns set up in the cavity. They also examined the effect of extending the slab using lossless materials like Teflon, which smooths the peaks and results in flatter power and temperature profiles. Jolly and Turner (1990) extended the above model to convective drying of porous materials. The material is assumed to be a solid matrix in which the liquid, vapor, and air reside. Darcy's law for gas and liquid was considered to account for the convective transport of the air and water. Fick's law was assumed to be applicable for the vapor transport. The electrical properties were considered as functions of both moisture and temperature. Their results showed that the drying times could be halved using a combination of convective and MW drying. The MW heating causes the moisture to migrate to the surface from the

Table 1.3 Summary of Microwave Heating/Drying Modeling

Model Characteristics	*Remarks*	*References*
Lambert's law to describe EM fields in the sample	Exponential decay of EM fields inside the material	
EM and heat transfer models without mass transfer	Modeling MW drying of low-moisture materials, such as wood	
1-D model with intermediate moisture contents	Mostly applied in MW drying of foods	
1-D model with phase change for high moisture contents	MW heating of high moisture-content foods	Ni et al. (1999)
2-D models with simultaneous heat and mass transfer		
Combined MW-convective drying		Sharma and Prasad (2002); Kumar (2009); Bingol et al. (2008)
3-D models without coupling of EM and heat transfer models	Models that do not include the variation in parameters due to temperature and moisture content	
3-D models without mass transfer	Models that do not include mass transfer	Dev et al. (2010); Klinbun et al. (2011)
3-D model completely coupled model with phase change for a multimode cavity	Pseudo steady state assumptions. No turntable or mode stirrers considered	Murthy (2003)
3-D model completely coupled model	Considering turntable and mode stirrers	Geedipalli et al. (2007)
Industrial systems		Clemens and Saltiel (1996); Chen (2008)

core of the sample. Sharma (2000) also showed that the drying times could be significantly reduced when a combination of MW and convective heating is used. The models in the context of the MW heating are mostly for the drying process, which is a low-power process, and the materials are often of low moisture content. Ni et al. (1999) demonstrated the importance of the liquid pumping effect for MW heating of high moisture materials. Liquid pumping was defined by the authors as the expulsion of the liquid water to the surface of the material without change in phase. The authors had considered a 1-D formulation of the problem, including the convective transport of the vapor, liquid, and air. They were also able to predict the pressures inside the sample, which could be used to study the material integrity and safety of the MW heated material. In another effort, Ramaswamy and Holm (1999) modeled intensive drying of paper and board. They modeled the liquid pumping under the effect of intensive heating using steam as used in the paper industry. They also concluded that there is a definite possibility to use this effect of liquid expulsion to the surface for accelerating the drying process in porous materials. Zhao and Turner (2000) developed a coupled (EM and heat transfer, mass transfer processes not included) computational model for the MW heating of wood. Murthy and Prasad (2005) developed a completely coupled 3-D model considering the phase change and the multimode nature of the MW cavity. The simulations indicated a complex pattern of electric fields inside the MW oven that is dependent on the wave guide, cavity, and sample dimensions and sample placement (Figures 1.3 through 1.5).

Complexity of the electric field patterns in multimode ovens can be observed from Figures 1.4 and 1.5.

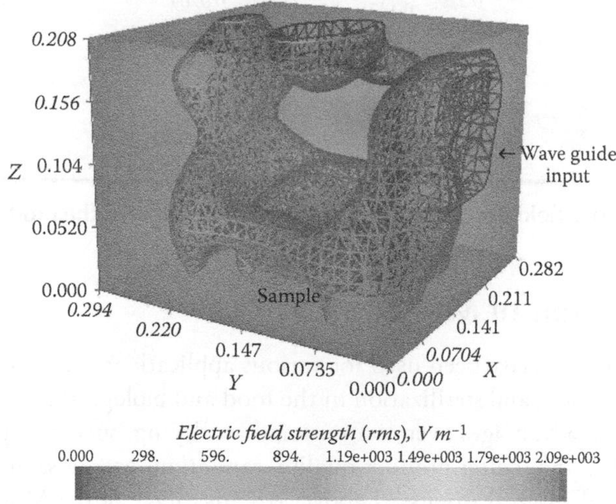

Figure 1.3 Electric field distribution inside the oven cavity for the mode TE 1:0:0.

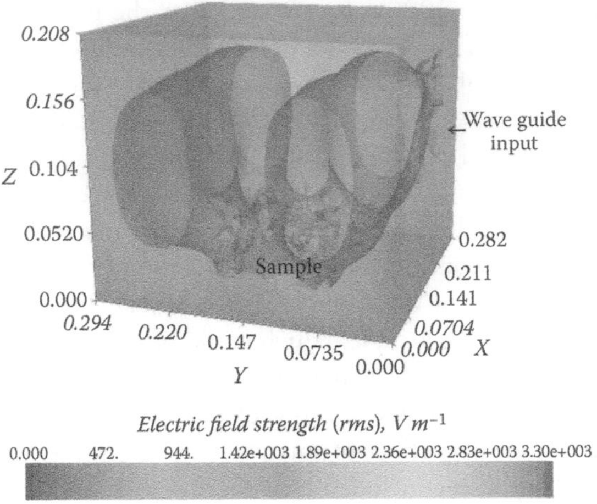

Figure 1.4 Electric field distribution inside the oven cavity for the mode TE 10:0:0.

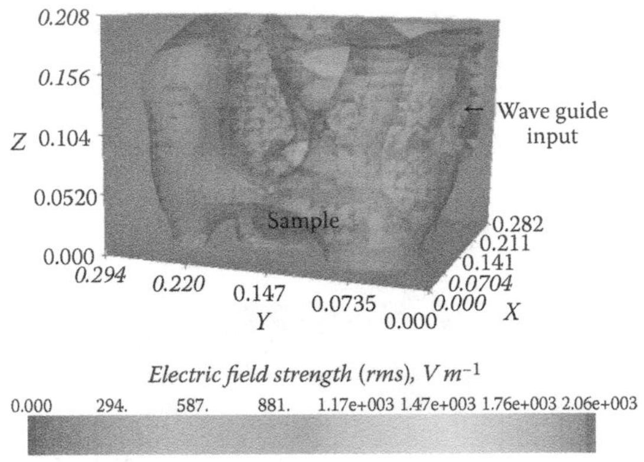

Figure 1.5 Electric field distribution inside the oven cavity for the mode TE 20:0:0.

1.6 Application of MW

MW energy processes have been used for various applications, such as processing, drying, pasteurization, and sterilization in the food and biological industries. MW drying has several advantages over conventional air drying, such as higher drying rates, uniform heating, better product quality, sanitation, energy saving, and efficient utilization of space as discussed above. An advantage of the MW heating is the possibility for precise control of temperature in the system as the MW power input to the system can be instantaneously switched on/off.

The MW application also has some limitations associated with it such as low efficiency of magnetron devices (~50%) and scattering of energy byproduct. Uneven heating of product, textural damage, and formation of hot spots due to nonuniform distribution of the EM field inside the cavity is a major problem in the application of MW (Zhang et al. 2006).

However, the MW drying technique has been found to be highly effective when it is used in combination with another drying method, such as hot air drying, vacuum drying, freeze drying, osmotic drying, etc. (Zhang et al. 2006). Another way to address some of these limitations is to apply MW energy in a pulsed manner. MW finish drying, defined as the use of MWs in the last stage (falling rate period), has also been suggested for better quality of food products (Funebo and Ohlsson 1998). During the initial stage of drying, food materials have very high moisture, which can be easily removed with hot air. However, in conventional drying in which about two thirds of the drying time is used to evaporate the last 33% of moisture, MW-assisted drying can significantly reduce the drying times (Wang and Sheng 2006).

In MW-convective drying, moisture migrates to the surface due to MW heat generation inside the material and is removed by hot air. To prevent the condensation of the driven moisture on the surface of the food, it is necessary to heat the air to increase its moisture-carrying capacity. The drying rate is enhanced in MW-convective drying, and heating is more uniform. Several researchers (Tulasidas 1994; Prabhanjan et al. 1995; Funebo and Ohlsson 1998; Sharma 2000; Maskan 2001; Dadali et al. 2007; Kumar 2009; Kumar et al. 2011) have reported MW-convective drying of foodstuffs and found considerable improvements in the drying speed and quality of the dried product. MW power can be applied in various ways during hot air drying: throughout the drying period, during the starting period to pump out the moisture, or in the last stage when most of the moisture is already removed and drying rates are very slow. A lab scale MW-assisted hot air drying system is shown in Figure 1.6.

The vacuum drying technique has been used in recent years to obtain higher quality dried heat-sensitive products and high-value fruits and vegetables. MWs have also been used to accelerate the vacuum drying process and improve the efficiency of the process while maintaining higher quality. Many studies have been conducted to study the effect of MW-vacuum drying on drying behavior and product quality of fruits and vegetables: bananas (Drouzas and Schubert 1996), cranberries (Yongsawatdigal and Gunasekaran 1996), carrots (Cui et al. 2004), grapes (Clary et al. 2005), and mushrooms (Giri and Prasad 2007). Giri and Prasad (2007) conducted studies on MW vacuum drying of mushrooms and found that rehydration of dried mushroom was better in the case of the MW-assisted processes than that of hot air drying.

Microwave power can be applied along with the application of fluidization (also known as MW-assisted fluidized bed or spouted bed drying), which can improve the heat and mass transfer and uniformity of heating due to agitation. Many studies

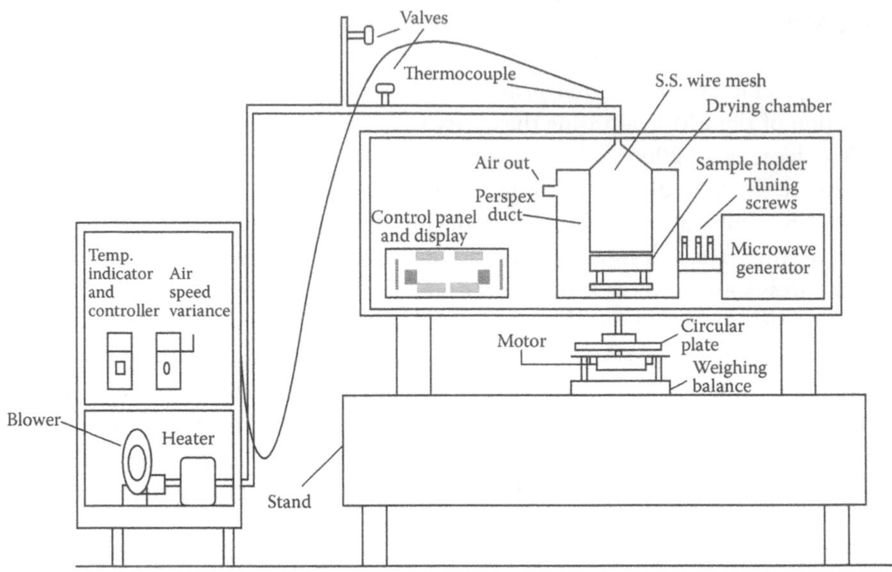

Figure 1.6 Schematic diagram of MW-convective drying system. (From Kumar, D., Studies on Microwave-Convective drying of Okra, M.Tech Thesis, Agricultural and Food Engineering Department, Indian Institute of Technology, Kharagpur, 2009; Kumar, D., Prasad, S., & Murthy, G.S., *Journal of Food Science and Technology*, 2011, DOI: 10.1007/s13197-011-0487-9.)

(Ang et al. 1977; Sunderland 1980; Duan et al. 2007) have been conducted on the application of MWs to assist freeze drying, a novel technique for dehydration of heat-sensitive food products and pharmaceutical materials. Because freeze drying is an expensive technique, application of MWs to decrease the drying time could result in a lower overall cost of dehydration (Zhang et al. 2006). Other than drying, application of MWs has been studied for blanching, thawing, pasteurizing, and sterilizing of food products (Sharma and Prasad 2002).

Use of MW power has also been investigated for biomass pretreatment and pyrolysis in the area of renewable energy. The energy production from biomass is an important alternative that can address these concerns of a limited supply of fossil fuels and global warming. Biomass can be processed using various technologies, such as direct combustion, gasification, pyrolysis, and fermentation to get direct energy or liquid fuels. The application of MW energy in these processing technologies can lead to many advantages: reducing the transportation and storage cost of biomass by removing moisture, better control of processes due to uniform and efficient heating, ease of control and decrease in processing time, etc.

Pyrolysis is thermochemical processing of biomass at high temperatures (400°C–500°C) and in absence of oxygen to produce charcoal, syngas, and bio-oils. Pyrolysis of biomass at moderate temperatures and short residence times, known as fast pyrolysis, produces high yields of bio-oils. The main advantage of using MWs in thermal processing of biomass is uniform internal heat generation due to large penetration depth, which can diminish the need for size reduction of biomass before processing, which itself is an energy intensive and costly process (Krieger-Brockett 1994; Yu et al. 2009). In conventional heating of large-size biomass, the heating rate is very slow due to low thermal conductivity and diffusion, which results in production of char and lower volatiles yield. This problem can be addressed to some extent by using MWs. The application of MWs in thermal processing of biomass has been investigated by many researchers (Krieger-Brockett 1994; Menendez et al. 2002; Miura et al. 2004; Dominguez et al. 2006; Yu et al. 2006, 2009; Wang et al. 2008; Wan et al. 2009). Biomass feed stocks, organic wastes, and coal do not absorb MW energy very effectively, so to achieve the higher temperature required for pyrolysis, receptors are mixed with samples. This process is known as MW-induced pyrolysis, and receptors are high dielectric substances that heat very rapidly under MW fields, such as metal oxides, char, non-stoichiometric oxides, etc. (Monsef-Mirzai et al. 1995; Menendez et al. 2002; Dominguez et al. 2006). Menendez et al. (2002) studied the MW-induced pyrolysis of sewage sludge by mixing char as a receptor and concluded that the MW pyrolysis process was fast, energy-efficient, and resulted in high volume reduction. Yu et al. (2009) studied the effect of MW power levels during pyrolysis of corn stover and observed increasing corn stover degradation and gas production with increasing power levels.

Bioethanol has been considered as an important alternative to liquid transportation fuels because of their compatibility with the current infrastructure. The recalcitrance structure of biomass and the presence of a complex matrix of cellulose, hemicellulose, and lignin limit the efficient conversion of biomass into sugars during the hydrolysis process. As mentioned earlier, MWs penetrate inside the material and produce internal heat due to the interaction of polar molecules, which results in a fast removal of water. Higher drying rates also cause structural changes in biomass, which affects the downstream processing process. Several researchers have investigated the potential use of MW/chemical pretreatment to improve the hydrolysis efficiency of lignocellulosic biomass for ethanol production (Ooshima et al. 1984; Azuma et al. 1984; Zhu et al. 2005a; Zhu et al. 2006; Keshwani et al. 2007; Keshwani and Cheng 2009; Gong et al. 2010). Zhu et al. (2005a), Zhu et al. (2005b), and Zhu et al. (2006) studied the use of MWs along with different chemicals (alkali, acid, H_2O_2) for pretreatment of rice straw. The authors observed higher water, xylose, and lignin removal and hydrolysis rates compared to conventional heating chemical pretreatment. The oscillations of polar molecules, the collision of moving ions due to application of MW energy produce heat and result in physical and chemical changes in the biomass. Acetic acid, a component of hemicellulose, is released due to thermal effects, which results in an acidic aqueous environment and enhances auto-hydrolysis (Keshwani et al. 2007; Keshwani and Cheng 2009). Keshwani et al. (2007) studied the effect of MW/chemical pretreatment on hydrolysis yields of switchgrass by varying the treatment time and concentration of chemicals (H_2SO_4 and NaOH). The hydrolysis yields were slightly less than that of switchgrass treated with conventional chemical pretreatments; however, time required was observed to be reduced by about six times.

1.7 Summary

MW and RF heating refers to the use of EM waves of certain frequencies to generate heat in a material. Typically, MW food processing uses two frequencies of 2450 and 925 MHz. Of these, the 2450 MHz frequency is used for household ovens, and both are used in industrial heating. An estimated 225 million domestic MW ovens in use around the world points to the importance of MW heating technology. While MW and RF heating technologies have many advantages, their application in combination with conventional drying techniques provides several advantages such as reduced drying times and increased energy efficiencies compared to conventional heating technologies. In addition to conventional applications in food processing, there is a potential for MW heating to be applied in novel areas, such as thermo-chemical and biochemical pretreatment of biomass for biofuel production.

Nomenclature

a, b, and d dimensions of the cavity in X, Y, and Z directions, respectively
c speed of light in vacuum (m s^{-1})
c_v, c_w, and c_a concentration of vapor, water, and air (kg m^{-3}), respectively
dS surface element
d_p penetration depth
f frequency of electric field (Hz)
f_{nml} TE_{nml} or TM_{nml} mode's resonant frequency
h_v, h_w, and h_a enthalpy of vapor, water, and air (kJ kg^{-1}), respectively
i rate of evaporation of water (kg m^{-3})
k_{eff} effective heat conductivity (W ms)
n, m, and l number of half sinusoidal variations in the standing wave pattern along X, Y, and Z axes, respectively
n_v, n_w, and n_a fluxes of vapor, water, and air (kg m^{-2} s), respectively
p polarization
$\dot{q}$ rate of heat generation (W m^{-3})
t time (s)
B magnetic flux density vector (Wb m^{-2})
C_p specific heat (J kg^{-1} °C^{-1})
D electric flux density vector (C m^{-2})
E electric field vector (V m^{-1})
$|E|$ modulus of electric field (V m^{-1})

E_0 RMS electric field intensity at the surface in which the wave is traveling
H magnetic field vector (A m^{-2})
P power dissipated (W m^{-3})
Q RF power density (W m^{-3})
T local temperature (°C)
V voltage between the two electrodes (V)
Z distance in Z direction

Greek Symbols

ε electrical permittivity (F m^{-1})
$(\varepsilon$-1$)$ electric susceptibility
ε_0 absolute permittivity of the free space (8.86×10^{-12} F m^{-1})
ε' dielectric constant
ε'' dielectric loss factor
μ magnetic permeability (H m^{-1})
σ electrical conductivity (mhos m^{-1})
ρ charge density (C m^{-3})
ρ_m density (kg m^{-3})
$\tan\delta$ loss tangent
α attenuation constant
α_1 thermal diffusivity (m s^{-2})
λ_0 wavelength in free space (12.24 at 2.45 GHz)

References

Ang, T.K., Ford, J.D. and Pei, D.C.T. (1977). Microwave freeze-drying of food: A theoretical investigation. *International Journal of Heat and Mass Transfer*, 20, 517–526.

Ayappa, K.G., Davis, H.T., Crapiste, G., Davis, E.A. and Gordon, J. (1991). Microwave heating: An evaluation of power formulations. *Chemical Engineering Science*, 46(4), 1005–1016.

Azuma, J.I., Tanaka, F. and Koshijima, T. (1984). Enhancement of enzymatic susceptibility of lignocellulosic wastes by microwave irradiation. *Journal of Fermentation Technology*, 62(4), 377–384.

Bingol, G.P.Z., Roberts, J.S., Devres, Y.O. and Balaban, M.O. (2008). Mathematical modelling of microwave-assisted convective heating and drying of grapes. *International Journal of Food Agricultural and Biology Engineering*, 1(2), 46–55.

Birla, S.L., Wang, S. and Tang, J. (2008). Computer simulation of radio frequency heating of model fruit immersed in water. *Journal of Food Engineering*, 84(2), 270–280.

Buffler, C.R. (1993). *Microwave Cooking and Processing: Engineering Fundamentals for a Food Scientist*. Van Nostran Reinhold, New York.

Chen, H. (2008). Simulation model for moving food packages in microwave heating processes using conformal FDTD method. *Journal of Food Engineering*, 88(3), 294–305.

Clary, C.D., Wang, S.J. and Petrucci, V.E. (2005). Fixed and incremental levels of microwave power application on drying grapes under vacuum. *Journal of Food Science*, 70(5), 344–349.

Clemens, J. and Saltiel, C. (1996). Numerical modeling of materials processing in the microwave furnaces. *International Journal of Heat and Mass Transfer*, 39(8), 1665–1675.

Cui, Z.W., Xu, S.Y. and Sun, D.W. (2004). Microwave–vacuum drying kinetics of carrot slices. *Journal of Food Engineering*, 65(2), 157–164.

Dadali, G., Apar, D.K. and Ozbek, B. (2007). Microwave drying kinetics of okra. *Drying Technology*, 25, 917–924.

Datta, A.K. and Ramaswamy, C.A. (2001). *Microwave Technology for Food Applications*, Marcel Dekker, New York.

Dev, S.R.S., Gariepy, Y., Orsat, V. and Raghavan, G.S.V. (2010). FDTC modeling and simulation of microwave heating of in-shell eggs. *Progress in Electromagnetics Research*, 13, 229–243.

Dincov, D.D., Parrott, K.A. and Pericleous, K.A. (2002). Coupled 3-D finite difference time domain and finite volume methods for solving microwave heating in porous media. *Lecture Notes in Computer Science*, 2329(1), 813–822.

Dominguez, A., Menendez, J.A., Inguenzo, M. and Pis, J.J. (2006). Production of bio-fuels by high temperature pyrolysis of sewage sludge using conventional and microwave heating. *Bioresource Technology*, 97, 1185–1193.

Drouzas, A.E. and Schubert, H. (1996). Microwave application in vacuum drying of fruits. *Journal of Food Engineering*, 28, 203–209.

Duan, X., Zhang, M. and Mujumdar, A.S. (2007). Studies on the microwave freeze drying technique and sterilization characteristics of cabbage. *Drying Technology*, 10(25), 1725–1731.

Fu, W.B. and Metaxas, A.C. (1994). A mathematical derivation of power penetration depth for thin lossy materials. *Journal of Microwave Power and Electromagnetic Energy*, 27, 217–222.

Funebo, T. and Ohlsson, T. (1998). Microwave-assisted air dehydration of apple and mushroom. *Journal of Food Engineering*, 38(3), 353–367.

Galema, S.A. (1997). Microwave chemistry. *Chemical Society Reviews*, 29(3), 233–238.

Geedipalli, S.S.R., Rakesh, V. and Datta, A.K. (2007). Modeling the heating uniformity contributed by a rotating turntable in microwave ovens. *Journal of Food Engineering*, 82, 359–368.

Giri, S.K. and Prasad, S. (2007). Drying kinetics and rehydration characteristics of microwave-vacuum and convective hot-air dried mushroom. *Journal of Food Engineering*, 78, 512–521.

Gong, G., Liu, D. and Huang, Y. (2010). Microwave-assisted organic acid pretreatment for enzymatic hydrolysis of rice straw. *Biosystems Engineering*, 107, 67–73.

Hill, J.M. and Marchant, T.R. (1996). Modeling microwave heating. *Applied Mathematical Modeling*, 20, 3–15.

Jolly, P.G. and Turner, I.W. (1990). Non-linear field solutions of one-dimensional microwave heating. *Journal of Microwave Power and Electromagnetic Energy*, 25, 3–15.

Keshwani, D.R. and Cheng, J.J. (2009). Modeling changes in biomass composition during microwave-based alkali pretreatment of switchgrass. *Biotechnology and Bioengineering*, 105(1), 88–97.

Keshwani, D.R., Cheng, J.J., Burns, J.C., Li, L. and Chiang, V. (2007). Microwave pretreatment of switchgrass to enhance enzymatic hydrolysis. 2007 ASABE Annual Meeting Paper No. 077127, Minneapolis, MN.

Klinbun, W., Rattanadecho, P. and Pakdee,W. (2011). Microwave heating of saturated packed bed using a rectangular waveguide (TE10Mode): Influence of particle size, sample dimension, frequency, and placement inside the guide. *International Journal of Heat and Mass Transfer*, 54, 1763–1774.

Krieger-Brockett, B. (1994). Microwave pyrolysis of biomass. *Research of Chemical Intermediates,* 20(1), 39–49.

Kumar, D. (2009). Studies on microwave-convective drying of okra. M. Tech Thesis, Agricultural and Food Engineering Department, Indian Institute of Technology, Kharagpur.

Kumar, D., Prasad, S. and Murthy, G.S. (2011). Optimization of microwave-assisted hot air drying conditions of okra using response surface methodology. *Journal of Food Science and Technology,* DOI: 10.1007/s13197-011-0487-9.

Liao, X. (2002). Dielectric properties and their application microwave-assisted organic chemical reactions. PhD Thesis, Departmental of Agriculture and Biosystem Engineering, McGill University, Macdonald Campus, Quebec, Canada.

Maskan, M. (2001). Kinetics of color change of kiwifruits during hot air and microwave drying. *Journal of Food Engineering,* 48, 169–175.

Menendez, J.A., Inguanzo, M. and Pis, J.J. (2002). Microwave-induced pyrolysis of sewage sludge. *Water Research,* 36, 3261–3264.

Miura, M., Kaga, H., Sakurai, A., Kakuchi, T. and Takahashi, K. (2004). Rapid pyrolysis of wood block by microwave heating. *Journal of Analytical and Applied Pyrolysis,* 71, 187–199.

Monsef-Mirzai, P., Ravindran, M., McWhinnie, W.R. and Burchill, P. (1995). Rapid microwave pyrolysis of coal. *Fuel,* 74(1), 20–27.

Mujumdar, A.S. (2000). *Drying Technology in Agriculture and Food Science.* Science Publishers, Enfield, NH.

Murthy, G.S. (2003). Modelling and simulation of microwave heating of food materials. M. Tech. Thesis, Agricultural and Food Engineering Department, Indian Institute of Technology, Kharagpur, India.

Murthy, G.S. and Prasad, S. (2005). A completely coupled model for microwave heating of foods in microwave oven. ASAE Paper No. 056062. ASAE, St. Joseph, MI.

Nelson, S.O. (1994). Measurement of microwave dielectric properties of particulate materials. *Journal of Food Engineering,* 21, 365–384.

Nelson, S.O. (1999). Dielectric properties measurement techniques and applications. *Transactions of the ASAE,* 42(2), 523–529.

Ni, H., Datta, A.K. and Torrance, K.E. (1999). Moisture transport in intensive microwave heating of biomaterials: A multiphase porous media model. *International Journal of Heat and Mass Transfer,* 42, 1501–1512.

Nyfors, E. and Vainikainen, P. (1989). *Industrial Microwave Sensors,* Artech House, Inc., Norwood, MA.

Ooshima, H., Aso, K., Harano, Y. and Yamamoto, T. (1984). Microwave treatment of cellulosic materials for their enzymatic hydrolysis. *Biotechnology Letters,* 6(5), 289–294.

Prabhanjan, D.G., Ramaswamy, H.S. and Raghavan, G.S.V. (1995). Microwave-assisted convective air drying of thin layer carrots. *Journal of Food Engineering,* 25, 283–293.

Ramaswamy, H. and Tang, J. (2008). Microwave and radio frequency heating. *Food Science and Technology International,* 14, 423–427.

Ramaswamy, S. and Holm, R.A. (1999). High intensity drying. *Drying Technology,* 17(1 & 2), 73–95.

Rowley, A.T. (2001). Radio frequency heating. In: Richardson, P., ed. *Thermal Technologies in Food Processing.* Woodhead Publishing, Cambridge, UK: pp. 163–177.

Ryynänen, S. (1995). The electromagnetic properties of food materials: A review of the basic principles. *Journal of Food Engineering,* 26, 409–429.

Sharma, G.P. (2000). Microwave convective drying of garlic cloves. PhD Thesis, Agricultural and Food Engineering Department, Indian Institute of Technology, Kharagpur, India.

Sharma, G.P. and Prasad, S. (2002). Dielectric properties of garlic (*Allium sativum L.*) at 2450 MHz as function of temperature and moisture content. *Journal of Food Engineering*, 52, 343–348.

Sunderland, J.E. (1980). Microwave freeze-drying. *Journal of Food Process Engineering*, 4, 195–212.

Tang, J., Ikediala, J.N., Wang, S., Hansen, J.D. and Cavalieri, R.P. (2000). High-temperature-short-time thermal quarantine methods. *Postharvest Biology and Technology*, 21,129–145.

Thostenson, E.T. and Chou, T.W. (1999). Microwave processing: Fundamentals and applications. *Composites, Part A: Applied Science and Manufacturing*, 30, 1055–1071.

Tulasidas, T.N. (1994). Combined convective and microwave drying of grapes. PhD Thesis, Department of Agriculture Engineering, McGill University, Macdonald Campus, Quebec, Canada.

Tulasidas, T.N., Raghavan, G.S.V., VandeVoort, F. and Girard, R. (1995). Dielectric properties of grapes and sugar solutions at 2.45 GHz. *Journal of Microwave Power and Electromagnetic Energy*, 30(2), 117–123.

Venkatesh, M.S. and Raghavan, G.S.V. (2004). An overview of microwave processing and dielectric properties of agri-food materials. *Biosystems Engineering*, 88(1), 1–18.

Wan, Y., Chen, P., Zhang, B., Yang, C., Liu, Y., Lin, X. and Ruan, R. (2009). Microwave-assisted pyrolysis of biomass: Catalyst to improve product selectivity. *Journal of Analytical and Applied Pyrolysis*, 86, 161–167.

Wang, J. and Sheng, K. (2006). Far-infrared and microwave drying of peach. *LWT-Food Science and Technology*, 39, 247–255.

Wang, X., Chen, H., Luo, K., Shao, J. and Yang, H. (2008). The influence of microwave drying on biomass pyrolysis. *Energy and Fuels*, 22(1), 67–74.

Wang, Y., Wig, T.D., Tang, J. and Hallberg, L.M. (2003). Dielectric properties of foods relevant to RF and microwave pasteurization and sterilization. *Journal of Food Engineering*, 57, 257–268.

Whitaker, S. (1977). Simultaneous heat, mass and momentum transfer in porous media: A theory of drying. *Advances in Heat Transfer*, 13, 119–203.

Yongsawatdigal, J. and Gunasekaran, S. (1996). Microwave-vacuum drying of cranberries: Part II. Quality evaluation. *Journal of Food Processing and Preservation*, 20(2), 145–156.

Yu, F., Ruan, R., Deng, S., Chen, P. and Lin, X. (2006). Microwave pyrolysis of biomass. 2006 ASABE Annual Meeting Paper No. 066051, Portland, OR.

Yu, F., Ruan, R. and Steele, P. (2009). Microwave pyrolysis of corn stover. *Transactions of ASABE*, 52(5), 1595–1601.

Zante, H.J.V. (1973). *The Microwave Oven*. Houghton Mifflin Company, Boston, USA.

Zhang, M., Tang, J., Mujumdar, A.S. and Wang, S. (2006). Trends in microwave-related drying of fruits and vegetables. *Trends in Food Science and Technology*, 17, 524–534.

Zhao, H. and Turner, I.W. (2000). The use of coupled computational model for studying the microwave heating of wood. *Applied Mathematical Modeling*, 24, 183–197.

Zhu, S., Wu, Y., Yu, Z., Liao, J. and Zhang, Y. (2005a). Pretreatment by microwave/alkali of rice straw and its enzymatic hydrolysis. *Process Biochemistry*, 40, 3082–3086.

Zhu, S., Ziniu, Y., Yuanxin, W., Xia, Z., Hui, L. and Ming, G. (2005b). Enhancing enzymatic hydrolysis of rice straw by microwave pretreatment. *Chemical Engineering Communications*, 192, 1559–1566.

Zhu, S., Wu, Y., Yu, Z., Wang, C., Yu, F., Jin, S., Ding, Y., Chi, R., Liao, J. and Zhang, Y. (2006). Comparison of three microwave/chemical pretreatment processes for enzymatic hydrolysis of rice straw. *Biosystems Engineering*, 93(3), 279–283.

Chapter 2

Emerging Drying Technologies for Agricultural Products

Chung Lim Law, Wan Ramli Wang Daud, and Arun S. Mujumdar

Contents

2.1 Introduction

Food products are diverse in physical, chemical, and bio-chemical properties. A large assortment of dryers has been developed to dehydrate and preserve food products to meet different quality requirements. More than 500 dryer types have been reported in the technical literature, although about 100 types are commercially available. Differences in dryer design are due to different physical attributes of product, different modes of heat input, different operating temperature and pressure, etc. Most conventional dryers use hot air as the drying medium, convection and conduction as the mode of heat transfer, operated at atmospheric pressure, etc.

New developments in dryers and emerging drying technologies can be classified into many categories, which include the following list. However, all new developments and emerging drying technologies must be cost-effective. This is needed to ensure market acceptance.

- New product or process not made or invented heretofore
- Higher capacities than current technology permits
- Better quality and quality control than currently feasible
- Reduced environmental impact
- Safer operation
- Better efficiency (resulting in lower cost)
- Lower cost (overall)
- Shorter processing time

Conventional dryers in the food industry are spray dryers, freeze dryers, vacuum dryers, tray dryers, drum dryers, fixed bed dryers, fluidized bed dryers, filter dryers, etc. Table 2.1 gives the general characteristics of the dryers. Although these dryers are classified as conventional, there are a great number of areas related to these dryers that require further improvement. This leads to evolutionary innovation in drying technology. Mujumdar (2007a) has identified various areas and aspects that need further research and development. The paper also discusses a great deal on some advancements and emerging drying technologies, such as drying of nanomaterials, superheated steam drying, pulse combustion drying, and heat pump assisted drying.

Some of the areas (including limitations) as stated in Table 2.1 have been addressed by researchers, and significant improvements as well as advancements have been made over the past few years, which led to new developments in drying.

Table 2.1 General Characteristics of Some Conventional Dryers and Areas for Further Improvement

Dryer Type	*General Characteristics*	*Areas for Further R&D*
Tray/ Rotary	• Materials are placed on trays and directly make contact with drying medium (typically hot air). • Heat transfer mode is typically convective. Conductive is possible by heating the trays. • A cylindrical drying chamber is rotating while materials are tumbling in the chamber. • Drying medium (typically hot air) is charged into the chamber and makes contact with the materials in cross flow. • Flights are used to agitated the material. • Internal heat exchanger is installed to allow conductive heat transfer.	• Uniformity of air flow profile. • Uniformity of final product quality and moisture content. • Hybrid mode by combining tray dryers with MW drying, etc. • Prediction of particle motion, particle residence time distribution, and uniformity of particle final moisture content. • Effect of poly-dispersity and cohesiveness of solids on drying kinetics and characteristics. • Design of flights, internal heat exchanger. • Effect of solids hold-up and hot air injection on drying kinetics/ characteristics.
Drum	• A drum covered with thin filter medium (typically filter cloth) is partially submerged into suspension. • While rotating, a thin layer of filter cake is formed and the cake is then subjected to drying medium (typically hot air). • Dried thin filter cake is then removed from the drum using a scraper.	• Noncircular shape of drying chamber. • Heat transfer to thin film of suspensions, including effects of crystallization, boiling, etc. • Enhancement of drying rate by radiant heat or jet impingement.

(continued)

Table 2.1 (Continued) General Characteristics of Some Conventional Dryers and Areas for Further Improvement

Dryer Type	*General Characteristics*	*Areas for Further R&D*
Flash	• Flash dryer is used to remove surface moisture, especially rigid and nonporous media. • Materials are charged into a fast moving drying medium stream, drying occurs while the drying medium carries the materials. • Cyclone is normally used to separate the drying medium and the material.	• Modeling of particle motion including effects of agglomeration, attrition, and geometry of dryer. • Use of pulse combustion exhaust, superheated steam, internal heat exchangers, variable cross section ducts, hot air injection. • Computational fluid dynamic simulation.
Spray	• Atomizer mounted on top of a drying chamber sprays liquid/suspension and forms droplets. • Drying medium (typically hot air) is supplied into the chamber concurrently or counter-currently. • Hot air exits the chamber at the chamber outlet and carries dried powder. • Separation of hot air and powder takes place in cyclone.	• Effects of types of atomizer on droplet flow pattern, product properties, agglomeration, size reduction. • Effect of chamber geometry. • Injection of supplementary air. • Use of superheated steam. • Computational fluid dynamic simulation on various dryer designs and types.
Fixed bed	• Materials are placed in a drying column and the materials form a bed of particulate solids. • Drying medium (typically hot air) is introduced into the column from the bottom and makes contact with the material in cross flow.	• Uniformity of air flow across the bed of particles. • Uniformity of product quality and final moisture content. • Design of internal heat exchanger, agitator, etc.

(continued)

Table 2.1 (Continued) General Characteristics of Some Conventional Dryers and Areas for Further Improvement

Dryer Type	*General Characteristics*	*Areas for Further R&D*
Fluidized bed	• Similar to fixed bed dryer but operating hot air velocity is higher to ensure the particles are suspended in the air stream. • Large contacting surface areas between the drying medium and the material compared with fixed bed dryer. • Conventional fluidized bed is not suitable for drying fine powders (due to channeling and slugging) and coarse particles (due to formation of big bubbles). • However, modified FBD, such as vibrating FBD, agitating FBD, etc., can be used to dry difficult-to-fluidize particles. • If the materials are polydispersed, the hot air stream may carry over some fine particles. • A cyclone is used to separate the fine particles from the gas stream.	• Effect of particle moisture content/poly-dispersity on fluidization hydrodynamics, agglomeration, heat, and mass transfer. • Effect of agitation, vibration, pulsation, acoustic, radiation on drying kinetics and characteristics. • Design of internal heat exchangers. • Classification of particle type based on fluidization quality at varying particle moisture content and stickiness. • Mathematical modeling on fluidization hydrodynamics, heat, and mass transfer by taking into account agitation, vibration, pulsation, internal heat exchanger, varying particle moisture content, etc.

(continued)

Table 2.1 (Continued) General Characteristics of Some Conventional Dryers and Areas for Further Improvement

Dryer Type	*General Characteristics*	*Areas for Further R&D*
Vacuum	• Need to maintain high vacuum. • Drying chamber is operated at reduced pressure or vacuum. • Boiling point of water/solvent is reduced, thus reducing the operating temperature. • However, absence of drying medium in the vacuum drying chamber disables convective heat transfer but enhances mass transfer at low temperatures.	• Combined mode of heat transfer, e.g., MW vacuum drying. • Hybrid drying, e.g., vacuum superheated steam drying, etc. • Use of internal heating media. • Enhancement in drying kinetics by incorporating radiant heat input, internal heating media, etc.
Freeze	• Vacuum freeze drying is expensive in terms of capital costs and operating costs due to very low vacuum required at very low temperature. • Drying times are long; most operated batch-wise. • Suitable only for very high value products like pharmaceutical products.	• Use of magnetic/electric/acoustic fields to control nucleation and crystal size of ice during freezing; permits better quality product. • Effects of intermittent/cyclic/variable heat inputs and variable operating profiles on drying kinetics and characteristics as well as product quality.
Batch dryer	• Not all dryers can operate in batch mode. • Good for low capacity needs. • Tray, rotary, drum, fixed bed, fluidized bed vacuum dryers, etc., can be operated batch-wise.	• Use of heat pump including chemical heat pump. • Reduction in labor costs.

Source: Modified from Mujumdar, A.S., An overview of innovation in industrial drying: Current status and R&D needs. *Transport Porous Media*, 66, 3–18. 2007a.

2.2 New Developments in Drying

The past few decades have seen many new developments in dryers to overcome operational difficulties and improve product quality. Recently developed dryers have some general attributes:

- Multimode heat input concurrently or sequentially to match drying kinetics without adverse effect on quality, e.g., convection followed by or simultaneously with conduction, radiation, or microwave heat input
- Time-dependent heat input for batch drying to match drying kinetics with heat input
- Superheated steam as drying medium at high, atmospheric, or sub-atmospheric pressure
- Low-temperature dehumidified air as drying medium at modified atmosphere, which eliminates the existence of oxygen
- New gas-particle contactors used as dryers, e.g., impinging streams (confined opposing jets, more than 20 variants reported)
- Multistage drying, e.g., spray drying followed by fluid bed/vibrated bed as second and/or third stage

New techniques have been developed based on different aspects of drying, e.g., how drying materials are handled in the dryers, how effective is the contact between drying materials and drying medium, type of drying medium, atmosphere of drying, mode of heat input, energy sources, mode of combustion, mode of heat transfer, number of stages, control of dryer, operating profiles, etc. Table 2.2 lists some of the new developments with reference to the drying aspects mentioned above.

Some of the new designs of dryers are the result of combination of the new developments mentioned in Table 2.2. For instance, microwave-assisted vacuum dryer, heat pump-assisted fluidized bed dryer, spray-fluidized bed dryer cooler, etc. Table 2.3 shows the combinations of new developments that led to a new design of dryers given in the examples stated above. The list of new dryer designs in Table 2.3 can be much longer if one combines the ideas/new developments given in Table 2.2.

Table 2.4 shows several new developments with reference to the feed type. Food products have different physical forms. Conventional dryers are only suitable for particulate products except freeze dryers and spray dryers for liquid and suspension. With the new developments listed in Table 2.4, new types of dryers can also be used to dry nonparticulate feed such as liquid suspension, paste, sludge, and continuous sheet type of material.

Table 2.2 New Developments with Reference to Different Drying Aspects

Drying Aspects	*Conventional*	*New Developments*
Mode of mass transfer	Internal moisture diffusion Convective transfer on surface	Edible solids—sorption medium
Mode of heat transfer	Convective/conductive/radiant/volumetric using microwave or RF	Volume heating—electromagnetic waves Inert solids conduction medium
Number of modes of heat transfer	Single	Combine mode/hybrid
Mode of heat input	Steady	Intermittent/multivariable
Number of modes of heat input	Single	Combine mode
Mode of gas flow	Constant	Variable/cyclic/periodic
Temperature profile	Constant	Cyclic/periodic/on-off/step-down
Pressure	Atmospheric	Vacuum/reduced pressure/cyclic
Number of stages	Single	Multistage
Number of types	Single	Multistage with different dryer types
Drying medium	Air/combustion gas	Superheated steam Low temperature dehumidified air
Contact between drying medium and material	Direct/indirect (slow) Co-current/counter current /mixed flow	Electromagnetic field High impact impingement
Pre-treatment method	Chemical	Chemical-mechanical
Energy source	Fossil fuel, biomass, waste oil, sun, electricity, waste heat recovery	Phase change materials Renewable energy sources when fossil fuel becomes expensive

(continued)

Table 2.2 (Continued) New Developments with Reference to Different Drying Aspects

Drying Aspects	*Conventional*	*New Developments*
Combustion of fossil fuel	Direct firing	Pulse combustion
Combine with other unit operation	No	Filtration/cooling/ classification
Drying medium	Hot air Flue gases	Superheated steam Low temperature dehumidified air Hot air + superheated steam mixture Hot air + cooling air
Number of stages	One (common) Two/three (same dryer type)	Multistage with different dryer types
Dryer control	Manual Automatic	Fuzzy logic Model-based control Artificial neural networks
Handling of material	Static Layered Fixed bed Packed bed Sprayed	Fluidized bed Spouted bed Jetting Vibrated Agitated
Atmosphere	Atmospheric air	Modified atmosphere with nitrogen/CO_2 Critical fluid drying Vacuum drying

Source: Modified from Mujumdar, A.S., Classification and selection of industrial dryers. In S. Devahastin (ed.), *Mujumdar's practical guide to industrial drying* (pp. 23–36). Exergex Corp., Quebec, 2000; Mujumdar, A.S., Classification and selection of industrial dryers. In A.S. Mujumdar (ed.), *Practical guide to industrial drying* (pp. 23–35). Color Publications Pvt. Ltd., Mumbai, 2004; Mujumdar, A.S., Principles, classification and selection of dyers. In A.S. Mujumbar (ed.), *Handbook to industrial dying* (pp. 3–32). CRC Press, New York, 2007b.

Table 2.3 Combination of New Ideas and New Developments of Selected New Dryer Designs

New Dryers	*Combination of New Ideas and New Developments*
MW-assisted vacuum dryer	• Vacuum to reduce operating temperature • MW to enable volume heating
Heat pump-assisted fluidized bed dryer	• Heat pump to generate low temperature drying medium, suitable for drying heat-sensitive products • Fluidized bed to increase the contacting efficiency
Spray-fluidized bed dryer/cooler	• Spray dryer to form liquid droplets and powders after being dried by drying medium • Fluidized bed dryer to remove internal moisture • Fluidized bed cooler to remove heat and cool the material to avoid condensation during packaging

Table 2.4 Selected New Developments with Reference to Feed Type

Particulars	*Dryer Type*	*New Developments*
Liquid suspension	Drum Spray	Spray dryer using inert particle in a fluid/spout/vibrated bed Spray-fluid bed dryer Vacuum belt dryers Pulse combustion dryers
Paste/sludge	Spray Drum Paddle	Spray dryer using inert particle in a fluid/spout/vibrated bed Fluid bed with solids back mixing Superheated steam fluid bed dryer with inert solids

(continued)

Table 2.4 (Continued) Selected New Developments with Reference to Feed Type

Particulars	*Dryer Type*	*New Developments*
Particle	Tray Rotary Flash Fluidized bed Conveyor Freeze Vacuum	Superheated steam FBD Vibrated bed (variable frequency/amplitude) Ring dryer Pulsated fluid bed Jet-zone dryer Impinging streams Yamato rotary dryer
Continuous sheet	Multicylinder contact dryers; impingement dryers	Combined impingement/radiation dryers Combined impingement and through dryers Impingement jets and MW/RF/radiation dryers

Source: Modified from Mujumdar, A.S., Classification and selection of industrial dryers. In S. Devahastin (ed.), *Mujumdar's practical guide to industrial drying* (pp. 23–36). Exergex Corp., Quebec, 2000; Mujumdar, A.S., Classification and selection of industrial dryers. In A.S. Mujumdar (ed.), *Practical guide to industrial drying* (pp. 23–35). Color Publications Pvt. Ltd., Mumbai, 2004; Mujumdar, A.S., Principles, classification and selection of dryers. In A.S. Mujumdar (ed.), *Handbook of industrial drying* (pp. 3–32). CRC Press, New York, 2007b.

2.3 Emerging Drying Technologies

Numerous new conceptual designs for dryers have been proposed and tested at laboratory, pilot, or even full scale. Due to the inherent risk involved in introducing radically different new technologies, it is not surprising that most of the adopted new technologies are incremental variants of established technologies. Often they are hybrids of known technologies, e.g., spray followed by fluid or vibro-fluid bed dryer, a fluid bed dryer followed by a packed (column) bed dryer, etc. Because most food products are heat sensitive, drying must be carried out without exceeding the damage temperature of the product. Low-humidity air drying, using a heat pump to dehumidify and heat the drying air, is one way to accomplish high drying rates at lower temperatures. Vacuum drying gives a higher quality product, which retains form, nutritional value, color, etc. In this section, we discuss some of the more common emerging drying technologies. Most of them have been commercialized but are still not as widespread as they should be.

It must be pointed out that use of renewable energies, e.g., solar and wind, should be looked at seriously as concerns about energy shortage and global climate change will likely result in legislative actions minimizing fossil fuel usage. A solar drying system, particularly for agro-products and marine products is viable already. In the future, larger systems could be designed utilizing solar thermal, photovoltaic panels combined with wind power. To minimize use of oil or gas, one could use biomass to provide back-up heating in the absence of insulation and wind. Furthermore, use of thermal energy storage in water pools, pebble beds, and/or in phase change materials will make use of intermittent energy sources, such as solar and wind energy, commercially viable.

2.3.1 Intermittent Drying

By varying airflow rate, temperature, humidity, or operating pressure, individually or in tandem, the operating condition of a drying process can be monitored in order to reduce the operating cost, e.g., thermal input and power input. Figure 2.1 shows various possible variations of operating conditions. This technique is known as intermittent drying. Various modes are possible, namely stepwise with equal intervals or with varying intervals, step-up and step-down profiles. The idea is to obtain high energy efficiency without subjecting the product beyond its permissible temperature limit and stress limit while maintaining a high moisture removal rate. In addition, variable intermittency profiles, such as stepwise change of operating conditions, can be applied to minimize energy loss. One can also vary the mode of heat input (e.g., convection, conduction, radiation, or microwave/radio frequency heating). Multiple heat inputs can be used to remove both surface and internal moisture simultaneously.

All drying processes undergo a falling rate at which the drying rate is decreasing with reference to drying time. As such, enhancing the external driving force or drying potential at the latter stage of drying does not enhance the moisture removal rate. This is due to the fact that the removal of moisture at this stage is governed by internal diffusion. Hence, gentle drying should be applied at the latter stage of the falling rate period. One of the strategies is to reduce the operating temperature or flow rate of the drying medium. In this regard, a step-down profile of temperature or flow rate of drying medium can be applied.

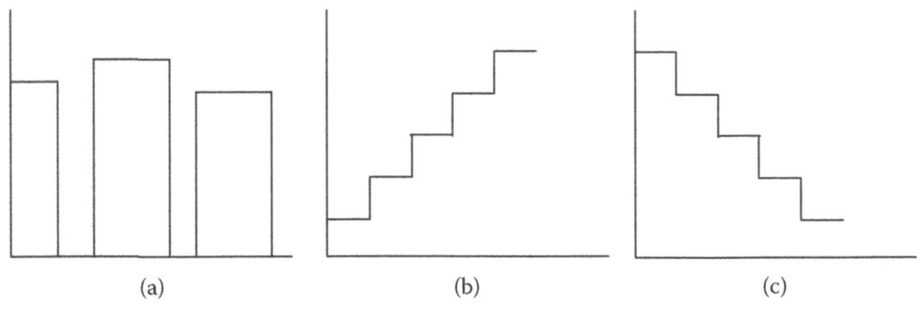

Figure 2.1 Intermittent mode, namely (a) step-wise, (b) step-up, and (c) step-down.

Some intermittent dryings have distinctive active and non-active drying. For instances, when heat input or the flow rate of processing air is stopped within a period of time, this period is known as non-active drying or tempering. During each tempering period, redistribution of internal moisture within the drying material occurs, and the moisture gradient within the material is reduced. Reduction of the moisture gradients has a beneficial effect on the mechanical property of the drying material because this prevents crack formation. Further, migration of internal moisture from the interior to the surface of the drying material during the non-active drying period allows the drying rate in the following active drying to be increased significantly.

Figure 2.2 shows the typical drying kinetics and drying curves for intermittent drying. Figure 2.2a shows that average moisture content reduces during active drying whereas it remains unchanged during the tempering period. However, during the tempering period, surface moisture content increases due to migration of internal moisture from the interior to the surface of the drying material. The availability of more surface moisture enhances the drying rate of the following active drying. Figure 2.2b shows that drying rates at the commencement of the active drying after the respective tempering period are increased noticeably. The phenomenon in which surface moisture is higher at the initial stage of each active drying period has been confirmed by nuclear magnetic resonance imaging. Xing et al. (2007) found that a wet layer near the pasta surface was observed after each tempering period. On the other hand, a dry layer was observed throughout continuous drying.

Chua et al. (2003) gave a comprehensive overview of intermittent drying for bio-products, and Law et al. (2008) discussed the latest advancements in intermittent drying for postharvest of bio-origin and agricultural products. Some recent research findings on intermittent drying of food products are summarized in Table 2.5. It is noteworthy that various modes of intermittency are possible. The modes of intermittency that one may consider includes cyclic temperature, stepwise temperature, cyclic pressure, and cyclic flow rate.

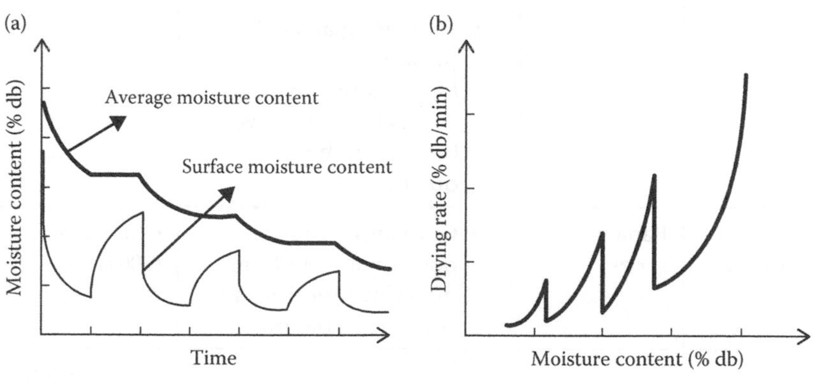

Figure 2.2 Typical drying kinetics and drying curve of intermittent drying.

Table 2.5 Research Findings of Intermittent Drying of Food Products Using Different Modes of Intermittency

Mode of Intermittency	*Product*	*Findings*	*References*
Cyclic temperature	Guava pieces	• Minimize degradation of ascorbic acid content.	Chua et al. (2000a)
Cyclic temperature	Pasta	• Longer tempering period at higher temperature was needed to ensure better product quality of pasta.	Xing et al. (2007)
Cyclic temperature	Solid byproducts from olive oil extraction	• Longer tempering period enables internal moisture to redistribute and level between the interior and surface, thus leading to higher removal of moisture.	Jumah et al. (2007)
Cyclic temperature	Rough rice	• Tempering at 45°C (tested within the range from 35°C to 55°C) for 2 h gave better milling recovery, head rice ratio, grain hardness, germination ratio, degree of whiteness, and cracked grains ratio. • Tempering at 60°C (tested at three temperatures 20°C, 50°C, and 60°C) gave better rough rice quality.	Madamba and Yabes (2005) Aquerreta et al. (2007)
Cyclic temperature in HPD	Banana slices	• Gave improved performance in terms of drying kinetics and color degradation.	Chua et al. (2001)

(continued)

Table 2.5 (Continued) Research Findings of Intermittent Drying of Food Products Using Different Modes of Intermittency

Mode of Intermittency	*Product*	*Findings*	*References*
Step-down temperature	Banana slices	• Reduced color degradation and effective drying time. • Induced higher drying rate.	Chua et al. (2001)
Tempering	Rough rice	• Reduced the number of fissured kernels. • Reduced effective drying time and increased drying rate.	Nguyen and Kunze (1984); Li et al. (1999); Iguaz et al. (2006) Iguaz et al. (2006); Cihan and Ece (2001)
Tempering in VFBD	Carrot cubes	• Reduced degradation of β-carotene and shortened effective drying time, while maintaining rehydration capacity.	Pan et al. (1999)
Pulsating in batch rotating SBD	Bio-material	• Gave better energy performance and produces better product quality.	Bon and Kudra (2007)
Intermittency of heat mode	Simulation	• Radiation-coupled convection gave higher drying rate and lower energy consumption.	Sun et al. (2005)
Intermittent infrared and continuous convection heating	Osmotically pretreated potato, carrot, and banana samples	• Produced premium quality dried products as compared with conventional drying.	Chua at el. (2004)

(continued)

Table 2.5 (Continued) Research Findings of Intermittent Drying of Food Products Using Different Modes of Intermittency

Mode of Intermittency	*Product*	*Findings*	*References*
Cyclic low pressure SS	Banana chips	• Shortened effective drying time. • Intermittent temperature gave higher retention of ascorbic acid. • Intermittent pressure led to greater degradation of ascorbic acid.	Thomkapanich et al. (2007)

Note: SBD: spouted bed dryer; SS: superheated steam; VFBD: vibrating fluidized bed dryer.

Intermittency can be applied to any direct dryer and batch dryer, such as a tray dryer, convective dryer, conveyor dryer, fluidized bed dryer, spouted bed dryer, etc. Many industrial dryers actually operate in the intermittent mode although they are not so labeled, for example, circulating bed dryers, spouted bed dryers, jet fluidized bed dryers, rotating fluidized bed dryers, etc. These dryers have two distinctive zones; one zone receives the drying medium and the other does not. Active drying occurs in the drying medium receiving zone whereas tempering occurs in the other zone.

Mujumdar (1991) identified and proposed for the first time the use of multiple modes of variable levels of heat input, simultaneous or consecutive, as well as cyclical variations in velocity or operating pressure as technologies of the future for batch and continuous heat pump drying processes. Using multiple modes of heat input, it is possible to speed up drying kinetics without adversely affecting the quality of dried products.

Experiments and mathematical modeling results have shown that intermittent application of heat is a good drying strategy for dehydrating heat-sensitive materials (Sun et al. 2005). This is due to the fact that internal moisture movement through the material is governed by diffusion, which cannot be enhanced by improving the contacting efficiency between the solid and gas phases. A period of tempering allows more moisture to travel from the interior to the surface where it can be carried away by the gas phase. In view of the range of possible parameter variations, including cycle time, a mathematical model is needed to develop optimal drying parameters for intermittent drying.

2.3.2 Heat Pump Drying

Low-temperature dehumidified air can be used to replace hot air in direct dryers. Although the temperature of this type of processing air is relatively lower than hot air, its relative humidity is normally low, e.g., 20% Rh at 20°C. This drying system incorporates a dehumidification cycle, in which condensation of dew allows the removal of water from the closed system of drying air circulation. Normally a heat pump system is used to perform the condensation and heating of the dehumidified air. The heat pump recovers the sensible and latent heats by condensing moisture from the drying air. Consequently, partial vapor pressure in the drying air decreases, which increases the driving potential of the drying air. The recovered heat is recycled back to the dryer through heating of the dehumidified drying air. Desiccant systems can also be used to dehumidify the air, but they tend to be expensive and, hence, are rarely used in drying applications.

Heat pump drying has been developed to accelerate the drying process at higher energy efficiency and preserve the quality of products that are sensitive to heat as well as to conserve overall energy. However, installation of a heat pump dryer normally requires a high capital cost, and its maintenance cost is high due to the need to maintain the compressor and refrigerant filters and refilling of refrigerant owing to leakage. Many advantages as well as limitations are associated with heat pump-assisted dryers. Some of these can be offset by incorporating heat pump drying with other types of drying methods in hybrid drying technologies. Table 2.6 shows the summary of the research findings on the advantages of heat pump drying over other conventional drying techniques.

To avoid oxidation of drying material, the drying air, which contains 21% oxygen, can be replaced with nitrogen or carbon dioxide. By eliminating the existence of oxygen, oxidation, and some undesirable reactions, which require oxygen, are thus avoided. This, in turn, reduces browning of products and improves the retention of bio-active ingredients. This strategy can also be applied to a heat pump, and it is known as modified atmosphere heat pump drying. In addition, researchers have also found that modified atmosphere heat pump drying increases the effective diffusivities of some food products. Table 2.7 shows some of the recent research findings on the modified atmosphere heat pump drying. Common inert gases used in these research works are carbon dioxide and nitrogen. Judging from the research findings given in Table 2.7, it reveals the great potential of modified atmospheric heat pump drying in the food drying industry.

Many conventional dryers, for example, tunnel, fluid bed, spouted bed, rotary, etc., that use convection as the primary mode of heat input can be modified and changed to a heat pump-assisted dryer. A generalized classification scheme for heat pump dryers has been reported by Kiang and Jon (2007). Numerous studies have been published on experimental studies and numerical simulations to predict the effects of using a heat pump in fluidized bed dryers (Alves-Filho and Strommen 1996; Strommen et al. 1999; Ogura et al. 2002, 2003), infrared dryers, microwave dryers, chemical heat

Table 2.6 Research Findings of Advantages of Heat Pump Drying

Base Case	*Findings*	*References*
Drying of onion in conventional dryer	• Onion slices dried in a heat pump dryer (HPD) led to energy savings on the order of 30% with better product quality.	Rossi et al. (1992)
Freeze drying of lactic acid bacteria	• Heat pump drying is more economical in capital and running cost.	Cardona et al. (2002)
Freeze drying of absorbable gelatin sponge	• The installation and running costs are 10 times and 4.5 times less compared with a freeze dryer. • The quality of heat pump-dried sponge is inferior in terms of reduced water intake capacity and conformability.	Waje et al. (2007)
Hot air drying of sapota pulp	• Drying time of heat pump drying is shorter than hot air drying. • Heat pump drying at higher temperatures decreased the sugar content and rehydration ratio of sapota powder.	Jangam et al. (2008)
Drying of Australian nectarines slices in cabinet and tunnel dryers	• Heat pump drying gave better product quality in terms of preserving the lactone and terpenoid contents.	Sunthonvit et al. (2007)
Drying of wood using conventional and combined conventional dehumidification drying	• Heat pump drying has lower energy consumption but longer drying time.	Zhang et al. (2007)
Heat pump drying of banana, guava, and potato	• Intermittent heat pump drying gave improvement in drying kinetics and color change.	Chua et al. (2000b)
Hot air drying of pine cones and pine pollen catkins	• Significant improvement in product quality and seed germination rate.	Chen et al. (2002)

Table 2.7 Research Findings of Modified Atmosphere Heat Pump Drying of Food Products

Atmosphere	*Product*	*Findings*	*References*
N_2, CO_2	Ginger	• Modified atmosphere HPD improved the drying characteristics as well as the retention of 6-gingerol. • The effective diffusivity was increased, resulting in better retention of flavor, even better than freeze drying.	Hawlader et al. (2006a)
Inert gas	Apple cubes	• Modified HPD drying resulted in more porous products and thus better rehydration.	Neill et al. (1998)
N_2	Apples	• Modified HPD produced excellent color and good retention of vitamin C.	Perera (2001)
N_2, CO_2	Apple, guava, and potato	• Modified HPD gave improvement of dried product quality not only in color but also in rehydration ability.	Hawlader et al. (2006b)
N_2, CO_2	Guava, papaya	• Effective diffusivity was 44% higher in guava and 16% higher in papaya compared to HPD, vacuum drying, and freeze drying. • Less browning, faster rehydration, and more vitamin C retention in the final products.	Hawlader et al. (2006c)

pump-assisted dryers (Alves-Filho 2002), spray-freeze dryers (Stapley and Rielly 2007), etc. Table 2.8 summarizes the research findings of some heat pump-assisted dryers.

Mujumdar (1991) proposed applying an intermittent drying strategy to heat pump drying some 20 years ago to save operating costs. This strategy is currently under active research. Furthermore, it is possible to use a smaller heat pump to service two or more drying chambers in cyclical mode, which may dry the same or different products (Mujumdar 2001). Chua et al. (2002a) have presented the

Table 2.8 Research Findings of Heat Pump-Assisted Drying

Hybrid Mode	*Product*	*Findings*	*References*
Subatmospheric spray-freeze dryer	Pharmaceuticals and biologicals and food products	• The authors claim that it offers scope for significant heat integration and thus energy saving by using heat from the condenser to supply the re-heater via a heat pump.	Stapley and Rielly (2007)
Fluidized bed	Mixture of puree/jam and inert agent	• Using CO_2 as the working fluid, the author claims that it requires considerably less energy for clean operation and produced high-quality dried products.	Alves-Filho (2002)
Tray dryer	Olive leaves	• High exergetic efficiency was obtained. • Gave good retention of total phenolic content and total antioxidant activity.	Erbay and Icier (2009)

effect of different temperature-time profiles on the quality of agricultural products in a tunnel HPD. Various configurations of a heat pump dryer are possible, for instance, single-stage heat pump for one-stage drying, single-stage heat pump for two-stage drying, and two-stage heat pump for two-stage drying. Li et al. (2009) compared the performance of various heat pump dryers using the configurations mentioned above and concluded that two-stage heat pump drying is better than one-stage. The performance of the two-stage heat pump for two-stage drying was satisfactory; however, it is not suitable for operation at high temperatures.

Review of progress made in heat pump drying has been presented by Chua et al. (2002b), which includes its operating and working principles, advantages and limitations, and opportunities for further R&D to achieve better product quality and energy efficiency. Klocker et al. (2002) discussed the design and construction of a batch type cabinet HPD using carbon dioxide as the working fluid.

Table 2.9 Research Findings of Advantages of Superheated Steam Drying

Base Case	*Findings*	*References*
No base case (drying of soybeans)	• Increasing the temperature in superheated steam drying of soybeans led to higher urease inactivation and brown pigment formation. • Increasing temperature as well as moisture content led to higher diffusion of moisture within the beans.	Prachayawarakorn et al. (2004)
Hot air drying of Asian noodle	• Superheated steam drying resulted in better product quality in terms of recovery, adhesiveness, and gumminess. • Superheated steam drying gave higher levels of starch gelatinization and retention of fiber content of spent grains. • Drying rates of superheated steam drying of sugar beet pulp was greater.	Pronyk et al. (2004)
Hot air drying of durian chips	• Superheated steam drying provided products with greater values of redness, yellowness, and rehydration capacity but lower values of lightness.	Jamradloedluk et al. (2007)
No base case drying of contaminated food products	• Superheated steam drying is beneficial for reducing the contamination of foods.	Cenkowski et al. (2007)
Hot air drying of basil leaves	• Low-pressure superheated steam drying gave better retention of aromatic compounds.	Barbieri et al. (2004)

2.3.3 Superheated Steam Dryers

Superheated steam can be used to replace heated air in almost all direct dryers. As it does not contain oxygen, oxidative or combustion reactions can be avoided. In addition, it also eliminates the risk of fire and explosion hazards. Because oxidative and combustion reactions are avoided, the quality of superheated, steam-dried

Table 2.10 Research Findings on Hybrid Superheated Steam Drying

Hybrid Mode	*Product*	*Findings*	*References*
Microwave and superheated steam drying	Sintered glass beads	• Higher kinetics than microwave nitrogen drying.	Shibata and Ide (2001)
Fluidization and superheated steam drying	Glass beads	• Fluidization enhanced the drying kinetics of porous material.	Tatemoto et al. (2007)

Table 2.11 Research Findings of Vacuum/Reduced Pressure Superheated Steam Drying

Product	*Findings*	*References*
Porous media immersed in glass beads	• Less drying time due to glass beads to transfer heat conductively.	Tatemoto et al. (2007)
Carrot cubes	• Better product quality in terms of color retention and rehydration ratio than vacuum drying. • However, the reduced pressure superheated steam drying gave poorer convective heat transfer and thus longer drying time.	Devahastin et al. (2004)
Carrot	• Better retention of β-carotene than hot air drying.	Suvarnakuta et al. (2007)

products tends to be better than the product quality when dried by conventional hot air. Superheated steam also allows pasteurization, sterilization, and deodorization of food products. This is particularly important for food and pharmaceutical products that require a high standard of hygienic processing. In addition, superheated steam drying also gives higher drying rates in both constant and falling rate periods. This, in turn, reduces the dryer size. Closed system superheated steam drying enables emitted odors, dust, or other hazardous components to be contained and thus mitigate the risks of these hazards. The pollutants are concentrated in the condensate of the effluent steam. On the other hand, desirable organic compounds can also be contained using the superheated steam drying method (Table 2.9).

Like many other hot air dryers, the performance of superheated steam drying can be enhanced by combining it with other drying techniques, such as microwave and fluidized bed drying. The hybrid mode allows different modes of heat transfer or heat input to be carried out in the same dryer. Furthermore, different ways of handling drying materials or use of an alternative drying medium can be achieved in a hybrid dryer. Table 2.10 gives the research findings reported by researchers on various hybrid modes of superheated steam drying.

Despite its many advantages, several limitations are present as well. Heat-sensitive materials are damaged at the saturation temperature of superheated steam corresponding to atmospheric or higher pressures. Thus, a reduced pressure superheated steam drying strategy was proposed by some researchers in order to reduce the damage or denaturation of heat-sensitive ingredients contained in the drying materials. Low-pressure superheated steam is, however, a poor heat transfer medium for convection, and hence, the drying rates are necessarily low. Use of supplementary heating by radiation or microwaves may be worthwhile to be considered to enhance the drying rate. Table 2.11 gives the research findings reported in the literature on vacuum or reduced pressure of superheated steam drying.

2.3.4 Osmotic Dehydration and Pretreatment

Thermal drying is an energy-extensive operation because it involves the phase transition of water. Osmotic dehydration may be applied to partially remove moisture of materials before the drying material is subjected to thermal drying. Mechanical pretreatments coupled with chemical pretreatment can also be applied prior to or during thermal drying to enhance the efficiency of moisture transport.

Mass transfer of solvent, typically diffusion, depends on various factors, such as operating conditions (pressure, concentration of the osmotic medium, treatment time, size and geometry, specific surface area of the material, and temperature), mode of phase contacting (solid-liquid phases), sample-to-solution ratio, composition of the solute, and the agitation level of the solution (Raoult-Walk 1994). As the mass transfer of solvent is rather slow, pretreatments are required to enhance the rate of diffusion. Pretreatment methods include freezing/thawing, vacuum treatment, exposure to ultra-high hydrostatic pressure, high-intensity electrical field pulses, microwave, ultrasound, application of centrifugal force, use of supercritical carbon dioxide, coating of edible layer, etc. and have been proven to enhance mass transfer of solvent for osmotic dehydration. Table 2.12 lists various methods of pretreatment for osmotic dehydration as well as the recent research findings.

The choice of osmotic solute controls the ratio of water loss to solute impregnation. Sugars (for fruits) and inorganic salts (for vegetables) are the most common solutes used in osmotic dehydration. Combination of both solutes, in ternary (sucrose-salt-water) solutions, can reduce solute impregnation and increase water loss (Sereno et al. 2001; Pereira et al. 2006; Medina-Vivanco et al. 2002).

Solute choice and concentration depend on several factors, namely the effect on sensorial properties, solute solubility, cell membrane permeability, its stabilizing effect, and cost (Qi et al. 1998). The common osmotic agents for dehydration of vegetables and fruits are table salt, sucrose, glucose, fructose, starch, corn syrup, glycerol, and plant hydrogenation colloid. Different osmotic agents behaves differently. Generally, a

Table 2.12 Application of Various Osmotic Pretreatment Methods on Different Products and Research Findings

Osmotic Pretreatment	*Product*	*Findings*	*References*
Freezing/ thawing	Pumpkin	• Did not affect the rate of water loss but greatly increases the rate of sugar gain.	Fito (1994)
Vacuum treatment	Apricot, strawberry, pineapple, porous materials	• Water transfer increased.	Shi et al. (1995); Lazarides and Mavroudis (1995)
Exposure to ultra-high hydrostatic pressure	Potato	• Moisture transfer and solid transfer increased owing to ultra-high hydrostatic damage to the cell structure, which, in turn, increases the permeability of the cell wall.	Rastogi et al. (2000, 2002)
Exposure to high-intensity electrical field pulses	Carrot Apple Pepper Mango	• Permeability of cell membrane was increased and resulted in higher water loss and lower sugar uptake.	Rastogi et al. (1999, 2002); Taiwo et al. (2003); Tedjo et al. (2002); Amami et al. (2005)
Exposure to microwave field	Apple	• Moisture loss was increased, and uptake of solids was restricted.	Li and Ramaswamy (2003)
Exposure to ultrasound	Apple Melon	• Ultrasound gives a rapid series of alternative compressions and expansions, which may increase the water effective diffusivity of the fruits, leading to faster air drying.	Simal et al. (1998); Rodrigues and Fernandes (2007); Duan et al. (2008); Xu et al. (2009); Rodrigues et al. (2009)
Application of centrifugal force	Potato Apple	• Slightly increases the water loss and lowers the solids uptake	Azuara et al. (1996)

(continued)

Table 2.12 (Continued) Application of Various Osmotic Pretreatment Methods on Different Products and Research Findings

Osmotic Pretreatment	*Product*	*Findings*	*References*
Use of supercritical CO_2 prior to osmotic dehydration	Mango	• Enhances mass transfer during the subsequent thermal drying.	Tedjo et al. (2002)
Coating of edible layer on the material surface	Strawberry Apple	• Limits solute uptake and increases dehydration efficiency.	Ishikawa and Nara (1993); Matuska et al. (2006); Emam-Djomeh et al. (2006)

low molecular weight osmotic agent performs better than an agent with high molecular weight. Mixtures of low concentrations of sucrose and additive salt are usually recommended judging from the point of view of health. Table 2.13 lists types of solute that can be used in osmotic dehydration as well as some recent findings.

Apart from using osmotic dehydration to perform pretreatment to partially remove moisture prior to thermal drying, non-chemical pretreatment methods can be employed as well to achieve this objective. The non-chemical pretreatment methods include blanching using hot water, chilling, freezing, abrasion, and drilling holes on the surface. Table 2.14 lists the methods as well as some recent research findings on this topic.

Research findings given in Table 2.14 show that drying performance of materials that possess waxy skin can be improved significantly by combining blanching and perforations. As the rate of water removal is improved, this contributes to the retention of color of some food material as the drying time is shortened.

2.3.5 Microwave Vacuum Drying

Microwave drying offers advantages in drying kinetics, energy saving, precise process control, fast start-up and shut-down times (Schiffmann 2007), cost of operation, quality of dried product (Gunasekaran 1999), compactness of microwave applicators (Puschner 2005), retarding microbial growth (Yaghmaee and Durance 2007), etc. In spite of the widespread use of domestic microwave ovens in household and offices, microwaves remain little used in industrial drying especially in the food industry. This is for various reasons, such as high start-up costs and requires sophisticated mechanical and electronic components (Erle 2005; Zhang et al. 2006), uneven heating resulted from focusing, corner, and edge heating, inhomogeneous electromagnetic field, and irregular shape and nonuniform composition of material (Kelen et al. 2006).

Table 2.13 Recent Findings on Osmotic Dehydration of Fruits and Vegetables

Solute	*Product*	*Findings/Remarks*	*References*
Glucose	Cabbage pieces	• Glucose and high-glucose syrup were found to be better dehydrating agents for increasing solids content whereas high-maltose syrup and maltodextrin were better agents for osmotic dehydration.	Hui et al. (2007)
Glucose and sucrose	Apple and apricot	• Mass transfer rate decreased during air drying after the removal of the free water acquired during the osmotic treatment.	Li and Ramaswamy (2003); Mandala et al. (2005)
Sucrose	Kiwi fruits	• Osmotic temperature affects the dehydration time. High temperature is detrimental to ascorbic acid. • Porous structure would release trapped air from the tissue structure at high temperature, resulting in more effective removal of water by osmotic pressure.	Cao et al. (2006); Hui et al. (2006)
Sucrose and EDTA	Banana slices	• Ethylenediamine tetraacetic acid (EDTA) presence in sucrose avoids undesirable decrease in brightness of the samples.	Hui et al. (2006)
Sucrose and chitin	Banana slices	• Chitin presence in sucrose produced samples in which brightness was less affected.	Waliszewski et al. (2002a)
Sucrose	Bananas and melons	• Osmotic dehydration (using sucrose solutions) followed by air drying reduced total drying time. Modification of solid soluble content was observed.	Waliszewski et al. (2002b); Oliveira et al. (2006)

(continued)

Table 2.13 (Continued) Recent Findings on Osmotic Dehydration of Fruits and Vegetables

Solute	*Product*	*Findings/Remarks*	*References*
Sucrose	Apple (cylinders)	• Microwave-assisted osmotic dehydration improved mass transfer rates and reduced solid gain rate. Higher diffusion rate of water transfer was obtained at lower solution temperatures. Application of microwave heating to the osmotic dehydration process limited uptake of solids and increased moisture loss.	Teles et al. (2006)
Sucrose	Potato and carrot	• Combination of intermittent infrared and continuous convection drying coupled with osmotic dehydration pretreatment was used. Intermittent infrared heating produced dried product with minimal color degradation.	Riva et al. (2005)
Sucrose	Blueberries	• Moisture loss and solid gain increased with increasing temperature and sucrose concentration.	Nsonzi and Ramaswamy (1998)
Ternary solution	Pumpkin cylinders (*Cucurbita pepo L.*)	• Using ternary osmotic solution with NaCl and sucrose gave appreciable water loss, sucrose gain, and NaCl gain in the final product. Efficiency of dehydration was improved when a small amount of NaCl was added to the solution.	Mayor et al. (2007)
Ternary solution	Apple, ginger, carrot, and pumpkin	• Drying time was reduced. It can be further shortened by using tempering. Carotene loss was noticeable during osmotic dehydration.	Pan et al. (2003)

Table 2.14 Application of Various Pretreatment Methods and Research Findings Reported in the Literature

Non-chemical Pretreatment	*Purpose*	*Findings/Remarks*	*References*
Blanching	To inactivate the activity of enzyme peroxidase, limit changes in pectin, and improve the rehydration characteristics of dry products.	• Blanching increased the drying rates of carrot particles (*Daucus carota L.*) at all temperatures in a spouted bed dryer. • Drying rate of whole paprika (without cutting) was improved after blanching was carried out for an appropriate period of time.	Zielinska and Markowski (2007); Ramesh et al. (2001)
Freezing	Ice crystals may damage the integrity of cellular compartments.	• Freezing and combined blanching and freezing pretreatment are very effective in improving the drying rate of peeled banana dried in a heat pump dryer.	Dandamrongrak et al. (2002)
Removal of fruit skin	Fruit skin/membrane limits the removal of internal moisture.	• Drying rate was greater in peeled and thinner sized sapota compared to unpeeled and whole fruits. • Removal of waxy layer of plums by abrasion improved the drying rate and maintained the quality of the dried plums.	Padmini et al. (2005); Cinquanta et al. (2002)
Drilling/ punching of holes on the fruit surface	Holes on the surface improves the rate of water removal.	• Drilled superficial holes (perforations of the skin) improved the drying rate of the fruits and roots significantly. Drying rate increased with increase of diameter and density of hole.	Yong et al. (2006)

Table 2.15 Research Findings of Microwave Vacuum Drying for Various Products

Product	*Findings*	*References*
Carrots Chive leaves	• Carotenoid retention of carrot slices and chlorophyll retention of chive leaves in MVD was similar to freeze drying and much better than those dried by conventional hot air. • Blanching was not necessary when using MVD.	Cui et al. (2004)
Potatoes	• MVD drying of potato at 60°C for 150 min gave uniform product quality, better color retention, and avoided scorch.	Setiady et al. (2007)
Garlic	• Garlic slices dried by MVD were slightly lighter and yellower than that of freeze-dried samples. • MVD–hot air drying gave similar flavor or pungency, color, texture, and rehydration ratio as FD garlic slices but better than hot air drying.	Li et al. (2007); Cui et al. (2003)
Mushrooms	• Drying rate of microwave vacuum drying of mushroom is better than hot air drying but slightly lower than microwave convective drying. • Rehydration ratio and color of dried mushrooms in microwave-vacuum drying was better than those of microwave-convection drying.	Sutar and Prasad (2007)
Strawberries	• The retention of ascorbic acid, phenolic compounds, and antioxidative capacity of dried strawberries in MVD can be further improved by optimizing the operating parameters.	Böhm et al. (2006)

(continued)

Table 2.15 (Continued) Research Findings of Microwave Vacuum Drying for Various Products

Product	*Findings*	*References*
Seafood	• Drying time was shortened. • Longer period of constant rate drying was observed.	Tsuruta and Hayashi (2007)
Honey	• MVD drying retained the contents of fructose, glucose, maltose, and sucrose in honey but slightly reduced the content of aroma compounds (volatile acids, alcohols, aldehydes, and esters).	Cui et al. (2008)
Aspirin Paracetamol	• MVD of pharmaceutical powders (aspirin, paracetamol, lactose, maize starch) using different solvents (water, ethanol, methanol, acetone) in reduced operating pressure gave shorter drying times.	Farrel et al. (2005)
Aspirin Paracetamol	• Decreasing pressure increased the drying rate and effective diffusivity.	McLoughlin et al. (2003)
Lactose Maize starch	• Reduced operating pressure gave shorter drying times.	Farrel et al. (2005)
Hydrochlorothiazide	• MVD is suitable for liquid evaporation. • Product temperature is always below solvent boiling point.	Berteli et al. (2007)
Ganoderma extract	• MVD-VD gave similar retention of polysaccharide and triterpenes to that of freeze-dried and better than VD. • Total drying time is reduced due to higher drying rate.	Cui et al. (2006)

(continued)

Table 2.15 (Continued) Research Findings of Microwave Vacuum Drying for Various Products

Product	*Findings*	*References*
Spinach (*Spinacia oleracea L.*) and okra	• Microwave drying caused more brown compound(s) (shown by BI index and *a* value). *L* and *b* values decreased also indicating the dried samples were darker compared to the fresh samples.	Dadali et al. (2007a, b)
Cranberries (*Vaccinium macrocarpon*)	• Microwave power had a more positive effect than vacuum on the color of dried product.	Sunjka et al. (2004)

Judging from its pros and cons, microwave drying is typically combined with other drying methods to overcome its operational problems mentioned above. Microwave vacuum drying is one of the drying techniques that has been developed recently to dehydrate food products. It has been shown to produce dried products with improved texture and color (Yongsawatdigul and Gunasekaran 2006). Microwaves allow volume heating in which thermal heat can be transferred to the inner core of material even in the initial stage of drying while a vacuum can reduce thermal stresses. A combination of microwaves and vacuum in drying results in improved color and texture of dried products over air-dried products (Qing-Guo et al. 2006). Reduction of drying times in microwaves is beneficial for color (Cui et al. 2008), porosity and pore size (Sundaram and Durance 2007), aroma (Cui et al. 2008), shrinkage (Giri and Prasad 2006), and improved rehydration (Giri and Prasad 2007a, b; Khraisheh et al. 2004). Microwave vacuum drying (MVD) has been applied for the drying of different biomaterials, which are listed in Table 2.15. Table 2.15 also gives the findings reported in the literature.

2.3.6 Novel Spray Drying

Conventional spray dryers encounter various limitations. Recent developments in spray dryers include horizontal spray dryers and spray-freeze dryers. A horizontal spray dryer can be considered as an alternative to a vertical spray dryer when there is a limitation on headspace. Kwamya (1984) first evaluated the idea of a horizontal spray dryer by using a mathematical model. Later, Huang and Mujumdar (2004) suggested various configurations of horizontal spray dryers and evaluated their drying performance using computer simulation. Simulations from a computer fluid dynamics model revealed that the location of the atomizer plays an important role in avoiding wall collision (Huang and Mujumdar 2005).

Dwell time in a one-stage horizontal spray dryer is normally too short to completely remove moisture from droplets. To overcome this problem, Huang and Mujumdar (2006) investigated a two-stage horizontal spray dryer in which a fluidized bed dryer was incorporated in their study. This ensures removal of surface moisture during the formation of powder in the spray drying chamber while removal of internal moisture is carried out in the fluidized bed dryer installed at the bottom of the spray dryer. This design allows longer residence of powder in the dryers as residence time in the fluidized bed dryer can be monitored.

Horizontal spray dryer has been reportedly applied for egg, albumin, and whole egg powder; cheese powder; skim milk; whey protein; a variety of dairy products; fish products; meat products, such as chicken or beef broth, beef blood plasma, etc.; and vegetable products, including soy milk, soy protein, chocolate, enzymes, glucose, etc.

Spray-freeze drying (SFD) is another new development in spray drying that has been proven to minimize protein degradation, loss of stability, and problems

Table 2.16 Research Findings on Spray-Freeze Drying for Various Products

Product	*Findings*	*References*
Liposomal ciprofloxacin	• Powder produced from SFD has an improved mass median aerodynamic diameter and fine particle fraction as compared with jet-milled powder.	Sweeney et al. (2005)
Protein inhalation powders	• SFD produced stable formulations with highly porous particles, which leads to improved aerosolization and dissolution properties compared with other methods.	Maa et al. (1999)
Encapsulated darbepoetin alfa in poly(lactide-co-glycolide)	• Spray drying produced larger microspheres than SFD. • SFD powder has higher values for relative bioavailability and yield than spray drying.	Burke et al. (2004); Nguyen et al. (2004)
Magnesium sulphate powders	• Open pores were found on the dried powder surface. • Highly porous, porosity was reported 87%–90% with pore size less than 100 nm. • Suitable for fine powders, e.g., calcium phosphate, tetragonal zirconia polycrystals.	Yokota et al. (2001)

associated with obtaining a sterile product, which commonly occur in conventional drying methods (Henczka et al. 2006; Maa et al. 1999; Shekunov et al. 2004; Chow et al. 2007). It can also be used to create highly dispersible porous particles as well as for encapsulation purposes (Weers et al. 2007). SFD is indeed a hybrid of spray drying and freeze drying. In a typical SFD, liquids are sprayed and droplets are frozen by liquid nitrogen. The frozen droplets are then dried in a freeze dryer. As the dehydration of products is carried out in a freeze dryer, products dried with SFD tend to have similar quality as freeze-dried products. As such, they are highly porous and stable in term of solid dispersions of active drug and carrier materials (Van Drooge et al. 2005). In addition, many heat sensitive materials, which are in the form of liquid, are suitable to be dried in a SFD. Table 2.16 shows some of the recent findings on spray-freeze drying for various products.

SFD can also be combined or incorporated with drying techniques in order to enhance drying kinetics and drying performance. SFD can be combined with a fluidized bed to increase the transfer rate and improve the uniformity of the product during lyophilization. It was reported that SFD fluidized bed drying of pharmaceutical solutions and liquid foods required drying time of 2 hr (Mumenthaler and Leuenberger 1991; Leuenberger 2002). When the operating pressure is further reduced, the drying time may be lowered further due to the fact that lower specific gas demand (gram of dry air to sublime 1 g of ice) is required at lower pressures. It was reported that SFD fluidized bed drying of whey protein only required 1 hr of drying time (Anandharamakrishnan et al. 2006). Furthermore, Leuenberger et al. (2006) constructed a prototype of this dryer and reported that the dryer gives short drying times and allows advanced control and produces uniform product particle shape and size as well as high solubility.

Furthermore, SFD combined with heat pump drying deceases the energy load whereas combining with fluidization conveying overcomes the operational problem of fluidizing and elutriating cohesive frozen powder from substrate. Wang et al. (2006) confirmed this phenomenon and reported that it also shortened drying time and gave better retention of bio-active ingredients.

2.3.7 Pulse Combustion Drying

Pulse combustors of various designs produce high-temperature exhaust that is highly turbulent and periodic with a frequency in the range of 100 Hz. This is in the audible range, so the SPL (sound pressure level) is high and makes this one of its operational limitations. However, for heat and mass transfer, the conditions are ideal. If a liquid stream is introduced in the exhaust of the tailpipe of a pulse combustor, the jet is atomized, and the resulting spray is dried within milliseconds to a fine powder. An interesting study using sand as feeding material in a pulse combustion drying experiment showed that the presence of sand in the flow region of pulse combusted flue gas slightly affected the shape of sonic waves, causing a small amplitude reduction and no frequency change of the pressure fluctuations. It was found that the temperature rise of sand from 20°C to 600°C could be accomplished within a short period of time. This, in turn, resulted in a 25.5% reduction of natural gas consumption in a continuous burner (Benali and Legros 2004). Another study on the effect of a pulse combustor on the atomization characteristics found that higher air flow oscillating frequency, lower liquid feed rate, and moderate liquid viscosity give better atomizing effects (Xiao et al. 2008; Wu and Liu 2002). Despite the very high temperature environment, the evaporative cooling keeps the particles at temperatures below 60°C. Thus, pilot tests have shown that a highly heat sensitive material like vitamins or yeast can be dried in a pulse combustion dryer without loss of bio-availability (Wu and Mujumdar 2006).

This process has found some applications in industry, namely fine chemical, specialty chemical agri-food and pharmaceutical products (Wu and Liu 2002). The application of pulse combustion drying in the industry is not widespread mainly due to its high noise level and difficult scale-up.

2.3.8 Atmospheric Freeze Drying

Freeze drying is one of the advanced drying technologies for drying highly heat-sensitive valuable food and pharmaceutical products. Due to its complexity and high fixed and operating costs, its applications are restricted to delicate, heat-sensitive materials of high value (Wolff and Gilbert 1990a, b). This is due to the fact that this process involves many energy-intensive operations in series, namely freezing of the drying materials, heating of the frozen materials at low temperature to induce sublimation, condensation of water vapor, and consumption of mechanical energy to maintain the vacuum.

Because a vacuum incurs additional operating costs, new methods suggest carrying out freeze drying at atmospheric pressure. With the absence of a vacuum, the freeze drying is simplified. Here fluidization can be used to improve the contacting efficiency between the drying material and the freezing air. In addition, absorbent materials can also be used to assist the dehydration process, thus improving the external heat and mass transfers.

Di Matteo et al. (2003) found that atmospheric freeze drying gave higher external heat and mass transfer coefficients compared with those obtained from vacuum freeze drying. Because the operation is operated at atmospheric pressure, the removal of internal moisture is dependent on the diffusion at the latter stage. Thus, improving the external operating condition did not help to improve the internal moisture removal. It was also found that product size is a key parameter that affects the drying kinetics.

More recently, Rahman and Mujumdar (2008a, b, 2009) worked on a laboratory scale system of atmospheric freeze drying (AFD) that utilizes a vibrated bed and mixing with an adsorbent to enhance the heat and mass transfer rates. They used a vortex tube to produce a cold stream of air from compressed air. The hot stream from the vortex tube can be used to supply the heat of sublimation needed

Table 2.17 Application of Various Combined/Hybrid Drying and Research Findings Reported in the Literature

Hybrid Drying	*Product*	*Findings/Remarks*	*References*
Osmotic microwave	Mango	• Reduced drying time and energy requirement. • Microwave power influences drying kinetics but may also produce charred pieces.	Andre's et al. (2007)
Pulse vacuum osmotic dehydration	Mango slices	• Temperature and solution concentration affected drying kinetics (positive effect). • Vacuum time affected solids gain and water loss effective diffusivity. • However, osmotic solution recirculation and vacuum pressure had no effect on drying kinetics and product quality.	Ito et al. (2007)
Vacuum microwave drying	Edamame	• Drying rate was accelerated. • Quality of dried samples was enhanced. • Produced a porous structure and improved retention of vitamin C and chlorophyll, color, and microstructural changes and rehydration capacity.	Hu et al. (2006)
Freeze drying – air drying	Strawberry	• Author claimed that it is better than vacuum freeze drying. • Similar product quality as freeze dried products in terms of color and bacterial count. • However, its total capital and operating costs are estimated to be about half of those of freeze drying.	Xu et al. (2006)
Microwave convective drying and microwave vacuum drying	Cranberries	• Color parameters of products produced by both methods were quite similar. • Microwave vacuum-dried cranberries had softer texture.	Sunjka et al. (2004)

for drying, although this was not done in their setup. They showed, using carrot, apple, and potato slices as test models, that the quality of AFD product is much better than that of heat-pump dried product, comparable to vacuum-dried product although inferior to vacuum freeze-dried one. The drying time is long relative to vacuum freeze drying, but the operating and capital costs are much lower. They also examined the effect of osmotic pretreatment. Generally, such treatment reduces initial moisture content, but the presence of sugar or salt crystals that develop during drying tends to slow down the overall drying rate substantially. This observation has also been made in microwave freeze drying of several fruits and vegetables.

2.3.9 Combined/Hybrid Drying

Two dryer types or two drying strategies can be combined so that the advantages of both dryer types or both drying strategies can be realized in one dryer. For instance, combined/hybrid drying can be employed to enable heat input in various modes to remove surface and internal moistures simultaneously, to alter the operating conditions of conventional drying in order to enhance drying kinetics, etc. Table 2.17 gives some of the research findings of combined/hybrid drying reported in the literature.

2.4 Drying of Heat-Sensitive Food Products

Typically, food products are heat-sensitive and hence need lower temperature operation. These products contain active ingredients that are sensitive to heat as well as oxidative reactions. Therefore, some reactions (e.g., Maillard, browning reactions) aided by oxygen and higher temperatures need to be avoided. It can be accomplished by pretreatment (chemical, thermal, and mechanical pretreatments). Vegetable products are sensitive to heat and must be carefully dried to keep color, nutritional content, sensory value, etc. Further to this, retention of vitamin C (ascorbic acid) and pro-vitamin A (beta-carotene) content is dependent on drying method and operating conditions.

The same dryers may be operated under milder conditions to process heat-sensitive bio-products. If susceptible to oxidative damage, drying may be accomplished in a vacuum, at subzero temperatures, or in an inert atmosphere, e.g., nitrogen, CO_2, or superheated steam. Highly heat-sensitive and high-value products are freeze-dried. Freeze drying is 10 times more expensive than spray drying, which itself is an expensive operation in terms of energy and capital. Cryoprotective chemicals are often added in freeze drying to avoid wall rupture. Freeze drying is almost always batch-processed due to very long drying times, typically with proper drying schedule (time varying drying conditions).

Liquid feeds may be dried using spray drying, drum drying, spin flash, or fluid beds of inert particles. Concurrent spray dryers are better suited for heat-sensitive materials. Inert gas, e.g., nitrogen may be used to avoid oxidation or avoid fire and explosion hazards when organic solvents are present. Multistage drying is beneficial for large production rates and when milder conditions are needed to remove internal moisture. For particulate or sludge-like feeds numerous choices exist, e.g., fluidized beds, vibrated beds, conveyer dryers, turbo dryers, tray dryers, etc.

For pasty or sludge-like materials, solids back mixing helps with drying kinetics. Low-temperature drying under vacuum conditions may be used for very highly temperature-sensitive materials. In general,

- Milder drying conditions are more expensive but give a better quality dried product.
- Blanched products (e.g., parsley) have better color and nutritional content upon drying.
- Rapid drying retains more vitamin C; slow drying in sun, solar dryers results in a greater loss of vitamin C.
- Freeze drying typically yields the best results but at the highest cost.
- Heat pump drying is a good option for low-temperature drying.
- Modified atmosphere heat pump drying has been shown to give better quality; it is more expensive.
- Low-pressure operation reduces the drying rate and hence increases the cost of drying.
- Increased cost of drying must be weighed against the premium the market can give for a higher quality product.

Example 2.1

A dryer processes 500 kg h^{-1} of wet material. Water content of the material is to be reduced from 0.4 to 0.04 (dry basis). The drying medium is air, initial temperature is 15°C, relative humidity (Rh) is 50%; the temperature is elevated to 120°C after going through a pre-heater and exits the dryer at 45°C, Rh 80%. The dryer has an internal heater that supplies additional heat to the drying medium. Determine (a) the amount of water removed and (b) the flow rate of the humid air.

Solution:

(a) Amount of water removed, W

$$\text{Amounted removed from the drying material} = W = \dot{M}_{db}(X_1 - X_2) =$$
$$\text{amount of water carried away by the drying medium} = W = L(H_2 - H_1)$$

where, $\dot{M}_{db}$ is the mass flow rate of dry material (kg h^{-1}), X is the moisture content of drying material ((kg H_2O) (kg dry material)$^{-1}$), L is the mass flow rate of dry air (kg h^{-1}), H is the absolute humidity of air ((kg H_2O) (kg dry air)$^{-1}$). Subscript '1' denotes dryer inlet, '2' denotes dryer outlet

$$W = \dot{M}_{db}(X_1 - X_2)$$
$$= 500(0.4 - 0.04) = 180 \text{ kg } H_2O \text{ h}^{-1}$$

(b) Flow rate of humid air, L'

Flow rate of humid air is related to the flow rate of drying by the following equation

$$L' = L(1 + H_o)$$

Subscript 'o' denotes pre-heater inlet. L can be obtained from the mass equation above

$$W = L(H_2 - H_1)$$

or

$$L = \frac{W}{(H_2 - H_1)}$$

From a psychrometric chart, air at 15°C and Rh of 50% has an absolute humidity of 0.005 (kg water) (kg dry air)$^{-1}$, and air at 45°C and Rh of 80% has an absolute humidity of 0.052 (kg H_2O) (kg dry air)$^{-1}$.

$$L = \frac{180}{(0.052 - 0.005)} = 3830 \text{ kg dry air h}^{-1}$$

Flow rate of humid air, L'

$$L' = L(1 + H_o)$$
$$= 3830\,(1 + 0.005) = 3849 \text{ kg humidity air h}^{-1}$$

2.5 Summary

New advances in the drying of food products have been achieved in recent years owing to the demand for improved operation efficiency and better product quality. Food products may appear in the form of wet solid, liquid, suspension, or paste, which require refrigeration for storage or drying for preservation. Dried foods are easier to transport and are less expensive to handle and process. As such, dehydration and drying are important operations to the food industry. Food products are heat sensitive. Conventional drying methods, such as hot air drying, tends to damage and denature the product, destroy active ingredients, cause case hardening and browning. In recent years, the introduction of new value-added food products has resulted in demand for effective and efficient drying technologies. This chapter discusses and compares conventional drying methods and includes some recent developments with reference to food drying. The new developments include heat pump-assisted drying with controlled atmosphere, intermittent batch drying, osmotic dehydration, various pretreatments, MW-vacuum drying, low pressure superheated steam drying, pulse combustion drying, novel spray dryer designs, atmospheric freeze drying, etc. The advantages as well as the limitations of these new emerging drying technologies are discussed and compared with reference to the conventional drying methods focusing on food processing industries.

References

Ade-Omowaye, B.I.O., Rastogi, N.K., Angersbach, A. and Knorr, D. (2003). Combined effects of pulsed electric field pre-treatment and partial osmotic dehydration on air drying behavior of red bell pepper. *Journal of Food Engineering*, 60, 89–98.

Alves-Filho, O. (2002). Combined innovation heat pump drying technologies and new cold extrusion techniques for production of instant foods. *Drying Technology*, 20, 1541–1557.

Alves-Filho, O. and Strommen, I. (1996). Performance and improvements in heat pump dryers. In *Proceedings IDS'96*, pp. 405–415.

Amami, E., Vorobiev, E. and Kechaou, N. (2005). Effect of pulsed electric field on the osmotic dehydration and mass transfer kinetics of apple tissue. *Drying Technology*, 23(3), 581–595.

Anandharamakrishnan, C., Khwanpruk, K., Rielly, C. and Stapley, A. (2006). Spray freeze-drying at sub-atmospheric pressures. In the *Proceedings of the IDS'06*, pp. 636–642.

Andre's, A., Fito, P., Heredia, A. and Rosa, E.M. (2007). Combined drying technologies for development of high-quality shelf-stable mango products. *Drying Technology*, 25, 1857–1866.

Aquerreta, J., Iguaz, A., Arroqui, C. and Vırseda, P. (2007). Effect of high temperature intermittent drying and tempering on rough rice quality. *Journal of Food Engineering*, 80, 611–618.

Azuara, E., Garcia, H.S. and Beristain, C.I. (1996). Effect of the centrifugal force on osmotic dehydration of potatoes and apples. *Food Research International*, 29(2), 195–199.

Barbieri, S., Elustondo, M. and Urbicain, M. (2004). Retention of aroma compounds in basil dried with low pressure superheated steam. *Journal of Food Engineering*, 65, 109–115.

Benali, M. and Legros, R. (2004). Thermal processing of particulate solids in a gas-fired pulse combustion system. *Drying Technology*, 22, 347–362.

Berteli, M.N., Marsaioli, Jr., A. and Rodier, E. (2007). Study of a microwave assisted vacuum drying process applied to the granulated pharmaceutical drug hydrochlorthiazide (HCT). *Journal of Microwave Power and Electromagnetic Energy*, 40, 241–250.

Böhm, V., Kühnert, S., Rohm, H. and Scholze, G. (2006). Improving the nutritional quality of microwave-vacuum dried strawberries: A preliminary study. *Food Science and Technology International*, 12, 67–75.

Bon, J. and Kudra, T. (2007). Enthalpy-driven optimization of intermittent drying. *Drying Technology*, 25, 523–532.

Burke, P.A., Klumb, L.A., Herberger, J.D., Nguyen, X.C., Harrell, R.A. and Zordich, M. (2004). Poly(lactide-co-glycolide) microsphere formulations of *Darbepoetin Alfa*: Spray drying is an alternative to encapsulation by spray-freeze drying. *Pharmaceutical Research*, 21, 500–506.

Cao, H., Zhang, M., Mujumdar, A.S., Du, W. and Sun, J. (2006). Optimization of osmotic dehydration of kiwifruits. *Drying Technology*, 24, 89–94.

Cardona, T.D., Driscoll, R.H., Peterson, J.L., Srzednicki, G.S. and Kim, W.S. (2002). Optimizing conditions for heat pump dehydration of lactic acid bacteria. *Drying Technology*, 20, 1611–1632.

Cenkowski, S., Pronyk, C., Zmidzinska, D. and Muir, W.E. (2007). Decontamination of food products with superheated steam. *Journal of Food Engineering*, 83, 68–75.

Chen, G., Banister, P., Carrington, C.G., Velde, P.T. and Burger, F.C. (2002). Design and application of dehumidifier dryer for drying pine cones and pollen catkins. *Drying Technology*, 20, 1633–1643.

Chow, A.H.L., Tong, H.H.Y., Chattopadhyay, P. and Shekunov, B.Y. (2007). Particle engineering for pulmonary drug delivery. *Pharmaceutical Research*, 24, 411–437.

Chua, K.J., Chou, S.K. Ho, J.C. Mujumdar, A.S. and Hawlader, M.N.A. (2000a). Cyclic air temperature drying of guava pieces: Effects on moisture and ascorbic acid contents. *Transactions IChemE*, 78, Part C, 72–78.

Chua, K.J., Mujumdar, A.S., Chou, S.K., Hawlader, M.N.A. and Ho, J.C. (2000b). Heat pump drying of banana, guava and potato pieces: Effect of cyclic variations of air temperature on convective drying kinetics and color change. *Drying Technology*, 18, 907–936.

Chua, K.J., Chou, S.K., Hawlader, M.N.A., Ho, J.C. and Mujumdar, A.S. (2002a). On the study of time-varying temperature drying–effect on drying kinetics and product quality. *Drying Technology*, 20, 1579–1610.

Chua, K.J., Chou, S.K., Mujumdar, A.S., Ho, J.C. and Hawlader, M.N.A. (2002b). Heat pump drying: Recent developments and future trends. *Drying Technology*, 20, 1559–1577.

Chua, K.J., Mujumdar, A.S. and Chou, S.K. (2003). Intermittent drying of bioproducts–an overview. *Bioresource Technology*, 90, 285–295.

Chua, K.J., Mujumdar, A.S., Hawlader, M.N.A., Chou, S.K. and Ho, J.C. (2001). Batch drying of banana pieces–effect of stepwise change in drying air temperature on drying kinetics and product colour. *Food Research International*, 34, 721–731.

Chua, K.J., Chou, S.K., Mujumdar, A.S., Ho, J.C. and Hon, C.K. (2004). Radiant convective drying of osmotic treated agro-products: Effect on drying kinetics and product quality. *Food Control*, 15, 145–158.

Cihan, A. and Ece, M.C. (2001). Liquid diffusion model for intermittent drying of rough rice. *Journal of Food Engineering*, 49, 327–331.

Cinquanta, L., Matteo, M.D. and Esti, M. (2002). Physical pre-treatment of plums (*Prunus domestica*). Part 2: Effect on the quality characteristics of different prune cultivars. *Food Chemistry*, 79, 233–238.

Cui, Z.W., Sun, L.J., Chen, W. and Sun, D.W. (2008). Preparation of dry honey by microwave-vacuum drying. *Journal of Food Engineering*, 84, 582–590.

Cui, Z.W., Xu, S.Y. and Sun, D.W. (2003). Dehydration of garlic slices by combined microwave-vacuum and air drying. *Drying Technology*, 21, 1173–1184.

Cui, Z.W., Xu, S.Y. and Sun, D.W. (2004). Effect of microwave-vacuum drying on the carotenoids retention of carrot slices and chlorophyll retention of Chinese chive leaves. *Drying Technology*, 22, 563–575.

Cui, Z.W., Xu, S.Y., Sun, D.W. and Chen, W. (2006). Dehydration of concentrated ganoderma lucidum extraction by combined microwave-vacuum and conventional vacuum drying. *Drying Technology*, 24, 595–599.

Dadali, G., Apar, D.K. and Ozbek, B. (2007a). Colour change kinetics of okra undergoing microwave drying. *Drying Technology*, 25, 925–936.

Dadali, G., Demirhan, E. and Ozbek, B. (2007b). Colour change kinetics of spinach undergoing microwave drying. *Drying Technology*, 25, 1713–1723.

Dandamrongrak, R., Young, G. and Mason, R. (2002). Evaluation of various pre-treatments for the dehydration of banana and selection of suitable drying models. *Journal of Food Engineering*, 55(2), 139–146.

Devahastin, S., Suvarnakuta, P., Soponronnarit, S. and Mujumdar, A.S. (2004). A comparative study of low-pressure superheated steam and vacuum drying of a heat-sensitive material. *Drying Technology*, 22, 1845–1867.

Di Matteo, P., Donsi, G. and Ferrari, G. (2003). The role of heat and mass transfer phenomena in atmospheric freeze-drying of foods in a fluidised bed. *Journal of Food Engineering*, 59, 267–275.

Duan, X., Zhang, M., Li, X.L. and Mujumdar, A.S. (2008). Ultrasonically enhanced osmotic pretreatment of sea cucumber prior to microwave freeze drying. *Drying Technology*, 26(4), 420–426.

Emam-Djomeh, Z., Dehghannya, J. and Sotudeh Gharabagh, R. (2006). Assessment of osmotic process in combination with coating on effective diffusivities during drying of apple slices. *Drying Technology*, 24, 1159–1164.

Erbay, Z. and Icier, F. (2009). Optimization of drying of olive leaves in a pilot-scale heat pump dryer. *Drying Technology*, 27, 416–427.

Erle, U. (2005). Drying using microwave processing. In H. Schubert and M. Regier (eds.), *The Microwave Processing of Foods*, pp.142–152. CRC Press, New York.

Farrel, G., McMinn, W.A.M. and Magee, T.R.A. (2005). Microwave-vacuum drying kinetics of pharmaceutical powders. *Drying Technology*, 23, 2131–2146.

Fito, P. (1994). Modeling of vacuum osmotic dehydration of food. *Journal of Food Engineering*, 22, 313–328.

Giri, S.K. and Prasad, S. (2006). Modeling shrinkage and density changes during microwave-vacuum drying of button mushroom. *International Journal of Food Properties*, 9, 409–419.

Giri, S.K. and Prasad, S. (2007a). Drying kinetics and rehydration characteristics of microwave-vacuum and convective hot-air dried mushrooms. *Journal of Food Engineering*, 78, 512–521.

Giri, S.K. and Prasad, S. (2007b). Optimization of microwave-vacuum drying of button mushrooms using response-surface methodology. *Drying Technology*, 25, 901–911.

Gunasekaran, S. (1999). Pulsed microwave-vacuum drying of food materials. *Drying Technology*, 17, 395–412.

Hawlader, M.N.A., Perera, C.O. and Tian, M. (2006a). Comparison of the retention of 6-gingerol in drying of ginger under modified atmosphere heat pump drying and other drying methods. *Drying Technology*, 24, 51–56.

Hawlader, M.N.A., Perera, C.O., Tian, M. and Chng, K.J. (2006b). Properties of modified atmosphere heat pump dried foods. *Journal of Food Engineering*, 74(3), 392–401.

Hawlader, M.N.A., Perera, C.O., Tian, M. and Yeo, K.L. (2006c). Drying of guava and papaya: Impact of different drying methods. *Drying Technology*, 24, 77–87.

Henczka, M., Bałdyga, J. and Shekunov, B.Y. (2006). Modelling of spray-freezing with compressed carbon dioxide. *Chemical Engineering Science*, 61, 2880–2887.

Hu, Q., Zhang M., Mujumdar, A.S., Du, W. and Sun, J. (2006). Effects of different drying methods on the quality changes of granular edamame. *Drying Technology*, 24, 1025–1032.

Huang, L. and Mujumdar, A.S. (2004). Spray drying technology – principles and practice'. In A.S. Mujumdar (ed.), *Guide to industrial drying: Principles, equipment and new developments*, pp. 143–173. Color Publications Pvt. Ltd., Mumbai, India.

Huang, L. and Mujumdar, A.S. (2006). Numerical study of two-stage horizontal spray dryers using computational fluid dynamics. *Drying Technology*, 24, 727–733.

Hui, C., Zhang, M., Mujumdar, A.S., Du, W. and Sun, J. (2006). Optimization of osmotic dehydration of kiwifruits. *Drying Technology*, 24, 89–94.

Hui, C., Zhang, M., Duan, X., Zhang, C. and Sun, J. (2007). Effect of sugar pretreatment on quality of dehydrated cabbage. *Drying Technology*, 25, 1545–1549.

Iguaz, A., Rodriguez, M. and Virseda, P. (2006). Influence of handling and processing of rough rice on fissures and head rice yields. *Journal of Food Engineering*, 77, 803–809.

Ishikawa, M. and Nara, H. (1993). Osmotic dehydration of food by semipermeable membrane coating. In R.P. Singh and H.A. Wirakartakusumah (eds.), *Advances in Food Engineering*, pp. 73–77. CRC Press, New York.

Ito, A.P., Tonon, R.V., Park, K.J. and Hubinger, M.M. (2007). Influence of process conditions on the mass transfer kinetics of pulsed vacuum osmotically dehydrated mango slices. *Drying Technology*, 25, 1769–1777.

Jamradloedluk, J., Nathakaranakule, A., Soponronnarit, S. and Prachayawarakorn, S. (2007). Influences of drying medium and temperature on drying kinetics and quality attributes of durian chip. *Journal of Food Engineering*, 78, 198–205.

Jangam, S.V., Joshi, V.S., Mujumdar, A.S. and Thorat, B.N. (2008). Studies on dehydration of sapota (*Achras zapota*). *Drying Technology*, 26, 369–377.

Jumah, R., Al-Kteimat, E., Al-Hamad, A. and Telfah, E. (2007). Constant and intermittent drying characteristics of olive cake. *Drying Technology*, 25, 1421–1426.

Kelen, A., Ress, S., Nagy, T., Pallai, E. and Pintye-Hodi, K. (2006). Mapping of temperature distribution in pharmaceutical microwave vacuum drying. *Powder Technology*, 162, 133–137.

Khraisheh, M.A.M., McMinn, W.A.M. and Magee, T.R.A. (2004). Quality and structural changes in starchy foods during microwave and convective drying. *Food Research International*, 37, 497–503.

Kiang, C.S. and Jon, C.K. (2007). Heat pump drying systems. In A.S. Mujumdar (ed.), *Handbook of industrial drying*, 3rd ed. pp. 1103–1131. CRC Press, New York.

Klocker, K., Schmidt, E.L. and Steimle, F. (2002). A drying heat pump using carbon dioxide as working fluid. *Drying Technology*, 20, 1659–1671.

Kudra, T. and Mujumdar, A.S. (2009). *Advanced drying technologies,* 2nd ed. CRC Press, New York.

Kwamya, M. (1984). M. Eng. Thesis, Department of Chemical Engineering. McGill University, Montreal, Canada.

Law, C.L., Waje, S., Thorat, B.N. and Mujumdar, A.S. (2008). Innovation and recent advancement in drying operation for postharvest processes. *Stewart Postharvest Review*, 4(1), 1–23.

Lazarides, H.N. and Mavroudis, N.E. (1995). Freeze/thaw effects on mass transfer rates during osmotic dehydration. *Journal of Food Science*, 60(4), 826–828.

Leuenberger, H. (2002). Spray-freeze-drying: The process of choice for low water soluble drugs? *Journal of Nanoparticle Research*, 4, 111–119.

Leuenberger, H., Plitzko, M. and Puchkov, M. (2006). Spray freeze drying in a fluidized bed at normal and low pressure. *Drying Technology*, 24, 711–719.

Li, H. and Ramaswamy, H.S. (2003). Continuous flow microwave-osmotic combination drying of apple slices. Poster 58–54 presented at IFT Annual Conference, Chicago, Il, 2003.

Li, M., Ma, Y., Gong, W. and Su, W. (2009). Analysis of CO_2 transcritical cycle heat pump dryers. *Drying Technology*, 27, 548–554.

Li, Y., Xu, S.Y. and Sun, D.W. (2007). Preparation of garlic powder with high allicin content by using combined microwave-vacuum and vacuum drying as well as micro-encapsulation. *Journal of Food Engineering*, 83, 76–83.

Li, Y.B., Cao, C.W., Yu, Q.L. and Zhong, Q.X. (1999). Study on rough rice fissuring during intermittent drying. *Drying Technology*, 17, 1779–1793.

Maa, Y.F., Nguyen, P.A., Sweeney, T., Shire, S.J. and Hsu, C.C. (1999). Protein inhalation powders: Spray drying vs spray freeze drying. *Pharmaceutical Research*, 16, 249–254.

Madamba, P.S. and Yabes, R.P. (2005). Determination of the optimum intermittent drying conditions for rough rice (*Oryza sativa*, L.). *Lebensm.-Wiss. u.-Technology*, 38, 157–165.

Mandala, I.G., Anagnostaras, E.F. and Oikonomou, C.K. (2005). Influence of osmotic dehydration conditions on apple air-drying kinetics and their quality characteristics. *Journal of Food Engineering*, 69, 307–316.

Matuska, M., Lenart, A. and Lazarides, H.N. (2006). On the use of edible coating to monitor osmotic dehydration kinetics for minimal solids uptake. *Journal of Food Engineering*, 72, 85–91.

Mayor, L., Moreira, R., Chenlo, F. and Sereno, A.M. (2007). Osmotic dehydration kinetics of pumpkin fruits using ternary solutions of sodium chloride and sucrose. *Drying Technology*, 25, 1749–1758.

McLoughlin, C.M., McMinn, W.A.M. and Magee, T.R.A. (2003). Microwave-vacuum drying of pharmaceutical powders. *Drying Technology*, 21, 1719–1733.

Medina-Vivanco, M., Sobral, P.J.A. and Hubinger, M.D. (2002). Osmotic dehydration of tilapia fillets in limited volume ternary solutions. *Chemical Engineering Journal*, 86, 199–205.

Mujumdar, A.S. (1991). Drying technologies of the future. *Drying Technology*, 9, 325–347.

Mujumdar, A.S. (2000). Classification and selection of industrial dryers. In S. Devahastin (ed.), *Mujumdar's practical guide to industrial drying* (pp. 23–36). Exergex Corp., Quebec.

Mujumdar, A.S. (2004). Classification and selection of industrial dryers. In A.S. Mujumdar (ed.), *Practical guide to industrial drying* (pp. 23–35). Color Publications Pvt. Ltd., Mumbai.

Mujumdar, A.S. (2006). Some recent developments in drying technologies appropriate for post-harvest processing. *International Journal of Postharvest Technology and Innovation*, 1, 76–92.

Mujumdar, A.S. (2007a). An overview of innovation in industrial drying: Current status and R&D needs. *Transport Porous Media*, 66, 3–18.

Mujumdar, A.S. (2007b). Principles, classification and selection of dryers. In A.S. Mujumdar (ed.), *Handbook of industrial drying* (pp. 3–32). CRC Press, New York.

Mumenthaler, M. and Leuenberger, H. (1991). Atmospheric spray-freeze drying: A suitable alternative in freeze-drying technology. *International Journal of Pharmaceutics*, 72, 97–110.

Neill, M.B., Rahman, M.S., Perera, C.O., Smith, B. and Melton, L.D. (1998). Colour and density of apple cubes dried in air and modified atmosphere. *International Journal of Food Properties*, 1, 197–205.

Nguyen, X.C., Herberger, J.D. and Burke, P.A. (2004). Protein powders for encapsulation: A comparison of spray-freeze drying and spray drying of Darbepoetin Alfa. *Pharmaceutical Research*, 21, 507–514.

Nguyen, C.N. and Kunze, O.R. (1984). Fissures related to post-drying treatments in rough rice. *Cereal Chemistry*, 61, 63–68.

Nsonzi, F. and Ramaswamy, H.S. (1998). Osmotic dehydration kinetics of blueberries. *Drying Technology*, 16, 725–741.

Ogura, H., Ishida, H., Kage, H. and Mujumdarm, A.S. (2003). Enhancement of energy efficiency of a chemical heat pump-assisted convective dryer. *Drying Technology*, 21, 279–292.

Ogura, H., Yamamoto, T., Kage, H., Matsuno, Y. and Mujumdar, A.S. (2002). Effect of heat exchange conditions on hot air production by a chemical heat pump dryer using CaO/ $Ca(OH)_2$ reaction. *Chemical Engineering Journal*, 86, 3–10.

Oliveira, I.M., Fernandes, F.A.M., Rodrigues, S., Sousa, P.H.M., Maia, G.A. and Figueiredo, R.W. (2006). Modeling and optimization of osmotic dehydration of banana followed by air drying. *Journal of Food Process Engineering*, 29, 400–413.

Padmini, T., Raghavan, G.S.V. and Thangaraj, T. (2005). Dehydration of sapota flakes. In *Proceedings of 4th Asia Pacific Drying Conference*, II, 1161–1165.

Pan, Y.K., Zhao, L.J., Dong, X., Mujumdar, A.S. and Kudra, T. (1999). Intermittent drying of carrot in a vibrated fluid bed: Effect on product quality. *Drying Technology*, 17, 2323–2340.

Pan, Y.K., Zhoa, L.J., Zhang, Y., Chen, G. and Mujumdar, A.S. (2003). Osmotic dehydration pretreatment in drying of fruits and vegetables. *Drying Technology*, 21, 1101–1114.

Pereira, L.M., Ferrari, C.C., Mastrantonio, S.D.S., Rodrigues, A.C.C. and Hubinger, M.D. (2006). Kinetic aspects, texture, and color evaluation of some tropical fruits during osmotic dehydration. *Drying Technology*, 24, 475–484.

Perera, C.O. 2001. Modified atmosphere heat pump drying of food products. In *Proceedings of the 2nd Asia-Oceania Drying Conference (ADC '01)* (pp. 469–476), Batu Ferringhi, Malaysia.

Prachayawarakorn, S., Prachayawasin, P. and Soponronnarit, S. (2004). Effective diffusivity and kinetics of urease inactivation and color change during processing of soybeans with superheated-steam fluidized bed. *Drying Technology*, 22, 2095–2118.

Pronyk, C., Cenkowski, S. and Muir, W.E. (2004). Drying foodstuffs with superheated steam. *Drying Technology*, 22, 899–916.

Puschner, P. (2005). Improved microwave process control. In H. Schubert and M. Regier (eds.) *The Microwave Processing of Foods*, pp. 264–291. CRC Press, New York.

Qi, H., LeMaguer, M. and Sharma, S.K. (1998). Design and selection of processing conditions of a pilot scale contactor for continuous osmotic dehydration of carrots. *Journal of Food Process Engineering*, 21(1), 75–88.

Qing-Guo, H., Min, Z., Mujumdar, A.S., Wei-Hua, D. and Jin-Cai, S. (2006). Effects of different drying methods on the quality changes of granular edamame. *Drying Technology*, 24, 1025–1032.

Rahman, S.M.A. and Mujumdar, A.S. (2008a). A novel atmospheric freeze-drying system using a vibro-fluidized bed with adsorbent and multimode heat input. *Drying Technology*, 26, 393–403.

Rahman, S.M.A. and Mujumdar, A.S. (2008b). Sublimation of ice in a novel atmospheric freeze drying system using vortex tube and multimode heat input: Simulation and experiment. *Asia-Pacific Journal of Chemical Engineering*, 3(4), 408–416.

Rahman, S.M.A. and Mujumdar, A.S. (2009). A novel atmospheric freeze drying system using a vortex tube and multimode heat supply. *International Journal of Postharvest Technology and Innovation*, 1(3), 249–266.

Ramesh, M.N., Wolf, W., Tevini, D. and Jung, G. (2001). Influence of processing parameters on the drying of spice parika. *Journal of Food Engineering*, 49, 63–72.

Raoult-Walk, A.L. (1994). Recent advances in the osmotic dehydration of foods. *Trends in Food Science and Technology*, 5, 255–260.

Rastogi, N.K., Angersbach, A. and Knorr, D. (2000). Synergistic effect of high hydrostatic pressure pretreatment and osmotic stress on mass transfer during osmotic dehydration. *Journal of Food Engineering*, 45, 25–31.

Rastogi, N.K., Eshtiaghi, M.N. and Knorr, D. (1999). Accelerated mass transfer during osmotic dehydration of high intensity electrical field pulse pretreated carrots. *Journal of Food Science*, 64(6), 1020–1023.

Rastogi, N.K., Raghavarao, K.S.M.S., Niranjan, K. and Knorr, D. (2002). Recent developments in osmotic dehydration: Methods to enhance mass transfer. *Trends in Food Science and Technology*, 13, 48–59.

Riva, M., Campolongo, S., Leva, A.A., Maestrelli, A. and Torreggiani, D. (2005). Structure-property relationships in osmo-air-dehydrated apricot cubes. *Food Research International*, 38, 533–542.

Rodrigues, S. and Fernandes, F.A.M. (2007). Use of ultrasound as pretreatment for dehydration of melons. *Drying Technology*, 25, 1791–1796.

Rodrigues, S., Oliveira, F.I.P., Gallão, M.I. and Fernandes, F.A.N. (2009). Effect of immersion time in osmosis and ultrasound on papaya cell structure during dehydration. *Drying Technology*, 27, 220–225.

Rossi, S.J., Neues, I.C. and Kicokbusch, T.G. (1992). Thermodynamics and energetic evaluation of a heat pump applied to drying of vegetables. In A.S. Mujumdar (ed.), *Drying '92* (pp. 1475–1483). Elsevier Science, Amsterdam.

Schiffmann, R.F. (2007). Microwave and dielectric drying. In A.S. Mujumdar (ed.), *Handbook of industrial drying*, 3rd ed. (pp. 285–305). CRC Press, New York.

Sereno, A.M., Moreira, R. and Martinez, E. (2001). Mass transfer coefficients during osmotic dehydration of apple in single and combined aqueous solutions of sugar and salt. *Journal of Food Engineering*, 47, 43–49.

Setiady, D., Clary, C., Younce, F. and Rasco, B.A. (2007). Optimizing drying conditions for microwave-vacuum (MIVAC®) drying of russet potatoes (*Solanum tuberosum*). *Drying Technology*, 25, 1483–1489.

Shekunov, B., Chattopadhyay, P., Seitzinger, J., Bałdyga, J. and Henczka, M. (2004). Optimization of spray-freezing for production of respirable particles. *The AAPS Journal*, 6, Abstract M1145.

Shi, X.Q., Fito, P. and Chiralt, A. (1995). Influence of vacuum treatment on mass transfer during osmotic dehydration of fruits. *Food Research International*, 28(5), 445–454.

Shibata, H. and Ide, M. (2001). Combined superheated steam and microwave drying of sintered glass beads: Drying rate curves. *Drying Technology*, 19, 2063–2079.

Simal, S., Benedito, J., Sanchez, E.S. and Rossello, C. (1998). Use of ultrasound to increase mass transfer rates during osmotic dehydration. *Journal of Food Engineering*, 36, 323–336.

Stapley, A. and Rielly, C. (2007). Advances in unconventional freeze drying technology. *Chemical Engineer*, London, 790, 33–35.

Strommen, I., Eikevik, T.M. and Alves-Filho, O. (1999). Optimum design and enhanced performance of heat pump dryers. In *Proceedings of the ADC'99*, pp. 66–80. Bandung, Indonesia.

Sundaram, J. and Durance, T.D. (2007). Influence of processing methods on mechanical and structural characteristics of vacuum microwave dried biopolymer foams. *Food and Bioproducts Processing*, 85, 264–272.

Sunjka, P.S., Rennie, T.J., Beaudry, C. and Raghavan, G.S.V. (2004). Microwave-convective and microwave-vacuum drying of cranberries: A comparative study. *Drying Technology*, 22, 1217–1231.

Sun, L., Islam, M.R., Ho, J.C. and Mujumdar, A.S. (2005). A diffusion model for drying of a heat sensitive solid under multiple heat input modes. *Bioresource Technology*, 96, 1551–1560.

Sunthonvit, N., Srzednicki, G. and Craske, J. (2007). Effects of drying treatments on the composition of volatile compounds in dried nectarines. *Drying Technology*, 25, 877–881.

Sutar, P.P. and Prasad, S. (2006). Modeling microwave vacuum drying kinetics and moisture diffusivity of carrot slices. *Drying Technology*, 25, 1695–1702.

Suvarnakuta, P., Devahastin, S. and Mujumdar, A.S. (2007). A mathematical model for low-pressure superheated steam drying of a biomaterial. *Chemical Engineering and Processing*, 46, 675–683.

Sweeney, L.G., Wang, Z., Loebenberg, R., Wong, J.P., Lange, C.F. and Finlay, W.H. (2005). Spray-freeze-dried liposomal ciprofloxacin powder for inhaled aerosol drug delivery. *International Journal of Pharmaceutics*, 305, 180–185.

Taiwo, K.A., Angersbach, A. and Knorr, D. (2003). Effects of pulsed electric field on quality factors and mass transfer during osmotic dehydration of apples. *Journal of Food Process Engineering*, 26, 31–48.

Tatemoto, Y., Yanoa, S., Mawatarib, Y., Nodaa, K. and Komatsuc, N. (2007). Drying characteristics of porous material immersed in a bed of glass beads fluidized by superheated steam under reduced pressure. *Chemical Engineering Science*, 62, 471–480.

Tedjo, W., Taiwo, K.A., Eshtiaghi, M.N. and Knorr, D. (2002). Comparison of pretreatment methods on water and solid diffusion kinetics of osmotically dehydrated mangos. *Journal of Food Engineering*, 53(2), 133–142.

Teles, U.M., Fernandes, F.A.M., Rodrigues, S., Lima, A.S., Maia, G.A. and Figueiredo, R.W. (2006). Optimization of osmotic dehydration of melons followed by air-drying. *International Journal of Food Science and Technology*, 41, 674–680.

Thomkapanich, O., Suvarnakuta, P. and Devahastin, S. (2007). Study of intermittent low-pressure superheated steam and vacuum drying of a heat-sensitive material. *Drying Technology*, 25, 205–223.

Tsuruta, T. and Hayashi, T. (2007). Internal resistance to water mobility in seafood during warm air drying and microwave-vacuum drying. *Drying Technology*, 25, 1393–1399.

Van Drooge, D.J., Hinrichs, W.L.J., Dickhoff, B.H.J., Elli, M.N.A., Visser, M.R., Zijlstra, G.S. and Frijlink, H.W. (2005). Spray freeze drying to produce a stable Δ^9-tetrahydrocannabinol containing insulin-based solid dispersion powder suitable for inhalation. *European Journal of Pharmaceutical Science*, 26, 231–240.

Waje, S.S., Phadake, J., Pandey, R., Malshe, V.C. and Thorat, B.N. (2007). Absorbable gelatine foam: Synthesis and quality analysis. In *Proceedings of the ADC '07* (pp. 1058–1063). Hong Kong.

Waliszewski, K.N., Pardio, V.T. and Ramirez, M. (2002a). Effect of EDTA on colour during osmotic dehydration of banana slices. *Drying Technology*, 20, 1291–1298.

Waliszewski, K.N., Pardio, V.T. and Ramirez, M. (2002b). Effect of chitin on colour during osmotic dehydration of banana slices. *Drying Technology*, 20, 719–726.

Wang, Z.L., Finlay, W.H., Peppler, M.S. and Sweeney, L.G. (2006). Powder formation by atmospheric spray-freeze drying. *Powder Technology*, 170, 15–52.

Weers, J.G., Tarara, T.E. and Clark, A.R. (2007). Design of fine particles for pulmonary drug delivery. *Expert Opinion on Drug Delivery*, 4, 1–17.

Wolff, E. and Gilbert, H. (1990a). Atmospheric freeze-drying part 1: Design, experimental investigation and energy saving advantages. *Drying Technology*, 8, 385–404.

Wolff, E. and Gibert, H. (1990b). Atmospheric freeze-drying part 2: Modelling drying kinetics using adsorption isotherms. *Drying Technology*, 8, 405–428.

Wu, Z. and Liu, X. (2002). Simulation of spray drying of a solution atomized in a pulsating flow. *Drying Technology*, 20, 1097–1117.

Wu, Z. and Mujumdar, A.S. (2006). R&D needs and opportunities in pulse combustion and pulse combustion drying. *Drying Technology*, 24, 1521–1523.

Xiao, Z., Xie, X., Yuan, Y. and Liu, X. (2008). Influence of atomizing parameters on droplet properties in a pulse combustion spray dryer. *Drying Technology*, 26, 427–432.

Xing H, Takhar P.S., Helms G. and He, B. (2007). NMR imaging of continuous and intermittent drying of pasta. *Journal of Food Engineering*, 78, 61–68.

Xu, H., Zhang, M., Duan, X., Mujumdar, A.S. and Sun, J. (2009). Effect of power ultrasound pretreatment on edamame prior to freeze drying. *Drying Technology*, 27, 186–193.

Xu, Y., Zhang, M., Mujumdar, A.S., Duan, X. and Sun, J. (2006). A two-stage vacuum freeze and convective air drying method for strawberries. *Drying Technology*, 24, 1019–1023.

Yaghmaee, P. and Durance, T. (2007). Efficacy of vacuum microwave drying in microbial decontamination of dried vegetables. *Drying Technology*, 25, 1099–1104.

Yokota, T., Takahata, Y., Katsuyama, T. and Matsuda, Y. (2001). A new technique for preparing ceramics for catalyst support exhibiting high porosity and high heat resistance. *Catalysis Today*, 69, 11–15.

Yong, C.K., Islam, M.R. and Mujumdar, A.S. (2006). Mechanical means of enhancing drying rates: Effect on drying kinetics and quality. *Drying Technology*, 24, 397–404.

Yongsawatdigul, J. and Gunasekaran, S. (1996). Microwave-vacuum drying of cranberries: Part II. Quality evaluation. *Journal of Food Processing and Preservation*, 20, 145–156.

Zhang, B., Zhou, Y., Ning, W. and Xie, D. (2007). Experimental study on energy consumption of combined conventional and dehumidification drying. *Drying Technology*, 25, 471–474.

Zhang, M., Tang, J., Mujumdar, A.S. and Wang, S. (2006). Trends in microwave related drying of fruits and vegetable. *Trends in Food Science and Technology*, 17, 524–534.

Zielinska, M. and Markowski, M. (2007). Drying behavior of carrots dried in a spout–fluidized bed dryer. *Drying Technology*, 25, 261–270.

Chapter 3

Osmo-Concentration of Foods

Sujata Jena and H. Das

Contents

3.1 Introduction

The process of contacting pieces of food with concentrated sugar, salt, or mixtures of solutes is an age-old concept of preservation of foods, especially horticultural products, such as fruits and vegetables. This process is scientifically termed as osmotic dehydration (OD). OD is regarded as a minimal processing dehydration method for production of intermediate moisture foods (IMF), which are not shelf stable. Consequently, the osmotically dehydrated product should further be processed by various other drying methods, viz. air drying, vacuum drying, freeze drying, and microwave drying, etc. to obtain a shelf-stable product. Hence, osmo-concentration can be treated as a pretreatment method for canning, freezing, and drying. This method also increases the sugar-to-acid ratio and improves the texture and stability of pigments during dehydration and storage. A large amount of energy is needed in drying operations for providing the latent heat of vaporization of water. But in osmo-concentration, about 50% of the water is removed without any phase change. This results in potential energy saving in a dehydration operation. Hence, OD is acknowledged as an energy-efficient mild partial dehydration method. Osmo-concentration of foods has tremendous potential in food processing industries. Minimization of water load by OD prior to drying makes it an efficient complementary processing step in the overall chain of integrated food processing.

3.2 General Principles

OD, sometimes termed as dewatering-impregnation soaking (DIS) (Raoult-Wack 1994), involves immersion of food solids in a hypertonic solution. During this process, two major counter-current flows take place: removal of water from the food to the osmotic solution and infusion of solutes from the solution to the food through semipermeable cell membranes (Figure 3.1). The driving force for water removal is the chemical potential between the solution and the intracellular fluid. For solutions, higher concentrations of solutions give a higher chemical potential and a

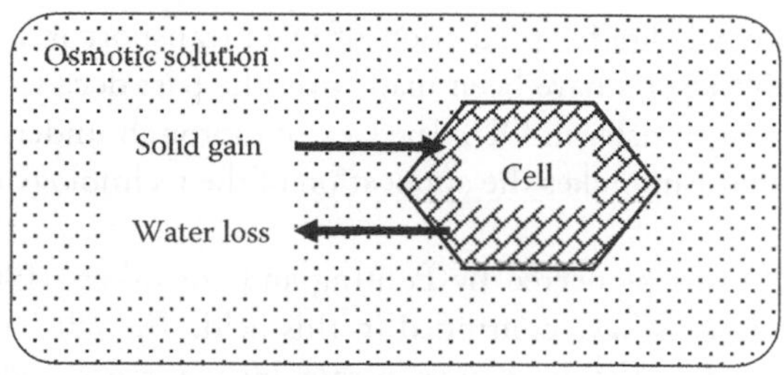

Figure 3.1 Counter-current flows during OD.

higher osmotic pressure (Π), which is a measure of this potential. In solutions of osmotically active solutes, viz. salt, sucrose, fructose, etc. (in the case of fruits and vegetables), the osmotic pressure can be determined using the van't Hoff model:

$$\Pi = cRT \tag{3.1}$$

where Π is the osmotic pressure, c is the molar concentration of all solutes, R is the universal gas constant, and T is the absolute temperature.

In the case of the OD process, two solutions of different osmotic pressures are separated by a semipermeable membrane (i.e., permeable to low molecular weight substances, such as water, and nonpermeable to high molecular weight solutes like sugars); a migration of low molecular weight components take place until equilibrium is reached between the osmotic pressure of the intracellular fluid and the surrounding osmotic solution. Theoretically, it is assumed that solute molecules, being large in size, do not migrate into the food from the syrup through the semipermeable cell membrane during osmo-concentration. However, due to the complex internal structure of plant cells, it is practically impossible to obtain a perfect semipermeable membrane. Hence, in reality, some food solutes viz. organic acids, vitamins, pigments, minerals, and sugars leach into the solution, and some solutes from the solution infuse the food during osmotic dehydration (Cohen and Yang 1995; Singh et al. 2007). These simultaneous water and solute diffusion processes take place due to water and solute activity gradients across the cell membrane (Tsamo et al. 2005; Torreggiani and Gianni 2001). Through the incorporation of solute in the food system, it is possible, to a certain extent, to change its nutritional and functional properties, achieving specific formulations of the product without modifying its integrity. Hence, a wide range of applications ranging from high dehydration with low solute impregnation to low dehydration with high solute uptake are possible.

The osmotic dehydration process is characterized by dynamic and equilibrium periods. In the dynamic period, the mass transfer rates are increased or decreased until equilibrium is reached. Equilibrium is the end point of the osmotic process, i.e., the net rate of mass transport is zero. The removal of water is mainly by diffusion and capillary flow, whereas solute uptake or leaching is only by diffusion.

The fundamental knowledge for prediction of mass transport is still a grey area, although considerable efforts have been made over the past decade to improve the understanding of mass transfer in OD. The lack of thorough understanding of the mass transport mechanism makes the application of the technology empirical even today.

Research on OD was pioneered by Ponting and coworkers (1966), and since then, steady investigations have continued in this field. The efficiency of the OD process is characterized by the mass transfer rate. Hence, most studies on OD are concentrated on establishing the process parameters, viz. temperature, time, agitation, type of solute, and solute concentration, etc. for enhancing the mass transfer rates. Three major indices are commonly used to characterize the OD process kinetics. These indices are water loss (WL), solid gain (SG), and weight reduction (WR). As a result of water and solute exchanges, the product loses weight and shrinks. The net WR is therefore the difference between water removal and SG. The following equations represent the formula for evaluation of the percentage of moisture loss, percentage of SG, and percentage of WR.

$$\text{WL} = \left(\frac{\text{weight of initial moisture} - \text{weight of final moisture}}{\text{initial weight of product}} \right) \times 100 \tag{3.2}$$

$$\text{SG} = \left(\frac{\text{weight of final solids} - \text{weight of initial solids}}{\text{initial weight of product}} \right) \times 100 \tag{3.3}$$

$$\text{WR} = \text{WL} - \text{SG} \tag{3.4}$$

In addition to the above indices, the ratio of WL to SG P_e is also sometimes used to measure the efficiency of the osmo-concentration process with respect to solute impregnation (Ramaswamy and Marcotte 2006). This index is shown in the following equation.

$$P_e = \frac{\text{WL}}{\text{SG}} \tag{3.5}$$

3.3 Modeling of Osmo-Concentration Process

3.3.1 Process Modeling and Diffusion Coefficients

Most of the models developed for the osmo-concentration process are based on the assumption that mass transfer is the rate-limiting step, and its rate can be approximately predicted by appropriate analytical solutions of a simplified Fick's second law of unsteady state diffusion. Fick's second law for unidirectional unsteady-state diffusion process is given by the following equation:

$$\left(\frac{\partial C}{\partial t}\right) = -D\left(\frac{\partial^2 C}{\partial x^2}\right) \tag{3.6}$$

If the diffusion is radial, Equation 3.6 takes the form of:

$$\left(\frac{\partial C}{\partial t}\right) = -D\left(\frac{\partial^2 C}{\partial x^2} + \frac{2}{x}\frac{\partial C}{\partial x}\right) \tag{3.7}$$

where t is the time, s, C is the concentration of diffusing substance, x is the distance parameter, m, and D is the diffusion coefficient, $m^2\ s^{-1}$.

The analytical solutions of these equations were developed by Crank (1975) with the following assumptions:

a. The diffusion process during osmo-concentration involves only two counter-current flows, viz. transfer of moisture out of the cell and flow of solids into the cell.
b. The process is isothermal.
c. Each sample is homogeneous and isotropic with uniform initial moisture and solid distribution.
d. The osmotic solution concentration is constant during the OD process.
e. The mass transfer coefficient between the osmotic solution and food sample is infinite.
f. The effect of shrinkage is negligible because the solid gain during OD is expected to compensate for product shrinkage.
g. The osmotic solution is well agitated.
h. Resistance to mass transfer at the minor surfaces is negligible in comparison to internal diffusion resistance.

Crank (1975) developed several analytical solutions for different shapes, viz. (1) infinite flat plate; (2) rectangular parallelepiped and cube; (3) infinite cylinder; and (4) sphere. Of these various solutions, models for some of the most commonly encountered shapes in food applications have been discussed below. These models are used to determine moisture and solid diffusion coefficients or diffusivity during OD.

3.3.1.1 Infinite Flat Plate

The OD process, in the case of infinite slab-shaped food materials, is governed by Equation 3.6 for unidirectional unsteady-state diffusion. The solution to this equation, as developed by Crank (1975), is in terms of moisture ratios (M_R) and solid ratios (S_R) based on the above assumptions. If the infinite slab is dehydrated from both the faces, the analytical solution of Equation 3.6 is written as Equation 3.8 for moisture ratios and Equation 3.9 for solid ratios for well-agitated unlimited volume of osmotic solution under the following boundary conditions:

$$C = C_0 \text{ at } t = 0, -l < x < +l \quad \text{and} \quad C = C_1 \text{ at } t > 0 \quad \text{and} \quad x = l$$

where C_0 and C_1 are initial and bulk concentrations, respectively.

$$M_R = \left(\frac{m_t x_t - m_e x_e}{m_0 x_0 - m_e x_e}\right) = \frac{8}{\pi^2} \sum_{n=0}^{\infty} \frac{1}{(2n+1)^2} \exp\left[-\left(n+\frac{1}{2}\right)^2 \pi^2 F_{ow}\right] \tag{3.8}$$

$$S_R = \left(\frac{m_e s_e - m_t s_t}{m_e s_e - m_0 s_0}\right) = \frac{8}{\pi^2} \sum_{n=0}^{\infty} \frac{1}{(2n+1)^2} \exp\left[-\left(n+\frac{1}{2}\right)^2 \pi^2 F_{os}\right] \tag{3.9}$$

For a well-agitated limited volume osmotic solution, the analytical solutions Equations 3.8 and 3.9 take the form of Equations 3.10 and 3.11, respectively.

$$M_R = \left(\frac{m_t x_t - m_e x_e}{m_0 x_0 - m_e x_e}\right) = 1 - \sum_{n=1}^{\infty} \frac{2\alpha(1+\alpha)}{\left(1+\alpha+\alpha^2 q_n^2\right)} \exp\left[-q_n^2 F_{ow}\right] \tag{3.10}$$

$$S_R = \left(\frac{m_e s_e - m_t s_t}{m_e s_e - m_0 s_0}\right) = 1 - \sum_{n=1}^{\infty} \frac{2\alpha(1+\alpha)}{\left(1+\alpha+\alpha^2 q_n^2\right)} \exp\left[-q_n^2 F_{os}\right] \tag{3.11}$$

where M_R is moisture ratio and S_R is solid ratio. m, x, and s represent the mass of the sample (kg), the moisture content of the sample (kg kg^{-1}), and the solid content of the sample (kg kg^{-1}). The subscripts 0, t, and e correspond to initial, any osmo-concentration time t, and equilibrium conditions, respectively. F_{ow} and F_{os} are Fourier numbers defined as $D_w t/l^2$ and $D_s t/l^2$, respectively, where D_w and D_s are effective moisture diffusivity and solid diffusivity (m^2 s^{-1}), respectively; t is the osmo-concentration time (s); and l is the half thickness of the infinite slab (m). q_n is the nonzero positive root of the equation $\tan q_n = -\alpha q_n$, and α is the ratio of volume of solution to that of the sample.

For estimation of moisture and solid diffusivity in the above case, Fourier numbers for moisture and solid diffusion can be obtained from Equations 3.8 or 3.10 and 3.9 or 3.11, respectively. Normally, the first three terms of Equations 3.8 or 3.10 and 3.9 or 3.11 are taken into consideration for calculation as the value of q_n

increases with values of n, and contribution of exponential terms will reduce with increased values of n (Das 2005). The values of Fourier numbers are then plotted against the corresponding values of osmo-concentration time, t. The values of moisture and solid diffusivity are estimated from the slopes of these plots.

3.3.1.2 Rectangular Parallelepiped and Cube

The solution to Fick's second law (Equation 3.6) for diffusion from a rectangular parallelepiped of dimensions 2a, 2b, and 2c (for cubes, all sides are taken as equal, i.e., 2a = 2b = 2c) is expressed in the form of moisture ratio and solute ratio as given by the following equations:

$$M_R = \left(\frac{m_t x_t - m_e x_e}{m_0 x_0 - m_e x_e}\right) = \sum_{n=1}^{\infty} C_n^3 \exp\left[-\frac{D_w t q_n^2}{A^2}\right] \tag{3.12}$$

$$S_R = \left(\frac{m_e s_e - m_t s_t}{m_e s_e - m_0 s_0}\right) = \sum_{n=1}^{\infty} C_n^3 \exp\left[-\frac{D_s t q_n^2}{A^2}\right] \tag{3.13}$$

where M_R is moisture ratio, and S_R is solid ratio. m, x, and s represent the mass of the sample (kg), the moisture content of the sample (kg kg^{-1}), and the solid content of the sample (kg kg^{-1}), respectively. The subscripts 0, t, and e correspond to initial, any osmo-concentration time, t, and equilibrium conditions, respectively. $C_n = \dfrac{2\alpha(+\alpha)}{\left(1+\alpha+\alpha^2 q_n^2\right)}$ and $1/A^2 = \left(\dfrac{1}{a^2}+\dfrac{1}{b^2}+\dfrac{1}{c^2}\right)$, D_w, and D_s are effective moisture diffusivity and solid diffusivity (m^2.s^{-1}), respectively; t is the osmo-concentration time (s). q_n is the nonzero positive root of the equation tan $q_n = -\alpha q_n$, and α is the ratio of volume of solution to that of the sample.

When the Fourier numbers for moisture $\left(F_{ow} = \dfrac{D_w t}{A^2}\right)$ and solute diffusion $\left(F_{os} = \dfrac{D_s t}{A^2}\right)$ are greater than 0.1, the higher terms involving $n > 1$ of Equations 3.12 and 3.13 can be neglected. These equations can be then rewritten in the form of the following equations:

$$-\ln\left(\frac{M_R}{C_n^3}\right) = \frac{D_w t q_n^2}{A^2} \tag{3.14}$$

$$-\ln\left(\frac{S_R}{C_n^3}\right) = \frac{D_s t q_n^2}{A^2} \tag{3.15}$$

Then, by plotting $-\ln\left(M_R/C_n^3\right)$ or $\left(S_R/C_n^3\right)$ against osmo-concentration time t, moisture and solid diffusivity can be estimated from the slopes of these plots.

3.3.1.3 Infinite Cylinder

$$M_R = \left(\frac{m_t x_t - m_e x_e}{m_0 x_0 - m_e x_e}\right) = 1 - \sum_{n=1}^{\infty} \frac{4}{a\alpha_n} \exp\left[-F_{ow}(a\alpha_n)^2\right] \tag{3.16}$$

$$S_R = \left(\frac{m_e s_e - m_t s_t}{m_e s_e - m_0 s_0}\right) = 1 - \sum_{n=1}^{\infty} \frac{4}{a\alpha_n} \exp\left[-F_{os}(a\alpha_n)^2\right] \tag{3.17}$$

where the $a\alpha_n$s are the roots of the equation $Jo(a\alpha_n) = 0$, a is the radius of the infinite cylinder, and $F_{ow} = \frac{D_w t}{A^2}$ and $F_{os} = \frac{D_s t}{A^2}$ are Fourier numbers for moisture and solute diffusion. For infinite cylinders, where length $l >> a$ (radius), $A = a$ (Rastogi et al. 1997).

Fourier numbers corresponding to different values of moisture ratio (Equation 3.16) and solid ratio (Equation 3.17) can be plotted against osmo-concentration time. The slopes of these plots are referred to as moisture and solid diffusivity, respectively.

3.3.1.4 Sphere

For spherical geometries, the analytical solution to Fick's equation for diffusion under constant OD process conditions is expressed in the form of the following equations (Crank 1975):

$$M_R = \frac{6}{\pi^2} \sum_{n=1}^{\infty} \frac{1}{n^2} \exp\left(-n^2\pi^2 D_w \frac{t}{r^2}\right) \tag{3.18}$$

$$S_R = \frac{6}{\pi^2} \sum_{n=1}^{\infty} \frac{1}{n^2} \exp\left(-n^2\pi^2 D_s \frac{t}{r^2}\right) \tag{3.19}$$

where M_R is moisture ratio and S_R is solid ratio. t, r, D_w, and D_s are osmo-concentration time (s), radius of sphere (m), effective moisture diffusivity, and solid diffusivity (m^2 s^{-1}), respectively.

When Fourier numbers for moisture $\left(F_{ow} = \frac{D_w t}{r^2}\right)$ and solute diffusion $\left(F_{os} = \frac{D_s t}{r^2}\right)$ are greater than 0.1, the higher terms involving $n > 1$ of Equations 3.16 and 3.17 can be neglected. Then, by plotting $-\ln(\pi^2 M_R/6)$ or $(\pi^2 S_R/6)$ against osmo-concentration time t, moisture and solid diffusivity can be estimated from the slopes of these plots.

3.3.2 Moisture Variation with Osmo-Concentration Time

Osmo-concentration is a simultaneous WL and SG process. Prediction of moisture content during osmo-concentration necessitates estimation of the above two parameters. A simple model using Fick's second law was developed by Jena and Das (2005) to predict the moisture content of apple and pineapple slices by estimating equilibrium WL, sugar diffusivity, and sugar gain from theoretical considerations. The model determines maximum WL at the attainment of osmotic equilibrium between the sugar solution and the fruit slice and then uses Fick's second law to find out the variation of WL with the time of osmo-concentration. The model also uses a constant called the "sugar–water diffusion quotient" for the first time to predict sugar diffusivity and sugar gain after a certain duration of osmo-concentration.

The model is based on the following assumptions:

a. There is no exchange of solutes between the fruit slices and sugar solution other than sugar and water.
b. All the water in the fruit slices is considered to be confined within the cell wall of the fruit through which water and sugar permeates.
c. Sugar gain is considered as equivalent to SG by the fruit slices.
d. At equilibrium, osmotic pressure of the solution confined within the cell wall of the fruit becomes equal to the average of the initial osmotic pressures of the solutions present inside and outside the cell wall.
e. The cell wall is considered as a hypothetical semipermeable membrane, which allows only sugar and water to permeate.

Major steps of the model are detailed below.

3.3.2.1 Water Loss and Sugar Gain during Osmo-Concentration

Water loss L_t ((kg water) (kg slice)$^{-1}$) and sugar gain S_t ((kg sucrose) (kg slice)$^{-1}$) after a time t during osmo-concentration can be given by Equations 3.20 and 3.21, respectively.

$$L_t = W_i X_i - W_t X_t \tag{3.20}$$

$$S_t = W_t (1 - X_t) - W_i (1 - X_i) \tag{3.21}$$

where W_i (kg) is the initial weight of the fruit slice, X_i ((kg water) (kg slice)$^{-1}$) is the initial moisture content of the slice, W_t (kg) is the final weight of the fruit slice, and X_t ((kg water) (kg slice)$^{-1}$) is the final moisture content of the slice.

3.3.2.2 Theoretical Maximum WL

Theoretical maximum WL L_Π ((kg water) (kg slice)$^{-1}$) from the fruit slice (i.e., equilibrium water loss) was obtained from the osmotic pressure equilibrium consideration as per assumption (d). Considering sucrose content of the fruit slice as constant during the process of WL, the following equation can be used to calculate L_Π:

$$\frac{1}{2}\left[\frac{\dfrac{W_iX_{si}}{M_s}}{\dfrac{W_iX_i}{\rho_w}}+\frac{\dfrac{W_{si}S_{si}}{M_s}}{\dfrac{W_{si}S_{wi}}{\rho_w}}\right]RT=\left[\frac{\dfrac{W_iX_{si}}{M_s}}{\dfrac{W_iX_i-L_\pi}{\rho_w}}\right]RT \tag{3.22}$$

where W_{si} (kg) is the initial weight of the sucrose solution, X_{si} ((kg sucrose) (kg slice)$^{-1}$) is the initial sucrose content of the fruit slice, S_{wi} ((kg water) (kg solution)$^{-1}$) is the initial moisture content of the sucrose solution, S_{si} ((kg sucrose) (kg solution)$^{-1}$) is the initial concentration of the sucrose solution, M_s (342 kg (kg mole)$^{-1}$) is the molecular weight of sucrose, M_w(18 kg (kg mole)$^{-1}$) is the molecular weight of the water, ρ_w (kg m^{-3}) is the density of water, and R (8314 J k mole^{-1} K^{-1}) is the universal gas constant. $\left[\frac{\dfrac{W_iX_{si}}{M_s}}{\dfrac{W_iX_i}{\rho_w}}\right]RT$ and $\left[\frac{\dfrac{W_{si}S_{si}}{M_s}}{\dfrac{W_{si}S_{wi}}{\rho_w}}\right]RT$ are the osmotic pressure of the sugar water solutions within and outside the fruit solid before osmo-concentration, respectively. The right-hand side of Equation 3.22 is the osmotic pressure of the sucrose solution within the fruit solid after the completion of osmo-concentration. On simplification, Equation 3.22 reduces to the following:

$$L_\pi = W_i\left[X_i-\frac{2X_{si}}{\left(\dfrac{X_{si}}{X_i}+\dfrac{S_{si}}{S_{wi}}\right)}\right] \tag{3.23}$$

3.3.2.3 Water Diffusivity

Water diffusivity D_w (m^2 s^{-1}) in fruit slices can be obtained by using the following solution of Fick's second law for unidirectional unsteady-state diffusion developed by Crank (1975) for an infinite slab allowing permeation of moisture from both sides in a well-agitated, limited volume of osmotic solution.

$$L_t = L_\pi\left[1-\sum_{n=1}^{\infty}\frac{2\propto(1+\propto)}{1+\propto+\propto^2\ q_n^2}\exp\left(-\frac{D_w q_n^2 t}{l^2}\right)\right] \tag{3.24}$$

where l (m) is the half thickness of the fruit slice. q_n, n = 1, 2, 3, ..., ∞ are the roots of the transcendental equation: tan $q_n = -\alpha q_n$ and α is the ratio of the weight of water in the sucrose solution to that in the fruit slice.

$$\alpha = \frac{W_{si}S_{wi}}{W_i X_i} \tag{3.25}$$

The first three roots of q_n were taken into consideration for a fairly accurate estimation of the value of water diffusivity D_w (m² s⁻¹). For a particular set of experimental data, a value of D_w is assumed, and the value of L_t at different times of osmo-concentration is calculated. The sum of squares of the difference between the actual and the calculated value of L_t is then estimated. The value of D_w at which the sum of squares is the minimum is taken as the actual value of water diffusivity during osmo-concentration.

3.3.2.4 Sugar Diffusivity and Sugar Gain

Diffusivities of sugar and water through the cell wall of fruit can be assumed to be proportional to $((T + 273)/M)^{1/p}$, where T (°C) is the absolute temperature of the fruit-sugar syrup system, M (kg (kg mole)$^{-1}$) is the molecular weight (which is denoted as M_s for sugar or M_w for water) and "p" is a constant designated as "sugar–water diffusion quotient." A similar type of relationship is used to relate Knudsen type diffusion of gases through solids (Treybal 1981). The value of "p" will depend on the type of fruit. It is calculated from the computational flow chart given in Figure 3.2. The value of sugar diffusivity D_s (m^2 s^{-1}) can thus be calculated from the value of water diffusivity D_w (m^2 s^{-1}) by using the following equation (Equation 3.26). No experimental sugar gain data is required for the prediction.

$$D_s = D_w\left(\frac{M_w}{M_s}\right)^{1/p} \tag{3.26}$$

The mass flux of sugar and water depend on the product of diffusivity and concentration gradient along the direction of moisture movement. During osmo-concentration, the mass flux of the water and sugar takes place under the same concentration gradients of water and sugar across the cell wall. Therefore, the

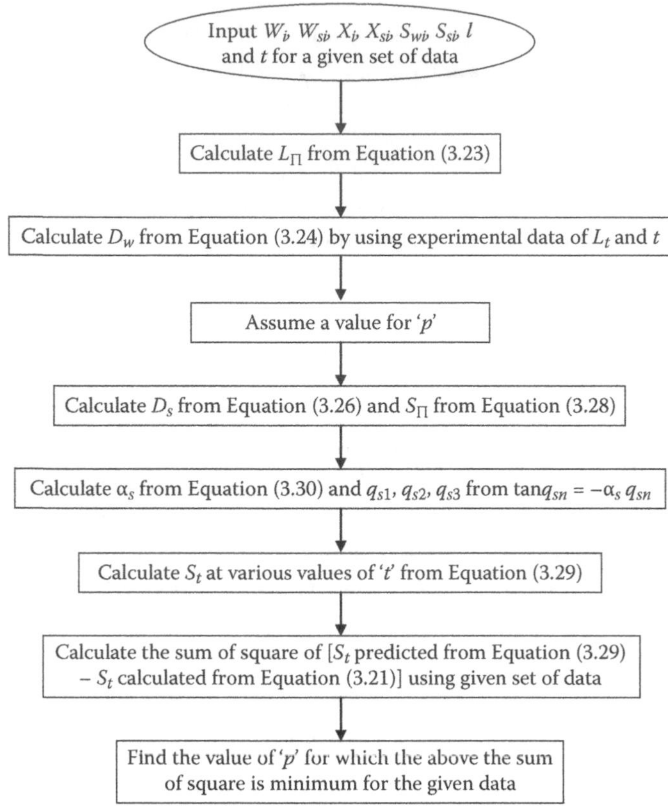

Figure 3.2 Computational flow chart for estimation of the value of "p."

theoretical maximum WL, L_Π ((kg water) (kg slice)$^{-1}$), and the corresponding amounts of maximum sugar gain, S_Π ((kg sucrose) (kg slice)$^{-1}$) will depend on the diffusivity of the cell wall against water and sugar, respectively, i.e.,

$$\frac{L_\pi}{S_\pi} = \frac{D_w}{D_s} \tag{3.27}$$

Using the relationships given in Equation 3.26 and 3.27, maximum sugar gain S_Π ((kg sucrose) (kg slice)$^{-1}$) can be estimated by:

$$S_\pi = L_\pi \left[\frac{M_w}{M_s}\right]^{1/p} \tag{3.28}$$

Sugar gain S_t ((kg sucrose) (kg slice)$^{-1}$) after a time t (s) of osmo-concentration can be predicted by using the following equation, which is similar to Equation 3.24:

$$S_t = S_\pi \left[1 - \sum_{n=1}^{\infty} \frac{2\alpha_s(1+\alpha_s)}{1+\alpha+\alpha_s^2 q_{sn}^2} \exp\left(-\frac{D_s q_{sn}^2 t}{l^2} \right) \right] \tag{3.29}$$

where q_{sn}, $n = 1, 2, 3, \ldots, \infty$ is the root of the transcendental equation $\tan q_{sn} = -\alpha_s q_{sn}$, α_s is the ratio of the weight of the sugar in the sucrose solution to that in the fruit slice, and

$$\alpha_s = \frac{W_{si} S_{si}}{W_i X_{si}} \tag{3.30}$$

3.3.2.5 Moisture Contents at Different Times of Osmo-Concentration

The moisture contents X_t ((kg water) (kg slice)$^{-1}$) in a fruit slice can be predicted from the values of WL, L_t (Equation 3.25) and sugar gain, S_t (Equation 3.29):

$$X_t = \frac{W_i X_i - L_t}{W_i + S_t - L_t} \tag{3.31}$$

where W_i (kg) is the initial weight of the fruit slice, X_i ((kg water) (kg slice)$^{-1}$) is the initial moisture content of the slice, L_t ((kg water) (kg slice)$^{-1}$) is the WL, and G_t ((kg sucrose) (kg slice)$^{-1}$) is the sugar gain after a time t during osmo-concentration.

Exercise 3.2 illustrates the use of the above modeling to predict the moisture content of apple rings during osmotic dehydration. A program developed in MATLAB® was used for solving the sample problem. In the problem, the percentage deviation error, E is used to compare the predicted and experimental values. E is calculated using Equation 3.32. An E value of less than 10% indicates a good fit (Lemon et al. 1985).

$$E = \frac{100}{N} \sum_{i=1}^{N} \left| \frac{pre - exp}{exp} \right| \tag{3.32}$$

where N is the number of experimental data, pre is the predicted values, and exp is the experimental values.

3.4 Factors Influencing Osmo-Concentration

3.4.1 Geometry of Food Materials

OD depends on the geometry of the sample due to a different specific surface area-to-volume ratio and varying diffusion lengths of water and solutes involved in mass transfer. Lerici et al. (1985) found that solute gain increased as the ratio of surface area to minimum diffusion length increased, and WL increased to maximum and then decreased. This decrease in WL may be due to a reduction in diffusion caused by high SG at the surface and consequent formation of a solute barrier layer. The size and shape of fruits and vegetables have a definite effect on the kinetics of OD. The commonly used shapes for osmo-concentration are finite disc, finite cylinder, infinite cylinder, cube, or sphere, etc. Agnelli et al. (2005) observed that WL and SG during OD were more in apple cubes of 1.5 cm as compared to 2 cm. A negative effect of size on the WL and SG were observed by Panagiotou et al. (1998) and Fernandes et al. (2006).

3.4.2 Physico-Chemical Properties of Food Materials

The chemical composition (protein, carbohydrate, fat, etc.) and physical structure (porosity, arrangement of cells, fiber orientation, and skin) have been found to influence the OD kinetics. The loss of membrane integrity due to heating was the cause of poor osmotic behavior (Islam and Flink 1982). It is a well-recognized fact that the osmo-concentration process will be influenced by the biological structure and compactness of the food material. Certain fruits, viz. citrus and tomatoes, were found not to be suitable due to their excessive juice loss during the process. Fruits that are porous in nature are better suited to vacuum treatment during OD.

3.4.3 Nature, Type, and Concentration of Osmotic Solution

Because osmotic pressure of a solution is dependent on the solute type, its properties, and its concentration, these parameters play a major role in determining the efficiency of the OD process. The most commonly used osmotic agents are sucrose for fruits and sodium chloride for vegetables, fish, and meat. Other osmotic agents include glucose, fructose, lactose, dextrose, maltose, polysaccharides, maltodextrin, corn starch syrup, sorbitol, and a combination of these osmotic agents. Sethi et al. (1999) reported that low molecular weight polysaccharides (glucose, fructose, sorbitol, etc.) favored sugar uptake because of the high velocity of penetration of the molecule. In such case, solid enrichment becomes the main effect instead of dehydration. A number of researchers have investigated the use of binary mixtures of solutes with sucrose as a means of reducing solute cost and improving the effectiveness of the OD process. For the mixed systems, WL was lower in comparison to using sodium chloride alone. Sodium chloride increases the driving force for dehydration owing to its water activity-lowering capacity, and its low molecular weight allows a higher rate of penetration in the material (Azoubel and Murr 2004). However, its use is limited because a salty taste is imparted to the food. In addition, sucrose allows the formation of a sugar surface layer, which becomes a barrier to the removal of water and the solute uptake. Typical WL and SG kinetics of ginger cubes using different types of osmotic agents are presented in Figures 3.3 and 3.4, respectively. The combination of sugar and salt solutions for osmotic dehydration has been reported by Singh et al. (2007) and Sacchetti et al. (2001). The ultimate choice of blends depends on many factors, viz. solute cost, toxicity, its reaction with food material, its capacity to lower water activity and organoleptic compatibility with the end product.

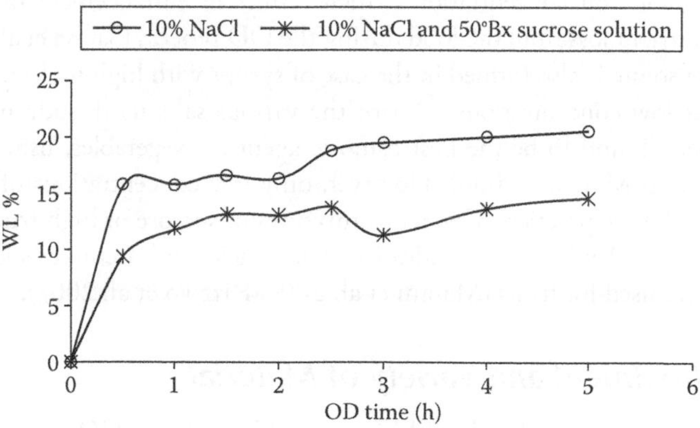

Figure 3.3 Effect of type of osmotic agents on WL kinetics of ginger.

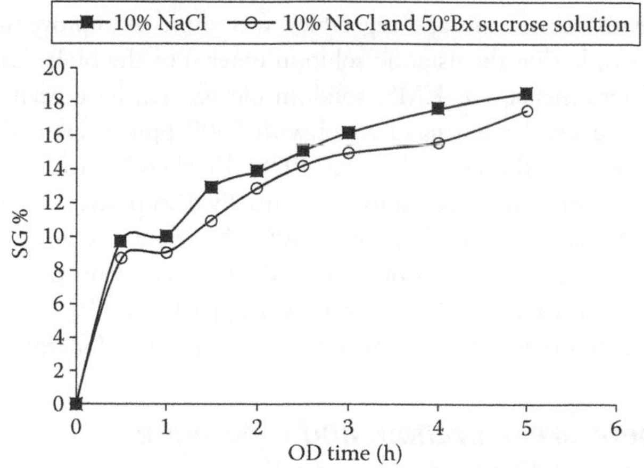

Figure 3.4 Effect of type of osmotic agents on SG kinetics of ginger.

The pH of osmotic syrup affects the OD efficiency. Acidification has been found to increase the rate of water loss by changes in tissue properties and consequential changes in the texture of fruits and vegetables (Moy et al. 1978). WL was maximum at pH 3.0 for apple rings using corn syrup (Contreras and Smyrl 1981). In a more acidic solution (pH < 2.0), the apple rings were found to be very soft whereas in a pH range of 3–6, firmness was maintained. The softening may be due to hydrolysis and depolymerization of pectin.

Mass transfer rate increases with an increase in osmotic solution concentration (Jena and Das 2005; Lombard et al. 2008; İspir and Toğrul 2009). This may be due to the fact that water activity of the solution decreases with an increase in osmotic solution concentration. A solute barrier layer at the surface of the product is formed with increased solution concentration, which results in enhancement of WL and reduction of nutrient loss from the food during the OD process (Saurel et al. 1994). A similar layer of solute is also formed in the case of syrups with high molecular weight solutes even at low concentrations. Out of the various salts used, sodium chloride (NaCl) has been found to be the best osmotic agent for vegetables, fish, and meat (Ramaswamy and Marcotte 2006). However, only low concentrations of salt solutions (0.5%–10%) are practical. Generally, mixtures of sucrose or high fructose corn syrup (40°Bx–70°Bx) with various additives (citric acid and ascorbic acid, lactose, sorbitol, etc.) are used for fruits (Marani et al. 2007; Rizzolo et al. 2007).

3.4.4 Pretreatment and Variety of Material

Common pretreatments, such as blanching, antioxidant dip, and KMS, are practiced as part of the OD process. However, blanching and freezing prior to OD has been found to be detrimental for plant materials as these pretreatments disrupt the plant cells and result in poor osmo-concentration. Physical treatments, e.g., punctures and abrasions on the surface of fruits and vegetables, was found to improve the efficiency of the OD process, letting the osmotic solution enter into the biological structure. Other chemical treatments, viz. KMS, solidum oleate, etc., have been reported to improve OD efficiency. Apple rings treated with 2500 ppm sulphur dioxide gave the best quality product (Ramamurthy et al. 1978). Biochemical treatments, such as the use of enzymes, could also be used to preferentially disrupt the skin or peel of the plant structure (Ramaswamy and Marcotte 2006). A variety or cultivar may greatly influence the osmo drying behavior of fruits and vegetables depending on the agro climatic conditions. The *Golden Delicious* variety of apple from a dry temperate region was found to be better than *Red Gold* for osmo-vacuum drying (Sharma et al. 1998).

3.4.5 Temperature, Duration, and Operating Pressure of Osmo-Concentration

Osmotic kinetics is greatly influenced by the temperature of an osmotic solution. WL increases with an increase in process temperature, whereas the positive effect on solid gain is less pronounced (Jena and Das 2005; Ramallo and Mascheroni 2005; İspir and Toğrul 2009). The effect of increasing the solution temperature of a highly concentrated solution was attributed to a decrease in the viscosity of osmotic solution, resulting in high diffusion rates for both water and solids (Singh et al. 2007). Too high temperatures result in possible damage to plant tissue structure and a "cooked taste" of the end product. Too low temperatures lead to increased processing time. The commonly employed temperature range for OD is between 40°C and 70°C. Increased temperature also resulted in poor retention of vitamin C and carotenoids in the case of bell peppers (Ade-Omowaye et al. 2002).

Most studies on OD are carried out in batch systems with highly concentrated solutions. An increased net weight loss was observed with an increase in process duration at a constant concentration of osmotic solution. Although the weight loss increases as a function of process duration, the rate of weight loss decreases. According to Tsamo et al. (2005), the kinetics of osmo-concentration of onion slices could be divided into three phases: (i) initial lag phase of 15 min during which rate of mass transfer is slow, (ii) an accelerated phase of 30–60 min showing peak mass transfer rate, and (iii) a decreasing or retarded phase in which the rate of water removal decreases.

In general, OD is carried out at atmospheric pressure. However, the effect of a vacuum on OD was also investigated by many researchers (Fito et al. 2001; Giraldo et al. 2006; Peiro-Mena et al. 2006; Atarés et al. 2008; Lombard et al. 2008). The application of a vacuum pulse at the beginning of OD has been reported to have beneficial effects on the structural changes of product and mass transfer kinetics leading to shorter processing time. Solid-liquid interfacial area and mass transfer between both phases could be increased using sub-atmospheric conditions. Application of a vacuum pulse facilitated WL especially at higher concentrations and temperatures (Lombard et al. 2008). Applying a high vacuum for long durations was found to have an adverse effect on fruit tissue structure (Mujica-Paz et al. 2003). Vacuum impregnation during osmotic dehydration also introduced controlled quantities of a solution in the porous structure of fruits and vegetables (Fito et al. 2001).

3.4.6 Sample-to-Solution Ratio and Agitation of Osmotic Solution

A sample-to-solution ratio is one of the major factors affecting the mass transfer rate during OD. The rate of osmosis (both WL and SG) increases with an increase in the ratio of sample to solution up to a certain level and, afterwards, levels off (Singh et al. 2007). The most employed sample-to-solution ratios were between 1:1 and 1:10. Higher ratios have also been used by many researchers to avoid significant dilution of medium during osmosis (Nieto et al. 2001). An optimum ratio of 1:4 for pineapple has been defined by Uddin and Islam (1985).

The OD process can be accelerated by agitation or circulation of the syrup around the sample with a low sample-to-solution ratio of 1:4 or 1:5 (Hawkes and

Flink 1978; Lenart and Flink 1984; Sethi et al. 1999; Matusek and Meresz 2002; Ramaswamy and Marcotte 2006). This is mainly due to the reduced mass transfer resistance at the surface of the sample, i.e., localized dilution is avoided. However, the improvement in mass transfer rate is so small that in some cases it might be more economical to use no agitation due to the extra agitation equipment needs and damage to the sample (Ponting et al. 1966).

3.5 Methods to Accelerate Mass Transfer during Osmo-Concentration

The rate of mass transfer during osmo-concentration is generally low. Various techniques have been tried by many investigators to improve the rate of WL and SG. These techniques include agitation, vacuum impregnation, application of ultra-high hydrostatic pressure (Rastogi and Niranjan 1998), applying high-intensity electrical field pulses (Rastogi et al. 1999) prior to OD, applying ultrasound (Simal et al. 1998; Mulet et al. 2003; Fernandes et al. 2009), γ-irradiation (Rastogi et al. 2006), and centrifugal force (Azuara et al. 1996) during OD. The influence of agitation and vacuum impregnation has already been discussed in previous sections.

Application of high pressures during osmo-concentration causes permeabilization of the cell structure by damaging cell wall structure and making the cell more permeable. This leads to significant changes in the tissue structure, resulting in increased mass transfer rates during OD. An increase in WL and SG was observed during osmo-concentration of potato treated at 400 Mpa (Rastogi et al. 2000). This was attributed to the synergistic effect of cell permeabilization and osmotic stress during the dehydration process.

Application of high-intensity electric pulse (HELP) treatment at room temperature with a peak electric strength of 0.5–2.5 kV cm^{-1}, number of pulses = 20, and a pulse duration of 400 μs showed enhanced mass transfer rates in bell peppers (Ade-Omowaye et al. 2002). The increase in effective diffusivities of moisture and solids with increased field strength in the case of carrots subjected to HELPs was attributed to increased cell permeability, facilitating the transport of water and solute (Rastogi et al. 1999). The combination of HELP with OD results in a product having higher vitamin C and carotenoid retention. According to traditional belief, it is necessary to have the cell membrane intact for achieving high mass transfer rates in case of OD. However, the recent findings of high pressure and HELP treatment in OD are in contradiction to the conventional theory, i.e., high mass transfer rates could be achieved through disintegrated cell structure (Raoult-Wack 1994; Rastogi et al. 1999; Ade-Omowaye et al. 2002). This fact is gaining importance because it is inevitable to work with food items whose cell structure is disintegrated during ripening or primary processing.

The applicability of sonication or ultrasound in OD has been reported by many researchers (Simal et al. 1998; Mulet et al. 2003; Fernandes et al. 2009). According to Fernandes et al. (2009), ultrasound induced disruption of cells and formation of microscopic channels in the fruit structure without breakdown of the cells. Consequently, ultrasound applications increased sugar loss and water diffusivity in pineapple cells because of the formation of microscopic channels, which offered lower resistance to water and sugar diffusion. Simal et al. (1998) recommended the use of a lower solution temperature to obtain high mass transfer rates during OD of apple cubes.

Azuara et al. (1996) worked on the application of centrifugal force at 64 xg at 30°C in osmo-concentration apple and potato slices (cylindrical shape) using a mixed solute of sucrose and salt. An enhanced WL by 15% and reduced solid uptake by 80% was observed. However, further investigations are still in place to understand the effect of rotational speed, temperature, and concentration of osmotic solution, type of solute, and solute geometry in osmotic dehydration with centrifugal force.

γ-Irradiation treatment prior to OD has been reported to increase mass transfer rate (Rastogi et al. 2006) in potato and carrot pieces. The increase in mass transfer was attributed to an increase in cell wall permeability. Softening of fruit tissues was also observed due to γ-irradiation exposure.

3.6 Industrial Applications of Osmo-Concentration

Industrial applications of OD are very limited due to the lack of development of the technology. However, dried fruits, candy, and dehydrated vegetables are some of the major applications of osmo-concentration. For the production of dried fruits, fresh fruit pieces are soaked in sucrose syrup in batch systems at normal atmospheric pressure and at temperatures in the range of 30°C–80°C. The osmo-concentrated fruits are then air dried to produce dried products of 15%–20% moisture content wet basis. Figure 3.5 shows a typical process flowchart for production of dehydrated pineapple rings.

Thailand is the major producing country of such dried fruits. Dried fruits are mainly prepared from mango, papaya, and pineapple. Jackfruit, banana, and ginger are also used to produce dry fruits using this technology. Dehydrated papaya (popularly known as "tutti-frutti") is being produced in small- and medium-scale industries in South Asian countries, including India. France is one of the European countries that has developed technologies for candy and other OD products. Bertuzzei, Italy, is one of the leading producer companies of such products.

Osmo-concentration has been applied in the field of concentration of fruit juices. Raoult-Wack (1994) designed a process for production of fruit and vegetable aromatic concentrates. In this process, the fruit pieces are first osmotically dehydrated, then crushed, refined, and pasteurized or frozen to produce concentrates. From the recent developments, it has also been possible to produce fruit juice concentrates using only osmo-concentration. However, by this process, it is not

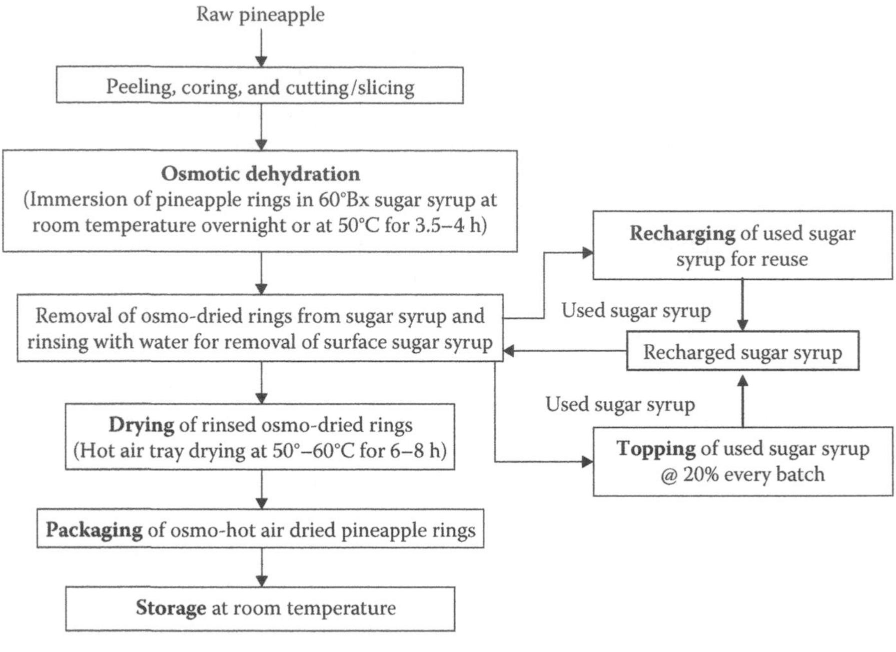

Figure 3.5 Process flowchart for production of dehydrated pineapple rings.

feasible to concentrate the fruit juice above a certain level (40°Bx–50°Bx) due to high osmotic pressure barriers. To overcome such difficulties, application of an osmotic gradient across a polymeric membrane (instead of pressure gradient as in the case of reverse osmosis/ultra filtration) can be employed. This process is termed "direct osmosis" or "osmotic membrane distillation." A hybrid process, ultrafiltration/reverse osmosis followed by direct osmosis has been successfully demonstrated on a pilot plant scale for concentration of fruit and vegetable juices in Mildura and Melbourne, Australia. The osmotic membrane distillation plant was designed and fabricated by Zenon Environmental, Burlington, Ontario.

Typical batch type equipment for OD consists of a mixing tank for the syrup with heating elements to maintain the temperature. Due to the difference in densities, the product tends to float in the syrup. Hence, the product is placed in a perforated basket to avoid floating of food pieces. To increase the contact between the product and the solution, arrangement for periodic agitation is also included in such batch systems. After the treatment, the product is removed by draining the osmotic solution. It is also possible to develop an osmotic process involving a series of 'batch systems' with sugar concentrations in increasing order to minimize osmotic shocks. However, scaling up of such processes is very difficult.

Semi-batch or semi-continuous units were also developed to overcome problems associated with batch systems. In these systems, the product is packed in a column, and the solution is pumped and recirculated. These units are named "batch-recirculation" and are similar to units commonly found in conventional solid-liquid extraction.

Development of continuous OD units is considered difficult as many factors have to be taken into consideration. Percolation units similar to solid-liquid solvent extraction have been developed for OD. In such units, the product is introduced continuously and immersed into the solution or the product is introduced and the solution is sprayed over the product. A rotating drum was used to circulate the fruit pieces immersed in the solution (Figure 3.6). A similar process has also been developed using a rotating drum, and the solution is atomized with a stream in a rotating drum containing fruit slices. But the major difficulty in this process was to control the residence time of the product.

In another design, a belt type osmotic-contractor was developed. In this design, food pieces were evenly distributed on a belt, and the osmotic solution was sprayed over it. Using a conveyor belt, this process can be made as a concurrent or counter-current type. This design was mainly developed for sugar infusion into cranberries. A pilot scale continuous OD unit has also been developed by some researchers for carrots and other fruits and vegetables. In this system, the main part is a perforated wheel divided into 24 compartments, 12 of which are submerged in osmotic solution. During rotation of the wheel, a certain amount of the fruit sample is subjected to osmotic treatment with the same temperature, concentration, and processing time. This design is particularly suitable for long process time and can be used instead of a conveyor belt-type immersion system, which requires a lot of floor space during installation. Both batch and continuous systems can be improved using

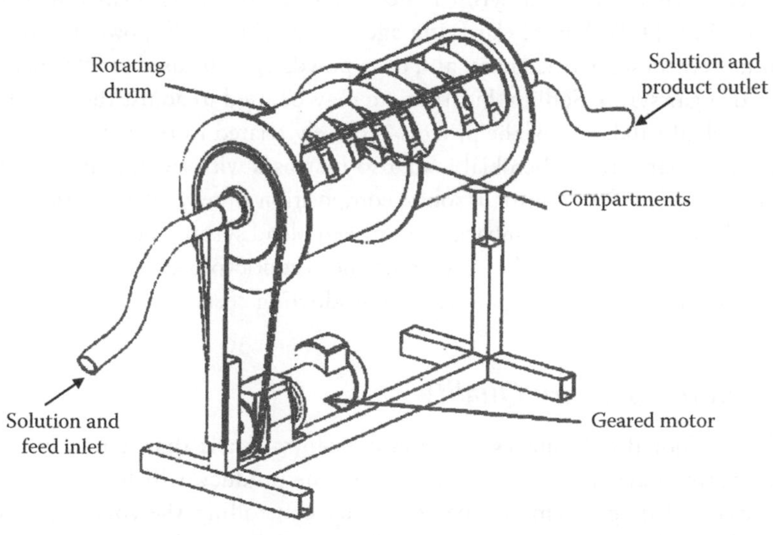

Figure 3.6 Rotating drum unit for osmo-concentration.

vibration to the conveyor. Use of fluidized bed drying in combination with OD is presently under investigation.

3.7 Constraints Associated with Osmo-Concentration

3.7.1 Management of Osmotic Solution

Management of osmotic solution poses the major challenge to make the process industrially viable on a large scale. Management of syrup includes syrup composition, syrup concentration, reuse of syrup, solute addition, and waste disposal. Dilute solution, if not reused, is considered as an industrial waste as it contains a large amount of organic matter. Dilution of osmotic solution, changes in pH, and color of sugar syrup are some of the concerns associated with osmo-concentration. Hence, management of solution is important not only because it is economical, but also because it will have a lower impact on the environment. Several researchers have worked on reconcentration of diluted sugar syrup. The methods practiced for reconcentration of syrup include evaporation at atmospheric pressure, vacuum evaporation, membrane filtration (reverse osmosis), and cryo-concentration. However, these operations are very costly. The use of diluted syrup as an ingredient in other foods, viz. juices, concentrates, tablet syrups, etc., is also an attractive alternative solution for small-scale industries.

During OD of fruits, especially pineapple, the pH of sugar syrup reduces from about 6.2 to around 4.0, and the turbidity of sugar syrup increases, which results in undesirable changes in the product in subsequent uses. Hence, clarification and pH balance of the used sugar syrup is necessary to ensure its quality in future applications in OD. Biological clarifying agents, viz. skim milk powder and okra in the form of fresh stem, dried okra, and okra powder, can be used for clarification of the used sugar syrup. Sodium bicarbonate may be used to adjust the pH of the syrup. Microbial validation of the process is closely related to the management of concentrated sugar syrup. Microbial load also increases with the number of times the syrup is recycled. The control of solute composition in recycling is easier in the case of single solute syrup in comparison to mixed blend syrup. The cost of osmotic syrup is also a key factor in the success of the osmotic process. The sample-to-solution ratio should be kept high to reduce production costs.

3.7.2 Product Sensory Quality

One of the major disadvantages of osmo-concentration is that it may increase saltiness or sweetness or decrease the acidity of the product, which is not desired in some cases. This problem can be solved by controlling the solute diffusion and optimizing the process parameters to improve sensory characteristics of the product. Edible semipermeable membrane coatings made up of lipids, resins, polysaccharides, and proteins can also be used to reduce solute uptake and improve WL (Lenart and Piotrowski 2001; Khin et al. 2006). Khin et al. (2006) used solutions of 1% sodium alginate and 2% low methoxy pectinate as coating materials of potato cubes for osmotic dehydration. The use of these hydrophilic coating materials resulted in effective control of solute uptake at high temperatures while maintaining the WL rate.

3.7.3 Process Control and Design

Inadequate and limited information about experiments available in the literature creates hurdles in the effective design and control of osmo-concentration process by the food industry. Both qualitative and quantitative predictions of processing factors are necessary for process design in industries. Online measurements of syrup properties can provide continuous control of the process. Fruit and vegetable pieces tend to float in syrup due to density difference. Moreover, the viscosity of the syrup also exerts considerable mass transfer resistance and causes difficulty in agitation and the syrup tends to adhere to the food material. Breakage of food material may occur during continuous process or by mechanical agitation during batch process. These are certain inherent problems associated with osmo-concentration that needs research attention.

3.8 Summary

Osmo-concentration is still regarded as a mild and energy efficient minimal processing methods to produce intermediate moisture foods. Due to its energy and quality related advantages, is also gaining popularity as a complimentary processing step in food processing. Prediction of osmotic mass transfer, though a complex area, can be effectively dealt with the help of different process models discussed in the chapter. In view of the continuously emerging new techniques, application of the discussed methods to enhance the mass transfer during osmo-concentration would be highly useful. The scaling up of osmo-concentration facility; a challenging task; has been addressed through citation of industrial applications. Nevertheless osmo-concentration process has still some blind spots that needs to be addressed to in future. A great amount of work on osmo-concentration has been carried out in the past. However, the studies were not sufficient to make the process industrially viable. Considerable research has been undergone for understanding the mass transfer phenomena of OD. Still, significant advances have to be made to enhance the mass transfer rates. There is ample scope for research and development in this area. At present, reconcentration of used syrup is being done by atmospheric or vacuum evaporation. Use of other reconcentration methods, viz. membrane processing and cryo-concentration, has tremendous potential for research and investigation. High solute concentration and microbial problems are less severe in the OD process. However, these findings have to be strengthened through systematic and in-depth studies. The challenge of clarification and pH balance of used sugar syrup using bio-cleansers and other suitable additives has been little investigated scientifically. Considerable research scope is still there to explore this possibility.

Special emphasis is needed for equipment design, especially for handling fragile food materials with simultaneous provision of automatic control and online measurements. The application of osmo-concentration to gel-type structures needs further investigation to benefit other foods, viz. egg or dairy products, and non-food applications.

Active research is going on around the globe to develop functional foods through addition of bioactive components to food materials. OD can play a major role in this area. Hence, new technologies and equipment could be designed to develop osmo-concentrated functional foods. Thorough research is needed to study the process parameters for various food products, either to extract or preserve the bioactivity during processing of functional foods. Although this type of study has already been applied in other fields of research, such as solid-liquid extraction, the challenge lies in applying this to the processing of biological materials and to minimize the environmental impact of these processes.

EXERCISE 3.1

OD of 1 kg of ginger cubes was carried out using 10% NaCl solution at a temperature of 45°C. The solution-to-sample ratio was kept at 5:1. The osmo-concentration process was carried out for 5 h. Using the following osmo-concentration data, determine percentage WL, SG, and WR at different time intervals using Equations 3.2, 3.3, and 3.4, respectively. Experimental data are presented in Table 3.1.

Table 3.1 Osmo-Concentration Data

Time, h	*Weight of Ginger, W (kg)*	*Moisture Content, X (% wb)*	*Solid Content, S (%)*
0	1.00	87.04	12.96
1	0.92	84.21	15.79
2	0.85	78.76	21.24
3	0.77	75.23	24.77
4	0.69	73.61	26.39
5	0.63	69.18	30.82

Solution

Weight of raw ginger $W_i = 1$ kg
Initial moisture content $X_i = 0.8704$ kg kg^{-1}
Final moisture content at 1 h osmo-concentration time $X_1 = 0.8421$ kg kg^{-1}
Initial amount of moisture present in ginger = $m_i = W_iX_i = 1*0.8704 = 0.8704$ kg
Final amount of moisture present after 1 h $m_1 = W_1X_1 = 0.92*0.8421 = 0.775$ kg
Using Equation 3.2, the percentage WL after 1 h

$$\text{WL} = \left(\frac{m_i - m_1}{W_i}\right)*100 = \left(\frac{0.87 - 0.775}{1}\right)*100 = 9.57\%$$

Amount of solid content present initially $s_i = W_i\,S_i = 1*0.1296 = 0.1296 \approx 0.13$ kg
Amount of solid content present after 1 h $s_1 = W_1S_1 = 0.92*0.1579 = 0.145$ kg
From Equation 3.3, the percentage SG after 1 h

$$\text{SG} = \left(\frac{s_1 - s_i}{W_i}\right)*100 = \left(\frac{0.145 - 0.13}{1}\right)*100 = 1.57\%$$

From Equation 3.4, the percentage of weight reduction after 1 h WR = 9.57 − 1.57 = 8%.

Proceeding in the same manner, WL, SG, and WR in ginger at different time intervals are:

t (h)	1	2	3	4	5
%WL	9.57	20.10	29.10	36.25	43.46
%SG	1.57	5.10	6.11	5.25	6.46
%WR	8.00	15.00	23.00	31	37

By this example, it can be comprehended that approximately 45% of moisture reduction with less than 10% solid impregnation can be achieved by this simple treatment of OD. The process also results in about 40% WR, which decreases the bulk weight of material resulting in convenience of handling and transport.

EXERCISE 3.2

Apple slices of 8 ± 0.1 mm thickness were immersed in sucrose solution of 50°Bx at a temperature of 50°C for OD. The OD was carried out for 5 h. The ratio of weight

Table 3.2 Experimental Data for OD of Apple Slices

Time, h	*Weight of Apple Slice, W_i (kg)*	*Moisture Content, X_i (% wb)*
0	0.01594	87.514
1	0.01254	75.314
2	0.01131	69.185
3	0.01275	67.543
4	0.01097	66.614
5	0.01086	65.897

of sucrose solution to apple slices was 4:1. Initial sucrose content of apple is 6.56%. Using the data presented in Table 3.2, characterize the moisture content profile of the apple slices during the dehydration process. Also compare predicted moisture contents with the experimental values at different OD times.

Solution

Initial weight of apple slice, $W_i = 0.01594$ kg
Weight of sucrose solution, $W_{si} = 4*W_i = 0.06376$ kg
Half thickness of slice, $l = 0.004$ m
Initial moisture content of slice, $X_i = 0.87514$ (kg water) (kg slice)$^{-1}$
Initial sucrose content of apple, $X_{si} = 0.0656$ kg (kg slice)$^{-1}$
Initial sucrose content of sucrose solution, $S_{si} = 0.6$ (kg sucrose) (kg solution)$^{-1}$
Initial moisture content of sucrose solution $S_{wi} = 1 - S_{si}0 = 0.4$ (kg sucrose) (kg solution)$^{-1}$
Molecular weight of water, $M_w = 18$ kg (kg mole)$^{-1}$
Molecular weight of sucrose, $M_s = 342$ kg (kg mole)$^{-1}$
Using Equation 3.23, maximum WL, $L_{\text{II}} = 0.01262$ (kg water) (kg slice)$^{-1}$

Using Equation 3.20, WL, L_t at 0, 1, 2, 3, 4, and 5 h is 0, 0.00451, 0.00613, 0.00534, 0.00664, and 0.00679 (kg water) (kg slice)$^{-1}$, respectively.

Using Equation 3.25, the value of $A = 1.828$

First three roots of transcendental equation from standard tables q_1, q_2, and q_3 are 1.857, 4.825, and 7.923, respectively.

A value of water diffusivity, $D_w = 1*10^{-11}$ was assumed, and L_t at different times was predicted from Equation 3.24. The value at which the sum of square of the difference between predicted and experimental L_t is minimum is taken to be the final water diffusivity value = $1.5*10^{-10}$ m^2 s^{-1}.

The predicted values of WL, L_t at 0, 1, 2, 3, 4, and 5 h using the above D_w value in Equation 3.20 are 0.0013, 0.00372, 0.0051, 0.0061, 0.0068, and 0.0075 (kg water) (kg slice)$^{-1}$, respectively.

Using Equation 3.30, the value of $\alpha_s = 36.535$

First three roots of transcendental equation from standard tables q_{s1}, q_{s2}, and q_{s3} are 1.588, 4.718, and 7.856, respectively.

S_t at 0, 1, 2, 3, 4, and 5 h using Equation 3.21 are 0, 0.00111, 0.0015, 0.00215, 0.00167, and 0.00171 (kg sucrose) (kg slice)$^{-1}$, respectively.

Following the procedure explained in Figure 3.2, the value of p based on minimum sum of square of the difference between predicted and experimental $S_t = 4.19$.

Putting the values of $L_{II} = 0.01262$ kg water slice^{-1}, M_w and M_s in Equation 3.28, maximum sucrose gain, $S_{II} = 0.00625$ (kg sucrose) (kg slice)$^{-1}$.

Using the calculated values of $D_w = 1.5*10^{-10}$ m^2 s^{-1} and $p = 4.19$ in Equation 3.26, the value of sugar diffusivity, $D_s = 7.421 \times 10^{-11}$ m^2 s^{-1}

Putting the calculated values of values of S_{II} and D_s in Equation 3.29, predicted sucrose gain S_t at 0, 1, 2, 3, 4, and 5 h are 0.00043, 0.00095, 0.00132, 0.0016, 0.0019, and 0.0021 (kg sucrose) (kg slice)$^{-1}$, respectively.

Using the predicted values of L_t and S_t and values of W_i and X_i in Equation 3.31, moisture contents at different osmo-concentration times are predicted to be 0.839, 0.777, 0.728, 0.687, 0.649, and 0.613 (kg water) (kg slice)$^{-1}$.

The developed model can be tested for accuracy by the percent deviation error, E (Equation 3.32) = 3.97%. It suggests a good fit. Figure 3.7 depicts the experimental and predicted moisture profile of the apple slices.

This numerical example illustrates that moisture content of fruit slices during osmo-concentration can be predicted accurately by using the model described.

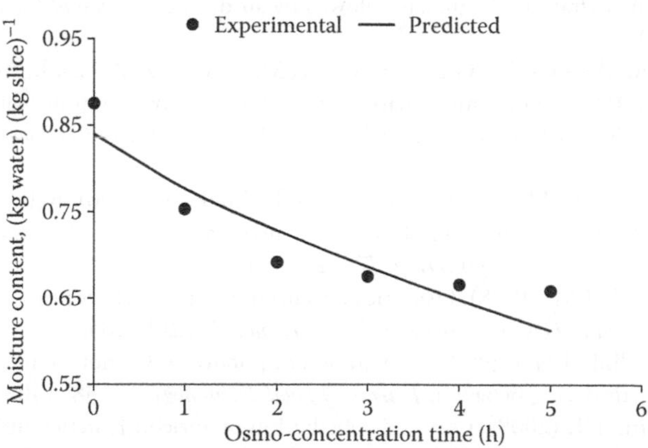

Figure 3.7 Moisture profile of apple slices during osmo-concentration.

References

Ade-Omowaye, B.I.O., Rastogi, N.K., Angerbach, A. and Knorr, D. (2002). Osmotic dehydration of bell peppers: Influence of high intensity electric field pulses and elevated temperature treatment. *Journal of Food Engineering*, 54, 35–43.

Agnelli, M.E., Marani, C.M. and Mascheroni, R.H. (2005). Modelling of heat and mass transfer during (osmo) dehydrofreezing of fruits. *Journal of Food Engineering*, 69, 415–424.

Angu, K. and Jena, S. (2010). Mass transfer studies on ginger during osmotic dehydration. Unpublished B. Tech. Project report, Central Agricultural University, Imphal, India.

Atarés, L., Chiralt, A. and González-Martínez, C. (2008). Effect of solute on osmotic dehydration and rehydration of vacuum impregnated apple cylinders (cv. Granny Smith). *Journal of Food Engineering*, 89(1), 49–56.

Azoubel, P.M. and Murr, E.X.F. (2004). Mass transfer kinetics of osmotic dehydration of cherry tomato. *Journal of Food Engineering*, 61, 291–295.

Azuara, E., Garcia, H.S. and Beristain, C.I. (1996). Effect of centrifugal force on osmotic dehydration of potatoes and apples. *Food Research International*, 29, 195–199.

Cohen, J.S. and Yang, C.S. (1995). Progress in food dehydration. *Trends in Food Science and Technology*, 6, 20–24.

Contreras, J.S. and Smyrl, T.G. (1981). An evaluation of osmotic concentration of apple rings using corn solids solutions. *Canadian Institute of Food Science and Technology Journal*, 14, 310–314.

Crank, J. (1975). *The mathematics of diffusion*. Oxford University Press, London.

Das, H. (2005). *Food processing operation analysis*. Asian Books Pvt Ltd, New Delhi, India.

Dhingra, D., Singh, J., Patil, R.T. and Uppal, D.S. (2008). Osmotic dehydration of fruits and vegetables: A review. *Journal of Food Science and Technology*, 45(3), 209–217.

Fernandes, F.A.N., Gallão, M.I. and Rodrigues, S. (2009). Effect of osmosis and ultrasound on pineapple cell tissue structure during dehydration. *Journal of Food Engineering*, 90(2), 186–190.

Fernandes, F.A.N., Rodrigues, S., Gaspareto, C.A. and Oliveira, E.L. (2006). Optimization of osmotic dehydration of bananas followed by air drying. *Journal of Food Engineering*, 77, 188–193.

Fito, P., Chiralt, A., Betroret, N., Grass, M., Chafe, M., Martinez-Monzo, J., Andres, A. and Vidal, I. (2001). Vacuum impregnation and osmotic dehydration in matrix engineering: Application in functional fresh food development. *Journal of Food Engineering*, 49, 175–183.

Giraldo, G., Vazquez, R., Martin-Esparza, M.E. and Chiralt, A. (2006). Rehydration kinetics and soluble solids lixiviation of candied mango fruit as affected by sucrose concentration. *Journal of Food Engineering*, 77, 825–834.

Hawkes, J. and Flink, J.M. (1978). Osmotic concentration of fruit slices prior to freeze dehydration. *Journal of Food Processing and Preservation*, 39, 265–284.

Islam, M.N. and Flink, J.N. (1982). Dehydration of potato. II. Osmotic concentration and its effect on air drying behavior. *Journal of Food Technology*, 17, 387–403.

İspir, A. and Toğrul, İ.T. (2009). Osmotic dehydration of apricot: Kinetics and the effect of process parameters. *Chemical Engineering Research and Design*, 87(2), 166–180.

Jena, S. and Das, H. (2005). Modeling for moisture variation during osmo-concentration in apple and pineapple. *Journal of Food Engineering*, 66(4), 425–432.

Khin, M.M., Zhou, W. and Perera, C.O. (2006). A study of the mass transfer in osmotic dehydration of coated potato cubes. *Journal of Food Engineering*, 77, 84–95.

Lemon, C.J., Bakshi, A.S. and Labuza, T.P. (1985). Evaluation of moisture sorption isotherm equation: Part 1, fruits, vegetables and meat products. *Lebensmittel Wissenschaft und Technologie*, 18, 111–115.

Lenart, A. and Flink, J.M. (1984). Osmotic concentration of potato. I. Criteria for the end-point of the osmosis process. *Journal of Food Technology*, 19, 45–63.

Lenart, A. and Piotrowski, D. (2001). Drying characteristics of osmotically dehydrated fruits coated with semipermeable edible films. *Drying Technology*, 19, 849–877.

Lerici, C.R., Pinnavaia, G., Rosa, M.D. and Bartolucci, L. (1985). Osmotic dehydration of fruit: Influence of osmotic agents on drying behavior and product quality. *Journal of Food Science*, 50, 1217–1219.

Lombard, G.E., Oliveira, J.C., Fito, P. and Andrés, A. (2008). Osmotic dehydration of pineapple as a pretreatment for further drying. *Journal of Food Engineering*, 85(2), 277–284.

Marani, C.M., Agnelli, M.E. and Mascheroni, R.H. (2007). Osmo-frozen fruits: Mass transfer and quality evaluation. *Journal of Food Engineering*, 79, 1122–1130.

Matusek, A. and Meresz, P. (2002). Modelling of sugar transfer during osmotic dehydration of carrots. *Periodica Polytechnica ser Chemical Engineering*, 46, 83–92.

Moy, J.H., Lau, N.B.H. and Dollar, A.M. (1978). Effects of sucrose and acids on osmovac-dehydration of tropical fruits. *Journal of Food Processing and Preservation*, 2, 131–135.

Mujica-Paz, H., Valdez-Fragoso, A., Lopez-Malo, A., Palou, E. and Welti-Chanes, J. (2003). Impregnation properties of some fruits at vacuum pressure. *Journal of Food Engineering*, 56, 307–314.

Mulet, A., Cárcel, J.A., Sanjuán, N. and Bon, J. (2003). New food drying technologies—use of ultrasound. *Food Science and Technology International*, 9(3), 215–221.

Nieto, A., Castro, M.A. and Alzamora, S.M. (2001). Kinetics of moisture transfer during air drying of blanched and/or osmotically dehydrated mango. *Journal of Food Engineering*, 50, 175–185.

Panagiotou, N.M., Karathanos, V.T. and Maroulis, Z.B. (1998). Mass transfer modeling of the osmotic dehydration of some fruits. *International Journal of Food Science and Technology*, 33, 267–284.

Peiro-Mena, R., Camacho, M.M. and Martinez-Navarrete, N. (2007). Compositional and physicochemical changes associated to successive osmodehydration cycles of pineapple (*Ananas comosus*). *Journal of Food Engineering*, 79, 842–849.

Peiro-Mena, R., Dias, V.M.C., Camacho, M.M. and Martinez-Navarrete, N. (2006). Micronutrient flow to the osmotic solution during grape fruit osmotic dehydration. *Journal of Food Engineering*, 74, 299–307.

Ponting, J.D., Watters, G.G., Forrey, R.R., Jackson, R. and Stanley, W.L. (1966). Osmotic dehydration of fruits. *Food Technology*, 20, 125–129.

Ramallo, L.A. and Mascheroni, R.H. (2005). Rate of water loss and sugar uptake during the osmotic dehydration of pineapple. *Brazilian Archives of Biology and Technology*, 485, 761–770.

Ramamurthy, M.S., Bongirwar, D.R. and Bandyopadhyay, C. (1978). Osmotic dehydration of fruits: Possible alternative to freeze-drying. *Indian Food Packer*, 32(1), 108–112.

Ramaswamy, H. and Marcotte, M. (eds) (2006). *Food Processing: Principles and Applications*, CRC Press, Boca Raton, FL, pp. 317–377.

Raoult-Wack, A.I. (1994). Recent advances in the osmotic dehydration of foods. *Trends in Food Science and Technology*, 5, 255–260.
Rastogi, N.K., Angersbach, A. and Knorr, D. (2000). Synergestic effect of high hydrostastic pressure pretreatment and osmotic stress on mass transfer during osmotic dehydration. *Journal of Food Engineering*, 45, 25–31.
Rastogi, N.K. and Niranjan, K. (1998). Enhanced mass transfer during osmotic dehydration of high pressure treated pineapple. *Journal of Food Science*, 63, 508–511.
Rastogi, N.K., Eshtiaghi, M.N. and Knorr, D. (1999). Accelerated mass transfer during osmotic dehydration of high intensity electrical filed pulse pretreated carrots. *Journal of Food Science*, 64, 1020–1023.
Rastogi, N.K., Raghavarao, K.S.M.S. and Niranjan, K. (1997). Mass transfer during osmotic dehydration of banana: Fickian diffusion in cylindrical configuration. *Journal of Food Engineering*, 31, 423–432.
Rastogi, N.K., Raghavarao, K.S.M.S., Niranjan, K. and Knorr, D. (2002). Recent developments in osmotic dehydration: Methods to enhance mass transfer. *Trends in Food Science and Technology*, 13(2), 48–59.
Rastogi, N.K., Suguna, K., Nayak, C.A. and Raghavarao, K.S.M.S. (2006). Combined effect of γ-irradiation and osmotic pretreatment on mass transfer during dehydration. *Journal of Food Engineering*, 77, 1059–1063.
Rizzolo, A., Geril, F., Prinzivalli, C., Buratti, S. and Torreggiani, D. (2007). Headspace volatile compounds during osmotic dehydration of strawberries (cv camarosa): Influence of osmotic solution composition and processing time. *LWT–Food Science and Technology*, 40, 529–535.
Sacchetti, G., Gianotti, A. and Rosa, M.D. (2001). Sucrose-salt combined effects on mass transfer kinetics and product acceptability: Study on apple pretreatment. *Journal of Food Engineering*, 49, 163–173.
Saurel, R., Raoult-Wack, A., Rios, G. and Guilbert, S. (1994). Mass transfer phenomenon during osmotic dehydration of apple. I. Frozen plant tissue. *International Journal of Food Science and Technology*, 29, 531–542.
Sethi, V., Sahni, C.K., Sharma, K.D. and Sen, N. (1999). Osmotic dehydration of tropical temperate fruits: A review. *Indian Food Packer*, Jan/Feb, 34–43.
Sharma, K.D., Sethi, V. and Maini, S.B. (1998). Osmotic dehydration in apple: Influence of variety, location and treatment on mass transfer and quality of dried rings. *Acta Alimentaria*, 27(3), 245–256.
Simal, S., Benedito, J., Sanchez, E.S. and Rosello, C. (1998). Use of ultrasound to increase mass transport rate during osmotic dehydration. *Journal of Food Engineering*, 36, 323–336.
Singh, B., Kumar, A. and Gupta, A.K. (2007). Study of mass transfer kinetics and effective diffusivity during osmotic dehydration of carrot cubes. *Journal of Food Engineering*, 79, 471–480.
Torreggiani, D. and Gianni, B. (2001). Osmotic pretreatments in fruit processing: Chemical, physical and structural effects. *Journal of Food Engineering*, 49, 247–253.
Treybal, R.E. (1981). *Mass Transfer Operations*. McGraw Hill Book Co., New York.
Tsamo, C.V.P., Bilame, A.F., Ndjouenkeu, R. and Nono, Y.J. (2005). Study of material transfer during osmotic dehydration of onion slices (*Allium cepa*) and tomato fruits (*Lycopersicon esculentum*). *LWT–Food Science and Technology*, 38, 495–500.
Uddin, M.B. and Islam, N. (1985). Development of shelf stable pineapple products by different methods of drying. *Journal of Institution of Engineers of Bangladesh*, 13, 5–13.
Waliszewski, K.N., Pardio, V.T. and Ramirez, M. (2002a) Effects of chitin on color during osmotic dehydration of banana slices. *Drying Technology*, 20, 1291–1298.

Chapter 4

Membrane Processing of Food Materials

Satyanarayan Dev, Abid Hussain, and Vijaya Raghavan

Contents

4.1 Introduction

Membrane processing is an expanding field in preservation, nutritional performance, and safety of foods. Membrane processing permits separation and concentration of food material without use of heat and, therefore, has been widely applied in the dairy, food, and beverage industries. In a membrane process, a membrane can be described as an interphase, usually heterogeneous, acting as a barrier to the flow of molecular and ionic species present in liquids and/or vapors contacting its two surfaces. The success of any membrane-based operation, such as demineralization, desalination, purification, fractionation, bioseparation, etc., depends on the membrane surface area. Membrane-based processes are gaining importance in biotechnology applications due to their potential efficiency for size- and/or charge-based protein separation with high purity and throughput. Electric or ultrasonic fields are now imposed simultaneously to improve the separation efficiency of the membranes.

Membrane processing in the food industry involves high energy saving, high yield, and superior quality products compared to the conventional processes of thermal pasteurization, evaporation, drying, and dehydration. The major advantages of membrane processing of food materials include the following:

- Membrane processing provides flexibility in equipment design and operation as the process is modular.
- Separation occurs without any phase change (except pervaporation) and at low pressure drop.
- Membrane processing is a gentle and mild operation and retains the chemical identity of the food. The process involves no thermal treatment, thereby retaining the original quality of foods.
- Membrane processing offers low operating and maintenance cost and requires minimum space for operation.
- Membrane processing allows separation of a wide range of molecular weight components in the process stream.
- Membrane processing is quite effective in processing dilute solutions. The process is resistant to chemicals and temperatures.

However, there is a paucity of information in membrane processing regarding various process variables, such as feed rate, cross-flow velocity, cleaning procedure, nature of membrane fouling, economics of membrane replacement, etc. Introduction of membrane processing also involves a high initial capital investment and adequate back-up support for replacement of membranes, technological expertise for plant operation, and training facilities.

4.2 Membrane Processes

A membrane can be defined as a region of discontinuity interposed between two phases or as a phase that acts as a barrier to prevent mass movement but allows restricted and/or regulated passage of one or more species through it. The driving forces that can lead to a significant flux of food materials are of practical significance for membrane separation processes. These driving forces may be hydrostatic pressure difference, concentration gradient, temperature gradient or electrical potential difference.

- A hydrostatic pressure difference between two phases separated by a membrane can lead to generation of a flux and to the separation of chemical species when the hydrodynamic permeability of the membrane is different for different components.
- A concentration difference between two phases separated by a membrane can lead to the transport of materials and to the separation of various chemical species when the diffusivity and the concentration of the various chemical species in the membrane are different for different components.
- A temperature gradient between two phases separated by a membrane can lead to the transport of materials and to the separation of various components when the thermal conductivities of the various components in the membrane are different for different components.
- A difference in electrical potential between two phases separated by a membrane can lead to the transport of materials and to the separation of various chemical species when the differently charged particles have different mobilities in the membrane.

The broad classification of membrane processing based on these driving forces is presented in Table 4.1. Various membrane processes and their major characteristics along with their applications in the food industry are summarized in Table 4.2.

Table 4.1 Broad Classification of Membrane Processing by Driving Force

Driving Force			
Pressure	*Concentration*	*Temperature*	*Electropotential*
Gas-vapor separation Reverse osmosis (RO) Nanofiltration (NF) Ultrafiltration (UF) Microfiltration (MF) Pervaporation	Membrane extraction Dialysis	Membrane distillation	Electrodialysis Electroosmosis

Table 4.2 Various Membrane Processes and Their Applications

Membrane Process	*Membrane Type*	*Driving Force*	*Method of Separation*	*Range of Application*
Microfiltration	Symmetric microporous membrane; 0.1 to 10 μm pore radius	Hydrostatic pressure difference 0.1 to 1 bar	Sieving mechanism due to pore radius and absorption	Sterile filtration clarification
Ultrafiltration	Asymmetric microporous membrane; 1 to 10 μm pore radius	Hydrostatic pressure difference 0.5 to 5 bar	Sieving mechanism	Separation of micromolecular solution
Reverse osmosis	Asymmetric skin type membrane	Hydrostatic pressure difference 20 to 100 bar	Solution diffusion mechanism	Separation of salt and microsolutes from solution
Dialysis	Symmetric microporous membrane; 0.1 to 10 μm pore radius	Concentration gradient	Diffusion in convention-free layer	Separation of salt and microsolutes from macromolecular solution
Electrodialysis	Cation and anion exchange membrane	Electrical potential gradient	Electrical charge of particle and size	Desalination of ionic solution
Gas separation	Homogeneous or porous polymer	Hydrostatic pressure concentration gradient	Solubility, diffusion	Separation of gas mixture

4.3 Membrane Types

4.3.1 Symmetric and Asymmetric Membranes

A membrane may be symmetric or asymmetric. Symmetric membranes are uniform and have relatively thick (50–500 μm) barriers. On the other hand, asymmetric membranes are composed of an ultrathin microporous layer (≤ 1 μm thick) supported by a 50–500 μm microporous sublayer. The sublayer acts as a supporting agent and does not influence the separation characteristics or the flux rate through the membrane. Asymmetric membranes are primarily used for pressure-driven membrane processes, such as ultrafiltration (UF) and reverse osmosis (RO). Because the flux rate in RO or UF processes is inversely proportional to the thickness of the actual barrier layer, asymmetric membranes exhibit much higher flux rates than symmetric structures of comparable thickness. In addition to a higher flux rate, asymmetric membranes retain all rejected materials at the membrane surface where they can be removed by the shear forces applied by the feed solution moving parallel to the membrane surface.

4.3.2 Microporous Membranes

The microporous media represent the simplest form of a membrane as far as mass transport properties and separation mode are concerned. The microporous membranes consist of a solid matrix with defined pores, which have diameters ranging from less than 10 nm to more than 50 pm. The separation of various components is achieved by a sieving mechanism. The microporous membranes can be made from various materials, such as metal oxides, graphite, metals, or polymers. The most common membranes are porous ceramics made from silica or aluminum oxide by molding and sintering. Equivalent membranes can be made using finely powdered polymers.

4.3.3 Homogeneous Membranes

A homogeneous membrane consists of a dense film through which a mixture of components is transported under pressure, concentration, or electrical potential. The separation of various components in a solution is directly related to their transport rate within the membrane phase, which is determined by their diffusivity and concentration of the membrane matrix. An important property of homogeneous membranes is that chemical species of similar size and identical diffusivities, may be separated when their concentration, i.e., their solubility in the film, differs significantly. A typical example of a homogeneous membrane is the silicone rubber film used to separate helium from other gases.

4.3.4 Ion Exchange Membranes

Ion exchange membranes consist of highly swollen gels or a microporous structure with the pore walls carrying fixed positive or negative charges. A membrane with fixed positive charges is referred to as an anion exchange membrane because it binds anions from the surrounding fluid, and a membrane containing fixed negative charges is called a cation exchange membrane. Separation in charged membranes is achieved by exclusion of co-ions, i.e., ions that bear the same charges as the fixed ions of the membrane structure. Separation properties of these membranes are determined by the charge and concentration of the ions in the surrounding solution and in the membrane structure. In an ideal ion exchange membrane, the charge density is in the range of 2 to 4 milliequivalents per gram of membrane material. The main application of ion exchange membranes is in electrodialysis for the desalination of aqueous solutions.

4.4 Membrane Module, Rating, and Performance

A membrane separation process to be utilized on a large scale requires not only a membrane with the desired separation characteristics, but also equipment that is compact, reliable, and inexpensive. A useful membrane process requires the development of a membrane module containing large surface areas of membrane. Therefore, membranes are integrated into different modules to accommodate large membrane areas in a small volume to withstand the applied pressure and the cross-flow velocity required to maintain a clean membrane surface. The common membrane module configurations include flat plate, tubular, hollow fiber, hollow fine fiber, and spiral. Polyethersulfone, polysulfone, cellulose acetate, regenerated cellulose, ceramic, and sintered metal membranes are commonly available membranes for commercial applications. Examples of various membranes with their module and membrane type are represented in Table 4.3. Besides economic considerations, chemical and engineering aspects are of prime importance for design of a membrane module. Because various membrane separation processes differ significantly in their operational concepts and applications, the membrane modules and system designs used for various processes are equally different.

In membrane processing, ultrafiltration (UF) and nanofiltration (NF) membranes are rated by molecular weight cut-off (MWCO). The unit of MWCO is Dalton (D) and expressed in kD (1 kD = 1000 D). UF membranes are available in the range of 1 to 1000 kD MWCOs whereas NF membranes range from 0.01 to 1 kD. For RO membranes, rating is commonly listed as percentage of salt retention. The percentage of retention depends on the type of salt and operating conditions in the membrane process.

The performance of a membrane is usually expressed in two terms: ability of the membrane to produce a large volume of filtrate in a short operation time and degree of purity of the filtrate with respect to solute concentration. The permeate flux and solute rejection factor are two terms used to define the performance of a membrane.

Table 4.3 Applications of Membranes with Their Module and Membrane Type

Product	*Application*	*Range*	*Module*	*Membrane*
Dairy	Milk concentration	RO	Spiral	CA, TFC
	Whey concentration	RO	Spiral	CA, TFC
	Whey fractionation	UF	Spiral	PS, PVDE, PES
	Lactose concentration	RO	Spiral	CA, TFC
	Milk pasteurization	MF	Tubular	Ceramic
	Desalination	NF	Spiral	TFC
Juice	Clarification	MF	Tubular	Ceramic
		MF	Hollow fiber	PS, PE
		UF	HF	PS
		MF/UF	Tubular	PVDF, PS
	Concentration	RO/NF	Spiral	TFC
		RO/NF	Tubular	TFC
		RO/NF	HFF	TFC
	Caustic recovery	MF	Tubular	Ceramic
Gelation	Concentration	UF	Spiral	PS
Corn sweetener	Dextrose clarification	MF	Spiral	PS/PES
Sugar	Clarification	MF/UF	Tubular	Ceramic
	Preconcentration	RO	Spiral	TFC

Note: CA: cellulose acetate; MF: microfiltration; NF: nanofiltration; PE: polyster; PES: polyestersulfone; PS: polysulfone; PVDF: polyvinylidine difluoride; RO: reverse osmosis; TFC: thin film composite; UF: ultrafiltration.

$$\text{Permeate flux} = \frac{\text{Permeate volume}}{\text{membrane area} \times \text{time}} \tag{4.1}$$

$$\text{Solute rejection factor} = \left(1 - \frac{\text{solute concentration in the permeate}}{\text{solute concentration in the feed}}\right) \tag{4.2}$$

4.5 Membrane Classification Based on Pressure Drive

Pressure-driven membrane processes, i.e., microfiltration (MF), ultrafiltration (UF), and nanofiltration (NF) involve separation mechanisms in porous membranes, and reverse osmosis (RO) and pervaporation make use of tight and dense membranes. MF and UF membranes separate on the basis of a simple sieving mechanism. The particle dimensions in relation to the pore size distribution of the membrane determines whether or not a particle can pass through the membrane. Pervaporation or vapor permeation is an alternative to RO in recovering aroma compounds.

4.5.1 *Microfiltration*

Microfiltration (MF) uses porous membranes with a cut-off pore-size of microns. MF is primarily used to remove suspended solids and colloids, fat and high molecular weight protein, as well as bacteria, and to retain color and taste compounds, alcohols, and sugars. A pressure gradient is maintained across the membrane to maintain the fluid flow through the MF membrane. The MF membrane is usually made from a thin polymer film with a uniform pore size and a high pore density, approximately 80%. In MF, the feed is applied perpendicular to the membrane surface, and the filtered particles are accumulated on the membrane surface in the form of filter cake. As the thickness of the cake increases with time, the permeation rate decreases correspondingly. To reduce the build-up of the cake on the membrane surface, an alternative cross flow or tangential operation can be used. The feed flows parallel to the membrane surface and the permeate pass through the membrane with proper driving force.

4.5.2 *Ultrafiltration*

Ultrafiltration (UF) is an ideal process for fractionation, concentration, and purification of liquid and removes particles in the range of 0.001 to 2 μm. Solvents and salts of low molecular weight will pass through while larger molecules are retained. UF requires pressure of 1 to 7 bar to overcome the viscous resistance of liquid permeation through the membrane. The UF process is most effective in removal of anti-nutritional factors such as oligosaccharides, phytic acids, and some trypsin inhibitors from vegetable proteins to produce purified protein isolates or concentrates with superior functional quality. UF is also used in separation and fractionation of individual milk proteins from lactose and minerals. The UF Permeate is significantly sterile due to the removal of microorganisms. Other industrial applications of UF include enzyme recovery, concentration of sucrose and tomato paste, treatment of still effluents in brewing and distilling, and removal of protein hazes from honey and syrups.

Diafiltration is a special type of UF process. The retentate is diluted with water and re-ultrafiltered to improve the recovery and purity of the retained stream.

Typical uses include recovery of antibiotics from the fermentation broth enhancing protein, lactose from whey protein, and recovery of fruit sugars at the end of juice clarification.

4.5.3 Nanofiltration

Nanofiltration (NF) membranes separate species using both charge and size exclusion. Therefore, models to describe this process must induce both steric and electrostatic effects on mass transport. Usually, NF is used for separation of components with molecular weights in the range of 200–1000 D. NF operates at lower pressure than RO (≈ 0.5 bar) but yields a higher flow rate of water. In the food industry, applications of NF are quite numerous. In the dairy sector, NF is used to concentrate whey and recycle clean-in-place solutions. In processing of sugar, dextrose syrup and sugar juice are concentrated by NF. NF is used for degumming of solutions in edible oil processing as well as production of continuous cheese and alternative sweeteners (Landhe and Kumar 2010; Cesar et al. 2009). The production of salt from natural brines uses NF as a purification process. The pharmaceutical and biotechnical industries allow the use of NF in their purification processes. In aqueous systems, NF uses hydrophilic polymeric materials, such as polyether-sulphone, polyamides, and cellulose derivatives. These materials, in contact with organic solvents, quickly lose their stability. Special membranes have, therefore, been developed to provide the same kind of performance as in aqueous systems, and they are now used for solvent exchange processes and removal of catalysts and heavy metals.

4.5.4 Reverse Osmosis

The reverse osmosis (RO) mechanism is based on size, shape, ionic charge, and membrane interactions. RO is usually employed either alone or in combination with other membrane separation processes, such as MF and UF. RO separates all organic and inorganic solutes from solution. The permeate in RO is pure water as the RO membrane rejects dissolved and suspended materials above 0.0001–0.001 μm at moderate to high pressure. The tight nature of RO membranes requires the highest operating pressure compared to other membrane separation processes. The energy requirement for RO is significantly lower than for mechanical vapor compression systems. Other advantages of RO include a high-quality product without heat damage and reduced waste treatment volume and, therefore, reduced cost involved in the overall process. However, in spite of the high selectivity and solute retention capacity, the major drawbacks of RO membranes include limited operating pressure range and membrane fouling. In most food applications, from an economical point of view, the RO process is used as a pre-concentration step with other technologies.

Applications of RO in food processing include concentration of whey for cheese manufacture and ice cream, as well as a pre-concentration stage prior to drying, as the process requires. RO is quite effective in concentration and purification of fruit juices, enzymes, and liquors in fermentation process and in the removal of monovalent and polyvalent ions, bacteria, and organic materials with a molecular weight greater than 100 D to produce pure water for beverage manufacture; in dealcoholization process to produce low-alcohol beers, ciders, and wine; and in recovery of proteins from distillation residues, dilute juices, and waste water from corn milling.

4.6 Membrane Classification Based on Electro-Potential Drive

4.6.1 Electrodialysis

Electrodialysis is an electro-chemical separation process in which a gradient in electric potential is used to separate ions. The interest in electrodialysis (ED) as a membrane process has been triggered by recent developments in membrane materials. These new materials offer better stability and performance, thus giving scope for new applications. In ED, charged (ion exchange) membranes are used to separate molecules or ions in an electrical field on the basis of differences in charge and transport velocity through the membrane. Often, these membranes have very narrow pores (1–2 nm wide) and charged sites. In an ED cell, a number of cation and anion exchange membranes are placed between an anode and a cathode. When a current is applied, positively charged ions migrate through the cation exchange membrane while the negatively charged ions migrate through the positively charged anion exchange membrane. Some ion exchange membranes can even discriminate between monovalent and multivalent ions, such as Mg^{2+} and Na^+.

Large-scale applications of ED can be found in the desalination of whey used in ice cream, bread, cake, sauces, and baby foods. ED is preferable to RO because it does not affect taste, color, or flavor. Also, the desalting of species of high molecular mass is very effective, and the process can be very precisely controlled. Major disadvantages of ED are the high operation costs and the susceptibility to fouling. ED is used to demineralize milk and whey, deacidify fruit juices, and de-ash sugar solutions, such as dextrose.

The energy consumption in an ED separation procedure can be expressed as

$$E = i^2 nRt \tag{4.3}$$

where E is the energy consumption, i is the electric current through the stack, R is the resistance of the cell, n is the number of cells in a stack, and t is time. The electric current needed to desalt a solution is directly proportional to the number of ions transferred through the ion exchange membranes from the feed stream to concentrated brine. It is given as

$$i = \frac{zFQ\Delta c}{\vartheta} \tag{4.4}$$

where F is the Faraday constant, z is the electrochemical valence, Q is the feed solution flow rate, Δc is the concentration difference between the feed solution and the diluate, and ϑ is the current utilization. The current utilization is directly proportional to the number of cells in a stack and is governed by the current efficiency.

The total current utilization can, therefore, be expressed by the following relation:

$$\vartheta = n\mu_s\mu_w\mu_m \tag{4.5}$$

where ϑ is the current utilization; n is the number of cells; and μ_s, μ_w, and μ_m are the current efficiencies due to incomplete permselectivity of the membrane, transport of water with solvated ions, and parallel currents through the stack manifold respectively. In ED, n, μ_w, and μ_m are always smaller than unity. Due to the Donnan potential of the membrane, μ_s decreases with the concentration of the feed solution. In solutions with salt concentrations in excess of 3–5 mol liter^{-1}, the current utilization reaches such low values that often the ED process becomes uneconomical. A combination of Equations 4.4 and 4.5 gives the energy consumption in ED as a function of the current applied in the process, the electrical resistance of the stack, i.e., the resistance of the membrane and the electrolyte solution in the cells, the current utilization, and the amount of salt removed from the feed solution.

$$E = \frac{inRt(zFQ\Delta c)^2}{\vartheta^2} \tag{4.6}$$

where E is the energy requirement, n is the number of cells in a stack, R is the resistance of a cell, t is the time, z is the electrochemical valence of the components to be removed, F is the Faraday constant, Q is the volume flow rate of the feed solution, Δc is the concentration difference between the feed solution and the diluate, and ϑ is the current utilization. Equation 4.6 indicates that the electrical energy required in ED is directly proportional to the amount of salts that have to be removed from a certain feed volume to achieve the desired product concentration. Current utilization decreases with increasing feed solution concentration. At very high feed solution concentration, the current utilization approaches zero, and the energy consumption reaches an infinitely high value. The required membrane area is also a function of the feed solution concentration, the desalting effect as a function of salts to be removed, the current density, and the current utilization.

$$A = \frac{zFnQ\Delta c}{I\vartheta} \tag{4.7}$$

where A is the membrane area, z is the electrochemical valence, F is the Faraday constant, Q is the feed solution flow rate, Δc is the concentration difference between the feed solution and the diluate, n is the number of cells in a stack, I is the current density, and ϑ is the current utilization. Thus, in ED desalination, the energy cost as well as the investment cost increase steeply with feed solution concentration and approach an infinitely high value when the concentration in the feed solution reaches a value that eliminates the permselectivity of the ion exchange membranes.

4.6.2 Electrodeionization

An evolution of ED is electrodeionization (EDI). It is best described as a combination of ED and ion exchange (IEX) and functions in a similar way to ED but allows deionization to a much lower level. EDI has been pursued more recently and has begun to find commercial applications almost always as part of a hybrid process for producing demineralized water.

4.6.3 Bipolar Electrodialysis

Bipolar electrodialysis operates by splitting water into H^+ and OH^- and, therefore, converting salts into the corresponding alkaline and acidic solutions. The technology has been designed to replace the ion exchange process because it does not have the disposal problems associated with the regeneration of chemicals (Bazinet et al. 1998).

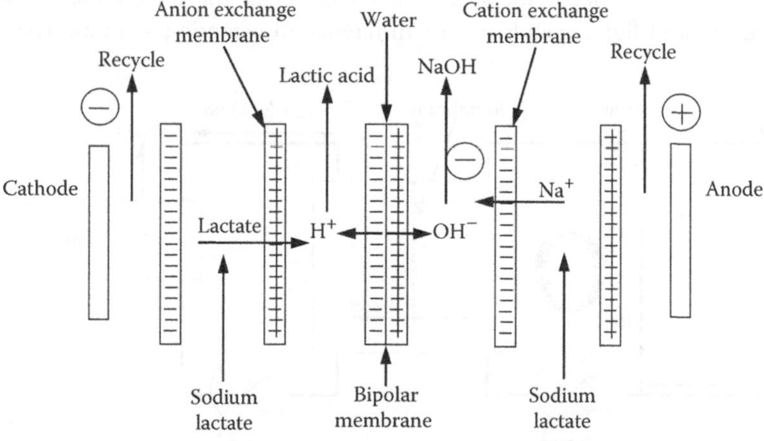

Figure 4.1 A schematic diagram of the use of electrodialysis and bipolar membrane to produce lactic acid from sodium lactate solution produced by a fermenter.

An example of applying a bipolar membrane in combination with electrodialysis is shown in Figure 4.1. A sodium lactate solution from a fermenter is fed into an electrodialysis-bipolar membrane unit. The negatively charged lactate migrates through the anion exchange membrane and recombines with the H^+ ion generated by the bipolar membrane. Presently, this technology is quite effective for production of vitamin C.

4.7 Membrane Classification Based on Concentration Drive

4.7.1 Direct Osmosis Concentration

Direct osmosis concentration (DOC) is capable of concentrating fruit juice at a lower temperature and lower pressure, thereby, maintaining the original flavor and color profile of the juices. In the DOC process, an osmotic agent solution is used to establish an osmotic pressure gradient across a semipermeable membrane and, thus, to remove water from products. An osmotic agent is generally a solid highly soluble in water; hygroscopic; nontoxic; inert toward the flavor, odor, and color of the food; and which does not pass through the membrane. Generally, the higher the concentration of the dissolved solids and the lower the molecular weight of the dissolved solids, the higher the osmotic pressure. The most frequently employed agents are sodium chloride, sucrose, glycerol, cane molasses, and corn syrup. The osmotic agent solutions should have an osmotic pressure greater than that of the concentrated fruit juice. For instance, the osmotic pressure of a 74°Brix high fructose corn syrup is about 270 bar that is greater than 90 bar for 42°Brix pulpy orange juice. A schematic diagram of a DOC process for concentration of fruit juices is shown in Figure 4.2.

Unlike the RO process, pressure differences across the membrane in DOC are negligible, and flux depends on the difference in osmotic potential. The only

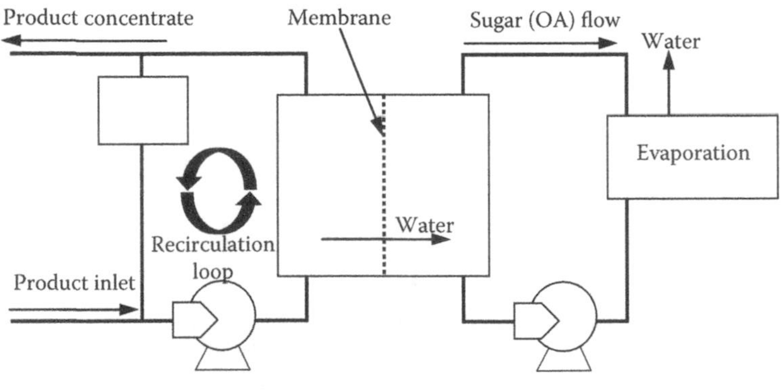

Figure 4.2 A simplified flow diagram of a DOC process. (From Jiao, B. et al., *Journal of Food Engineering*, 63, 303–324, 2004.)

hydraulic pressure (≈ 2 bar) is necessary to pump the juice and the osmotic agent solution over the membrane surfaces. Juices containing large amounts of both dissolved and suspended solids can be concentrated with minimum fouling because the solids are not forced against the membrane.

4.7.2 Dialysis

A dialyzer is a device in which one or more solutes are transferred from one fluid to another through a membrane as a result of a concentration gradient. Generally, microporous membranes such as cellophane tubing are used as the separation barrier in dialysis. Under optimum operating conditions, the fluids are circulated. The solution to be depleted of solute may be called the feed, and the fluid receiving the solute may be termed the dialysate. The overall efficiency of a dialyzer is governed by two independent factors: the ratio of the flow rates of the two fluids and the rate constant for solute transport between the fluids, which is determined by the properties of the membrane, the fluid channel geometry, and the local fluid velocities. The dialyzer flow parameters are shown schematically in Figure 4.3. In the diagram, Q represents the volumetric flow rates, and C represents the solute concentrations. The subcripts B and D refer to feed and dialysate, respectively; the subscripts i and o designate inlet and outlet conditions, respectively.

Making an overall solute transport balance over the dialyzer:

$$N = Q_B\,(CB_i - C_{Bo}) = Q_D\,(C_{Di} - C_{Do}) \tag{4.8}$$

where N is the overall solute transport rate, which can also be expressed as follows:

$$N = k^*A\Delta C \tag{4.9}$$

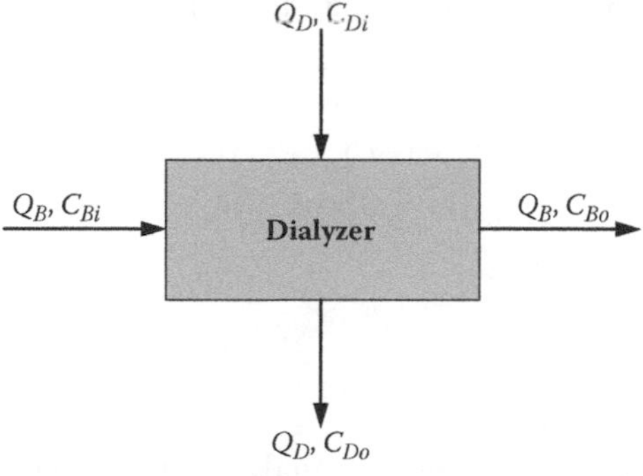

Figure 4.3 Flow parameters of a dialyzer.

where ΔC is the average solute concentration difference between the fluids, A is the membrane area, and k^* is an overall rate constant describing the transfer of solutes from the feed solution to the dialysate. The average solute concentration difference is the logarithmic mean of the inlet and outlet differences. The three most common cases are parallel flow of both fluids, countercurrent flow, and flow with the dialysate completely mixed. The efficiency of a dialyzer is expressed in terms of its dialysance D_B, which is defined as

$$D_B = \frac{N}{C_{Bi} - C_{Di}} \tag{4.10}$$

The dimensionless ratio D_B/Q_B can be regarded as a convenient efficiency parameter because it represents the fraction of maximum attainable solute depletion in the feed that can actually be achieved in the device.

By simultaneous solution of Equations 4.9 and 4.10 and use of the logarithmic mean concentration driving force, the dialysance ratio D_B/Q_B can be expressed for various flow geometries as follows:

(i) Parallel flow (the feed and dialysate solutions flow in the dame direction)

$$\frac{D_B}{Q_B} = \frac{1}{1+Z}[1-(1+Z)e^{-R}] \tag{4.11}$$

(ii) Mixed dialysate flow (the dialysate is mixed by stirring)

$$\frac{D_B}{Q_B} = \frac{1-(1+Z)e^{-R}}{Z-(1+Z)e^{-R}} \tag{4.12}$$

(iii) Countercurrent flow (the feed and dialysate solutions flow in opposite directions)

$$\frac{D_B}{Q_B} = \frac{1-e^{-R}}{1+Z(1-e^{-R})} \tag{4.13}$$

where, $Z = \frac{Q_B}{Q_D}$ and $R = \frac{k^* A}{Q_B}$.

The above relationships can be used to analyze dialyzer performance. It should be pointed out that the overall rate constant k^*, which determines the transfer of solutes from a feed solution to the dialysate is not only influenced by the properties of the membrane, but also by the fluid boundary layers on each side of the membrane. Due to the resistance of the boundary layer to the transport of solute, a depletion of solutes is obtained at the feed solution membrane interface, and a high concentration of solutes is obtained at the dialysate membrane interface in relation to the concentration of the bulk solutions. The resistance of the boundary layer is directly proportional to its thickness, which again is determined by the degree of turbulence in the bulk solutions. This has to be considered when designing a dialysis process or process equipment. Applications of dialysis include removal of salts and other low molecular weight solutes from a mixture containing macromolecular components.

4.8 Membrane Classification Based on Temperature Drive

4.8.1 Membrane Distillation

In membrane distillation (MD), two aqueous solutions at different temperatures are separated by a microporous hydrophobic membrane. In these conditions, a net pure water flux from the warm side to the cold side occurs. The process takes place at atmospheric pressure and at a temperature that may be much lower than the boiling point of the solutions. The driving force is the vapor pressure difference between the two solutions. The phenomenon can be described as a three-phase sequence: formation of a vapor gap at the warm solution–membrane interface; transport of the vapor phase through the microporous system; and condensation of vapor at the cold side membrane–solution interface. The most suitable materials for MD membranes include polyvinyldifluoride, polytetrafluroethylene, and polypropylene. The size of micropores ranges between 0.2 and 1.0 lm. The porosity of the membrane ranges from 60% to 80% of the volume and the overall thickness from 80 to 250 lm, depending on the absence or presence of support. In general, the thinner the membrane and greater the porosity of the membrane, the greater the flux rate.

In MD, the permeate flux decreases with an increase in concentration of feed solution. This phenomenon can be attributed to the reduction of the driving force due to the decrease in vapor pressure of the feed solution and to the increase in viscosity of the juice solution. At higher concentration ratios, fluxes are higher in the MD process than in the RO. The MD flux gradually increases with an increase of the temperature difference between feed juice and the cooling water. At a lower temperature gradient, a decreasing of the permeate flux is observed due to the lower vapor pressure gradient between the two sides of the membrane.

4.8.2 Osmotic Distillation

Osmotic distillation (OD) has been successfully applied to the concentration of liquid foods, such as milk, fruit and vegetable juice, instant coffee and tea, and various non-food aqueous solutions. The process can be used to extract water from aqueous solutions under atmospheric pressure and at room temperature, thus avoiding thermal degradation of the solution. It is, therefore, particularly adapted in concentration of heat-sensitive products, such as fruit juices. As compared to RO and MD, the OD process has a potential advantage, which might overcome the drawbacks of RO and MD for concentrating fruit juice, because the RO process suffers from high osmotic pressure limitation, and in MD some loss of volatile components and heat degradation may occur due to the heat requirement for the feed stream in order to maintain the water vapor pressure gradient. OD, on the other hand, does not suffer from any of the problems mentioned above when operated at room temperature.

4.9 Permeate Flux

The performance of the membrane processing is dependent on the effectiveness of the pressure, concentration, electropotential, or temperature gradients on the membrane surface. In fact, all these parameters in the process are governed by certain phenomena through the membrane surface. The basic fundamental equations influencing these driving forces through the membrane are summarized in Table 4.4.

Table 4.4 Relationships Between Various Fluxes and Corresponding Driving Forces

Relationship	*Flux*	*Driving Force*	*Constant Proportionality*
Fick's law $J = -D\Delta c$	Mass (J)	Concentration difference (Δc)	Diffusion coefficient (D)
Ohm's law $i = \frac{\Delta U}{R}$	Electricity (i)	Electrical potential difference (ΔU)	Electrical resistance (R)
Fourier's law $Q = k\Delta T$	Heat (Q)	Temperature difference (ΔT)	Heat conductivity (k)
Hagen-Poiseuille's law $V = h_d\Delta P$	Volume (V)	Pressure difference (ΔP)	Hydrodyanamic permeability (h_D)

4.9.1 Permeate Flux and Pore Size

A solvent flowing through the pores of a membrane is a function of pore diameter d_p, number of pores N, porosity ε, applied pressure P_T, viscosity of the solvent μ, and thickness of the membrane Δx. The model most frequently used for describing flow through pores is based on the Hagen Poiseseille model for laminar flow through the pore.

$$J = \frac{\varepsilon d_p^2 P_T}{32 \Delta x \mu} \tag{4.14}$$

where J is the permeate flux (m s^{-1}).

4.9.2 Concentration Polarization

Permeate flux is the rate at which permeate is produced. This is usually expressed as mass or volumetric flow rate of permeate per unit area of membrane. The osmotic pressure difference across the membrane and the resistance provided by the membrane is given as

$$N = \frac{\Delta P - \Delta \pi}{\mu \sum R} \tag{4.15}$$

Figure 4.4 presents the concentration, pressure, and osmotic pressure on each side of the membrane. The permeate flux is directly proportional to the difference between the transmembrane pressure ΔP and the osmotic pressure difference $\Delta \pi$ in

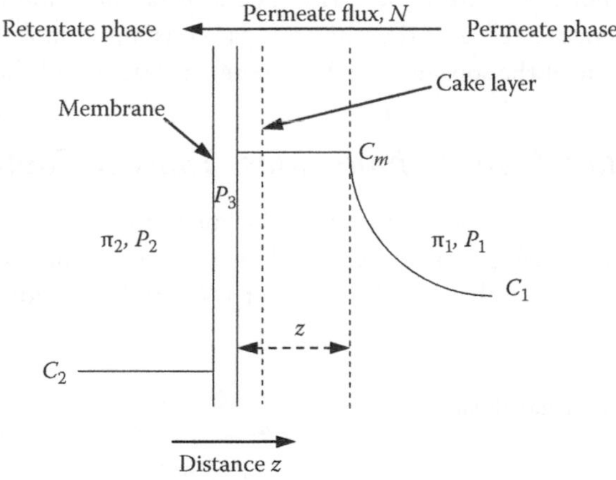

Figure 4.4 Concentration polarization.

which $\Delta P = P_1 - P_2$ and $\Delta\pi = \pi_1 - \pi_2$. $\sum R$ is the sum of the mass transfer resistances of the membrane itself and fouling layers due to deposits from the feed materials.

Equation 4.15 may be expanded into

$$N = \frac{\Delta P - \Delta\pi}{\mu(R_m + R_f + R_p)} \tag{4.16}$$

where R_m, R_f, and R_p are the resistance to the flow of permeate by the membrane, the fouling layer, and the concentration polarization layer, respectively.

4.10 Estimation of Transport Model Parameters

4.10.1 Osmotic Pressure

Osmotic pressure in dilute solutions can be determined by the van't Hoff model, and Gibb's model can be used at higher concentrations. Osmotic pressure variation with concentration of solute is approximately linear in the low-concentration range.

van't Hoff model:

$$\pi = CRT \tag{4.17}$$

Gibb's model:

$$\pi = -\frac{RT}{V}\ln X \tag{4.18}$$

where π is the osmotic pressure (Pa), C is the concentration solute (kg m^{-3}), T is the absolute temperature (K), R is the universal gas constant (J mol^{-1} K^{-1}), V is the partial molar volume of the solvent, and X is the molar fraction of the solvent.

4.10.2 Boundary Layer Osmotic Mass Transfer Coefficients

The boundary layer osmotic mass transfer coefficient h_{os} (m s^{-1}) can be determined for simple geometries using non-dimensional mass-transfer correlations relating Sherwood number (N_{sh}) with Reynolds number (N_{re}) and Schmidt number (N_{sc}) (Kulkani et al. 1992).

(i) Developing laminar flow

$$N_{sh} = 0.664 N_{re}^{0.5} N_{sc}^{033} \left[\frac{d_h}{L}\right]^{0.5} \tag{4.19}$$

(ii) Fully developed laminar flow

$$N_{sh} = 1.62 N_{re}^{0.33} N_{sc}^{0.33} \left[\frac{d_h}{L}\right]^{0.33} \tag{4.20}$$

(iii) Turbulent flow

$$N_{sh} = 0.023 N_{re}^{0.8} N_{sc}^{0.33} \tag{4.21}$$

where Sherwood number $N_{sh} = \frac{kd_h}{D}$, Reynolds number $N_{re} = \frac{vd_h}{\mu}$, and Schmidt number $N_{sh} = \frac{\mu}{D}$ k denotes the Boltzmann's constant (1.381 × 10^{-23} J K^{-1}), d_h denotes the hydraulic diameter (m) of the membrane pore space, D denotes the diffusivity (m^2 s^{-1}) of fluid food, v denotes the cross flow velocity (m s^{-1}) of the liquid food, and μ denotes the viscosity (Pa s) of the fluid food.

4.10.3 Membrane Resistance

The membrane resistance r_M (Pa s m^{-1}) can be determined by conducting experiments using pure water and calculating the slope of the flux-pressure relationship. It can also be predicted using the Hagen-Poiseuille equation for the capillary flow.

$$r_M = \frac{128\mu b}{\pi \sum \left[n_p d_p^4\right]} \tag{4.22}$$

where μ is the dynamic viscosity (Pa s) of the fluid food, b is the membrane thickness (m), n_p is the number of particles/pores, and d_p diameter (m) of pores/particles.

4.10.4 Hydraulic Resistance of Cake

The hydraulic resistance r_{HR} (Pa s m^{-1}) of the boundary layer to the solvent flow in the MF process can be determined when the cake is incompressible by using the Carmen-Cozeny relationship.

$$r_{HR} = \frac{k(1-\varepsilon)^3 A_{SC}^2 \delta_c}{\varepsilon^3} \tag{4.23}$$

The void fraction ε of the randomly packed cake layer is about 0.4, and the specific surface area of solids A_{sc} (m^{-1}) is $3/r_s$ for rigid spheres, where r_s is the radius (m) of the sphere. The coefficient k has a value of about 5 in most applications, and δ_c is the cake layer thickness (m).

4.10.5 Membrane Based Mass Transfer Coefficient

The mass transfer coefficient h_M (m s^{-1}) in the boundary layer can be determined using the Leveque solution for analogous heat transfer problem in laminar flow in tubes (Zydney and Colton 1986).

$$h_M = 0.807\left(\frac{\gamma_w D^2}{L}\right)^{1/3} \tag{4.24}$$

where γ_w is the shear rate at the wall (s^{-1}) of the fluid food, D is the diffusivity (m^2 s^{-1}) of solute in the solution, and L is the tube length (m). The diffusivity D (m^2 s^{-1}) of the solute is estimated by using the Zydney and Colton (1986) equation as follows:

$$D = 0.0075 d_p^2 \gamma_w \tag{4.25}$$

where d_p is the diameter (m) of the particles/membrane pores, and γ_w is the shear rate at the wall (s^{-1}) of the fluid food.

4.10.6 UF Membrane Resistance

The membrane resistance γ_{mUF} (Pa s m^{-1}) of membrane to solvent flow in UF can be determined by conducting experiments using pure water or be predicted using the Hagen-Poiseuille equation for the capillary flow as in the Equation 4.22.

4.10.7 Hydraulic Resistance of Cake in UF

The hydraulic resistance r_{HR} (Pa s m^{-1}) of the boundary layer to solvent flow in UF process for micro-molecular compounds is best found by experimentation. However, Equation 4.26 can be rearranged to produce a linear relationship between $\frac{1}{J}$ and $\frac{1}{(P_1 - P_3)}$, which can be used to determine ϕ, which relates r_{HR} to the pressure difference.

$$J = \frac{(P_1 - P_2)}{r_{HR} + \phi(P_1 - P_2)} \tag{4.26}$$

where J is the permeate flux rate (m^3 m^{-2} s^{-1}); P_1, P_2, and P_3 are the pressures (Pa) in the retentate phase at the boundary layer between the cake and membrane surface and permeate phase, respectively; and ϕ is the proportionality coefficient (s m^{-1}) between cake layer resistance and pressure.

4.10.8 UF Based Mass Transfer Coefficient

Boundary layer osmotic mass transfer coefficient h_{uf}(m s^{-1}) can be determined by using Equations 4.19, 4.20, and 4.21, depending upon the flow regimen. In the equations, the diffusivity D of the fluid food is estimated by the Brownian diffusivity expressed by the Stokes-Einstein relationship as given by:

$$D = \frac{kT}{\mu \grave{\imath} d_p} \tag{4.27}$$

where k is the Boltzmann's constant (1.381 × 10^{-23} J K^{-1}), T is the absolute temperature (K), μ is the dynamic viscosity (Pa s) of the fluid food, and d_p is the diameter (m) of the membrane pores/particles.

4.11 Electro-Ultrafiltration

Electro-ultrafiltration (EUF) is an effective method to decrease gel layer formation on the membrane surface and to increase the flux rate. The EUF combines forced flow electrophoresis with the UF to control or eliminate the gel polarization layers. Suspended and colloidal particles have electrophoretic mobility measured by zeta potential. Most naturally occurring suspension solids, emulsions, and proteins are negatively charged. Placing an electric field across an UF membrane facilitates transport of the retained species away from the surface membrane. Thus, the retention of practically rejected solutes can be improved.

The basic principle of EUF is presented in Figure 4.5. The mechanism is based on both electro-phoretic and electro-osmosis effects. The process is a combination of a number of mechanisms, including ion association, ion adsorption, or ion dissolution. In the process, two electrodes are positioned on either side parallel to the UF membrane and used to apply the electric gradient. The field vector perpendicular to the membrane provokes a displacement of colloid species toward the electrode with the opposite sign. When an electric field is applied to the membrane, permeate is either concentrated or diluted by differences in the charge of solute and the direction of the electric potential gradient. The accumulation of the solutes on the membrane surface is limited by the imposed electro-phoretic force. In addition, the filtration rate through the filter cake is dramatically enhanced due to electro-osmosis as a secondary electro-kinetic phenomenon.

This method is best suited for the separation of protein because its surface charge changes according to the solution pH. As biological products such as proteins and peptides are sensitive to shear stress and temperature, the coupled effects of electric field and pressure served as an additional driving force for the separation, which is an interesting way to improve the membrane permeates flux without increasing the shear stress. Because proteins carry a net electrical charge, an electrical field may be

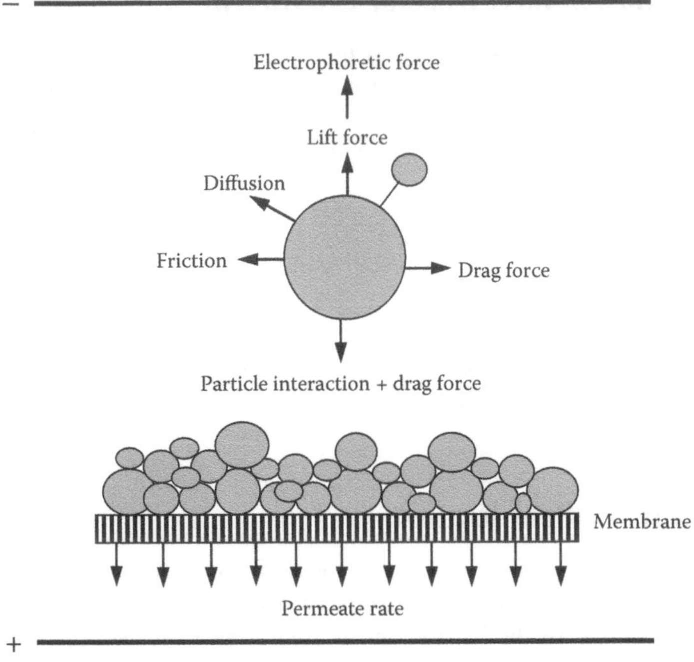

Figure 4.5 Principle of electro-ultrafiltration.

used to reduce the influence of the polarized layer. By applying a suitable external DC electric field, the protein molecules can be transmitted through the membrane due to the electrostatic attraction and concentration polarization can be reduced. The electrochemical properties of the membrane surface and the solutes have a great effect on the nature and magnitude of the interactions between the membrane and the substances being used and their separation characteristics.

4.12 Membrane Emulsification

An emulsion is a dispersion of droplets of one immiscible liquid within another liquid. An emulsifier may be added to stabilize the dispersion. An emulsifier process is a molecule consisting of a hydrophilic and a hydrophobic part and concentrates at the interface between the immiscible liquids, where they form interface films. The hydrophilic part of the emulsifier may consist of a fatty acid whereas the hydrophilic part of the emulsifier may consist of glycerol, possibly esterified with acetic, citric, or tetraric acid. A simple membrane emulsification process is shown in Figure 4.8. The dispersed phase is pressed through the pores of a microporous membrane, and the continuous phase flows along the membrane surface. Droplets grow at pore outlets until, upon reaching a certain size, they detach. This is determined by the balance between the drag force on the droplet from the flowing continuous phase, the buoyancy of the droplet, the interfacial tension forces, and the driving pressure.

The droplet at a pore tends to form a spherical shape under the action of interfacial tension, but some distortion may occur depending on the flow rate of continuous phase and the contact angle between the droplet and membrane surface. The final droplet size and size distribution are not only determined by the pore size and size distribution of the membrane but also by the degree of coalescence, both at the membrane surface and in the bulk solution (Figures 4.6 and 4.7).

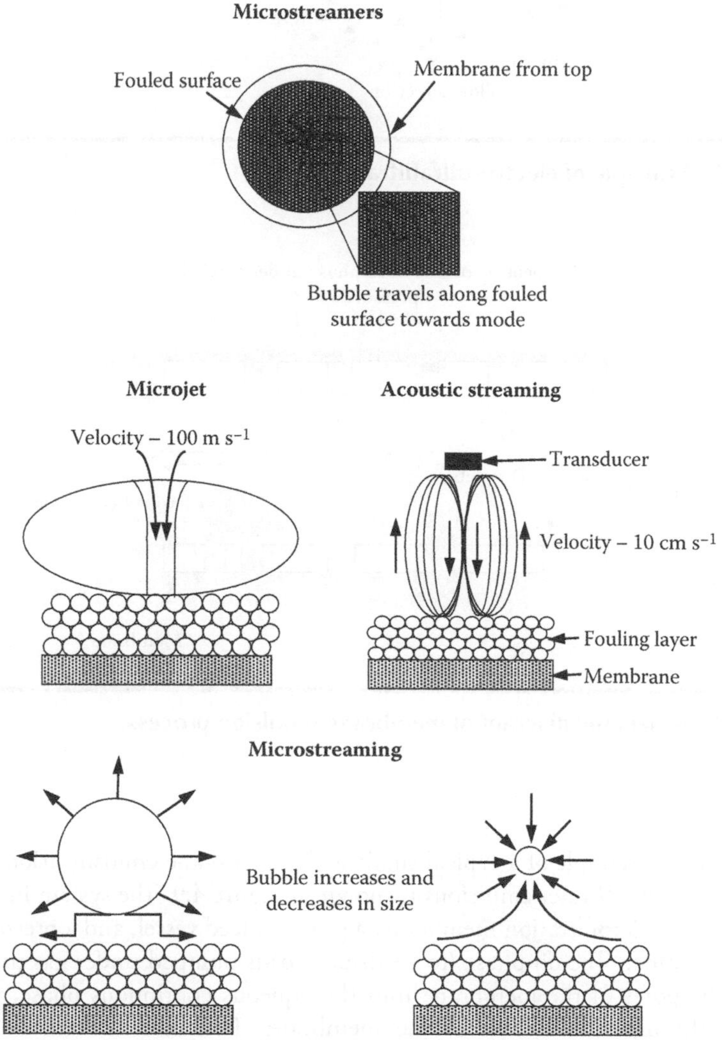

Figure 4.6 Possible mechanisms for removal/detachment observed with ultrasound cleaning.

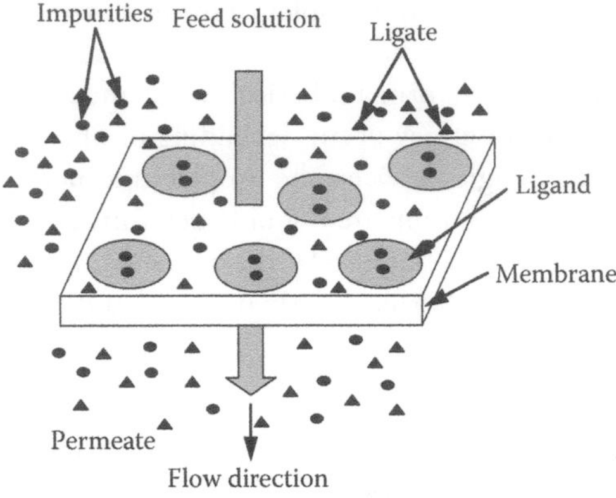

Figure 4.7 Principle of electro-ultrafiltration.

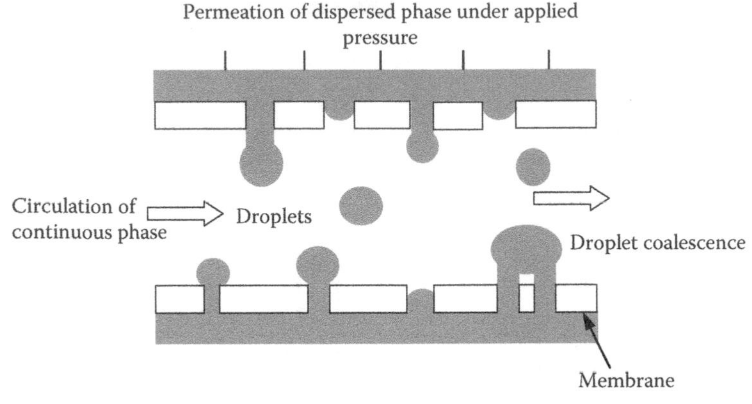

Figure 4.8 Schematic diagram of membrane emulsion process.

A schematic example of a typical small-scale, membrane emulsification apparatus for making oil/water emulsions is shown in Figure 4.9. The system incorporates a tubular microfiltration membrane, a pump, a feed vessel, and a pressurized (N2) oil container. The oil phase (to be dispersed) is pumped under gas pressure through the pores of the membrane into the aqueous continuous phase, which circulates through the middle of the membrane. Pore size and distribution, membrane porosity, membrane surface type, emulsifier type and concentration, dispersed phase flux, velocity of the continuous phase, and transmembrane pressure are some of the important process controlling parameters in membrane

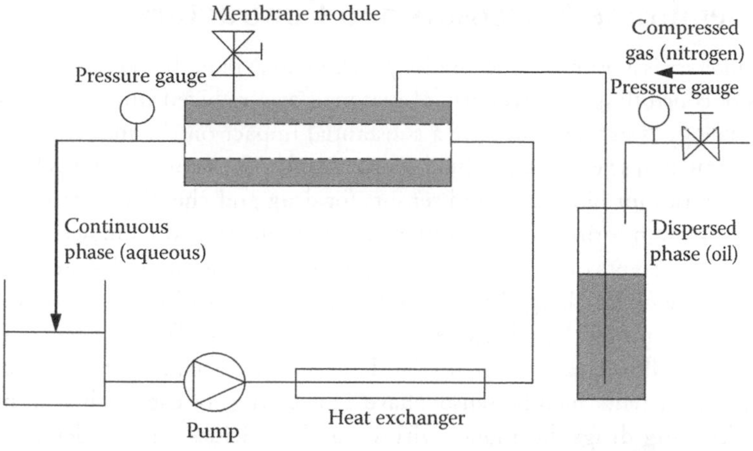

Figure 4.9 Schematic diagram of a simple small-scale membrane emulsification system.

emulsification. In the membrane emulsification process, the transmembrane pressure ΔP is defined as

$$\Delta P = \frac{P_d - (P_{c1} + P_{c2})}{2} \tag{4.28}$$

where P_d is the pressure of the dispersed phase outside the membrane. P_{c1} and P_{c2} are the pressures at both ends of the membrane module. The minimum emulsification pressure, i.e., the pressure at which the dispersed phase outside the membrane must be at to just begin to permeate through the membrane, can be calculated from the equation for capillary pressure:

$$P_c = \frac{4\gamma \cos\theta}{d_p} \tag{4.29}$$

where d_p is the pore diameter, P_c is the oil/water interfacial tension, and θ is the contact angle between the dispersed phase and the membrane surface. P_c is also referred to as the critical pressure.

The critical factors that influence the effectiveness of the membrane emulsification include membrane pore size, distribution, porosity and surface type, velocity of continuous phase, emulsifier, emulsification pressure and flux of dispersed phase, temperature, viscosity, and pH.

4.13 Membrane Biosensors and Contractors

Nanotechnology is a promising technology seeking to exploit distinct technological advances of controlling the structure of materials at a reduced dimensional scale. Nanotechnology is expected to have a substantial impact on the modern research and development of science and technology now and in the future. Nanotechnology offers possibilities for biosensors and sensors for drug and chemical screening and environmental monitoring. When a cell membrane detects a target molecule, it can turn electrical currents on or off by opening or closing molecular channels. When the channels are open, charged ions can pass in to or out of the cell. These ions would not pass through the otherwise insulating membrane. When these channels are open, the ion flow creates a potential difference across the membrane, which, in turn, creates a current. Such biosensors have a huge range of uses such as in medicine, for detecting drugs, hormones, viruses, and pesticides and to identify gene sequences for diagnosing genetic disorders. Even a simple biosensor design offers valuable advances in low-cost sensing for clinical medicine, food, and the health care industry (Figure 4.9).

In membrane contractors, the membrane functions as an interface between two phases, one of which must be a fluid. The contractor does not control the passage of permeate across the membrane. Essentially, microporous membranes of hollow fiber membranes/modules are most commonly used in the membrane contractors. Membrane contractors allow the creation of one immobilized phase interface between two phases participating in separation though the porous membrane. This immobilized phase interface in two phase configurations may include (i) fluid phase membrane contractors (two phases in contact: gas–gas, liquid–liquid, vapor–liquid, or supercritical fluid–fluid); (ii) solid–fluid phase membrane contactors (one fluid phase in contact with one solid phase: liquid–solid, gas/vapor–solid, supercritical fluid–solid); and (iii) multiphase interphase-based membrane contractors (the membrane system has two immobilized interfaces; each interface is between two immiscible fluid phases: gas–liquid and liquid–gas, liquid–liquid and liquid–liquid). Delivery or recovery of gases from liquids is the largest application of a contractor. One example is the blood oxygenator used during surgery in which a patient's lungs cannot function normally. Membrane contractors have also some industrial applications, most commonly to deoxygenate ultrapure water from electrolysis or boiler feed water and to adjust carbonation levels in beverages.

4.14 Dynamic and Vibratory Membrane Filter Systems

Eliminating membrane fouling requires the prevention of accumulation of material in the gel layer. The dynamic membrane filter (DMF) system and vibratory membrane filtration (VMF) system can be used for this purpose. The DMF system,

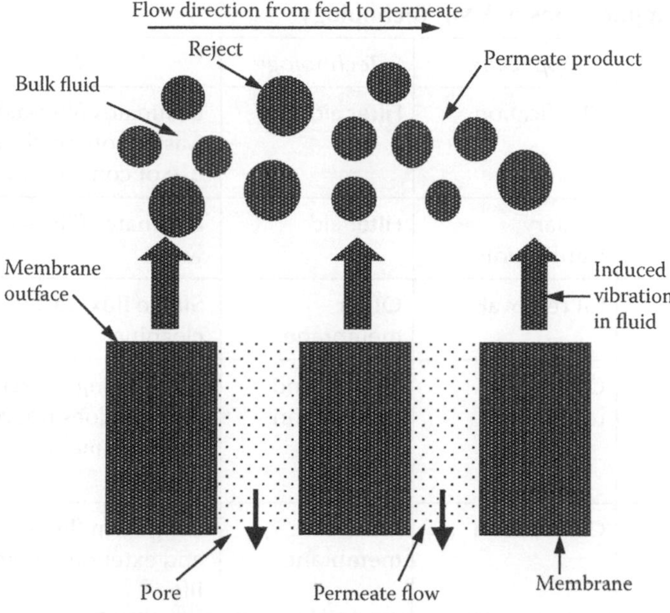

Figure 4.10 VMF keeps rejected species away from the membrane surface, allowing free passage of permeate.

in which a disc spinning in close proximity to the membrane surface generates wall shear rates, provides extremely high transmission of solutes in the filtrate and, accordingly, can be used for bioburden reduction in whey protein concentrate or the filtration of infant formula fractions. However, the DMF system is affected by the solid content and viscosity of the feed solution. A VMF system, on the other hand, is unaffected by solids content or viscosity; the only criterion being that the material should be pumpable. In a VMF system, the energy generating the wall shear rate is directed into the membrane itself. Energy usage efficiency is very high with shear waves being propagated a short distance from the membrane surface (Figure 4.10). The various applications of VMF system are summarized in Table 4.5.

4.15 Membrane Fouling

Fouling is a commonly encountered problem in membrane processing. Membrane fouling can result from the accumulation of multivalent ions, organic matter, and microbial cells in suspension. This accumulation leads to a decrease in the process efficiency and to an increase in the production cost. Membrane fouling is known to depend on several parameters, such as composition of feed, physicochemical

Table 4.5 Applications of VMF Technology

Product	*Application*	*Technology*	*Note*
Beef tallow	Clarification	Filter aid	Eliminates filter-aid waste, potentially allows use of concentrate
Gelation	Primary clarification	Filter aid	Eliminates filter-aid waste
Skim milk	Fat removal	Other membrane	Stable flux, easier cleaning
Casein	Off-flavor removal	Settling and decantation	Time saving—direct process does not require settling time or use of cyclone
Sugar	Clarification	Other membrane	Long-term flux stability and extended membrane life
Soy sauce	Crystalline deposit removal	Filter aid	No filter-aid waste, high flux rate
Xanthan gum	Concentration prior to extraction	None	Concentration (up to 9%–10%) prior to extraction improves speed of process, reduces solvent usage, increases downstream capacity of process
Tea	Removal of leaf solids	Centrifuge	Improved clarification
Coffee	Clarification	Centrifuge	Improved clarification
Apple juice	Clarification	Other membrane	Stable flux, membrane cleanability unaffected by fruit sugars that may be a problem to cross-flow systems
Corn mud	Removal of oil and protein	Centrifuge, filter aid	One step clarification, eliminates waste

(continued)

Table 4.5 (Continued) Applications of VMF Technology

Product	*Application*	*Technology*	*Note*
Gluten	Concentration of protein	Other membrane	Can be run up to limits of pumpability failure
Hydrogenated oil	Removal of catalyst	Plate filter	Catalyst and product separate
Barm beer	Separation of yeast, beer recovery	Centrifuge	Low bioburden filtrate can be directly blended into production beer, high solids achievable in excess of those typical from a centrifuge, no cell damage, concentrate can be sold as animal feed, single pass operation can be used
Yeast extract	Concentrate and wash extract	Centrifuge	Contained single step process—no need for multiple centrifuge cascade
Egg yolk	Bioburden reduction	Pasteurization	Lower energy, no protein degradation

properties of membranes, and also the hydrodynamic conditions. Irreversible adherence and growth of microbial cells on the membrane surface leading to biofilm development, referred to as biofouling, is a major component of membrane fouling in water separation applications. In drinking water plants, biofouling leads to higher operating pressures, increases the frequency of chemical cleanings, and leads to membrane deterioration. Moreover, biofilms and other materials accumulated on the membrane surface, which cannot be removed by cross-flow, backflushing, or backpulsing, can result in permanent permeability loss.

Many authors have studied the fundamental mechanisms involved in membrane fouling by feed materials, which may be as follows:

- The formation of a gel layer due to concentration polarization
- Adsorption of species on the membrane surface and inside the pore structure
- Deposition and pore blocking after the formation of species aggregates due to denaturation

The factors affecting membrane fouling can be classified into four groups: membrane statistics, membrane physicochemical characteristics, feed characteristics, and operating conditions. The complex interactions between these parameters complicate the understanding of membrane fouling. For a given membrane processing, the fouling behavior is directly determined by sludge characteristics and hydrodynamic conditions. But operating conditions and input characteristics have indirect influences on the membrane fouling by modifying sludge characteristics. Table 4.6 gives the relationship between various fouling factors and membrane fouling. There are various models developed based on the various parameters, such

Table 4.6 Fouling Control Strategies Based on Various Fouling Factors

Control Strategy	*Control Item and its Effect on Membrane Fouling Factor*
Hydraulic control	HRT reduces; sludge viscosity increases; EPS increases
	Aeration increases; permeability fiber movement increases; cake-removing efficiency increases; cake resistance decreases
	Periodical backwashing; flux increases; operation period increases; total resistance decreases
	Sub-critical/low flux operation; sustainable operation
Chemical control	Powdered activated carbon; EPS decreases; irremovable fouling decreases
	Membrane fouling reducer; cake porosity increases; soluble EPS decreases
	Flocculation/coagulation; organic matter decreases
	Chemically enhanced backwashing/remove fouling
Biological control	SRT increases; bound EPS decreases
	MLSS/viscosity increases; permeate flux decreases; cake; fouling decreases
	F/M ratio decreases; fouling resistance decreases
	Filamentous bacteria decreases; bound EPS decreases

Note: EPS: extracellular polymeric substance; F: food; HRT: hydraulic retention time; M: microorganism; MLSS: mixed liquid suspended solid; SRT: solids retention time.

as membrane parameters, feed parameters, operating conditions, etc. to predict the fouling characteristics in the membrane processing.

4.15.1 Pore Model

In the pore model, pores are represented as a bundle of N cylindrical non-intersecting, homogeneously distributed capillaries of uniform radius r_p and uniform length l_p. Assuming an incompressible Newtonian fluid and time independent, fully developed laminar flow through the membrane pores, the pore model can be derived from a momentum balance using cylindrical coordinates (Cheryan 1998).

$$J = \frac{\pi N r_p^4 P}{8A\mu l_p} \tag{4.30}$$

This model relates transmembrane pressure P, permeate viscosity μ, and membrane characteristics to permeate flux. Mass transport is controlled by membrane resistance and is directly proportional to the applied transmembrane pressure. Operating below critical pressure (i.e., in absence of membrane fouling and at minimal concentration polarization), which means very low pressure and feed concentration and high feed velocity, the pore model provides an adequate description of the membrane filtration.

4.15.2 Film Theory Model

The film theory model can be derived from a mass balance for a dissolved component across the boundary layer near the membrane surface. At steady state, convective transport of the species to the membrane surface is just equal to diffusive transport of the solute from the membrane into the bulk solution and convective transport of the solute from the membrane surface to the permeate side, which leads to the next equation (Andres et al. 1991):

$$J = \frac{D_s}{\delta}\left(\frac{C_w - C_{pe}}{C_b - C_{pe}}\right) \tag{4.31}$$

where D_s is the diffusion coefficient of the solute, and δ is the boundary layer thickness. This model describes a mass transfer controlled flux and is valid only in the presence of concentration polarization without fouling. If transmembrane pressure exceeds a critical pressure, the solute concentration at the membrane surface will be high enough and the boundary layer will exhibit gel-like properties. Thus, the gel concentration will be the upper limit of the solute concentration at the membrane surface. This results in a constant limiting flux, which is independent of pressure-driven force and membrane permeability.

4.15.3 Resistance in Series Model

Another approach to predict permeate flux is the resistance-in-series model based on the flow of solvent through several transport layers. The membrane is a selective barrier, in which resistance R_m depends upon the mechanical and chemical structure as well as on membrane thickness. Separation of a solute by the membrane gives rise to an increased solute concentration in the boundary layer at the membrane surface and an additional resistance due to concentration polarization R_{cp}. Adsorption and deposition of matter from the process feed within the membrane pores and on the membrane surface give rise to a fouling layer with an extra resistance R_f to solvent flow. A series resistance of the membrane, boundary, and fouling layers is used to relate permeate flux to the applied transmembrane pressure (Choi et al. 2000):

$$J = \frac{P}{\mu(R_m + R_{cp} + R_f)} \tag{4.32}$$

This model can be used for the filtration of dilute solutions or suspensions containing a wide range of solutes, colloids, or particles through a porous membrane at constant pressure considering polarization and fouling phenomena.

4.15.4 Osmotic Pressure Model

The osmotic pressure model assumes a deviation from pure solvent flux due to concentration polarization and membrane fouling. When the concentration difference between both sides of the membrane is high enough, an osmotic pressure P arises and is opposite to the transmembrane pressure P that reduces the net driving force. When a layer of deposited matter is formed, frictional drag due to permeation through this fouling layer leads to an additional hydraulic resistance R_f in series with the membrane resistance R_m. Combining both phenomena, the permeate flux is given by Bacchin et al. (2002) as:

$$J = \frac{P - \pi}{\mu(R_m + R_f)} \tag{4.33}$$

This model can be applied to feed streams of small or macromolecular species without colloids.

4.15.5 Standard Blocking Model

The standard blocking model, just like the pore model, assumes a membrane composed of a bundle of regular cylindrical pores. Consequently, the pore volume decreases proportionally to permeate volume due to the adsorption and deposition of solutes on the pore walls. Pores get plugged up, and the membrane resistance increases as a consequence of pore size reduction. In this case, the flow rate is given by Hermia (1982) as:

$$Q = \frac{Q_0}{(1 + 0.5k_{sb}Q_0t)^2} \tag{4.34}$$

During the processing of concentrated solutions of permeable solutes, such as sugars and salts, internal pore plugging of UF membranes may predominate.

4.15.6 Complete Blocking Model

Herein, after reaching the membrane, each solute participates in blocking by means of pore sealing without superposition of solutes. The pore fraction, which is completely clogged by solutes, is proportional to the amount of permeate flow through the membrane. An exponential decrease in flow rate is found (Hermia 1982):

$$Q = Q_0 \exp(-k_{cb}t) \tag{4.35}$$

The complete blocking model may describe membrane fouling by means of a very dilute feed stream.

4.15.7 Intermediate Blocking Model

In the intermediate blocking model, each solute can reach the membrane at any point, i.e., it can settle on the membrane surface or on another previously arrived solute or it can seal a pore. The probability of a solute to block a pore can be evaluated and leads to the following equation for flow rate (Hermia 1982):

$$Q = \frac{Q_0}{(1 + k_{ib}\, Q_0\, t)^2} \tag{4.36}$$

This model will be valid for dilute solutions with solutes and pores of about the same size.

4.15.8 Cake Filtration Model

In the cake filtration model, macromolecules, colloids, or aggregates deposit over the membrane surface, forming a cake or fouling layer, increasing hydraulic flow resistance due to foulant deposition. The decrease in the permeate volumetric flow rate can be described as follows (Hermia 1982):

$$Q = \frac{Q_0}{\sqrt{\left(1 + 2k_{ef}\, Q_0^2 t\right)}} \tag{4.37}$$

This model may be applied for a more concentrated feed when the solute size is larger than the pore size.

4.16 Integrated Membrane Technology

Development of efficient bioseparation methods is important for a broad range of business areas including pharmaceuticals, nutrition and health products, biobased materials, and crop production chemicals. Depending on the value of the end product and the scale of production, the processing required varies significantly. Key factors that have an impact on the choice of separation strategy include process throughput, particle size of the product, impurities, and desired concentration of product.

Membrane systems are extensively used throughout the various biological and chemical industries to control a variety of products. These membrane processes are successful because they are effective and economically implemented on a large scale, which is required for industrial applications. A combination of different membrane processes gives interesting benefits, which cannot be achieved by a single-membrane operation. The possibility of redesigning overall industrial production by the integration of various already-developed membrane operations is becoming of particular interest. This synergy means that better results can be achieved. Integration membrane technology also shows the simplicity of the units and the possibility of advanced levels of autoimmunization. Membrane hybrid technology works best when developed as one concept. This clearly lies in the influence that processes have on each other. The rationalization of industrial production by use of these technologies permits low environmental impacts, low energy consumption, and higher quality of final products.

4.17 Membrane-Assisted Food Processing

In the food and beverage industries, membrane filtration is state-of-the-art technology for clarification, concentration, fractionation, desalting, and purification of a variety of beverages. It is also applied to improving the food safety of products while avoiding heat treatment. Some examples of final products using this technique are fruit and vegetable juices, such as apple or carrot; cheeses, such as ricotta; ice cream; butter; some fermented milks; skimmed or low-lactose dairy products; microfiltered milk; non-alcoholic beers; wines; ciders; etc.

4.17.1 Dairy Industry

Ultrafiltration of milk offers substantial advantages to both manufacturers and consumers. During the cheese-making process, some of the nutrients found in milk are lost in the whey (e.g., carbohydrates, soluble vitamins and minerals). These losses have a considerable impact on the economics of the processing operation. Ultrafiltration is an effective means of recovering the byproducts, which can be used for further food formulations. At the same time, the result is cheese products of higher nutritional value at a better price. Another application in cheese is the use of microfiltration to remove undesirable microorganisms from the milk used in the production of raw milk cheeses (Rosenberg 1995).

Classical techniques used to improve milk's shelf life and safety are based on heat treatments, such as pasteurization and sterilization. But those techniques modify some sensory properties of milk, for example, its taste. Microfiltration constitutes an alternative to heat treatment to reduce the presence of bacteria and improve the microbiological safety of dairy products while preserving the taste. Fresh microfiltered milk has a longer shelf life than traditionally pasteurized fresh milk. There is also a new development in membrane technology manufacture, which leads to a similar hygienic safety as "thermization" of skimmed milk at 50°C. This will allow the commercialization of new milk, which can be stored at room temperature for six months and with a taste similar to fresh pasteurized milk.

The removal of bacteria and spores from milk to extend its shelf life by MF is an alternative way to ultra-pasteurization. In this approach, the organoleptic and chemical properties of the milk are unaltered. The first commercial system of this so-called Bactocatch was developed by Alfa Laval, Inc., and marketed by Tetra Pak under the name Tetra Alcross Bactocatch. In this process, the raw milk is separated into skim milk and cream. The resulting skim milk is microfiltered using ceramic membranes with a pore size of 1.4 mm at constant transmembrane pressure (TMP). Thus, the retentate contains nearly all the bacteria and spores while the bacterial concentration in the permeate is less than 0.5% of the original value in milk.

The retentate is then mixed with a standardized quantity of cream. Subsequently, this mix is subjected to a conventional high heat treatment at 130°C for 4 s and reintroduced into the permeate, and the mixture is then pasteurized. Because less than 10% of the milk is heat treated at the high temperature, the sensory quality of

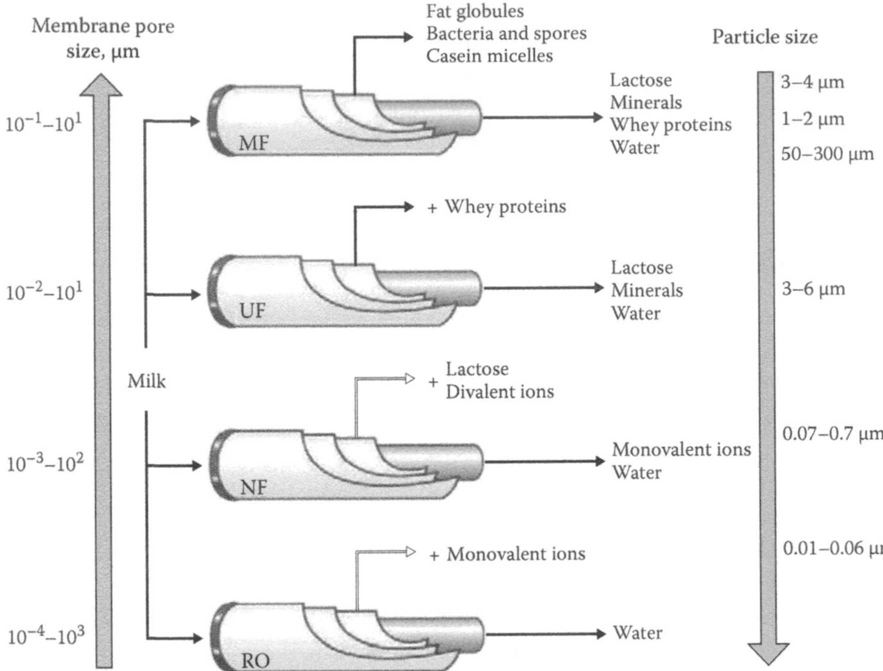

Figure 4.11 Membrane technology in milk processing.

the milk is significantly improved. Figure 4.11 gives a diagrammatic representation of the technology behind membrane processing in the dairy industry.

4.17.2 Fermented Food Products

In the production of fermented food products, such as beer, wine, and vinegar, membranes have been used for clarification after fermentation. MF and clarification along with dealcoholization by RO are some of the applications in beer manufacture. Use of MF/UF in wine making reduces the number of steps in clarification, stabilization, and filtration in one step. RO increases the concentration of tannins and organoleptic factors in the must by water removal.

Microfiltration (MF) and ultrafiltration (UF) have been proven to be effective in refining soy sauce at high yield and ensuring quality products with a long shelf life. Compared with conventional sedimentation and filter-aid filtration, membrane filtration can improve the recovery of soy sauce, eliminate the need for filter medium (thus eliminating additional solid waste), and require less downtime for cleaning. Since the 1990s, MF or UF has become a more attractive technology for sterilization and clarification of soy sauce in China and other Asian countries. The need for high-end products with low-salt content and light-colored soy sauce provided new opportunities for application of membrane technologies, such as reverse osmosis (RO) and nanofiltration (NF), to the production of these kinds of soy sauce. MF and UF was found suitable. UF of soy sauce brought about complete removal of bacteria besides clarification, improving color, retaining flavor, and reducing sodium chloride.

Cross-flow membrane processes were introduced to downstream processing of fermentation products in the 1970s for enzymes, antibiotics, organic acids, vitamins, amino acids, yeast, and biopolymers and, since then, became a standard unit of operation for the recovery and purification of fermentation products (Frank 2010a). UF with diafiltration is the state-of-the-art process in the bulk pharmaceutical industry to separate solutions containing high and low molecular weight solutes (Frank 2010b). The key advantages of this process compared to conventional processes include: high product purity and process yield and elimination of filter aids and wetting agents. MF can be used as pretreatment before enzyme concentration stages to remove impurities. The enzymes can then be either used directly or further purified and concentrated by crystallization, precipitation, adsorption, or UF/NF. By UF, the enzyme strength can typically be concentrated 25 times with hardly any loss of enzyme activity. The initial concentration can be done by UF/NF. The purification effect can be further enhanced by using diafiltration, thus increasing the purity by reducing color and endotoxins. The diafiltration/washing step is also used to remove excess salt when the enzyme is recovered by salt extraction. The enzymes are then typically standardized and either directly used as liquid bulk product or spray dried with an optional preconcentration step by evaporation to be used as a powder. Further, RO can be used to recover purified water from the UF permeate and from the evaporator condensate.

4.17.3 Clarification of Vinegar

The vinegar fining by UF can be applied for a wide range of vinegar types and results in a vinegar product on the permeate side that has similar color and organoleptic qualities to the original vinegar but no turbidity. Additionally, proteins, pectins, yeast, fungi, bacteria, and colloids are removed, and thus the filtration/sedimentation and the clarification are substituted and the storage time reduced.

4.17.4 Fruit Juices

The UF process removes the suspended solids and other high molecular solids, and the filtered juice obtains good clarity and excellent quality. In order to achieve high yield, high capacity, and excellent quality, the juice is pretreated with enzyme and filtered before the UF system is utilized. A newer effective technique combines a high-speed separator with spiral-wound UF modules. Concentration by RO removes 50% of the water and hastens subsequent evaporation, economizing on energy and saving volatile flavor compounds (Merry 2010).

4.17.5 Oilseed Proteins

Research on use of membrane technology is focused on protein extraction and functionality. Commercial food-grade proteins from oilseeds, such as flax and canola, have been developed. Purified plant proteins may be blended with other proteins for use as functional ingredients in bakery foods, beverages, meat protein replacers, and nutritious supplements.

4.17.6 Full-Fat Soy Protein Concentrates (Soy Milk)

A purified protein-fat concentrate or a "soy milk" devoid of oligosaccharides and with lower trypsin inhibitor, reduced off-flavors, and low in phytic acid was possible with UF. Large-diameter tubular membranes were used to ultrafilter oligosaccharides from the water extracts of soya. Soya protein isolate and concentrate could be obtained from an optimum sequence of UF, continuous diafiltration and UF. Several plant proteins have been isolated, namely, corn, sunflower, and dry beans (Wan et al. 2010).

4.17.7 Egg Products

Whole-egg concentration using UF can produce liquid whole egg with up to 40%–44% total solids (TS), and some of the low molecular weight components (e.g., salts and sugars) are removed with permeate. Egg-white concentration by UF can produce liquid egg white with up to 20%–21% TS by removing salts, glucose, and other low molecular components with permeate. RO concentration of egg white can produce egg white up to 24% TS approximately.

4.17.8 Filtration and Dry Degumming of Vegetable Oil

The advantages of using membranes in vegetable oil processing is mainly to reduce the energy consumption, thereby reducing the cost associated with it, improving the product quality, and increasing the yield of vegetable oils. Cross-flow UF membranes soaked in solvents of varying polarity were effective in the degumming of vegetable oil with minimum oil loss and producing lecithin byproduct. UF membranes have been applied with a relatively better success rate for degumming of oil obtained by screw press (Ladhe and Kumar 2010).

4.18 Future Trends

Pervaporation involves the mass transfer of components through a commonly nonporous polymeric or zeolite membrane combined with a phase change from liquid to vapor. The driving force of pervaporation is an activity difference between the feed and permeate side, and the mass transfer can be described based on the solution diffusion model. Prospective applications of this concept are being investigated for dealcoholization by using hydrophilic membranes instead of RO for aroma recovery from fruit juices, wine, and herbal and flower extracts. Moreover, the use of UF membranes also has potential for deacidification, dewaxing, removal of coloring matter, and recovery of proteins and other value added concentrates, which is currently under study.

References

Andres, J.L., Stanley, K., Cheifetz, S., and Massagué, J. (1991). Membrane anchored and soluble forms of betaglycan, a polymorphic proteoglycan that binds transforming growth factor-β. *Journal of Cell Biology*, 109 (1989), 3137–3145.

Bacchin, P., Si-Hassen, D., Starov, V., Clifton, M.J., and Aimar, P. (2002). A unifying model for concentration polarization, gel-layer formation and particle deposition in cross-flow membrane filtration of colloidal suspensions. *Chemical Engineering Science*, 57, 77–91.

Bazinet, L., Lamarche, F., and Ippersie, D. (1998). Bipolar-membrane electrodialysis: Applications of electrodialysis in the food industry. *Trends in Food Science and Technology*, 9 (3), 107–113.

Cesar de, M.C., Ming, C., Rodrigo, C.B., Ana, P.B.R., Lireny, A.G.G., and Auiz, A.V. (2009). State of art of the application of membrane technology to vegetable oils: A review. *Food Research International*, 42 (5–6), 536–550.

Cheryan, M. (1998). *Ultrafiltration and microfiltration handbook*. CRC Press, Boca Raton, FL.

Choi, K.H., Peck, D.H., Kim, C.S., Shin, D.R., and Lee, T.H. (2000). Water transport in polymer membranes for PEMFC. *Journal of Power Sources*, 86 (1–2), 197–201.

Frank, L. (2010a). Cross flow membrane applications in the food industry. *Membrane technology, vol. 3: Membranes for food applications*. P. Klaus-Viktor, P. Suzana, and G. Lidietta, (eds.), Wiley-VCH, Weinheim.

Frank, L. (2010b). Membrane processes for the production of bulk fermentation products. In *Membrane technology: A practical guide to membrane technology and applications in food and bioprocessing*. Z.F. Cui and H.S. Muralidhara (eds.), Elsevier, Oxford, pp. 123–115.

Hermia, J. (1982). Constant pressure blocking filtration laws: application to power law non-Newtonian fluids. *Trans IChemE*, 60, 183–187.

Jiao, B., Cassano, A., and Drioli, E. (2004). Recent advances in membrane processes for the concentration of fruit juices: A review. *Journal of Food Engineering*, 63, 303–324.

Kulkani, S.S., Funk, E.W., and Li, N.N. (1992). Ultrafiltration: Theory and mechanistic concepts. In *Membrane handbook*. W.S.W. Ho and K.K. Sirkar, (eds.), Van Nostrand Reinhold, New York, p. 398.

Ladhe, A.R., and Kumar, N.S.K. (2010). Application of membrane technology in vegetable oil processing. In *Membrane technology: A practical guide to membrane technology and applications in food and bioprocessing*. Z.F. Cui and H.S. Muralidhara (eds.), Elsevier, Ltd., pp. 63–78.

Merry, A. (2010). Membrane processes in fruit juice processing. In *Membrane technology: A practical guide to membrane technology and applications in food and bioprocessing*. Z.F. Cui and H.S. Muralidhara (eds.), Elsevier, Oxford, pp. 33–43.

Rosenberg, M. (1995). Current and future applications of membrane technology in the dairy industries. *Trends in Food Science and Technology*, 6, 12–1.

Wan, Y., Luo, J., and Cui, Z. (2010). Membrane application in soy sauce processing, In *Membrane technology: A practical guide to membrane technology and applications in food and bioprocessing*. Z.F. Cui and H.S. Muralidhara (eds.), Elsevier, Oxford, pp. 45–62.

Zydney, A.L., and Colton, C.K. (1986). A concentration polarization model for filtrate flux in cross flow micro-filtration of particulate suspensions. *Chemical Engineering Communication*, 47, 1–21.

Chapter 5

High Hydrostatic Pressure Processing of Food Materials

P. Srinivasa Rao, S. Chakraborty, N. Kaushik, B. Pal Kaur, and N.R. Swami Hulle

Contents

5.1 Introduction

Today's consumer demand for safe, minimally processed, additive-free, and shelf-stable foods that retain appearance, natural flavor, and texture has been a driving force for commercial application of non-thermal food processing methods. High hydrostatic pressure (HHP) processing is one such novel technology, which has gained popularity in food industries for the past two decades due to its potential to meet all the above requirements. It is also termed as hyperbaric pressure processing or ultra-high pressure or high-pressure processing (HPP) or pascalization.

In HPP, a food is subjected to high-pressure in the range of 100 to 1000 MPa with or without heat in which pressure is instantaneously and uniformly transmitted throughout the product irrespective of its size and geometry. To be clear, 10,000 kg weight balancing on a 1 cm^2 area will produce a pressure of 1000 MPa. Recently, HPP systems are available for treatment up to 1400 MPa, but the volume of sample treated at that pressure is very small (up to 30 mL). The typical advantage of this technology is the ability to inactivate both food spoilage microorganisms as well as enzymes at low or ambient temperature along with the minimal effect on flavor and nutritional attributes of the products (Sequeira-Munoz et al. 2006). However, a complete or larger extent of inactivation of specific enzymes may require thermal assistance along with high-pressure application. Following initial successes with fruit juices and jams, the HPP technology has been applied to a wider range of food products, including milk and milk products, meat, smoothies, ham, guacamole, salsa, rice products, fish, and shellfish (Murchie et al. 2005).

Future advances are expected from the synergistic effect of high pressure and temperature combinations in rapidly evolving pressure-assisted thermal processing (PATP) technology. PATP, which is yet to be commercialized, will require more complex safety validation procedures than HPP, particularly for low-acid foods (pH < 4.5). PATP conditions are sufficiently severe to achieve the inactivation of bacterial spores, and recent studies suggest that pressure application can lower the degradation rate of product quality caused by high temperature treatments (Vazquez-Landaverde et al. 2007).

5.2 Principles of High-Pressure Processing

5.2.1 Le Chatelier's Principle

Chemical reactions that result in a decrease in total volume (negative activation volume) are enhanced by pressure. Conversely, reactions resulting in an increased total volume (positive activation volume) are slowed down by pressure. HPP involves the application of high-pressure, which reduces the availability of molecular space favoring the chain interactions and finally inducing negative volume change. The effects of pressure on protein stabilization are also governed by this principle, i.e., the negative changes in volume with an increase in pressure cause an equilibrium shift toward bond formation. The breaking of ions is also enhanced by high-pressure as this leads to a volume decrease due to the electrostriction of water. Consequently, high-pressure can disrupt large molecules of microbial cell structures, such as enzymes, proteins, lipids, and cell membranes, leaving small molecules, such as vitamins and flavor components, unaffected (Linton and Patterson 2000).

5.2.2 Isostatic Rule

The effects of high-pressure are uniform as its distribution follows isostatic rule which is nearly instantaneous throughout the food irrespective of its geometry and equipment size. This unique property of HPP in comparison with other preservation technologies has facilitated the scale-up laboratory findings to full-scale production.

5.2.3 Compression Energy

Energy input during the high-pressure process is very small compared to thermal processes. Therefore chemical reactions involving covalent bonds remain unaffected.

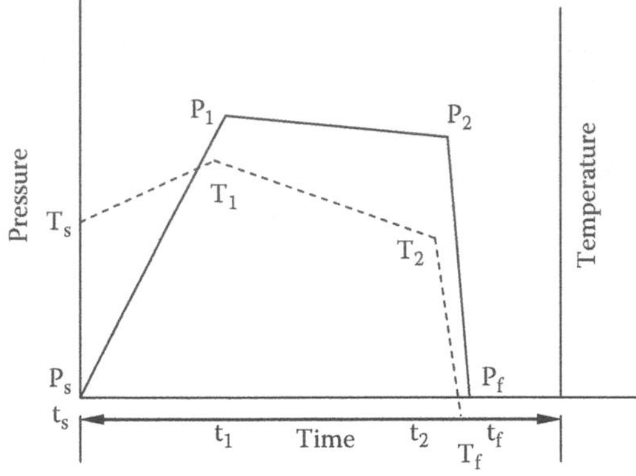

Figure 5.1 Variation of pressure and temperature in a non-insulated high-pressure vessel.

5.2.4 Heat of Compression

Pressurization from an initial pressure P_s to P_1 (Figure 5.1) is accompanied by a uniform temperature increase through adiabatic heating. Each material has its own specific heat of compression values, for example, water ~3°C/100 MPa, fats and oils ~6°C/100–8°C/100 MPa (Balasubramanian and Balasubramaniam 2003). However, with 30% aqueous monopropylene glycol (MPG), the average rise in temperature is about 2°C per 100 MPa. The magnitude of temperature increase, in part, also depends upon the initial temperature, material compressibility and specific heat, and the target pressure. The maximum product temperature at process pressure is independent of the compression rate as long as heat transfer to the surroundings is negligible. During the pressure holding time (P_1 to P_2) the temperature of the product decreases from T_1 to T_2 due to heat loss through the pressure vessel. Foods normally cool down to their original temperature on decompression if no heat transfer occurs through the walls of the pressure vessel during the holding time. Thus, high-pressure offers a unique way to increase the temperature of the product only during the treatment.

5.3 High-Pressure Processing System

A typical HPP system consists of a high-pressure vessel and its closure, a pressure generation system, a temperature controller, and a material handling system. The key components among these are the pressure vessels and the high hydrostatic pressure generating pumps or pressure intensifiers.

5.3.1 Pressure Vessel

The heart of a high-pressure processing system is the pressure vessel, which, in many cases, is simply a forged monolithic, cylindrical vessel constructed using a low-alloy steel of high tensile strength. The wall thickness is determined by the maximum working pressure, vessel diameter, and number of cycles for which the vessel is designed. For any high-pressure system, the working pressure is a very important parameter for determining the working life of the equipment as the number of failures increases with the magnitude of working pressure (Otero and Sanz 2000). The required wall thickness can be reduced by using multilayer, wire-wound, or other prestressed designs (Figure 5.2).

Prestressing by wire-winding and other technologies is preferred over monoblock for safe and reliable commercial-size vessels operating at higher pressures in excess of 400 MPa. Wire winding increases equipment costs and is thus preferable for operations requiring ~600 MPa, such as guacamole salsa production, and uneconomical for operations, such as oyster shucking requiring 200–400 MPa. However, to inactivate bacterial spores, a pressure vessel capable of operating at 700 MPa and temperature higher than 100°C is essential.

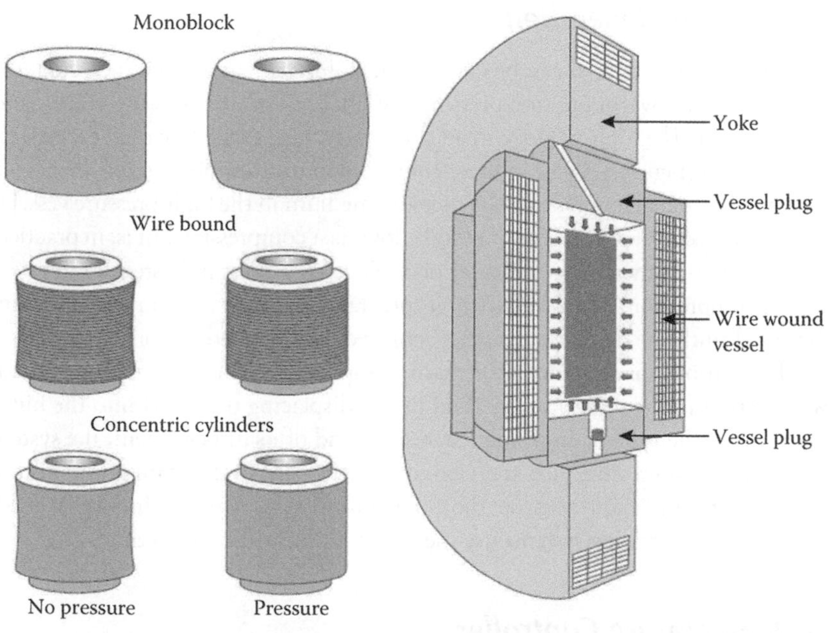

Figure 5.2 High-pressure vessel technologies. (From Torres, J.A., and Velazquez, G., *Journal of Food Engineering*, 67, 95–112, 2005.)

5.3.2 Closures

In general, two types of closure mechanisms are used, viz. multi-threaded and yoke-held top and bottom caps. Most fast-cycling CIP systems use interrupted threaded closures, allowing very fast opening and closing of the vessel and hence, minimizing vessel down time for loading and unloading. The threaded closures are self-centering and can be automatically opened and closed by means of a hoist device, guiding the closure without any thread friction. The yoke is a heavy metal frame, which is capable of holding top and bottom caps in position while high-pressure is being applied for treatment of foods. The wire-winding technology is also used for the yoke holding the top and bottom seals.

5.3.3 Pressure-Transmitting Medium

Pressure is generated by compressing any food-grade fluid. In most current cold applications, pressure-transmitting medium (PTM) is simply purified water mixed with a small percentage of soluble oil for lubrication and anticorrosion purposes. Other fluids used are aqueous solution of mono-propylene glycol (MPG) for high temperature and isopropyl alcohol (IPA) for low temperature HPP, etc.

5.3.4 Pressure Generation

The pressurization of material can be carried out in four different ways, i.e., cold isostatic pressing (CIP), warm isostatic pressing (WIP), hot isostatic pressing (HIP), and chemical reaction. The high-pressure in a HPP system is generated by direct compression (piston type) (Figure 5.3a) or indirect compression (pump and/or intensifier type) (Figure 5.3b). In direct compression, the pressure medium in the high-pressure vessel is directly pressurized by a piston. This method allows fast compression but is, in practice, restricted to small-diameter, laboratory or pilot plant high-pressure systems. However, the indirect compression method uses a high-pressure intensifier to pump the medium into the vessel until the desired pressure is achieved. In this method, oil at 20 MPa is fed into the high-pressure oil side of the main pump piston, which has an area ratio of 30:1 with respect to the high-pressure fluid piston displacing the PTM into the high-pressure vessel. When the main piston reaches the end of its displacement, the system is reversed, and high-pressure oil is then fed to the other side of the main pump piston, and the high-pressure fluid exits on the other pump side. Most of industrial cold, warm, and isostatic pressing systems use the indirect pressurization method.

5.3.5 Temperature Controller

Temperature of the product and pressure medium inside the high-pressure vessel can be controlled by heating/cooling of the entire pressure vessel or by internal heating/cooling. In the latter case, the temperature-control device is placed inside

(a) Direct pressure generator

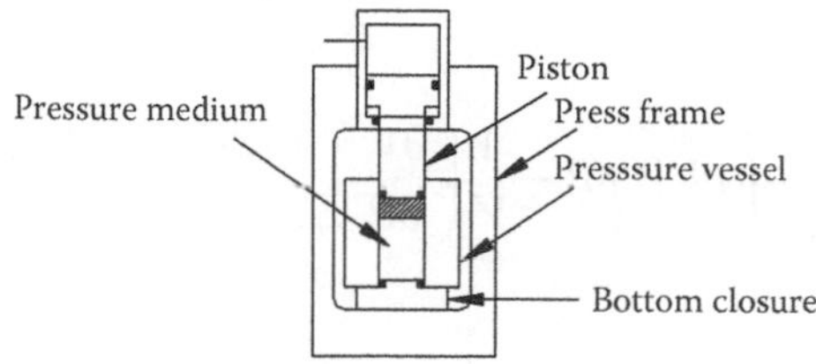

(b) Indirect pressure generator

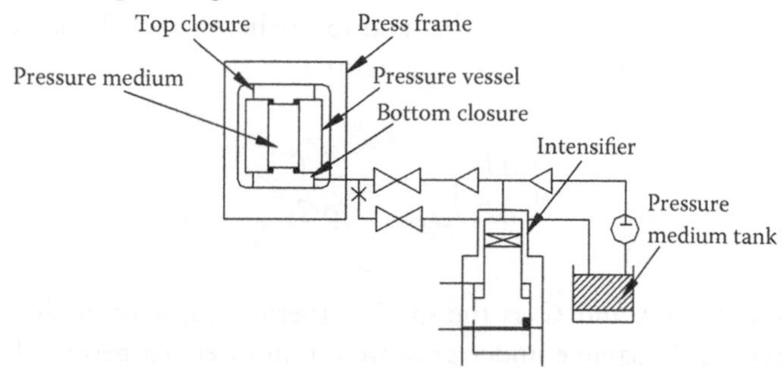

Figure 5.3 Methods for the generation of high isostatic pressure. (From Kanda, T., Recent trend of industrial high-pressure equipment and its application to food processing. In R. Hayashi (Ed.), *Pressure-Processed Food Research and Development* (p. 341). Japan, 1990.)

the vessel. The simplest execution of external heating uses electric heater bands wound around the high-pressure vessel.

5.4 Work Input for High-Pressure Processing

The required work to increase the pressure to the desired pressure level can be estimated if the compressibility of the pressure-transmitting medium (PTM) is known. Assuming the pressure (P) increases following the equilibrium path, the element of work (W) to increase the pressure (or decrease the volume (V)) of a system can be written:

$$dW = -PdV \tag{5.1}$$

The variation of V with P at constant T defines the compressibility of a substance (β):

$$\beta = -\frac{1}{V}\left(\frac{\partial V}{\partial P}\right)_T \tag{5.2}$$

Introducing Equation 5.1 to Equation 5.2 yields

$$W = \int PV\beta \, dP - \frac{1}{V}\left(\frac{\partial V}{\partial P}\right)_T \tag{5.3}$$

The temperature within the pressure vessel does not remain constant during pressure build-up. The temperature within the pressure system will increase as the pressure increases. Equation 5.3 can be used to predict this adiabatic heat generation during pressure build-up (Pfister et al. 2001).

$$\left(\frac{\partial T}{\partial P}\right)_{\text{adiabatic}} = \frac{\beta T}{\rho C_P} \tag{5.4}$$

where ρ is the density, and C_p is the specific thermal capacity of the substance. However, keeping the sample under pressure for an extended period of time does not require any additional energy (Cheftel and Culioli 1997), meaning no work input is required during the pressure hold period.

5.5 Modes of Operation

Industrial high-pressure treatment is either a batch or semi-continuous process. The equipment selection for HPP depends on the kind of food product to be processed. Solid food products or foods with large solid particles can be treated only in a batch mode process. Liquid, slurry, and other pumpable products have the additional option of semi-continuous production (Ting and Marshall 2002).

5.5.1 Batch Processing

Batch-type processing has been a preferred method for high-pressure treatment of packaged foods. Advantages of adopting batch processing are that it eliminates any risk that the food may be contaminated by lubricants or wear particles from machinery and the equipment does not need cleaning between product changes, thus eliminating any danger of cross contamination by food particles. The vessels may be operated in a vertical, horizontal, or tilting mode. However, handling, drying, and storage of packages lengthen the overall processing cycle and, hence, increase the overall cost of the process. The food is prepared and aseptically filled/sealed in flexible packaging, then placed in a pressure chamber for pressurizing, using suitable PTM. Pressurization is done either by pumping the medium into the vessel or by reducing the volume of the pressure chamber by using a piston.

Compressibility of PTM is an important factor e.g., water is compressed by up to 15% of volume at pressure above 600 MPa. Once the target pressure is reached, the pump or piston is stopped, the valves are closed, and the pressure is maintained without further energy input. Immediately after target dwell time, the system is depressurized followed by opening of the vessel and product unloading. With the advent of technology, the process control through valves and pumps operation has become more accurate; hence pressures obtained are very much with close hysteresis. The system is again reloaded with product, either manually or by using hydraulic mechanisms, depending on the degree of automation possible. The total time for pressurization, holding, and depressurization is referred to as the "cycle time." The cycle time and the loading factor (i.e., the percentage of the vessel volume actually used for holding packed product, primarily a factor of package shape) determine the throughput of the system. In the commercial high-pressure batch process, generally, throughput is maximized by shortening the dwell time. Batch-type high-pressure vessels are available as laboratory units with volumes of 0.035 to 2 L. Pilot plant vessels have capacities of 5 to 25 L, and batch production pressure vessels can be supplied with volumes of several hundred liters. In batch operation, two or more pressure vessels can be driven by a single intensifier. However, the compression time is a function of pump capacity (hp). Therefore, the compression energy required in high-pressure vessels depends upon the vessel volume. Table 5.1 shows the examples of the compression energy contained in high-pressure vessels filled with water at 400 kPa.

Numerous batch mode high-pressure vessels are available for chemical and food applications. Depending on versatility in applications of HPP, the vessels have various sizes and operating pressures as shown in Table 5.2. Capacity of the system depends upon operating pressure, shape, and size of the vessel; vessel loading and unloading; and the shape and size of food packages.

5.5.2 Semi-Continuous Processing

Direct introduction of food into a high-pressure chamber is an alternative process compared to the batch one. This has been achieved industrially, so far, only in a semi-continuous mode. Pumpable products can be pumped in and out of the processing vessel through special high-pressure transfer valves and isolators. To

Table 5.1 Compression Energy Contained in High-Pressure Vessels Filled with Water at 400 kPa

Internal vessel volume (*L*)	10	50	100	250	1000
Energy (*kJ*)	192	960	1920	4800	19,200

Table 5.2 Some of the Existing High-Pressure Vessels with Specifications

Maximum Pressure (MPa)	*Diameter (mm)*	*Length (mm)*	*Internal Volume (L)*	*Manufacturer*
196	2000	3000	9400	Kobe Steel
100	1700	4000	9000	Engineered Pressure Systems
200	1000	4000	3150	Engineered Pressure Systems
410	600	4500	1250	Engineered Pressure Systems
310	540	3000	687	Avure Technologies
900	500	2500	500	ABB Autoclaves
600	190	1220	35	Avure Technologies
800	110	625	5	Stansted Fluid Power
800	110	375	3.5	Stansted Fluid Power
900	100	250	2	Stansted Fluid Power
1030	100	1000	8.5	Engineered Pressure Systems
1380	90	550	3.5	Engineered Pressure Systems
1400	22	100	0.035	Stansted Fluid Power

package the product after treatment, additional aseptic filling systems are required. Current semi-continuous HPP systems are used for treating liquid food in the pressure vessel with a free piston for the compression (Figure 5.4). The free piston is displaced after filling up the vessel by a low-pressure food pump. When filled, the inlet port is closed, and PTM is introduced behind the free piston to compress the liquid food. After the desired holding time, decompression of the medium releases the pressure within the vessel. The treated liquid is discharged from the pressure vessel to a sterile hold tank through a discharge port. A low-pressure pump is used to move the free piston toward the discharge port. A semi-continuous system with a processing capacity of 600 L h^{-1} of liquid food and a maximum operating pressure of 400 MPa is used commercially to process grape fruit juice in Japan. Multiple units

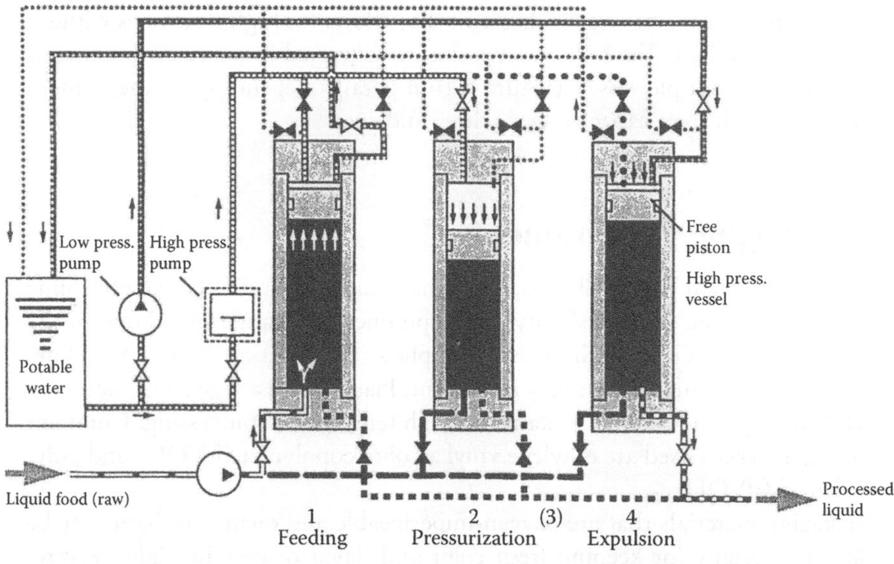

Figure 5.4 Multi-vessel arrangement for semi-continuous high-pressure system. (From Singh, R.P., Technical elements of new and emerging non-thermal food technologies, FAO bulletin, 2001.)

can be sequenced so that while one unit is being filled others are in various stages of operation (Palou et al. 2002).

5.5.3 Bulk Processing

Bulk processing is applicable only to pumpable foods, and it offers the advantage that the system is much less significant than in the case of the in-container process. However, the bulk process requires an aseptic design of special high-pressure components. In fact, there is no established rule to use in choosing between an in-container and a bulk process apart from the basic one of whether the food product is pumpable or not. Moreover, if the batch and semi-continuous processes are compared, there is an economic advantage of about 25% for the semi-continuous system. The bulk process requires a specific system to fill the product into the vessel without having any influence on the quality and the constitution of the food. Also, the maximum amount of the product must fill the vessel, and the presence of air must be avoided. Furthermore, the food remains in contact with the vessel surfaces, which should be of non-corrosive and food-grade type. The high-pressure pumping unit pressurizes PTM that has to be separated from the food product by a reliable system. The pressurized food is discharged from the vessel through a suitably designed aseptic high-pressure valving unit that allows for high flow rates without damaging the product. The bulk process can be converted into a semi-continuous one by utilizing multiple vessels (Figure 5.4) in parallel depending on the number of vessels required, their capacity, cycle time, and cost.

5.6 Packaging Requirements

HPP requires airtight and flexible packages that can withstand a change in volume corresponding to the compressibility of the product, as change in volume in the food matrix is a function of the pressure applied (Hugas et al. 2002). Therefore, selection of packing materials is very important. Plastic films are generally accepted for HPP although they are not suitable for high temperature processing. Common packaging materials used are ethylene vinyl alcohol copolymer (EVOH) and polyvinyl alcohol (PVOH).

Packaging materials that are oxygen-impermeable and opaque to light may be developed especially for keeping fresh color and flavor of certain high-pressure-treated foods (Hayashi 1995). On the other hand, metal cans and glassware are not suitable for high-pressure treatment. In production, the use of flexible pouches can achieve a high packing ratio; the use of semi-rigid trays is also possible; vacuum-packed products are identically suited for HPP. Because the size and shape of the product will have major effects on the stacking effectiveness of the product carrier, they must be optimized for the most cost-effective process; it is obviously uneconomical to treat empty space. A critical control point in the food industry is the handling of certain products after HPP; improper handling can lead to the product becoming re-contaminated before packaging.

5.7 Effects of High-Pressure Processing on Food Quality

5.7.1 Effect of High-Pressure on Microorganisms

Microorganisms, a heterogeneous group of organisms, are different members capable of growing at temperatures from well below freezing (extreme psychrophiles) to temperatures above 100°C (extreme thermophiles). However, each species has a particular range in which it can grow best; this range is determined largely by the influence of temperature on cell membranes and enzymes, and growth is restricted to those temperatures at which cellular enzyme and membranes can function. As with temperature, large differences in pressure resistance can be apparent among various strains of the same species.

When a food is subjected to high-pressure, the compression is instantly transmitted through the hydrostatic media to the microbes present in the food.

Compression appears to affect microbial inactivation by altering the proteins responsible for replication, integrity, and metabolism. High-pressure alters hydrogen and ionic bonds responsible for holding proteins in their native forms. Thus, observed microbial inactivation kinetics can be postulated to be the result of the irreversible denaturation of one or more critical proteins in the microbes. Because the extent of irreversible protein denaturation is a function of its structure, a wide range of pressure resistances must be expected among vegetative microbes. This indicates that a critical protein that seems to be denatured can be repaired as the repair proteins are not damaged in that condition.

High-pressure treatment is also known to cause sublethal injury to microbes, which is a particularly important consideration for any preservation method. Given favorable conditions, such as prolonged storage in a suitable substrate or medium, sublethally injured cells may be able to recover. On the other hand, cell death is associated with irreversible damage to cell components, which are essential for cell growth and reproduction. Cell repair can be affected by the food composition. Acids in foods may inhibit the repair of damaged cell proteins, making the microbe more sensitive to pressure or temperature.

Pressure treatment at 400 MPa for 15 min or at 500 MPa for 3 min at room temperature achieves satisfactory microbial reduction similar to thermal pasteurization, but the high processing cost of HPP restricts its commercial viability. Taking an example of processing of milk, HPP treatments at 586 MPa for 3 and 5 min at moderate temperature (55°C) extend the refrigerated shelf life of milk to more than 45 days, retaining its volatile profiles similar to those observed after conventional HTST treatments. On the other hand, ultra-high temperature (UHT) processing (135°C–150°C for 3–5 s) yields milk that is stable at room temperature for 6 months, but inducing strong "cooked" off-flavor notes in the product during processing limits consumer acceptance of UHT (Steely 1994).

5.7.1.1 Applications of High-Pressure Processing for Microbial Inactivation

HPP has been used with hundreds of products to inactivate spoilage-causing organisms and enzymes, germinate or inactivate some bacterial spores, extend shelf life, and reduce the potential for food-borne illness. The cell membrane is the primary site of pressure-induced cell damage in case of microbes. However, pressure sensitivity can differ widely even within strains of one species (Table 5.3). Exponentially growing cells tend to be more sensitive to pressure-induced cell damage than stationary phase cells. In addition, Gram-positive bacteria are generally more resistant to pressure than Gram-negative. Bacterial spores are highly pressure resistant because pressures exceeding 1200 MPa may be needed for their inactivation (Knorr 1995).

Some of the recent findings exploring the effect of HPP on microbial inactivation in mango, lychee, and black tiger shrimp have been presented below as case studies.

Table 5.3 Examples of Application of High-Pressure Technology for Food Preservation

Product	*Processing Conditions (MPa/°C/min)*	*Process and Quality Attributes*	*References*
Orange juice	500/35/5	Improved shelf life, better consistency, low acid loss	Polydera et al. (2003)
Sausage	500/65/5	Better texture, improved taste, more juicy, less firm, no loss in color	Mor-Mor and Yuste (2003)
Cheese	400/20/20	Higher yield, increased microbial content, less crumbly, no color change	Sandra et al. (2004)
Salmon	200/20/10	Lighter color, increased tissue firmness, shelf life extended	Lakshmanan et al. (2003)
Pear	400/20/30	Browning, firm texture	Prestamo and Arroyo (2000)
Oyster	100–800/10/20	Oyster muscles get detached from the shells, resulting in shucking, but the recovered tissue has good shape and is more voluminous and juicy than that of untreated oysters, lower microbial load	Cruz et al. (2004)

Case Study 5.1: Mango Pulp

Fresh mango pulp was treated at 100 to 600 MPa and 29°C ± 5°C for different holding times at 400 MPa min^{-1} ramp rate of pressurization and analysis was performed for viable microbial counts, viz. aerobic mesophilic count (AMC), yeast and mold count (YMC), lactic acid bacteria count (LAB), total coliforms (TC), and psychrotrophic count (PC).

Both pressurization and hold time had significant effect on lowering the microbial counts ($p < 0.05$). Increased pressure intensity and longer hold duration both resulted in reduced viability of microorganisms. Psychrotrophic microbes were found to be most resistant to single pulse pressurization (2.87 log reduction at 600 MPa for 1 s) whereas coliforms were found to be most sensitive (5.23 log reduction at 600 MPa for 1 s). Also, yeast and mold were found to be most barotolerant to static pressurization (4.31 log reduction at 300 MPa for 15 min) whereas coliforms were found to be most barosensitive (4.02 log reduction at 300 MPa for 5 min) for the same. Detection limit of the microbiological assay was 10 cfu g^{-1}. Reduction of all the studied microbial counts to < 10 cfu g^{-1} in the sample was reported for 300 MPa for 20 min and 400 MPa for 15 min treatments (> 5.9 log reduction). Microbial destruction effects of high-pressure have been hypothesized to protein denaturation, cell membrane rupture, and destabilization of cytoplasmic contents due to increased membrane permeability, which proves fatal for microbes (Meersman and Heremans 2008). Based on this study, it was concluded that high-pressure pasteurization processes can be designed for mango pulp considering psychrotrophs as a target group for single pulse pressurization and yeast and mold as target group for static pressurization processes.

Case Study 5.2: Lychee Juice

Lychee juice was pressurized at 332–668 MPa pressure for 7–23 min of hold times at 215–385 MPa min^{-1} ramp rate of pressurization (process temperature: 29°C ± 5°C) and analyzed for viable microbial counts, viz. AMC, YMC, LAB, TC, and PC.

All the studied microbial counts were reduced to ≤ 1 cfu ml^{-1} on application of > 500 MPa pressure for > 10 min, giving more than a 3 log cycle reduction in aerobic mesophiles and yeast and mold counts. More than a 2 log cycle reduction was achieved for coliforms and psychrotrophic bacteria. Aerobic mesophiles (AMC) were found to be most resistant and coliforms were found to be least resistant. Increasing intensity of pressure, dwell time, and ramp rate decreased the viability of all the microbes studied. Thus, within the studied range of variables, pressure of > 500 MPa for > 10 min will be agreeable for production of microbiologically safe lychee juice.

Case Study 5.3: Black Tiger Shrimp (*Penaeus monodon*)

Shrimp samples treated at 100, 270, and 435 MPa for 5 min at room temperature (25°C ± 2°C) were evaluated for total plate count (TPC), *E. coli* (Gram-negative), and *S. aureus* (Gram-positive) count. HPP significantly reduced the individual microflora present in shrimp, and this decrease was greater at higher pressures (Figure 5.5). Pressure treatments of 270 and 435 MPa significantly ($p < 0.05$)

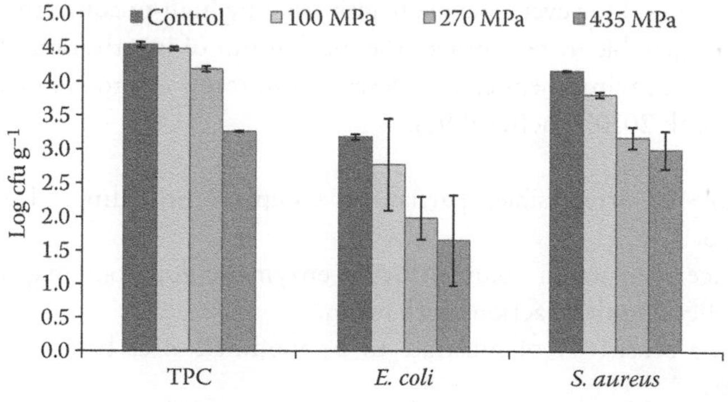

Figure 5.5 Microbiological quality of high-pressure-treated black tiger shrimp.

reduced TPC by 0.37 and 1.29 log cycles; *S. aureus* by 0.98 and 1.16 log cycles; and *E. coli* by 1.20 and 1.53 log cycles (Kaur et al. 2013). Gram-negative bacteria are less resistant to HPP than Gram-positive bacteria (Gram and Huss 2000); this has been explained as a result of the complexity of Gram-positive cell membranes. The present results agreed with this Gram-negative susceptibility to high-pressure treatment. The study concluded HPP could be a useful technology for extending the shelf life of the valuable catch and enhancing the market potential of black tiger shrimp.

5.7.2 Effect of High-Pressure Processing on Enzymes

5.7.2.1 Enzyme Structure and Activity

An enzyme is a special molecular device that determines the pattern of biochemical transformations. Usually it is a globular protein that possesses the functional characteristics of binding one or more substrate molecules. The interesting properties behind its uniqueness are its catalytic power and specificity. Both properties are maintained by a three-dimensional hydrophobic cleft bonded with a substrate called an active site. The structure of any enzyme is characterized by four distinct aspects: primary, secondary, tertiary, and quaternary structures. The primary structure refers to the linear number and sequence of amino acids present in it. The local spatial arrangements and array of amino acids confer the secondary structure. The tertiary structure refers to the three-dimensional compact folding and coiling in α-helix manner. The quaternary structure is governed by several structural subunits. The transition between the native conformations (generally tertiary or quaternary structures) of the enzyme is dependent on the interactions between the surrounding solvent molecules and the molecules present near the surface. Disturbing this balance by any means may change its conformational structure and finally results in the loss of its activity.

5.7.2.2 Effect of High-Pressure on Enzyme Activity

It is well known that high-pressure has a significant impact on the activity and stability of enzymes. However, enzyme inactivation by high-pressure and temperature are not comparable to each other. The mechanism of inactivation of enzymes in a high-pressure environment can be described in terms of protein denaturation (Ludikhuyze et al. 2010; Cheftel 1992).

- Reversible or irreversible, partial or complete unfolding of enzymatic structure.
- Difference in reaction volume during enzymatic reactions may be altered finally affecting the reaction mechanism.
- Increasing sensitivity of substrate to be modified after being unfolded by pressure.
- Enzyme-substrate bonding may become stronger by the release of an intracellular enzyme.

In general, the primary structure of the enzyme remains unaffected by the application of high-pressure (Heremans 1993; Mozhaev et al. 1994). High-pressure affects the tertiary and quaternary structure of an enzyme by modifying the electrostatic and hydrophobic interactions as well as hydrogen bonding. The secondary structure may be affected at higher pressure (above 700 MPa). Formation of hydrogen bonds during pressurization may change in the molar volume as well as global volume (Balny and Masson 1993). The quaternary structure, which is the most pressure-sensitive structure of the enzyme, tends to be closely packed by application of high-pressure. Large alteration in volume of hydration influences the denaturation of the enzyme under a high-pressure environment (Mozhaev et al. 1994).

High-pressure inactivation of enzymes can be influenced by water activity (Mozhaev et al. 1996), several osmolytes, such as amino acid, sugar, and polyols (Timasheff 2002), and also by total soluble solid (TSS) in the medium (Cano et al. 1997).

5.7.2.3 Thermodynamics of High-Pressure Inactivation of Enzyme

Thermodynamically, the effect of pressure on any chemical system is generally characterized by the variation in molar volume (ΔV) of the system. In case of pressure-induced denaturation of enzyme, change in molar volume between native state and denatured state at constant temperature can be considered as ΔV. According to Le Chatelier's principle, with the increase in pressure, positive and negative ΔV will try to shift the equilibrium toward bond breaking and formation, respectively. However, according to Heremans (1982), molecular ordering within the system has to be increased with pressure. Considering the conversion of enzymes from native state to denatured state as a chemical reaction having the equilibrium constant as K, Gibbs free energy (ΔG) can be related by the following equation:

$$\Delta G = -RT \ln K \tag{5.5}$$

The temperature and pressure dependency of ΔG can be expressed in terms of entropy change (ΔS) and ΔV:

$$\frac{d\Delta G}{dT} = -\Delta S \tag{5.6}$$

$$\frac{d\Delta G}{dP} = \Delta V \tag{5.7}$$

According to Planck, change in volume can be expressed in terms of pressure gradient of equilibrium constant (K) given by

$$-\Delta V = \frac{RTd \ln K}{dP} \tag{5.8}$$

As the magnitude of ΔG for a conversion from the native to denatured state of an enzyme is in the order of several thousand calories, threshold pressure to induce denaturation in the enzymatic structure should be in the order of several MPa.

5.7.2.4 Applications

5.7.2.4.1 α-Amylase

α-Amylase (α-1,4 glucan glucanohydrolase EC. 3.2.1.1) is capable of hydrolyzing the 1,4-α-D-glucosidic linkages in polysaccharides. It contains a calcium ion to maintain the structural integrity of the enzyme (Vallee et al. 1959; Bush et al. 1989). In the case of high-pressure and thermal stability of α-Amylase, it is specific toward its source. Three different *Bacillus species* in Tris-buffer (pH 8.6) produce α-amylases having a wide range of thermo- and pressure stability, and among those, *B. licheniformis* produced the most resistant α-amylase (Weemaes et al. 1996). Pressure and temperature showed both synergistic and antagonistic effect on the inactivation of α-amylase from *B. subtilis* in Tris buffer (Ludikhuyze et al. 1997). α-Amylase from *B. amyloliquefacien* showed significant stability up to 400 MPa and 25°C (Raabe and Knorr 1996) whereas stability of porcine pancreatic α-amylase was hindered at 300 MPa (Matsumoto et al. 1997).

5.7.2.4.2 Lipoxygenase

Lipoxygenase (LOX; EC 1.13.1.13) accelerates the formation of odorous scission hydroperoxide products by oxygenation of cis,cis-1,δ4-pentadiene fatty acids. Soybean LOX in Tris buffer (pH 8.3) showed first-order inactivation in the domain up to 750 MPa and 75°C (Heinisch et al. 1995). High-pressure inactivation of soy-LOX was faster in lower pH range (pH 3.5–4). At elevated pressure (above 500 MPa), soy-LOX was found to be stable at room temperature (30°C) but deviation on either side of that temperature led to its increased inactivation (Indrawati et al. 1999). Green bean LOX extract followed first-order inactivation kinetics over a wide range of pressure up to 700 MPa, having maximum inactivation at room temperature (Indrawati et al. 2000).

5.7.2.4.3 Peroxidase

Peroxidase (POD; EC 1.11.1.7) is responsible for producing off-flavor in the food product. Thermal stability of this enzyme is well known, but its pressure sensitivity varies widely depending upon the source as well as treatment conditions.

For significant inactivation of POD in green bean, higher pressure of 900 MPa was needed at room temperature whereas combined pressure-temperature treatment at 600 MPa was found to be effective for the same sample (Quaglia et al. 1996). Similar pressure resistivity up to 800 MPa was reported in the case of POD from carrot (Anese et al. 1995) and pH was affecting POD activity significantly. Significant POD inactivation was reported at 600 MPa. In other studies, activity of POD was found to be increased by the application of high-pressure. POD activity was increased in strawberry puree beyond 300 MPa at 20°C (Cano et al. 1997). Increased POD activity was also found out in tomato puree when it was treated below 350 MPa at room temperature (Hernandez and Cano 1998) in non-syruped lychee at pressure up to 200 MPa at 40°C (Phunchaisri and Apichartsrangkoon 2005).

5.7.2.4.4 Polyphenoloxidase

Polyphenoloxidase (PPO; EC 1.14.18.1) catalyzes both the oxidation of monophenol and diphenol into o-quinones, which finally results in tissue discoloration and loss of nutrients from fruits and vegetables (Whitaker 1994). High-prcssurc inactivation of PPO is highly dependent on its source. Partially purified mushroom PPO was found to be more pressure resistant than commercial purified mushroom PPO (Gomes and Ledward 1996). Strawberry PPO was stable (23% inactivation) even after treatment at 690 MPa, 90°C for 30 min (Terefe et al. 2010). On the other hand, strawberry PPO was activated in the pressure range of 285–400 MPa (Cano et al. 1997). Activation of PPO was also found in the case of Bartlett pears at 400–500 MPa for 10 min (Asaka and Hayashi 1991). PPO from red raspberries was found to be stable at 400–800 MPa (Garcia-Palazon et al. 2004). Similarly, stability of apricot PPO at 25°C was also reported even up to 900 MPa (Rovere et al. 1994). A stabilizing effect of grape PPO was observed at 45°C within the pressure range of 100–600 MPa (Rapeanu et al. 2005).

5.7.2.4.5 Pectic Enzymes

Pectinmethylesterase (PME; EC 3.1.1.11) de-esterifies the carboxyl groups in pectin molecules into galacturonic acid and methanol. Galacturonic units, the de-esterified product in pectin molecules are the substrates for polygalacturonase (PG; EC 3.2.1.15), which acts on the hydrolytic cleavage of α-1,δ-4 glycoside bonds between galacturonic acid residues. It finally results in tissue softening of fruits and vegetables. Pressure stability of PME and PG varies widely depending upon the specific fruits and vegetables. Presence of both pressure labile and pressure stable fractions of PME were reported in the case of oranges (Nienaber and Shellhammer 2001). Pressure labile fraction was inactivated within the range of 400–600 MPa and 25°C–50°C, whereas another fraction was found to be stable even at 900 MPa and 25°C. Similarly, the presence of different fraction of PME with varying pressure stability was reported in the case of purified carrot (Ly-Nguyen et al. 2002c), banana (Ly-Nguyen et al. 2002b), and strawberry (Ly-Nguyen et al. 2002a). Tomato PME was found to be stable at 600–800 MPa (Shook et al. 2001).

Inactivation of tomato PG was reported at 600 MPa and 30°C for 5 min (Crelier et al. 2001) and within 350 to 500 MPa (Fachin et al. 2003; Rodrigo et al. 2006). The presence of two isoforms of PG having varying temperature and pressure sensitivity was reported, but they were found to be inactivated at 500 MPa, 25°C for 15 min (Peeters et al. 2004).

Some of the recent findings exploring the effect of HPP on enzyme inactivation in mango, lychee, and *aloe vera* are presented below as case studies.

Case Study 5.4: Mango Pulp (*Mangifera indica* cv. Amrapali)

PPO, POD, and PME activities were investigated in high-pressure-treated mango pulp. Both pressure and treatment time had a positive effect on the activity of all three enzymes. High-pressure has been known to decrease activity potential of enzymes by denaturing the enzyme protein and thus limiting its ability to act as biocatalysts. PPO, POD, and PME were inactivated to maximum extent of 82%, 65%, and 75%, respectively, when the sample was treated at 600 MPa for 120 min (Figure 5.6). Pressure-assisted enzyme inactivation followed second-order inactivation model.

POD was found to be most baroresistant in mango pulp followed by PME whereas PPO was found to be most pressure sensitive. A second-order rate showed that beyond a certain pressure level–dwell time application, further reduction in enzyme activity is not possible. This may be due to the presence of pressure-resistant fraction in the enzyme, which showed residual activity even after extreme treatment condition. Complete inactivation of the enzyme needs another hurdle, such as temperature-assisted HPP or addition of solutes, such as sugar, lowering of pH, etc., which can act as a synergist with pressure.

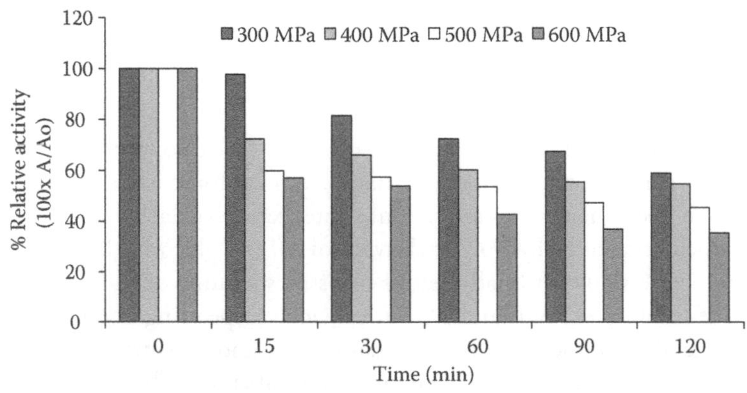

Figure 5.6 Relative activities of POD enzyme in high-pressure-processed mango pulp.

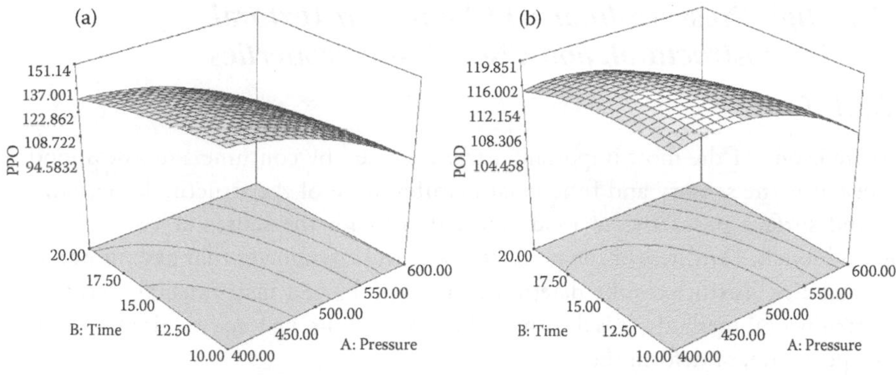

Figure 5.7 Effect of high-pressure and dwell time on enzyme activites of (a) PPO and (b) POD in lychee juice.

Case Study 5.5: Lychee Juice (*Litchi chinensis* cv. Bombai)

PPO and POD activity in high-pressure-treated lychee juice were investigated. PPO activity was significantly affected positively by all the process parameters. With the increase in each parameter, increased inactivation was found in the case of PPO (Figure 5.7). The effect of dwell time and ramp rate were less compared to pressure in the case of PPO activity. High-pressure has been known to induce denaturation of the enzyme, thus limiting its ability to act as a biocatalyst. POD activity was significantly affected by pressure followed by dwell time and ramp rate. However, similar to PPO, increased pressure intensity showed reduced POD activity in lychee juice samples. Maximum PPO and POD inactivations were found to be 50% and 25%, respectively. This study reveals that POD is more resistant than PPO toward pressure inactivation in lychee juice. These results are supported by the findings of Phunchaisri and Apichartsrangkoon (2005) in lychee fruits when POD was reported to be more resistant than PPO in high-pressure environment.

Case Study 5.6: *Aloe Vera* Juice (*Aloe barbadensis* Miller)

PME activity of high-pressure-treated *aloe vera* juice was studied, varying pressure level (63–736 MPa), dwell time (3–37 min) and pH (2.3–5.7). Pressure level affected the PME activity significantly ($p < 0.05$), dwell time, and changes in pH had no significant effect on PME activity ($p > 0.05$) within the range studied. A fresh sample without treatment had PME activity of 0.053 unit ml^{-1} and maximum inactivation up to 30.19% was achieved at 736 MPa for a dwell time of 20 min. A small increase of 5%–13% in PME activity was observed for samples treated at 200 MPa; this increase might be due to effective extraction of enzymes due to damage of cell wall membranes (Dornenburg and Knorr 1998). Pressure and dwell time had a negative effect, indicating a decrease in response whereas PME activity increased with an increase in pH. The quadratic terms of pressure showed significant effect ($p < 0.05$) followed by dwell time and pH. Interacting terms pressure–dwell time and dwell time–pH had a significant effect on PME activity ($p > 0.05$). Similar results for PME activity were reported by Basak and Ramaswamy (1996) for orange juice; they reported a linear relationship for pressure inactivation of PME.

5.7.3 High-Pressure-Induced Changes in Textural, Microstructural, and Viscoelastic Properties

5.7.3.1 Texture

Texture is one of the most important attributes used by consumers to assess food quality. It is the sensory and functional manifestation of the structural, mechanical, and surface properties of food detected through the senses of vision, hearing, and touch. Hardness of the product is a critical parameter that determines its marketability. Texture can be determined by sensory and instrumental methods. In instrumental methods, it is determined by measuring the force required to compress, penetrate, or deform the product.

5.7.3.2 Food Microstructure

The organization of several chemical components, that is, proteins, carbohydrates, fats, etc., forms the microstructure of the food. The structure and functionality of any food are governed by macroscopic properties of the material which mainly depend on the changes of its microstructure. Microstructure investigation can help in quantifying product changes during processing and may also improve the understanding of mechanisms and changes in quality factors, especially the changes in food texture. Few structural elements in food are recognized by the naked eye while most are discernible only with the aid of microscope and other physical methods (e.g., MRI, light diffraction).

5.7.3.3 Viscoelastic Properties

The science of rheology has many applications in the fields of food acceptability, food processing, and food handling. Rheology concerns the flow and deformation of substances particularly to their behavior in the transient area between solid and fluid. Moreover, it attempts to define a relationship between the stress acting on a given material and the resulting deformation and/or flow that takes place. Rheological properties are determined by measuring force and deformation as a function of time.

Some of the recent findings depicting the rheological behavior of high-pressure-treated mango pulp, *aloe vera* juice, and black tiger shrimp have been presented below as case studies.

Case Study 5.7: Mango Pulp (*Mangifera indica* cv. Amrapali)

Rheological characteristics of mango pulp were studied using a rheometer (RVDV-III Ultra Rheometer, Brookfield Engineering, USA) with a small sample adapter (spindle SC4-21). The sample (15 mL) was loaded into the cylinder and allowed to equilibrate at 25°C. Sample temperature was controlled by a circulating water bath

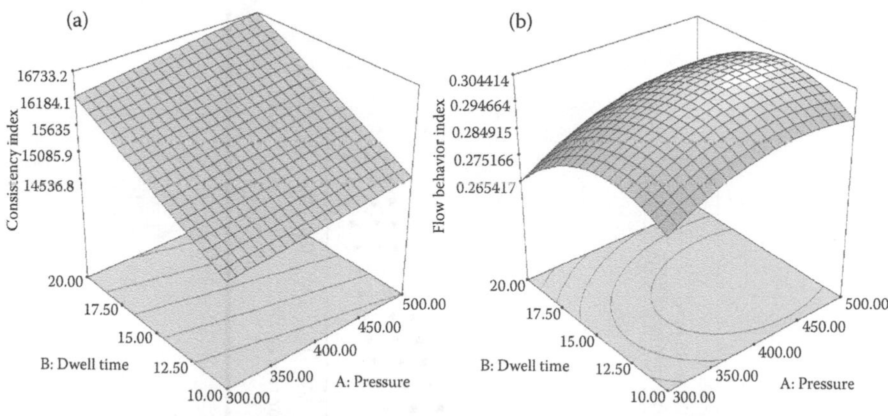

Figure 5.8 Response surfaces for (a) consistency index and (b) flow behavior index of high-pressure-treated mango pulp.

attached to the adapter with an accuracy of ± 0.2°C throughout the experiment. For steady shear measurements, the rheometer was programmed with shear rate changing from 0.1 to 50 s^{-1}, and the time interval for each measurement was fixed at 30 s. The viscosity data obtained from the experiments were analyzed by fitting different mathematical models using Rheocalc software (Brookfield Engineering, USA). Consistency index (CI), flow behavior index (FI), yield stress (YS), and viscosity (V) values obtained from the measurement were used to characterize the effect of independent parameters on rheological behavior of high-pressure-treated mango pulp within the pressure range of 300–500 MPa.

Rheological behavior of mango pulp was significantly affected by high-pressure processing. Increased consistency was obtained both with hold time and level of pressure applied (Figure 5.8). However, effect of dwell time was more significant. Pressure level affects the flow index but was not affected by dwell time. No significant effect of pressure application was noted on the viscosity of mango pulp.

Case Study 5.8: *Aloe Vera* Juice (*Aloe barbadensis* Miller)

Changes in rheological behavior of untreated, high-pressure-treated (500 MPa for 15 min), and thermally treated (95°C for 15 min) *aloe vera* juice were compared. Figure 5.9 shows the effect of different processing parameters on the flow characteristics of *aloe vera* juice. It clearly indicated that apparent viscosity decreased after both the treatments; however, this change was minimal in the case of high-pressure-treated juice compared to thermal treatment. The viscous behavior of *aloe vera* juice is mainly due to the presence of high molecular weight polysaccharides composed of a mixture of acetylated glucomannan, and it loses its activity after extraction rapidly due to enzymatic degradation (Yaron et al. 1992; Rodriguez et al. 2010). Also, molecular interactions are disrupted due to high temperatures during thermal treatment. The experimental results were analyzed using Casson and Power law models. High-pressure-treated *aloe vera* juice showed higher consistency index, flow behavior index, and casson yield stress values followed by thermal treated and control samples. This change in consistency might be attributed to the

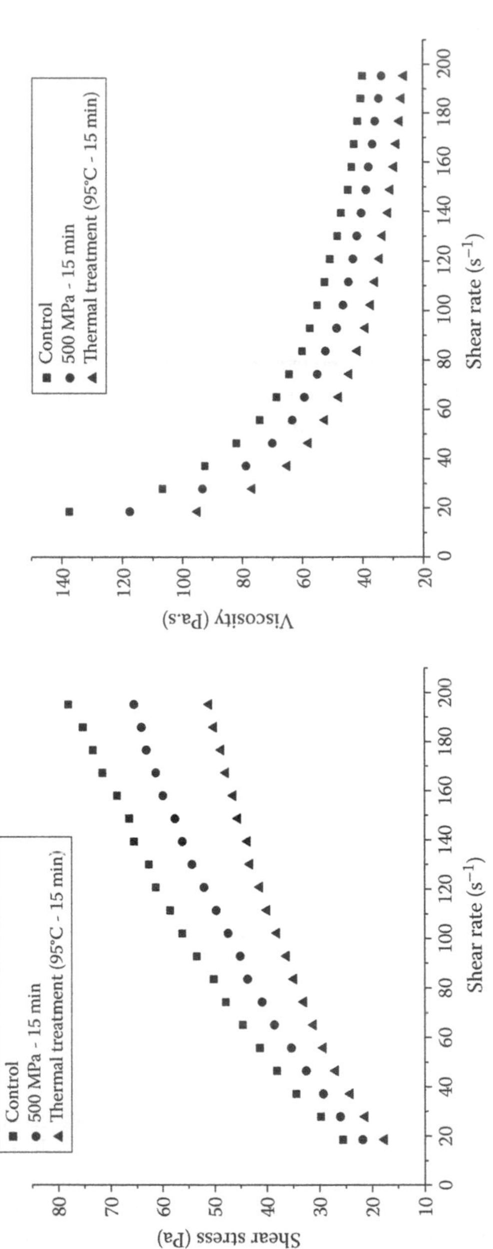

Figure 5.9 Flow behavior of treated *aloe vera* juice.

fact that high-pressure causes a change in hydration capacity of peptic substances present in juice creating compact structures (Krebbers et al. 2003). HPP affects the solid-gel transition of the polysaccharide forming gels, which might be responsible for the increase in the yield stress of *aloe vera* juice (Mozhaev et al. 1994).

Case Study 5.9: Black Tiger Shrimp

Black tiger shrimp was vacuum packaged and high-pressure-treated at selected pressure levels of 100, 270, and 435 MPa for 5 min at room temperature (25°C ± 2°C). Texture was investigated using a texture analyzer. HPP samples showed significantly ($p < 0.05$) higher hardness values than untreated ones. Increased hardness was obtained with increased pressure intensity. The effect of high-pressure on hardness may be explained by myofibrillar protein denaturation and aggregation. This also led to increased whiteness in treated samples (Figure 5.10). High-pressure-induced denaturation of myosin led to the formation of structures that contained hydrogen bonds and were additionally stabilized by disulfide bonds. These disulphide bonds contribute to the hardness. During chilled storage (2°C), the hardness values of both treated and untreated samples decreased significantly ($p < 0.05$), but treated samples were firmer than the control (Figure 5.11). Because

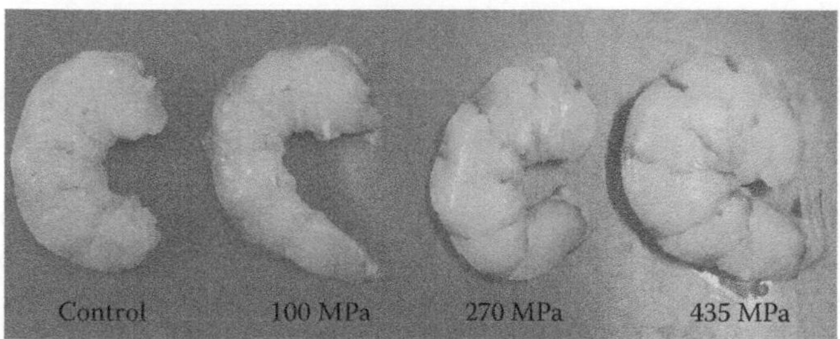

Figure 5.10 Change in appearance of high-pressure-treated black tiger shrimp.

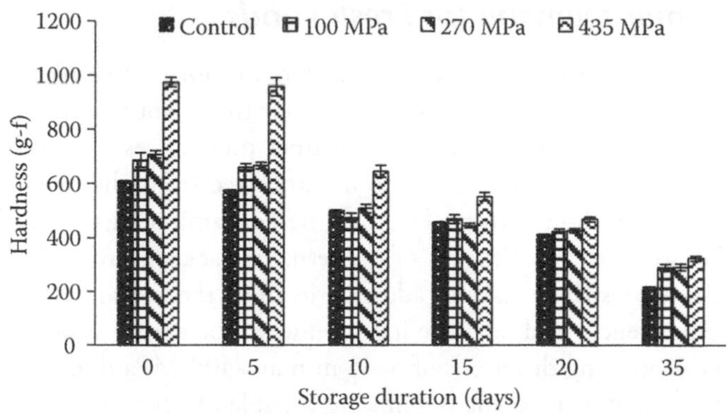

Figure 5.11 Changes in hardness of high-pressure-treated black tiger shrimp during chilled storage.

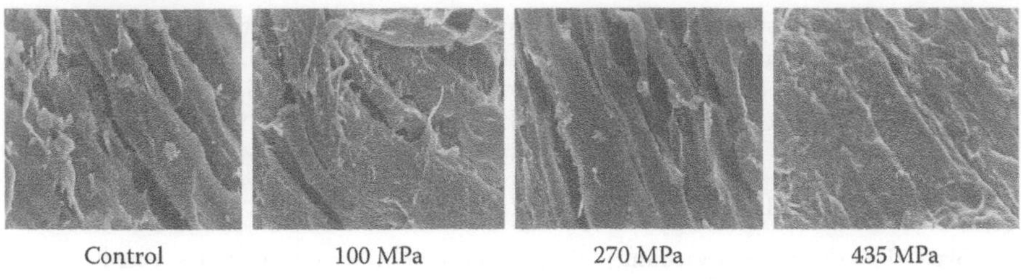

Figure 5.12 Micrographs of high-pressure-treated black tiger shrimp.

HPP inactivates enzymes and reduces the microbial load, thus the spoilage rate was decreased in processed samples during storage.

Micrographs were obtained using scanning electron microscopy. Compactness of muscles of treated samples remained intact immediately after pressurization (Figure 5.12). Higher pressure resulted in a denser muscle fiber network, which finally led to an increased firmness. The extracellular space decreases with the increase in pressure resulting from compaction of muscle and protein gel network formation.

5.8 Opportunities for High-Pressure Processing

The successful introduction of a new technology demands the identification of specific competitive advantages over existing ones. Technologically, high-pressure has proved its merit, however, to tap the maximum potential and economize the equipment cost, it can be applied on high-value commodities when both food quality and safety are the main concerns.

5.8.1 Consumer Demand for Fresh Foods

The classical examples of satisfying consumer demand for a fresh-like product is the high-pressure-processed avocado, which requires treatment under 650 MPa for ~1 min. It enjoys a huge demand due to consumer preferences for products with acceptable shelf life that are convenient to use and free from chemical additives. Prior to processing, fresh-cut fruit salads may be contaminated with *Salmonella, E. coli* O157:H7 and other pathogens of concern. In-package processing adopted during HPP eliminates these risks in addition to meet the consumer demand for healthiness, convenience, and labor-saving products. The earlier systems available required longer processing durations of 5–15 min at ~400 MPa due to limitations of the technology; however, recent systems are capable of generating much higher pressures in conjunction with moderate temperature applications, hence reducing the processing times considerably (1–3 min), making commercial sterilization possible.

5.8.2 Commercial High-Pressure Food Processing

The first high-pressure-processed foods were introduced to the Japanese market in 1990 by Meidi-ya, marketing a line of jams, jellies, and sauces packaged and processed without application of heat (Thakur and Nelson 1998). Other available products include fruit preparations, fruit juices, rice cakes, and raw squid in Japan; fruit juices, especially apple and orange juice, in France and Portugal; and guacamole and oysters in the USA (Hugas et al. 2002). Shellfish, such as oysters and crustaceans, are commercially pressure treated commodities in countries such as the USA, Canada, New Zealand, Australia, South Korea, and Greece (Leadley 2009). High-pressure is very effective for shucking raw meat from the rigid shell of crustaceans and mollusks without cooking.

5.9 Advantages and Disadvantages of High-Pressure Processing

5.9.1 Key Advantages

- HPP reduces microbiological loads and inactivates the enzymes, ensuring high quality and shelf-stable foods.
- Vitamins, essential nutrients, and the sensory profile of foods undergo minimal or no damage unlike thermal processing.
- Intensity of HPP treatment is independent of mass, thus reducing the cycle time as compared to thermal processing.
- It enables instant transmittance of pressure throughout the system, irrespective of size and geometry, thereby making size reduction optional, which can be a great advantage.
- Pasteurization can be done by the application of high-pressure even under chilled conditions, which retains the nutritional and functional qualities of foods.
- HPP can be used to retard the browning reaction in foods. The condensation in a Maillard reaction does not show any acceleration by high-pressure treatment due to suppression of free stable radicals derived from melanoidin.
- Protein gels induced by high-pressure treatment are glossy and transparent because of rearrangement of water molecules surrounding the amino acid residues in a denatured state.

5.9.2 Disadvantages

- High capital cost restricts the use of HPP for a commercial scale.
- Implementation of comprehensive quality assurance of high-pressure-treated product in terms of microbial safety finally leads to higher processing costs.
- Machinery required for HPP system is very complex and requires extremely high precision in its construction, use, and maintenance.

5.10 Future Trends

Allergenicity is a key concern in the safety assessment of novel foods. The incidence of food allergies is rapidly increasing as is their severity and the number of foods involved. New studies on the putative allergenicity of high-pressure-processed foods may be needed. Information regarding the combination of minimum temperature and time requirements for high-pressure-processed food is yet to be developed, and it is important to establish microbiological criteria for safe production of high-pressure-treated food.

Problem 5.1

Efficiency of a high-pressure treatment can be expressed as the ratio of logarithmic reduction of microbial load ($\log_{10}(N/N_0)$) to barometric power (MPa s) of the treatment. For a single pulsed high-pressure system, pressurizing and depressurizing rates were 40 MPa s^{-1} and 80 MPa s^{-1}, respectively. At 30°C, in a strawberry puree sample, D value was found to be 2 min at 400.1 MPa. Find out the efficiency of 2 min isobaric treatment at 400.1 MPa.

Solution:

Barometric power is the area under the curve in P vs. t plot.

$$\text{Treatment efficiency} = -\log(N/N_0)/\int P\,dt \tag{5.9}$$

$$\begin{aligned}\int P\,dt &= \Delta\text{ABC} + \text{rectangle BCDE} + \Delta\text{DEF}\\ &= 0.5\times(400\div 40)\times 400 + 400\times 120 + 0.5\times(400\div 80)\times 400\\ &= 51000 \text{ MPa s}\end{aligned} \tag{5.10}$$

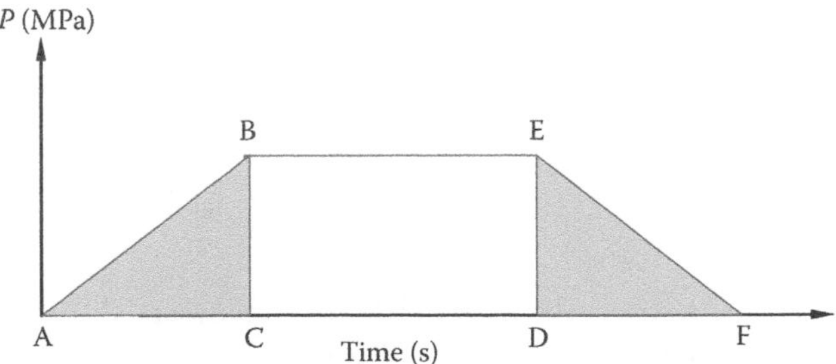

For D value, $(N/N_0) = 0.1$

So, treatment efficiency = $-[\log(0.1)]/51000 = 1.96 \times 10^{-5}$

Problem 5.2

Orange pectinmethylesterase (PME) follows a first-order inactivation model in the range of 400–600 MPa. From a high-pressure treatment at 18°C, inactivation rate constants of PME at 400, 500, and 600 MPa were 0.024, 0.107, and 0.478 min^{-1}, respectively taking atmospheric pressure as reference. What is the activation volume for PME in that experimental domain?

Solution:

$$k = k_{ref} \exp\left[\frac{-V_a}{RT}(P - P_{ref})\right] \tag{5.11}$$

Taking natural log (ln) on both sides and putting the values,

$$R = 8.314 \text{ cc MPa K}^{-1} \text{ mol}^{-1}$$

$$T = 273.15 + 18 = 291.15 \text{ K}$$

For obtaining V_a in cc mol^{-1},

$$\ln(0.024 \div 60) = \ln k_{ref} - (V_a \times (400 - 0.1) \div (8.314 \times 291.15))$$

$$\ln(0.107 \div 60) = \ln k_{ref} - (V_a \times (500 - 0.1) \div (8.314 \times 291.15))$$

Therefore, by solving the above two equations, activation volume (V_a) = –36.198 cc mol^{-1}

Problem 5.3

High-pressure inactivation of *S. cerevisiae* ascospores in orange juice at 25°C showed different D values at different pressures given below:

Pressure (MPa)	*D value (min)*
500	0.18
400	0.97
300	5.23

Find out the z_p value of *S. cerevisiae* at 25°C.

Solution:

$$D = D_1 10^{\left(\frac{P_1 - P}{z_P}\right)} \tag{5.12}$$

Putting $D = 0.18$ min, $P = 500$ MPa, $D_1 = 0.97$ min, and $P_1 = 400$ MPa in Equation 5.12, $z_P = 136.7$ MPa.

Problem 5.4

In an inactivation model, ln k (k = kinetic rate constant in min^{-1}) of *Z. bailii* in Tris buffer (pH 6.5) follows an elliptic model of pressure and temperature dependency without interaction terms of P and T within the domain of 120–300 MPa and 0°C–50°C. P and T used in the model were measured with reference to the midpoint of the experimental domain. Different treatments of *Z. bailii* showed the following values of k:

Pressure (MPa)	Temp (°C)	k value (min^{-1})
160	25	0.013
220	25	0.065
210	20	0.025
210	30	0.063
210	25	0.031

Find out the D value at 200 MPa and 40°C.

Solution:

We know that the elliptic model for enzyme inactivation without interaction terms is

$$\ln k = a + b\,(P - P_{ref}) + c(P - P_{ref})^2 + d(T - T_{ref}) + e(T - T_{ref})^2 \tag{5.13}$$

$$T_{ref} = (0 + 50) \div 2 = 25°\text{C and } P_{ref} = (120 + 300) \div 2 = 210 \text{ MPa.}$$

Putting all given values in Equation 5.13,

$$\ln 0.031 = a = -3.473$$

$$\ln 0.065 = -3.473 + b(220 - 210) + c(220 - 210)^2 \tag{5.14}$$

$$\ln 0.013 = -3.473 + b(160 - 210) + c(160 - 210)^2 \tag{5.15}$$

Solving Equations 5.14 and 5.15, $b = 0.0645$ and $c = 0.00094278$.

Also,

$$\ln 0.025 = -3.473 + d(20 - 25) + e(20 - 25)^2 \quad (5.16)$$

$$\ln 0.063 = -3.473 + d(30 - 25) + e(30 - 25)^2 \quad (5.17)$$

Solving Equations 5.16 and 5.17, $d = 0.09241$ and $e = 0.00985$.

Now substituting the values of a, b, c, d, and e in Equation 5.13 at $P = 200$ MPa and $T = 40°C$:

$$\begin{aligned} \ln k &= -3.473 + 0.0645(200 - 210) + 9.4278 \times 10^{-4}(200 - 210)^2 \\ &\quad + 0.09241(40 - 25) + 0.00985(40 - 25)^2 \\ k &= 0.6561\,\text{min}^{-1} \end{aligned}$$

Therefore, $D = 2.303/k = 2.303 \div 0.6561 = 3.51$ min.

Acknowledgment

The authors gratefully acknowledge the financial assistance provided by the National Agricultural Innovation Project (NAIP) under World Bank Support for initiating high-pressure research in India.

References

Anese, M., Nicoli, M.C., Dall'aglio, G., and Lerici, C.R. (1995). Effect of high pressure treatments on peroxidase and polyphenoloxidase activities. *Journal of Food Biochemistry*, 18, 285–293.

Asaka, M., and Hayashi, R. (1991). Activation of polyphenoloxidase in pear fruits by high pressure treatment. *Journal of Agricultural and Biological Chemistry*, 55(9), 2439–2440.

Balasubramanian, S., and Balasubramaniam, V.M. (2003). Compression heating influence of pressure transmitting fluids on bacteria inactivation during high pressure processing. *Food Research International*, 36(7), 661–668.

Balny, C., and Masson, P. (1993). Effects of high pressure on proteins. *Food Reviews International*, 9, 611–628.

Basak, S., and Ramaswamy, H.S. (1996). Ultra high pressure treatment of orange juice: A kinetic study on inactivation of pectin methyl esterase. *Food Research International*, 29(7), 601–607.

Bush, D.B., Sticher, L., van Huystee, R., Wagner, D., and Jones, R.L. (1989). The calcium requirement for stability and enzymatic activity of two isoforms of barley aleurone α-amylase. *Journal of Biological Chemistry*, 264, 19392–19398.

Cano, M.P., Hernandez, A., and Ancos, B.D. (1997). High pressure and temperature effects on enzyme inactivation in strawberry and orange products. *Journal of Food Science*, 62(1), 85–88.

Cheftel, J.C. (1992). Effects of high hydrostatic pressure on food constituents: An overview. In C. Balny, R. Hayashi, K. Heremans, and P. Masson (Eds.), *High Pressure and Biotechnology*, (pp. 195–209). Montrouge: John Libbey.

Cheftel, J.C., and Culioli, J. (1997). Effects of high pressure on meat: A review. *Meat Science*, 46(3), 211–236.

Crelier, S., Robert, M.C., Claude, J., and Juillerat, M.A. (2001). Tomato (*Lycopersicon esculentum*) pectinmethylesterase and polygalacturonase behaviors regarding heat and pressure induced inactivation. *Journal of Agricultural and Food Chemistry*, 49, 5566–5575.

Cruz, M.R., Smiddy, M., Hill, C., Kerry, J.P., and Kelly, A.L. (2004). Effects of high pressure treatment on physicochemical characteristics of fresh oysters (*Crassostrea gigas*). *Innovative Food Science and Emerging Technologies*, 5(2), 161–169.

Dornenburg, H., and Knorr, D. (1998). Monitoring the impact of high-pressure processing on the biosynthesis of plant metabolites using plant cell cultures. *Trends in Food Science & Technology*, 9(10), 355–361.

Fachin, D., Van Loey, A.M., Ly Nguyen, B., Verlent, I., Indrawati, and Hendrickx, M.E. (2003). Inactivation kinetics of polygalacturonase in tomato juice. *Innovative Food Science and Emerging Technology*, 4, 135–142.

Garcia-Palazon, A., Suthanthangjai, W., Kajda, P., and Zaaetakis, I. (2004). The effects of high hydrostatic pressure on β-glucosidase, peroxidase and polyphenoloxidase in red raspberry (*Rubus idaeus*) and strawberry (*Fragaria* x *ananassa*). *Journal of Food Chemistry*, 88, 7–10.

Gomes, M.R.A., and Ledward, D.A. (1996). Effect of high-pressure treatments on the activity of some polyphenoloxidases. *Journal of Food Chemistry*, 56, 1–5.

Gram, L., and Huss, H.H. (2000). Fresh and processed fish and shellfish. In: B.M. Lund, T.C. Baird-Parker, and G.W. Gould (Eds.), *The Microbiological Safety and Quality of Food*, vol. 1 (pp. 472–506). Gaithersburg, MD: Aspen Publishers.

Hayashi, R. (1995). Advances in high pressure food processing technology in Japan. In A.G. Gaonkar (Ed.), *Food Processing: Recent Developments*, (pp. 185–195). Amsterdam: Elsevier.

Heinisch, O., Kowalski, E., Goossens, K., Frank, J., Heremans, K., Ludwig, H., and Tauscher, B. (1995). Pressure effects on the stability of lipoxygenase: Fourier transform-infrared spectroscopy (FT-IR) and enzyme activity studies. *Zeitschrift für Lebensmittel-Untersuchung und -Forschung*, 201, 562–565.

Heremans, K. (1982). High pressure effects on proteins and other biomolecules. *Annual Review of Biophysics and Bioengineering*, 11, 1–21.

Heremans, K. (1993). The behaviour of proteins under pressure. In R. Winter, and J. Jonas, (Eds.), *High Pressure Chemistry, Biochemistry and Materials Science*, (pp. 443–469). Dordrecht: Kluwer Academic Publishers.

Hernandez, A., and Cano, M.P. (1998). High-pressure and temperature effects on enzyme inactivation in tomato puree. *Journal of Agricultural and Food Chemistry*, 46, 266–270.

Hugas, M., Garriga, M., and Monfort, J.M. (2002). New mild technologies in meat processing: High pressure as a model technology. *Meat Science*, 62(3), 359–371.

Indrawati, Van Loey, A.M., Ludikhuyze, L.R., and Hendrickx, M.E. (1999). Soybean lipoxygenase inactivation by pressure at subzero and elevated temperatures. *Journal of Agricultural and Food Chemistry*, 47, 2468–2474.

Indrawati, Van Loey, A., Ludikhuyze, L., and Hendrickx, M. (2000). Kinetics of pressure inactivation at subzero and elevated temperature of lipoxygenase in crude green beans (*Phaseolus vulgaris* L.) extract. *Journal of Biotechnology Progress*, 16, 109–115.

Kanda, T. (1990). Recent trend of industrial high pressure equipment and its application to food processing. In R. Hayashi (Ed.), *Pressure-Processed Food Research and Development*, (p. 341). San-Ei Shuppan, Kyoto.

Kaur, B.P., Kaushik, N., Rao, P.S., and Chauhan, O.P. (2013). Effect of high-pressure processing on physical, biochemical, and microbiological characteristics of black tiger shrimp (*Penaeus monodon*). *Food and Bioprocess Technology*, 6(6), 1390–1400.

Knorr, D. (1995). Hydrostatic pressure treatment of food: Microbiology. In G.W. Gould (Ed.), *New Methods for Food Preservation*, (pp. 159–175). Glasgow: Blackie Academic and Professional.

Krebbers, B., Matser, A.M., Hoogerwerf, S.W., Moezelaar, R., Tomassen, M.M.M., and van den Berg, R.W. (2003). Combined high-pressure and thermal treatments for processing of tomato puree: Evaluation of microbial inactivation and quality parameters. *Innovative Food Science & Emerging Technologies*, 4(4), 377–385.

Lakshmanan, R., Piggott, J.R., and Paterson, A. (2003). Potential applications of high pressure for improvement in salmon quality. *Trends in Food Science & Technology*, 14, 354–363.

Leadley, C. (2009). High pressure processing of fish and shellfish. *Seafish*, Fact sheet 09.

Linton, M., and Patterson, M.F. (2000). High pressure processing of foods for microbiological safety and quality. *Acta Microbiologica et Immunologica Hungarica*, 47(2–3), 175–182.

Ludikhuyze, L., Van den Broeck, I., Weemaes, C., Herremans, C., Van Impe, J., Hendrickx, M., and Tobback, P. (1997). Kinetics for isobaric-isothermal inactivation of *Bacillus subtilis* α-amylase. *Journal of Biotechnology Progress*, 13, 532–538.

Ludikhuyze, L., Van Loey, A., Denys, S. and Hendrickx, M. (2010). Effects of high pressure on enzymes related to food quality, In M.E.G. Hendrickx, and D. Knorr (Eds.), *Ultra-high Pressure Treatments of Foods* (pp. 115–166). New York: Kluwer Academic/ Plenum Publisher.

Ly-Nguyen, B., Van Loey, A.M., Fachin, D., Verlent, I., Duvetter, T., Vu, S.T., Smout, C., and Hendrickx, M.E. (2002a). Strawberry pectinmethylesterase (PME): Purification, characterization, thermal and high-pressure inactivation. *Journal of Biotechnology Progress*, 18, 1447–1450.

Ly-Nguyen, B., Van Loey, A.M., Fachin, D., Verlent, I., and Hendrickx, M.E. (2002b). Purification, characterization, thermal, and high-pressure inactivation of pectin methylesterase from bananas (cv *cavendish*). *Journal of Biotechnology Bioengineering*, 78, 683–691.

Ly-Nguyen, B., Van Loey, A.M., Fachin, D., Verlent, I., Indrawati, and Hendrickx, M.E. (2002c). Partial purification, characterization, and thermal and high-pressure inactivation of pectin methylesterase from carrots (*Daucus carrota* L.). *Journal of Agricultural and Food Chemistry*, 50, 5437–5444.

Matsumoto, T., Makimoto, S., and Taniguchi, Y. (1997). Effect of pressure on the mechanism of hydrolysis of maltotetraose, maltopentaose, and maltohexose catalyzed by porcine pancreatic α-amylase. *Biochimica et Biophysica Acta*, 1343, 243–250.

Meersman, F., and Heremans, K. (2008). High hydrostatic pressure effects in the biosphere: From molecules to microbiology. In C. Michiels, D.H. Bartlett, and A. Aertsen (Eds.), *High-Pressure Microbiology*, (pp. 1–17). Washington, DC: ASM Press.

Mor-Mor, M., and Yuste, J. (2003). High pressure processing applied to cooked sausage manufacture: Physical properties and sensory analysis. *Meat Science*, 65(3), 1187–1191.

Mozhaev, V.V., Heremans, K., Frank, J., Masson, P., and Balny, C. (1994). Exploiting the effects of high hydrostatic pressure in biotechnological applications. *Trends in Biotechnology*, 12(12), 493–501.

Mozhaev, V.V., Heremans, K., Frank, J., Masson, P., and Balny, C. (1996). High pressure effects on protein structure and function. *Journal of Proteins*, 24, 81–91.

Murchie, L.W., Cruz-Romero, M., Kerry, J.P., Linton, M., Patterson, M.F., and Smiddy, M. (2005). High pressure processing of shellfish: A review of microbiological and other quality aspects. *Innovative Food Science & Emerging Technologies*, 6(3), 257–270.

Nienaber, U., and Shellhammer, T.H. (2001). High-pressure processing of orange juice: Kinetics of pectinmethylesterase inactivation. *Journal of Food Science*, 66, 328–331.

Otero, L., and Sanz, P.D. (2000). High-pressure shift freezing. Part 1: Amount of ice instantaneously formed in the process. *Biotechnology Progress*, 16(6), 1030–1036.

Palou, E., Lopez-Malo, A., and Welti-Chanes, J. (2002). Innovative fruit preservation using high pressure. In J. Welti-Chanes, G.V. Barbosa-Canovas, and J.M. Aguilera (Eds.), *Engineering and Food for the 21st Century*, (pp. 715–726). Food Preservation Technology Series, Boca Raton, FL: CRC Press.

Peeters, L., Fachin, D., Smout, C., Loey, A.V., and Hendrickx, M.E. (2004). Influence of β-subunit on thermal and high-pressure process stability of tomato polygalacturonase. *Journal of Biotechnology Bioengineering*, 86, 5443–5449.

Pfister, M.K.H., Butz, P., Heinz, V., Dehne, L.I., Knorr, D. and Tauscher, B. (2001). *Influence of high pressure treatment on chemical alterations in foods: A literature review*, (pp. 10–11). BgVV-Heft, Berlin.

Phunchaisri, C., and Apichartsrangkoon, A. (2005). Effects of ultra-high pressure on biochemical and physical modification of lychee (*Litchi chinensis* Sonn.). *Food Chemistry*, 93(1), 57–64.

Polydera, A.C., Stoforos, N.G., and Taoukis, P.S. (2003). Comparative shelf life study and vitamin C loss kinetics in pasteurised and high pressure processed reconstituted orange juice. *Journal of Food Engineering*, 60, 21–29.

Prestamo, G., and Arroyo, G. (2000). Preparation of preserves with fruits treated by high pressure. *Alimentaria*, 318, 25–30.

Quaglia, G.B., Gravina, R., Paperi, R., and Paoletti, F. (1996). Effect of high pressure treatments on peroxidase activity, ascorbic acid content and texture in green peas. *LWT-Food Science of Technology*, 29, 552–555.

Raabe, E., and Knorr, D. (1996). Kinetics of starch hydrolysis with *Bacillus amyloliquefaciens* α-amylase under high hydrostatic pressure. *Starch-Starke*, 48(11–12), 409–414.

Rapeanu, G., Van Loey, A., Smout, C., and Hendrickx, M. (2005). Effect of pH on thermal and/or pressure inactivation of Victoria grape (*Vitis vinifera sativa*) polyphenoloxidase: A kinetic study. *Journal of Food Science*, 70, 301–307.

Rodrigo, D., Cortés, C., Clynen, E., Schoofs, L., Van Loey, A., and Hendrickx, M. (2006). Thermal and high-pressure stability of purified polygalacturonase and pectinmethylesterase from four different tomato processing varieties. *Journal of Food Research International*, 39(4), 440–448.

Rodriguez, E.R., Martin, J.D., and Romero, C.D. (2010). Aloe vera as a functional ingredient in foods. *Critical Reviews in Food Science and Nutrition*, 50(4), 305–326.

Rovere, P., Carpi, G., Maggi, A., Gola, S., and Dall'Aglio, G. (1994). Stabilization of apricot puree by means of high pressure treatments. *Prehrambeno-Tehnološka i Biotehnološka revija* (*Food technology and biotechnology review*), 32(4), 145–150.

Sandra, S., Stanford, M.A., and Meunier Goddik, L. (2004). The use of high-pressure processing in the production of queso fresco cheese. *Journal of Food Science*, 69(4), 153–158.

Sequeira-Munoz, A., Chevalier, D., LeBail, A., Ramaswamy, H.S., and Simpson, B.K. (2006). Physicochemical changes induced in carp (*Cyprinus carpio*) fillets by high pressure processing at low temperature. *Innovative Food Science & Emerging Technologies*, 7(1–2), 13–18.

Shook, C.M., Shellhammer, T.H., and Schwartz, S.J. (2001). Polygalacturonase, pectinesterase, and lipoxygenase activities in high-pressure-processed diced tomatoes. *Journal of Agricultural and Food Chemistry*, 49, 664–668.

Singh, R.P. (2001). Technical elements of new and emerging non-thermal food technologies, FAO bulletin.

Steely, J.S. (1994). Chemiluminiscence detection of sulfur compounds in cooked milk. In: C.J. Mussinan, and M.E. Keelan, (Eds.), *Sulfur Compounds in Foods* (pp. 22–35). Chicago, IL: American Chemical Society.

Terefe, N.S., Yang, Y.H., Knoezer, K., Buckow, R., and Versteeg, C. (2010). High pressure and thermal inactivation kinetics of polyphenoloxidase and peroxidase in strawberry puree. *Innovative Food Science and Emerging Technologies*, 11, 52–60.

Thakur, B.R., and Nelson, P.E. (1998). High pressure processing and preservation of foods. *Food Reviews International*, 14(4), 427–447.

Timasheff, S.N. (2002). Protein hydration, thermodynamic binding and preferential hydration. *Journal of Biochemistry*, 41, 13473–13482.

Ting, E.Y., and Marshall, R.G. (2002). Production issues related to UHP food. In J. Welti-Chanes, G.V. Barbosa-Canovas, and J.M. Aguilera (Eds.), *Engineering and Food for the 21st Century* (pp. 727–738). Food Preservation Technology Series, Boca Raton, FL: CRC Press.

Torres, J.A., and Velazquez, G. (2005). Commercial opportunities and research challenges in the high pressure processing of foods. *Journal of Food Engineering*, 67, 95–112.

Vallee, B.L., Stein, E.A., Sumerwell, W.N., and Fischer, E.H. (1959). Metal content of α-amylases of various origins. *Journal of Biological Chemistry*, 234, 2901–2905.

Vazquez-Landaverde, P.A., Qian, M.C., and Torres, J.A. (2007). Kinetic analysis of volatile formation in milk subjected to pressure-assisted thermal treatments. *Journal of Food Science*, 72(7), E389–E398.

Weemaes, C., De Cordt, S., Goossens, K., Ludikhuyze, L., Hendrickx, M., Heremans, K., and Tobback, P. (1996). High pressure, thermal, and combined pressure-temperature stabilities of α-amylases from *Bacillus* species. *Journal of Biotechnology and Bioengineering*, 50, 49–56.

Whitaker, J.R. (1994). *Principles of Enzymology for the Food Sciences* (2nd ed., pp. 543–556). New York: Marcel Dekker.

Yaron, A., Cohen, E., and Arad, S.M. (1992). Stabilization of aloe vera gel by interaction with sulfated polysaccharides from red microalgae and with xanthan gum. *Journal of Agricultural and Food Chemistry*, 40(8), 1316–1320.

Chapter 6

High-Voltage Pulsed Electric Field Processing of Foods

Jatindra K. Sahu

Contents

There are many well-recognized food-processing technologies being commercialized worldwide nowadays for conversion of raw materials into various food products with desirable functional, structural, chemical, physical, and sensory qualities. These include drying, dehydration, evaporation, pasteurization, sterilization, baking, roasting, separation, size reduction, coagulation, freezing, thawing, homogenization, milling, storage, mixing, and packaging. These technologies have been used for many years in the food industries and have gained the consumers' acceptance worldwide for their potential applications. More than ever, consumers' growing appetite for fresh; safe; chemical-, additive-, or preservative-free; and minimally processed foods has forced the food industry to develop many non-thermal food processing technologies, which cause minimum nutritional and sensory quality impairment in the processed foods. Today, non-thermal technologies such as pulse electric field processing, high-pressure processing, food irradiation, ultraviolet light, ultrasound, arc discharge, oscillating magnetic field, pulsed light processing, pulsed x-ray processing, plasma, chemicals (ozone, carbon dioxide), and a combination of these technologies with thermal or without thermal treatments to create novel food products are the major areas of research in food science and technology.

6.1 Pulse Electric Field (PEF) Processing

Pulsed electric field processing of foods involves the application of pulses of high voltage in the range of 20–80 kV cm^{-1} in between two electrodes. PEF technology, when first introduced in the food industry, produced thermal inactivation of some selected microorganisms. However, later on, use of higher voltages (3000–4000 V) showed that, in addition to the thermal inactivation, there is enhanced microbial destruction by the electricity itself. Although PEF processing of foods has not yet been introduced on a commercial scale, the process includes relatively low energy consumption and low processing temperature, which results in maximum retention of sensory and nutritional qualities of a food.

Today, PEF processing is gaining popularity in the food industry as an alternative method for food pasteurization and sterilization. However, the technology is not suitable for solid foods containing air bubbles. The presence of air bubbles inside a PEF system induces non-uniform treatment of the products. The technology is also restricted to foods that can withstand high electric fields. Homogeneous liquids with low electrical conductivity provide ideal conditions for continuous treatment in PEF processing. Foods with high electrical conductivity reduce the resistance of the treatment chamber and consequently, require more energy to achieve a specific electrical field. Particle size in the liquid food should also be smaller than the gap between the two electrodes in the treatment chamber for proper functioning of the PEF system.

6.2 PEF-Generating System

An electrically conductive food product placed in between a high voltage and a grounded electrode develops an electric field. The developing electrical field can be predicted from the Laplace equation $\nabla^2\varphi = 0$, where ∇ denotes the Laplace operator, and φ denotes the electrical potential. The most important parameters, which attract more attention in PEF processing of foods, are the process parameters, i.e., electric field strength, pulse width, number of pulses, and the design of the treatment chamber. The electric field strength is dependent upon the magnitude of pulse strength, and the number of pulses depends upon the duration of processing time. To get the required electric field strength, which induces inactivation of microbial cells, it is necessary to control the treatment time and to regulate the level of pulse in the treatment chamber (Putri et al. 2010).

Figure 6.1 shows a schematic diagram of a high pulse generator system. The major components of a high PEF system are a high-voltage power supply, an energy storage capacitor, and a discharge switch. A high voltage generator produces a high voltage charge, which supercharges the energy storage capacitor. The capacitor is then discharged through the food material by a switch releasing a pulse of duration within microsecond to millisecond between the two parallel electrodes in the treatment chamber.

The energy stored in the capacitor is given by the following expression:

$$Q = 0.5C_oV^2 \tag{6.1}$$

where V is the charging voltage (V), and C_o is defined as the capacitance (F) of the energy storage capacitor. The value of C_o can be expressed as

$$C_o = \frac{t}{R} = \frac{t\sigma A}{d} \tag{6.2}$$

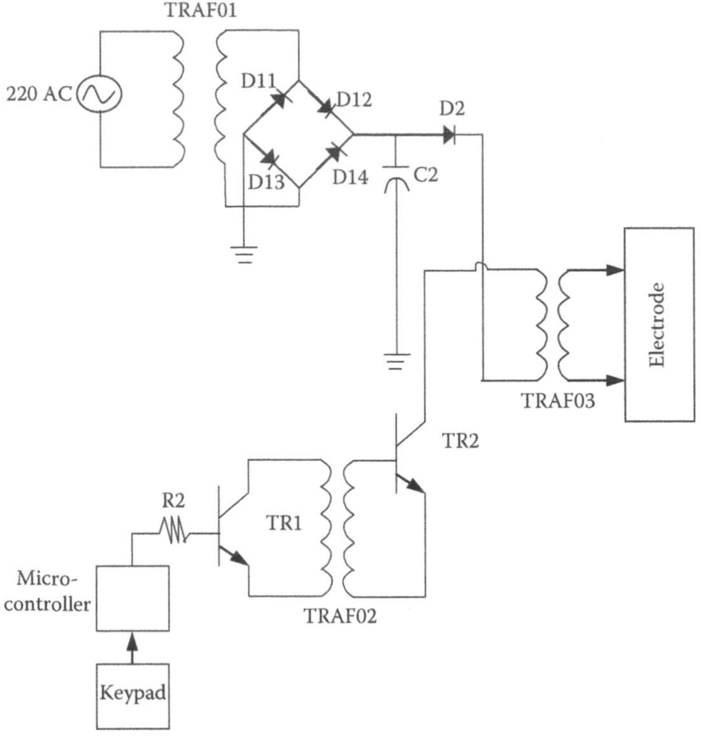

Figure 6.1 Schematic diagram of a high-voltage pulse generator.

where t is the pulse duration (s^{-1}), R is the resistance (Ω), σ is the electrical conductivity ($S\ m^{-1}$) of the food, A is the area (m^2) of electrode surface, and d is the gap (m) between two electrodes. The energy stored in the capacitor can be discharged almost instantaneously at very high levels of power.

To generate a high-intensity PEF with a food, a large flux of current is required to flow in a very short duration of time. The generation of a pulse involves slow charging and rapid discharging of the capacitor. The discharge is accomplished using high-voltage switches that operate reliably at high power and repetition rate. The treatment chamber is used to transfer high-intensity pulses to food materials. The electric field strength E ($V\ m^{-1}$) field, generated between the two parallel electrodes placed in the treatment chamber when a high voltage from the storage capacitor is applied is given as

$$E = \frac{V}{d} \tag{6.3}$$

where V is the voltage (V), and d is the gap (m) between two electrodes. Although a high PEF system is intended to operate at ambient temperature, the high electric field treatment may cause a rise in temperature of the material, depending on the field strength, pulse frequency, and number of pulses. Therefore, the whole system is fitted with refrigeration coils to control the temperature rise in the food product.

6.3 Pulse Wave Shape

Treatment time in a PEF system is defined as the number of pulses multiplied by the pulse width, which is a function of pulse waveform. High electric pulse is generated through some circuit-dependent waveforms. In order to increase the performance of a PEF system, it requires a high-voltage pulse generator with high effectiveness, better energy efficiency, and flexible circuit design (Love 1998). In a high pulse generator system, a resistor–inductor–capacitor (RLC) circuit gives higher efficiency than a resistor–capacitor (RC) circuit. Maximum efficiency of a RLC circuit lies between 40% and 50%, while an RC circuit gives only 38% (De Haan and Willcock 2002).

The electric pulses in a PEF system may be applied in various forms, i.e., exponential decaying, square wave, oscillatory, bipolar, or instant reverse charges. Oscillatory pulses are the least efficient for microbial inactivation whereas square-wave pulses are more energy and lethally efficient than exponential decaying pulses. Bipolar pulses are more lethal than monopolar pulses because the alternating changes in the movement of charged molecules with bipolar pulses cause a stress in the cell membrane and enhance its electric breakdown. Bipolar pulses also offer the advantages of minimum energy utilization, reduced deposition of solids on the electrode surface, and decreased food electrolysis. The pulse discharge may also be in the form of a square wave. In a square wave, the voltage increases quickly from zero to the maximum value at which it remains for the given time period and then decreases instantly to zero. This form of wave is used when the aim is to maximize energy absorption, which minimizes the heat transfer. An exponential decay voltage wave is a unidirectional voltage that rises rapidly to the maximum value and decays slowly to zero. The specific energy input over the pulse duration for an exponential decay pulse can be calculated as

$$Q = \frac{n[2ktE_{\max}]}{10\rho} \tag{6.4}$$

where E_{max} is the peak electric field strength (V m^{-1}), k is the electrical conductivity (S m^{-1}), ρ is the density (kg m^{-3}) of the product, t is the pulse duration (s), and n is the number of pulses. Figure 6.2 shows the circuit diagrams of generation of exponential decay, square wave, bipolar, and instant reversal pulse waveforms.

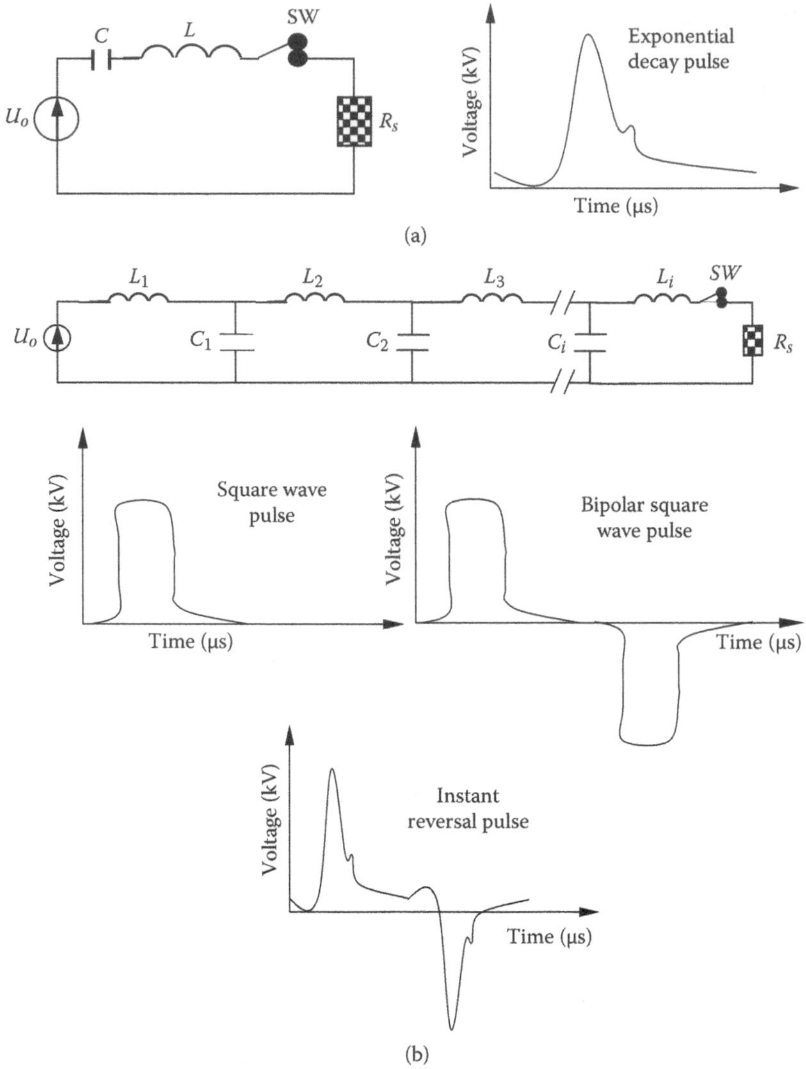

Figure 6.2 Types of electrical pulses most used in PEF processing and corresponding pulse forming networks (PFN). U_o is the charging voltage of the power supply, R_s the treatment chamber resistance, and SW is the switch. (a) Simple RC circuit, C is the capacitor, and L is the parasite inductor; (b) a more complex PFN, C_i are the capacitors, and L_i the inductors.

6.4 Treatment Chamber

The PEF process involves the application of a short pulse of a high-voltage electric field to a food, flowing through or placed in between two electrodes, which constitutes a treatment chamber. A PEF treatment chamber consists of at least two electrodes and insulation that forms a volume called a treatment zone in which the food receives the high electric pulses. The electrodes are made of inert materials, such as titanium. The size of the treatment chamber determines the electric field strength to be generated by a high-voltage generator. The two most important criteria that are considered in designing a treatment chamber are that the treatment chamber should impart a uniform electric field to the food with a minimum increase in the product temperature and the electrodes should minimize the effect of electrolysis (Zhang et al. 1995). A treatment chamber can be modeled as a parallel combination of capacitors and resistors. If the shape of the chamber is in the form of a tube, the expressions for calculating resistance R (Ω) and capacitance C (F) are as follows:

$$R = \frac{\ln(r_{LV}/r_{HV})}{2\pi\delta l} \tag{6.5}$$

$$C = \frac{2\varepsilon_r \varepsilon_0 l}{\ln(r_{LV}/r_{HV})} \tag{6.6}$$

where l is length (m) of electrode, δ is the electrical conductivity (S m^{-1}) of the food, r_{LV} is the diameter (m) of the low voltage electrode, r_{HV} is the diameter (m) of the high-voltage electrode, ε_r is the relative permittivity (F m^{-1}) of the food, and $\varepsilon_0 = 8.8 \times 10^{-12}$ F m^{-1}.

The processing of food materials using a PEF system may be either static or continuous. In a static PEF system, discrete portions of a fluid foodstuff are treated as a unit by subjecting the entire foodstuff to a PEF treatment chamber, in which a uniform electric field is applied to all elements of the foodstuff. In a continuous PEF system, the foodstuff to be treated flows into and discharges from the PEF treatment system continuously by a pump.

EXERCISE 6.1

During PEF treatment of apple juice, a voltage of 65 kV was applied in exponentially decaying form to create an electric field strength of 35 kV cm^{-1} at ambient temperature. If resistance of the PEF treatment chamber, pulse duration, and the surface area of the electrodes is 50 Ω, 20 μs, and 0.325 cm^2, respectively, calculate the gap between the two electrodes. The electrical conductivity of the apple juice is 0.2200 S m^{-1} at the processing temperature.

Solution:

With the given data, the capacitance of the energy storage capacitor is estimated as

$$C_o = \frac{Q}{0.5V^2} = \frac{35000}{0.5*(65000)^2} = 1.657*10^{-5} \text{ F}$$

Therefore, the gap between the two electrodes in the treatment chamber is

$$d = \frac{t\sigma A}{C_o} = \frac{20*10^{-6}*0.2200*0.325}{1.657*10^{-5}} = 0.086 \text{ cm}$$

EXERCISE 6.2

The height and diameter of a PEF treatment chamber is 45 cm and 7 cm, respectively. The chamber can accommodate 1.7 L of fresh apple juice. The chamber is to be modeled as a parallel combination of capacitors and resistors. The electrical conductivity and permittivity of the apple juice is 0.2200 S m^{-1} and 8.8×10^{-12} F m^{-1}, respectively. During the treatment, the temperature of the apple juice is set at 20°C, and the relative permittivity is 75 F m^{-1}. If the diameter of high and low voltage electrodes is 0.1 cm and 7 cm, respectively, determine the value of the resistance and capacitance of the treatment chamber.

Solution:

The resistance R of the treatment chamber can be estimated using Equation 6.5.

$$R = \frac{\ln(r_{LV}/r_{HV})}{2\pi\delta l}$$

$$= \frac{\ln(7/0.1)}{2\pi*0.2200*0.45} = 6.83\ \Omega$$

Similarly, the capacitance C can be calculated using Equation 6.6.

$$C = \frac{2\varepsilon_r\varepsilon_0 l}{\ln(r_{LV}/r_{HV})}$$

$$= \frac{2*28.8*10^{-12}*75*0.45}{\ln(7/0.1)} = 0.457 \text{ nF}$$

6.4.1 Static Treatment Chamber

The continuous treatment chambers are more efficient for large-scale operations. There have been many static treatment chambers developed by the various research groups, and their subsequent modification or development have made the PEF processing more easy, convenient, reliable, and cost-effective. Some of the static PEF treatment chamber designs that are available commercially in the market include U-shape type, parallel plate type, disk-shape type, wire cylinder type, rod-shape type, and sealed type. All the treatment chambers are very typical and specific in their design, i.e., electric field strength, treatment time, treatment temperature, pulse wave shape, and product type.

6.4.2 Continuous Flow Treatment Chamber

Development of continuous flow treatment chambers are based on the principles of the static chamber design. The efficiency of the microbial inactivation by a PEF is determined mainly by three factors such as process parameters (i.e., electric field intensity, treatment time, pulse wave shape, treatment temperature), product parameters (i.e., product conductivity, pH and ionic strength, food type, hurdle approach), and microbial parameters (i.e., type, concentration, and growth stage of microorganisms). Of these factors, only process parameters are modulated in the treatment chamber. Turbulent flow through the treatment chamber is more effective than the laminar flow because residence time distribution in turbulent flow is more homogeneous than in laminar flow (Barbosa-Canovas et al. 1999).

6.4.2.1 Electric Field Strength

Electric field strength applied to a PEF system determines the extent of microbial inactivation in the food. Increasing voltage and treatment time increases the number of pulses and electric field strength. The microbial inactivation increases with an increase in the electric field intensity, above the critical transmembrane potential (TMP). Bacteria have proven to be more resistant to a pulse electric field than yeast under high electric fields although spores are more resistant microbial entities. The field strength is determined as the force per unit charge. The voltage between the electrodes is proportional to the charge that moves between them. Therefore, the electric field is a function of the voltage in the treatment chamber. The electric field strength generated by a high voltage pulse circuit depends upon the amount of high voltage-pulse supplied to the treatment chamber and also the size of the treatment chamber. The higher the electric field strength, the shorter can be the processing time.

Calculation of electric field strength E_f(V m^{-1}) in a cylindrical chamber can be expressed by the following equation:

$$E_f = \frac{V_o}{r\left(\ln\dfrac{r_{LV}}{r_{HV}}\right)} r_{HV} \leq r \leq r_{LV} \tag{6.7}$$

where V_o is the average voltage (V) of the treatment chamber, r is the radius (m) of the treatment chamber, r_{LV} is the diameter (m) of the low-voltage electrode, and r_{HV} is the diameter (m) of high-voltage electrode. Table 6.1 presents some of the typical values of electric field strength used in different PEF systems.

The power requirement in a PEF system can be reduced by reducing the frequency, pulse width, or voltage. The electric field strength has a pronounced effect on pulse width on the inactivation rate. Therefore, energy efficiency can be maximized by increasing electric field strength and reducing the pulse width. The pulse width τ is given by

$$\tau = RC \tag{6.8}$$

Depending on the resistance of the treatment chamber, the pulse width varies. If the frequency is reduced, the number of pulses received by the flowing liquid food is restricted as given by

$$f = \frac{nF}{v} \tag{6.9}$$

where n is the number of pulses, F is the volumetric flow rate ($m^3\ s^{-1}$), and v is the volume of the treatment chamber (m^3). The continuous power rating of an exponential pulse generator is given by

$$P = \frac{f\tau V^2}{2R} = \frac{fCV^2}{2} \tag{6.10}$$

where P is the power (W), f is the frequency of pulses (Hz), τ is the pulse width, V is the peak voltage (V), C is the capacitance of the capacitor (F), and R is the resistance of the treatment chamber (Ω). In calculation of input specific energy, which is required during PEF application, the following equation is used:

$$W_{PEF} = \frac{V^2 t}{R_c v_c} \tag{6.11}$$

Table 6.1 Typical Values of Electric Field Strength in PEF Processing

Field Strength (kV cm⁻¹)	*Purpose*	*Reference*
25 to 40	Microbial inactivation	Sensoy et al. (1997)
5 to 25	Microbial inactivation	Dunn and Pearlman (1987)
35 or 70	Microbial inactivation	WSU PEF research group

where V (kV) is the peak tension that is given during the PEF processing, t is the total processing time (s), and R_c and v_c are resistance (Ω) and volume (cm^3) of the treatment chamber, respectively.

EXERCISE 6.3

During pasteurization of milk using a PEF system, the resistance of the treatment chamber is 0.22 Ω, and the volume in the treatment chamber is 0.95 L. Calculate the input-specific energy per liter of milk required during 6 s processing of the milk.

Solution:

The specific input energy W_{PEF} required by the PEF system for pasteurization of milk can be calculated by using Equation 6.11 as below:

$$W_{PEF} = \frac{(29.22)^2 * 6}{0.22 * (0.95 * 10^{-3}) * (100)^3} = 24.511 \text{ kJ L}^{-1}$$

6.4.2.2 *Treatment Time*

In PEF processing, as the treatment time increases, the inactivation rate of microbial spores increases rapidly and then gently, gradually flattens, and finally, no significant change occurs even treated with more time (Barbosa-Canovas et al. 1999). The pulse width influences microbial reduction by affecting the critical electric field (electric field intensity below which inactivation does not occur). Pulses with longer width decrease the value of the critical electric field, which results in a higher rate of inactivation; however, an increase in pulse duration may also result in an undesirable increase in product temperature. Critical treatment time also depends on the electric field intensity applied. Above the critical electric field, the critical treatment time reduces with the higher electric field.

6.4.2.3 *Treatment Temperature*

PEF treatment may be carried out at ambient, sub-ambient, or slightly above ambient temperature. However, mild temperature enhances the effectiveness of a PEF system as a preservation method, but the effect of the temperature on inactivation is complicated. On the contrary, higher temperature induces damage in sensory and physical qualities of foods. It is observed that increasing the inlet temperature from 22°C to 50°C would lead to a higher microbial inactivation rate (Calderon-Miranda et al. 1999). With constant electric field strength, the inactivation increases with an increase in temperature because the electric field intensity causes some increase in the temperature of foods. Therefore, proper cooling is necessary to maintain product temperature far below those generated by thermal pasteurization.

6.4.2.4 Cooling System

The change in temperature during the PEF processing should be monitored to achieve a non-thermal operation because the unavoidable difference between the inlet and outlet temperature is due to heat dissipation in the product as a result of ohmic heating (Alkhafaji and Farid 2007). High-voltage pulses produced by storage capacitor discharge contain a finite amount of energy Q_{pulse} that reaches the treatment chamber as defined by the following equation (Sepulveda et al. 2005):

$$Q_{pulse} = \left(\frac{R_{Ch}}{R_T}\right)\frac{CV^2}{2} \tag{6.12}$$

where C is the capacitance (F) of the discharging capacitor, V is the charging voltage (V), R_{Ch} is the electrical resistance (Ω) of the treatment chamber, and R_T is the total electrical resistance (Ω) of the system through which the capacitor is being discharged. Repetitive application of high-voltage pulses causes heating of the treated product as energy (ΔT) and is released into the treated product (Sepulveda et al. 2005). This energy is calculated by the following equation:

$$\Delta T = \frac{fQ_{pulse}}{F\rho C_p} \tag{6.13}$$

where f is the pulsing frequency (s^{-1}), F is the flow rate ($m\ s^{-3}$) of the product flowing through the treatment chamber, ρ is the density ($kg\ m^{-3}$) of the product, and C_p is its specific heat ($J\ kg^{-1}\ K^{-1}$).

6.5 Temperature Rise

The application of PEF to a fluid may be a batch or continuous type. A continuous-type operation may be single pass or recirculation type.

6.5.1 Single-Pass Operation

The energy E (J) dissipated during the release of discharge voltage V (V) is given by Equation 6.1. Taking into account the energy frequency f (s^{-1}) of the pulses, the energy flow dissipated Q ($J\ s^{-1}$) is given as

$$E = \frac{fCV^2}{2} \tag{6.14}$$

where C (μF) is the capacitance of the capacitor. In the process, one part of the energy ϕ will heat the food that flows through the PFE treatment chamber. This ratio must be less than one and depends on the electrical conductivity of the food placed in the treatment chamber. Taking an energy balance at the stationary state on the system,

$$q\rho C_p (T_f - T_i) = \phi Q \tag{6.15}$$

where q is the flow rate (1 s^{-1}), ρ is the density (kg m^{-3}) of the food flowing through the treatment chamber, and C_p is the specific heat (J kg °C^{-1}) of the food. Therefore, the increase in temperature ($T_f - T_i$) of the fluid food can be written as

$$(T_f - T_i) = \frac{0.5\phi fCV^2}{q\rho C_p} \tag{6.16}$$

6.5.2 Multi-Pass Operation

In multi-pass (recirculation type) operation, similar to the continuous operation, a part ϕ of the energy dissipated during the discharge of the capacitor will increase the temperature of the food. In the recirculation operation, the energy balance in the PEF system is the same as that of the single pass operation, which is given by Equation 6.15. Assuming non-stationary conditions, perfect mixing, and adiabatic conditions, the energy balance in the system leads to the following linear relationship between the temperature and time:

$$T_{sf} = T_{so} + \frac{rQ}{V_s \rho C_p} t \tag{6.17}$$

where T_{sf} is the initial system temperature (°C) after treatment time t (s), T_{so} is the initial temperature (°C) of the system, r is the part of energy released to the product (fraction), Q is the discharge of the capacitor (J s^{-1}), V_s is the volume (L) of the system, ρ is the density (kg L^{-1}) of the product, and C_p is the specific heat (kH kg^{-1} °C) of the product.

6.6 Microbial Inactivation

Two theories are mostly accepted by the various researchers as the result of application of PEF treatment on inactivation of microorganisms.

6.6.1 Dielectric Rupture Theory

Usually, cell membranes are considered as capacitors filled with dielectric materials whose dielectric constant is on the order of two (Zimmerman 1986). Most foods have a dielectric constant in the range of 60 to 80. When a pulsed electric field is applied to a microbial cell to cause electroporation of the cell membrane, ions inside the cells move along the field until they are held by the membrane. This is believed to be the primary event that leads to pore-formation or electroporation in a cell (Figure 6.3). As a result, free charges accumulate at both membrane surfaces. The accumulation of charges on the membrane surfaces increases the electromechanical stress or transmembrane potential (TMP), which is quite higher than the applied electric field. Due to the attraction of opposite charges induced on the inner or outer surface of the cell membrane, compression pressures occur, resulting in a decrease in the membrane thickness. Pore formation occurs when a certain threshold value of electric field strength is exceeded. The decrease in resistance further leads to irreversible breakdown, resulting in the inactivation of the cell membranes. The TMP induced during the pore formation is given as

$$\mathrm{TMP}_m = 1.5rE\cos\theta \tag{6.18}$$

where TMP_m is the transmembrane potential (V), r is the radius of the cell (mm), E is the applied electric field strength (V mm^{-1}), θ is the angle (degree) between a given membrane site and the field direction (Figure 6.3). In general, an increase in the value of TMP leads to a reduction in the membrane thickness.

The number and size of the pores depends on the field strength and the treatment time. The electric field intensity at which membrane breakdown occurs is

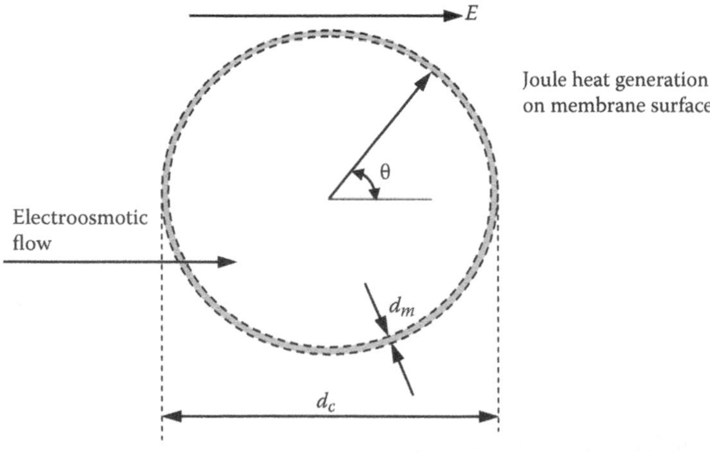

Figure 6.3 Biological cell in the external electric field. (Tsong, T.Y., *Biophysics Journal*, 60: 297–306, 1991.)

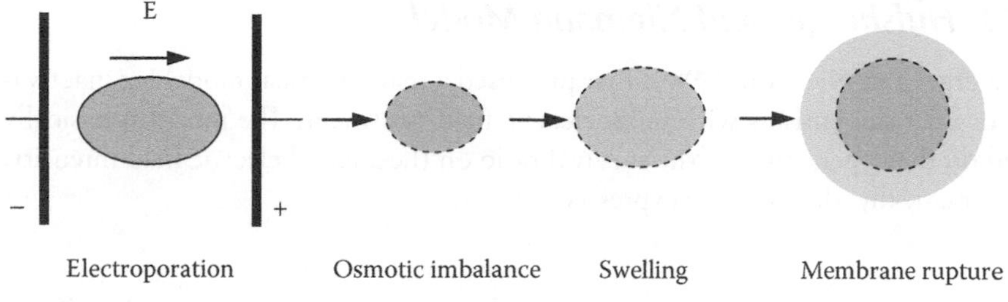

Figure 6.4 Electroporation process of cell membrane. (Vega-Mercado, H. et al. *Trends in Food Science and Technology,* 8(5): 151–157, 1997.)

called the threshold or critical electric field. When the electric field applied reaches a value close to the critical electric field or when the treatment time is short, the number and size of the generated pores is low. In this condition, the permeabilization of the membrane is reversible because the cell membrane restores its structure and functionality when PEF ceases. However, when more intense PEF treatment is applied, the number of pores and their size increases resulting in irreversible permeabilization or mechanical disruption of the cell (Zimmerman 1986).

6.6.2 Electroporation Theory

Electroporation is explained as a phenomenon in which a cell is exposed to a high-voltage electric field that temporarily destabilizes the lipid bilayer and protein of the cell membrane (Tsong 1991; Castro et al. 1993). The opening and closing of protein channels in a cell membrane depends on transmembrane potential. The potential of protein channels in the cell membranes is on the order of 50 mV. When an electric field is applied, many protein channels that are voltage-sensitive open up. The protein channels, once open, experience current that is much larger than the current normally experienced by the protein channels during metabolic activities (Dornenburg and Knorr 1993). As a result, protein channels are irreversibly denatured by Joule heating or electric modification of many functional groups, such as the hydroxyl, carboxyl, sulfhydryl, or amino group. The cell membrane electroporation process is shown in Figure 6.4.

6.7 Prediction of Microbial Inactivation

Some empirical mathematical models have been proposed to describe the relationship between the electric field intensity and microbial inactivation. The models were established for the calculation of survival fractions of the PEF-treated cells based on the electric field strength and total treatment time.

6.7.1 Hulsherger and Niemann Model

Hulsherger and Niemann (1980) first proposed a mathematical model for inactivation of micro-organisms with pulse electric field treatment. The model is basically based on the dependency of the survival ratio on the applied electric field intensity. Mathematically, the model is expressed as:

$$s = e^{-\frac{E-E_c}{k_c}} \tag{6.19}$$

where s is the survival fraction, E is the applied electric field strength (kV cm^{-1}), E_c is the critical electric field strength (kV cm^{-1}), and k_c is the constant factor (kV cm^{-1}). The same authors proposed another model correlating survival fraction with treatment time as below.

$$s = e^{\frac{t_c-t}{k_t}} \tag{6.20}$$

where s is survival fraction, t is total treatment time (μs), t_c is critical treatment time (μs), and k_t is a constant factor (μs).

6.7.2 Peleg Model

Peleg (1995) described a sigmoid shape microbial inactivation model by pulse electric field treatment. The model relates the percentage of surviving micro-organisms as a function of the electric fields and number of pulses applied. The expression is described as:

$$s = \frac{1}{\left\{1+e^{\frac{E-E_c}{k_e}}\right\}} \tag{6.21}$$

where, $k_e(t) = k_{eo}e^{k_1 t}$ and $k_c(t) = k_{co}e^{-k_2 t}$; where, s is the survival fraction, E is the applied electric field strength (kV cm^{-1}), E_c is critical electric field strength (kV cm^{-1}) and k_c, k_{eo}, E_{co}, k_1 (μs^{-1}), k_2 (μs^{-1}) are model constants.

6.7.3 Hulsherger and Others Model

Hulsheger et al. (1981) proposed an inactivation kinetic model which correlates microbial survival fraction with pulse electric field treatment time. The model is expression as:

$$s = \left(\frac{t}{t_c}\right)^{\frac{E_c-E}{k'}} \tag{6.22}$$

where *s* is the survival fraction, *t* is the total treatment time (μs), t_c is critical treatment time (μs), *E* is the applied electric field strength (kV cm^{-1}), E_c is the critical electric field strength (kV cm^{-1}) for the targeted microorganisms, and *k′* is a constant (kV cm^{-1}).

6.8 Microbial Shape Factor

PEF-induced microbial inactivation models relate the cell survival fraction after treatment and the applied electric field strength or the treatment time. However, the microbial cell shape forms one of the important factors to determine the induced transmembrane potential (TMP) in the cell membrane. The TMP for a spherical shape cell (Figure 6.3) depends on the angle θ between the external electric field direction and the radius-vector on the membrane surface. The induced TMP for a spherical cell can be determined by solving the Maxwell's equations in spherical coordinates (Neumann 1996). The final expression is given by:

$$\mathrm{TMP}_m = -\frac{3}{2} f(k) rE\cos\theta \tag{6.23}$$

where TMP_m is the transmembrane potential induced by an external electric field strength E_c, *r* is the cell radius, and *f*(*k*) is the shape factor. For calculation of TMP_m at a particular location at the membrane, the angle θ of the radial direction vector has to be specified. Per definition, θ is zero (and cos θ = 1) when the vector coincides with the direction of the electrical field. Hence, the highest membrane potential differences are assumed to occur at the two poles of the cell in field direction. For a very small relative membrane width $d_m \ll d_c$, the shape factor *f*(*k*) can be estimated by the following expression (Kotnik and Miklavcic 2000):

$$f(k) \approx \left(1 + \frac{\sigma_c d_c}{4\,\sigma_m d_m}\left(2 + \frac{\sigma_c}{\sigma_e}\right)\right)^{-1} \tag{6.24}$$

where σ_m, σ_c, and σ_e are the electrical conductivities of the membrane, intracellular and intercellular media, respectively.

Neumann (1989) suggested that the shape factor *f*(*k*) is an explicit function of the electrical conductivities of the suspending medium, the plasma k_i, the cell membrane k_M, and the ratio of the membrane thickness and the cell radius. The author reported the shape factor by the following equation:

$$f(k) = \frac{1}{1 + \frac{\left(k_M\left(2 + \frac{k_i}{k}\right)\right)}{\frac{2k_iD}{r}}} \tag{6.25}$$

However, TMP_m being in the range of nS m^{-1} for intact membranes, it is readily seen from Equation 6.23 that $f(k)$ can be approximated by 1, independently from the electrical conductivity of the suspending medium.

For cells with non-spherical shape, the TMP_m can be obtained by solving the Maxwell's equations in ellipsoidal coordinates. In these cases, Equation 6.26 is applied to calculate the shape factor.

$$TMP_m = -f(k)A_FE \tag{6.26}$$

Equation 6.26 yields the local membrane potential difference at the distance A_F from the center in the direction of the external electrical field. The shape factor $f(k)$ is a function of the three semi-axes (A_1, A_2, A_3) of elliptical cells.

$$f(k) = \frac{2}{2 - A_1A_2A_3 \int_0^{\infty} \frac{1}{\left(\left(s + A_F^2\right)\left(\sum_{n=1}^{3} \sqrt{s + A_n^2}\right)\right)} ds} \tag{6.27}$$

Zimmermann et al. (1974) derived a mathematical equation to calculate the membrane potential TMP_m for nonspherical cells. The equation is based on the assumption that the cell shape consists of a cylinder with two hemispheres at each end. In this case, the shape factor $f(k)$ for a rod-shaped microorganism is given by

$$f(k) = L\,(1 - 0.33\,D) \tag{6.28}$$

where L is the length of the particle, and D is the diameter of the cylinder. Therefore, the induced potential for the rod-shaped microbial cells is given by

$$V_m = \left(\frac{L}{1 - 0.33}\right)DE_c \tag{6.29}$$

where E_c is external electric field strength applied to inactivate the targeted microorganism, and D is the diameter of the cell. Table 6.2 shows the shape and size of some of the microbial cells.

Table 6.2 Geometry and Mean Size of Selected Microbial Cells

Microorganisms	*Shape*	*Size (μm)*
Listeria monocytogenes	Short rods	0.4–0.5 × 0.5–2
Yersinia enterocolitica	Straight rod to cocobacilli	0.5–0.8 × 1–3
Lactobacillus brevis	Rod with rounded ends	0.7–1.0 × 2–4
Bacillus subtilis (vegetative cells)	Rods with rounded or squared ends	Small: 0.5 × 1.2 large: 2.5 × 10
Lactobacillus plantarum	Rod with rounded ends	0.9–1.2 × 3–8
Salmonella senftenberg	Straight rods	0.7–1.5 × 2–5
Escherichia coli	Straight rods	1–1.5 × 2–6
Saccharomyces cerevisiae	Ellipsoidal shape	3–15 × 2–8

Source: *Bergey's Manual of Systematic Bacteriology*. Williams and Wilkins, Baltimore, 1986.

EXERCISE 6.4

A fruit juice is required to undergo PEF treatment to reduce its microbial contamination. If the critical electric field strength is 22.85 kV cm^{-1} and the survival fraction of the microorganisms is 0.152 after treatment, determine the electric field strength required for the inactivation. Assume the value of k_c = 0.26 kV cm^{-1}.

Solution:

Electrical field strength dependent survival fraction s is given as

$$s = e^{-\frac{E-E_c}{k_c}}$$

Substituting the corresponding values of s = 0.152, E_c = 22.85 kV cm^{-1}, and k_c = 0.26 kV cm^{-1}, $0.152 = e^{-\frac{E-22.85}{0.26}}$.

So, solving for the electrical field strength E = 23.34 kV cm^{-1}.

6.9 Joule Overheating

The electric current flowing in a conductive media generates Joule heat and warms up the medium. The increase in medium temperature is proportional to the treatment time t and the current density through the media.

$$\Delta T = \frac{(\sigma E)^2 t}{\sigma \rho C_p} \tag{6.30}$$

where $(\sigma E)^2$ is the current density; and σ, ρ, and C_p are the mean specific electrical conductivity, density, and specific heat of the medium, respectively.

In the absence of any thermal diffusion, the ratio of the local temperature rise in a membrane ΔT_m to the surrounding media ΔT_c is

$$\frac{\Delta T_m}{\Delta T_c} = \frac{\sigma_c \rho_c C_{pc}}{\sigma_m \rho_m C_{pm}} = \frac{\sigma_c}{\sigma_m} = 10^5 \tag{6.31}$$

where subscripts m and c correspond to the membrane and to surrounding cellular media, respectively. From the above equation, it can be concluded that membranes are the main elements for the Joule heating in a cellular system, and the local temperature rise on a membrane is very high.

$$\Delta T_m = \frac{(\sigma E)^2 t}{\sigma_m \rho_m C_{pm}} \tag{6.32}$$

The external electric field may cause the thermal damage to a membrane. But the heat diffusion effectively abates the Joule overheating of the membrane surface. The total temperature rise on a membrane during the heating averages over the distance $d_T \approx \sqrt{X^t}$, where $X \approx 10^{-7}$ m^2 s^{-1} is the thermal diffusivity. Hence, the actual temperature rise on a membrane is given by

$$\overline{\Delta T_m} = \frac{d_m \Delta T_m}{d_T} \tag{6.33}$$

The value of Equation 6.33 is much lesser than ΔT_m. This indicates that the temperature rise in PEF treatment is much less or negligible.

EXERCISE 6.5

During PEF processing of a tissue, the applied electric field strength is 500 V cm^{-1}, and treatment duration is 0.01 s. The mean specific electrical conductivity in the tissue and the density and specific heat of the medium are 0.1 Ω^{-1} m^{-1}, 10^3 kg m^{-3}, and 2.5 kJ kg^{-1} K^{-1}, respectively. If the specific electrical conductivity and diameter of the tissue membrane are 0.1 Ω^{-1} m^{-1} and 5×10^{-9} m, respectively, calculate the temperature rise during processing. The thermal diffusivity is 10^{-7} m^2 s^{-1}.

Solution:

The temperature rise on the membrane surface after the PEF treatment can be calculated from Equations 6.32 and 6.33.

$$\Delta T_m = \frac{(0.1 * 500)^2 * 0.01}{0.1 * 1000 * 2.5} = 2 * 10^5 \text{ K}$$

So, the temperature rise on the membrane surface is given by

$$\overline{\Delta T_m} = \frac{5 \times 10^{-9} * 5 \times 10^5}{\sqrt{10^{-7}}} = 26 \text{ K}$$

6.10 Electroosmosis Transport through Membrane

Electroosmosis is the phenomenon of the directed electrolyte flow near the charged surface in a pore under the external electric field intensity.

$$V = \mu E \tag{6.34}$$

where E is the external electric field intensity, $\mu = \varepsilon\varepsilon_0\zeta/\eta$ is the electrophoretic mobility, ζ is the zeta potential of the pore surface, $\varepsilon\varepsilon_0$ is the permittivity, and η is the electrolyte viscosity inside the pore. Substituting corresponding values of the parameters $\varepsilon\varepsilon_0 \approx 80\varepsilon_0 \approx 7.1 \times 10^{-10}$ Fm^{-1} and $\eta \approx 0.001$ Pa and $\zeta \approx 10$ mV, the value of $\mu \approx 10^{-8}$ m^{2}V^{-1}s^{-1}. The electroosmosis phenomenon can be used in practical applications for moisture transport in water-saturated biological materials, their dehydration, and dewatering. At PEF treatment of the biological materials, the critical electrical field intensity generated inside the membrane is rather high $E_m \approx 10^8$ V m^{-1}, and the velocity of the electroosmotic flow through the membrane pore is on order high ($V \approx 1$ m s^{-1}). Therefore, electroosmosis can be the dominant mechanism, which can control the fast transmembrane exchange, causes chemical imbalance, and enhances lysis of the microbial cells.

6.11 Applications of PEF

PEF processing of foods is superior compared to the traditional heat treatment because the process greatly reduces the detrimental changes to the sensory and physical properties of the foods. PEF has been applied to various foods to preserve the physical, chemical, and sensory qualities by inactivating the spoilage microorganisms, to extend the shelf life of bread, to enhance the process of extraction, and to accelerate the process of drying and dehydration. Different applications of PEF technology in cold processing of foods are discussed in more detail in Chapter 7.

6.12 Summary

High-voltage pulse electric field is a novel technique for processing of foods in the modern research of food science and technology. The technology is gaining popularity throughout the world due to its potential applications in various food processing operations. In this chapter, an attempt has been made to give an insight into the pulse electric field; a concept for system design; and the effects of pulse electric variables on physical, chemical, and biological factors of foods. Various applications of the technology have been described systematically. However, in order to have a good market for the high pulsed electric field foods, research, such as packaging material development, low cost system design, and appropriate storage and preservation conditions, are the need of the hour.

References

Alkhafaji, S., and Farid, M.M. (2007). An investigation on pulsed electric fields technology using new treatment chamber design. *Innovative Food Science and Emerging Technology*, 8: 205–212.

Barbosa-Canovas, G.V., Gongora-Nieto, M.M., Pothakamury, U.R., and Swanson, B.G. (1999). *Preservation of Foods with Pulsed Electric Fields*. Academic Press, London.

Bergey's Manual of Systematic Bacteriology. (1986). Williams and Wilkins, Baltimore, MD.

Calderon-Miranda, M.L., Barbosa-Canovas, G.V., and Swanson, B.G. (1999). Inactivation of *Listeria innocua* in liquid whole egg by pulsed electric fields and nisin. *International Journal of Food Microbiology*, 51(1): 7–17.

Castro, A.J., Barbosa-Canovas, G.V., and Swanson, B.G. (1993). Microbial inactivation of foods by pulsed electric fields. *Journal of Food Processing and Preservation*, 17: 47–73.

De Haan, S.W.H., and Willcock, P.R. (2002). Comparison of energy performance of pulse generation circuits for PEF. *Innovative Food Science and Emerging Technology*, 3: 349–356.

Dornenburg, H., and Knorr, D. (1993). Cellular permealibilization of cultured plant tissue by high electric field pulse and ultra high pressure for recovery of secondary metabolites. *Food Biotechnology*, 7: 35–48.

Dunn, J.E., and Pearlman, J.S. (1987). Methods and apparatus for extending the shelf-life of fluid food products. US Patent No. 4,695,472.

Hulsheger, H., and Niemann, E.G. (1980). Lethal effects of high voltage pulses on *E. coli* K12. *Radiation and Environmental Biophysics*, 18(4): 281–288.

Hulsheger, H., Potel, J., and Niemann, E.G. (1981). Killing of bacteria with electric pulses of high electric field strength. *Radiation and Environmental Biophysics*, 20: 53–65.

Kotnik T., and Miklavacic, D. (2000). Second order model of membrane electric field induced by alternating electric fields. *IEEE Transaction of Biomedical Engineering*, 47: 1074–1081.

Love, P. (1998). Correlation of Fourier transforms of pulsed electric field waveform and microorganism inactivation. *IEEE Transactions on Dielectrics and Electrical Insulation*, 5(1): 142–147.

Neumann, E. (1989). The relaxation hysteresis of membrane electroporation. In: E. Neumann, A.E. Sowers, and C. Jordan (eds.), *Electroporation and electrofusion in cell biology* (pp. 61–82). Plenum Press, New York.

Neumann, E. (1996). Gene delivery by membrane electroporation. In: P.T. Lynch and M.R. Davey (eds.), *Electrical manipulation of cells* (pp. 157–184). Chapman and Hall, New York.

Peleg, M. (1995). A model of microbial survival after exposure to pulsed electric fields. *Journal of Science, Food and Agriculture*, 67: 93–99.

Putri, R.I., Syamsiana, I.N., and Hawa, L.C. (2010). Design of high voltage pulse generator for pasteurization by pulse electric field (PEF). *International Journal of Computer and Electrical Engineering*, 2(5): 916–923.

Sensoy, I., Zhang, Q.H., and Sastry, S.K. (1997). Inactivation kinetics of *Salmonella dublin* by pulsed electric field. *Journal of Food Process Engineering*, 20: 367–381.

Sepulveda, D.R., Gongora-Nieto, M.M., San-Martin, M.F., and Barbosa-Canovas, G.V. (2005). Influence of treatment temperature on the inactivation of *Listeria innocua* by pulsed electric fields. *Lebensmittel-Wissenschaft und -Tech*, 38(2): 167–172.

Tsong, T.Y. (1991). Mini review, Electroporation of cell membranes. *Biophysics Journal*, 60: 297–306.

Vega-Mercado, H., Martin-Belloso, O., Qin, B.-L., Chang, F.-J., Gongora-Nieto, M.M., Barbosa-Canovas, G.V., and Swanson, B.G. (1997). Non-thermal food preservation: Pulsed electric fields. *Trends in Food Science and Technology*, 8(5): 151–157.

Zhang, Q.H., Barbosa-Canovas, G.V., and Swanson, B.G. (1995). Engineering aspects of pulsed electric field pasteurization. *Journal of Food Engineering*, 25(2): 261–281.

Zimmermann, U. (1986). Electric breakdown, electropermeabilization and electrofusion. *Reviews of Physical, Biochemical and Pharmaceuticals*, 105: 176–256.

Zimmermann, U., Pilwat, G., and Riemann, F. (1974). Dielectric breakdown of cell membranes. *Biophysics Journal*, 14: 881–899.

Chapter 7

Cold Pasteurization of Fruit Juices Using Pulsed Electric Fields

Faiyaz Ahmed

Contents

7.1 Introduction

Thermal processing of fruit juices is widely practiced today due to the exceptional ability to sterilize numerous spoilage and pathogenic bacteria and inherent enzymes. However, the thermal processes can alter the nutritional and sensory qualities of a food. The quest for energy conservation by the manufacturers to reduce the carbon footprint of the processes involved in food processing and preservation and the increasing consumer demand for fresh-like, quality foods have given rise to the development of innovative non-thermal food processing technologies, including ionizing radiation, high-intensity light pulses, high isobaric pressure, electric or magnetic fields, antimicrobial chemicals, polycationic polymers, lytic enzymes, and pulsed electric fields.

The pulsed electric field (PEF) technology can be considered as a potential alternative to traditional thermal processing of foods, wherein a high-intensity electric field generated between two electrodes creates a large flux of electrical current that flows through a food without a significant change in nutrients and the sensory characteristics of the food (Dunn 2001). In the PEF system, pasteurization of foods is achieved within microseconds, and thus, the technology has attracted much attention of researchers worldwide resulting in numerous publications in various application zones of PEF. Apart from microorganisms, enzymes play a major role in the shelf life of plant products, and therefore, food preservation techniques target inactivation of both the microorganisms and the inherent enzymes. PEF also exerts its food preservation ability by inactivation of microorganisms and enzymes.

There is a growing interest in the application of PEF in food processing. Generally, applications of PEF in food processing have been directed to two main categories: microbial inactivation and preservation of liquid foods and enhancement of mass transfer and texture in solids and liquids.

A large portion of works on PEF have been focused on reducing microbial load in liquid or semi-solid foods in order to extend their shelf life and ensure their safety. Frequently studied liquid products are orange juice, apple juice, tomato juice, milk, and liquid egg (Hermawan et al. 2004; Evrendilek and Zhang 2005; Amiali et al. 2006). It has been evidenced that PEF caused inactivation of pathogenic and food spoilage microorganisms as well as selected enzymes, resulting in better retention of flavors and nutrients and fresher taste compared to heat pasteurized products (Espachs-Barroso et al. 2003; Sepulveda et al. 2005). Juice extraction and dehydration are the other areas in which PEF has potential applications in future. In view of the above, the present chapter presents a summary of pulsed electric field applications in extending the shelf life of some important fruit juices, including orange, apple, grape, cranberry, and tomato.

7.2 Microbial Inactivation

A number of mechanisms have been put forth to explain the effect of PEF on microorganisms, including structural damage to the cell membrane leading to ion leakage, loss of metabolites and protein, and DNA damage. In Chapter 5, different theories behind microbial inactivation by PEF are discussed. In summary, the microbial cells develop structural fatigue due to induced membrane potential and mechanical stress, resulting in material flow after the loss of integrity of the cellular membrane. This results in an imbalance in the osmotic pressure between the cytosol and the external medium resulting in swelling/shrinking and disruption of the microbial cells.

The electric potential causes an electrostatic charge separation in the membrane of microbial cells due to the dipole nature of the molecules of the membrane. The cell membrane is regarded as an insulator shell to the cytoplasm due to its electrical conductivity, which is six to eight times weaker than the cytoplasm (Barbosa-Cánovas et al. 1999). Electrical charges are accumulated in cell membranes when microbial cells are exposed to electric fields. The accumulation of negative and positive charges in cell membranes forms transmembrane potential. The charges attract each other and generate compression pressure, which causes the membrane thickness to decrease. A further increase in the electric field strength beyond a critical membrane potential leads to pore formation (electroporation). Hamilton and Sale (1967) reported that cell lysis with the loss of membrane integrity occurred when transmembrane potential was approximately 1 V. This critical electrical potential varies depending on the pulse duration time, number of pulses, and PEF treatment temperature. This phenomenon has been substantiated by the findings of Harrison et al. (1997) wherein disruption of cell organelles and lack of ribosomes was observed in *Saccharomyces cerevisiae* in PEF-treated apple juice. The researchers proposed that damaged organelles and a lack of ribosomes is an alternative inactivation mechanism of PEF.

Further, the microbial inactivation capacity is dependent on a number of factors, such as strength of the electric field, exposure time, pulse width and shape, and the treatment temperature (Knorr et al. 1994). Apart from these parameters, the microbial inactivation also depends on the electric conductivity, density, viscosity, pH, and water activity of the substance that is being treated. It is also observed that the microbial inactivation by PEF is directly proportional to the strength of the electric field, exposure time, pulse width and shape, and the treatment temperature but inversely proportional to the electric conductivity of the product (Wouters et al. 2001). It is also interesting to note that bacteria are generally more resistant to PEF than yeasts (Barbosa-Cánovas et al. 1999), and among bacteria, Gram-positive bacteria are more resistant than Gram-negative bacteria (Hulsheger et al. 1983).

7.3 Enzyme Inactivation

The very definition of the enzymes as proteinaceous catalysts that change the rate of a biochemical reaction without undergoing any qualitative and quantitative change provides an answer for the effect of PEF on enzymes. It is well established that proteins carry an electric charge, and the application of an electric field induces association or dissociation of functional groups, movements of charged chains, and changes in alignment of helices, and this change in conformation results in enzyme inactivation. This mechanism is supported by a number of studies, wherein papain was inactivated by the loss of helical structure; alkaline phosphatase was inactivated by polarization followed by aggregation and loss of native structure (Castro et al. 2001). Similar to microbial inactivation, enzyme inactivation also depends on parameters such as strength of the electric field, exposure time, pulse width, treatment temperature, and structure of the enzyme and product characteristics (Yeom and Zhang 2001).

7.4 Applications of PEF

7.4.1 (In)activation of Microorganisms

The PEF was first introduced in the food industry as an alternative method for pasteurization of milk and later on various liquid foods. However, the effect of PEF has a great effect on inactivation of several microorganisms in various foods. The effect of PEF treatment on inactivation of the microorganisms is a function of field strength and treatment time. The electric field causes conformational changes in the membrane structure, resulting in impairment of semiperishable properties of the microorganisms. In general, it has been demonstrated that bacterial cells in the stationary growth phase are more resistant to PEF than cells in the exponential growth stage (Woulters et al. 1999). The different process parameters of some of the studies conducted to inactivate microorganisms are summarized in Table 7.1.

7.4.2 (In)activation of Enzymes

In general, enzymes are more resistant to PEF treatment. The three-dimensional molecular structure of globular protein is stabilized by hydrophobic interactions, hydrogen bonding, van der Waal interactions, ion pairing, electrostatic forces, and steric constraints. High PEF treatment influences the conformational state of the protein through charge, dipole, or induced dipole chemical reactions. The charged groups and structure are highly susceptible to various types of electric field perturbations. Association and dissociation of ionizable groups, movement of charged side chains, changes in structure, and alignment of helices and, thus, the overall shape of the protein may all be influenced by external electric fields. Because high PEF can be used to inactivate enzymes, which are responsible for development of oxidative off-flavor and color in the foods, the quality of the foods treated with PEF can be preserved. The reaction kinetics of the enzyme activation follows the first-order reaction kinetics.

$$EA = EA_o e^{-(k_1 t)} \tag{7.1}$$

Table 7.1 Summary of the Studies Conducted on Fruit Juices Using PEF

Juice	*PEF Conditions*	*Microorganism*	*Reference*
Apple	12 kV cm^{-1}, 20 pulses, exponential decay, <30°C	*Saccharomyces cerevisiae, Escherichia coli*	Qin et al., 1994
Apple	25 kV cm^{-1}, 558 J, exponential decay, <25°C	*Saccharomyces cerevisiae*	Zhang et al., 1994b
Apple	50 kV cm^{-1}, square wave, 29.6°C	*Saccharomyces cerevisiae*	Qin et al., 1995
Apple	40 kV cm^{-1}, 64 pulses, exponential decay, 15°C	*Saccharomyces cerevisiae*	Harrison et al., 1997
Apple	29 kV cm^{-1}, square wave 34 kV cm^{-1}, 166 μs of treatment time, 1.5 mL s^{-1}, 800 pps	*Escherichia coli* O157:H7	Evrendilek et al., 2000
Apple	35 kV cm^{-1}, 94 μs of treatment time, 85 L h^{-1}, 952 Hz	Aerobic microorganisms, yeasts, and molds	Evrendilek et al., 2000
Apple	20 kV cm^{-1}, 10.4 pulses, square wave	*Saccharomyces cerevisiae*	Cserhalmi et al., 2002
Apple	32.3 kV cm^{-1}	*Zygosaccharomyces balii* ascospores, Vegetative cells (V), ascospores (A)	Raso et al., 1998a
Apple	50, 58, and 66 kV cm^{-1} and number of pulses 2, 4, 8, and 16	Aerobic microorganisms, yeasts, and molds	Ortega-Rivas et al., 1998
Cranberry	36.5 kV cm^{-1}, 22°C	*Byssochlamys fulva* Canidiospores	Raso et al., 1998b
Cranberry	40 kV cm^{-1}, 150 μs treatment time, square wave	Aerobic microorganisms, yeasts, and molds	Jin and Zhang, 1999
Cranberry	36.5 kV cm^{-1}	*Zygosaccharomyces balii* ascospores, Vegetative cells (V), ascospores (A)	Raso et al., 1998a

(continued)

Table 7.1 (Continued) Summary of the Studies Conducted on Fruit Juices Using PEF

Juice	*PEF Conditions*	*Microorganism*	*Reference*
Grape	35.0 kV cm^{-1}	*Zygosaccharomyces balii* ascospores, Vegetative cells (V), ascospores (A)	Raso et al., 1998a
Grape	65 kV cm^{-1}, 20 pulses, 50°C	Aerobic microorganisms, yeasts, and molds	Wu et al., 2005
Orange	29.5 kV cm^{-1}, 60 µs treatment time, square wave	Aerobic microorganisms	Qiu et al., 1998
Orange	30 kV cm^{-1}, 240 µs treatment time, 2 µs pulse width, 1000 Hz, 2 mL s^{-1}	Aerobic microorganisms, yeasts, and molds	Jia et al., 1999
Orange	30 kV cm^{-1} or 50 kV cm^{-1}, 100 L h^{-1}	*Listeria mesenteroides, Escherichia coli, Listeria innocua, Saccharomyces cerevisiae* ascospore	McDonald et al., 2000
Orange	35 kV cm^{-1}, 59 µs treatment time, 1.4 µs pulse width, 600 pps, 98 L h^{-1}	Aerobic microorganisms, yeasts, and molds	Yeom et al., 2000b
Orange	40 kV cm^{-1}, 97 µs treatment time, 2.6 µs pulse width, 1000 pps, 500 L h^{-1}	Aerobic microorganisms, yeasts, and molds	Min and Zhang, 2002b
Orange	34.3 kV cm^{-1}	*Zygosaccharomyces balii* ascospores, Vegetative cells (V), ascospores (A)	Raso et al., 1998a
Orange	32 kV cm^{-1}, 92 µs treatment time, 3.3 µs pulse width, 800 Hz, 79 L h^{-1}	Aerobic microorganisms, yeasts, and molds	Sharma et al., 1998

(continued)

Table 7.1 (Continued) Summary of the Studies Conducted on Fruit Juices Using PEF

Juice	*PEF Conditions*	*Microorganism*	*Reference*
Orange	90 kV cm^{-1}, 20 pulses, 45°C	*Salmonella typhimurium*	Liang et al., 2002
Orange	40 kV cm^{-1} for 150 μs, 55°C	Aerobic microorganisms, yeasts, and molds	Walkling-Ribeiro et al., 2009
Orange	35 kV cm^{-1} for 59 μs	Aerobic microorganisms, yeasts, and molds	Ayhan et al., 2003
Pineapple	33.0 kV cm^{-1}	*Zygosaccharomyces balii* ascospores, Vegetative cells (V), ascospores (A)	Raso et al., 1998a
Tomato	40 kV cm^{-1}, 57 μs treatment time, 2 μs pulse width, 1000 pps, 500 L h^{-1}	Aerobic microorganisms, yeasts, and molds	Min and Zhang, 2002a
Tomato	80 kV cm^{-1}, 20 pulses, 50°C	Aerobic microorganisms, yeasts, and molds	Nguyen and Mittal, 2007

where EA = % residual enzyme activity, and $t = n\,\tau$. For instance, in the case of peach PPO, the first order constant of inactivation k_1 varied from 9 (at E = 3 kV cm^{-1}) to 234 μs^{-1} (at 24 kV cm^{-1}) (Kotnic and Miklavacic 2000). The value of k_1 is expressed as function of the electric field strength E and can be expressed as

$$k_1 = k_{o1}e^{-(\omega E)} \tag{7.2}$$

where k_{o1} and ω are constants. The value of EA can be expressed as a function of the total energy density W_t supplied as follows:

$$EA = EA_o e^{-(K\omega_t)} \tag{7.3}$$

where K is a constant whose value varies depending upon the shape of the pulse, i.e., bipolar exponential decay pulses, mono- or bipolar pulses. In general, bipolar pulses are more effective than monopolar pulses.

7.4.3 Drying and Dehydration

High pulse field treatment increases the permeability of plant cells, thereby increasing the drying rate. Angersbach et al. (1997) observed that mass transfer during fluidized bed drying of PEF-treated potato cubes increases, thereby reducing the drying time of the cubes to one-third. PEF pretreatment also accelerates osmotic dehydration of fruits. For example, in PEF processing of carrot, applied energy in the range of 0.004 to 2.25 kJ kg^{-1} increases the cell disintegration index between 0.09 and 0.84 with a rise in product temperature less than 1°C. The high PEF treatment increases the water loss during osmotic dehydration of apple slices (Taiwo et al. 2001).

7.4.4 Rehydration

High pulse electric field and osmotically treated samples enhance the rehydration capacity of sugar-rich fruit slices because of absorption of sugar. High PEF treatment alone results in the least rehydration capacity, probably due to greater shrinkage from the faster water loss during air dehydration as a result of increased membrane permeabilization behavior of many dehydrated foods.

7.4.5 Extraction of Fruit Juices and Beverages

Orange: Orange juice is one of the most popular and widely consumed juices in the world and has attracted much scrutiny related to its quality and shelf life by scientists worldwide. Yeom et al. (2000a) evaluated the effect of pulsed electric field at 35 kV cm^{-1} for 59 μs and heat pasteurization at 94.6°C for 30 s on the quality of orange juice and found that PEF treatment prevented the growth of microorganisms for 112 days, inactivated 88% of pectin methyl esterase activity, and retained higher amounts of vitamin C and flavor compounds with lower browning index, higher whiteness (*L*), and higher hue angle (θ) values than the heat-pasteurized orange juice. Similar observations were reported by Ayhan et al. (2001), wherein orange juice treated with pulsed electric fields (35 kV cm^{-1} for 59 μs) stored in glass and polyethylene terephthalate bottles exhibited more than 16 weeks of shelf life with higher retention of flavor compounds, color, and vitamin C. In another study, PEF (35 kV cm^{-1} for 59 μs) processing resulted in significant increase in the hydrocarbons D-limonene, α-pinene, myrecene, and valencene in the orange juice during storage for 112 days at 4°C and 22°C. The microorganisms in PEF-processed orange juice, along with the flavor and color of the juice, remained stable at 4°C for 112 days (Ayhan et al. 2002).

Sánchez-Moreno et al. (2005) studied the effect of various processing technologies, such as high pressure (400 MPa/40°C/1 min), pulsed electric fields (35 kV cm^{-1}/750 μs), low pasteurization (70°C/30 s), and high pasteurization (90°C/1 min), on the retention of bioactive components (vitamin C, carotenoids, and flavanones) in orange juice. Of the technologies tested, PEF and HP were found to be effective in preserving the bioactive compounds and radical scavenging activity of freshly squeezed orange juice. Earlier, Jia et al. (1999) had reported similar observations, wherein PEF-treated orange juice (30 kV cm^{-1} for 240 or 480 μs) retained higher amounts of flavor compounds compared to heat-treated (90°C for 1 min). Also a greater decrease in total plate count, yeast, and mold counts was observed compared to untreated juice. However, heat treatment exhibited a higher reduction in the microbial load compared to PEF treatment. These observations are in good agreement with another study, wherein application of PEF at intensities of 25 kV cm^{-1}, 280 μs, and 330 μs did not affect pH, °Brix, total acidity, turbidity, hydroxymethylfurfural (HMF), and color to a significant extent compared to conventional HTST treatment (98°C, 21 s). It was also noted that sensory characteristics of the PEF-treated juice were more similar to the untreated juice than the HTST–pasteurized juice, but heat pasteurization was more efficient in inactivating microbial flora and pectinmethylesterase (PME) and preventing the growth of microbial flora and reactivation of PME at 2°C and 12°C for 10 weeks. However, the shelf life of the PEF-treated juice was established as 4 weeks at 2°C, which was considered to be a reasonable shelf life for this type of foodstuff (Rivas et al. 2006).

Application of PEF (50 pulses at 28 kV cm^{-1}) in a continuous system to grapefruit, lemon, orange, and tangerine did not induce any significant changes in pH, °Brix, electric conductivity, viscosity, color, organic acid content, and volatile flavor compounds, indicating the need for in-depth work regarding the mechanisms of changes during PEF treatment and subsequent storage (Cserhalmi et al. 2006). Thermosonication (55°C for 10 min) followed by continuous PEF (40 kV cm^{-1} for 150 μs) of the orange juice resulted in safe storage of the juice for 168 days. Although the microbial counts were within safe limits, thermal pasteurization resulted in significantly lower microbial counts compared to PEF treatment. It was also observed that PEF-treated juice retained more color compared to thermally pasteurized juice during storage (Walkling-Ribeiro et al. 2009).

Cortes et al. (2006) reported that high-intensity pulsed electric field processing of orange juice with different electric field intensities (25, 30, 35, and 40 kV cm^{-1}) and different treatment times (30–340 μs) retains more carotenoid content compared to a conventional heat treatment (90°C, 20 s). It was concluded that PEF processing of orange juice is an alternative to the thermal treatment of pasteurization provided that it is kept refrigerated.

Elez-Martínez et al. (2006) evaluated the inactivation of orange juice peroxidase by high-intensity pulsed electric fields (5 to 35 kV cm^{-1} for up to 1500 μs using square-wave pulses in mono- and bipolar mode). Results indicated that the inhibition of orange juice peroxidase was proportional to the increase in the electric field strength, treatment time, pulse frequency, and the pulse width. Monopolar pulses were more effective than bipolar pulses. Orange juice POD activity decreased with electric energy density input. Liang et al. (2002) studied the effect of combinations of moderately high temperatures (<60°C), antimicrobial compounds, and pulsed electric field (PEF) treatments to reduce *Salmonella* in pasteurized and freshly squeezed orange juice. *Salmonella typhimurium* were found to decrease with an increase in pulse number and treatment temperature. At a field strength of 90 kV cm^{-1}, a pulse number of 20, and a temperature of above 46°C, cell death and injury were greatly increased, and the *Salmonella* numbers were reduced by 5.9 log cycles in freshly squeezed orange juice (without pulp) treated at 90 kV cm^{-1}, 50 pulses, and 55°C. Further, the combination of nisin and lysozyme had a pronounced bactericidal effect.

Tomato: Min et al. (2003) studied the effects of a commercial-scale pulsed electric field (40 kV cm^{-1} for 57 μs) on the quality of tomato juice in comparison with the thermal processing (92°C for 90 s). The results indicated that, both thermally and PEF-processed juice showed microbial shelf life at 4°C for 112 days, and the lipoxygenase activities of thermally and PEF-processed juices were 0% and 47%, respectively. Although no significant differences were observed in the concentration of lycopene, °Brix, pH, or viscosity of juices during storage, PEF processed juice contained significantly higher ascorbic acid, and the flavor and overall acceptability was higher than thermally processed juice. In another study, Min and Zhang (2002a) evaluated the effects of PEF (40 kV cm^{-1} for 57 μs) on the flavor and color of tomato juice during storage at 4°C for 112 days and compared with thermal processing (92°C for 90 s). The PEF-processed tomato juice retained more flavor compounds of *trans*-2-hexenal, 2-isobutylthiazole, *cis*-3-hexanol and showed lower nonenzymatic browning and higher redness than thermally processed juice and was preferred by the panelists over thermally processed juice.

Nguyen and Mittal (2007) evaluated the effect of combination of moderately high temperatures (<50°C), antimicrobial compounds, and pulsed electric field (PEF) to reduce naturally occurring microbes in tomato juice. The results indicated that the microbial count decreased with the increase in pulse number and treatment temperature at constant field strength to an extent of 4.4 log. As reported earlier, no reduction of vitamin C was observed, and the juiced stayed microbiologically safe at 4°C for 28 days. Although polygalacturonase was unaffected by PEF, the activity of pectin methyl esterase was reduced by 55%.

Odriozola-Serrano et al. (2008) investigated the effect of a high-intensity pulsed electric field (35 kV cm^{-1} for 1500 μs in bipolar 4-μs pulses at 100 Hz) on the bioactive compounds and antioxidant capacity of tomato juice. PEF-treated juice contained significantly higher amounts of lycopene and vitamin C compared to thermal processing (90°C for 1 min or 30 s). However, these values were lower than that of fresh juice. No significant differences were observed in the phenolic compound contents of the fresh, PEF-treated, and thermally processed juices. The authors opined that a high-intensity pulsed electric field may be appropriate to achieve nutritious and fresh-like tomato juice. In another study, the processing of tomato juice using high-intensity pulsed electric fields enhanced lycopene, β-carotene, and phytofluene and the red color of juices with no significant changes in phenolic compounds, pH, and soluble solids compared to heat pasteurized and untreated juices. It was also observed that PEF-processed tomato juices contained a higher content of lycopene, neurosporene, γ-carotene, and quercetin through the storage time than heat-pasteurized and untreated juices. The authors opined that PEF may be appropriate to achieve safety and nutritional wholesomeness of tomato juices.

Grape: Wu et al. (2005) explored the effect of a combination of different obstacles, such as moderately high temperatures (<50°C), antimicrobial compounds and pulsed electric field (PEF) to reduce naturally occurring microbes in red and white grape juice. Results indicated that application of PEF caused a significant decrease in microbial count when a constant number of pulses were applied. Further, vitamin C content remained unchanged. Praporscic et al. (2007) reported that application of a pulsed electric field ($E = 250–1000$ V cm^{-1}) at constant pressure ($P = 5$ bar) resulted in significant improvement of juice yield and quality of different white grapes (Muscadelle, Sauvignon, and Semillon).

Apple: Treatment of fresh apple juice inoculated with *E. coli* O157:H7 and *E. coli* 8739 with PEF (30, 26, 22, and 18 kV cm^{-1} and total treatment time 172, 144, 115, and 86 μs) resulted in a 5-log reduction in both cultures. Results showed no difference in the sensitivities of *E. coli* O157:H7 and *E. coli* 8739 against PEF treatment, and the authors concluded that PEF is a promising technology for the inactivation of *E. coli* O157:H7 and *E. coli* 8739 in apple juice (Evrendilek et al. 1999). Zhang et al. (1994b) treated *Saccharomyces cerevisiae* suspended in apple juice with exponential-decay and square-wave pulsed electric fields and found that both waveforms are effective in the microbial inactivation. However, inactivation with square-wave pulses was greater than with exponential-decay pulses. The authors concluded that square-wave pulsed electric fields resulted in significant energy savings compared to exponential-decay pulses in food pasteurization.

Apple juice subjected to PEF treatment (50, 58, and 66 kV cm^{-1} and number of pulses 2, 4, 8, and 16) reduced aerobic plate count, yeasts and molds, and aciduric bacteria, and no perceptible changes in the quality attributes, such as pH, acidity, and soluble solids was observed. However, fading of juice color was observed (Ortega-Rivas et al. 1998). Earlier, Qin et al. (1995) achieved significant inactivation of *Saccharomyces cerevisiae* in apple juice. Using high-intensity pulsed electric fields in a continuous system, PEF-treated apple juice was found to have a shelf life of 3 weeks at both 4°C and 25°C storage conditions. Schilling et al. (2008) observed a synergistic effect of heat and PEF in inactivating apple polyphenoloxidase, retaining the wholesomeness of the juice. It was concluded that PEF can be utilized with moderate heat treatment for fresh-like juice.

Cranberry: Application of high-voltage pulsed electric field (20 and 40 kV cm^{-1} for 50°C and 150°C) to cranberry juice significantly reduced the viable microbial cells without affecting the volatile and flavor profile of the juice as compared to thermal processing (90°C for 90 s). PEF treatment resulted in no growth of molds and yeasts during storage at 22°C and 4°C and no growth of aerobic bacteria during storage at 4°C. The authors opined that PEF could be used an alternative process to thermal pasteurization for cranberry juice (Jin and Zhang 1999).

Miscellaneous products: Walkling-Ribeiro et al. (2010) evaluated the effect of combining moderate heat and PEF as an alternative technique toward achieving microbiological safety and shelf life of a smoothie-type beverage. The results indicated that the combined technique achieved better stability of Brix and viscosity and superior microbiological shelf stability. It was concluded that a combined process of moderate heat and PEF is a feasible processing alternative for a fruit smoothie. Bouzrara and Vorobiev (2000) reported that a process consisting of a combined pressing and pulsed electric field (PEF) treatment increases the efficiency of juice extraction from sugar beet cossettes by causing pore formation and destruction of the semipermeable barrier of the cell membrane. It was also reported that mechanical pressing associated with PEF treatment increased juice yield threefold. It was concluded that PEF could be used as an alternative process to the standard thermal and mechanical techniques for extracting cellular material.

7.5 Summary

Consumption of juice has increased manifold due to the nutrition and the convenience it offers, and thermal treatment is evidenced to be the best processing method to keep them microbiologically safe. However, thermal treatment tends to alter the sensory qualities and also accelerates the loss of heat labile bioactive compounds. On the other hand, a pulsed electric field offers similar benefits without affecting the product quality parameters and the bioactive components. Considering this, there is great potential to exploit the utility of pulsed electric fields in commercial processing of fruit juices and other beverages.

References

Amiali, M., Ngadi, M.O., Raghavan, G.S.V. and Smith, J.P. (2006). Inactivation of *Escherichia coli* O157:H7 and Salmonella enteritidis in liquid egg white using pulsed electric field. *Journal of Food Science*, 71, 88–94.

Angersbach, A., Heinz, V. and Knorr, D. (1997). Electrishce Leifahigkeit als Maβ des Zellaufschluβgrades von Zellalaren Materialien durch Verarbeitunnsprozesse. *Lebenspittel – und Verpackungstechnik*, 42, 195–200.

Ayhan, Z., Yeom, H.W., Zhang, H.Q. and Min, D.B. (2001). Flavor, color, and vitamin C retention of pulsed electric field processed orange juice in different packaging materials. *Journal of Agricultural and Food Chemistry*, 49, 669–674.

Ayhan, Z., Zhang, Q.H. and Min, D.B. (2002). Effects of pulsed electric field processing and storage on the quality and stability of single-strength orange juice. *Journal of Food Protection*, 65, 1623–1627.

Barbosa-Canovas, G.V., Gongora-Nieto, M.M., Pothakamury U.R. and Swanson, B.G. (1999). *Preservation of Foods with Pulsed Electric Fields*. Academic Press Ltd. London.

Bouzrara, H. and Vorobiev, E. (2000). Beet juice extraction by pressing and pulsed electric fields. *International Sugar Journal*, 102, 194–200.

Castro, A.J., Swanson, B.G., Barbosa-Cánovas, G.V. and Dunker, A.K. (2001). Pulsed electric field modification of milk alkaline phosphatase activity. In: Barbosa-Cánovas, G.V., and Zhang, Q.H. (eds.). *Pulsed electric fields in food processing: fundamental aspects and applications*. Lancaster, PA: Technomic Publishing, pp. 83–103.

Cortés, C., Esteve, M.J., Rodrigo, D., Torregrosa, F. and Frígola, A. (2006). Changes of color and carotenoids contents during high intensity pulsed electric field treatment in orange juices. *Food and Chemical Toxicology*, 44, 1932–1939.

Cserhalmi, Z., Sass-Kiss, A., Tóth-Markus, M. and Lechner, N. (2006). Study of pulsed electric field treated citrus juices. *Innovative Food Science and Emerging Technologies*, 7, 49–54.

Cserhalmi, Z., Vidacs, I., Beczner, J. and Czukor, B. (2002). Inactivation of *Saccharomyces cerevisiae* and *Bacillus cereus* by pulsed electric fields technology. *Innovative Food Science and Emerging Technologies*, 3, 41–45.

Dunn, J.E. (2001). Pulsed electric field processing: An overview. In Barbosa-Cánovas, G.V., and Zhang, Q.H., (eds.). *Pulsed electric fields in food processing: Fundamental aspects and applications*. Lancaster, PA: Technomic Publishing, pp. 1–30.

Elez-Martínez, P., Aguiló-Aguayo, I. and Martín-Belloso, O. (2006). Inactivation of orange juice peroxidase by high-intensity pulsed electric fields as influenced by process parameters. *Journal of the Science of Food and Agriculture*, 88, 71–81.

Espachs-Barroso, A., Barbosa-Cánovas, G.V. and Martín-Belloso, O. (2003). Microbial and enzymatic changes in fruit juice induced by high-intensity pulsed electric fields. *Food Reviews International*, 19, 253–273.

Evrendilek, G.A., Jin, Z.T., Ruhlman, K.T., Qiu, X., Zhang, Q.H. and Richter, E.R. (2000). Microbial safety and shelf-life of apple juice and cider processed by bench and pilot scale PEF systems. *Innovative Food Science and Emerging Technologies*, 1, 77–86.

Evrendilek, G.A. and Zhang, Q.H. (2005). Effects of pulse polarity and pulse delaying time on pulsed electric fields-induced pasteurization of *E. coli* O157:H7. *Journal of Food Engineering*, 68, 271–276.

Hamilton, W.A. and Sale, A.J. (1967). Effects of high electric fields on microorganisms. *Biochimica et Biophysica Acta*, 148, 789–800.

Harrison, S.L., Barbosa-Cánovas, G.V. and Swanson, B.G. (1997). *Saccharomyces cerevisiae* structural changes induced by pulsed electric field treatment. *Lebensm-Wiss u-Technol*, 30, 236–240.

Hermawan, G.A., Evrendilek, W.R., Zhang, Q.H. and Richter, E.R. (2004). Pulsed electric field treatment of liquid whole egg inoculated with *Salmonella enteritidis*. *Journal of Food Safety*, 24, 1–85.

Hulsheger, H., Potel, J. and Niemann, E.G. (1983). Electric field effects on bacteria and yeast cells. *Radiation and Environmental Biophysics*, 22, 149–162.

Jia, M., Zhang, Q.H. and Min, D.B. (1999). Pulsed electric field processing effects on flavor compounds and microorganisms of orange juice. *Food Chemistry*, 65, 445–451.

Jin, T.Z. and Zhang, H.Q. (1999). Pulsed electric field inactivation of microorganisms and preservation of quality of cranberry juice. *Journal of Food Processing and Preservation*, 23, 481–497.

Kotnik, T. and Miklavacic, D. (2000). Second order model of membrane electric field induced by alternating electric fields. *IEEE Transaction of Biomedical Engineering*, 47, 1074–1081.

Knorr, D., Geulen, M., Grahl, T. and Sitzmann, W. (1994). Food application of high electric fields pulses. *Trends in Food Science and Technology*, 5, 71–75.

Liang, Z., Mittal, G.S. and Griffiths, M.W. (2002). Inactivation of *Salmonella typhimurium* in orange juice containing antimicrobial agents by pulsed electric field. *Journal of Food Protection*, 65, 1081–1087.

McDonald, C.J., Lloyd, S.W., Vitale, M.A., Petersson, K. and Innings, F. (2000). Effects of pulsed electric field on microorganisms in orange juice using electric field strengths of 30 and 50 kV/cm. *Journal of Food Science*, 65, 984–989.

Min, S., Jin, T.Z. and Zhang, H.Q. (2003). Commercial scale pulsed electric field processing of tomato juice. *Journal of Agricultural and Food Chemistry*, 51, 3338–3344.

Min, S. and Zhang, Q.H. (2002a). Effects of commercial scale pulsed electric field processing on flavor and color of tomato juice. *Journal of Food Science*, 68, 1600–1606.

Min, S. and Zhang, Q.H. (2002b). Inactivation kinetics of tomato juice lipoxygenase by pulsed electric fields. *Journal of Food Science*, 68, 1995–2001.

Nguyen, P. and Mittal, G.S. (2007). Inactivation of naturally occurring microorganisms in tomato juice using pulsed electric field (PEF) with and without antimicrobials. *Chemical Engineering and Processing: Process Intensification*, 46, 360–365.

Odriozola-Serrano, I., Soliva-Fortuny, R. and Martín-Belloso, O. (2008). Changes of health-related compounds throughout cold storage of tomato juice stabilized by thermal or high intensity pulsed electric field treatments. *Innovative Food Science and Emerging Technologies*, 9, 272–279.

Ortega-Rivas, E., Zárate-Rodríguez and Barbosa-Cánovas, G.V. (1998). Apple juice pasteurization using ultrafiltration and pulsed electric fields. *Food and Bioproducts Processing*, 76, 193–198.

Praporscic, L., Lebovka, N., Vorobiev, E. and Mietton-Peuchot, M. (2007). Pulsed electric field enhanced expression and juice quality of white grapes. *Separation and Purification Technology*, 52, 520–526.

Qin, B.-L., Chang, F.-J., Barbosa-Cánovas, G.V. and Swanson, B.G. (1995). Non-thermal inactivation of *Saccharomyces cerevisiae* in apple juice using pulsed electric fields. *LWT - Food Science and Technology*, 28, 564–568.

Qin, B.L., Zhang, Q., Barbosa-Cánovas, G.V., Swanson, B.G. and Pedrow, P.D. (1994). Inactivation of microorganisms by pulsed electric fields with different voltage waveforms. *IEEE Transactions on Dielectrics and Electrical Insulation*, 1, 1047–1057.

Qiu, X., Sharma, S., Tuhela, L., Jia, M. and Zhang, Q.H. (1998). An integrated PEF pilot plant for continuous non-thermal pasteurization of fresh orange juice. *Transactions of the American Society of Agricultural Engineers*, 41, 1069–1074.

Raso, J., Calderon, M.L., Gongora, M., Barbosa-Cánovas, G. and Swanson, B.G. (1998a). Inactivation of *Zygosaccharomyces Bailii* in fruit juices by heat, high hydrostatic pressure and pulsed electric fields. *Journal of Food Science*, 63, 1042–1044.

Raso, J., Calderon, M.L., Gongora, M., Barbosa-Cánovas, G.V. and Swanson, B.G. (1998b). Inactivation of mold ascospores and conidiospores suspended in fruit juices by pulsed electric fields. *Lebensm Wiss u Technol*, 668–672.

Rivas, A., Rodrigo, D., Martínez, A., Barbosa-Cánovas, G.V. and Rodrigo, M. (2006). Effect of PEF and heat pasteurization on the physical-chemical characteristics of blended orange and carrot juice. *LWT - Food Science and Technology*, 39, 1163–1170.

Sánchez-Moreno, C., Plaza, L., Elez-Martínez, P., Ancos, B.D., Martín-Belloso, O. and Cano, P.M. (2005). Impact of high pressure and pulsed electric fields on bioactive compounds and antioxidant activity of orange juice in comparison with traditional thermal processing. *Journal of Agricultural and Food Chemistry*, 53, 4403–4409.

Schilling, S., Schmid, S., Jaäger, H., Ludwig, M., Dietrich, H., Toepfl, S., Knorr, D., Neidhart, S., Schieber, A. and Carle, R. (2008). Comparative study of pulsed electric field and thermal processing of apple juice with particular consideration of juice quality and enzyme deactivation. *Journal of Agricultural and Food Chemistry*, 56, 4545–4554.

Sepulveda, D.R., Gongora-Nieto, M.M., San-Martin, M.F. and Barbosa-Cánovas, G.V. (2005). Influence of treatment temperature on the inactivation of *Listeria innocua* by pulsed electric fields. *LWT - Food Science and Technology*, 38, 167–172.

Sharma, S.K., Zhang, Q.H. and Chism, G.W. (1998). Development of a protein fortified fruit beverage and its quality when processed with pulsed electric field treatment. *Journal Food Quality*, 21, 459–473.

Taiwo, K.A., Angersbach, A. and Knorr, D. (2001). Influence of high intensity electric field pulses and osmotic dehydration on the rehydration characteristics of apple slices at different temperatures. *Journal of Food Engineering*, 52, 185–192.

Walkling-Ribeiro, M., Noci, F., Cronin, D.A., Lyng, J.G. and Morgan, D.J. (2009). Shelf life and sensory evaluation of orange juice after exposure to thermosonication and pulsed electric fields. *Food and Bioproducts Processing*, 87, 102–107.

Walkling-Ribeiro, M., Noci, F., Cronin, D.A., Lyng, J.G. and Morgan, D.J. (2010). Shelf-life and sensory attributes of a fruit smoothie-type beverage processed with moderate heat and pulsed electric fields. *LWT - Food Science and Technology*, 43, 1067–1073.

Woulters, P.C., Dutreux, N., Smelt, J.P.P.M. and Lelieveld, H.L.M. (1999). Effect of pulsed electric fields on inactivation kinetics of *Listeria innocua. Applied Environmental Microbiology*, 65, 5364–5371.

Wouters, P.C., Alvarez, I. and Raso, J. (2001). Critical factors determining inactivation kinetics by pulsed electric field food processing. *Trends in Food Science and Technology*, 12, 112–121.

Wu, Y., Mittal, G.S. and Griffiths, M.W. (2005). Effect of pulsed electric field on the inactivation of microorganisms in grape juices with and without antimicrobials. *Biosystems Engineering*, 90, 1–7.

Yeom, H.W., Streaker, C.B., Zhang, H.Q. and Min, D.B. (2000a). Effects of pulsed electric fields on the quality of orange juice and comparison with heat pasteurization. *Journal of Agricultural and Food Chemistry*, 48, 4597–4605.

Yeom, H.W., Streaker, C.B., Zhang, Q.H. and Min, D.B. (2000b). Effects of pulsed electric fields in the activity of microorganisms and pectin methyl esterase in orange juice. *Journal of Food Science*, 65, 1359–1363.

Yeom, H.W. and Zhang, Q.H. (2001). Enzymatic inactivation by pulsed electric fields: A review. In: Barbosa-Cánovas, G.V. and Zhang, Q.H. (eds.). *Pulsed electric fields in food processing: Fundamental aspects and applications*. Lancaster, PA: Technomic Publishing, pp. 57–63.

Zhang, Q., Chang, F.J., Barbosa-Cánovas, G.V. and Swanson, B.G. (1994a). Inactivation of microorganisms in semisolid foods using high voltage pulsed electric fields. *Lebensm Wiss u Technology*, 27, 538–543.

Zhang, Q., Monsalve-González, A., Qin, B.-L., Barbosa-Cánovas, G.V. and Swanson, B.G. (1994b). Inactivation of *Saccharomyces cerevisiae* in apple juice by square-wave and exponential-decay pulsed electric fields. *Journal of Food Process Engineering*, 17, 469–478.

Chapter 8

Ultrasonic System for Food Processing

Balunkeswar Nayak

Contents

8.1 Introduction

Consumers of foods are more health concerned than ever and have developed awareness on the food quality, safety, and value addition of the products they consume. It is a challenge for the food industry, regulatory agencies, and scientific research community to provide highly nutritious food that is safe from any contamination or adulteration. While food quality and safety researchers continuously thrive for better food processing equipment and techniques to retain the fresh-like properties and characteristics of foods, regulatory agencies are looking for better detection methods to check contamination and adulteration of foreign materials in the food products. Everyday new food processing and detection technologies have been developed and introduced to the agricultural and seafood markets to assess the quality and safety of food products. However, one technology has some advantage over another in terms of principle, processing time, quality of finished product, environmental impact, and financial cost. Conventionally, heat treatments have been applied, either for sterilization or pasteurization, to control or destroy microorganisms that are responsible for food deterioration and causing illness in human health. However, food products, such as fruit and vegetables, fat and oils, sugar, dairy, meat, coffee and cocoa, meal, and flours, are complex mixtures of vitamins, sugars, proteins and lipids, fibers, aromas, pigments, antioxidants, and other organic and mineral compounds. Processing effectiveness of the food products is dependent on the intensity of treatment temperature and time. Most of the time, focus on providing a safe food by killing all the microorganisms with intense heat for longer duration leads to loss of important food nutrients and alteration of physical, chemical, and sensorial attributes developing undesirable flavors, colors, and taste of food. As a result, new minimal thermally processed and nonthermal technologies are of great interest to food producers, processors, researchers, and regulators. While a number of nonthermal food processing technologies, viz., high hydrostatic pressure (HHP), pulsed electric field (PEF), cold plasma, and supercritical

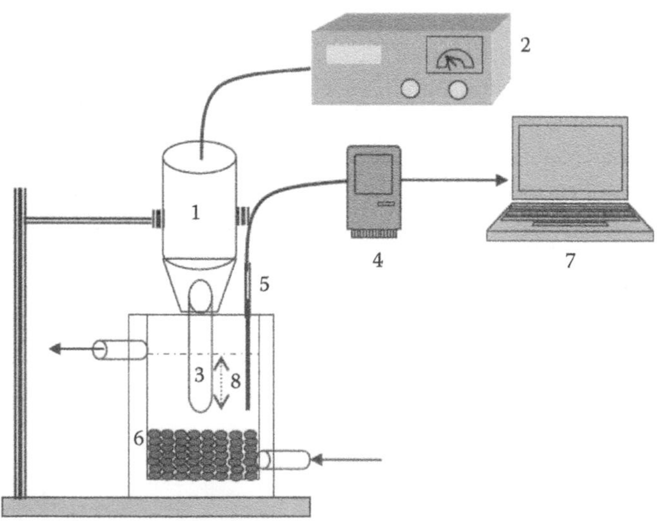

Figure 8.1 Ultrasound setup for the experimental case study. (1) Ultrasound transducer, (2) ultrasonic generator, (3) ultrasound probe (19 mm), (4) data logger, (5) temperature probe, (6) jacketed glass beaker, (7) computer, and (8) depth of probe into the water (2.5 cm). (Adapted from Rawson, A. et al., *Ultrasonics Sonochemistry*, 18, pp. 1172–1179, 2011.)

carbon dioxide, are available in the market, ultrasound has emerged as an important processing and detection technology for its unique application in minimizing processing, maximizing quality, and ensuring the safety of food products.

While application of ultrasound in food processing is relatively new, it has been proved that high-intensity ultrasonic waves can rupture plant/animal cells impacting organelles and composition and low-intensity ultrasound can modify metabolism of cells and detect alterations in physical and chemical properties. In combination with heat, pressure, and electromagnetic waves, ultrasonication can accelerate the rate of sterilization of foods, thus lessening both the duration and intensity of thermal treatment and the resultant damage. Ultrasound is applied to impart positive effects in food processing such as improvement in mass transfer affecting the kinetics, preservation for better food quality, assistance of thermal treatments, and manipulation of texture and food analysis (Knorr et al. 2011). A complete setup of ultrasound technology used for different purpose is given in Figure 8.1. The advantages of ultrasound over heat sterilization include the minimization of flavor loss, greater homogeneity, and significant energy savings.

8.2 Basic Principles

Acoustic waves are mechanical waves that need a material or medium to propagate. In general, acoustic waves are classified by considering human audible frequency as reference with a range of 20 Hz to 20 kHz. Lower frequencies are classified as infrasound, and higher frequencies as ultrasound. While travelling through a medium, the ultrasound waves change their properties and characteristics (frequency, velocity, attenuation, etc.) depending on physical properties such as the density of the medium. The study of these variations is used in diagnosis applications in the range of megahertz (MHz) to characterize the medium, and the power applied is not higher than 1 W cm^{-2} (Patist and Bates 2008). When the applied power of ultrasound is higher, the acoustic waves could affect the medium, which has enormous applications in industry and food regulation. This use of ultrasonic technology is known as "power ultrasound" or "high intensity ultrasound," and the main objective is to induce changes in products or processes.

Ultrasound produces acoustic waves, which are mechanical waves that travel through a medium, resulting in a series of compression and rarefaction. At high power levels, the rarefaction exceeds the attractive forces between molecules in a liquid phase and forms bubbles due to cavitation. Interference of each bubble with its neighbor causes collapse, and release of energy increases temperature and pressure in the medium. The cavitation collapse in aqueous medium also generates shear forces that can produce mechanical and chemical effects. With low intensities (or high frequencies), acoustic streaming is the main mechanism (Leighton 2007). Acoustic streaming is the motion and mixing within the fluid without formation of bubbles (Alzamora et al. 2011). Higher intensities (low frequencies) induce acoustic cavitation (Povey 1998) due to the generation, growth, and collapse of large bubbles, which causes the liberation of higher energies.

Application of ultrasound for food processing is based on the amount of energy generated by the transducer and is characterized by acoustic power (W), acoustic intensity (W m^2), or acoustic energy density (W cm^3 or W mL^{-1}) (Knorr et al. 2004). With the help of a calorimeter, ultrasonic intensity or acoustic energy density can be determined as the following equations (Tiwari et al. 2009):

$$\text{ultasonic intensity} = \frac{4P}{\pi d^2} \tag{8.1}$$

$$\text{acoustic energy density} = \frac{P}{V} \tag{8.2}$$

where P is the absolute ultrasonic power, and it is defined as $mC_p\left(\frac{dT}{dt}\right)_{t=0}$, where C_p is the specific heat capacity, and $\frac{dT}{dt}$ is the rate of change of temperature during sonication. The sonication treatment and the cavitation activity in a treatment chamber may vary if the sample volume and probe location change even for the same ultrasound intensity.

8.2.1 Low-Power Ultrasound

Low-intensity, high-frequency ultrasound typically with less than 1 W cm^{-2} and at more than 100 kHz is used for detection and analysis of physical properties of foods. At low intensities, the power levels are so small that they do not alter the properties of the material through which the ultrasound propagates. Low power ultrasound has been used to evaluate the composition of raw and fermented meat products, fish and poultry, cheese during processing, commercial cooking oils, bread and cereal products, bulk and emulsified fat-based food products, food gels, and aerated and frozen foods. It is also used for the detection of adulteration of honey and modification in the state, size, and type of protein (Awad et al. 2012).

Ultrasonic velocity (v) is determined by density (ρ) and elastic modulus (E) of the medium, according to the Newton–Laplace equation (Blitz 1963):

$$v = \sqrt{\frac{E}{\rho}} \tag{8.3}$$

Composition and physical properties impact on the moduli and density of a material. Therefore, Equation 8.3 implies that the ultrasound velocity of the solid form of a material is larger than that of its liquid form. For instance, the velocity will be higher in a solid food than that of its viscoelastic or viscous form. The interaction of sound waves with matter alters both the velocity and attenuation of the sound waves via absorption and/or scattering mechanisms (McClements 1995). Attenuation is caused the energy loss in comparison and decompression in ultrasonic waves due to both absorption and scattering contributions. Attenuation coefficient is expressed as a measure of decrease in amplitude of an ultrasonic wave as it travels within a material. Response and sensitivity of ultrasound velocity or attenuation coefficient to molecular organization and intermolecular interactions of compounds in food with different composition make ultrasound velocity measurements (UVMs) suitable for physical state and molecular processes (Buckin et al. 2003). With this principle of ultrasound that could change viscosity, compressibility, wall material, and scattering and adsorption effects (Povey 1997), foreign and adulterated materials and defects in processed and packaged food can be detected (Leemans and Destain 2009). The basic principle will also help in studying phase transition and crystallization kinetics in bulk fats, emulsions, and solid lipid nanoparticles (Povey et al. 2009). It is known that the absorption contribution of attenuation is associated with homogeneous materials, whereas the scattering only exists in heterogeneous ones.

Acoustic impedance is a fundamental physical characteristic, which depends on the composition and microstructure of a material. It is the product of density and sound velocity passing through the boundary of different materials, which affects the reflection coefficient. Materials with different densities will have different acoustic impedances, which results in reflections from the boundary between two materials with different acoustic impedances. Attenuation (A) and acoustic impedance (z) are expressed by the following relationships (McClements 1995):

$$A = A_0 e^{-ax} \tag{8.4}$$

$$R = \frac{A_T}{A_t} = \frac{Z_1 - Z_2}{Z_1 + Z_2} \tag{8.5}$$

where A_0 is the initial (unattenuated) amplitude of the wave, x is the distance traveled, R is the ratio of the amplitude of reflected wave (A_T) to the incident wave (A_t) reflection coefficient, and z_1 and z_2 are the acoustic impedances of two materials. From Equation 8.5, a small amount of ultrasound will be reflected from the surface of a material that has very similar acoustic impedance to its surrounding, whereas a high percentage of ultrasound is reflected when two materials have different acoustic impedances.

8.2.2 High-Power Ultrasound

At high intensities, an ultrasonic wave generates intense pressure, shear, and temperature gradients within a material, which can physically disrupt its structure.

High-energy (high-power, high-intensity) ultrasound uses intensities higher than 1 W cm^{-2} at frequencies between 20 and 500 kHz, which are disruptive and induce effects on the physical, mechanical, or chemical/biochemical properties of foods. These effects have positive impacts in food processing, preservation, and safety. Among various applications of this technology as an alternative to conventional food processing operations controlling microstructure and modifying textural characteristics of fat products (sonocrystallization), emulsification, defoaming, modifying the functional properties of different food proteins, inactivation or acceleration of enzymatic activity to enhance shelf life and quality of food products, and microbial inactivation are common (Awad et al. 2012). The main applications of ultrasound in food processes are linked to the effects it has on heat or mass transfer operations. Ultrasound has been applied in osmotic dehydration, brining, freezing, extraction, and enhancement of heat transfer in heat exchangers.

Major parameters that affect power ultrasound are energy, intensity, pressure, velocity, and temperature. High-power ultrasound can be described by the following equation:

$$P_a = P_{max} \sin 2\pi ft \tag{8.6}$$

where P_a is the acoustic pressure (a sinusoidal wave), which is dependent on time (t), frequency (f), and the maximum pressure amplitude of the wave. P_{amax} is related to the power input or intensity (I) of the transducer:

$$I = \frac{P_{amax}}{2\rho v} \tag{8.7}$$

where ρ is the density of the medium, and v is the sound velocity in the medium.

8.3 Applications of Ultrasound

8.3.1 Food Quality

The food industry is becoming increasingly aware of the importance of developing new analytical techniques to study complex food materials and to monitor food adulteration, unit operations, and other processes for evaluating the processed or final food products. The possibility of using low-intensity ultrasound to characterize foods was first realized over 60 years ago; however, it is only recently that the full potential of the technique has been apprehended. Rapid advances in microelectronics have led to the development of relatively inexpensive ultrasonic instrumentation that is now commercially available. Modern ultrasound instruments can be fully automated, can make rapid and precise measurements, are nondestructive and noninvasive, can easily be adapted for online applications, and can be used to analyze systems that are optically opaque. Scattering and reflection of low-power/intensity, high-frequency acoustic pulses sent from a transducer and received by either the same or a different transducer are used for food quality assessment. Based on this basic principle for analysis of foods using ultrasound technology, a relationship between measurable ultrasonic parameters (velocity, attenuation coefficient, and impedance) and physicochemical properties of food materials (composition, structure, and physical state) can be established. This relationship can either be empirical, by preparing a calibration curve that relates the property of interest to the measured ultrasonic parameters, or theoretical, by using equations that describe the propagation of ultrasound through materials. During the past 50 years, ultrasound has been used to analyze foods for quality control. A plethora of studies given in Tables 8.1 through 8.3 provide brief descriptions of the use of lower power ultrasound for quality analysis and control in various foods including meat products and plant materials. Other important applications of low-power ultrasound include analyzing deterioration of edible oil during frying, determining solid fat and droplet size distribution in homogenized milk, and assessing volume fraction of some components in foods, such as fruit juices, syrups, and alcoholic beverages and mechanical properties of cheeses.

Low-power ultrasound has been used for food quality control based on assessing physical properties such as texture, rheology, and composition such as water and fat content. For instance, ultrasound parameters (velocity, attenuation) correlated with the rheological properties of dough prepared from flours of a range of quality, showing that ultrasound (at ~100 kHz) may be a suitable measurement method to discriminate types of flour for different applications (Álava et al. 2007). In a separate study, researchers observed that acoustic impedance of cake batter provides a direct measurement of the gas content in the batter (Gómez et al. 2008). Some researchers combined the results of both the attenuation and the velocity of the sound at 54 kHz to get information on the porosity, cell size, and shape to characterize the structure of breadcrumb (Elmehdi et al. 2003). A noncontact ultrasonic technique to characterize the porous structure of breadcrumb and the phase velocity and attenuation measurements were used to assess structural differences between two different bread types in the frequency range of 40 kHz to 1 MHz (Lagrain et al. 2006). Antonova et al. (2003) used mechanical and ultrasonic techniques to determine crispness in breaded fried chicken nuggets under different storage conditions. The above investigators used a pair of dry-coupling 250 kHz ultrasonic transducers to perform the ultrasonic transmission through the fried chicken nuggets.

Investigating on the effects of ultrasonic velocity (at 4 MHz) on the Atlantic cod fillets, Ghaedian et al. (1998) observed that the velocity is decreasing linearly over the range 1575–1595 m s^{-1} with increasing moisture content between 78% and 82%. Using a shear reflectivity method of ultrasound at 10 MHz, two honeys with a moisture content over the range 15%–20% were distinguished. Application of this

Table 8.1 Applications of Low-Power Ultrasound in Analysis and Quality Control of Meat Products

Meat Product	*Measurements*	*Parameter*	*Advantage*
Livestock, beef cattle carcass, sheep carcass, carcass traits of Bali bulls, growing lambs	Fat and muscle accretion and body composition, intramuscular fat (IMF) percentage, carcass traits, degree of muscle development	Vibration—reflection images of tissues and internal organs in live animals	Enhance genetic improvement programs for livestock; quality control of meat
Pigs	Characterize and classify back fat from animals of different breeds and feeding regimes	V	Quality control; improve meat quality traits in breeding animals
Atlantic mackerel	Fat content (solids and nonfat content)	V	Easy, rapid, and nondestructive method
Atlantic cod fillets	Structural deterioration of tissue during post mortem	A	
Chicken	Composition of chicken analogs; solid fat content	V	Nondestructive and rapid detection; alternative to x-ray methods
Skinless poultry breast	Defects and internal objects	V	
Raw meat mixtures	Composition	V	Quality control
Fermented meat	Quality determination	V	
Dairy products (cheese)	Composition, defects, and internal objects, rheology	V	Inline quality control

Source: Awad, T.S. et al., *Food Research International*, 48, pp. 410–427, 2012. With permission.

Note: A, attenuation coefficient; V, ultrasonic velocity.

Table 8.2 Applications of Low-Power Ultrasound for Analysis and Quality Control of Plant Food Resources

Application	*Measurements*	*Parameters*	*Advantages*
Fruits and vegetables	Firmness, mealiness, dry weight, oil contents, soluble solids, and acidity	V, A	Indirect assessment proper harvesting time, storage, and shelf life
Reconstituted orange juice	Sugar content and viscosity	V	Quality control
Fruit peels	Ripeness	A	Quality control
Plums and tomato	Maturity and sugar content	A	Correlated well with firmness
Potatoes	Defects	A	Quality control
Oils and fat-based products	Density, impedance, celerity, absorption losses coefficient, dynamic viscosity, and compressibility modulus	V, A	Nondestructive, noninvasive, simple pulse-echo method
	Composition, purity, quality, density, solid fat, phase transition, polymorphism	V	Authentication of food fat contents, improving real-time quality control
Cereal product (bread dough)	Extent of mixing and rheological properties	V, A	Online dough quality control
	Rheological properties, kinetics of bread dough fermentation	V, relative delay, signal amplitude	Nondestructive, quality control of bread dough
Batters	Monitor specific quality of batters as it is mixed; consistency	Acoustic impedance	Quality control of sensorial properties
Biscuits and cereal products	Sensory crispness	V, wave amplitude	Quality control of sensorial properties

Source: Awad, T.S. et al., *Food Research International*, 48, pp. 410–427, 2012. With permission.

Note: A, attenuation coefficient; V, ultrasonic velocity.

Table 8.3 Applications of Low-Power Ultrasound for Analysis and Quality Control of Other Food Products

Application	*Measurements*	*Parameters*	*Advantages*
Food oil-in-water (O/W) emulsions	Disperse phase volume fraction, solid fat content, droplet size and size distribution, sedimentation, creaming, coalescence, flocculation, composition, crystallization and melting temperatures, crystallization kinetics and stability	V, A	Quality control and assurance, help optimizing formulations, extending shelf life and long-term storage stability, and controlling physicochemical properties of food emulsions and emulsion-based delivery systems
Aerated food products (ice cream, whipped cream, confectionery, bread dough, and desserts)	Dispersed gas phase, bubble morphology, mean bubble size and uniformity	V, A	Quality control of aerated food systems
Honey	Physical and mechanical properties, adulteration, high frequency dynamic shear rheology, viscosity, and moisture content	V	Quality assurance, measure continuously the rheology of samples flowing through a pipe without disturbing them; measure the rheology of a sample packed in a container without having to open the container

Food gels			
Tofu	To identify aggregation and the ripening processes/textural or gelation	V, A	Quality control allows to sensitively differentiate between carrageenan types
Carrageenan		V, A	
Food protein	Hydration, solubility, foaming capacity, flexibility, changes in conformation	V	Understanding and controlling the functionality of protein in complex food systems
	Size and concentration of soluble proteins and casein micelle in skimmed milk	A	
	Isoelectric point and precipitation	V, A	
Food freezing			
Gelatin, chicken, and beef	Temperature of frozen food and ice content	Time of flight of ultrasonic pulses, V	Quality control, extending the shelf life and preserving the quality of many food products

Source: Awad, T.S. et al., *Food Research International*, 48, pp. 410–427, 2012. With permission.

Note: A, attenuation coefficient; V, ultrasonic velocity.

technique could be used for potential honey quality control. The variations of ultrasonic velocity and attenuation with storage time and elastic modulus of apples are given in Figures 8.2 and 8.3. Benedito et al. (2001) studied the chemical composition of fermented sausages with varying protein (2%–21%), moisture (7%–76%), and fat (2%–90%) contents at two temperatures (4°C and 25°C) at 1 MHz of ultrasound velocity. Mörlein et al. (2005) showed that animal carcasses can be classified based on intramuscular fat content by spectral analysis of ultrasound echo signals (at ~3.5 MHz). Similarly, meat yield in pigs was predicted by the back fat thickness and longissimus muscle using ultrasound scanning at 3 MHz (Cisneros et al. 1996).

Fermentation processes in bread and alcoholic drinks were assessed using low-power ultrasound. Online ultrasonic velocity measurements at 1, 2, and 4 MHz were used to monitor alcoholic fermentation and the contributions of potential of gas production, microorganism growth, polysaccharide hydrolysis, and monosaccharide catabolism in synthetic broths (glucose, fructose, and sucrose) (Resa et al. 2009). In a recent study, a polynomial approach was established to simultaneously determine the sugar and ethanol concentration in industrial fermentation fluids based on measurement of acoustic velocity using the pulse echo method (center frequency ~2 MHz) (Schöck and Becker 2010). Skaf et al. (2009) designed a nondestructive acoustic sensor suitable to monitor the kinetics of bread dough fermentation by using ultrasound at 4–146 kHz. The device was also sensitive to the

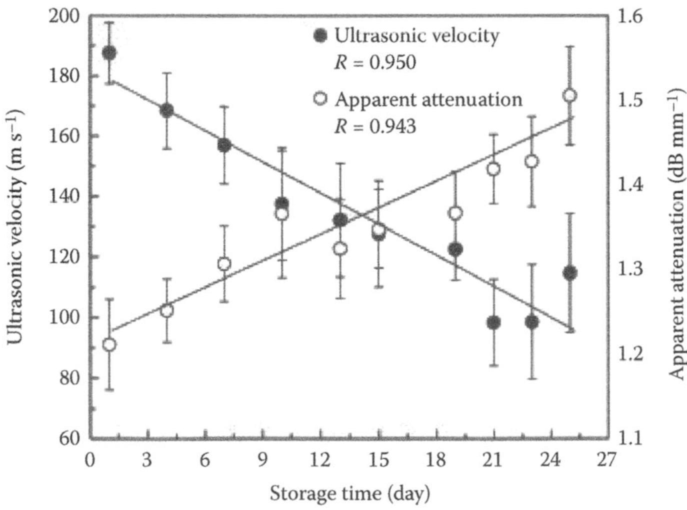

Figure 8.2 Variations of the mean value of ultrasonic velocity and attenuation with storage time for apples. Each point represents the mean value of 20–30 apples. Straight lines are data fitted lines. Vertical lines represent the 95% confidence intervals. (From Kim, K.B. et al., *Postharvest Biological Technology*, 52, pp. 44–48, 2009. With permission.)

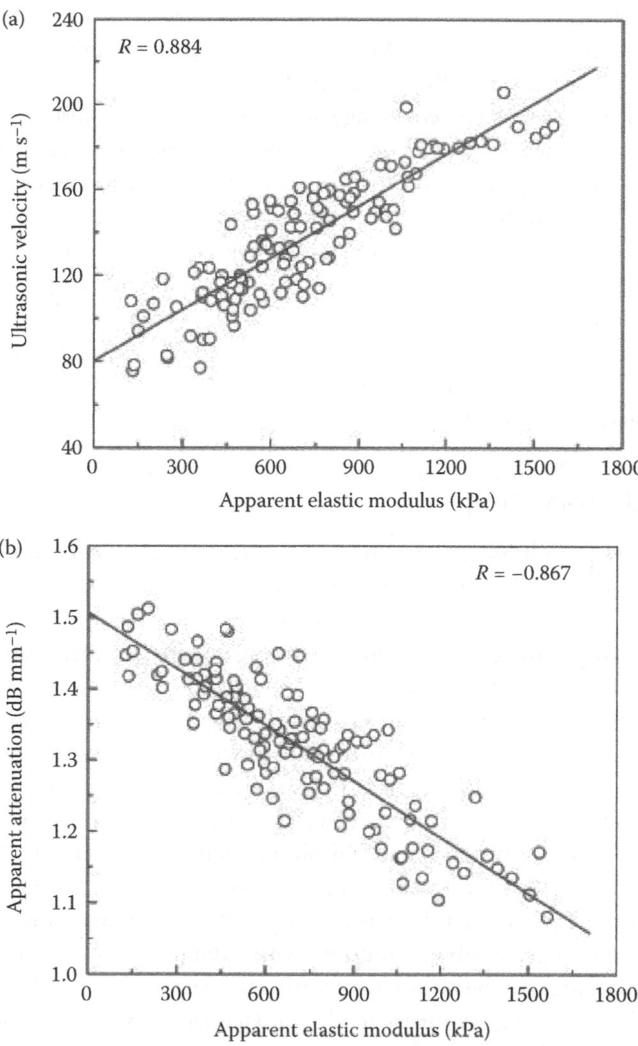

Figure 8.3 Variations in (a) ultrasonic velocity and (b) apparent attenuation with the apparent elastic modulus for apples. (From Kim, K.B. et al., *Postharvest Biological Technology*, 52, pp. 44–48, 2009. With permission.)

influence of the technological parameters of temperature and nature of the ingredients. In a study, the propagation of ultrasound waves at 1 MHz through tofu has been used to monitor the development of tofu texture during gelation, allowing the assessment of the quality of tofu during its manufacture (Ting et al. 2009). Similarly, ultrasonic parameters (attenuation, velocity) were used to determine the optimal renneting time, by relating ultrasonic variables to enzyme activity in milk during coagulation, and to indicate the maturation degree of cheeses (Benedito et al. 2002).

Recently, a method to estimate the fat and moisture content of fresh and blended cheeses using acoustic velocity at 1 MHz and temperatures ranging from 3°C to 29°C has been proposed (Telis-Romero et al. 2011).

Ultrasound is a promising technology to detect foreign materials (e.g., glass, metal, plastic), suspended particles (e.g., microorganisms), or internal structural defects (e.g., holes, cracks) in foods due to its ability to detect changes in acoustic impedance (Z) between different regions within a given volume. Ultrasonic pulse compression (UPC) has been used to detect foreign objects in contained foods, consistency of some liquids, and discrepancies in liquid levels in polymer-based beverage bottles (Bermúdez-Aguirre et al. 2011). Similarly, Knorr et al. (2004) used ultrasonic signals in a time-frequency analysis to detect and identify standard foreign materials, as well as raw material contaminants (kernels imbedded in cherry flesh), in various food products.

8.3.2 Food Processing

High-intensity power ultrasound can induce chemical and physical changes in food systems due to its acoustic and hydrodynamic cavitations. Acoustic and hydrodynamic cavitations are able to generate intensities required to induce chemical and physical changes in the composition of food materials. Ultrasound-induced formation, growth, and implosive collapse of cavities (gas bubbles) in liquids release high amounts of highly localized energy. The collapse of bubbles has different impacts on the surfaces depending on the location of released energy. Near the solid surfaces, it forms asymmetrical microjets that help cleaning of surfaces from contaminants, whereas microjets generated near the interface between two immiscible liquids (oil-in-water or water-in-oil) facilitate emulsification (Thompson and Doraiswamy 1999). Free hydroxyl radicals formed from the dissociation of water molecules in aqueous solutions as a result of the high temperature and pressure of the collapsing gas bubbles associated with cavitation (sonolysis) (Riesz and Kondo 1992) can enhance the rate of mass transport reactions due to the generation of local turbulence and liquid microcirculation (acoustic streaming) (Gogate and Pandit 2011). Potential oxidative damage associated with free radicals is a concern to human health.

High-power ultrasound may be used for various food processing applications, such as extraction of polyphenolic compounds, flavors, degassing, emulsification, destruction of foams, enhancement of crystallization and modifying polymorphism, and prevention of pathogenic bacterial contamination from the food processing surfaces. Application of this technology in various food processing operations is illustrated in Table 8.4. Besides, the technology has also been used for various unit operations such as blanching, drying, and extraction.

Rawson et al. (2011) investigated on the effect of ultrasound as pretreatment with conventional blanching treatment in the drying of carrots and studied various physical and chemical changes in it. The study was conducted from the viewpoint that the quality of a dehydrated product depends not only on the drying conditions but also on the pretreatments employed before drying. Blanching is a pretreatment that involves heating a product to a high temperature for a short period and is normally carried out prior to dehydration to inactivate enzymes, which may otherwise lead to formation of unacceptable color and flavors (Baloch et al. 1997). Blanching pretreatments are also used to improve the final texture and color after rehydration (García-Reverter et al. 1994). Blanching has also been reported to reduce drying times, though thermal degradation of the product may occur as a result of the relatively high temperatures used. In common with other thermal processes, blanching has been shown to affect the concentration of some bioactive compounds in vegetables including polyphenols, ascorbic acid, carotenoids, and polyacetylenes (PAs) (S. Hansen et al. 2003). Given the possible detrimental effect of blanching on the nutritional quality of some products, there is a need to develop alternative pretreatment methods that have minimal impact on the nutritional and organoleptic properties of food. Power ultrasound is an emerging and promising alternative technology for food processing applications, which has been identified as a possible pretreatment to replace blanching.

Rawson et al. (2011) studied the effect of ultrasound-assisted hot air drying (UPHD) of carrot disks and compared it with conventional blanched hot air dried (BHD) carrot disks for the retention of PAs. The authors reported that BHD led to significant decrease in the levels of bioactive PAs, viz., falcarindiol (FaDOH), falcarindiol-3-acetate (FaDOAc), and falcarinol (FaOH), compared to control, which was attributed to the relatively high temperatures required for blanching treatment (80°C, 3 min) leading to oxidative and thermal degradation. In contrast, UPHD samples sonicated at the highest treatment conditions (61 mm and 10 min) had higher retention levels for all three PAs compared to BHD and hot air dried samples only. In the case of FaOH, the most biologically active of the three PAs, pretreatment with ultrasound at amplitude levels of 42.7 and 61.0 mm resulted in significantly higher levels of retention compared to BHD samples: 60% and 64.4% higher at 3 min treatment time, whereas it was 37.2% and 86.2% higher at 10 min treatment time. The authors concluded that higher ultrasound amplitudes, that is, high intensity and longer holding times, may lead to improved retention of PAs compared to hot air dried samples with no pretreatment. It is possible that ultrasound treatment will increase the extractability of PAs by providing the activation energy required to dissociate PAs from pectin-rich cell walls. In addition, the greater retention of PAs at higher amplitudes and holding times could be related to improved denaturation of enzymes, which are responsible for their degradation. Improved extraction efficiency following sonication has been attributed to the propagation of ultrasound pressure waves, with induced cavitation and high shear forces resulting in increased mass transfer (Vilkhu et al. 2008). Simal et al. (1998) suggested that the degassing effect observed under sonication may be similar to that observed under vacuum treatment, which can enhance diffusion into pores on the surface and may explain the enhanced extractability.

Table 8.4 Applications and Characteristics of High-Power Ultrasound in Some Food Processes

Application	*Effect/Mechanism*	*Parameters*	*Advantages*
Sonocrystallization			
Crystallization kinetics of model triglycerides (tripalmitin and trilaurin)	Cavitation induces formation of nucleation active sites and creates smaller crystals with modified properties.	20 kHz and 100 W for 2 s	• HPU decreased crystallization induction times, increased nucleation rate, and modified polymorphic crystallization, microstructure, texture, and melting behavior.
Functional properties of anhydrous milk, palm kernel oil, shortening		20 kHz for 10 s	• Tunable by varying sonication time, power, duration of the acoustic pulse, and crystallization temperature.
Emulsification			
Edible nanoemulsions	Collapse of cavitation forms high energy microjets near interfaces.	Irradiation time and power, oil viscosity, and interfacial tension	• Facilitate the formation of small (40 nm) nanoemulsions. • Decreased amount of surfactants. • More stable droplets. • High loading.

Defoaming			
To prevent decay and oxidation, enhance freshness and quality, and extend shelf life; maximize production and avoid problems in process control and equipment operation	Dissolved gas/oxygen moves toward cavitation bubbles, which grow in size by coalescence then rise releasing the entrapped gas to the environment.	20 kHz in pulsed operation (1 s/1 s)	• Effective procedure to remove foam and dissolved oxygen (80% of foam reduction) with very low energy consumption 40 kJ L^{-1}) in supersaturated milk. • Control of excess foam produced during the filling operation of bottles and cans on high-speed canning lines and in fermenting vessels and other reactors of great dimensions.
Food Proteins			
Whey protein	Cavitation	Power (20, 40, 500 kHz) and time	• Increased protein solubility and foaming ability.
Whey protein isolate	Cavitation	20 kHz, 15 min	• Increased solubility.
Whey protein concentrate			• Significant increase in apparent viscosity.
Soy protein isolates (SPI); soy protein concentrate (SPC)	Cavitational forces of ultrasound treatment with probe, and microstreaming and turbulent forces after treatment with baths	20 kHz probe and ultrasound bath (40 and 500 kHz)	• Significant changes in conductivity and rheological properties, increased solubility for SPC, and increased specific surface area. • Less energy and shorter time compared to traditional and current technology.

Source: Awad, T.S. et al., *Food Research International*, 48, pp. 410–427, 2012. With permission.

Cavitation during high-power ultrasound helped in achieving a faster rate of water diffusion during ultrasound-assisted drying. Fernandes and Rodrigues (2008) observed an increase in the effective diffusivity of water in the fruit after the application of ultrasound, which reduced air-drying time. In a separate study, effective water diffusivity in strawberries increased while treating for osmotic dehydration combined with ultrasonic energy that also reduced total processing time compared to osmotic dehydration (Garcia-Noguera et al. 2010). The action of high-power ultrasound is due to cavitation, which generates high shear forces and microbubbles that enhance surface erosion, fragmentation, and mass transfer resulting in high yield of extracted materials and fast rate of extraction. For instance, application of high-power ultrasound has improved the extraction process of a variety of food components (e.g., herbal, oil, protein, polysaccharides) as well as bioactive ingredients (e.g., antioxidants) from plant and animal resources (Vilkhu et al. 2008).

8.3.3 Food Safety

8.3.3.1 Microbial Safety

Conventional thermal pasteurization and sterilization are the most common techniques to inactivate microorganisms including pathogenic bacteria and enzymes in food products. However, many times the intensities of processing time and temperature have adverse effects on the amount of nutrient, development of undesirable flavors, and deterioration of functional properties of food products. Ultrasound is one of the new preservation techniques that could inactivate or control microbial activity by virtue of its cavitation effects. Acoustic cavitation can be classified as either transient or stable. Transient cavitation occurs when the cavitation bubbles, filled with gas or vapor, undergo irregular oscillations and finally implode producing high local temperatures and pressures that would disintegrate biological cells and denature enzymes. In addition, the imploding bubble also produces high shear forces and liquid jets in the solvent used that may have sufficient energy to physically damage the cell wall or cell membrane. On the other hand, stable cavitation refers to bubbles that oscillate in a regular fashion for many acoustic cycles inducing microstreaming in the surrounding liquid, which can also induce stress in the microbiological species. The inactivation effect of ultrasound has also been attributed to the generation of intracellular cavitation, and these mechanical shocks can disrupt cellular structural and functional components up to the point of cell lysis. Spores appear to be more resistant than vegetative forms, whereas enzymes are reported to be inactivated by ultrasound due to a depolymerization effect. For inactivation of microorganisms, critical processing factors are the nature of the ultrasonic waves, the exposure time with microorganisms, the type of microorganisms, the volume of food processed, the composition of food, temperature, and interactions between these parameters.

If the ultrasound is applied alone at ambient temperature, intense power and long contact times are required to inactivate microorganisms, especially for microbial inactivation in real food systems. The decimal reduction time value (*D* value) of vegetative *Staphylococcus aureus* treated with ultrasound (150 W, 20 kHz) took a longer time (187 min) at ~14°C when the microorganism was suspended in UHT milk (Ordóñez et al. 1987) in comparison to 36.5 min at 11°C. Figures 8.4 and 8.5 show the influence of power output on the survival rate of various microorganisms. In another study, Bermúdez-Aguirre and Barbosa-Cánovas (2008) observed that the increase in butter fat content of milk (fat-free, 1%, 2%, and whole milk) has increased the resistance of *Listeria innocua* against ultrasound (400 W, 24 kHz) performed at 63°C for 30 min. The resistance of *L. innocua* could be due to the bacteria that adhere to the surface of the newly formed milk fat globules. The bacteria were protected either by the rough surface of the fat globules or by concealment within the globules that were disturbed with ultrasound, providing a protective fat layer against the heat and cavitation generated with the sonication. In general, spores of bacteria (e.g., *Bacillus* and *Clostridium* spp.) are more resistant to cavitational effects than vegetative cells, which are in growth phase. Fungi are generally more resistant than vegetative microorganisms, aerobes are more resistant than anaerobes, and

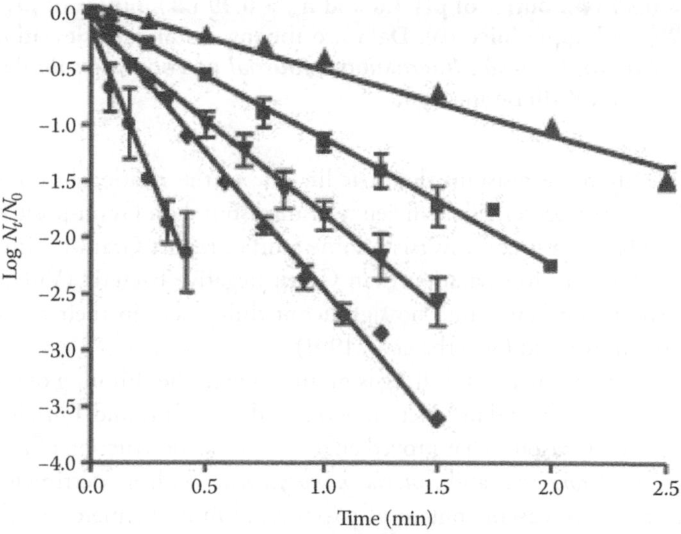

Figure 8.4 Survival curves of *Enterococcus faecium* CECT 410 (▲), *Listeria monocytogenes* CECT 4031 (■), *Salmonella enterica* serovar *enteritidis* CECT 4300 (▼), *Cronobacter sakazakii* CECT 858 (◆), and *Yersinia enterocolitica* CECT 4315 (●) to ultrasound under pressure (35°C, 117 µm, 200 kPa) treated in citrate–phosphate buffer of pH 7.0. Data are means ± standard deviation (error bars). (From Arroyo, C. et al., *International Journal of Food Microbiology*, 144, pp. 446–454, 2011. With permission.)

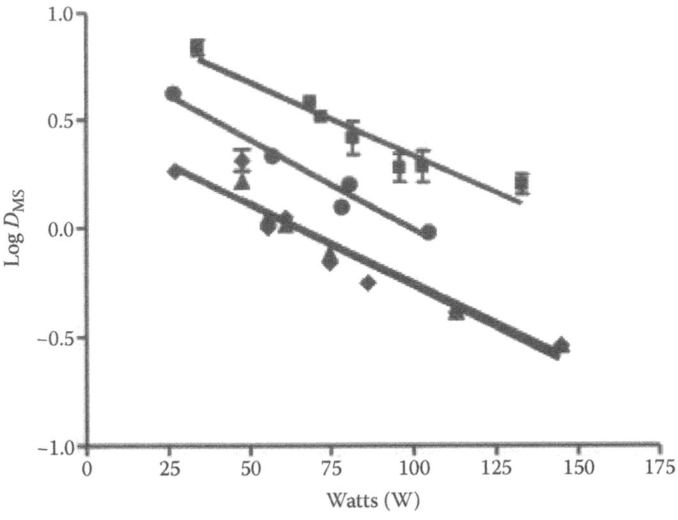

Figure 8.5 Influence of power input (watts) on the inactivation rate (log D_{MS}) of *C. sakazakii* CECT 858 by ultrasound treated in four different media: buffer of pH 7.0 and a_w > 0.99 (◆), buffer of pH 4.0 and a_w > 0.99 (▲), buffer of pH 7.0 and a_w = 0.94 (■), and apple juice (●). Data are means ± standard deviation (error bars). (From Arroyo, C. et al., *International Journal of Food Microbiology*, 144, pp. 446–454, 2011. With permission.)

cocci are typically more resistant than bacilli, due to the relationship of cell surface and volume. The bactericidal efficiency of ultrasound on Gram-positive versus Gram-negative bacteria is controversial. Some studies report Gram-positive bacteria to be more resistant to cavitation than Gram-negative bacteria (Villamiel and Jong 2000), whereas others found no significant differences in their resistance to inactivation by ultrasound (Scherba et al. 1991).

The mechanism of microbial killing is mainly due to the thinning of cell membranes, localized heating, and production of free radicals (Butz and Tauscher 2002). Use of high-power ultrasound has proved effective for the destruction of *Escherichia coli*, *Pseudomonas fluorescens*, and *Listeria monocytogenes* with no detrimental effect on the total protein or casein content of pasteurized milk (Cameron et al. 2009). Although high-power ultrasound alone is known to disrupt biological cells, it can accelerate the rate of sterilization of foods when combined with heat or pressure or both treatments. When ultrasound is applied in combination with pressure, it is known as manosonication (MS). Similarly, combination of ultrasound and heat is known as thermosonication (TS), and combination of ultrasound, pressure, and heat is known as manothermosonication (MTS). Combination of ultrasound with heat and/or pressure reduces both the duration and intensity of the treatment and the resultant damage compared to the individual treatments alone (Knorr et al.

2004). Bermúdez-Aguirre et al. (2009) compared conventional batch pasteurization (63°C, 0–30 min) with TS (400 W, 24 kHz, 63°C, 0–30 min) inactivation of *L. innocua* and mesophilic bacteria in raw whole milk. Thermal pasteurization resulted in 0.69 log and 5.5 log reductions for 10 and 30 min treatments, respectively. With a combination of ultrasound (60%, 90%, or 100% of 400 W) and heat (63°C), a 5 log reduction was achieved after just 10 min, and a synergistic rather than additive effect was observed. Besides, the thermosonicated milk had improved whiteness and similar physicochemical properties compared to the conventionally pasteurized milk. Czank et al. (2010) found that TS treatment was considerably more effective (*E. coli* D_{45} °C = 1.74 min, D_{50} °C = 0.89 min; *Saccharomyces epidermidis* D_{45} °C = 2.08 min, D_{50} °C = 0.94 min) than ultrasound (150 W, 20 kHz) treatment alone (*E. coli* D_4 °C = 5.94, *S. epidermidis* D_4 °C = 16.01 min) for inactivation of *E. coli* and *S. epidermidis* in human milk. Cabeza et al. (2004) reported that combined ultrasound (24 kHz, 400 W) and heat (52–58°C) enhanced the killing rate of a *Salmonella enterica* strain on intact egg shells compared to heat treatment alone. The treatment process yielded a product that not only reduced the salmonellae to a safe level but also had negligible damage to the thermolabile egg components. Haughton et al. (2012) found that *Campylobacter jejuni* was more susceptible to TS (24 kHz, 53°C) than to sonication or thermal treatment alone, with respective mean inactivation of 4.7, 3.2, and 1.5 $\log_{10}$ colony forming units per milliliter. In contrast, the effectiveness of sonication and TS generally decreased when similar treatments were applied to inoculated poultry products (Haughton et al. 2012). In another study, TS (24 kHz, 24.4, 42.7, and 61 μm amplitude, 2–10 min, 25°C–45°C) of watermelon juice significantly decreased the ascorbic acid, lycopene, and total phenolic content at increasing amplitude levels and maximum processing time (Rawson et al. 2011). Thus, optimization of the TS conditions is essential to ensure microbial food safety without compromising the quality of the food. Some studies suggest that osmosonication that is a combination of ultrasound treatment of a medium and its subsequent storage at high osmotic pressure may be a potential alternative to thermal stabilization processes for producing high-quality juice concentrates. Wong et al. (2012) reported that osmosonication enhanced membrane damage and viability loss of *Salmonella* spp. in orange juice. Sonication of orange juice (50 W, 20 kHz, 25°C), combined with its subsequent concentration and storage (48 h) at high osmotic pressure (10.9 MPa), reduced *Salmonella* spp. by 5 log. Moreover, sonication did not affect the key physicochemical and functional attributes of the juice.

The pathogens *Bacillus cereus* and *Bacillus licheniformis* spores have been found to be resistant to ultrasonic treatment alone (Burgos et al. 1972). Combining the effect of a 20 kHz ultrasonic probe on samples under static pressure indicated that an increase in static pressure resulted in an increased level of bacillus spore inactivation (Raso et al. 1998). MS treatment at 500 kPa for 12 min inactivated over 99% of the spores. Increasing the amplitude of ultrasonic vibration of the transducer, that is, the acoustic power entering the system, increased the level of inactivation, for example, a 20 kHz probe at 300 kPa, 12 min sonication at 90 lm amplitude inactivated 75% of the spores. Raising the amplitude to 150 lm resulted in 99.5% spore inactivation. Finally, increasing the thermal temperature of the treatment resulted in greater rates of inactivation certainly at 300 kPa compared to thermal treatment alone. Huang et al. (2006) reported that a combination of ultrasound (34.6 W, 20 kHz, 55°C, 5 min) and HHP (2 × 2–4 cycles at 138 MPa) resulted in the higher inactivation (3.2 log) of *Salmonella enteritidis* in liquid whole egg over PEF, HHP, and ultrasound alone.

Recently, TS has been combined with other emerging technologies for improved microbial inactivation while minimizing nutrient loss in the food products. For instance, Noci et al. (2009) reported on the inactivation of *L. innocua* in low-fat UHT milk preheated to 55°C (>60 s) by combined TS (24 kHz, 400 W, 80 s) and PEF (40 kV cm^{-1}). Cell death number by TS-PEF was similar to that achieved by conventional pasteurization. In addition, the TS treatment substantially decreased the severity of the temperature/time exposure over thermal treatment alone. Similar findings were reported by Walkling-Ribeiro et al. (2009a) when orange juice was treated using TS-PEF (30 kHz, 55°C, 10 min). In another study, the effect of TS (10 min at 55°C) in combination with PEFs (40 kV cm^{-1}) on the inactivation of *S. aureus* (SST 2.4) in orange juice was investigated and then compared with conventional pasteurization (94°C for 26 s). The investigators reported that no significant effect was found on the pH, conductivity, °Brix, juice color, and the nonenzymatic browning index using TS/PEF than thermal treatment (Walkling-Ribeiro et al. 2009b). Moreover, the TS/PEF combination has proven the storage life of orange juice without affecting the sensory acceptability by consumers (Walkling-Ribeiro et al. 2009a).

8.3.3.2 Enzymatic Activity

Ultrasound has mixed effects of inactivation and activation on enzyme activity. According to the literature, inactivation of monomeric enzymes generally involves either defragmentation of the enzyme or formation into aggregates (Mawson et al. 2011), whereas polymeric enzymes tend to fragment into monomeric subunits during ultrasonication. Inactivation of enzymes by ultrasound is primarily attributed to cavitation phenomenon. Indeed, the cavitation effects generated by bubble collapse (mechanical, thermal, chemical) may be sufficient to cause irreversible destruction and deactivation of enzymes (Mawson et al. 2011). In addition, the extreme agitation created by microstreaming could disrupt van der Waals interactions and hydrogen bonds in the polypeptide, causing protein denaturation (Tian et al. 2004). Most likely, more than one mechanism is operative. For instance, free radical mediated deactivation of lipoxygenase (LOX) by MTS has been suggested (Lopez and Burgos 1995) and possibly protein denaturation. Splitting of the heme group from peroxidase by MTS was reported to inactivate the enzyme (López et al. 1994), whereas the inactivation of trypsin has been partly attributed to the large interfacial area created by ultrasound, which disrupts hydrophobic interactions and hydrogen bonds (Tian et al. 2004). Generally, ultrasonication in combination with other treatments is more effective in enhancing the enzyme inactivation efficacy. In particular, simultaneous application of low-intensity ultrasound and mild heat (TS) and/or pressure (MS, MTS) is reported to increase the effectiveness of inactivation of various enzymes in food. Nevertheless, the sensitivity can vary between enzymes. For instance, Terefe et al. (2009) reported a synergistic effect of combined ultrasound (20 kHz, 65 μm amplitude) and heat (50°C–75°C) on inactivation of pectin methylesterase (PME) and polygalacturonase (PG) in tomato juice. TS increased the inactivation rate of PME by 1.5–6 fold and that of PG by 2.3–4 fold over 60°C–75°C, with the highest increase corresponding to the lowest temperature (Terefe et al. 2009). Wu et al. (2008) observed a synergistic effect of TS (24 kHz, 60°C–65°C) on the rate of inactivation of PME in tomato juice compared to thermal treatment alone. However, at 70°C, the ultrasound was not effective, and this was attributed to impairment of cavitation at this temperature (Wu et al. 2008). Ganjloo et al. (2008) compared ultrasonic blanching (20 kHz, 25% power, 80°C–95°C) with hot water blanching and found that the combined treatment provided a more rapid and effective inactivation of seedless guava peroxidase at comparable temperature and time.

8.3.3.3 Chemical Safety/Food Allergy

While high-intensity ultrasound waves interact with the food system, a part of the wave energy is absorbed and converted into mechanical and thermal energy to change the chemical and physical properties of the food system by cyclic generation and collapse of cavities by increasing pressure and temperature (Mason et al. 1992; McClements 1995). This phenomenon resulted in reversible unfolding, loss of secondary structure, formation of new intramolecular/intermolecular interactions, and rearrangements of disulfide bonds (Davis and Williams 1998). Based on this principle, it is highly possible that high-power ultrasound can induce protein structure changes affecting the allergenicity of food products. Reaction to food allergens in children and adults is prevalent in almost all countries of the world. Ultrasound could be one of the emerging technologies to assess and reduce allergenicity of major food allergens in peanut, milk, fish and shellfish, crustacean, wheat, tree nuts, and soybean. Although there are not many studies on the effects of ultrasound on food allergens, Li et al. (2006) reported that high-intensity ultrasound at 50°C produces an important effect on the integrity and structure of shrimp proteins, which corresponds to a relevant decrease (20%) on the overall allergenicity as compared to raw shrimp extracts.

Alteration in protein structure and the sequences for their allergenicity could also depend on the condition of ultrasonic treatment (e.g., ultrasonic treatment combined with heat) and ultrasound intensity. Villamiel and Jong (2000b) demonstrated that denaturation of α-lactalbumin and β-lactoglobulin in milk was higher when high-intensity ultrasound treatment was applied in combination with heat. Synergism between heat and ultrasound was explained by the reduction in viscosity of the heated milk resulting in a better penetration of ultrasound into the liquid. Better penetration resulted in better absorption in the food system. Some researchers reported that heat labile birch pollen-related allergens of hazelnuts such as Cor a 1 and Cor a 2 lost 90% immunoreactivity after being heated (K.S. Hansen et al. 2003).

8.4 Summary

As a nonthermal processing technology, ultrasound has emerged in food science and technology with diversified applications in the areas of food analysis and quality control, processing, and food safety. While the application of low-power (high-frequency) ultrasound provides a noninvasive and simple technique that can be used for estimating the food composition (fish, eggs, dairy, etc.), monitoring physicochemical and structural properties (emulsions, dairy products, and juices), and detecting contamination by metals and other foreign materials (canned food, dairy foods, etc.), high-power (low-frequency) ultrasound modifies the food properties by inducing mechanical, physical, and chemical/biochemical changes through cavitation, which reduces reaction time and increases reaction yield under mild conditions compared to the conventional route. Scattered, transmitted, and reflected acoustic pulses are used in food quality assurance. Enzymatic inactivation for quality preservation is a requisite for stabilization of some food materials. The chemical (free radical production, hot spots) and physical forces (microstreaming) generated by acoustic cavitation promote severe damage to the cell wall, resulting in the inactivation of the microorganisms. In addition to the cavitation effects, the bactericidal effect of ultrasound in liquid foods is also attributed to intracellular cavitation, which disrupts the structural and functional components up to the point of cell lysis. Over the decades, researchers were able to optimize many ultrasound applications either for the testing or processing of food products. Efforts are continued to integrate fully automated ultrasound systems to the food production lines, which will help reduce cost, save energy, and ensure the production of high value and safe food products.

An understanding of the physicochemical properties and functional properties of a specific food should guide in the selection of the appropriate ultrasound sensing or processing system in terms of probe design, geometry, and characteristics (e.g., frequency), as well as operation conditions that provide optimum results for each individual application (Knorr et al. 2011). The advantages of the technology are versatile and profitable to the food industry, yet more research efforts are still needed to design and develop efficient power ultrasonic systems that support large-scale operations and that can be adapted to various processes. These technologies can be integrated to other unit operations, such as blanching, drying, osmotic dehydration, rehydration, frying, extraction, freezing, and thawing, for improving process efficiency. However, there are many challenges in upscaling the technology for industrial purposes. Most of the studies are based on the frequency of ultrasound available commercially (20 or 40 kHz). The technology needs to be rigorously tested and proved to be safe while being commercially viable. Use of different frequencies in ultrasound along with varying parameters such as treatment temperature, time, and acoustic power should be studied. Detailed information on the properties such as densities, compressibilities, heat capacities, and thermal conductivities of a material, is needed in order to make theoretical predictions of its ultrasonic properties. Nevertheless, ultrasound can fulfill its potential through its unique use for various purposes in food safety and quality control in the coming years.

References

Álava, J.M., Sahi, S.S., García-Álvarez, J., Turó, A., Chávez, J.A., García M.J. et al. (2007). Use of ultrasound for the determination of flour quality. *Ultrasonics*, 46, pp. 270–276.

Alzamora, S.M., Guerrero, S.N., Schenk, M., Raffellini, S., López-Malo, A. (2011). Inactivation of microorganisms. In H. Feng, G. Barbosa-Canovas, J. Weiss (Eds.), *Ultrasound Technologies for Food and Bioprocessing*. New York: Springer, pp. 321–343.

Antonova, I., Mallikarjunan, P., Duncan, S.E. (2003). Correlating objective measurements of crispness in breaded fried chicken nuggets with sensory crispness. *Journal of Food Science*, 68, pp. 1308–1315.

Arroyo, C., Cebrián, G., Pagán, R., Condón, S. (2011). Inactivation of *Cronobacter sakazakii* by ultrasonic waves under pressure in buffer and foods. *International Journal of Food Microbiology*, 144, pp. 446–454.

Awad, T.S., Moharram, H.A., Shaltout, O.E., Asker, D., Youssef, M.M. (2012). Applications of ultrasound in analysis, processing and quality control of food: A review. *Food Research International*, 48, pp. 410–427.

Baloch, A.K., Buckle, K.A., Edwards, R.A. (1997). Effect of processing variables on the quality of dehydrated carrot. *Journal of Food Technology*, 12, pp. 295–307.

Benedito, J., Carcel, J.A., Gisbert, M., Mulet, A. (2001). Quality control of cheese maturation and defects using ultrasonics. *Journal of Food Science*, 66 (1), pp. 100–104.

Benedito, J., Carcel, J.A., Gonzalez, R., Mulet, A. (2002). Application of low intensity ultrasonics to cheese manufacturing processes. *Ultrasonics*, 40, pp. 19–23.

Bermúdez-Aguirre, D., Barbosa-Cánovas, G.V. (2008). Study of butter fat content in milk on the inactivation of *Listeria innocua* ATCC 51742 by thermo-sonication. *Innovative Food Science and Emerging Technology*, 9, pp. 176–185.

Bermúdez-Aguirre, D., Mawson, R., Versteeg, K., Barbosa-Canovas, G.V. (2009). Composition properties, physicochemical characteristics and shelf life of whole milk after thermal and thermo-sonication treatments. *Journal of Food Quality*, 32, pp. 283–302.

Bermúdez-Aguirre, D., Mobbs, T., Barbosa-Cánovas, G.V. (2011). Ultrasound applications in food processing. In H. Feng, G.V. Barbosa-Cánovas, J. Weiss (Eds.), *Ultrasound Technologies for Food and Bioprocessing*. New York: Springer, pp. 65–105.

Blitz, J. (1963). *Fundamentals of Ultrasonics*. London: Butterworths.

Buckin, V., O'Driscoll, B., Smyth, C. (2003). Ultrasonic spectroscopy for materials analysis: Recent advances. *Spectroscopy Europe*, 15(1), pp. 20–25, 321–343.

Burgos, J., Ordonez, J.A., Sala, F. (1972). Effect of ultrasonic waves on the heat resistance of *Bacillus cereus* and *Bacillus licheniformis* spores. *Applied Microbiology*, 24, pp. 497–498.

Butz, P., Tauscher, B. (2002). Emerging technologies: Chemical aspects. *Food Research International*, 35, pp. 279–284.

Cabeza, M.C., Ordóñez, J.A., Cambero, M.I., Hoz, L., García, M.L. (2004). Effect of thermo-ultrasonication on *Salmonella enterica* serovar *enteritidis* in distilled water and intact egg shells. *Journal of Food Protection*, 67(9), pp. 1886–1891.

Cameron, M., McMaster, L.D., Britz, T.J. (2009). Impact of ultrasound on dairy spoilage microbes and milk components. *Dairy Science and Technology*, 89, pp. 83–98.

Cisneros, F., Ellis, M., Miller, K.D., Novakofski, J., Wilson, E.R., McKeith, F.K. (1996). Comparison of transverse and longitudinal real-time ultrasound scans for prediction of lean cut yields and fat-free lean content in live pigs. *Journal of Animal Science*, 74, pp. 2566–2576.

Czank, C., Simmer, K., Hartmann, P.E. (2010). Simultaneous pasteurization and homogenization of human milk by combining heat and ultrasound: Effect on milk quality. *Journal of Dairy Research*, 77, pp. 183–189.

Davis, P.J., Williams, S.C. (1998). Protein modification by thermal processing. *Allergy*, 53, pp. 102–105.

Elmehdi, H.M., Page, J.H., Scanlon, M.G. (2003). Using ultrasound to investigate the cellular structure of bread crumb. *Journal of Cereal Science*, 38, pp. 33–42.

Fernandes, F.A.N., Oliveira, F.I.P., Rodrigues, S. (2008). Use of ultrasound for dehydration of papayas. *Food and Bioprocess Technology*, 1, pp. 339–345.

Fernandes, F.A.N., Rodrigues, S. (2008). Application of ultrasound and ultrasound-assisted osmotic dehydration in drying of fruits. *Dry Technology* 26, pp. 1509–1516.

Ganjloo, A., Rahman, R.A., Bakar, J., Osman, A., Bimakr, M. (2008). Feasibility of high-intensity ultrasonic blanching combined with heating for peroxidase inactivation of seedless guava (*Psidium guajava* L.). *Proceedings of the 18th National Congress in Food Technology*, Mashhad, Iran.

Garcia-Noguera, J., Oliveira, F.I.P., Gallão, M.I., Weller, C.L., Rodrigues, S., Fernandes, F.A.N. (2010). Ultrasound-assisted osmotic dehydration of strawberries: Effect of pretreatment time and ultrasonic frequency. *Drying Technology*, 28, pp. 294–303.

García-Reverter, J., Bourne, M.C., Mulet, A. (1994). Temperature blanching affects firmness and dehydration of dried cauliflower florets. *Journal of Food Science*, 59, pp. 1181–1183.

Ghaedian, R., Coupland, J., Decker, E.A., McClements, D.J. (1998). Ultrasonic determination of fish composition. *Journal of Food Engineering*, 35, pp. 323–337.

Gogate, P.R., Pandit, A.B. (2011). Sonocrystallization and its application in food and bioprocessing. In H. Feng, G. Barbosa-Canovas, J. Weiss (Eds.), *Ultrasound Technologies for Food and Bioprocessing*. New York: Springer, pp. 467–493.

Gómez, M., Oliete, B., García-Álvarez, J., Ronda, F. Salazar, J. (2008). Characterization of cake batters by ultrasound measurements. *Journal of Food Engineering*, 89, pp. 408–413.

Hansen, K.S., Ballmer-Weber, B.K., Luttkopf, D., Skov, P.S., Wuthrich, B., Bindslev-Jensen, C., Vieths, S., Poulsen, L.K. (2003). Roasted hazelnuts allergenic activity evaluated by doubleblind, placebo-controlled food challenge. *Allergy*, 58, pp. 132–138.

Hansen, S., Purup, S., Christensen, L. (2003). Bioactivity of falcarinol and the influence of processing and storage on its content in carrots (*Daucus carota* L). *Journal of Science of Food and Agriculture*, 83, pp. 1010–1017.

Haughton, P.N., Lyng, J.G., Morgan, D.J., Cronin, D.A., Noci, F., Fanning, S. et al. (2012). An evaluation of the potential of high-intensity ultrasound for improving the microbial safety of poultry. *Food and Bioprocess Technology*, 5(3), pp. 992–998.

Huang, E., Mittal, G.S., Griffiths, M.W. (2006). Inactivation of *Salmonella enteritidis* in liquid whole egg using combination treatments of pulsed electric field, high pressure, and ultrasound. *Biosystem Engineering*, 94(3), pp. 403–413.

Leemans, V., Destain, M.F. (2009). Ultrasonic internal defect detection in cheese. *Journal of Food Engineering*, 90(3), pp. 333–340.

Leighton, T. (2007). What is ultrasound? *Progress in Biophysics and Molecular Biology*, 93, pp. 3–83.

Li, Z., Hong, L., Li-min, C., Khalid, J. (2006). Effect of high intensity ultrasound on the allergenicity of shrimp. *Zhejiang University. Science B*, 7(4), pp. 251–256.

Kim, K.B., Lee, S., Kim, M.S., Cho, B.K. (2009). Determination of apple firmness by non-destructive ultrasonic measurement. *Postharvest Biological Technology*, 52, pp. 44–48.

Knorr, D., Froehling, A., Jaeger, H., Reineke, K., Schlueter, O., Schoessler, K. (2011). Emerging technologies in food processing. *Annual Review of Food Science and Technology*, 2, pp. 203–235.

Knorr, D., Zenker, M., Heinz, V., Lee, D.U. (2004). Applications and potential of ultrasonics in food processing. *Trends in Food Science and Technology*, 15, pp. 261–266.

Lagrain, B., Boeckx, L., Wilderjans, E., Delcour, J.A., Lauriks, W. (2006). Non-contact ultrasound characterization of bread crumb: Application of the Biot-Allard model. *Food Research International*, 39, pp. 1067–1075.

Lopez, P., Burgos, J. (1995). Lipoxygenase inactivation by manothermosonication: Effects of sonication physical parameters, pH, KCl, sugars, glycerol, and enzyme concentration. *Journal of Agricultural & Food Chemistry*, 43, pp. 620–625.

López, P., Sala, F.J., Fuente, J.L., Condón, S., Raso, J., Burgos, J. (1994). Inactivation of peroxidase, lipoxygenase, and polyphenol oxidase by manothermosonication. *Journal of Agricultural & Food Chemistry*, 42, pp. 252–256.

Mason, T.J., Lorimer, J.P., Bates, D.M. (1992). Quantifying sonochemistry: Casting some light on a 'black art'. *Ultrasonics* 30(1), pp. 40–42.

Mawson, R., Gamage, M., Terefe, M.S., Knoerzer, K. (2011). Ultrasound in enzyme activation and inactivation. In H. Feng, G.V. Barbosa-Cánovas, J. Weiss (Eds.), *Ultrasound Technologies for Food and Bioprocessing*. New York: Springer, pp. 369–404.

McClements, D.J. (1995). Advances in the application of ultrasound in food mushrooms, Brussels sprouts and cauliflower by applying power ultrasound and its rehydration properties. *Journal of Food Engineering*, 81, pp. 88–97.

Mörlein, D., Rosner, F., Brand, S., Jenderka, K.V., Wicke, M. (2005). Non-destructive estimation of the intramuscular fat content of the longissimus muscle of pigs by means of spectral analysis of ultrasound echo signals. *Meat Science*, 69, pp. 187–199.

Noci, F., Walkling-Ribeiro, M., Cronin, D.A., Morgan, D.J., Lyng, J.G. (2009). Effect of thermosonication, pulsed electric field and their combination on inactivation of *Listeria innocua* in milk. *International Dairy Journal*, 19, pp. 30–35.

Ordóñez, J.A., Aguilera, M.A., Garcia, M.L., Sanz, B. (1987). Effect of combined ultrasonic and heat treatment (thermoultrasonication) on the survival of a strain of *Staphylococcus aureus*. *Journal of Dairy Research*, 54, pp. 61–67.

Patist, A., Bates, D. (2008). Ultrasonic innovations in the food industry: From the laboratory to commercial production. *Innovative Food Science and Emerging Technologies*, 9(2), pp. 147–154.

Povey, M.J.W. (1997). *Ultrasonic Techniques for Fluids Characterization*. San Diego, CA: Academic Press.

Povey, M.J.W., Awad, T.S., Huo, R., Ding, Y.L. (2009). Quasi-isothermal crystallization kinetics, non-classical nucleation and surfactant-dependent crystallisation of emulsions. *European Journal of Lipid Science and Technology*, 111(3), pp. 236–242.

Povey, M.J.W., Mason, T.J. (Eds.). (1998). *Ultrasound in Food Processing*. London: Blackie Academic & Professional, pp. 183–192.

Raso, J., Pagán, R., Condón, S., Sala, F.J. (1998). Influence of temperature on the lethality of ultrasound. *Applied Environmental Microbiology*, 64, pp. 465–471.

Rawson, A., Tiwari, B.K., Tuohy, M.G., O'Donnell, C.P., Brunton, N. (2011). Effect of ultrasound and blanching pretreatments on polyacetylene and carotenoid content of hot air and freeze dried carrot discs. *Ultrasonics Sonochemistry*, 18, pp. 1172–1179.

Resa, P., Elvira, L., de Espinosa, F.M., González, R., Barcenilla, J. (2009). On-line ultrasonic velocity monitoring of alcoholic fermentation kinetics. *Bioprocessing Biosystem Engineering*, 32, pp. 321–331.

Riesz, P., Kondo, T. (1992). Free radical formation induced by ultrasound and its biological implications. *Free Radical Biology and Medicine*, 13(3), pp. 247–270.

Scherba, G., Weigel, R.M., O'Brien, W.D. (1991). Quantitative assessment of the germicidal efficacy of ultrasonic energy. *Applied Environmental Microbiology*, 57, pp. 2079–2084.

Schöck, T., Becker, T. (2010). Sensor array for the combined analysis of water–sugar–ethanol mixtures in yeast fermentations by ultrasound. *Food Control*, 21(4), pp. 362–369.

Simal, S., Benedito, J., Sanchez, E.S., Rossello, C. (1998). Use of ultrasound to increase mass transfer rates during osmotic dehydration. *Journal of Food Engineering*, 36, pp. 323–336.

Skaf, A., Nassar, G., Lefebvre, F., Nongaillard, B. (2009). A new acoustic technique to monitor bread dough during the fermentation phase. *Journal of Food Engineering*, 93, pp. 365–378.

Telis-Romero, J., Váquiro, H.A., Bon, J., Benedito, J. (2011). Ultrasonic assessment of fresh cheese composition. *Journal of Food Engineering*, 103, pp. 137–146.

Terefe, N.S., Gamage, M., Vilkhu, K., Simons, L., Mawson, R., Versteeg, C. (2009). The kinetics of inactivation of pectin methylesterase and polygalacturonase in tomato juice by thermosonication. *Food Chemistry*, 117, pp. 20–27.

Thompson, L.H., Doraiswamy, L.K. (1999). Sonochemistry: Science and Engineering. *Industrial and Engineering Chemistry Research*, 38(4), pp. 1215–1249.

Tian, Z.M., Wan, M.X., Wang, S.P., Kang, J.Q. (2004). Effects of ultrasound and additives on the function and structure of trypsin. *Ultrasonics Sonochemistry*, 11(6), pp. 399–404.

Ting, C.H., Kuo, F.J., Lien, C.C., Sheng, C.T. (2009). Use of ultrasound for characterising the gelation process in heat induced $CaSO_4 \cdot 2H_2O$ tofu curd. *Journal of Food Engineering*, 93, pp. 101–107.

Tiwari, B.K., Muthukumarappan, K., O'Donnell, C.P., Cullen, P.J. (2009). Inactivation kinetics of pectin methylesterase and cloud retention in sonicated orange juice. *Innovative Food Science and Emerging Technology*, 10, pp. 166–171.

Vilkhu, K., Mawson, R., Simons, L., Bates, D. (2008). Applications and opportunities for ultrasound assisted extraction in the food industry—A review. *Innovative Food Science and Emerging Technology*, 9, pp. 161–169.

Villamiel, M., Jong, P. (2000a). Inactivation of *Pseudomonas fluorescens* and *Streptococcus thermophilus* in Trypticase® soy broth and total bacteria in milk by continuous-flow ultrasonic treatment and conventional heating. *Journal of Food Engineering*, 45, pp. 171–179.

Villamiel, M., Jong, P. (2000b). Influence of high-intensity ultrasound and heat treatment in continuous flow on fat proteins, and native enzymes of milk. *Journal of Agricultural and Food Chemistry*, 48, pp. 472–478.

Walkling-Ribeiro, M., Noci, F., Cronin, D.A., Lyng, J.G., Morgan, D.J. (2009a). Shelf life and sensory evaluation of orange juice after exposure to thermosonication and pulsed electric fields. *Food and Bioproducts Processing*, 87, pp. 102–107.

Walkling-Ribeiro, M., Noci, F., Riener, J., Cronin, D.A., Lyng, D.G., Morgan, D.J. (2009b). The impact of thermosonication and pulsed electric fields on *Staphylococcus aureus* inactivation and selected quality parameters in orange juice. *Food and Bioprocess Technology*, 4(2), pp. 422–430.

Wong, E., Vaillant-Barka, F., Chaves-Olarte, E. (2012). Synergistic effect of sonication and high osmotic pressure enhances membrane damage and viability loss of salmonella in orange juice. *Food Research International*, 45(2), pp. 1072–1079.

Wu, J., Gamage, T.V., Vilkhu, K.S., Simons, L.K., Mawson, R. (2008). Effect of thermosonication on quality improvement of tomato juice. *Innovative Food Science and Emerging Technology*, 9, pp. 186–195.

Chapter 9

CA and MA Storage of Fruits and Vegetables

T.K. Goswami and S. Mangaraj

Contents

9.1 Introduction

Fruits and vegetables are good sources of carbohydrates, vitamins, minerals, and dietary fiber. Being a living biological entity, fruits and vegetables respire and transpire. Before harvest, when they are attached to the parent plant, losses due to respiration and transpiration are replenished by water, photosynthates, and minerals from the plant. After harvest, losses of respirable substrates and moisture are not replenished, and therefore, deterioration occurs soon followed by senescence or total death. The physiological and biochemical changes during respiration and transpiration are influenced by environmental factors such as temperature, ethylene, O_2, and CO_2 concentration, and in general, these biological activities cause a decline in the quality of the produce and limit its shelf life.

Controlled atmosphere (CA) and modified atmosphere (MA) storage systems involve altering and maintaining an atmospheric composition different from normal air composition. This method must be considered to supplement the maintenance of optimal conditions of temperature and relative humidity for the commodity. The underlying principle of these methods lies in reducing the respiration rate and ethylene synthesis during storage.

Respiration is a metabolic process by which organic material in living cells is continuously broken down, evolving CO_2, H_2O, and energy. The respiration rate, expressed as either rate of oxygen consumption or CO_2 evolution is dependent—besides factors such as temperature and the state of maturity of the fruits and vegetables—on the gas composition of the environment in which they are stored. In CA and MA, the fruits are stored in a depleted O_2 and elevated CO_2 environment. In CA, the gas composition is "controlled" in the sense that it is closely monitored and controlled, and MA refers to the gas exchange between the fruit and its microenvironment within a suitable enclosure. A recent approach has been made to modify both the micro- and macro-environments of storage by combining both MA and CA storage. Because there is an inverse relationship between respiration rates and post-harvest life of fresh commodities, the effects of CA and MA are beneficial.

At the commercial level, CA is most widely applied during the storage and transport of apples and pears. It is also applied, to a lesser extent, on kiwifruits, avocados, persimmons, pomegranates, nuts, and dried fruits. Atmospheric modification during long-distance transport is used for apples, avocados, bananas, blueberries, cherries, figs, kiwifruits, mangoes, nectarines, peaches, pears, plums, raspberries,

Table 9.1 Comparison Between CA and MA Storage of Fruits and Vegetables

CA Storage	*MA Storage*
• In CA storage, the atmosphere is modified, and its composition is precisely controlled according to the specific requirements of the food product throughout the storage period. The optimum level of O_2 and CO_2 concentration, temperature, and relative humidity lowers the rate of respiration and the ethylene production rate, reduces ethylene action, delays ripening and senescence, retards the growth of decay-causing pathogens, and controls insects.	• MA packaging (MAP) of fresh produce refers to the technique of sealing actively respiring produce in polymeric film packages to modify the O_2 and CO_2 concentration levels within the package atmosphere. The reduced O_2 and increased CO_2 concentration can potentially reduce respiration rate, ethylene sensitivity and production, compositional changes, decay, and physiological changes, namely, oxidation.
• In a CA storage facility, the gases are continuously monitored and adjusted artificially to maintain the optimal concentration along with temperature and relative humidity for the entire storage period. It needs a highly sophisticated machinery system, technical skill, and operative measure to establish the conditions for such a long period to achieve the goal.	• In MA, the modification of the atmosphere inside the package is achieved by the natural interplay between the two processes, the respiration of the products, and the permeation of gases through the packaging. MA involves the exposure of produce to the atmosphere generated in a package by the interaction of the produce, the package, and the external atmosphere. The initial atmosphere may be either an air or a gas mixture.
• The aim of CA storage is to artificially monitor and control the storage environment precisely to extend the storage life for a longer period preserving the natural quality of the commodity. CA takes less time to establish the desired atmospheric conditions.	• The goal of MAP is to achieve the equilibrium concentration of O_2 and CO_2 within the package by the natural diffusion process for extending shelf life and maintaining freshness. MAP is used when the composition of the atmosphere is not closely controlled.

(*continued*)

Table 9.1 (Continued) Comparison Between CA and MA Storage of Fruits and Vegetables

CA Storage	*MA Storage*
• The application of CA is for long-term storage and transportation while preserving the original quality.	• The application of MA is mainly for short-term storage, transportation, distribution, and retailing.
• This high degree of atmospheric regulation associated with CA is capital intensive and expensive to operate and thus is more appropriate for commodities that are amenable to long-term storage.	• MA has a great advantage in developing countries because it can economically be done by saving the high cost of machineries. Additionally, the need therefore for such a technique is much greater because of the dearth of refrigerated storage.
• The success of CA storage is based on the degree of control employed over the atmospheric composition. CA implies a greater degree of precision than MA in maintaining specific levels of O_2, CO_2, and other gases.	• The success of MA depends mainly on the maintenance of temperature and proper selection of the film for gas permeation. No further control is exerted over the initial composition, and the gas composition is likely to change with time owing to the diffusion of gases into and out of the product, the permeation of gases into and out of the pack, and the effects of product and microbial metabolism.
• The shelf life is extended for a longer period.	• The shelf life is lesser compared to CA storage.

and strawberries. Technological developments geared toward providing CA during transport and storage at reasonable cost (positive benefit/cost ratio) are essential if the application of this technology to fresh fruits and vegetables is to be expanded.

Although CA and MA storage have both been shown to be effective in extending the post-harvest life of many commodities, their commercial applications have been limited by the relatively high cost of these technologies. There are, however, a few cases in which a positive return on investment (cost/benefit ratio) can be demonstrated. In a comparison of losses due to decay during retail marketing of strawberries shipped in air and those shipped in an environment consisting of 15%

CO_2-enriched air (MA within pallet cover), the use of MA was observed to reduce losses by 50% (an average of 20% losses was sustained in strawberries shipped in air versus 10% losses in those shipped by MA). The economic loss of 10% value (US$50–75 per pallet) was much greater than the cost of using MA (US$15–25 per pallet). Use of CA during marine transportation can extend the post-harvest life of those fruits and vegetables that would normally have a short post-harvest life potential, thereby allowing the use of marine transportation instead of air transport for the shipment of such produce. In terms of cost and benefit, savings realized with the use of marine transportation are much greater than is the added cost of CA service (Table 9.1).

Many green vegetables and most horticultural produce are quite sensitive to ethylene damage. Their exposure to ethylene must therefore be minimized. Ethylene contamination in ripening rooms can be minimized by (i) using ethylene levels of 100 ppm instead of the higher levels often used in commercial ripening operations, (ii) venting ripening rooms to the outside on completion of exposure to ethylene, (iii) at least once per day ventilating the area around the ripening rooms or installing an ethylene scrubber, and (iv) use of battery-powered forklifts instead of engine-driven units in ripening areas (Table 9.2).

Ethylene-producing commodities should not be mixed with ethylene-sensitive commodities during storage and transport. Potassium permanganate, an effective oxidizer of ethylene, is commercially used as an ethylene scrubber. Scrubbing units based on the catalytic oxidation of ethylene are used to a limited extent in some commercial storage facilities.

Table 9.2 Classification of Horticultural Crops According to Their CA Storage Potential at Optimum Temperatures and RH

Storage Period (Month)	*Commodities*
> 12	Almond, Brazil nut, cashew, filbert, macadamia, pecan, pistachio, walnut, and dried fruits and vegetables
6 to 12	Some cultivars of apples and European pears
3 to 6	Cabbage, Chinese cabbage, kiwifruit, persimmon, pomegranate, and some cultivars of Asian pears
1 to 3	Avocado; banana; cherry; grape (no O_2); mango; olive; onion (sweet cultivars); some cultivars of nectarine, peach, and plum; and tomato (mature green)
< 1	Asparagus, broccoli, cane berries, fig, lettuce, muskmelons, papaya, pineapple, strawberry, sweetcorn, fresh-cut fruits and vegetables, and some cut flowers

9.2 Ripening Pattern of Fruits and Vegetables

Basically there are two distinct patterns of ripening that can be identified, and these are termed as climacteric and non-climacteric. The difference between the two ripening patterns is shown in Figure 9.1.

In climacteric fruits, the respiration rate shows a decreasing trend to the lowest value termed the pre-climacteric minimum followed by a sharp rise in respiration rate to the climacteric peak. This sudden upsurge is called respiratory climacteric, and such fruits are known as climacteric fruits. Subsequently, the respiration rate decreases. Therefore, respiratory climacteric is defined as a period in the ontogeny of fruit during which a series of biochemical changes are initiated by the autocatalytic production of ethylene, making the change from growth to senescence and involving an increase in respiration rate leading to ripening of the fruit (Biale 1964). Non-climacteric fruits, as depicted in Figure 9.1, show neither a rise in respiration rate nor an associated production of ethylene during the ripening process. Usually climacteric fruits are harvested in a mature but unripe condition, and then they are allowed to ripen either naturally or by using certain chemicals.

The climacteric and non-climacteric fruits are distinguishable by their response to treatment with ethylene. In immature climacteric fruits, ethylene treatment, as low as 0.1 to 10 ppm, reportedly hastens the onset of climacteric, and the associated ripening changes without appreciably altering the pattern or the magnitude of respiratory change. However, in non-climacteric fruits, it led to an increase in ethylene production and respiration with no correlation with the inception of ripening. In climacteric fruits, ethylene application is effective only during the pre-climacteric stage whereas, in non-climacteric fruits, stimulation in the rate of respiration has

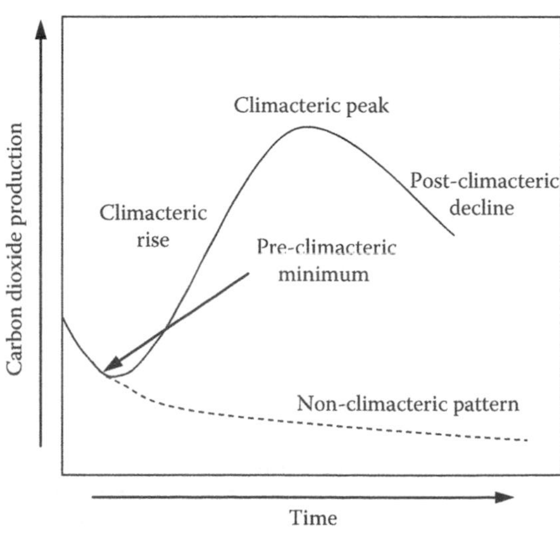

Figure 9.1 Ripening patterns for fruits and vegetables.

been observed at any stage. Once climacteric fruits start to ripen, there is very little that can be done, except to market them for immediate consumption (Ryall and Pentzer 1982). Hence, it is recommended that fruits should be stored before the onset of climacteric rise. A list of common climacteric and non-climacteric fruits is given in Table 9.3.

Table 9.3 Classification of Fruits According to Respiratory Behavior During Ripening

Climacteric Fruits		*Non-climacteric Fruits*	
Apple	Papaya	Blackberry	Loquat
Apricot	Passion fruit	Blueberry	Lychee
Avocado	Peas	Cacao	Okra
Banana	Pear	Caju	Olive
Biriba	Persimmon	Carambola	Orange
Blueberry	Plantain	Cashew apple	Peas
Breadfruit	Plum	Cherry	Pepper
Cherimoya	Quince	Cucumber	Pineapple
Fig	Rambutain	Date	Pomegranate
Guava	Sapodilla	Eggplant	Prickly pear
Jackfruit	Sapota	Grape	Raspberry
Kiwifruit	Soursop	Grapefruit	Strawberry
Mango	Tomato	Jujube	Summer squash
Muskmelon	Watermelon	Lemon	Tamarillo
Nectarine		Lime	Tangerine
		Longan	

Source: Adapted from Kays, S.J., Metabolic processes in harvested products respiration. In *Post Harvest Physiology of Perishable Plant Products*, Van Nostrand Reinhold Publication, New York, 1991; Salunke, D.D. and Kadam, S.S. (eds.) *Handbook of Fruit Science and Technology*, Marcel Dekker Inc., New York, p. 550, 1995. Saltveit, M.E., *Commercial Storage of Fruits, Vegetables and Florist and Nursery Crops*, Postharvest Technology Centre RIC, Department of Plant Science, University of California, USA. Irtwange, S.V., *Agricultural Engineering International: the CIGR E-journal*. 4 (8), 2006.

9.3 Controlled and Modified Atmospheric Storage

The positive influence of CA and MA storage on the shelf life of an agricultural commodity also stems from its influence on the various enzymatic changes taking place in the fruit during ripening (Kader 1987). Elevated carbon dioxide atmospheres inhibit activity of 1-aminocyclopropane 1-carboxylic acid synthase (key regulatory site for ethylene synthesis) while ACC oxidase activity is stimulated at low carbon dioxide and inhibited at elevated carbon dioxide and/or low oxygen levels. Elevated carbon dioxide atmospheres, thus, inhibit ethylene action. Optimum atmospheric compositions retard biosynthesis of carotenoids and anthocyanins and biosyntheis and oxidation of phenolic compounds. CA also slows down the activity of cell wall-degrading enzymes, influencing softening of fruit tissue. Low oxygen and/or elevated carbon dioxide levels are also known to influence flavor by reducing loss of acidity, starch-to-sugar conversion, sugar inter-conversion, and biosynthesis of flavor volatiles. A word of caution, however, needs to be exercised while choosing the exact limits for oxygen and carbon dioxide levels. Levels below 1% oxygen and more than 15% carbon dioxide have been found to result in irreversible detrimental effects in the commodity quality. The potential benefit of CA and MA storage on fruits and vegetables are summarized in Tables 9.4 and 9.5.

- Produce in CA/MA storage maintains freshness and extends shelf life from several days to several weeks, compared to conventional storage.
- Reduction of oxygen and the increment of carbon dioxide environment suppresses the respiration rate of the commodity, thereby slowing vital processes and prolonging the maintenance of post-harvest quality.
- Reduction of respiration rate, loss of moisture, production of metabolic heat, yellowing, browning decay, and ethylene sensitivity and production takes place.
- Delay of ripening takes place.
- Reduction of physiological injury, disorder, and pathological deterioration takes place.
- Quality advantages, such as color, moisture, flavors, and maturity retention, occur.
- Reduction of fungal growth and diseases is common.
- Increased shelf life allowing less frequent loading of retail display on shelves occurs.
- In CA/MA storage, little or no chemical preservatives are used.
- Reduction of handling and distribution of unwanted or low-grade produce.
- Quality advantages transferred to the consumer.

Table 9.4 Potential Benefit of CA and MA Storage for Fresh Fruits

Commodity	*Temperature (°C)*	*MA/CA*		*Potential for Benefit*
		% O_2	*% CO_2*	
Apple	0–3	1–3	1–5	Excellent
Apricot	0–5	2–3	2–3	Fair
Avocado	5–13	2–5	3–10	Good
Banana	12–15	2–5	2–5	Excellent
Fig	0–5	5–10	15–20	Good
Grape	0–2	2–5	1–3	Fair
Guava	10–15	2–5	2–5	Good
Kiwifruit	0–5	1–2	3–5	Excellent
Lemon	10–15	5–10	0–10	Good
Lime	10–15	5–10	0–10	Good
Lychee	0–2	2–3	2–5	Good
Nectarine	0–5	1–2	3–5	Good
Olive	5–10	2–3	0–1	Fair
Orange	5–10	5–10	0–5	Fair
Mango	10–15	3–5	5–10	Fair
Papaya	10–15	3–5	5–10	Fair
Peas	0–5	1–2	3–5	Good
Pear, Asian	0–5	2–4	0–1	Good
Pear, European	0–5	1–3	0–3	Excellent
Persimmon	0–5	3–5	5–8	Good
Pineapple	8–13	2–5	5–10	Fair
Plum and prune	0–5	1–2	0–5	Good
Raspberry	0–3	5–10	15–20	Excellent
Strawberry	0–2	5–10	15–20	Excellent
Sweet cherry	0–2	3–10	10–15	Good
Nuts and dried fruits	0–25	0–1	0–100	Excellent

Source: Adapted from Parry, R.T., Principles and Application of Modified Atmosphere Packaging of Food. R.T. Parry (ed.) Blackie (Chapman and Hall), UK, pp. 1–18, 1993; Kader, A.A., Proceedings of the 7th International Controlled Atmosphere Research Conference. A.A. Kader (ed.). Postharvest Hort. Series 17 (2): 1–36, University California, Davis, CA, USA, 3, 1–34, 1997; Mahajan, P.V., Studies on controlled atmosphere storage for apple and litchi using liquid nitrogen. Unpublished PhD Thesis. IIT Kharagpur. India, 2001; Irtwange, S.V., *Agricultural Engineering International: the CIGR E-journal*. 4 (8), 2006.

Table 9.5 Potential Benefit of CA and MA Storage for Fresh Vegetables

Commodity	*Temperature (°C)*	*MA/CA Storage*		*Potential for Benefit*
		% O_2	*% CO_2*	
Artichokes	0–5	2–3	2–3	Good
Asparagus	0–5	15–20	5–10	Excellent
Beans	5–10	2–3	4–7	Fair
Beets	0–5	2–5	2–5	Fair
Broccoli	0–3	1–2	5–10	Excellent
Brussels sprouts	0–5	1–2	5–7	Good
Cabbage	0–5	2–3	3–7	Excellent
Cantaloupes	3–7	3–5	10–15	Good
Carrots	0–5	3–5	2–5	Fair
Cauliflower	0–2	2–3	2–5	Fair
Celery	0–5	1–1	0–5	Good
Corn, sweet	0–5	2–4	5–10	Good
Cucumbers	8–12	3–5	0–2	Fair
Honeydews	10–12	3–5	0–2	Fair
Leeks	0–5	1–2	3–5	Good
Lettuce	0–5	1–3	0–3	Good
Mushroom	0–3	Air	10–15	Fair
Okra	8–12	3–5	0–2	Fair
Onions, dry	0–5	1–2	0–5	Good
Onions, green	0–5	1–2	10–20	Fair
Peppers, bell	8–12	3–5	0–2	Fair
Peppers, chili	8–12	3–5	0–3	Fair
Potatoes	4–10	2–3	2–5	Fair
Radish	0–5	1–5	2–3	Fair
Spinach	0–5	18–21	10–20	Good
Tomato	15–20	3–5	0–3	Good

Source: Adapted from Parry, 1993; Irtwange, S.V., *Agricultural Engineering International: the CIGR E-journal*. 4 (8), 2006.

9.4 Principle of CA Storage

In CA storage, the temperature needs to be reduced to the desired level without delay. Rapid cooling is required to reduce the respiration rate and arrest the physiological changes that take place in the fruit, thus enhancing its storage life. Rapid cooling is difficult to accomplish with a conventional refrigeration system, necessitating more refrigeration capacity (Watkins et al. 1997). Once the commodity is cooled, only about 10% of the installed refrigeration capacity is utilized to maintain the storage temperature while the rest of the evaporator capacity remains idle during this maintenance phase (Waelti 1994). Hence, it would be very useful to have a system wherein the initial cool down of the storage facility is achieved rapidly and the use of cryogenic fluids, such as liquid nitrogen, is a viable option in this respect (Mahajan and Goswami 1997).

Liquid nitrogen is a colorless, odorless cryogenic liquid with a boiling point of –195.6°C and a latent heat of 199.58 $kJkg^{-1}$ (ASHRAE Handbook 1990). Cooling with liquid nitrogen is obtained by simply allowing the cryogenic fluid to expand at atmospheric pressure. The sensible heat obtained from the expanded cold nitrogen gas has excellent potential to be used as a cooling medium. Moreover, due to its inertness and high expansion ratio of 646 between liquid and gaseous nitrogen (ASHRAE Handbook 1990), the vaporized nitrogen gas displaces the oxygen gas from the storage space. Thus, the utilization of liquid nitrogen for CA storage seems to have multiple benefits because alongside the prospect of rapid cool down of the storage space, liquid nitrogen also accomplishes a pulling down of the oxygen concentration of the storage space.

Exposure of fruits and vegetables to harsh temperatures leads to poor ripening and loss of flavor and texture of the fruit (Nair et al. 2003). Because liquid nitrogen is a refrigerant of very low temperature, while using a system with liquid nitrogen for the cooling of a commodity, it is necessary to ensure that the fruit is not overtly stressed and liable to chilling injury.

Extreme conditions of oxygen and carbon dioxide have been reported to affect the biochemical and physiological changes in the commodity during CA storage. Also, the level of damage caused depends on the plant material and varies between the cultivar of the same crop. No one atmosphere is best for all produce, and a suitable atmosphere needs to be determined for the specific crop variety. The main consideration while identifying a storage protocol for any commodity would be quality and acceptability at the end of storage. Hence, to identify the optimal storage gas composition, it is essential to understand the influence the gases have on the quality of the commodity. Also, uniformity of gas composition and temperature distribution within the storage facility needs to be ensured for the optimal design of the storage process.

9.5 Principle of MA Storage

MA storage is a technique used for prolonging the shelf life of fresh or minimally processed foods by changing the composition of the air surrounding the food in the package. The storage life of food products is considerably extended by lowering oxygen concentration and enhancing the carbon dioxide concentration surrounding the food, which reduces the respiration rate of food products and activity of insects or microorganisms in food (Jayas and Jeyamkondan 2002). The terms MA and CA storage are used interchangeably. They differ based on the degree of control exerted over the atmospheric composition. In MA storage, the gas composition is modified initially, and it changes dynamically depending on the respiration rate of the food product and permeability of film or storage structure surrounding the food product. In CA storage, the gas atmosphere is continuously controlled throughout the storage period (Gorris and Tauscher 1999; Saltveit 2003). MA packaging involves the exposure of produce to the atmosphere generated in a package by the interaction of the produce, the package, and the external atmosphere. The initial atmosphere may be either air or a gas mixture. Different additives that may affect the atmosphere may be introduced into the package before it is sealed.

The aim of MA storage is to achieve the equilibrium concentration of oxygen and carbon dioxide within the package within the shortest possible time because of the interaction of the produce, the package, and the external atmosphere, and these concentrations lie within the desired level required, which remains nearly constant throughout the period of storage to continue the metabolic process at a minimum possible rate for maintaining freshness and maximum shelf life of stored commodity. The more complex forms of MA storage aim at achieving a closely specified atmosphere of the package by carefully selecting many relevant parameters and orchestrating them harmoniously to rapidly achieve the desired gas composition. The parameters should be selected so that the atmosphere is maintained for as long a period as the packaged commodity requires. Thus, this type of MA storage is more carefully controlled than CA storage and makes much greater demands because all the controls have to be programmed into the package before it is sealed.

MA storage is a dynamic system during which respiration and permeation occur simultaneously. Factors affecting both respiration and permeation must be considered when designing a package. The commodity mass kept inside the package, storage temperature, oxygen, and carbon dioxide partial pressures and stage of maturity are known to influence respiration in a package (Kader et al. 1989; Das 2005). Thickness, surface area of the packaging film that is exposed to atmosphere and across which permeation of oxygen and carbon dioxide takes place, and volume of void space present inside the package as well as temperature, relative humidity, and gradient of oxygen and carbon dioxide partial pressures across the film are known determinants of permeation (Das 2005). In a MA storage system, fresh fruits are sealed in permselective polymeric film packages. Due to respiration of the packaged fruits, oxygen starts depleting, and carbon dioxide starts accumulating within the package because of the consumption of oxygen and the production of carbon dioxide in the respiration process. Consequently, respiration begins to decrease while oxygen and carbon dioxide concentration gradients between package and ambient atmosphere begin to develop. The development of concentration gradients induces ingress of oxygen and egress of carbon dioxide through the polymeric films. Thus, the cyclic process continues until a steady state is established. In a properly designed MA storage system, after a period of transient state, an equilibrium state is established. At equilibrium, the amount of oxygen entering into the package and that of carbon dioxide permeating out of the package become equal to the amount of oxygen consumed and that of carbon dioxide evolved by the packaged commodity, respectively. The package atmosphere is then considered to be in dynamic equilibrium with the external atmosphere.

9.6 Respiration of Fruits and Vegetables

Respiration is an important physiological process of fruits and vegetables and involves oxidative breakdown of organic reserves to simpler molecules, including carbon dioxide and water with the release of energy (Kays 1991). Oxygen required for the respiration reaction comes from the surrounding air. It diffuses into the fruit and is dissolved in the cell solution where the reaction takes place. Carbon dioxide formed increases carbon dioxide concentration of the cell solution, and the water formed becomes part of the cell water (Ryall and Pentzer 1982). The organic substrates broken down in this process may include carbohydrates, lipids, and organic acids. The process consumes oxygen in a series of enzymatic reactions. Glycolysis, the tricarboxilic acid cycle, and the electron transport system are the metabolic pathways of aerobic respiration.

Respiratory gas exchange depends on the type of substrate present in the produce, and the dynamics of gas exchange and the end products of respiration also depend on the organic reserve available for metabolism (Tables 9.6 and 9.7). Depending upon the organic reserve being oxidized, the RQ values for fresh produce normally range from 0.7 to 1.3 (Fonseca et al. 2002). Very high values of RQ or a sudden shift in RQ value indicates a shift in the respiration cycle to the anaerobic cycle (Saltveit 2004). Respiration rate is commonly expressed as rate of oxygen consumption and/or carbon dioxide production per unit weight of the commodity. Phan et al. (1975) reported that the high rate of respiration is usually associated with short life, and the study of factors affecting respiration rate is of great significance from the standpoint of handling and storage.

Do and Salunkhe (1975) pointed out that CA can influence respiration, one of the important metabolic processes of fruits, on three levels: (a) aerobic respiration, (b) anaerobic respiration, and (c) a combination of the two. The rate of respiration and loss of carbohydrate considerably get reduced in a CA atmosphere than conventional refrigerated storage. In some cases, the average rate of loss of carbohydrates

Table 9.6 Optimum Conditions of CA/MA Storage for Some Fruits and Their Shelf Life

Commodity	*Storage Temperature (°C)*	*Optimum MA/CA*		*Injuries Atmosphere*		*Marketable Life (Days)*		*Major Benefit under MA/CA Storage*	*Commercial Potential*
		% O_2	*% CO_2*	*% O_2*	*% CO_2*	*RA Storage*	*CA Storage*		
Apple	0–3	3	3	2	10	200	300	Maintains firmness and acidity	Excellent
Avocado	7	2–5	3–10	1	15	12	56	Delays softening	Good
Banana	12–15	2	5	1	8	21	60	Suppression of climacteric pattern	Excellent
Grapes	0–2	3–5	1–3	1	10	40	90–100	Disease control	Fair
Guava	12–15	2–5	2–5	2	12	15–20	45	Delays ripening and chilling injury	

Lemon	15	3–5	0–5	1	6	130	220	Green color retention	Good
Lychee	0–5	3–5	3–5	2	14	20–30	2230	Delay in ripening	Good
Mango	13	3–5	5–8	2	8	14–28	21–45	Delayed ripening	Good
Orange	5–10	10	5	5	5	42	84	Maintenance of firmness	Fair
Papaya	13	3–5	5–8	2	8	14–28	21–35	Less decay	Fair
Pears	0–1	2–3	0–1	1	2	200	300	Delays the flesh and core browning	Excellent
Pineapple	10–15	2–5	10	2	10	12	10–15	Reduces chilling injury	Fair
Strawberry	0	4–10	15–20	1	12	7	7–15	Less decay	Excellent

Source: Adapted from Kader, *Proceedings of the 7th International Controlled Atmosphere Research Conference*. Postharvest Hort. Series 17 (2): 1–36, 1997; University California, Davis, CA, USA, 3, 1–34; Mahajan, P.V., Studies on controlled atmosphere storage for apple and litchi using liquid nitrogen. Unpublished PhD Thesis. IIT Kharagpur. India, 2001; Irtwange, S.V., *Agricultural Engineering International: the CIGR E-journal*. 4 (8), 2006.

Table 9.7 Various Types of Respiratory Substances Involved in Respiration Equation

Substrate	*Respiration Reaction*	*Example*
Palmitic acid	$C_{16}H_{32}O_2 + 11O_2 \rightarrow C_{12}H_{22}O_{11} + 4CO_2 + 5H_2O$	Oilseeds
Malic acid	$C_4H_6O_5 + 3O_2 \rightarrow 4CO_2 + 3H_2O$	Apple
Glucose	$C_6H_{12}O_6 + 6O_2 \rightarrow 6CO_2 + 6H_2O$	Mango

by respiration was between 1.2 and 1.4 times in normal air than that in 10% oxygen and between 1.35 to 1.55 times in the absence of carbon dioxide than that in the presence of 10% carbon dioxide (Do and Salunkhe 1975). Metlitskii et al. (1983) revealed that concentration of oxygen in the atmosphere substantially affects the process of respiration in fruits. The intensity of respiration has a hyperbolic relationship with oxygen concentration in some fruits.

An oxygen concentration of 5% considerably delays the climacteric rise in respiration compared to fresh air for most of the fruits. The decrease in respiration rate at low oxygen is due to a reduced oxygen supply for respiratory reactions. Talasila et al. (1992) concluded that the oxygen consumption rate of strawberries increased exponentially with an increase in the temperature in the range of 5°C to 20°C and decreased with any decrease in oxygen concentrations from 20% to 1%.

9.7 Heat of Respiration

Toledo et al. (1969) have recommended that the heat of respiration (MJ/ton-day) of fresh produce exposed to CA be estimated as 30% of its normal air composition at the same temperature. A large amount of heat is released during respiration as latent and sensible heat. The produce temperature tends to increase logarithmically if it is not properly ventilated or chilled as a result of sensible heat generated in respiration (Ciobanu et al. 1976). Raghavan et al. (1984) found that the heat of respiration could be lowered by 30% in a silicone membrane type CA system, resulting in a reduced load on the refrigeration unit. They noted 10% overall energy conservation in CA storage over conventional refrigeration techniques.

Living fruits and vegetables evolve heat owing to respiration. This tends to increase the product's surface temperature and thereby acts as a driving force for moisture transfer and deterioration. Besides maintaining field-fresh quality and increasing the storage life of many perishables, CA storage reduces the amount of heat produced by the product. Kang and Lee (1998) proposed a model for predicting the transpiration rate based on heat and mass balance and a model for heat of respiration based on the oxidation of glucose. Hoang et al. (2001) validated an equation, in which the heat of respiration is a function of the temperature of storage as in Equation 9.1.

$$Q_R = 0.32\ T_p^2 - 159.25\ T_p + 19757 \tag{9.1}$$

where Q_R is the respiration heat (W kg^{-1}), and T_p is the product temperature (K).

9.7.1 Measurement of Respiration Rate

Determination of respiration rate is useful when investigating the physiology of many plant products. Changes in respiration rate may indicate the stage of ripening in climacteric fruit. A continuously high rate of respiration is often associated with a shortened shelf life. Respiration measurements may also be used to calculate the respiratory quotient (RQ), which is the ratio of carbon dioxide produced to oxygen consumed. Respiration rate can be determined by measuring the consumption of oxygen and/or evolution of carbon dioxide. The respiratory gas exchange rates have been expressed as the amount of gas consumed or evolved per unit time on the basis of fresh weight of tissue. Amount of gases have been expressed both in terms of volume as well as weight (Kays 1991).

The usual methods of respiration rate determination are closed or static system and flow-through or flushed system. In a closed system, a gas-tight container of known volume is filled with the product and flushed with air of known composition (usually ambient), and all the inlet and outlet valves are closed (Figure 9.2). After an interval of time, gas samples from the container are analyzed for oxygen and carbon dioxide concentration. The respiration rate is calculated from the concentration

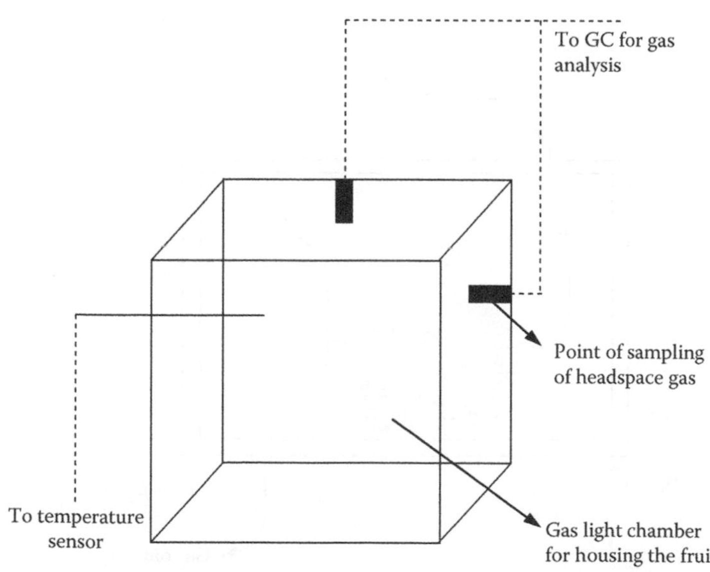

Figure 9.2 Schematic presentation of a closed system respirometer.

difference, weight of produce, and free volume of chamber. The respiration rates in terms of oxygen and carbon dioxide at a given temperature are calculated using the following equations as given by Kays (1991):

$$R_{O_2} = \left[\frac{(Y_{O_2})_t - (Y_{O_1})_{t+1}}{\Delta t}\right]\frac{V_f}{W} \tag{9.2}$$

$$R_{CO_2} = \left[\frac{(Z_{CO_2})_t - (Z_{CO_1})_{t+1}}{\Delta t}\right]\frac{V_f}{W} \tag{9.3}$$

where R_{O_2} is the respiration rate (ml), [O_2] kg^{-1} h^{-1}; R_{CO_2} is the respiration rate (ml), [CO_2] kg^{-1} h^{-1}; Y_{O_2} and Z_{CO_2} are the gas concentrations (%) for oxygen and carbon dioxide, respectively; t is the storage time (h); Δt is the time difference between two gas measurements; V_f is the free volume of the respiration chamber (ml); and W is the weight of the fruit (kg).

In the flow-through system (Figure 9.3), the product is enclosed in an impermeable container through which a gas mixture flows at a constant rate. The respiration rates are calculated from the absolute differences in gas concentrations between the outlet and the inlet when the system reached a steady state.

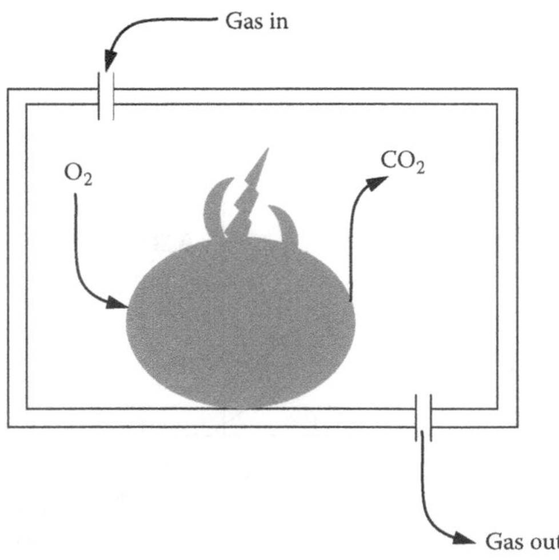

Figure 9.3 Flow-through system with a commodity enclosed in a respirometer.

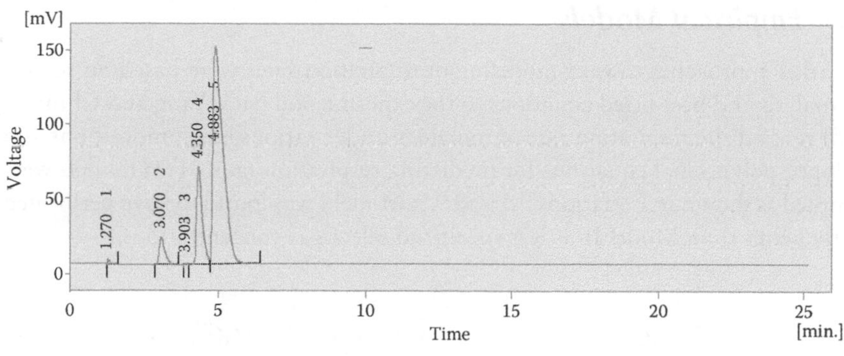

Figure 9.4 GC–TCD chromatogram for CO_2, O_2, and N_2, respectively.

9.7.2 Analysis of Respiratory Gases

Gas chromatography is the commonly reported method for quantitative and qualitative estimation of carbon dioxide and oxygen in a given sample. For measurement of respiratory gases, the columns used are a molecular sieve for oxygen and Porapak for carbon dioxide estimation. In recent times, digital analyzers have also been employed for gas analysis. Such systems employ sensors for detection of gases in a sample. The principle of detection for oxygen and carbon dioxide is as given below:

Carbon Dioxide: Sensing is performed by passing the gas through an enclosed sensor that utilizes an infrared energy source, an infrared filter, and thermopile detector. The thermopile is a voltage-generating device that outputs a signal representing the carbon dioxide level present.

Oxygen: The basis of the oxygen-measuring subsystem is a solid-state oxygen ion–conduction material called zirconium oxide. Due to oxygen vacancies in the ceramic lattice structure of this material, oxygen ions are able to move in the solid material at an elevated temperature. This property enables the measurement of oxygen in a gas of unknown composition. The chromatogram for carbon dioxide and oxygen during gas analysis using GC is shown in Figure 9.4.

9.8 Modeling of Respiration Rate

Various mathematical models for prediction of respiration rates of fruits and vegetables have been developed. These models incorporate those storage parameters, which influence respiration significantly. The models are used to predict respiratory behavior of the fruit under various storage conditions.

9.8.1 Empirical Models

The earlier approaches toward modeling of respiration rates were based on regression analysis and best-fitted equations to the experimental data. Yang and Chinnan (1988) studied the respiration rate of tomatoes under various gas compositions and developed polynomial equations for predicting respiration rates. Two models were attempted as shown in Equations 9.4 and 9.5. Model I was found to have performed slightly better than Model II as it is specific to select gas concentrations.

$$\begin{aligned} RR = \alpha_0 + \alpha_1 G_{O_2} + \alpha_2 G_{CO_2} + \alpha_3 t + \alpha_4 G_{O_2}^2 + \alpha_5 G_{CO_2}^2 \\ + \alpha_6 t^2 + \alpha_7 G_{O_2} G_{CO_2} + \alpha_8 G_{O_2} t + \alpha_9 G_{CO_2} t \end{aligned} \tag{9.4}$$

$$RR\left(G_{O_2}, G_{CO_2}\right) = \beta_0 + \beta_1 t + \beta_2 t^2 \tag{9.5}$$

where t is the storage time (h), $\alpha_{i,\, i=0 \text{ to } 9}$ are parameters, $\beta_{i,\, i=0 \text{ to } 2}$ are also parameters, and R is the respiration rate (ml kg^{-1} h^{-1}). Cameron et al. (1989) used Equation 9.6 to fit the data of oxygen depletion with respect to time. The first derivative of oxygen depletion rate was determined as in Equation 9.7. The rate of respiration was then calculated using Equation 9.8.

$$G_{O_2} = a\left[l - e^{-(b+ct)^a}\right] \tag{9.6}$$

$$\frac{dG_{O_2}}{dt} = acd\left[(b+ct)^{d-l}\right]e^{-(b+ct)^a} \tag{9.7}$$

$$RR = \frac{dG_{O_2}}{dt} V_f W^{-1} \tag{9.8}$$

where a, b, c, and d are constants; V_f is the void volume of jar (ml); and W is the weight of fruit (kg).

Talasila et al. (1992) studied the effect of temperature, oxygen concentration, and carbon dioxide concentration on the respiration rate of strawberries and developed the following empirical model for predicting respiration rate:

$$RR = c^{kT}\left[a_0 + a_1 G_{O_2} + a_2 G_{O_2}^2 + a_3 G_{CO_2}^2 + a_4 G_{O_2} G_{CO_2}\right] \tag{9.9}$$

where T is the temperature of the product (°C), $a_{i,\, i = 0 \text{ to } 4}$ are model parameters, and k is also a model parameter.

9.8.2 Models Based on Theoretical Approach

Makino et al. (1996a) proposed the following model based on the Langmuir theory of adsorption:

$$RR = \frac{a \times b \times G_{O_2}}{l + \left(a \times G_{O_2}\right)} \tag{9.10}$$

where R is the oxygen consumption rate of the produce (mmol kg^{-1} h^{-1}), b is the maximum oxygen consumption rate (mmol kg^{-1} h^{-1}), G_{O_2} is the oxygen concentration (kPa) and a is the rate parameter (kPa^{-1}).

The model was found to be suitable for describing the oxygen consumption rate of several kinds of fresh produce and valid for the design of the MAP systems. However, the model was derived for an atmosphere containing a negligible amount of carbon dioxide. Kader (1986) reported that carbon dioxide gas depresses the respiration rate of fresh produce. This effect was duly taken care of, and a respiration model considering the depressing effect of carbon dioxide gas was proposed as follows (Makino et al. 1996b):

$$RR = \frac{a \times b \times G_{O_2}}{l + \left(a \times G_{O_2}\right) + \left(a \times i \times G_{O_2} \times G_{CO_2}\right)} \tag{9.11}$$

where G_{CO_2} is the carbon dioxide concentration (kPa), and i is the rate parameter (kPa^{-1}).

9.8.3 Regression Models

9.8.3.1 Time-Dependent Model

Instantaneous respiration rates may theoretically be obtained from plotting gas concentrations versus time on a graph paper and measuring the slopes. This method, however, is not recommended because many data sets have large experimental variations. Instead, a regression function is often used to fit the data of gas concentration versus time, and the respiration rate at a given time is determined from the first derivative of the regression function (Hagger et al. 1992; Mahajan and Goswami 2001). By using the generated experimental respiration data, a nonlinear regression analysis was done to fit oxygen concentration and carbon dioxide concentration at different storage periods. The resultant regression equations for oxygen consumption and carbon dioxide evolution are shown in Equations 9.12 and 9.13 to determine the values of the coefficients a and b.

$$Y_{O_2} = 0.21 - \left[\frac{t}{a \times t + b}\right] \tag{9.12}$$

$$Z_{CO_2} = \left[\frac{t}{a \times t + b}\right] \tag{9.13}$$

where a and b are the regression coefficients, t is storage period in h, and Y_{O_2} and Z_{CO_2} are the gas concentrations for oxygen and carbon dioxide, respectively. The first derivative of the regression functions were used to determine the rate of change of gas concentration as outlined in Equations 9.14 and 9.15.

$$\frac{d(Y_{O_2})}{dt} = a \times t(a \times t + b)^{-2} - (a \times t + b)^{-1} \tag{9.14}$$

$$\frac{d(Z_{CO_2})}{dt} = -a \times t(a \times t + b)^{-2} + (a \times t + b)^{-1} \tag{9.15}$$

The respiration rate of the sample at any given time was then calculated by substituting the values of dY_{O_2}/dt and dZ_{CO_2}/dt obtained from Equations 9.14 and 9.15 in Equations 9.16 and 9.17, respectively.

$$R_{O_2} = -\frac{d[(Y_{O_2})]}{dt}\frac{V_f}{W} \tag{9.16}$$

$$R_{CO_2} = \frac{d[Z_{CO_2}]}{dt}\frac{V_f}{W} \tag{9.17}$$

The temperature dependence of the model coefficients a and b were estimated by linear interpolation between the two temperatures.

9.8.3.2 Two, Three, and Four Parameters Model

Mahajan (2001) studied the respiration rate of apple and lychee under various gas compositions and temperatures, and Das (2005) developed two parameter regression models for predicting respiration rates for apple and lychee. These models take into account the changes in oxygen concentration and carbon dioxide concentration with storage time at a specified temperature. The respiration rates at any combination of oxygen and carbon dioxide concentrations at a specified temperature can be predicted by using the values of the regression constants. The value of R_{O_2} and R_{CO_2} can be expressed as a function of oxygen and carbon dioxide concentrations by a second-order regression equation of the following type:

$$R_{O_2} = b_0 + b_1 Y_{O_2} + b_2 Z_{CO_2} + b_3 Y_{O_2}^2 + b_4 Z_{CO_2}^2 + b_5 Y_{O_2} Z_{CO_2} \quad (9.18)$$

$$R_{CO_2} = c_0 + c_1 Y_{O_2} + c_2 Z_{CO_2} + c_3 Y_{O_2}^2 + c_4 Z_{CO_2}^2 + c_5 Y_{O_2} Z_{CO_2} \quad (9.19)$$

where b_0, b_1, b_2, b_3, b_4, b_5, and c_0, c_1, c_2, c_3, c_4, and c_5 are the regression constants or regression coefficients. Values of constants b_0, b_1, b_2, b_3, b_4, b_5, and c_0, c_1, c_2, c_3, c_4, and c_5 at different temperatures can be obtained by using the least squares method. For some intermediate temperatures, values of the constants can be obtained through interpolation.

Prasad (1995), Mahajan (2001), and Das (2005) have developed a regression model for predicting the respiration rates of apple. The respiration rate was represented as a function of oxygen concentration (Y_{O_2}), carbon dioxide concentrations (Z_{CO_2}), and storage temperature (T). The value of R_{O_2} and R_{CO_2} can be expressed as a function of oxygen and carbon dioxide concentrations and temperature by a second-order regression equation of the following type:

$$\begin{aligned} R_{O_2} = {} & b_0 + b_1 T + b_2 Y_{O_2} + b_3 Z_{O_2} + b_4 T^2 + b_5 Y_{O_2}^2 \\ & + b_6 Z_{CO_2}^2 + b_7 T Y_{O_2} + b_8 T Z_{CO_2} + b_9 Y_{O_2} Z_{CO_2} \end{aligned} \quad (9.20)$$

$$\begin{aligned} R_{CO_2} = {} & c_0 + c_1 T + c_2 Y_{O_2} + c_3 Z_{CO_2} + c_4 T^2 + c_5 Y_{O_2}^2 \\ & + c_6 Z_{CO_2}^2 + c_7 T Y_{O_2} + c_8 T Z_{CO_2} + c_9 Y_{O_2} Z_{CO_2} \end{aligned} \quad (9.21)$$

where b_0, b_1, b_2, b_3, b_4, b_5, b_6, b_7, b_8, b_9, and c_0, c_1, c_2, c_3, c_4, c_5, c_6, c_7, c_8, and c_9 are the regression constants or regression coefficients.

Considering that the respiration rate is a function of the four factors (i = 4), namely oxygen and carbon dioxide gas concentrations, time, and temperature of storage influence the respiratory kinetics; a third model based on a second-degree polynomial function was proposed as shown in Equation 9.22:

$$R_{O_2} = b_0 + \sum_{i=1}^{4} b_i X_i + \sum_{i=1}^{4} b_{ii} X_i^2 + \sum_{i=1}^{3} \sum_{j=i+1}^{4} b_{ij} X_i X_j \quad (9.22)$$

where R_{O_2} is the respiration rate expressed as oxygen consumption, b_i is a constant coefficient, and X_i (i = 1 to 4) is the coded independent variable linearly related to temperature, storage time, and oxygen and carbon dioxide gas concentrations, respectively. A similar relationship was also used to express the respiration rate in terms of carbon dioxide evolution. The complete set of experimental data pooled from all the temperatures considered for this study was analyzed by multiple regression analysis using SYSTAT 8.0 to determine the value of the model coefficients.

9.9 Enzyme Kinetics

Lee et al. (1991) proposed a model based on the principle of enzymatic kinetics with a Michaelis-Menten type equation, which agreed well with experimental data on cut broccoli and has been widely used to model respiration rates of fresh produce

Table 9.8 Types of Inhibition and Resulting Equations for Respiration Rate Using Principles of Enzyme Kinetics

Types of Inhibition	*Description*	*Resulting Equation*	*Eq. no.*
No inhibition	Simple Michaelis-Menten relationship between O_2 concentration and consumption	$\frac{V*O_2}{K_m+O_2}$	(9.23)
Competitive	CO_2 competes with substrate O_2 for same active site of enzyme The maximum respiration rate is lower in high CO_2 concentrations	$\frac{V*O_2}{K_m*\left[1+\frac{CO_2}{K_i}\right]+O_2}$	(9.24)
Uncompetitive	CO_2 reacts with enzyme substrate complex The maximum respiration rate is not much influenced at high CO_2 concentrations	$\frac{V*O_2}{K_m+\left[1+\frac{CO_2}{K_i}\right]*O_2}$	(9.25)
Non-competitive	A one-to-one combination of competitive and uncompetitive	$\frac{V*O_2}{(K_m+O_2)*\left[1+\frac{CO_2}{K_i}\right]}$	(9.26)

Source: Adapted from Peppelonobes and Leven, Postharvest Biology and Technology. 7. 27–40 1996; Hertog, M.L.A.T.M. et al., *Postharvest Biology Technology*. 14. pp. 335–349, 1998.

Note: K_m = Michaelis-Menten constant (% O_2), K_i = inhibition constant (% CO_2) and V_m = maximum respiration rate (ml kg^{-1} h^{-1})

(Mahajan and Goswami 2001). This model is a simplification that tends to fit the experimental data very well, being based on one limiting enzymatic reaction in which the substrate is O_2. Another reason for its use is the similarity with microbial respiration for which this equation is widely used. The different types of inhibition and the resulting influence on the equation have been summarized in Table 9.8.

Further analysis of the model parameters, V_m, K_m, and K_i showed that with an increase in temperature, both V_m and K_m increased as shown in Equations 9.27 and 9.28, respectively (Lakakul et al. 1999).

$$V_m = 1.67^{-10} \times e^{0.069T} - 1.06^{-10} \tag{9.27}$$

$$K_m = 50\,T + 660 \tag{9.28}$$

Hertog et al. (1998) introduced temperature dependence in respiration using Arrhenius equations and validated the model with experimental data on tomatoes, apples, and chicory.

9.10 Temperature Dependency of Model Parameters

The temperature dependence of the model parameters of the Michaelis-Menten equations were quantified using an Arhhenius-type equation (Mahajan and Goswami 2001) using Equation 9.29, which was linearized as in Equation 9.30.

$$R_m = R_p \exp\left[\frac{E_a}{RT_{abs}}\right] \tag{9.29}$$

$$\ln R_m = \frac{-E_a}{R}\left(\frac{1}{T_{abs}}\right) + \ln R_p \tag{9.30}$$

where R_m is the model parameter of enzyme kinetics, R_p is the respiration pre-exponential factor, E_a is the activation energy (kJ g^{-1} mole^{-1}), T_{abs} is the storage temperature (K), and R is the universal gas constant (8.314 kJ kg-mole^{-1} K^{-1}).

9.11 Modeling of Gaseous Exchange in MA Storage Systems

Various MA storage systems have been developed and investigated for increasing the shelf life of a fresh commodity. Common gases used in MA storage are carbon dioxide, oxygen, and nitrogen. Carbon dioxide is bacteriostatic. Nitrogen is an inert gas; it does not possess any bacteriostatic effect. It is used as a filler gas in the MA gas mixture. The inhibitory effect of carbon dioxide increases with a decrease in temperature.

When the fresh commodity is sealed in a selected polymeric film package, it constitutes a dynamic system in which respiration of the product and the gas permeation through the film take place simultaneously. The theory is that the plastic film serves as the regulator of oxygen flow into the package and the flow of carbon dioxide out of the package. For a particular transient period and at a given temperature, the rate of oxygen consumption and the rate of carbon dioxide evolution of the packaged commodity depend greatly on oxygen concentration and carbon dioxide concentration. Considering that there is no gas stratification inside the packages and that the total pressure is constant, the differential mass balance equations that describe the oxygen concentration changes in a package containing respiring product are

Rate of oxygen entry into package space − rate of oxygen consumed by product = rate of oxygen accumulation inside package space

That is,

$$\mathrm{A_p P_{O_2}}\left(Y_{O_2}^{a} - Y_{O_2}^{i}\right) - W_p R_{O_2} = V_{fp}\left[\frac{\mathrm{d}Y_{O_2}^{i}}{\mathrm{d}t}\right] \tag{9.31}$$

or

$$\left(\frac{\mathrm{d}Y_{O_2}^{i}}{\mathrm{d}t}\right) = -\left(\frac{W_p}{V_{fp}}\right)R_{O_2} + \left(\frac{\mathrm{A_p P_{O_2}}}{V_{fp}}\right)\left(Y_{O_2}^{a} - Y_{O_2}^{i}\right) \tag{9.32}$$

Similarly, the carbon dioxide concentration changes in a package can be written as

Rate of carbon dioxide generated by the fruits − rate of carbon dioxide leaving out of the package space = rate of carbon dioxide accumulation inside package space

That is,

$$W_p R_{CO_2} - A_p P_{CO_2}\left(Z^{i}_{CO_2} - Z^{a}_{CO_2}\right) = V_{fp}\left(\frac{dZ^{i}_{CO_2}}{dt}\right) \tag{9.33}$$

$$\left(\frac{dZ^{i}_{CO_2}}{dt}\right) = \left(\frac{W_p}{V_{fp}}\right) R_{CO_2} - \left(\frac{A_p P_{CO_2}}{V_{fp}}\right)\left(Z^{i}_{CO_2} - Z^{a}_{CO_2}\right) \tag{9.34}$$

where A_p is the area of the package through which the oxygen and carbon dioxide permeates (m^2), $Y^{a}_{O_2}$ is oxygen concentration in the atmospheric air (cm^3 per cm^3 of air), $Y^{i}_{O_2}$ is the oxygen concentration inside the package (cm^3 per cm^3 of air), $Z^{a}_{CO_2}$ is the carbon dioxide concentration in the atmospheric air (cm^3 per cm^3 of air), $Z^{i}_{CO_2}$ is the carbon dioxide concentration inside the package (cm^3 per cm^3 of air), W_p is the weight of the fruit kept inside the package (kg), V_{fp} is the free volume in the package (cm^3), t is the storage time (h), P_{CO_2} is the carbon dioxide permeability of packaging material, ($cm^3\ m^{-2}\ h^{-1}$ [concentration difference of carbon dioxide in volume fraction]$^{-1}$), R_{CO_2} is the respiration rate for carbon dioxide evolution by the fruits ($cm^3\ kg^{-1}\ h^{-1}$), and $\frac{dY^{i}_{CO_2}}{dt}$ is the rate of change of carbon dioxide concentration $Y^{i}_{CO_2}$ within the package at time t of storage (cm^3 per m^3 of air h^{-1}).

9.12 Case Study: CA Storage of Red Delicious Apples

For the red delicious variety of apple, the lowest respiration rate is reported to be at 2% carbon dioxide concentration, 1%–1.5% oxygen concentration and at a storage temperature of 1°C (Kupferman 1997). For a study conducted by Mahajan (2001) on a laboratory scale model, a lower limit and an upper limit were set for regulating oxygen and carbon dioxide concentrations of the CA chamber. Above 1% oxygen and below 2% carbon dioxide levels are not harmful for apple, but below 1% oxygen and above 2% carbon dioxide levels may lead to undesirable changes in the apple. Hence, the storage gas regimes and temperature were selected as shown in Table 9.6. These storage gas regimes are also found to be optimum for red delicious varieties of apple (Table 9.9). The RH of the CA chamber was maintained at 92%–95% by using the saturated solution of potassium nitrate.

9.12.1 Control Strategy

In control strategy, the initially established CA conditions, i.e., oxygen, carbon dioxide, and temperature, would have to be maintained within the set limits

Table 9.9 CA Conditions and Other Parameters Used for Evaluating the Control Strategy for Red Delicious Apples

Parameter	*Value*
O_2 concentration, %	
Upper limit of O_2 conc. in the chamber	2
Lower limit of O_2 conc. in the chamber	1
CO_2 concentration, %	
Upper limit of CO_2 conc. in the chamber	2
Lower limit of CO_2 conc. in the chamber	1
Initial N_2 concentration of the chamber, %	97
Set temperature (T), °C	1
Temperature tolerance limit, ± °C	0.2
Apple weight W, kg	10.85
Free volume of the CA chamber, m^3	0.0474

throughout the storage period. The oxygen and carbon dioxide concentrations are maintained by purging the ambient air and liquid nitrogen, respectively, inside the CA chamber. Liquid nitrogen purging was started whenever the carbon dioxide concentration reached the upper limit, i.e., 2%, and continued until it dropped to the lower limit, i.e., 1%. The oxygen concentration of the CA chamber was increased to its upper limit, i.e., 2% by the addition of ambient air through a pump whenever oxygen concentration drops to the lower limit, i.e., 1%. The refrigeration load arising out of heat in leak, heat of respiration, and fan heat was taken care of by the mechanical vapor compression refrigeration system fitted with the CA chamber.

9.12.2 Model Development

The various steps involved in developing the CFD model that describes the heat and mass transfer phenomena in the CA storage system are as follows.

9.12.2.1 Physical Model

The laboratory RCA storage cabinet was considered as the computational domain to simulate the transport processes during CA storage of the fruit. The frontal portion of the chamber had a door the full width and height of the chamber for introducing and removing the fruits during experimentation. The chamber on all sides was insulated using thermocole sheets, and hence, the chamber was assumed to be fully insulated from its surroundings. The full perimeter of the door was made leak-proof using a neoprene gasket. A fan was placed in the chamber for uniform distribution of the temperature and gases inside the chamber (Figure 9.5). The flow inside the chamber was hence deemed forced convection during CFD analysis. The evaporator coils of the vapor compression refrigeration unit, which were made of copper, coupled with the RCA chamber, were located on one side of the chamber. A fruit-loading tray with an overhead protection hood was placed centrally inside the chamber. The fruits, being a living entity, participated in gas exchange phenomena with their surroundings and heat generation due to their respiration. Therefore, the fruit trays were modeled as a heat and mass-generating porous media with permeable boundaries. Because the storage cabinet was firmly symmetrical with the vertical plane in the middle of the fruit tray, for saving computational resources and time, the three-dimensional problem was translated to a two-dimensional heat mass and flow issue for the present analysis. A numerical experiment was carried out to check the grid independency of the solution with three levels of grid sizes. The selected grid size for the computational domain was considered to be 2.5 cm in both directions. Corresponding to this grid size used, the computational time was affordable with sufficient accuracy in the solution.

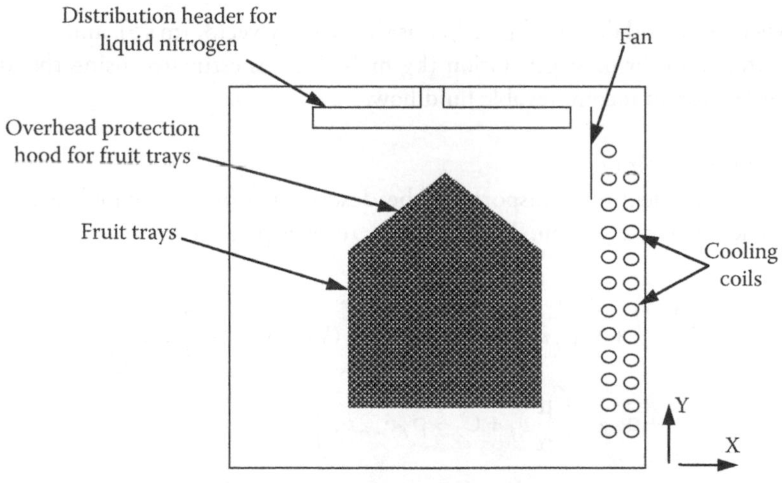

Figure 9.5 The physical domain.

9.12.2.2 Mathematical Model

The mathematical description of the transport processes outlined above led to a system of partial differential equations. However, to simplify the solution process, some assumptions were made as follows:

- Airflow in the porous medium was described by means of a Darcy–Forchheimer equation for porous media flow.
- The product was represented by a spherical shape, homogeneous and with constant properties throughout.
- Density, specific heat, and thermal conductivity did not vary within the temperature range of this application.
- The internal temperature and concentration gradients in a single fruit were neglected as the Biot numbers N_{Bi} for heat and mass transfer were found to be less than 0.1 for the situation under consideration.
- An incompressible airflow was assumed.

Governing equations

The following set of equations that governed the mass, momentum, heat, and species transport in the porous medium are as follows:

Continuity equation

$$\frac{\partial \rho_f}{\partial t} + \nabla(\rho_f \vec{v}) = S_M \tag{9.35}$$

where ρ_f is fluid density (kg m^{-3}), $\vec{v}$ is the velocity vector (m s^{-1}), and S_M is the source term for mass generation (kg m^{-3}s^{-1}). ρ_f is estimated using the ideal gas law for an incompressible fluid flow.

Momentum equation

The momentum transport can be described using a Reynolds-averaged Navier–Stokes equation. The equation can be expressed as

$$\begin{aligned}\frac{\partial(\rho_f \vec{v})}{\partial t} + \nabla(\rho_f \vec{v}\vec{v}) &= -\nabla p + \nabla\left(\mu\left[(\nabla\vec{v} + \nabla\vec{v}^T) - \frac{2}{3}\nabla\vec{v}I\right]\right) \\ &+ \rho_f \vec{g} - \left(\frac{\mu}{\alpha}v_j + C_2\frac{1}{2}\rho_f v_{mag} v_j\right)\end{aligned} \tag{9.36}$$

where p is the static pressure (Pa), $\bar{\bar{\tau}}$ is the stress tensor (Pa), $\rho_f \vec{v}$ is the gravitational force (N m^{-3}), μ is the dynamic viscosity (Pa s), and I is the unit tensor.

The permeability α (m^2) and internal resistance factor C_2 (m^{-1}) describing the resistance of the porous medium in the fruit trays to the air flow governed by the Ergun law and Darcy Forchheimer equation is estimated from the following equations (Xu and Burfoot 1999a):

$$\alpha = \frac{d_p^2 \varepsilon^3}{150(1-\varepsilon)^2} \tag{9.37}$$

and,

$$C_2 = \frac{3.5(1-\varepsilon)}{d_p \varepsilon^3} \tag{9.38}$$

where ε is the porosity of the medium, and d_p is mean product diameter (m).

Energy equation

The following standard equation was used to describe the energy transfer during storage:

$$\frac{\partial}{\partial t}\left(\varepsilon \rho_f E_f + (1-\varepsilon)\rho_p E_p\right) + \nabla\left(\vec{v}(\rho_f E_f + p)\right) = \nabla\left[k_{eff} \nabla T - \left(\sum_i h_i \vec{J_j}\right)\right] + S_f^h \tag{9.39}$$

where E_f is fluid energy (J kg^{-1}), E_p is product energy (J kg^{-1}), ρ_p is product density (kg m^3), k_{eff} is effective thermal conductivity of porous medium (W m^{-1} K^{-1}), h_i is enthalpy of species '*i*' (J kg^{-1}), J_j is diffusion flux (kg m^{-2} s^{-1}), and S_f^h is enthalpy (W m^{3}).

The effective thermal conductivity of porous medium k_{eff}, is estimated using the inbuilt function in the Fluent solver as

$$k_{eff} = (k_f \varepsilon) + [k_p(1 - \varepsilon)] \tag{9.40}$$

The first three terms on the right-hand side of Equation 9.39 represent energy transfer due to conduction, species diffusion, and viscous dissipation, respectively. The transport of enthalpy due to species diffusion can be neglected if the Lewis number N_{Le} is far away from unity. In the present case, however, N_{Le} was found to vary between 0.35 and 0.81. Hence, the species diffusion term was considered in the energy equation.

S_f^h is a positive heat source due to the metabolic heat of respiration, which is defined as a user-defined function of the following form (Kang and Lee 1998):

$$Q_R = \frac{2816 \times \rho_p}{6 \times 3600} \frac{\left(R_{O_2} + R_{CO_2}\right)}{2} \tag{9.41}$$

where Q_R is the heat of respiration (J kg^{-1} h^{-1}), ρ_p is the density of fruit (kg m^{-3}), R_{O_2} is the respiration rate of oxygen consumption as per Equation 9.12, and R_{CO_2} is the respiration rate of carbon dioxide generation as per Equation 9.13.

Species transport equation

$$\frac{\partial(\rho_f Y_i)}{\partial t} + \nabla(\rho_f \vec{v} Y_i) = -\nabla \vec{J}_i + S_i \tag{9.42}$$

where ρ_f is the density of air-vapor mixture (kg m^{-3}), Y_i is the mass fraction of species 'i'; J_i is the diffusion flux of i (kg m^{-2} s^{-1}) arising due to concentration gradients, and S_i is the source term (kg m^{-3} s^{-1}).

The mass source terms for oxygen consumption and carbon dioxide generation were incorporated into the model as a user-defined function as described by Michaelis-Menten type equation. The source term for water loss from product is defined by the following equation (Kang and Lee 1998):

$$L = \frac{Q_R W + hA(T_a - T_p)}{3600\lambda} \tag{9.43}$$

where L is the rate of water loss (kg s^{-1}), W is the weight of produce stored (kg), h is the convective heat transfer coefficient (W m^{-2} K^{-1}), T_a is the temperature of storage atmosphere (K), T_p is the product temperature (K), and λ is the latent heat of moisture evaporation (kJ kg^{-1}).

9.13 Summary

Consumer demand for more natural, minimally processed, and fresh foods is increasing day by day. Modified atmosphere packaging is a well-proven method for preserving the natural quality of fruits and vegetables in addition to extending the storage life. The method is gaining acknowledgment throughout the world due to strict regulations on the use of other chemical preservation methods. The method is commercially successful for preserving certain fruits and vegetables. With the vast basic and fundamental knowledge available on the method, the research in this area is taking a new dimension to suit the new consumer trends and demands. There is new interest in applying this technique to consumer-ready products. This technique can be integrated with active or interactive packaging to improve the control over the package atmosphere to achieve superior product quality and safety. Time–temperature indicators on the packages to show the remaining storage life of the food product would improve food safety and inventory control.

References

ASHRAE Handbook (1990). Refrigeration systems and applications, American Society of Heating, refrigeration and Air-conditioning Engineers. 1971. Tullie circle, NE Atlanta, GA 30329.

Biale, J.B. (1964). Growth, maturation and senescence in fruits. *Science*, 146, 880–888.

Cameron, A.C., Boylan-Pett, W., and Lee, J. (1989). Design of modified atmosphere-packaging systems: Modelling oxygen concentrations within sealed packages of tomato fruits. *Journal of Food Science*, 54 (6), 1413–1415.

Ciobanu, A., Lascu, G., Bercescu, V., and Niculescu, L. (1976). Effects of low temperature on foods. In *Cooling Technology in the Food Industry*. Abacus Press, Kent, England.

Das, H. (2005). *Food Processing Operations Analysis*. Asian Books Pvt Ltd, New Delhi.

Do, J.Y. and Salunkhe, D.K. (1975). Controlled atmosphere storage: Basic considerations. In *Postharvest physiology, handling and utilization of tropical and sub-tropical fruits and vegetables*. E.B. Pantastico (ed.), AVI Publishing Co., Inc., Westport, Connecticut.

Fonseca, S.C., Oliveira, F.A.R., and Brecht, J.K. (2002). Modeling respiration rate of fresh fruits and vegetables for modified atmosphere packages: A review. *Journal of Food Engineering*, 52, 99–101.

Gorris, L. and Tauscher, B. (1999). Quality and safety aspects of novel minimal processing technology. In *Processing of Foods: Quality Optimization and Process Assessment*, F.A.R. Oliveira and J.C. Oliveira (Eds.), CRC Press, Boca Raton, FL, pp. 325–339.

Granado Lorencio, F., Olmedilla Alonso, B., Herrero-Barbudo, C., Sánchez-Moreno, C., de Ancos, B., Martínez, J.A., Pérez-Sacristán, B., and Blanco-Navarro, I. (2008). Modified-atmosphere packaging (MAP) does not affect the bioavailability of tocopherols and carotenoids from broccoli in humans: A cross-over study. *Food Chemistry*, 106, 1070–1076.

Hagger, P.E., Lee, D.S., and Yam, K.L. (1992). Application of an enzyme kinetic based respiration model to closed system experiments for fresh produce. *Journal of Food Processing Engineering*, 15, 143–157.

Hertog, M.L.A.T.M., Peppelenbos, H.W., Evelo, R.G., and Tijskens, L.M. (1998). A dynamic and generic model on the gas exchange of respiring produce: The effects of oxygen, carbon dioxide and temperature. *Postharvest Biology Technology*, 14, 335–349.

Hoang, M.L., Verboven, P., Baelmans, M., and Nicolaï, B.M. (2001). Effect of process, box and product properties on heat and mass transfer during cooling of horticultural products in pallet boxes. *ASAE annual meeting*, Paper Number. 13014, Sacramento, July 28–31, 2001.

Irtwange, S.V. (2006). Application of modified atmosphere packaging and related technology in postharvest handling of fresh fruits and vegetables. *Agricultural Engineering International: The CIGR E-journal*, 4 (8).

Jayas, D.S. and Jeyamkondan, S. (2002). Modified atmosphere storage of grain meat fruit and vegetables. *Biosystems Engineering*, 82 (3), 235–251.

Kader, A.A. (1986). Biochemical and physiological basis for effects of controlled and modified atmospheres on fruits and vegetables. *Food Technology*, 40 (5), 99–104.

Kader, A.A. (1987). *Postharvest physiology of vegetables*. Marcel Dekker, New York.

Kader, A.A. (1997). A summary of CA requirements and recommendations for fruits other than apples and pears. In *Proceedings of the 7th International Controlled Atmosphere Research Conference,* A.A. Kader, Postharvest Hort. Series 17 (2): 1–36, University California, Davis, CA, USA, 3, 1–34.

Kader, A.A., Zagory, D., and Kerbel, E.L. (1989). Modified atmosphere packaging of fruits and vegetables. *CRC Critical Reviews in Food Science and Nutrition*, 28, 1–30.

Kang, J.S. and Lee, D.S. (1998). A kinetic model for transpiration of fresh produce in a controlled atmosphere. *Journal of Food Engineering*, 35, 65–73.

Kays, S.J. (1991). Metabolic processes in harvested products respiration. In *Post Harvest Physiology of Perishable Plant Products*, Van Nostrand Reinhold, New York.

Kupferman, E. (1997). Observations on storage regimes for apples and pears. *Tree Post Harvest Journal*, 8 (3), 3–5.

Lakakul, R., Beaudry, R.M., and Hernandez, R.J. (1999). Modeling respiration of apple slices in modified-atmosphere packages. *Journal of Food Science*, 64, 105–110.

Lee, D.S., Hagger, P.E., Lee, J., and Yam, K.L. (1991). Model for fresh produce respiration in modified atmosphere based on principles of enzyme kinetics. *Journal of Food Science*, 56 (6), 1580–1585.

Mahajan, P.V. (2001). Studies on controlled atmosphere storage for apple and litchi using liquid nitrogen. Unpublished PhD Thesis, IIT Kharagpur, India.

Mahajan, P.V. and Goswami, T.K. (1997). Controlled atmosphere storage of horticultural crops. *Indian Journal of Cryogenic*, 22 (4), 123–136.

Mahajan, P.V. and Goswami, T.K. (2001). Enzyme kinetics based modelling of respiration rate for apple. *Journal of Agricultural Engineering Research*, 79 (4), 399–406.

Makino, Y., Iwasaki, K., and Hirata, T. (1996a). A theoretical model for oxygen consumption in fresh produce under an atmosphere with carbon dioxide. *Journal of Agricultural Engineering Research*, 65, 193–203.

Makino, Y., Iwasaki, K., and Hirata, T. (1996b). Oxygen consumption model for fresh produce on the basis of adsorption theory. *Transactions of the ASAE*, 39, 1067–1073.

Metlitskii, L.V., Salkova, E.G., Volkind, N.L., Bondarev, V.I., and Yanyuk, V.Y.A. (1983). Engineering equipment for creating and regulating controlled atmosphere. In *Controlled Atmosphere Storage of Fruits* Amerind Publishing, India, pp. 68–94.

Nair, S., Singh, Z., and Tan, S.C. (2003). Aroma volatiles emission in relation to chilling injury in 'Kensington Pride' mango fruits. *The Journal of Horticultural Science and Biotechnology*, 78 (6), 866–873.

Parry, R.T. (1993). *Principles and application of modified atmosphere packaging of food.* R.T. Parry (ed.) Blackie (Chapman and Hall), UK, pp. 1–18.

Peppelenbos, H.W. and Leven, J. (1996). Evaluation of four types of inhibition for modelling the influence of carbon dioxide on oxygen consumption fruits and vegetables. *Postharvest Biology and Technology*, 7, 27–40.

Phan, C.T., Pantastico, E.B., Ogata, K., and Chachin, K. (1975). Respiration and respiratory climacteric. In *Postharvest physiology, handling and utilization of tropical and subtropical fruits and vegetables*. E.B. Pantastico (ed.), pp. 86–103.

Prasad, M. (1995). Development of modified atmosphere packaging system with permselective films for storage of red delicious apples. Unpublished PhD Thesis, Department of Agriculture and Food Engineering, Indian Institute of Technology, Kharagpur, India.

Raghavan, G.S.V., Gariépy, Y., Thériault, R., Phan, C.T., and Lanson, A. (1984). System for controlled atmosphere long-term cabbage storage. *International Journal of Refrigeration*, 7 (1), 66–71.

Rangana, S. (1995). *Handbook of analysis and quality control for fruits and vegetable products*. Tata McGraw-Hill, New Delhi, India.

Ryall, A.L. and Pentzer, W.T. (1982). *Handling, transportation and storage of fruits and vegetables*. Vol. II. AVI Publishing, Westport, CT.

Saltveit, M.E. (2003). Respiratory Metabolism. In *The Commercial Storage of Fruits, Vegetables, and Florist and Nursery Stocks*. Agriculture Handbook, Number 66. USDA. ARS.

Saltveit, M.E. (2004). Respiratory metabolism. In *The Commercial Storage of Fruits, Vegetables, and Florist and Nursery Stocks*. Agriculture Handbook Number 66. USDA. ARS.

Saltveit, M.E. (2005). *Commercial storage of fruits, vegetables and florist and nursery crops*, Postharvest Technology Centre RIC, Department of Plant Science, University of California, USA.

Salunke, D.D. and Kadam, S.S. (eds.) (1995). *Handbook of fruit science and technology*, Marcel Dekker Inc., New York, pp. 550.

Talasila, P.C., Chau, K.V., and Brecht, J.K. (1992). Effects of gas concentrations and temperature on O_2 consumption of strawberries. *Transactions of the ASAE*, 35 (1), 221–224.

Toledo, R., Steinberg, M.P., and Nelson, A.I. (1969). Heat of respiration of fresh produce as affected by controlled atmosphere. *Journal of Food Science*, 34 (3), 261–264.

Watkins, C., David, R.A., and Bartsch, J.A. (1997). Items of Interest for Storage Operators in New York and beyond. In *Cornell fruit handling and storage newsletter*. Cornell University, Ithaca, NY.

Waelti, H. (1994). Energy conservation in CA storages–program 101, FY–Briefing Report, Cooperative Extension, Department of Biological Systems Engineering, WSU.

Xu, Y. and Burfoot, D. (1999a). Simulation of the bulk storage of foodstuffs. *Journal of Food Engineering*, 39, 23–29.

Yang, C.C. and Chinnan, M.S. (1988). Modeling the effect of O_2 and CO_2 on respiration and quality of stored tomatoes. *Transactions of the ASAE*, 30 (3), 920–925.

Chapter 10

Innovation in Food Packaging

Mohammad Shafiur Rahman

Contents

10.1 Introduction

Packaging is one of the important unit operations in food processing and preservation. In addition to the direct approaches to food preservation, such as drying and freezing, other measures, such as packaging and quality management tools, need to be implemented in the process to avoid contamination and recontamination. Recently, tremendous progress has been made in the development of diversified packaging materials, packaging design, packaging equipment, and in achieving environmental safety and sustainability. In general, packaging performs five major functions: product containment, preservation and quality, presentation and convenience, protection, and providing storage history (Rahman 2007). Significant development has been achieved in each function. This chapter presents recent developments and progress in the area of food packaging, including innovative packaging design, new packaging materials, and addresses the issues of environmental concern due to food packaging.

10.2 Variations in Shapes and Design

It is important to display the product in an attractive manner to the potential buyer. A cleverly designed and beautifully produced package can help to sell a product, which is an essential ingredient of an effective marketing campaign. Consumer preferences and innovative technologies have always been the major forces that change the food industry. The designs based on marketing include apparent size, attention-drawing power, an impression of quality, and clear readability of the brand name (Ahmed and Alam 2012).

10.2.1 Shapes

Semi-rigid packaging containers are shifting gradually toward thin, light, and flexible plastics. These are stand-up, flexible pouches, stick packs for unit portion sizes, and easy-open and reclose options (Brody 2010b). The use of microwaves is increasing in food and meal preparation. There is a real need for the packaging industry to confront the particular problems in designing packages that deliver microwaved products to the dinner table. The critical aspect of microwaveable ready-to-eat meals is the safe release of internal steam buildup in packages during preparation. Microwaveable prepared foods are engineered to be reheated in achieving serving temperatures in microwave ovens. The microwave energy penetrates the food and generates steam, which, if under pressure during opening, could injure the consumer. Conversely, steam under pressure could expand and explode the package in the microwave oven. This could create, at the least, a mess to be cleaned up and, at the other extreme, a hazardous situation. Internal pressure may be relieved by a weak heat seal that fractures from excess steam pressure during microwave steam generation or the incorporation of shrink-film-covered vent valves that melt or otherwise fail as a result of steam pressure. More recently, there is a development of laser scored or perforated weaknesses in the film, which fail from internal steam-pressure and release the steam in the microwave oven. Laser processing of flexible film is sufficiently precise to be able to produce micro-perforations or partial-depth cuts that remain intact during distribution and early heating and fail only under steam pressure during cooking (Brody 2010b).

Many materials act as a microwave reflector and microwave absorbent. This could be used in designing microwaveable packaged foods (Yam and Lai 2006). Aluminum is often used to selectively shield microwaves from certain areas or locations of food. For example, a multi-component meal may consist of food items that needs to be heated at different rates in the microwave oven. The more microwave-sensitive food items can be shielded so that the entire meal can be heated more evenly. Aluminum is also used as an electromagnetic field modifier to redirect microwave energy (Bohrer and Brown 2001). It can intensify the microwave energy locally or redirect it to places in a package, for example, microwave energy can be redirected from the edges to the center for frozen food products, such as lasagna. Precautions are necessary to prevent arcing, which can occur between foil packages and oven walls and between two packages. Arcing can be prevented by following several design rules, for example, any foil components should be receded from the edge of the package to avoid arcing with oven walls (Russell 1999). In addition, it is necessary to test the packaged product to ensure that the package is safe to use (Yam and Lai 2006). Microwave-absorbent materials, commonly known as susceptors, are also used to generate localized heating. The major purpose of susceptors is to generate surface heating to mimic the browning and crisping ability of the conventional oven. Susceptors are available in different forms, such as flat pads, sleeves, and pouches, and in various patterns, such as portions of metallized layers that are deactivated. The patterns are designed to provide more control of heating (Yam and Lai 2006).

10.2.2 Closures

Closures are a key part of any rigid and semi-rigid package and can give a real competitive advantage. Lots of innovative closures have been developed over the last decade. Introducing a new closure is a sure way of rejuvenating an existing package (Robertson 2010). Paper cartons for milk or juice with an easy-open and easy-pour cap, thus can also increase consumption. The zipper and, more recently, the slide and pressure-sensitive adhesive flap have added to consumer convenience (Brody 2010b).

10.2.3 Time Temperature Indicator

The most common example of intelligent packaging is the time temperature indicator (TTI). TTIs are tags that can be applied to individual packages or shipping cartons to visually indicate whether a product has been exposed to time and temperature conditions (i.e., integrate or accumulate the exposure to temperature over time) that adversely affect the product quality. TTI could be used in chilled foods to identify the temperature abuse during storage, display, and distribution. It helps in ensuring proper handling and provides a gauge of product quality, such as microbial conditions and ripening status for sensitive products in which temperature control is imperative to efficacy and/or safety. TTIs are successfully applied as wireless devices to determine the influence of process or storage parameters on the products (Lo and Argin-Soysal 2006). According to the response mechanisms, TTIs can be divided into three groups: (i) biological, (ii) chemical, and (iii) physical systems. The tags are available in a one-dot version and a three-dot version with the three dots changing color at different rates. The change of color of the dot indicates the exposed time and temperature of the product (Mermelstein 1998). This could be used for continuous monitoring, such as individual packages and full cartons, machine readability, archival data recording (i.e., it does not reset if pack is re-chilled), shipping tracking, and easy traceability and recall (Robertson 2010).

There is considerable potential for use of TTIs in the food distribution chain, but there are two issues to be considered. One is the economics. When using a TTI for a relatively low-cost product, such as lettuce, the indicator also has to be relatively low in cost. This should be considered or addressed by the manufacturer of the indicator. The other issue is knowledge of the food products' quality deterioration kinetics (i.e., quantitative and precise measurement). The food processor must know the degradation kinetics of this product in order to select the appropriate indicator, i.e., how the quality characteristics of his product are changing with time and temperature exposure (Mermelstein 1998).

The intelligent ripeness indicator (such as RipeSense™) responds to the aroma released as fruit ripens, giving consumers a better way to determine the ripeness of the fruit before opening the package and eating, thus helping to reduce waste (Robertson 2010). Gas concentration indicators could create tremendous assistance, especially in modified atmosphere packaging. The most common oxygen indicator is pink when the ambient oxygen concentration is ≤0.1%, turning blue when the oxygen concentration is ≥0.5%. The presence of oxygen is indicated in 5 min or less, although the change from blue to pink may take 3 h or more. The "intelligent ink" has been developed by using light-sensitive nanoparticles, such as titania (TiO_2) that can only detect oxygen (Mills and Hazafy 2008). The ink is blue in air and ambient room light, but the color changes to white if irradiated with a pulse of UV light. The color can revert to blue under normal room light. In an oxygen-free atmosphere the ink remained colorless after the UV pulse. This type of ink could be inexpensive and could also be used to indicate if the original modified atmosphere inside a package is changed. A similar novel carbon dioxide intelligent pigment can be incorporated into a thermoplastic polymer to create a long-lived carbon dioxide-sensitive plastic film (Mills et al. 2010).

10.2.4 Smart Barcode

A radio frequency identification (RFID) tag increases convenience and efficiency in supply chain management and traceability. It can be read via radio meaning without any specific orientation and at greater distances with respect to a reading device than is required for a barcode. While a barcode indicates the type of item it is printed on, an RFID tag indicates not only the type of object it is attached to, but also a unique serial number, and thus it can distinguish a given package from every other one in the world (Robertson 2010). RFID uses radio waves to track items wirelessly. It makes use of tags or transponders (data carriers), readers (receivers), and computer systems (software, hardware, networking, and a database) (Ahmed and Alam 2012). The working principles of an RFID system are (Brody et al. 2008): (i) data stored in a tag are activated by a reader when an object with an embedded tag enters the electromagnetic zone of a reader, (ii) the data are transmitted to a reader for decoding, and (iii) the decoded data are transferred to a computer system for further processing.

2-D bar codes are capable of being scanned by smart phones to deliver additional information to users. The ScanLife™ solution goes a step beyond the ability to gather data and analyze by using a camera phone and an all-in-one 2-D barcode reader to quickly access websites for product pricing and more. This type of technology could also be used for marketing campaigns. SnapTags™ deliver interactive functionality and traceability to mass marketing by making any logo a gateway to mobile marketing (Robertson 2010).

10.2.5 Convenience

In many instances, consumers have limited time to shop, chop, cook, and clean, and they prefer prepared or ready-to-eat meals that taste just like homemade or need a minimum of preparation and are a good value for the money (Hollingsworth 2001). Eating styles, such as ready-to-eat meals, snacks, and microwaveable ready meals, have changed over the years, requiring innovation in packaging. Consumers prefer packages that are stable, rigid, and re-sealable. Convenience takes different forms (Robertson 2010): (i) easy to hold, use, open, and close and hand-held consumption, such as in the car or at your desk, enabling the eater to perform multiple tasks, and (ii) easy to prepare, cook, and reheat and ready-to-eat microwaveable meals, enabling an easy lifestyle. Other conveniences could be smaller portions, re-closable, and tamper-proof options. Numerous convenient, healthy, shelf-stable products are discussed by Kuhn (2011).

Value-added packaging allows in-package cooking and facilitates on-the-go consumption. Self-heating containers are also being developed for the convenience of consumers, who do not need to reheat the product during consumption. In a system described by Webb (1999), an exothermic reaction takes place with crushed limestone, and the heating process begins. In this system, users feel the heat of the container 2–3 min after pushing a button on the bottom surface of the can, and the container heats fully within 5 min.

10.2.6 Tamper Evidence

Consumers want tamper-evident closures to avoid packaging being opened unnoticed. Consumers prefer tamper-proof packaging with an easy opening option, yet consumers want to introduce tamper-evident closures to avoid packaging being opened unnoticed. In general, tamper-proof packaging makes products more difficult to open, so there is clearly a need to balance safety with consumer accessibility. The tamper-resistant package is to alert the consumer if tampering has taken place and to provide visible evidence of any tampering. In many cases, consumers are ready to pay more for tamper-resistant packaging. Tampering can be classified into four categories: two for kinds of tampering and two for location of tampering. Casual tampering or grazing happens in the store. The tamperer wants to taste or smell or change price by changing caps, but probably does not intend to do harm. Malicious, surreptitious tampering occurs outside the store at home or in a workshop. The tampered package may be returned to the store shelves. In the case of the normal route of entry, one could open the package and re-close it using the cap or the tear strip or the tear out easy-open end. Casually, one would do this, but a malicious tamperer might not do so. In the case of an evasive route of entry, the tamperer opens and re-closes the package by any means other than the cap or the tear strip. The intended route of entry and therefore the tamper-resistant feature is left undisturbed. Holograms are good security providers for labels, and they can guard against forgery because they cannot be copied successfully and are difficult to duplicate. A generic self-adhesive hologram tape with tamper-evident properties may be used on various substrates to protect, secure, and authenticate products because hologram tape can easily disclose whether or not cartons or boxes have been tampered with. Holographic shrink sleeves are reverse printed and attached with a tamper-evident holographic stripe inside the sleeves (Robertson 2010).

Covert taggants are unique, hard-to-replicate particles that can be added to inks, paper, films, and other materials, and they can be detected with a specialized reader. Beyond barcodes and optical authentication systems, covert taggants are used to include additional security features to applications from passports to packaging. In the food area, they are used on premium wines and spirits. Other high-tech covert taggants use surface enhanced raman scattering (SERS), a nanoscale phenomenon, to generate robust, unique, secure optical fingerprints. SERS materials provide a unique signal that can be detected using specialized readers and therefore provide highly counterfeit-resistant security solutions for a broad range of applications, including brand security (Robertson 2010).

Cases of extortion or sabotage have also been reported. In the mid 1970s, child-resistant packaging became an issue, leading to the development of child-proof lids for poisonous products. "Tamper-resistant" refers to the ability of the packaging to resist tampering (or opening), for example, for child protection, whereas "tamper-evident" refers to the ability of the packaging to reveal that it has been opened.

10.3 Progress in Developing New Materials

Earlier food packaging materials were used to provide only barrier and protective functions. Now various types of active substances can be incorporated into the packaging material to improve its functionality and to give new or extra functions. Bottani et al. (2011) conducted a packaging material survey considering food products within the solid, semi-solid, and liquid categories. They observed that polypropylene is by far the most widely used material, followed by glass, metal, PET, paper, tetra brick, polyethylene, and polystyrene.

10.3.1 Conventional Packaging Materials: Metals and Glass

Steel, tin, and aluminum are used mainly for canned foods and beverages. The most common use of metals for packaging is in tin-coated steel and aluminum cans. The principal advantages of metal (i.e., steel-based) cans are their strength, providing its mechanical stability; effective barrier properties (i.e., moisture, gas, and light); and resistance to high temperatures, providing stability during processing. The disadvantages of metal cans are their heavy mass, high cost, and tendency to interact with foods and the environment and their high energy requirement during manufacture. The main progress in metal cans happens in developing diversified shapes and different types of lacquering in order to avoid interactions with foods.

The huge expansion of the use of aluminum occurs in developing multilayered packaging materials and its use in making different types of lids, tubes, and utensils. Aluminum foil is difficult to use on modern fast packaging equipment because of creases, tearing, and marking effects. Foil may be used for formed or semi-rigid containers. Many instant meals are packed in cooking and eating trays made of aluminum with different compartments commonly formed in the tray to separate the meal components, especially with frozen foods.

No single film can satisfy all packaging requirements. Lamination is a technique for bonding films together to give a film with the properties of multiple constituents. By combining the qualities of choice from raw material films, a laminate can be tailor-made for its particular application. Each layer in the resulting laminate may exhibit different properties from its free state, such as mutual layer reinforcement in which cracks in a brittle layer are prevented from propagating by a high elongation (elastic) layer. The effect depends on good adhesion between the layers. Coatings are often used to enhance plastic film properties, such as printability. Aqueous and solvent coatings are applied to the substrate through water dispersions or emulsions, solvent solutions, or waxing. Aluminum metallized films are extensively used in food packaging applications and compared with films containing aluminum foil. Metallization has the following advantages: (i) lower environmental impact due to a significant reduction in the amount of raw material used and the recyclability of metalized film scrap as part of the base material, (ii) greater flexibility and resistance to flexion, and (iii) impressive presentation (Driscoll and Rahman 2007). One of the new technologies expected to enhance plastic film barriers to gases and vapors is the glass (silica)-coated plastics or the use of other ultrathin coatings, such as diamond, or other chemically treated surfaces (TFT 1997). Barrier properties can be incorporated by vapor-deposited silicon oxide, aluminum oxide, polyvinyl alcohol, and nanoclays (Brody 2010b).

Glass containers used to be and still are considered as a prestigious means of packaging and serve for the most expensive wines, liqueurs, perfumes, and cosmetics. It is highly inert, impermeable to gases and vapors, and amenable to the most diverse shaping. In its normal state, it has the advantage of transparency, but where required, it can be given different desired colors. It has complete as well as selective light protection properties. Its main disadvantages are its fragility, heavy mass, and high energy requirement during manufacturing. In addition to its marketing strength, glass has other advantages that give its muscle in today's marketplace. It is an excellent oxygen barrier and completely neutral in contact with foods. Glass also fits well into the modern recycling society because it can be recycled indefinitely. Glass packaging technology has developed to the extent that strength, minimal mass, color, and shape all have been improved. While glass won't supplant metal and plastic in volume, it is finding an increasingly strong niche at the high end of the food spectrum.

10.3.2 Plastics

Plastic films can be formed by lamination, co-extrusion, or impregnation. Polymers are the fastest growing group of materials in food packaging. Plastics are relatively cheap, light, easily processed and shaped, and easy to seal. It is less than half of the density of glass or aluminum or about one eighth of the density of steel. Plastics do not shatter like glass or buckle like metals. Two major drawbacks are their permeability to gases and vapors and the possibility of their interacting with the product. The main interactions are (Gnanasekharan and Floros 1997): (i) migration of volatile and nonvolatile compounds from packaging materials to the packaged food, including unreacted monomers or additives present in the polymerized packaging material; (ii) sorption of components from the food or from the environment into the packaging materials; and (iii) permeation of volatile compounds from the food through the packaging. The common plastics include polyethylene, polypropylene, polyvinyl chloride, polyvinylidene chloride, polytetrafluoroethylene, polystyrene, polysters, polyethylene terephthalate (PET), and cellulose (Komolprasert 2006; Selke 2006). Recent safety and environmental concerns are pushing the industry to develop alternatives biopolymers.

10.3.3 Biopolymers

The bio-based materials are usually called "environmentally friendly," "biodegradable," or "earth-friendly." Biodegradable plastics, under appropriate conditions of moisture, temperature, and oxygen availability, lead to fragmentation or disintegration of the plastics with no toxic or environmentally harmful residue (Chandra and Rustgi 1998). Biodegradable polymers can be classified according to their source as (Sorrentino et al. 2007):

1. Group 1 (extracted polymers): directly extracted or removed from biomass (i.e., polysaccharides, proteins, polypeptides, polynucleotides).
2. Group 2 (synthesized polymers): produced by classical chemical synthesis using renewable bio-based monomers or mixed sources of biomass and petroleum (i.e., polylactic acid or bio-polyester).
3. Group 3 (microbiologically transformed polymers): produced by microorganism or genetically modified bacteria (polyhydroxybutyrate, bacterial cellulose, xanthan, curdian, pullan).

Europe is far ahead of the United States, accounting for nearly 60% of the market for bio-based packaging (Robertson 2010). Biopolymers are based on renewable raw materials, which can be processed by established plastic processing technologies, such as injection and blow molding, blown or cast film extrusion, and extrusion. Components, e.g., cellulose, starch, or oils from plant-based biomaterials, such as corn, rapeseed, and soybean, can be extracted. Similarly, gelatin from animal skin and whey protein from milk can also be extracted (Robertson 2011). The compostable polymers are biodegraded in an industrial composting environment in fewer than 180 days. It means a defined temperature of about 60°C, a defined humidity, and the presence of microorganisms. These do not leave fragments longer than approximately 12 weeks in the residue and do not contain metals or toxins and do support plant life (Robertson 2011). One of the recently developed ones is the compostable polymer polylactide (polylactic acid, PLA), which is typically manufactured from cornstarch. Polylactides are water resistant and can be formed by injection molding, blowing, and vacuum forming and at room temperature.

Polyalkylene carbonate copolymers are a unique family of innovative thermoplastics representing a true breakthrough in polymer technology. These materials are produced through the copolymerization of carbon dioxide with one or more epoxides, and these are amorphous, clear, readily processed, and possess long-term mechanical stability. In addition, these are considered more environmentally friendly because they consume 50% fewer petrochemicals as compared to 100% petrochemically based polymers and exhibit biodegradable properties (Robertson 2010).

Polypropylene carbonate (PPC) is biodegradable aliphatic polyester reinforced with starch (Ge et al. 2004) and mango puree nanocomposite reinforced with cellulose nanofibers (Azeredo et al. 2009). Incorporation of low-cost and biodegradable cornstarch into PPC provides a practical way to produce a completely biodegradable and cost-competitive composite with good mechanical properties (Ge et al. 2004). Improvement in the thermal properties of PPC has been reported by mixing PPC with montmorillonite clay to form a bio-nanocomposite (Shi and Gan 2007).

10.3.4 Nanocomposites

Nanotechnology presents the packaging industry with opportunities to develop an array of exciting new products with enhanced or fundamentally different performance properties (Robertson 2010; Magnuson et al. 2011). In addition, it exhibits biocompatibility and biodegradability and, in some cases, controlled release capacity of antimicrobial and antioxidant in the cases of active antimicrobial food packaging. It contains a naturally occurring polymer (biopolymer) in combination with an inorganic moiety to be functionalized, including at least one dimension on the nanometer scale (Rhim and Ng 2007). Challenges remain in increasing the compatibility between clays and polymers and reaching complete dispersion of nanoplates (exfoliation). The most promising nanoscale-size fillers are montmorillonite and kaolinite clays and crumpled graphite nanosheets (Compton et al. 2010). Cellulose biofibers have their highly crystalline building nano-blocks and food contact complying non-MMT (non-montmorillonite) nanoclays (Sanchez-Garcia et al. 2010). The reviewed papers point out that these nanocomposites showed improved physical properties, for example, mechanical strength, thermal stability, gas barrier, physico-chemical, and recyclability (Sorrentino et al. 2007; Arora and Padua 2010). Azeredo et al. (2010) described use of cellulose nanofibers and glycerol as a plasticizer to improve the mechanical and water vapor barrier properties of edible chitosan films. In general, there is a need to develop more compatible filler-polymer systems, better processing technologies, and a systems approach to the design of polymer-plasticizer-filler (Magnuson 2011). Morris (2010) has expressed concerns over the long-term fate and disposal of these materials, which might then lead to the release of nanoparticles into the environment and back into the food chain, raising debate on the labeling, approval, traceability, and regulation of all these smart or intelligent biomaterials.

10.3.5 Smart or Intelligent Packaging

Smart packaging consists of active and intelligent packaging. Active packaging technologies involve interactions between the food, the packaging material, and the internal gaseous atmosphere (Labuza and Breene 1988). Examples are oxygen and carbon dioxide scavengers, ethylene scavengers, antimicrobial component releasers, ethanol emitters, moisture absorbers, and flavor or odor absorbers (Robertson 2006). In some cases, the new concept of active or life packaging material allows a one-way transfer of gases away from the product or the absorption of gases detrimental to the product. If there is gas exchange across the packaging plastic, the benefits of the effect are diminished (Birkbeck 1998).

Many packaged foods contain active enzymes and other materials that could perform or simulate a living system, such as respiration. Indeed, the benefits of controlled atmospheres with less oxygen and more carbon dioxide result in part from slowing down the effects of these enzyme systems. Ethylene affects the physiological processes of plants. As a plant hormone, ethylene regulates many aspects of growth development and senescence and is physiologically active in trace amounts (<0.1 ppm). It is a natural product of plant metabolism and is produced by all tissues of higher plants and by some microorganisms. Moreover, non-ethylene and non-respiratory organic volatiles may also have physiological and quality parameters. Active packaging can also play a part either by absorbing ethylene (or other volatiles) or preferential transmission of this gas. Other functions are oxygen scavenging (absorbing oxygen gas in the package and preventing rancidity), which are being developed as forms of sachets or polymer additives; moisture scavenging; ethylene scavenging; and ethanol emitting.

Oxygen removal is not always easy. Oxygen removal can be mechanical, but air in the packaging materials or residual oxygen cannot be effectively removed in many cases. Conventional oxygen scavenging is too slow to retard the changes in many products. An oxygen-scavenging system can also be incorporated in plastic packages, thus forming an integral part of its structure. There are significant technical and commercial advantages in not having to insert sachets or attach labels to trays or bags. It has the advantage of being activated just prior to use. The package can be manufactured and stored under standard conditions, then triggered to an activated state prior to filling. Oxygen scavengers can also be added in the cap of a glass container.

A recent oxygen-scavenging invention is based on surface modified phyllosilicate clay that is functionalized with active iron to create a naturally sourced and highly efficient oxygen scavenger (Harrington 2010). The intricate multistep process produces a clay-iron composite, which acts as a performance-enhancing carrier of the oxygen-scavenging iron. Oxygen is removed from the package by migrating through the packaging material and reacting with the dispersed active iron to produce iron oxide, which remains within the packaging material with the clay working as a barrier to any migration. The advantage of this material is to avoid the inconvenience of adding a foreign element (i.e., sachets of iron powder) to the package (Robertson 2010). Other intelligent packaging could prevent or respond to spoilage, for example, polymer opal films that change color to indicate spoilage or DNA biochip nanosensors that detect toxins, contaminants, and pathogens, and polymers that repel water and dirt (Magnuson et al. 2011).

Recently, there is a new concept of bioactive packaging technologies. The active packaging primarily deals with maintaining or increasing the quality and safety of packaged foods (active role) while bioactive packaging has a direct impact on the health of the consumer (bioactive role) by generating healthier packaged foods (Lopez-Rubio and Gavara 2006). The techniques mainly consist of micro- and nano-encapsulation and enzyme incorporation in packaging materials. Among the biomaterials already studied for the immobilization of enzymes and with potential interest in functional bioactive packaging, carrageenan, chitosan, gelatin, polylactic acid (PLA), polyglycolic acid (PGA), and alginate are very promising materials (Lopez-Rubio and Gavara 2006). The active enzymes are commonly lactase, glucose oxidase, invertase, glucoamylase, lysozyme, α-amylase, glucoamylase, naringinase, catalase, and lactase (Fernandez et al. 2008). These are mainly used for antimicrobial or oxygen scavenging.

10.3.6 Antimicrobial Packaging

Antimicrobial packaging is a promising form of active food packaging (Han 2000). Smart packaging that is antimicrobial adds a dimension of safety (Katz 1999). When antimicrobial agents are incorporated into a polymer, it limits or prevents microbial growth. Food packaging materials may obtain antimicrobial activity by common antimicrobial substances, radiation, or gas emission/flushing. Radiation methods may include the use of radioactive materials, laser-excited materials, UV-exposed films, or far-infrared-emitting ceramic powders. This application could be used for foods effectively, not only in the form of films, but also as containers and utensils. The incorporation of antimicrobial agents with polymeric packaging provides an economic and labor-free way to solve the food surface contamination problems (Weng et al. 1999).

An extensive list of antimicrobial agents used in packaging materials is given by Han (2000). Joerger (2007) reviewed 125 peer-reviewed articles on antimicrobial packaging. He pointed out that little good information was included on the experimental protocols for comparative or replicative purposes. The technology of incorporating the antimicrobial system into the package structure looms large as a challenge. He also concluded that no "magic bullet" exists among the many antimicrobial package compounds reported and cited. Much research was reported with essential oils whose two major limitations are effectiveness (typically on 2-D) and inherent color and taste (Robertson 2010). Antimicrobial films still face limitations and are perhaps still best viewed as part of a hurdle strategy to provide safe foods. It was reported that the inhibition provided by solutions of cinnamon and oregano essential oils was related not to the total amount of active compounds released but to the amount that reached the surface at a critical time; thus the rate of release (i.e., kinetics) is critically important (Gutierrez et al. 2010).

A bacteriocin (i.e., nisin) was the antimicrobial most commonly incorporated into films, followed by food-grade acids and salts, chitosan, plant extracts, and the enzymes lysozyme and lactoperoxidase (Joerger 2007). Nano-silver (NS), comprising silver nanoparticles (clusters of silver atoms that range in diameter from 1 to 100 nm) is attracting interest for a range of biomedical and food-surface applications owing to its potent antibacterial activity. The key to its broad-acting and potent antibacterial activity is the multifaceted mechanism by which NS acts on microbes (Chaloupka et al. 2010). The clothing industry has incorporated nano-silver into fabrics, such as socks, and exploited the antibacterial activity for the neutralization of odor-forming bacteria. NS also has been integrated into various food-contact materials, such as plastics used to fabricate food containers, refrigerator surfaces, bags, and chopping boards, and under the pretext of preserving foods longer by inhibiting microbial growth. The key is its broad-acting and potent antibacterial activity through multifaceted mechanisms of action on the microbes (Chaloupka et al. 2010). Several concerns about toxicity remain with the widespread adoption of nano-silver and need to be addressed (Robertson 2010). Other nano-antimicrobial compounds examined are chitosan/silver, nanoclays, fullernols, and nanoglass (Magnuson et al. 2011). Silver ions loaded in isolated soya film showed bacteriostatic and bactericidal effects, depending on their concentration (Sun et al. 2011).

Volatile antimicrobial compounds are also used in films. In this case, a direct contact of the antimicrobial material with the food surface is not necessary, and the antimicrobial agent can act through the intra-packaging atmosphere. Carvacrol, a volatile aroma compound extracted from thyme and oregano essential oils, is well known for its antimicrobial activity. Mascheroni et al. (2011) studied the release of carvacrol from wheat gluten/montmorillonite-coated papers induced by relative humidity.

10.3.7 Edible Packaging

Edible films or coatings applied to fresh, minimally processed, and processed fruits and vegetables are effective in extending their shelf life and maintaining their microbiological, sensory, and nutritional qualities. It mainly offers the opportunity to provide high concentrations of antimicrobials at the food surface and to create a modified atmosphere inside the food products. The most important properties to be evaluated in an edible coating are its microbiological stability, adhesion, cohesion, wettability, solubility, transparency, mechanical sensory properties, and permeability to water vapor and gases (Falguera et al. 2011). A recent dimension of the edible coating is to develop active envelopes, which include oil consumption reduction in deep-fat fried products, transport of bioactive compounds, and shelf-life extension of highly perishable products (Falguera et al. 2011). Enzyme-generated biofilms for packaging fresh food can be developed from cross-links of proteins (Rogers et al. 1999).

10.4 Diffusion Process in Packaging

The mechanical, thermal, optical, and mass transfer properties need to be considered for selecting packaging materials to determine the shelf life of a packaged food.

It is important to know the barrier properties (moisture, oxygen, carbon dioxide, nitrogen, and other low-molecular weight compounds), controlled released kinetics, leaching of packaging components, and absorption of aroma and flavor by the packaging materials in order to select a packaging material, especially polymeric plastics. In the case of diffusional mass transfer, the driving force is a difference in concentration or in partial pressures. In the case of a steady-state process, mass flow rate can be estimated from Fick's law of diffusion as:

$$\frac{J}{A} = \frac{D_e(C_1 - C_2)}{L} \tag{10.1}$$

where J is the mass flow rate through the packaging material (kg s^{-1}), A is the surface area (m^2), D_e is the effective mass diffusivity (m^2 s^{-1}), L is the thickness (m), and C_1 and C_2 are the concentrations for side 1 and 2 (kg m^{-3}), respectively. The concentrations of gas at the film surfaces are more difficult to measure than partial pressures. In this case, the concentration can be converted to partial pressures by using Henry's law (Singh and Heldman 2001):

$$C = S\,p \tag{10.2}$$

where S is the solubility (moles m^{-3} Pa), and p is the partial pressure (Pa), respectively. In this case, Equation 10.1 can be written as:

$$\frac{J}{A} = \frac{D_e S(p_1 - p_2)}{L} \tag{10.3}$$

where S is the solubility (moles m^{-3} Pa), p is partial pressure of gas (Pa), and $(D_e S)$ is known as the permeability coefficient [(mole m)/(s m^2 Pa)]. The diffusivity, solubility, and permeability coefficient of different polymers as a function of permeant are compiled by Singh and Heldman (2001).

In the case of an unsteady state, the concentration in the materials changes with time. In this case, the diffusion equation can be written as:

$$\frac{\delta C}{\delta t} = D_e\left(\frac{\delta^2 C}{\delta x^2}\right) \tag{10.4}$$

where x is the distance coordinate (m), and t is time (s). Crank (1975) provided the most complete array of a solution with different geometries and boundary conditions needed to describe the conditions at the surface of the object. Unsteady state mass transfer charts are also developed, and the procedures to determine mass average concentration as a function of time are provided by Singh and Heldman (2001).

The chart presents concentration ratio versus dimensionless ratios $D_e t/d_c^2$ for three standard geometries: infinite plate, infinite cylinder, and sphere (d_c is the characteristic dimension, half thickness for infinite plate, radius of infinite cylinder and sphere). The concentration ratio $[(C_{ma} - C_m)/(C_i - C_m)]$ contains the mass average concentration (C_{ma}) at any time t, the concentration of the diffusing component in the medium surrounding the food (C_m), and the initial concentration of the diffusing component with the food (C_i). At any concentration ratio, a dimensionless ratio $D_e t/d_c^2$ can be determined, which can be used to estimate the time to reach that concentration. An alternative approach to the chart could be used based on the experimental data and using an analogy to the heat transfer. The basic diffusion rate equation can be written as (Singh and Heldman 2001):

$$\log(C_m - C) = -\frac{t}{f} + \log[j(C_m - C_i)] \tag{10.5}$$

where f is the diffusion rate constant representing time required for a one log-cycle change in the concentration gradient, and j is the lag coefficient describing the region of nonlinearity in the relationship between the concentration gradient and time during the initial stages of diffusion.

10.5 Case Studies: Shelf-Life Prediction Based on Water Vapor Permeability

The moisture migration in a packaged food is controlled by two factors: (i) the permeance of the food itself and (ii) the permeance of the packaging film. The pseudo-steady state equation can be written as:

$$m_s \frac{dM}{dt} = \frac{\kappa}{x} A(p_o - p_p) \tag{10.6}$$

where dM/dt is the rate of moisture transfer (kg water)(kg dry solids s)$^{-1}$, m_s is the mass of dry solids in food (kg), A is the surface area of package (m^2), (κ/x) is the packaging film permeance to moisture (kg water)(s m^{2}Pa)$^{-1}$, and p_p and p_o are the partial pressures of water vapor inside and outside the package (Pa), respectively. In terms of water activity, Equation 10.6 can be written as:

$$\frac{dM}{p_w(a_o - a_p)} = \left(\frac{\kappa}{x}\right)\left(\frac{A}{m_s}\right) dt \tag{10.7}$$

where a_o and a_p are the relative humidity outside the package and water activity of food, and p_w is the vapor pressure of water at the temperature of packaging and its outside (Pa), respectively. The linear relationship of moisture and water activity can be written as:

$$M = b\,a + c \tag{10.8}$$

where b and c are the model parameters. The analytical solution of Equation 10.7, using a linear food isotherm, can be written as (Karel and Labuza 1969):

$$\ln\left(\frac{M_o - M_{pi}}{M_o - M_p}\right) = \left(\frac{\kappa}{x}\right)\left(\frac{A}{m_s}\right)\left(\frac{p_w}{b}\right)t \tag{10.9}$$

where M_o is the moisture content of food at the relative humidity outside the package, M_{pi} is the initial moisture content of food inside the package, M_p is the moisture content of food inside the package at any time (kg water)(kg dry solids)$^{-1}$, and t is time (s). Because linear isotherm is not valid over the entire region, so the use of the Guggenheim, Anderson and De Boer (GAB) model is preferred for practical purposes. The predicting equation can be written as:

$$\int_{a_o}^{a_p} \frac{1 + K^2(Y-1)a_p^2}{(a_o - a_p)(1 - Ka_p + YKa_p)^2}\, da_p = \left(\frac{p_w}{M_g YK}\right)\left(\frac{k}{x}\right)\left(\frac{A}{m_s}\right)t \tag{10.10}$$

where M_g is the GAB monolayer moisture content (kg)(kg dry solids)$^{-1}$, and K and Y are the GAB model parameters.

EXERCISE 10.1

Dried potato chips containing moisture 3(kg)(kg dry solids)$^{-1}$ are packaged in a polyethylenc bag. (κ/x) and (A/m_s) are given as 102*10^{-7}(kg water)(m h mm Hg)$^{-2}$ and 0.3. The linear moisture isotherm is given as $M = 0.088a + 0.02$. Calculate the shelf life of the product if stored at 0.75 relative humidity and 35°C, considering allowable moisture content of the product is 5(kg)(kg dry solids)$^{-1}$. The vapor pressure of water can be estimated from the following equation as (Labuza 1984):

$$\ln p_w = -\frac{5321.66}{T} + 21.03 \tag{10.11}$$

where p_w is in mm Hg, and T is in K.

Solution:

The vapor pressure of water (p_w) can be estimated from Equation 10.11 at 35°C as 42.6 mm Hg, and moisture content at relative humidity 0.75 can be estimated as 0.086. The time to reach moisture content 5(kg water)(kg dry solids)$^{-1}$ can be estimated from Equation 10.9 as

$$\ln\left(\frac{0.086-0.03}{0.086-0.05}\right)=(102*10^{-7})(0.3)\left(\frac{42.60}{0.088}\right)t$$

Solving the above equation for t, we can determine time as 2933.33 h or 12.22 days.

10.6 Environmental Concerns

Recently, concerns about the ecological dimensions of packaging are emerging. This means that packaging has to satisfy the required physical, chemical, and biological criteria during its life cycle as packaging (active protection function), and once the original function has been fulfilled, the packaging should decay without polluting the environment (passive protection function). The presence of plastics in the habitat of wildlife on both land and sea has created issues, which are being vigorously exploited by the environmental lobby to demand solutions from the plastics industry (Huang et al. 1990). Landfills are an important outlet for disposal of packaging waste. Currently, modern facilities are designed to address two environmental concerns (Almenar et al. 2012). The first one is to control the leachate (toxic polluting compounds) to the ground and surface water, and the second one is the control and use of landfill gases (carbon dioxide, methane, a small amount of nitrogen and oxygen, and a wide range of other gases). These gases are produced during the biodegradation of waste from bacterial action, and these may be toxic or explosive. The composition of the mixture depends on the composition, temperature, moisture content, and age of the waste. Most modern landfills use a gas capture technology in which the landfill gas is either flared to convert methane into carbon dioxide or collected and used as a substitute fuel or to generate energy (i.e., energy recovery) (Almenar et al. 2012). Thus, overall environmental safety is demanded from packaging material disposal; thus it needs to develop easily reusable, recyclable, easily disposable, or environmentally friendly packaging.

10.6.1 Reduce

An important consideration could be the reduction of the amount of packaging used for foods. In many cases, the efficient design of the preservation and distribution chain could reduce the amount of excessive packaging requirements for food. It is important to use packaging at an optimum level in addition to targeting only reuse and recycle. For example, the deterioration of product quality in bakery products is due to the crystallization of starch granules, causing staling; moisture uptake, causing a reduction in crispness; mold growth, as bakery products typically have a high moisture content; and rancidity from exposure of lipids to oxygen, drying out of the interior (crumb) of a loaf of bread. The main forms of packaging for bakery products have distribution rather than protection as their primary objective. Product quality is ensured by specifying a short shelf life. For example, bread should be consumed as quickly as possible. It is difficult to specify a packaging material for bread that could meet the competing requirements of keeping the crust dry and the crumb moist. Because the shelf life is short, the goal of bread packaging is partially moisture control, but its main purpose is to allow the product to be distributed safely and hygienically. The material used should be inexpensive as bread is a staple and basic. Materials, such as waxed paper, cellophane, and polyethylene, are commonly used in bread packaging. For many bakery products, even this level of packaging is excessive. Cakes and pastries are often distributed in cardboard boxes that are not airtight, made possible with a waxed paperboard support. Packages containing pies must have air holes so that the packages can "breath" moisture during heating or cooling, preventing crust uptake of moisture. There is a constant effort in the packaging industry to reduce the amount of material used for packaging; glass containers now are on average 30% lighter than in 1980, the weight of cans is now approximately 40% less than in 1970, and 2-liter polyethylene terephthalate (PET) soft-drink bottles are 25% lighter than in 1977 (Almenar et al. 2012).

10.6.2 Reuse

Repeat refers to the use of the same material again and again for some indefinite time. Reuse is the application of a structure for an extended period of time or until it wears out (Brody 2010a). Some packages, such as glass bottles and jars can return back to the original supply chain and be reused (Almenar et al. 2012).

10.6.3 Recycle

Recycle means to recover and reprocess into the same or similar or something useful so that the molecules are not lost in space. Recover is the act of identifying, finding, and retrieving materials that could be returned to the sender for eventual recycling (Brody 2010a). The present and future focus is to use materials that can be either recycled or burned without producing noxious fumes and the use of printers' ink that does not contain heavy metals or biodegradable inks (Brown 1993; Robertson 2010).

Contamination is the main concern in the case of recycling and reuse of postconsumer plastics. One of the many difficulties with recycling is that there is no control over how the consumer uses the container (e.g., for pesticides and chemicals), and the container may become contaminated. The ink on labels is also difficult to remove during recycling (Driscoll and Paterson 1999). In the case of recycling, lamination could be utilized whereby the food contact surface is a virgin layer placed over the recycled material, which means the virgin layer free of contaminants acts as a functional barrier. Other solutions for recycling are incineration, composting, and environmental degradation. However, this needs to develop biodegradable materials. In many cases, the energy requirement during recycling is also an important factor. PET bottles were 25% more energy efficient than glass and 65% more efficient than aluminum and had less impact on resources at all levels of possible recycling (Driscoll and Paterson 1999). In many cases, collection costs can be far greater than the potential market of these materials. Auditing recycling can provide an ounce of prevention that can help in avoiding the complications of improper disposal and the resulting liabilities.

10.6.4 Biodegradable Packaging

The environmentally friendly bio-based and/or biodegradable packaging has huge potential to grow due to consumer and retailer awareness. Packaging has to serve two different functions in two different directions; when used as packaging, it needs to be completely integrated, and after completion of its purpose, it should be disintegrated within a short period of time. It is a challenge to achieve these purposes completely in two dimensions in different directions. In reality, edible packaging could eliminate the environmental impact from waste; thus, in the future, edible packaging with safe antimicrobial, anti-fouling, and replant characteristics could emerge in the future.

10.7 Summary

Post and pre-processing of food materials accounts for a major portion of food losses in a food processing chain, and it is the major concern for many countries. In order to ensure international food security and nutrition, food safety and management with adoption of appropriate technologies for storage, packaging, and distribution play vital roles in a food processing and preservation system. Proper packaging and storage systems not only enhance the shelf life of various foods, but they also reduce losses due to microbial spoilage; handling and transportation; and development of off-flavor, color, and odor. In this regard, in this chapter, various new packaging design concepts, such as shapes, closures, TTIs, barcodes, and various novel packaging materials, such as nanocomposites, smart or intelligent, antimicrobial, and edible packaging, have been discussed. Most importantly, the effect of these packaging materials on the environment, which forms a key consideration in packaging design nowadays, has been elaborately discussed.

References

Ahmed, J. and Alam, T. (2012). An overview of food packaging: Material selection and the future of packaging. In: *Handbook of Food Process Design*. Ahmed, J. and Rahman, M.S. (eds.). Wiley-Academics, Oxford, pp. 1237–1283.

Almenar, E., Siddiq, M. and Merkel, C. (2012). Packaging for processed food and the environment. In: *Handbook of Food Process Design*. Ahmed, J. and Rahman, M.S. (eds.). Wiley-Academics, Oxford, pp. 1369–1405.

Arora, A. and Padua, G.W. (2010). Review: Nano-composites in food packaging. *Journal of Food Science*, 75(1), R43–R49.

Azeredo, H.M.C., Mattoso, L.H.C., Avena-Bustillos, R.F., Filho, G.C., Munford, M.L., Wood, D. and McHugh, T.H. (2010). Nanocellulose reinforced chitosan composite films as affected by nanofiller loading and plasticizer content. *Journal of Food Science*, 75(1), N1–N7.

Azeredo, H.M.C., Mattoso, L.H.C., Wood, D., Williams, T.G., Avena-Bustillos, R.J. and McHugh, T.H. (2009). Nano-composite edible films from mango puree reinforced with cellulose nanofibers. *Journal of Food Science*, 74, N31–N35.

Birkbeck, J. (1998). Does packaging affect nutritional quality? *The Food Technologists*, 28(4), 156–157.

Bohrer, T.H. and Brown, R.K. (2001). Packaging techniques for microwaveable foods. In: *Handbook of Microwave Technology for Food Applications*. Datta, A.K. and Anantheswaran, R.C. (eds.). Marcel Dekker, New York, pp. 397–469.

Bottani, E., Montanari, R., Vigali, G. and Guerra, L. (2011). A survey on packaging materials and technologies for commercial food products. *International Journal of Food Engineering*, 7(1), 12.

Brody, A.L. (2010a). Reusable food packaging-reverse distribution. *Food Technology*, 64(1), 69–71.

Brody, A.L. (2010b). Thin, light and flexible: Optimizing food packaging. *Food Technology*, 64(11), 73–75.

Brody, A.L., Bugusu, B., Han, J.H., Sand, C.K. and McHugh, T.H. (2008). Innovative food packaging solutions. *Journal of Food Science*, 73, R107–R116.

Brown, D. (1993). Plastics packaging of food products: The environmental dimension. *Trends in Food Science and Technology*, 4, 294–300.

Chaloupka, K., Malam, Y. and Seifalian, A.M. (2010). Nanosilver as a new generation of nanoproduct in biomedical applications. *Trends in Biotechnology*, 28, 580–589.

Chandra, R. and Rustgi, R. (1998). Biodegradable polymers. *Progress in Polymer Science*, 23, 1273–1335.

Compton, O.C., Kim, S., Pierre, C., Torkelson, J.M. and Nguyen, S.T. (2010). Crumpled graphene nanosheets as highly effective barrier property enhancers. *Advanced Materials*, 22, 4759–4763.

Crank, J. (1975). *The Mathematics of Diffusion*, 2nd edn. Oxford University Press, London.

Driscoll, R.H. and Paterson, J.L. (1999). Packaging and food preservation. In: *Handbook of Food Preservation*. Rahman, M.S. (ed.). Marcel Dekker, New York.

Driscoll, R.H. and Rahman, M.S. (2007). Types of packaging materials used for foods. In: *Handbook of Food Preservation*. Rahman, M.S. (ed.). CRC Press, Boca Raton, FL, pp. 917–938.

Falguera, V., Quintero, P.J., Jimenez, A., Munoz, J.A. and Ibarz, A. (2011). Edible films and coatings: Structures, active functions and trends in their use. *Trends in Food Science and Technology*, 22, 292–303.

Fernandez, A., Cava, D., Ocio, M.J. and Lagaron, J.M. (2008). Perspectives for biocatalysts in food packaging. *Trends in Food Science and Technology*, 19(4), 198–206.

Ge, X.C., Li, X.H., Zhu, Q., Li, L. and Meng, Y.Z. (2004). Preparation and properties of biodegradable poly(propylene carbonate)/starch composites. *Polymer Engineering Science*, 44, 2134–2140.

Gnanasekharan, V. and Floros, J.D. (1997). Migration and sorption phenomena in packaged foods. *Critical Reviews in Food Science and Nutrition*, 37, 519–559.

Gutierrez, L., Batelle, R., Sanchez, C. and Nerm, C. (2010). New approach to study the mechanism of antimicrobial protection of an active packaging. *Foodborne Pathogens and Disease*, 7, 1063–1069.

Han, J.H. (2000). Antimicrobial food packaging. *Food Technology*, 54(3), 56–65.

Harrington, R. (2010). 'Unique' oxygen scavenging method cheaper, more efficient, says company. *Food Production Daily*, November 10.

Hollingsworth, P. (2001). Convenience is key to adding value. *Food Technology*, 55(5), 20.

Huang, J., Shetty, A.S. and Wang, M. (1990). Biodegradable plastics: A review. *Advances in Polymer Technology*, 10(1), 23–30.

Joerger, R.D. (2007). Antimicrobial films for food applications: A quantitative analysis of their effectiveness. *Packaging Technology and Science*, 20, 231–273.

Karel, M. and Labuza, T.P. (1969). Optimization of protective packaging of space food. Aerospace Medical School, San Antonio, Texas.

Katz, F. (1999). Smart packaging adds a dimension to safety. *Food Technology*, 53(11), 106.

Komolprasert, V. (2006). Food packaging: New technology. In: *Handbook of Food Science, Technology, and Engineering*. Hui, Y.H. (ed.). CRC Press, Boca Raton, FL, pp. 130.1–130.10.

Kuhn, M.E. (2011). Serving up convenience. *Food Technology*, 65, 29–36.

Labuza, T.P. (1984). *Moisture sorptions: Practical aspects of isotherm measurement and use*. American Association of Cereal Chemists, St Paul, MN.

Labuza, T.P. and Breene, W.M. (1988). Applications of 'active packaging' for improvement of shelf-life and nutritional quality of fresh and extended shelf-life foods. *Journal of Food Processing and Preservation*, 13, 1–69.

Lo, Y.M. and Argin-Soysal, S. (2006). Units of operations. In: *Handbook of Food Science, Technology, and Engineering*. Hui, Y.H. (ed.). CRC Press, Boca Raton, FL, pp. 102.1–102.16.

Lopez-Rubio, A.R. and Gavara, R.J.M. (2006). Bioactive packaging: Turning foods into healthier foods through biomaterials. *Trends in Food Science and Technology*, 17(10), 567–575.

Magnuson, B.A., Jonaitis, T.S. and Card, J.W. (2011). A brief review of the occurrence, use, and safety of food-related nanomaterials. *Journal of Food Science*, 76, R126–R133.

Mascheroni, E., Guillard, V., Gastaldi, E., Gontard, N. and Chalier, P. (2011). Anti-microbial effectiveness of relative humidity-controlled carvacrol release from wheat gluten/montmorillonite coated papers. *Food Control*, 22, 1582–1591.

Mermelstein, N.H. (1998). Enzyme developments. *Food Technology*, 52(8), 122–124.

Mills, A. and Hazafy, D. (2008). A solvent-based intelligence ink for oxygen. *The Analyst,* 133, 213–218.

Mills, A., Skinner, G.A. and Grosshans, P. (2010). Intelligent pigments and plastics for carbon dioxide detection. *Journal of Materials Chemistry,* 20, 5008–5010.

Morris, V. (2010). Nanotechnology and food. IUFoST Scientific Information Bulletin, August 2010 Update.

Rahman, M.S. (2007). Packaging as a preservation technique. In: *Handbook of Food Preservation.* Rahman, M.S. (ed.). CRC Press, Boca Raton, FL, pp. 907–915.

Rhim, J.W. and Ng, P.K. (2007). Natural biopolymer-based nanocomposite films for packaging applications. *Critical Review in Food Science and Nutrition,* 47, 411–433.

Robertson, G.L. (2006). *Food Packaging Principles and Practice.* CRC Press, Boca Raton, FL.

Robertson, G.L. (2010). Innovations in food packaging technology. *Proceedings of the International Symposium: New Technologies for the Food Processing Industry,* December 19–21, pp. 51–62.

Robertson, T. (2011). Biopolymers and compostable polymers. *Food New Zealand,* 10(7), 20.

Rogers, P.J., Batzloff, M.R., Negus, S.M. and Tonissen, K.F. (1999). Enzyme generated biofilms for packaging fresh food. *Food Australia,* 51(7), 313–315.

Russell, A. (1999). Design considerations for success in microwave active packaging development. Future-Pak 99 conference proceedings, George O. Schroeder Associates.

Sanchez-Garcia, M.D., Lopez-Rubio, A. and Lagaron, J.M. (2010). Natural micro and nano-biocomposites with enhanced barrier properties and novel functionalities for food biopackaging applications. *Trends in Food Science and Technology,* 21, 528–536.

Selke, S.E.M. (2006). Food packaging: Plastics. In: *Handbook of Food Science, Technology, and Engineering.* Hui, Y.H. (ed.). CRC Press, Boca Raton, FL, pp. 131.1–131.5.

Shi, X. and Gan, Z. (2007). Preparation and characterization of poly(propylene)/montmorillonitenano composites by solution intercalation. *European Polymer Journal,* 43, 4852–4858.

Singh, R.P. and Heldman, D.R. (2001). *Introduction to Food Engineering,* 3rd edn. Academic Press, London.

Sorrentino, A., Gorrasi, G. and Vittoria, V. (2007). Potential perspectives of bio-nanocomposites for food packaging applications. *Trends in Food Science Technology,* 18, 84–95.

Sun, Q., Li, X., Wang, P., Du, Y., Han, D., Wang, F., Liu, X., Li, P. and Fu, H. (2011). Characterization and evaluation of the Ag^{+}-loaded soy protein isolate-based bactericidal film-forming dispersion and films. *Journal of Food Science,* 76, 438–443.

TFT. (1997). Intelligent packaging for better food. *The Food Technologist,* 27(3), 90–91.

Webb, V. (1999). The Ontro self-heating container. *Resource,* 6, 9–10.

Weng, Y., Chen, M. and Chen, W. (1999). Antimicrobial food packaging materials from poly(ethylene-co-methacrylic acid). *Food Science and Technology,* 32, 191–195.

Yam, K.L. and Lai, C.C. (2006). Microwavable frozen food or meals. In: *Handbook of Food Science, Technology, and Engineering.* Hui, Y.H. (ed.). CRC Press, Boca Raton, FL, pp. 133.1–133.7.

Chapter 11

Nanotechnology in Food Processing

Sanjib K. Paul and Jatindra K. Sahu

Contents

11.1 Introduction

Nanotechnology is an emerging field, which is expected to have a substantial impact on the modern research and development (R&D) of science, technology, and engineering now and in the future. Nanotechnology is acknowledged to represent a new frontier area in various dimensions of scientific innovations in the 21st century. The last 15 years have demonstrated advancements in every direction of nanotechnology, such as nanoparticles and powders, nanolayers and nanocoatings, electrical, optical and mechanical nanodevices, and nanostructured biological materials are developed and applied in various scientific innovations. Engineered nanomaterials are designed at the macromolecular (nanometer) level to avail the advantages of their small dimension and novel properties, which are generally not observed in their conventional, bulk state. Nowadays, nanotechnology is seeking to exploit the distinct technological advancement of controlling the structure of materials at a nanoscale approaching individual molecules at their molecular or macromolecular levels. This field of technology is a pure example of interdisciplinary activity comprising physical science, biological science, medical science and their engineering aspects.

Application of nanotechnology in the agri-food industry was first addressesd by the United States Department of Agriculture (USDA) roadmap published in September 2003. According to the roadmap, it was predicted that nanotechnology will transform the entire food industry, changing the ways foods are being produced, processed, stored, packaged, transported, marketed, and consumed (Joseph and Morrison 2006). The novel properties of nanomaterials has shown their potential to open up many new opportunities for the food industry. The major focus of new applications so far appears to be on food packaging, biosensor development, medicine, food processing ingredients, and health foods with only a few known examples in the mainstream of the food and beverage industries. Present market status reveals that food packaging applications hold the largest share of the current and short-term predicted market for nano-food industries. The most promising growth areas identified for the near future include active and smart or intelligent packaging, which is an area of thrust for the global researchers in this field. It is expected that nanotechnology will lead to fundamental changes in the way the materials, devices, and systems are understood and created in various sectors of science and technology including food industry.

11.2 Nanomaterials

Development of novel materials to meet the need of the hour in a sustainable way nowadays is attracting research attention all over the world. R&D attempts are being made by tailoring at their macromolecular state to improve the properties of materials in order to find alternative precursors that can give desirable features to the materials in a hazardless way. A nanometer scale is a billionth of a meter, i.e., about 1/80,000 of the diameter of a human hair or 10 times the diameter of a hydrogen atom. For comparison of diversity in size, the wavelength of visible light is between 100 and 700 nm. A leucocyte has a size of 10^4 nm; bacteria and viruses range in size between 10^3 to 10^4 and 75 to 100 nm respectively; protein is 5–50 nm; DNA is about 2 nm (width); and an atom is approximately 0.1 nm. The dimension of nanomaterials and devices are generally in the range of 1 to 100 nm (10^{-9} to 10^{-7} m) at least in one side, which is quite smaller than viruses and bacteria at least in one dimension while larger than most atoms and molecules (Kernes and Macnaghten 2006). Nanoscale structures permit the control and extensive enhancement of fundamental features of materials without altering the materials' chemical status at the molecular and sub-molecular levels. These emergent properties have the potential to bring revolutionary changes in various dimensions of food processing industries with its novel technological innovations.

In food systems, nanomaterials occur naturally or undergo various structural, dimensional, and functional changes within the nanoscale during processing. For instance, raw materials of several foods (protein, starch, and fat) show sudden and drastic changes in the material behavior as they undergo various modifications at nano- and micro-scale ranges during normal processing. But the specific interest of all workers has already concentrated on engineered nanomaterials with controlled shape and size which are either already being used or in trial experimentations in many products and process technologies (Morris and Parker 2008). Nanomaterials are already applied and tremendous outcomes are obtained in various day-to-day products, such as sunscreens, cosmetics, sporting goods, clothing, tires, and electronic and non-electronic devices as well as food and fodder. They are also used in medicines for the purpose of diagnosis, imaging, and drug delivery. The most interesting features of nanomaterials are that they have a much greater surface area-to-volume ratio than their conventional bulk forms and new quantum effects, which lead to greater chemical reactivity and, thus, increases their potential applicability in different sectors in general. This is due to the improvement of various material properties, such as electrcal, optical, mechanical, magnetic, catalytic, and barrier properties, as well as surface biocides, biodegradability, and intelligent functionality (Bradley et al. 2011; Lagaron and Lopez-Rubio 2011).

11.3 Types of Nanomaterials

Nanomaterials can be within 100 nm in one dimension (e.g., nano films), two dimensions (e.g., nano strands or fibers), or three dimensions (e.g., nano particles). They can exist in single, fused, aggregated, or agglomerated forms with spherical, ovoid, tubular, and irregular shapes. The most common types of nanomaterials include nanoparticles, nanotubes, dendrimers, quantum dots, and fullerenes. These different nanostructures can be synthesized and engineered from a wide range of materials by following appropriate processing techniques by maintaining suitable

Table 11.1 Classification of Nanomaterials

Criteria	*Classes*	*Examples*
Length of dimension(s)	One dimension < 100 nm Two dimensions < 100 nm Three dimensions < 100 nm	Films, coatings Tubes, weir, fibers, strand Particles, quantum dots
Manufacturing processes	Gas phase reaction Liquid phase reaction Mechanical processes	Flame synthesis, condensation, CVD Precipitation, sol-gel Ball milling, plastic deformation
Phase composition	Single-phase solid Multiphase solid Multiphase system	Crystalline, amorphous particles and layers Matrix composites, coated particles Colloids, aerogels

process parameters. Generally, materials that are used widely in commercial fields are gold, carbon, silver, titanium, metaoxides, and alloys (Azeredo et al. 2009). Some of them, including silver, titanium, gold, zinc, and silicon, have shown their potential applicability in food processing and packaging (Chaudhry et al. 2010). There are several approaches to classification of nanomaterials based on different criteria, such as dimension, phase composition, and manufacturing processes. Table 11.1 summarizes the classification of nanomaterials on the basis of these criteria.

11.4 Synthesis of Nanomaterials

There is an immense interest in the synthesis of nanostructures that have at least one dimension between 1 and 100 nm. The synthesis of nanomaterials can be achieved in both bottom-up and top-down approaches, i.e., either to assemble a few atoms together or dissociate (break or disassemble) bulk solids into finer pieces until they are constituted of only a few atoms or molecules with a desired size and shape (Sonalika et al. 2008). Advances in the process for producing nanostructured materials coupled with appropriate formulation strategies have enabled the production and stabilization of nanomaterials that have potential applications in food and allied industries.

There are different established methods of creating nanostructures, but definite structures, such as nanofiber or strand, nanoparticles, rods, buckyballs, and nanotubes, can be synthesized artificially for certain materials. They can also be arranged by methods based on equilibrium or near-equilibrium thermodynamics, such as methods of self-organization and self-assembly, sometimes also called bio-mimetic processes. Synthesized materials can be arranged into useful shapes by standardizing process parameters, so that, finally, the material can be applied to a certain specific application. The methods used for synthesis of nanoparticles include reverse micelle synthesis, gas phase synthesis, sputtered plasma processing, microwave plasma processing, and laser ablation. Along with these, recently nonhazardous and green synthesis method has also developed specifically for biological applications. Details of the synthesis and post-synthesis processing of nanomaterials and controlling of their size and shape may be found in from any standard textbook on nanotechnology.

11.5 Properties of Nanomaterials

Most sub-micro and microstructured materials have similar properties to the corresponding bulk materials. Nanomaterials have the structural dimensions just above the atoms or molecules, which show significant distinctions in terms of properties from that of bulk due to a large fraction of surface atoms, high surface energy, spatial confinement, and reduced imperfections, which do not exist in the corresponding bulk materials even at their micro and sub-microstructure level. This is due to their nanoscale dimensions, which have an extremely large surface area-to-volume ratio and make a large available surface of interfacial atoms. This provides more surface properties to the materials for the reactivity.

The energy-band structure and charge carrier density in the materials can be modified quite differently from their bulk and, in turn, will modify the electronic and optical properties of the materials along with some other features. For example, lasers, light-emitting diodes (LED), high density information storage, etc. using quantum dots and quantum wires are very promising in the future of optoelections and smart technologies.

Nanostructures and nanomaterials favor a self-purification process in that the impurities and intrinsic material defects will move to near the surface upon thermal annealing. This increases mechanical and structural properties along with material perfection. For example, the chemical stability for a certain nanomaterial may be enhanced, or the mechanical properties of the nanomaterial will be better than its bulk counterpart. The superior mechanical properties of carbon nanotubes are well known.

One of the most important properties of nanomaterials is the optical properties, which depend upon the size, shape, surface characteristics, and other variables, including doping and interaction with other nanostructures or the surrounding environment. Major applications of this property comprises the domain of optical detector, laser, sensor, imaging, phosphor, display, solar cell, photocatalysis, photoelectrochemistry, biomedicine, and disease diagnosis.

Electrical conductivity in nanotubes and nanorods, photoconductivity of nanorods, and electrical conductivity of nanocomposites are the major examples of electrical properties when materials reach the nanoscale range.

Mechanical properties that deal with bulk metallic and ceramic materials influence the porosity, grain size, superplasticity, polymer composites, particle-filled polymers, polymer-based nanocomposites filled with platelets, and carbon nanotube-based composites. The discussion of mechanical properties of nanomaterials is, to some extent, only of quite basic interest; the reason is that it is problematic to produce macroscopic bodies with a high density and a grain size in the range of less than 100 nm. However, it is possible to increase the mechanical properties of some other materials by introducing nanostructures into the base. For example, the mechanical properties have promisingly increased in edible films and coatings when introduced with certain nanoparticles. It might be the best possible alternative of nondegradable packagings with some other specific advantages.

In the nano-range, non-magnetic bulk materials show magnetic properties. For example, bulk gold and Pt are non-magnetic, but at the nano size, they exihibit magnetic properties. In the case of Pt and Pd, the ferromagnetism arises from the structural changes associated with size effects. However, gold nanoparticles become ferromagnetic when they are capped with appropriate molecules, i.e., the charge localized at the particle surface gives rise to ferromagnetic-like behavior in nanoprticles.

Some other properties, such as antimicrobial, antioxidant, barrier properties, catalytic, and reactivity, are also of great importance in the food processing industries. Antimicrobial properties are vital in the field of medicine, food preservation, packaging, and storage as well as certain specific purposes of biomedical studies. The silver nanoparticle is a good example of an antimicrobial material, which is presently in extensive use. Antioxidant nanomaterials have a great potentiality to prevent the oxidation of their surroundings, thus preventing them from oxidative reaction. Nanoparticles obtained from antioxidant materials are showing unexpected results in food processing and medicine as well as biochemical analysis.

11.6 Characterization of Nanosystems

Characterization of a nanosystem or nanomaterials includes a broad area of discussion, including its size, shape, morphology, surface characteristics, crystallinity, mechanical properties, conductivity, magnetic features, changes in molecular bonding, antimicrobial properties, and also the chemical and biochemical features of the material. From the juvenile stage of nanoscience and technology, a number of instruments and machinery devices are being developed for better understanding of the nanostructures. The basic characterization parameters and detection of nanostructures include (i) size, shape, and surface characteristic analysis; (ii) atomic structure and chemical feature analysis of nanomaterials; and (iii) determination of nanoparticles in aerosols and biological tissues.

Morphological features of nanomaterials, such as size, shape, surface characteristics, etc., are analyzed by various high-magnification and high-resolution microscopes and different modern optical and electrochemical techniques, such as electron microscopy (SEM, TEM, field-emission SEM, high-resolution TEM, etc.), epiphaniometry, laser granulometries, zeta potentials, and elliptically polarized light scattering. Various spectroscopic methods, such as vibrational spectroscopy (e.g., FTIR), nuclear magnetic resonance (NMR), X-ray, and UV spectroscopy, etc., X-ray diffraction (XRD) and neutron diffraction are used to analyze the atomic structure and chemical status of nanostructured materials. In addition, for detecting and determining nanomaterials and particles in aerosols and biological tissues, various specialized techniques, including use of a differential mobility analyzer and condensation particle counter, is precisely used by researchers and technologists. For the detail process of instrumentation, analysis, and data interpretation, a standard textbook on nanotechnology or material science can be referred.

11.7 Nano in Food Domain

The most promising applications of nanotechnology in various aspects of agriculture and food and their challenges for human and environmental health have been systematically reported by Chen and Yada (2011). Nanotechnology in animal health and production (livestock, livestock products, poultry, and aquaculture) started providing highly nutritious foods (meat, fish, eggs, milk, and their processed products) to the global society, which have been and will continue to be an important and integral part of human consumption. There are a number of significant challenges in animal and agricultural production in terms of food and environmental safety and security, which is the most important concern of the hour. Nanotechnology may offer effective and novel solutions to these challenges. Nowadays, applications of nanomaterials having desired functional characteristics are being analyzed and applied in food processing operations, such as packaging, storage, nutrient delivery, medicines, sensors, and in the improvement of various processing additives or ingredients (Chau and Yen 2008). Nanotechnology may bring a new era in food processing sector through development of novel foods in terms of quality and quantity to meet the present day and future requirement. In this arena, nanotechnology may provide improved nutritional, functional, and sensory properties to food components with enhancing quantitative production. Nanoparticles, nanoemulsions, nanocapsules, and various other specialized nanostructures are being designed to enhance the availability and dispersion of nutrients, antioxidants, and nutraceuticals in foods.

Applications of nanotechnology have shown improving feeding and nutritional efficiency of agricultural animals minimizing losses from animal diseases, including zoonoses (Morris 2009); improving reproductibility and fertility of livestock (Narducci 2007); ensuring animal product quality and safety (Bauman et al. 2008; Branton et al. 2008); and transforming animal waste and processing byproducts having environmental concerns into value-added products for human utility (Davis et al. 2009; Soghomonian and Heremans 2009). Nanotechnology improves water

Table 11.2 Nanomaterials Used for Different Applications According to Their Properties

Property	*Nanomaterial Used*	*Example of Applications*	*Reference*
Mechanical, tensile, and barrier property	Starch	Film and coatings	Mathew and Dufresne (2002)
	Beta-hydroxy octanoate	Film and coatings	Mathew and Dufresne (2002)
	Plant oil-clay hybrid nanomaterial	Packaging and film	Park et al. (2003)
	Nano-clay	Additives in plastic polymer	Chaudhry (2010)
	Nano-titanium nitride	Food packaging and processing aids	Chaudhry (2010)
Antioxidant	Nanoparticulate lycopene	Efficient nutrient delivery	Jochen et al. (2006)
	Nanoparticulate carotenoids	Efficient nutrient delivery for vitamin A precursor	Jochen et al. (2006)
	Vitamin E	Active materials and colorant in biopolymer matrix	Sabliov and Astete (2007); Sabliov (2009)
	Beta-carotene	Active materials and colorant in biopolymer matrix	Sabliov and Astete (2007); Sabliov (2009)
	Pt	Food supplement	FSAI (2008)
	Zn	Food supplement	FSAI (2008)

Antimicrobial	Ag, Fe, Ca, Mg, Ce, silica	Taste and flavor enhancer, salt effect in chips/crisps with minimum use and enhanced bioavailability of minerals	Chaudhury (2010)
	Nano-silver and nano-zinc oxide	Additives in food packaging	Chaudhry (2010); WWICS (2008)
	Nano-silver	Antibacterial in wheat flour	Park (2005)
	Nano-silver	Chitosan-based active-edible coating	Rhim et al. (2006)
	Carbon nanotubes	Antimicrobial food packaging and preservation against *E. coli*	Lanzon et al. (2009)
	Biogenic silver	Different filter systems for drinking water hygienization	DeGusseme et al. (2010); Sintubin et al. (2009)
	Zn	Food supplement/colorant	FSAI (2008)
	Vitamin E, itraconazole, and beta-carotene	Active materials and colorant in biopolymer matrix	Sabliov and Astete (2007); Sabliov (2009)
	Gold nanoparticles	Biosensors to detect *Salmonella spp.* in pork	Gong-Jun et al. (2009)
	Gold nanoparticles	Biosensors to detect *E. coli* O157:H7 in milk	Lin et al. (2008)

(*continued*)

Table 11.2 (Continued) Nanomaterials Used for Different Applications According to Their Properties

Property	*Nanomaterial Used*	*Example of Applications*	*Reference*
Antimicrobial	Cu & Au nanoparticles	Biosensors to detect *E. coli* in surface water	Zhang et al. (2009)
	Gold nanoparticles	Detection of *Salmonella typhi* in phosphate buffer solution	Dungchaia et al. (2008)
	Gold nanoparticles	Biosensor for the detection of *Mycobacterium avium subsp. Paratuberculosis* in milk	Yakes et al. (2008)
	Gold nanorods	Detection of *Pseudomonas aeruginosa* with 0.85% sodium chloride solution	Norman et al. (2008)
	Glyco-nanoparticles	Detection of *E. coli* with phosphate buffer solution	El-Boubbou et al. (2007)
	Silica particles (200 nm) coated with silver shells (ca. 20 nm)	Biosensor to detect *E. coli* in water	Kalele et al. (2006)
	Quantum dots	Sensor for *Salmonella typhi* in chicken carcass wash water	Yang and Li (2005)
Thickener/viscosity enhancer/texture improver	Cellulose	Ice-cream and food application	Hwang and Yeh (2010)
	Wheat and rice flour	Ice-cream and food application	Tsukamoto et al. (2010)
	Mayonnaise (water nano droplet emulsion)	Applied in low-fat products to make its texture as full-fat equivalent	Clegg et al. (2009)

Taste improver and additives/antiodorant	Nano-silver	Additives in food packaging	Chaudhry (2010)
	Nano-zinc oxide	Additives in food packaging	WWICS (2008)
	Liposomes, micelles, and protein	To mask the undesirable taste of certain additives and supplements or to protect them from degradation during processing	Chaudhry (2010)
Protection against UV and visible light wavelength and unique reaction characteristics	Nano-titanium dioxide	Additives in plastic polymer film and food packaging material, nano coatings	Chaudhry (2010)
	Gold, polystyrene, silica, quantum dots, carbon nanotubes, branched nano-particles and dendrimers	Biosensors and food authenticity	Gomez-Hens et al. (2008); Chaudhry et al. (2010)
Hydrophobicity	Nano-silica	Hydrophobic surface coating, food contact surfaces, packaging, clarification of beer and wines	Chaudhry (2010)
Catalytic property	Paladium	Dehalogenation of chlorinated solvents/clean and metal-free effluent	Hennebel et al. (2009a, b, 2010)

(continued)

Table 11.2 (Continued) Nanomaterials Used for Different Applications According to Their Properties

Property	*Nanomaterial Used*	*Example of Applications*	*Reference*
Carrier vehicle for nutrient delivery/ micronutrient/health supplement	Liposomes	Encapsulation and targeted delivery of food components	Mozafari et al. (2008)
	Iron	Food supplement	FSAI (2008)
	Nano-selenium	Additive to a tea product in China	Chaudhry (2010)
Gelation, heat stability, and thickener	Protein	Re-micellized calcium caseinate from dairy protein Increased functionality	FSAI (2008)
	Globular proteins (Fiber/fibrils)	Thermal stability, increased shelf life. Formation of transparent gel network for use as thickening agent	FSAI (2008)
Emulsions	Oil in water (O/W)	Stabilization of biologically active ingredients; delivery of active compounds; extended shelf life; flavor release; low fat products	Weiss et al. (2006)
Pesticide/herbicide/ insecticide detection	Single-walled carbon nanotubes	Detection of methyl parathion and chlorpyrifos (insecticides) in 0.002 M phosphate buffer solution (pH 7.0)	Viswanathan et al. (2009)
	Multiwalled carbon nanotubes	Biosensor to detect paraoxon (insecticide) in 0.05 M phosphate buffer (pH 7.4)	Prakash-Deo et al. (2005)

	CdTe QD (cadmium telluride quantum dot) nanoparticle	Sensing device for 2,4 D (herbicide) detection in 50 mM phosphate buffer solution (pH 7.4)	Vinayaka et al. (2009)
	(CdSe)ZnS core shell QDs	Detection of paraoxon (insecticide) in 50% CH_3OH/H_2O (v/v) solvent	Ji et al. (2005)
	Gold nanoparticles	Used in a biosensor to detect paraoxon (insecticide) in 20 mM glycine buffer (pH 9.0)	Simoniana et al. (2005)
Sugar detection	xGnPs (exfoliated graphite nanoplatelets) (thickness of 10 nm) decorated with Pt and Pd nanoparticles	Used in the detection of glucose in 50 mM phosphate buffer solution (pH 7.4)	Lu et al. (2007)
	ZnO:Co nanoclusters (5 nm)	Used in the detection of glucose in 0.1 M phosphate buffer solution (pH 7.4)	Zhao et al. (2007)
	Gold nanoparticles	Detection of glucose in fruit juice	Ozdemir et al. (2010)
	Multiwalled carbon nanotubes	Detection of fructose 0.1 M phosphate buffer solution (pH 5.0)	Tominaga et al. (2009)
	Single-walled carbon nanotubes	Used in honey to detect fructose	Antiochia et al. (2004)
	Fe_3O_4 magnetic nanoparticles	Detection and estimation of glucose in 0.2 M acetate buffer solution (pH 4.0)	Wei and Wang (2008)
	CdTe QDs	Glucose detection in 20 mM phosphate buffer solution (pH 7.4)	Li et al. (2009b)

quality and safety, preventing contamination, and supporting recycling and conservation. Some materials at their nano-range has a great efficiency in microbial disinfection (Li et al. 2009a; Nangmenyi and Economy 2009), desalination (Hoek and Ghosh 2009), elimination of heavy metals (Diallo 2009; Farmen 2009), water conservation and reutilization in agricultural production (Savage et al. 2009), and developing clean energy (Ciesielski et al. 2010). Table 11.2 provides a nonexhaustive list of current applications of nanomaterials with different functional properties in the food processing industry.

11.7.1 Post-Harvest Management

Methods that are being used to preserve the wholesomeness of agricultural produce on farms and during storage and marketing are generally based on cooling or refrigeration with or without control of composition of the atmosphere. Some alternative storage and packaging techniques, such as MAP, CAP, and use of plastic film and pesticides, have become quite useful for the purpose. However, drawbacks of these techniques have created innovations for the development of alternative storage management and packaging techniques, such as edible film coatings. A number of studies have been conducted to develop coating materials that would coat fruits and vegetables to provide them with an internal modified atmosphere (Park et al. 1994). However, development of an edible film coating incorporating nanoparticles will be an excellent innovation in order to reduce the post-harvest losses of fruits and vegetables. At the same time, the nano-coatings incorporated with biomaterials retain the nutritional qualities of fruits and vegetables developing a synergistic effect between biomaterial and nanomaterials and thus, improving the functional as well as barrier properties of the coating materials, reducing loss of moisture, volatile compounds, and also mechanical and biochemical degradation of fruits and vegetables during storage and transportation.

11.7.2 Food Packaging

Nanotechnology has the potential to influence the packaging by delaying oxidation and controlling moisture migration, microbial growth, respiration rate, and retention of volatile flavor and aroma (Brody et al. 2008). Applications of nanotechnology in food packaging and coating are already started to become a commercial reality at the R&D level (Chaudhry et al. 2010). It is expected that nanotechnology-derived food packaging will make up to 19% of the share of nanotechnology-enabled products and applications in the consumer industry by 2015 worldwide (Nanoposts Report 2008). Active nano-coatings with high antimicrobial functionality are being developed for hygienic and aseptic food contact surfaces and materials and hydrophobic nano-coatings for self-cleaning surfaces. For instance, nano-packaging with self-cleaning abilities or nanoscale filtration techniques has already shown the complete elimination of bacteria and other pathogens from milk or water without boiling (Maftoum 2009) and is now become available commercially. The coatings have been reported to be very efficient at keeping out or providing much less available oxygen and retaining carbon dioxide and also UV protection, thus controlling respiration rate leading to slow ripening and delayed spoilage. As an example, Ecology Coatings, Inc. (Auburn Hills, USA) has developed UV-cured coatings specifically to address opportunities within the paper and packaging industry, reducing the high cost of the existing technology for the same. Functional active agents, such as nano-antioxidants and/or nano antimicrobial materials can be loaded into the matrix of the coating to enhance the shelf life, retaining nutritional and associated quality attributes of food materials. The discovery of antimicrobial properties of nanozinc oxide and nanomagnesium oxide may provide more affordable materials for such applications in food packaging and coating. Edible nano-coatings can be used on meat, cheese, fruit and vegetables, confectionery, bakery products, and fast food and can provide a barrier to moisture and gas exchange; act as a vehicle to deliver color, flavor, antioxidants, various nutrients, enzymes, and anti-browning agents; and can also increase the shelf life of processed foods, even after the packaging is opened or broken (Azeredo et al. 2009). Incorporation of nanomaterials into the matrix of plastic polymers has led to development of novel food packaging materials, with extensive desirable properties and it is expected that it will occupy the greater share in the market within the near future.

Different nanostructures, including nano wheels, nanofibers, and nanotubes from different materials are investigated as an effective tool to improve the properties of food packages (McHugh 2008). Another potential application is being made with the nanoparticles in food packaging for the degradation of ripening gas, viz., ethylene, which further extends the shelf life by delaying the ripening and thus senescence and spoilage is also subsequently delayed (Hu and Fu 2003). Examples are plastic polymers with nano-clay as gas barrier, nano-silver and nano-zinc oxide for antimicrobial action, nano-titanium dioxide for UV protection, nano-titanium nitride for mechanical strength, and nano-silica for surface coating, which can ensure better protection and safety of foods, that include reduction in permeability of film and foils, reduction in deodorization, protection against UV light, improvement in mechanical and thermo-resistance properties, and acting as barriers against pathogens and non-pathogenic bacteria or fungi. Nanoform TiO_2 is transparent; it retains its UV resistance and is used as filler material in foils and plastic containers. Nylon nano-composites provide an excellent barrier to oxygen and carbon dioxide, keeping the product fresh and/or preventing off-flavor and rancidity due to oxidation. Use of nano-composite and specialized nanostructures with biopolymers as a base is expected to rise as they offer the possibility for carbon-neutral biodegradable materials for packaging to solve or bring an alternative for the synthetic polymers (Chaudhry and Castle 2011). The nanocomposite structure minimizes loss of carbon dioxide from beer and the influx of oxygen to bottles, keeping beer fresher and providing it up to a six-month shelf life (Joseph and Morrison 2006) and it can also be possible to apply for any carbonated beverages. Recent advances in preparation of natural biopolymer-based nano films and their potential use in packaging are reported by Rhim and Ng (2007). A concept to introduce active nanoparticles and nanostructures into polymer matrices could lead to manifold advantages to improve the performance of a food packaging material and to impart to it additional functionalities, such as antimicrobial, antioxidant, and scavenger and, thus, extending the shelf life, retaining nutritional and associated qualities of the packaged foods (Sorrentino et al. 2007). Chitosan-based nanocomposite films, especially silver-containing ones, showed a promising range of antimicrobial activity (Rhim et al. 2006). Silver nanoparticles can also be introduced in polymeric materials, such as PVC, PE, and PET, while polymerization occurs. Silver nanoparticles kill pathogens, bacteria, viruses, and fungi and are recognized as good and safe for packaging material. Such nano-based packaging materials are 100 times more secure than the normal bulk ones for storage of juices, milk, and other agri-products. This offers opportunities for utilization of agricultural and forest resources, byproducts, and waste and development of biopolymer nanocomposites. Phytoglycogen octenyl succinate nanoparticles embedded in ε-polylysine significantly increased the shelf life of various products (Scheffler et al. 2010). This nanoparticle provides an effective barrier against oxygen, free radicals, and metal ions that cause lipid oxidation and rancidity. Nanotechnology can also innovate in the intelligence of the packaging material to ensure a higher degree of food safety and quality by making the packaging materials smarter that can indicate the consumers about the internal condition of the food product rapidly and accurately. Research activities are ongoing for further development of intelligent packaging that will release a preservative if the food is about to start spoilage. This "release on command" preservative packaging operates by using a bio-switch developed through nanotechnological tools. Smart food packaging will produce an alert signal when oxygen has penetrated inside or if a food is going off. Such packaging is already in use in brewing and dairy industries and consists of nanofilters that can eliminate microorganisms and even viruses (Bhupinder 2010).

11.7.3 Nanobiosensors

Application of nanobiosensors in the food industry could lead to immense improvement in quality control, safety, and traceability with higher sensitivity, stability, and selectability. Scientists have focused insights into the potential advantages of nanotechnology in food safety by detecting the hazardous components (Farhang 2009). Advantages of nanobiosensors can go ahead with their use from raw material preparation, food processing, quality control, monitoring storage conditions, and preservation up to the quality status during storage just before consumption. Increasing use of tools and techniques developed by nanotechnology to detect carcinogenic components, including pathogens and biosensors for the preparation and consumption of safe foods have been reported (Shrivastava and Dash 2009). Nanosensors can also detect allergen proteins to prevent adverse reactions in consumers after taking food, such as peanuts, tree nuts, and gluten (Doyle 2006). Nanomaterials, such as carbon nanotubes, metal nanoparticles, nanowires, and nanocomposite, and specialized engineered nanostructurated materials of specific dimension and shape are playing an increasing role in designing, sensing, and biodetection systems with interest for applications in food analysis (Perez-Lopez and Merkoci 2011).

Presently, an immense interest is seen among R&D workers in modifying electrode surfaces with novel nanomaterials so as to achieve a faster electron transfer of biomolecules with a higher specificity. This advantage has inspired research in coupling nanomaterial-based biosensors with biomolecules. These biosensors have been applied in different areas, such as food quality, determination of food processing parameters and automation, clinical analysis, and environmental control. Integration of enzymes, proteins, or DNA sequences with nanomaterials, e.g., gold nanoparticles (Au NPs), single-wall carbon nanotubes (SWCNTs), multiwalled carbon nanotubes (MWCNTs), cadmium telluride quantum dot nanoparticles (CdTe QDs), and many others, essentially provide novel hybrid systems for food applications combining the conductive or semiconductive properties of nanomaterials with recognition or catalytic properties of biomaterials besides providing binding and electron transfer with biomolecules.

Most of the nanomaterial-based biosensors can be divided into two important classes. Electrochemical biosensors are developed to analyze various food samples, especially pork, milk, water, fruit juice, and honey for sugar, and pesticides as well as different microbes, such as *Salmonella* sp., *E. coli*, *Lactobacillus* sp., including pathogens and spoilage index (Gong-Jun et al. 2009; Zhang et al. 2009). Pesticide and insecticide residues are also analyzed by recently developed biosensors using nanomaterials, viz., SWCNT, MWCNT (multi-walled carbon nanotubes), and ZnO_2 (Wang and Li 2008; Viswanathan et al. 2009) Pt, Pd, Au, SWCNT, and MWCNT nanomaterial-based biosensors are being developed to determine sugar concentrations, such as glucose, fructose, and maltose in specific food samples (Zhao et al. 2007; Ozdemir et al. 2010).

Optical biosensors using nanoparticles of Au, Cd, Fe, and/or their oxides are used to determine the presence of pesticides and insecticide residues (Vinayaka et al. 2009) and microbes, including pathogens and sugar compounds (Wei and Wang 2008; Li et al. 2009a). In agricultural farms, networks of wireless nanosensors positioned across cultivated fields provide essential data, such as optimal times for planting and harvesting crops, time and level of water, fertilizers, pesticides, herbicides, and other treatments that need to be administered for specific plant physiology, pathology, and environmental conditions, leading to best agronomic intelligence with the aim to minimize resource inputs and maximize yield (Scott and Chen 2003).

11.7.4 Nanotechnology in Medicine

The present practice to deliver a drug in the form of chemicals in a body is to dump it into either the bloodstream or stomach and let it spread throughout the body.

For some chemicals, such as insulin, it is acceptable, but for others, such as chemotherapy drugs and some antibiotics, it is best to keep them as local as possible to minimize their damaging effects on surrounding tissues and harmless microbes. Systematically analyzed strategic nanotechniques can help humanity and may put drug-delivery devices in a way for supplying the drug at the appropriate location where they are needed. Nano-medicines and nanotechnology machines have created the possibility of diagnosing, treating, and preventing disease with the use of smart drugs and equipment that can detect as well as target the specific organs, cells, or pathogens and, thus, minimize the damage or adverse affect to healthy cells in the body. Nanosized feed supplements and feed additives, such as the nanoform of a biopolymer derived from a yeast cell wall that can bind mycotoxins to protect animals against mycotoxicosis and an aflatoxin-binding nano-additive for animal feed derived from modified nano-clay (Shi et al. 2005), are good present-day examples of this. Another example is a polystyrene nanoparticle with a polyethylene glycol linker and mannose targeting biomolecule that can potentially bind and eliminate food-borne pathogens in animal feed (Qu et al. 2005).

Some other strategies involve the use of manufactured nano-robots to repair specific cells. Nano crystalline silver is already being used as an antimicrobial agent in the treatment of wounds and external infections. Other useful devices using nanotechnology and other nanoproducts under development include the following:

- Quantum dots: Can identify the exact location of cells (e.g., cancer) in a body
- Nanoparticles: Can deliver drugs directly to cells to minimize damage to surrounding healthy cells of the body
- Nanoshells: Can focus heat from infrared light to destroy cancer cells with minimal damage to surrounding healthy cells
- Nanotubes: Can repair broken bones by providing a structure for new bone materials to grow

These nano devices will almost be the size of viruses or in between viruses and bacteria. They will be able to travel through and enter tissues, such as white blood cells and viruses. They will also be able to open and close cell membranes with the care and precision of a surgeon. Scientists are also developing a technique of nano encapsulation of probiotics to trigger the immune response of the individual, which can act as a *de novo* vaccine (Vidhyalakshmi et al. 2009).

11.7.5 Nutrient Delivery

Nutrients and nutraceuticals present in food as well as in the form of medicine have to pass through a number of adverse environments to reach the destiny of absorbtion. On its way, a substantial percentage get destroyed or goes to waste due to the destructive actions of the system. To make the nutrients maximally available for absorption into the body, a number of attempts, including microencapsulation are being made. The recent innovation in encapsulation and controlled-release technologies as well as the design of novel food delivery systems has been discussed (Huang et al. 2009). A detailed description of food nano-delivery systems based on lipids, proteins, and/or polysaccharides and current analytical techniques used for identification and characterization of these delivery systems in food products has been reported (Luykx et al. 2008). Design and application of nanoderived assemblies and nanomachinery systems as tools for improved delivery and bioavailability of bioactive nutrients and health improvers have been systematically analysed (Shimoni 2009).

The main achievements, such as harnessing casein micelles, a natural nanovehicle of nutrients for delivering hydrophobic bioactives; discovering unique nanotubes based on enzymatic hydrolysis of α-lactalbumin; introduction of novel encapsulation techniques based on cold-set gelation for delivering heat-sensitive bioactives, including probiotics; development and use of Maillard reaction-based conjugates of milk proteins and polysaccharides for encapsulating bioactives; introduction of β-lactoglobulin–pectin nanocomplexes for delivery of hydrophobic nutraceuticals in clear acidic beverages; development of core shell nanoparticles made of heat-aggregated β-lactoglobulin, nanocoated by beet pectin, for bioactive delivery; synergizing the surface properties of whey proteins with stabilization properties of polysaccharides in advanced W/O/W and O/W/O double emulsions; application of milk proteins for drug targeting, including lactoferrin or bovine serum albumin conjugated nanoparticles for effective *in vivo* drug delivery across the blood–brain barrier; β-casein nanoparticles for targeting gastric cancer; fatty acid-coated bovine serum albumin nanoparticles for intestinal delivery; and resistant starch for colon targeting have been reported (Zimet et al. 2011). Micelles are tiny spheres of oil or fat coated with a thin layer of bipolar molecules of which one end is soluble in fat and the other in water. The micelles are suspended in water or, conversely, water is encapsulated in micelles and suspended in oil. Such nano capsules can, for example, contain healthy ω-3 fish oil, which has a strong and unpleasant flavor, and only release it after reaching the stomach. Nanotechnology is showing great potentiality and scope in nutrient management and their availability. For example, the health benefits of curcumin could be enhanced by encapsulation in nanoemulsions (Wang et al. 2008). In the area of nanolaminated coatings on the bioavailability of encapsulated lipids, bioactive fat-liking compounds could be integrated into foods or beverages (Chaudhury and Castle 2011), which may enhance the ingredient's stability, delectableness, appeal, and bioactivity (McClements 2009). For example, nanoemulsions accelerate stability and oral bioavailability of epigallocatechin gallate and curcumin (Wang et al. 2007). DNA microarrays, micro-electro-mechanical systems, and microfluidics technologies is going to open the door for realization of the potential of nanotechnology in molecular approaches of food applications. Researchers examined the encapsulation and controlled release of active food ingredients using nanotechnological approaches and revealed its suitability and potentiality (Huang et al. 2010).

Solid-lipid nanoparticles are formed by controlled crystallization of food nanoemulsions and have been reported for delivery of bioactives and nutraceuticals, such as lycopene and carotenoids (Weiss et al. 2008). Findings also suggest that for targeted neutraceuticals and functional nutrients, such as lycopene, β-carotene, ω-3 fatty acids, phytosterols, and isoflavones as well as nanosized vehicles having at least one dimensional size of about 30 nm, can be used in even clear beverages without any phase separation (Garti and Aserin 2007; Garti 2008). From the results of the research on nutrient delivery with nanomaterials, several products are already entered into the market domain and it is expected to occupy a larger space throughout the globe within near future.

11.7.6 Food Additives and Ingredients

Food additives with nanoingredients are being sold in the United States and Germany, and it is claimed that these foods offer improved taste, flavor, texture, and consistency due to the effect of nanostructured materials. A number of oxides, such as titanium dioxide and silicon dioxide, are being used as additives in food compositions. The former has typically been used as a colorant, and the latter as a flow agent in foods (Ottaway 2009). The additives are mainly applied targeting at the sports and health food markets and contain minerals with a nano-formulation, such as silicon dioxide, magnesium, calcium, etc. The particle size of these minerals is claimed to be smaller than 100 nm, so they can pass through the stomach and GI wall and into the bloodstream more quickly than ordinary minerals with larger particle sizes. Use of silver nanostructures as antimicrobial, antiodorant, and health supplements (proclaimed) is being shown its extraordinary utilization potentiality and is expected to surpass all other nanomaterials used in different sectors in near future (WWICS 2008). Nanosilver can also be applied as an effective and ideal additive in antibacterial wheat flour as suggested by recent patent works (Park 2005).

Nano-additives can be incorporated in micelles or capsules of proteins or other natural food ingredients or food formulations to achieve desired functionality of the nanofood.

Nano-additives are also being potentially utilized by different researchers and food processors to improve the quality and to reduce the degradability and spoilage of food products. The US-based Oilfresh Corporation has marketed a new nanoceramic product, which reduces oil use in restaurants and fast food shops by half. As a result of its large surface area, the product prevents the oxidation and agglomeration of fats in deep fat fryers, thus extending the useful lifespan of the oil. An additional benefit is that oil heats up more quickly, reducing the energy required for cooking. Nanoparticles of different materials, including silver, gold, copper, platinum, palladium, iridium, titanium, and zinc are also used in production of various food supplements, viz., Mesosilver, Mesogold, Mesocopper, Mesoplatinum, Mesopalladium, Mesoiridium, Mesotitanium, and Mesozinc (FSAI 2008).

11.8 Risk of Nanotechnology in Food Processing

Nanotechnology promises big benefits for safety, security, quality, and shelf life of food products in food processing industries, provided the challenges it brings can be overcome. However, it is well established that the nanoparticles equipped with new chemical and physical properties that vary from normal macroparticles of the same composition may interact with the living systems, thereby causing unexpected hazard (Das et al. 2009). It also brings the prospect of consumer exposure to some insoluble and possibly biopersistent nanoparticles (hard nanomaterials) through consumption of food and drinks as well as fodder through the food chain and web. The major issues among people is that once in the body, the nanoparticles with large reactive surfaces may cross biological barriers to reach those parts of the body that are otherwise protected from entry of (larger) particulate materials. Although most foods processed at nanoscale should not raise any special health concerns, there is a huge gap of knowledge in our current understanding of the properties, behavior, and effects of hard nanomaterials, which may be used in food processing industries. Such knowledge gaps make it difficult to assess the risk to a consumer although a careful consideration of the nature of materials and applications can provide a basis for a conceptual risk categorization on a case-by-case basis.

Risk of migration of nanomaterials from the packaging into a food and subsequently to the body system is also being advocated (Bradley et al. 2011). Toxicological risk of nanomaterials is also of a major concern, but toxicology research and risk assessments in nanotechnologies are practically yet to be done properly, especially in the food sector, and few have proved to be valuable in terms of their use in assessing toxicity (Card and Magnuson 2010). Some materials exhibit toxicity at nanoscale and not at macroscale. For example, Cui et al. (2005) showed that single-walled carbon nanotubes inhibited human embryo kidney cell proliferation and negatively impacted on cell growth and cell turnover. Nanoadditives incorporated in foods are readily reactive with the other food compounds and stomach acids as well as enzymes of the gastrointestinal tract and thus will change the internal environment of the digestive system of an individual.

Although the risk potentiality of the nanomaterials are in the research stage, another important issue of concern to risk is that nanomaterials are extending their range of application in a unexpected progression, but still there is not a single regulation specifically for the application and analysis of nanomaterials in foods. General regulatory rules, such as European Union, Codex Alimentarius, USDA, USFDA, and PFA, are still followed for this purpose, but these regulations are concerned with general food safety, food contact surface, food hazard, general food adulteration, etc. But nanomaterials have certain extraordinary properties other than those of general bulk materials.

The initiative in concern regarding nanomaterials is observed to differ from country to country worldwide. In Taiwan, authorities have introduced a quality assurance system, viz., "Nano Mark System," for consumers, which certifies that a product uses a genuine nanotechnology, but food is not included as a category to which this symbol is assigned. In Australia, like in Europe, nanotechnologies are regulated by horizontal legislation of the European Commission (Lyons and Whelan 2010). NICNAS, which regulates chemicals for the protection of human health and the environment, have recently introduced new administrative processes to address nanotechnology (NICNAS 2010). In the United States, multiple federal agencies regulate products associated with nanotechnologies and nanomaterials, but there is no regulatory framework that provides consistent and comprehensive screening and protections for consumers. So it is transparent that a regulatory gap exists between commercial developments and public expectations about regulatory protections for the application of nanotechnological principles in food processing and consumption and its need is to be fulfilled globally by taking a rigid initiative within near future.

11.9 Summary

Science and technology of nanomaterials is at a juvenile stage, but it shows a great potential in the food processing sector. Efforts to facilitate international collaboration and information exchange are underway to ensure acceptance and utilization of the many benefits of nanotechnology. Many ideas about future applications of nanotechnology in food do not closely resemble the currently available food and processing technologies. Even though such futuristic speculations are probably not the aim of present-day research, the fact that they are suggested in public media influences the public awareness about food based on nanotechnology. The future of nanotechnology in food is expected to be extraordinary, which will change the entire scenario of food processing. However, extensive research can only bring many speculations into reality.

References

Antiochia, R., Lavagnini, I., and Magno, F. (2004). Amperometric mediated carbon nanotube paste biosensor for fructose determination. *Analytical Letters*, 37(8), 1657–1669.

Azeredo, H.C., Mattoso, L.H., Wood, D.F., Williams, T.G., Avena-Bustillos, R.D., and McHugh, T.H. (2009). Nanocomposite edible films from mango puree reinforced with cellulose nanofibers. *Journal of Food Science*, 74(5), 31–35.

Bauman, D.E., Perfield, II, J.W., Harvatine, K.J., and Baumgard, L.H. (2008). Regulation of fat synthesis by conjugated linoleic acid: Lactation and the ruminant model. *Journal of Nutrition*, 138, 403–409.

Bhupinder, S.S. (2010). Food nanotechnology – an overview. *Nanotechnology, Science and Applications*, 3, 1–15.

Bradley, E.L., Castle, L., and Chaudhry, Q. (2011). Applications of nanomaterials in food packaging with a consideration of opportunities for developing countries. *Trends in Food Science and Technology*, 22, 604–610.

Branton, D., Deamer, D.W., Marziali, A., Bayley, H., Benner, S.A., and Butler, T. (2008). The potential and challenges of nanopores sequencing. *Nature Biotechnology*, 26(10), 1146–1153.

Brody, A.I., Bugusu, B., Han, J.H., Sand, C.K., and McHugh, T.H. (2008). Innovative food packaging solutions. *Journal of Food Science*, 73(8), 107–116.

Card, J.W., and Magnuson, B.A. (2010). A method to assess the quality of studies that examine the toxicity of engineered nanomaterials. *International Journal of Toxicology*, 29(4), 402–410.

Chau, C.F., Wu, S.H., and Yen, G.C. (2007). The development of regulations for food nanotechnology. *Trends in Food Science and Technology*, 18(5), 269–280.

Chau, C.F., and Yen, G.C. (2008). A general introduction to food nanotechnology. *Food Info Online Features*. Available at http://www. foodsciencecentral.com/fsc/ixid15445. Accessed February 2010.

Chaudhry, Q. (2010). Applications of nanotechnology in food chain: Food application of nanotechnologies – An overview of potential benefits and risks. International symposium – Nanotechnology in the food chain: Opportunities and risk, November 24, Brussels, Belgium.

Chaudhry, Q., and Castle, L. (2011). Food applications of nanotechnologies: An overview of opportunities and challenges for developing countries. *Trends in Food Science and Technology*, 22, 595–603.

Chaudhry, Q., Castle, L., and Watkins, R. (Eds.). (2010). *Nanotechnologies in food*. Cambridge: Royal Society of Chemistry.

Chen, H., and Yada, R. (2011). Nanotechnologies in agriculture: New tools for sustainable development. *Trends in Food Science and Technology*, 22, 585–594.

Ciesielski, P.N., Hijazi, F.M., Scott, A.M., Faulkner, C.J., Beard, L., and Emmett, K. (2010). Photosystem-based biohybrid photoelectrochemical cells. *Bioresource Technology*, 101(9), 3047–3053.

Clegg, S.M., Knight, A.I., Beeren, C.J.M., and Wilde, P.J. (2009). Fat reduction whilst maintaining the sensory characteristics of fat using multiple emulsions. *5th International Symposium on Food Rheology and Structure (ISFRS, 2009)*, 238–241.

Cui, D., Tian, F., Ozkan, C.S., Wang, M., and Gao, H. (2005). Effect of single wall carbon nanotubes on human HEK293 cells. *Toxicology Letters*, 155(1), 73–85.

Das, M., Saxena, N., and Dwivedi, P.D. (2009). Emerging trends of nanoparticles application in food technology: Safety paradigms. *Nanotoxicology*, 3(1), 10–18.

Davis, R., Guliants, V.V., Huber, G., Lobo, R.F., Miller, J.T., and Neurock, M. (2009). An international assessment of research in catalysis by nanostructure and materials. Baltimore, MD: WTEC. Available at http://www.wtec.org/catalysis/WTEC-CatalysisReport-6Feb2009-color-hi-res.pdf. Accessed March 2010.

DeGusseme, B., Sintubin, L., Baert, L., Thibo, E., Hennebel, T., and Vermeulen, G. (2010). Biogenic silver nanoparticles for disinfection of water contaminated with viruses. *Applied and Environmental Microbiology*, 76, 1082–1087.

Diallo, M. (2009). Water treatment by dendrimer-enhanced filtration: Principles and applications. In N. Savage, M. Diallo, J. Duncan, A. Street, and R. Sustich (Eds.). *Nanotechnology Applications for Clean Water*, Norwich, NY: William Andrew, 143–155.

Doyle, M.E. (2006). Nanotechnology: A brief literature review. Food Research Institute Briefings, June. Available at http://www.wisc.edu/fri/briefs/FRIBrief_Nano-tech_Lit_Rev.pdf. Accessed August 2008.

Dungchia, W., Siangproh, W., Chaicumpa, W., Tongtawe, P., and Chailapakul, O. (2008). *Salmonella typhi* determination using voltammetric amplification of nanoparticles: A highly sensitive strategy for metalloimmuno assay based on a copper-enhanced gold label. *Talanta*, 77, 727–732.

El-Boubbou, K., Gruden, C., and Huang, X. (2007). Magnetic glyconanoparticles: A unique tool for rapid pathogen detection, decontamination, and strain differentiation. *Journal of the American Chemical Society*, 129, 13392–13393.

Farhang, B. (2009). Nanotechnology and applications in food safety. In G. Barbosa-Cánovas, A. Mortimer, D. Lineback, W. Spiess, K. Buckle and P. Colonna (Eds.). *Globals Issues in Food Science and Technology*, London: Academic Press, 401–410.

Farmen, L. (2009). Commercialization of nanotechnology for removal of heavy metals in drinking water. In N. Savage, M. Diallo, J. Duncan, A. Street, and R. Sustich (Eds.), *Nanotechnology Applications for Clean Water*, Norwich, NY: William Andrew, 115–130.

FSAI (Food Safety Authority of Ireland). (2008). *The relevance for food safety of applications of nanotechnology in the food and feed industries*. Dublin: FSAI.

Garti, N. (2008). *Delivery and Controlled Release of Bioactives in Foods and Nutraceuticals*. Cambridge: Woodhead Publishing.

Garti, N., and Aserin, A. (2007). Understanding and controlling the microstructure of complex foods. In M.D. Julian (Ed.), *Nanoscale liquid self-assembled dispersions in foods and the delivery of functional ingredients*, Cambridge: Woodhead Publishing, 504–553.

Gomez-Hens, A., Fernandez-Romero, J.M., and Aguilar-Caballos, M.P. (2008). Nanostructures as analytical tools in bioassays. *Trends in Analytical Chemistry*, 27(5), 394–406.

Gong-Jun, Y., Jin-Lin, H., Wen-Jing, M., Ming, S., and Xin-An, J. (2009). A reusable capacitive immunosensor for detection of Salmonella spp. based on grafted ethylene diamine and self-assembled gold nanoparticle monolayers. *Analytica Chimica Acta*, 647, 159–166.

Hennebel, T., De Corte, S., Vanhaecke, L., Vanherck, K., Forrez, I., and De Gussseme, B. (2010). Removal of diatrizoate with catalytically active membranes incorporating microbially produced palladium nanoparticles. *Water Research*, 44, 1498–1506.

Hennebel, T., Simoen, H., DeWindt, W., Verloo, M., Boon, N., and Verstraete, W. (2009a). Biocatalytic dechlorination of trichloroethylene by bio-precipitated and encapsulated palladium nanoparticles in a fixed bed reactor. *Biotechnology and Bioengineering*, 102, 995–1002.

Hennebel, T., Verhagen, P., De Gusseme, B., Simoen, H., Vlaeminck, S.E., and Boon, N. (2009b). Removal of trichloroethylene with bio-Pd in a pilot-scale membrane reactor. *Chemosphere*, 76, 1221–1225.

Hoek, E.M.V., and Ghosh, A.K. (2009). Nanotechnology based membranes for water purification. In N. Savage, M. Diallo, J. Duncan, A. Street, and R. Sustich (Eds.), *Nanotechnology Applications for Clean Water*, Norwich, NY: William Andrew, 47–58.

Hu, A.W., and Fu, Z.H. (2003). Nanotechnology and its application in packaging and packaging machinery. *Packaging Engineering*, 24, 22–24.

Huang, Q., Given, P., and Qian, M. (2009). *Micro/Nano Encapsulation of Active Food Ingredients*. Oxford: Oxford University Press.

Huang, Q., Yu, H., and Ru, Q. (2010). Bioavailability and delivery of nutraceuticals using nanotechnology. *Journal of Food Science*, 75, 50–57.

Hwang, L.S., and Yeh, A.I. (2010). Applying nanotechnology in food in Taiwan. Paper presented at the International Conference on Food Applications of Nanoscale Science *(ICOFANS)*, Tokyo, Japan, June 9–11.

Ji, X., Zheng, J., Xu, J., Rastogi, V.K., Cheng, T.C., and DeFrank, J.J. (2005). (CdSe)ZnS quantum dots and organophosphorus hydrolase bioconjugate as biosensors for detection of paraoxon. *Journal of Physical Chemistry*, 109, 3793–3799.

Jochen, W., Paul, T., and McClements, D.J. (2006). Functional materials in food nanotechnology. *Journal of Food Science*, 71(9), R107–R116.

Joseph, T., and Morrison, M. (2006). Nanotechnology in Agriculture and Food. Available at ftp://ftp.cordis.europa.eu/pub/nanotechnology/docs/nanotechnology_in_agriculture_and_food.pdf. Accessed November 2013.

Kalele, S.A., Kundu, A.A., Gosavi, S.W., Deobagkar, D.N., Deobagkar, D.D., and Kulkarni, S.K. (2006). Rapid detection of *Escherichia coli* by using antibody-conjugated silver nanoshells. *Small*, 2(3), 335–338.

Lagaron, J.M., and Lopez-Rubio, A. (2011). Nanotechnology for bioplastics: Opportunities, challenges and strategies. *Trends in Food Science and Technology*, 22, 611–617.

Lanzon, N., Kahl, J., and Ploeger, A. (2009). Nanotechnology in the context of organic food processing. Wissenschaftstagung Ökologischer Landbau, Zürich, February 11–13. Available at http://orgprints. org/14180/1/Lanzon_14180.pdf. Accessed February 2010.

Li, Q., Wu, P., and Shang, J.K. (2009a). Nanostructured visible-light photocatalysts for water purification. In N. Savage, M. Diallo, J. Duncan, A. Street, and R. Sustich (Eds.), *Nanotechnology Applications for Clean Water*, Norwich, NY: William Andrew, 17–37.

Li, X., Zhou, Y., Zheng, Z., Yue, X., Dai, Z., and Liu, S. (2009b). Glucose biosensor based on nanocomposite films of CdTe quantum dots and glucose oxidase. *Langmuir*, 25(11), 6580–6586.

Li, X.X., Cao, C., Han, S., Jong, S., and Sang, J. (2009). Detection of pathogen based on the catalytic growth of gold nanocrystals. *Water Research*, 43, 1425–1431.

Lin, Y.H., Chen, S.H., Chuang, Y.C., Lu, Y.C., Shen, T.Y., and Chang, C.A. (2008). Disposable amperometric immunosensing strips fabricated by Au nanoparticles-modified screen printed carbon electrodes for the detection of food borne pathogen *E. coli* O157:H7. *Biosensors and Bioelectronics*, 23, 1832–1837.

Lu, J., Do, I., Drzal, L.T., Worden, R.M., and Lee, I. (2007). Nanometal decorated exfoliated graphite nanoplatelet based glucose biosensors with high sensitivity and fast response. *Biosensors and Bioelectronics*, 23, 1825–1832.

Luykx, D.M., Peters, R.J.B., van Ruth, S.M., and Bouwmeester, H. (2008). A review of analytical methods for the identification and characterization of nano delivery systems in food. *Journal of Agriculture and Food Chemistry*, 56(18), 8231–8247.

Maftoum, N. (2009). Healthylicious. Available at http://nicolemaftoum.blogspot.com. Accessed February 2010.

Mathew, A.P., and Dufresne, A. (2002). Morphological investigation of nanocomposites from sorbitol plasticized starch and tunicin whiskers. *Biomacromolecules*, 3(3), 609–617.

McClements, D.J. (2009). Design of nano-laminated coatings to control bioavailability of lipophilic food components. *Journal of Food Science*, 166, 1–19.

McHugh, T.H. (2008). The world of food science. *Food Nanotechnology-Food Packaging Applications*, 4, 1–3.

Morris, K. (2009). Nanotechnology crucial in fighting infectious disease. *Lancet Infectious Diseases*, 9, 215.

Morris, V.J., and Parker, R. (2008). Natural and designed self-assembled nanostructures in foods. *The World of Food Science: Food Nanotechnology*. Available at www.worldfood science.org. Accessed February 2010.

Mozafari, M.R., Johnson, C., Hatziantoniou, S., and Demetzos, C. (2008). Nanoliposomes and their applications in food nanotechnology. *Journal of Liposome Research*, 18(4), 309–327.

Nangmenyi, G., and Economy, J. (2009). Nonmetallic particles for oligodynamic microbial disinfection. In N. Savage, M. Diallo, J. Duncan, A. Street, and R. Sustich (Eds.), *Nanotechnology Applications for Clean Water*, Norwich, NY: William Andrew 3–15.

Nanoposts Report (2008). Nanotechnology and consumer goods e-Market and applications to 2015. Available at www.nanoposts.com. Accessed February 2010.

Narducci, D. (2007). An introduction to nanotechnologies: What's in it for us? *Veterinary Research Communications*, 31(Suppl. 1), 131–137.

NICNAS. (2010). National industrial chemicals notification and assessment scheme. Australian Government. Gazette.

Norman, R.S., Stone, J.W., Gole, A., Murphy, C.J., and Sabo-Attwood, T.L. (2008). Targeted photothermal lysis of the pathogenic bacteria, *Pseudomonas aeruginosa* with gold nanorods. *Nano Letters*, 8, 302–306.

Ottaway, P.B. (2009). Nanotechnology in supplements and foods – EU concerns. *Nutraceuticals International*, 1.

Ozdemir, C., Yeni, F., Odaci, D., and Timur, S. (2010). Electrochemical glucose biosensing by pyranose oxidase immobilized in gold nanoparticle-polyaniline/AgCl/gelatin nanocomposite matrix. *Food Chemistry*, 119, 380–385.

Park, H.J., Chinnan, M.S., and Shewfelt, R.L. (1994). Edible coating effects on storage life and quality of tomatoes. *Journal of Food Science*, 59(3), 568–570.

Park, H.M., Lee, W.K., Park, C.Y., Cho, W.J., and Ha, C.S. (2003). Environmentally friendly polymer hybrids: Part I mechanical, thermal, and barrier properties of the thermoplastic starch/clay nanocomposites. *Journal of Material Science*, 38, 909–915.

Park, K.H. (2005). Preparation method antibacterial wheat flour by using silver nanoparticles. Korean Intellectual Property Office (KIPO), Pub. No/date 1020050101529A/24.10.2005.

Perez-Lopez, B., and Merkoci, A. (2011). Nanomaterials based biosensors for food analysis applications. *Trends in Food Science and Technology*, 22, 625–639.

Prakash-Deo, R., Wang, J., Block, I., Mulchandani, A., Joshi, K.A., and Trojanowicz, M. (2005). Determination of organophosphate pesticides at a carbon nanotube/organophosphorus hydrolase electrochemical biosensor. *Analytica Chimica Acta*, 530, 185–189.

Qu, L., Luo, P.G., Taylor, S., Lin, Y., Huang, W., and Tzeng, T.R.J. (2005). Visualizing adhesion-induced agglutination of *Escherichia coli* with mannosylated nanoparticles. *Journal of Nanoscience and Nanotechnology*, 5, 319–322.

Rhim, J.W., Hong, S.I., Park, H.M., and Ng, P.K.W. (2006). Preparation and characterization of chitosan-based nanocomposite films with antimicrobial activity. *Journal of Culture and Food Chemistry*, 54(16), 5814–5822.

Rhim, J.W., and Ng, P.K. (2007). Natural biopolymer-based nanocomposite films for packaging applications. *Critical Reviews in Food Science and Nutrition*, 47(4), 411–433.

Sabliov, C. (2009). New and emerging food applications of polymeric nanoparticles for improved health. *IFT International Food Nanoscience Conference 2009*, June 6. Available at http://members.ift.org. Accessed February 2010.

Sabliov, C.M., and Astete, C.E. (2007). Controlled release technologies for targeted nutrition: Encapsulation and controlled release of antioxidants and vitamins via polymeric nanoparticles. In N. Garti (Ed.), *Delivery and Controlled Release of Bioactives in Foods and Nutraceuticals.* Cambridge: Woodhead Publishing.

Savage, N., Diallo, M., Duncan, J., Street, A., and Sustich, R. (2009). *Nanotechnology Applications for Clean Water*. Norwich, NY: William Andrew.

Scheffler, S.L., Wang, X., Huang, L., Gonzales, F.S., and Yao, Y. (2010). Phytoglycogen octenyl succinate, an amphilic carbohydrate nanoparticle and epsilonpolylysine to improve lipid oxidative stability of emulsions. *Journal of Agricultural and Food Chemistry*, 58(1), 660–667.

Scott, N.R., and Chen, H. (2003). Nanoscale science and engineering or agriculture and food systems. In Roadmap Report of National Planning Workshop. Washington D.C. November 18–19. Available at www.nseafs.cornell.edu. Accessed February 2009.

Shi, Y.H., Xu, Z.R., Feng, J.L., Hu, C.H., and Xia, M.S. (2005). In vitro adsorption of aflatoxin adsorbing nano-additive for aflatoxin B1, B2, G1, G2. *Scientia Agricultura Sinica*, 38(5), 1069–1072.

Shimoni, E. (2009). Nanotechnology for foods: focus on delivering health. In G. Barbosa-Cánovas, A. Mortimer, D. Lineback, W. Spiess, K. Buckle, and P. Colonna (Eds.), *Global Issues in Food Science and Technology.* London: Academic Press, 411–424.

Shrivastava, S., and Dash, D. (2009). Agrifood nanotechnology: A tiny revolution in food and agriculture. *Journal of Nano Research*, 6, 1–14.

Simoniana, A.L., Good, T.A., Wang, S.S., and Wild, J.R. (2005). Nanoparticle-based optical biosensors for the direct detection of organophosphate chemical warfare agents and pesticides. *Analytica Chimica Acta*, 534, 69–77.

Sintubin, L., De Windt, W., Dick, J., Mast, J., Van der Ha, D., and Serstraete, W. (2009). Lactic acid bacteria as reducing and capping agent for the fast and efficient production of silver nanoparticles. *Applied Microbiology and Biotechnology,* 84, 741–749.

Soghomonian, V., and Heremans, J.J. (2009). Characterization of electrical conductivity in a zeolite like material. *Applied Physics Letters*, 95(15), 152112.

Sonalika et al. 2008.

Sorrentino, A., Gorrasi, G., and Vittoria, V. (2007). Potential prospective of bio-nanocomposites for food packaging applications. *Trends in Food Science and Technology*, 18, 84–95.

Tominaga, M., Nomura, S., and Taniguchi, I. (2009). D-Fructose detection based on the direct heterogeneous electron transfer reaction of fructose dehydrogenase adsorbed onto multi-walled carbon nanotubes synthesized on platinum electrode. *Biosensors and Bioelectronics*, 24, 1184–1188.

Tsukamoto, K., Wakayama, J., and Sugiyama, S. (2010). Nanobiotechnology approach for food and food related fields. Poster presented at the International Conference on Food Applications of Nanoscale Science (*ICOFANS)*, Tokyo, Japan, June 9–11.

Vidhyalakshmi, R., Bhakyaraj, R., and Subhasree, R.S. (2009). Encapsulation "The future of probiotics": A review. *Advanced Biological Research*, (3–4), 6–103.

Vinayaka, A.C., Basheer, S., and Thakur, M.S. (2009). Bioconjugation of CdTe quantum dot for the detection of 2,4-dichlorophenoxyacetic acid by competitive fluoroimmunoassay based biosensor. *Biosensors and Bioelectronics*, 24, 1615–1620.

Viswanathan, S., Radecka, H., and Radecki, J. (2009). Electrochemical biosensor for pesticides based on acetylcholinesterase immobilized on polyaniline deposited on vertically assembled carbon nanotubes wrapped with ssDNA. *Biosensors and Bioelectronics*, 24, 2772–2777.

Wang, M., and Li, Z. (2008). Nano-composite ZrO_2/Au film electrode for voltammetric detection of parathion. *Sensors and Actuators B*, 133, 607–612.

Wang, X., Jiang, Y., Wang, Y.W., and Huang, Q. (2007). Enhancing stability and oral bioavailability of polyphenols using nanoemulsions. *ACS Symposium Series, Vol. 1007, Micro/Nano-encapsulation of Active Food Ingredients*, 198–212.

Wang, X., Jiang, Y., Wang, Y.W., Huang, M.T., Ho, C.T., and Huang, Q. (2008). Enhancing antiinflammation activity of curcumin through O/W nanoemulsions. *Food Chemistry*, 108(2), 419–424.

Wei, H., and Wang, E. (2008). Fe_3O_4 magnetic nanoparticles as peroxidase mimetics and their applications in H_2O_2 and glucose detection. *Analytical Chemistry*, 80, 2250–2254.

Weiss, J., Decker, E.A., McClements, J., Kristbergsson, K., Helgason, T., and Awad, T.S. (2008). Solid lipid nanoparticles as delivery systems for bioactive food components. *Food Biophysics*, 3(2), 146–154.

Weiss, J., Takhistov, P., and McClements, D.J. (2006). Functional materials in food nanotechnology. *Journal of Food Science*, 71, R107–R116.

WWICS (Woodrow Wilson International Center for Scholars). (2008). The nanotechnology consumer inventory. Available at www.nanotechproject.org. Accessed July 2010.

Yakes, B.J., Lipert, R.J., Bannantine, J.P., and Porter, M.D. (2008). Detection of *Mycobacterium avium* subsp. Paratuberculosis by a sonicate immunoassay based on surface-enhanced Raman scattering. *Clinical and Vaccine Immunology*, 227–234.

Yang, L.J., and Li, Y.B. (2005). Quantum dots as fluorescent labels for quantitative detection of *Salmonella typhimurium* in chicken carcass wash water. *Journal of Food Protection*, 68(6), 1241–1245.

Zhang, L., Jiang, Y., Ding, Y., Povey, M., and York, D. (2007). Investigation into the antibacterial behaviour of suspensions of ZnO nanoparticles (ZnO nanofluids). *Journal of Nanoparticle Research*, 9(3), 479–489.

Zhao, Z.W., Chen, X.J., Tay, B.K., Chen, J.S., Han, Z.J., and Khor, K.A. (2007). A novel amperometric biosensor based on ZnO:Co nanoclusters for biosensing glucose. *Biosensors and Bioelectronics*, 23, 135–139.

Zimet, P., Rosenberg, D., and Livney, Y.D. (2011). Re-assembled casein micelles and casein nanoparticles as nano-vehicles for ω-3 polyunsaturated fatty acids. *Food Hydrocolloids*, 25, 1270–1276.

Chapter 12

Computational Fluid Dynamics in Food Processing

Tomás Norton and Brijesh Tiwari

Contents

12.1 Introduction

During food processing, thermo-fluid dynamics govern many processes and are involved in almost every process. Due to the complexity of the physical mechanisms involved in food processing, quantifying the phenomena must involve sophisticated design and analysis tools. Physical experimentation may allow for direct measurement; however, a comprehensive analysis would not only necessitate expensive equipment and require considerable time and expertise, but could also be intrusive and therefore affect the quality of the results (Verboven et al. 2004). Fortunately, ubiquitous processes, such as blanching, cooking, drying, sterilization, and pasteurization, which rely on food exchange to raise a product's food center to a prespecified temperature, are highly amenable to CFD modeling and have unarguably contributed to its exponential take-up witnessed over recent years. Coupled with other technological advancements, CFD has led to an improvement in both food quality and safety alongside reducing energy consumption in industrial processes and has reduced the amount of empiricism associated with a design process (Norton and Sun 2007).

The application of the principles of fluid motion and heat transfer to design problems in the food industry has undergone remarkable development in the last couple of decades. Problems involving heat and mass transfer, phase change, chemical reactions, and complex geometry, which once required either highly expensive experimental rigs or over-simplified computations, can now be modeled with a high level of spatial and temporal accuracy on personal computers. This progression is due to the development of advanced computer design and analysis tools, such as computational fluid dynamics (CFD), as these tackle complex problems in fluid mechanics and heat transfer and many other physical processes with important industrial applications. CFD is based on numerical methods that predict the governing transport mechanisms over a multidimensional domain of interest, and its physical basis is rooted in classical fluid mechanics. Since the first computer implementation of CFD in the 1950s, it has continued to be developed contemporaneously with the digital computer (Norton and Sun 2007). In its present-day form, CFD can be used to quantify many complex food-hydraulic phenomena and, as a result, has developed into a multifaceted industry, generating billions of Euros worldwide within a vast range of specializations (Xia and Sun 2002).

In light of the rapid developments in both food processing and CFD technology that has taken place over recent years, it is necessary to provide a critical and comprehensive account of the latest advances made through successful applications in research. First, the need for CFD to model food processing will be presented. The main equations governing the physical mechanisms encountered in food processing will then be discussed, followed by an extensive overview of the conventional processes, which have been modeled with CFD. A focus of the CFD modeling studies of emerging food processing technologies will then be highlighted. Finally, an overview of the challenges that are still encountered in the CFD modeling of food processes will be outlined.

12.2 Need for CFD Modeling in Food Processing

Owing to the complex thermo-physical properties of food, heat transfer to and from foods can stimulate complex chemical and physical alterations in them. Nowadays, with both food preservation and safety being equally important objectives of processing, it is necessary to promote quality characteristics of food while eradicating the threat of spoilage. For this to happen efficiently, the appropriate temperature and duration of heating must be known. Mathematical models permit the representation of a physical process from the analysis of measured data or physical properties over a range of experimental conditions. During food processing, heat transfer occurs due to one or more of three mechanisms: conduction, convection, and radiation, and depending on the food process, one mechanism is usually dominant (Wang and Sun 2003). As the food exchange phenomena are either distributed in space or in space and time, their fundamental representation is by means of partial differential equations (PDEs). These transport equations can be solved analytically for a small number of boundary and initial conditions (Nicolaï et al. 2001). With this in mind, CFD codes have been developed around numerical algorithms that can efficiently solve the PDEs governing all fluid flow, heat transfer, and many other physical phenomena. CFD techniques can be used to build distributed parameter models that are spatially and temporally representative of the physical system, thereby permitting the achievement of a solution with a high level of physical realism. With recent versatility, adaptability, and efficiency gained from progress in geometrical, physical, and numerical modeling, CFD can easily be applied to conventional and novel food processes ranging from experimental models to the scaled-up systems.

12.3 Numerical Representation of Food Processes

12.3.1 Numerical Techniques

CFD code developers have a choice of many different numerical techniques to discretize the transport equations. The most important of these include finite difference, finite elements, and finite volume. The finite difference technique is the oldest one used, and many examples of its application in the food industry exist. However, due to difficulties in coping with irregular geometry, finite difference is not commercially implemented. Furthermore, the current trend of commercial CFD coding is aimed toward developing unstructured meshing technology capable of handling the complex three-dimensional geometries encountered in the industry. Therefore, the prospects of finite difference being used in industrial CFD applications seem limited.

Finite element methods have historically been used in structural analysis when the equilibrium of the solution must be satisfied at the node of each element. Nicolaï et al. (2001) provided a short introduction to the use of the finite element method in conduction heat transfer modeling, and it will not be discussed here. Suffice to say that as a result of the weighting functions used by this method, obtaining a three-dimensional CFD solution with a large number of cells is impractical at present. Therefore, finite elements are not generally used by commercial CFD developers, especially as many of these CFD codes are marketed toward solving aerodynamic problems. Nonetheless, finite element methods have enjoyed use in the modeling of electromagnetic heating in microwave ovens (Geedipalli et al. 2007; Verboven et al. 2007), vacuum microwave drying (Ressing et al. 2007), radio frequency heating of food (Marra et al. 2007, 2009, 2010); conduction and mass transport during drying (Aversa et al. 2007).

With finite volume techniques, the integral transport equations governing the physical process are expressed in conservation form (divergence of fluxes), and the

Table 12.1 List of Current Available Software Packages

Company	*CFD Code*	*Features*
CHAM, Ltd., www.cham.co.uk	PHOENICS 2009	LEVL, SG, FV
ANSYS, Inc., www.ansys.com	ANSYS Fluent V12 ANSYS CFX V12	USG, LAG + PT, MPH + IPH USG, FV, FE CREM
CD Adapco Group, www.cd-adapco.com	STAR-CD V4.12 STAR-CCM+ V5.02	STAR-CD: USG, LAG + PT, MPH + IPH STAR-CCM+: Intuitive user interface, USG, LAG + PT, MPH + IPH
Flow Science, Inc., www.flow3d.com	FLOW-3D 8.2	Advanced moving obstacle capabilities; SG, LAG + PT, MPH + IPH
ADINA, Inc., www.adina.com	ANDINA-F	FE + FV, SG, ALE

volume integrals are then converted to surface integrals using Gauss's divergence theorem. This is a direct extension of the control volume analysis that many engineers use in thermodynamics and heat transfer applications, etc., so it can be easily interpreted. Thus, expressing the equation system through finite volumes forms a physically intuitive method of achieving a systematic account of the changes in mass, momentum, and energy as fluid crosses the boundaries of discrete spatial volumes within the computational domain. Also, finite volume techniques yield algebraic equations that promote solver robustness. Table 12.1 lists the currently available software packages.

12.3.2 Generic Equation and Its CFD Approximation

The basic requirement of CFD is to obtain a solution to a set of governing PDEs of the transported variable. As discussed above, the transport phenomena are generally described by commercial CFD developers using the finite volume method. The generic conservation convection-diffusion equation, after application of Gauss's theorem to obtain the surface integrals, can be described as follows:

$$\int_V \frac{\partial \rho\phi}{\partial t} + \int_A \mathrm{n}\,(\rho U \phi)\,dA = \int_A \mathrm{n}\,(\Gamma_\phi\, grad\ \phi)\,dA + \int_V S_\phi \tag{12.1}$$

The conservation principle is explicit in Equation 12.1, i.e., the rate of increase of ϕ in the control volume plus the net rate of decrease of ϕ due to convection is equal to the increase in ϕ due to diffusion and an increase in ϕ due to the sources (Versteeg and Malalasekera 1995). For a numerical solution, both the surface and volume integrals need to be solved on a discrete level, which means numerical interpolation schemes are required. As the convection term, the only nonlinear term in the equation needs to be approximated at each mesh element face, and it presents the greatest challenge in allowing the numerical scheme to preserve properties such as the stability, transportiveness, boundedness, and accuracy of a solution. For the same reason, numerical schemes are often called convection schemes as the accurate and stable representation of the convection term is a major requirement (Patankar 1980). The reader can refer to standard CFD textbooks, e.g., those of Patankar (1980) and Versteeg and Malalasekera (1995) for a complete discussion on the various properties of convection schemes.

12.3.3 Numerical Mesh

The volume mesh in a simulation is a mathematical description of the space or the geometry to be solved. One of the major advances to occur in meshing technology over recent years was the ability for tetrahedral, hexahedral hybrid, and even polyhedral meshes to be incorporated into commercial codes. This has allowed mesh to be fit to any arbitrary geometry, thereby enhancing the attainment of solutions for many industrial applications. In addition, some modern commercial CFD codes promote very little interaction between the user and individual mesh elements with regard to specifying the physics of the problem. Through this decoupling of the physics from the mesh, the user is allowed to concentrate more on the details of the geometry, and the transfer of simulation properties and solutions from one mesh to another is easier when the mesh independence of a simulation needs to be studied or when extra resolution is required.

As a result of unstructured meshing, local mesh refinement without creating badly distorted cells is achievable. However, as with structured meshes, mesh quality is still an important consideration, as poor quality can affect the accuracy of the calculated convective and diffusive fluxes. A common measure of quality is the skewness angle, which determines whether the mesh elements permit the computation of bounded diffusion quantities. Code developers may use the actual angle or an index between zero and one as the metric for skewness. When a skewness angle of 0° is obtained, the vector connecting the center of two adjacent elements is orthogonal to the face separating the elements; this is the optimum value. With skewness angles of 90° or greater, problems in terms of accuracy are caused; the solver may even divide by zero. Other than this, the versatility of these meshes has led to an increased take-up by the CFD community, and their uses are finding accurate solutions in many applications within the food industry.

12.4 Equations Governing Food Processing

12.4.1 Navier-Stokes Equations

The mathematical formulation of fluid motion has been complete for almost 200 years since the emergence of three very important scientists in the field of fluid mechanics. The first one is the Swiss mathematician and physicist Leonhard Euler (1707–1783) who formulated the Euler equations, which describe the motion of an inviscid fluid based on the conservation laws of physics, now defined as classical physics, namely the conservation of mass, momentum, and energy. The French engineer and physicist Claude-Louis Navier (1785–1836) and the Irish mathematician and physicist George Gabriel Stokes (1819–1903) later introduced viscous transport into the Euler equations by relating the stress tensor to fluid motion. The resulting set of equations, now termed the Navier-Stokes equations for Newtonian fluids, have formed the basis of modern day CFD (Anon. 2007).

$$\nabla \vec{v} = 0 \tag{12.2}$$

$$\rho \frac{\partial v_i}{\partial t} + \rho\, \vec{v}\, \nabla \vec{v}_i = -\nabla p + \mu \nabla^2 \vec{v}_i + \rho g \tag{12.3}$$

Equation 12.2 is a mathematical formulation of the law of conservation of mass (which is also called the continuity equation) for a fluid element where $\vec{v}$ consists of the components of $\vec{v}_i$; the solution for each requires a separate equation. The conservation of mass states that the mass flows entering a fluid element must balance exactly with those leaving for an incompressible fluid. Equation 12.3 is the conservation of momentum for a fluid element, i.e., Newton's second law of motion, which states that the sum of the external forces acting on the fluid particle is equal to its rate of change of linear momentum.

Any fluid that does not obey the Newtonian relationship between the shear stress and shear rate is called a non-Newtonian fluid. The shear stress in a Newtonian fluid is represented by the second term on the right hand side of Equation 12.3. Many food processing media have non-Newtonian characteristics, and the shear thinning or shear thickening behavior of these fluids greatly affects their food-hydraulic performance (Fernandes et al. 2006). Over recent years, CFD has provided better understanding of the mixing, heating, cooling, and transport processes of non-Newtonian substances. Of the several constitutive formulas that describe the rheological behavior of substances, which include the Newtonian model, power law model, Bingham model, and the Herschel Bulkley model, the power law is the most commonly used in food engineering applications (Welti-Chanes et al. 2005). However, there are some circumstances in which modeling the viscosity can be avoided as low velocities permit the non-Newtonian fluid to be considered Newtonian, e.g., as shown by Abdul Ghani et al. (2001).

12.4.2 Heat Transfer Equation

The modeling of food processes requires that the energy equation governing the heat transfer within a fluid system be solved. This equation can be written as follows:

$$\rho\frac{\partial(c_p\vec{v}_i)}{\partial t}+\nabla(\vec{v}\,T)=\lambda\nabla^2T+s_T \tag{12.4}$$

The transport of heat in a solid structure can also be considered in CFD simulations and becomes especially important when conjugate heat transfer is under investigation, in which case it is important to maintain continuity of food exchange across the fluid–solid interface (Verboven et al. 2004). The Fourier equation that governs heat transfer in an isotropic solid can be written as

$$\frac{\partial(\rho c_p\vec{v}_i)}{\partial t}=\lambda\nabla^2T+s_T \tag{12.5}$$

The main difference between the two equations is that the Fourier equation lacks a convective mixing term for temperature, which is incorporated into Equation 12.4. For a conjugate heat transfer situation in which evaporation at the food surface is considered and in which the heat transfer coefficient is known, the boundary condition for Equation 12.5 may be written as

$$h(T_{bf}-T_s)+\varepsilon\sigma\left(T_{bf}^4-T_s^4\right)=-\lambda\frac{\partial T}{\partial n}-\alpha\,N_s \tag{12.6}$$

where ε is the emission factor coefficient, and σ is the Stefan-Boltzmann constant. Heat transported by radiation and convection from air to food raises the sample temperature and also goes toward evaporating the free water at the surface (Aversa et al. 2007). The solution of Equation 12.6 can be used on the food surface to calculate the local heat transfer coefficients (Verboven et al. 2003).

$$h=\frac{-\lambda\left.\frac{\partial T}{\partial n}\right|_{surface}}{(T_{bf}-T_s)} \tag{12.7}$$

Equation 12.7 can be used provided the surface temperature is assumed independent of the coefficient during calculations.

The equation of state relates the density of the fluid to its thermodynamic state, i.e., its temperature and pressure. The profiles of food processing variables, i.e., temperature, water concentration, and fluid velocity, are functions of density variations caused by the heating and cooling of fluids. Because fluid flows encountered in food processing can be regarded as incompressible, there are two means of modeling the density variations that occur due to buoyancy. The first is the well-known Boussinesq approximation (Ferziger and Peric 2002). This has been used successfully in many CFD applications (Abdul Ghani et al. 2001):

$$\rho = \rho_{ref}[1 - \beta(T - T_{ref})] \tag{12.8}$$

The approximation assumes that the density differentials of the flow are only required in the buoyancy term of the momentum equations. In addition, a linear relationship between temperature and density with all other extensive fluid properties being constant is also assumed. This relationship only considers a single-component fluid medium; however, by using Taylor's expansion theorem, the density variation for a multi-component fluid medium can also be derived.

Unfortunately, the Boussinesq approximation is not sufficiently accurate at large temperature differentials (Ferziger and Peric 2002). Therefore, in such cases, another method of achieving the coupling of the temperature and velocity fields is necessary. This can be done by expressing the density difference by means of the ideal gas equation:

$$\rho = \frac{p_{ref} M}{RT} \tag{12.9}$$

This method can model density variations in weakly compressible flows, meaning that the density of the fluid is dependent on temperature and composition, but small pressure fluctuations have no influence.

12.4.3 Equation for Mass Transfer

In general, mass transfer in food products depends on local water concentration and is governed by Fick's law of diffusion of the form

$$\frac{\partial C}{\partial t} = \vartheta \nabla^2 C + s_c \tag{12.10}$$

where C is water concentration in food (ppm or mol m^{-3} etc.), and ϑ is the effective diffusion coefficient of water in food (m^2 s^{-1}). If the food is highly porous and dehydrated, then water vapor diffusion may be significant. However, for products with a void fraction lower than 0.3 (May and Perré 2002), vapor diffusion can be neglected. For mass transfer in the fluid medium, a passive scalar can represent the diffusion of mass and be written as

$$\frac{\partial C}{\partial t} + \nabla\ (\vec{v}C) = \vartheta \nabla^2 C + s_c \tag{12.11}$$

In flows involving gases mixing with air and chemicals in water, diffusion coefficients can be found in the literature, but for liquid foods, a reasonable approach is to make an educated guess and conduct a sensitivity analysis (Verboven et al. 2004). Using a passive scalar is only valid in low concentrations and becomes invalid when particulate sizes of about 1 μm are present, which can influence the flow properties.

12.4.4 Turbulence Models

Food processes are usually associated with turbulent motion, primarily due to the involvement of high flow rates and heat transfer interactions. Presently, the Navier-Stokes equations can be solved directly for laminar flows. For turbulent flow regimes, there are many turbulence models available, each being successful in different applications. It should be noted, however, that none of the existing turbulence models are complete, i.e., their prediction performance is highly reliant on turbulent flow conditions and geometry. In the following, some of the best performing turbulence models are discussed.

12.4.4.1 Eddy Viscosity Models

In turbulent flow regimes, engineers are generally content with a statistical probability that processing variables (such as velocity, temperature, and concentration) will exhibit a particular value in order to undertake suitable design strategies. Such information is afforded by the Reynolds averaged Navier-Stokes equations (RANS), which determine the effect of turbulence on the mean flow field through time averaging. By averaging in this way, the stochastic properties of turbulent flow are essentially disregarded, and six additional stresses (Reynolds stresses) result, which need to be modeled by a physically well-posed equation system to obtain closure that is consistent with the requirements of the study.

The eddy viscosity hypothesis (Boussinesq relationship) states that an increase in turbulence can be represented by an increase in effective fluid viscosity and that the Reynolds stresses are proportional to the mean velocity gradients via this viscosity (Ferziger and Peric 2002). For a k–ε type turbulence model, the following representation of eddy viscosity can be written as

$$\mu_t = \rho C_\mu \frac{k^2}{\varepsilon} \tag{12.12}$$

For the k-ω type turbulence without the low Reynolds number modifications, the eddy viscosity can be represented by (Wilcox 1993)

$$\mu_t = \rho \frac{k}{\omega} \tag{12.13}$$

This hypothesis forms the foundation for many of today's most widely used turbulence models, ranging from the simple models based on empirical relationships to variants of the two-equation k–ε model, which describe eddy viscosity through turbulence production and destruction (Versteeg and Malalasekera 1995). All eddy viscosity models have relative merits with respect to simulating food processes.

The standard k–ε model (Launder and Spalding 1974), which is based on the transport equations for the turbulent kinetic energy k and its dissipation rate ε, is semi-empirical and assumes isotropic turbulence. Although it has been successful in numerous applications and is still considered an industrial standard, the standard k–ε model is limited in some respects. A major weakness of this model is that it assumes an equilibrium condition for turbulence, i.e., the turbulent energy generated by the large eddies is distributed equally throughout the energy spectrum. However, in real life, energy transfer in turbulent regimes is not automatic, and a considerable length of time may exist between the production and the dissipation of turbulence.

The renormalization group (RNG) k–ε model (Choudhury 1993) is similar in form to the standard k–ε model, but owing to the RNG methods from which it has been analytically derived, it includes additional terms for dissipation rate development and different constants from those in the standard k–ε model. As a result, the solution accuracy for highly strained flows has been significantly improved. The calculation of the turbulent viscosity also takes into account the low Reynolds number if such a condition is encountered in a simulation. The effect of swirl on turbulence is included in the k–ε RNG model, thereby enhancing accuracy for recirculating flows.

In the realizable k–ε model (Shih et al. 1995), C_μ is expressed as a function of mean flow and turbulence properties instead of being assumed constant as in the case of the standard k–ε model. As a result, it satisfies certain mathematical constraints on the Reynolds stress tensor that are consistent with the physics of turbulent flows (for example, the normal Reynolds stress terms must always be positive). Also, a new model for the dissipation rate is used.

The k–ω model is based on modeled transport equations, which are solved for the turbulent kinetic energy k and the specific dissipation rate ω, i.e., the dissipation rate per unit turbulent kinetic. An advantage that the k–ω model has over the k–ε model is that its performance is improved for boundary layers under adverse pressure gradients as the model can be applied to the wall boundary without using empirical log-law wall functions. A modification was then made to the linear constitutive equation of the k–ω model to account for the principal turbulence shear stress. This model is called the SST (shear-stress transport) k–ω model and provides enhanced resolution of boundary layer of viscous flows (Menter 1994).

12.4.4.2 Reynolds Stress Model

The Reynolds stress closure model (RSM) generally consists of six transport equations for the Reynolds stresses: three transport equations for the turbulent fluxes of each scalar property and one transport equation for the dissipation rate of turbulence energy. RSMs have exhibited far superior predictions for flows in confined spaces in which adverse pressure gradients occur. Terms accounting for anisotropic turbulence, which are included in the transport equations for the Reynolds stresses, means that these models provide a rigorous approach to solving complex engineering flows. However, storage and execution time can be expensive for three-dimensional flows.

12.4.4.3 Large Eddy Simulation and Direct Eddy Simulation

Large eddy simulation (LES) forms a solution given the fact that large turbulent eddies are highly anisotropic and dependent on both the mean velocity gradients and geometry of the flow domain. With the advent of more powerful computers, LES offers a way of alleviating the errors caused by the use of RANS turbulence models. However, the lengthy time involved in arriving at a solution means that it is an expensive technique of solving the flow (Turnbull and Thompson 2005). LES provides a solution to large-scale eddy motion in methods akin to those employed for direct numerical simulation. It also acts as spatial filtering; thus only the turbulent fluctuations below the filter size are modeled. More recently, a methodology has been proposed by which the user specifies a region in which the LES should be performed with RANS modeling completing the rest of the solution; this technique is known as DES and has found to increase the solution rate by up to four times (Turnbull and Thompson 2005).

12.4.4.4 Near-Wall Treatment

An important feature of many two-equation turbulence models is the near-wall treatment of turbulent flow. Low Reynolds number turbulence models solve the governing equations all the way to the wall. Consequently, a high degree of mesh refinement in the boundary layer is required to satisfactorily represent the flow regime, i.e., $y^+ \leq 1$. Conversely, high Reynolds number k–ε models use empirical relationships arising from the log-law condition that describe the flow regime in the boundary layer of a wall. This means that the mesh does not have to extend into this region; thus, the number of cells involved in a solution is reduced. The use of this method requires $30 < y^+ < 500$ (Versteeg and Malalasekera 1995).

12.4.5 Radiation Models

In recent years, the number of radiation models incorporated into commercial CFD codes has increased, and some have been designed by the code developers themselves. The most common radiation models used in food processing simulations include discrete ordinate (DO) or surface-to-surface (S2S) models. The DO model takes into account media participation (Modest 1992). The S2S model considers the radiation heat exchange between two surfaces only (Siegel and Howell 1992), and the amount of radiation received and emitted by each surface is defined by the surface's view factors and the food boundary conditions. In contrast to a S2S solution, the solution for the DO model is coupled to the flow solution, and energy is exchanged between the fluid and the radiation field. Therefore, solution times for the S2S model can be almost half those of the DO model (Mistry et al. 2006). More recently, Chhanwal et al. (2010) compared three radiation models, namely the discrete transfer radiation model (DTRM), surface to surface (S2S), and discrete ordinates (DO) for an electric heating oven, and found that they all predicted similar temperature evolution in the baked product.

12.5 Optimizing Food Processes with CFD

Over the years, many of the conventional process, such as sterilization, drying, and cooking, have been optimized with CFD. Some of the studies conducted thus far are reviewed in this section.

12.5.1 Sterilization and Pasteurization

12.5.1.1 Plate Heat Exchangers for Milk Processing

The development of CFD models in recent years has contributed to the significant progress made in understanding of the food-hydraulics of heat exchangers (Jun and Puri 2005). It has been shown that plate geometry can influence fouling rates, and so CFD models of PHE food-hydraulics can bring about significant benefits for system optimization (Park et al. 2004). Many CFD studies of PHEs exist and have presented different techniques for geometry optimization, i.e., corrugation shape, or the optimization of other process parameters, i.e., inlet and outlet positions and PHE-product temperature differences (Kenneth 2004; Grijspeerdt et al. 2005). Grijspeerdt et al. (2005) investigated the effect of large temperature differences between product and PHE and noted that the larger the difference is, the greater the opportunity for fouling. Nema and Datta (2006) proposed a model to predict the fouling thickness and the milk outlet temperature in a helical triple tube heat exchanger as a function of temperature and shear stress on the surface of the heat exchanger. For this analogy to work, the milk outlet temperature was determined from the simulations, and the fouling thickness was determined based on the enthalpy balance and assuming that a constant heat flux across the heat exchanger wall leads to a dimensionless fouling factor in the form of a Biot number.

12.5.1.2 Plate Heat Exchangers for Yogurt Processing

Fernandes et al. (2006) studied the cooling of stirred yogurt in PHEs with CFD simulations in order to investigate the food-hydraulic phenomena involved in the problem. They modeled the rheological behavior of yogurt via a Herschel-Bulkley model with temperature influence on viscosity being accounted for through Arrhenius-type behavior. As well as accounting for this rheological behavior, they also provided a high level of precision in the PHE geometrical design and the imposed boundary conditions. During the course of these studies, it was found that due to the higher Prandtl numbers and shear thinning effects provided by the yogurt, the Nusselt number of the fully developed flows were found to be more than 10 times higher than those of water. This result presented a substantial food-hydraulic performance enhancement in comparison with that from Newtonian fluids (Maia et al. 2007). Furthermore, it was shown that PHEs with high corrugation angles may provide better opportunities for the gel structure breakdown desired during the production stage of stirred yogurt. Regarding stirred yogurt flow behavior, some other studies have been published in which both experimental and simulation results were reported (Fernandes et al. 2006; Mullineux and Simmons 2007). Afonso et al. (2003) conducted a CFD study to obtain correlations for the determination of convective heat transfer coefficients of stirred yogurt during the cooling stage in a plate heat exchanger, taking into account its rheological features. Despite modeling simplifications, good agreement between measurements and predictions owing to the prevailing effects of the high viscosity of stirred yogurt, i.e., the corrugated surfaces did not seem to cause significant perturbations in the flow development in their particular case. Recently, Fernandes et al. (2008) conducted simulations of laminar flows of Newtonian and power-law fluids through cross-corrugated chevron-type plate heat exchangers (PHEs) as a function of the geometry of the channels. Due to the geometrical complexity of the cross-corrugated chevron-type PHE passages, the authors found it difficult to predict the apparent viscosity observed during the flow of shear-thinning fluids unless the generalized viscosity concept was used. Using this concept, the authors showed that the flow index behavior influences the velocity profiles and the magnitude of the average interstitial velocity.

12.5.1.3 Canned Foods

During the sterilization process, rapid and uniform heating is desirable to achieve a suitable level of sterility with minimum destruction of the color, texture, and nutrients of food products (Tattiyakul et al. 2001). CFD has been used to investigate the heat and mass transfer phenomena occurring in canned products, such as soup, carboxyl-methyl cellulose (CMC), or cornstarch undergoing sterilization (Figure 12.1).

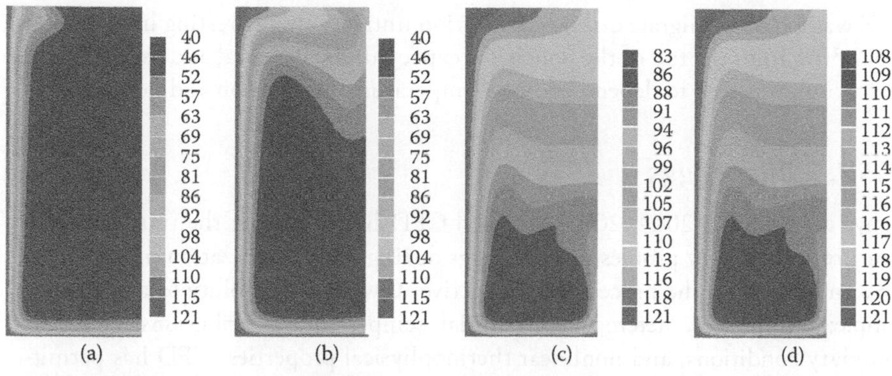

Figure 12.1 Temperature profiles in a can filled with CMC and heated by condensing steam (top insulated) after periods of (a) 54 s, (b) 180 s, (c) 1157 s, and (d) 2574 s. The right-hand side of each figure is centerline. (From Abdul Ghani, A.G., and Farid, M.M., Food sterilisation of foods using computational fluid dynamics. Chapter 13 in *Computational Fluid Dynamics in Food Processing*, D.-W. Sun (Ed.), 28 chapters, CRC Press, USA, pp. 347–381, 2007.)

In these investigations, CFD has highlighted the transient nature of the slowest heated zone (SHZ), quantified the amount of time needed for heat to be transferred fully throughout food as well as illustrating the sharp heterogeneity in temperature profile of the product when no agitation is applied to the can (Abdul Ghani et al. 1999). Abdul Ghani et al. (2002) conducted CFD studies of both natural and forced convection (via can rotation) sterilization processes of viscous soup and showed that forced convection was about four times more efficient. Further simulation studies of a starch solution undergoing transient gelatinization showed that uniform heating could be obtained by rotating the can intermittently during the sterilization process (Tattiyakul et al. 2001). CFD simulations have also been used to generate the data required for the development of a correlation to predict sterilization time (Farid and Abdul Ghani 2003). A more recent study simulated the sterilisation process of a solid-fluid mixture in a can, and showed that the position of the food in the can has a large influence on the sterilisation times experienced (Abdul Ghani and Farid 2006).

12.5.1.4 Foods in Pouches

In recent years, CFD has provided a rigorous analysis of the sterilization of three-dimensional pouches containing liquid foods (Abdul Ghani et al. 2002). Coupling first-order bacteria and vitamin inactivation models with the fluid flow has allowed transient temperature, velocity, and concentration profiles of both bacteria and ascorbic acid to be predicted during natural convection. The concentrations of bacteria and ascorbic acid after heat treatment of pouches filled with the liquid food were measured and found close agreement with the numerical predictions. The

SHZ was found to migrate during sterilization until eventually resting in a position about 30% from the top of the pouch. As expected, the bacterial and ascorbic acid destruction was seen to depend on both temperature distribution and flow pattern.

12.5.1.5 Intact Eggs

Denys et al. (2003, 2004, 2005) has used CFD has to predict the transient temperature and velocity profiles in intact eggs during the pasteurization process with the aim of making the process more effective. Owing to its ability to account for complex geometries, heterogeneous initial temperature distributions, transient boundary conditions, and nonlinear thermophysical properties, CFD has permitted a comprehensive understanding of this food process (Denys et al. 2003). Such an analysis has allowed the gap in the knowledge of this area to be filled as, up until recently, little information on the correct processing temperatures and times for safe pasteurization without loss of functional properties was available (Denys et al. 2004). In the series of papers published on this topic by Denys et al. (2003, 2004, 2005), a procedure to determine the surface heat transfer coefficient using CFD simulations of eggs filled with a conductive material of known food properties

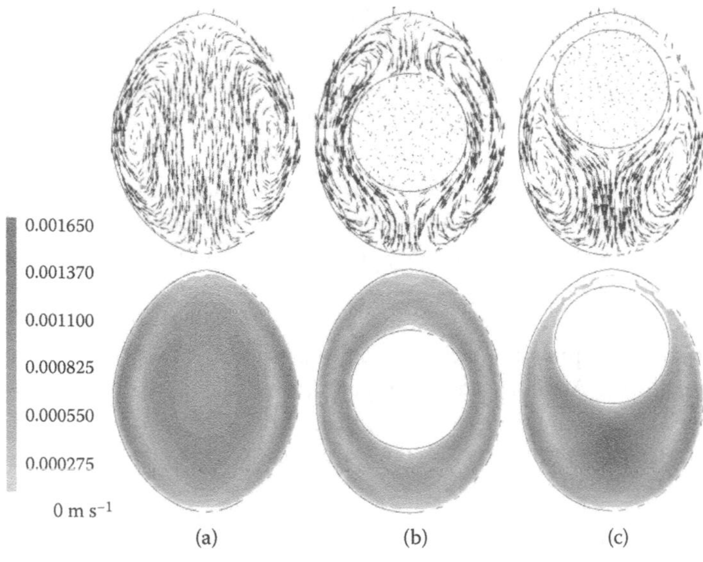

Figure 12.2 CFD-calculated velocity vector plots and velocity contours in a plane of symmetry of an egg filled with CMC and heated in a water bath after 30 s processing: (a) no yolk present, (b) conductive heating yolk present in the egg center, (c) yolk shifted toward the top of the egg. Process conditions for all cases were: T_i = (24.5 ± 0.2)°C; T_∞ = (59.4 ± 0.2)°C. (From Denys, S. et al., CFD analysis of food processing of eggs. Chapter 14 in *Computational Fluid Dynamics in Food Processing*, D.-W. Sun (Ed.), 28 Chapters, CRC Press, USA, pp. 347–381, 2007.)

was first developed. After this, conductive and convective heating processes in the egg were modeled as shown in Figure 12.2 (Denys et al. 2004). From this, it was revealed that, similar to the phenomena noted by Abdul Ghani et al. (1999) in canned food, the cold spot was found to move during the process toward the bottom of the egg. Moreover, again similar to the findings of Abdul Ghani et al. (1999), a cold zone as opposed to a cold spot was predicted. The location of the cold zone in the yolk was predicted to lay below its geometrical center even for the case in which the yolk was positioned at the top of the egg. It was concluded that no convective heating takes place in the egg yolk during processing.

12.5.2 Dehydration

12.5.2.1 Fluidized Bed Drying

Spouted beds are a type of fluidized bed dryer commonly found in the food industry in which heat is transferred from a gas to the fluidized particle. These beds are suitable for industrial unit operations that handle heavy, coarse, sticky, or irregularly shaped particles through a circulatory flow pattern (Mathur and Epstein 1974). Spouted beds have had many applications in the drying of foods, such as for drying of grains (Madhiyanon et al. 2002), diced apples (Feng and Tang 1998), etc. Because of the complex interactions that occur, empirical correlations are only valid for a certain range of conditions, and CFD simulations have been the only means of providing accurate information on the flow phenomena. However, similar to other drying applications, difficulties exist in modeling the interactions between the solid and liquid phases as well as limitations in computing power, and consequently, only a limited number of CFD simulations exist. To simulate the drying of agricultural products, various different approaches can be taken, i.e., a single-grain approach, a Eulerian-Eulerian approach, or a discrete-element particle (DEM) approach. In the DEM model, two techniques can be used to simulate the drying of granular foods, i.e., a soft sphere model and a hard sphere model. In the soft sphere model, the interaction forces using multiple particle-particle or particle-wall contacts can be predicted whereas, in the hard sphere model, it is only possible to compute the instantaneous change in particle velocity through particle collision (Li et al. 2010). However, the computational demand rises strongly with the number of traced particles, which limits its applicability.

The single-grain approach is when the drying of a representative grain is modeled. Using this approach, Markowski et al. (2009) undertook an experimental and theoretical investigation to determine the drying characteristics of barley grain dried in a spouted-bed dryer, in which the influence of the shape of a solid applied during modeling on the determined diffusivity of barley was studied. The finite element method was used to solve the equations applied in the barley drying process. The optimization techniques were applied to determine moisture diffusivity using an inverse modeling approach. Their results showed that when modeling the drying of a grain, the assumed geometry of a solid is of a fundamental importance as it has a direct influence on moisture diffusivity. From this, it was also shown that using spherical geometry for modeling the drying processes of grain is faulty if high accuracy of the results is expected from the modeling.

12.5.2.2 Spray Drying

The use of CFD in spray drying has been comprehensively reviewed in recent years, and for interested readers, the articles of Langrish and Fletcher (2003) and Fletcher et al. (2006) provide a good understanding of the topic. It is important to note that most CFD of spray dryers use the Eulerian-Lagrangian approach, which, as discussed above, demands a large amount of storage if the numbers of particles/droplets being modeled are great. Moreover, when large-scale systems need to be modeled, enhancing efficiency by parallel processing via domain decomposition would be difficult as particles will not remain in the one domain (Fletcher et al. 2006). Even though such difficulties are obviated with the Eulerian-Eulerian approach, many disadvantages exist, such as the loss of the time history of individual particles, the difficulty of modeling turbulent dispersion, and the inability to model interacting jets with some of these problems sharing common ground with fluidized bed modeling. However, in phenomena unrelated to fluidized bed applications, the Eulerian-Lagrangian approach permits reasonable predictions of drying rate, particle-wall depositions, and particle agglomeration whereas comparable applications using the Eulerian-Eulerian approach are limited. Lagrangian particle tracking will remain a key feature of spray dryer modeling into the future as it allows the modeler to gain understanding of particle histories, such as velocity, temperature, residence time, and the particle impact position, are important to design and operating spray drying, which is important given that the final product quality is dependent on these particle histories.

12.5.2.3 Forced-Convection Drying

Because of the complex geometry usually encountered in drying applications, theoretical studies are often not applicable, and to obtain the spatial distribution of transfer coefficients with reasonable accuracy it is necessary to solve the Navier-Stokes equations in the product surroundings. From the distribution of transfer coefficients, the correct temperature and moisture profiles within the product can be predicted so that the drying process can be optimized. Kaya et al. (2007) used this approach with CFD to determine the transfer coefficients, and then the heat and mass transfer within the food was simulated with an external program. In a later study, Kaya et al. (2007) used a similar CFD approach to find out the optimum inlet and outlet configurations and locations for a dryer containing a moist object. This approach was also recently used in order to investigate the effects of confined flow on the hydrodynamic, food, and mass transfer characteristics of a circular cylinder, and the interrelation of these mechanisms are investigated numerically (Ozalpa and

Dincer 2010). However, it is also possible to do this within the CFD package by determining the heat transfer coefficient with Equation 12.7 and the mass transfer with the Lewis relationship once a transient solution is performed. Such a technique was used by Suresh et al. (2001) in CFD simulations of the conjugate heat and mass transfer through a block immersed in a boundary layer flow. Conjugate heat and mass flux can also be determined with the help of empirical correlations of the transfer coefficients; however, these may not always be adequate (Zheng et al. 2007).

12.5.3 Cooking

12.5.3.1 Natural Convection Ovens

Electric ovens are commonly used household appliances that rely on conjugate food exchange to produce the desired cooking effect in a foodstuff. For that reason, CFD is an appropriate tool to quantify the internal food field and mass transfer, both of which are important for robust design and performance. Natural convection is dominant in these ovens when it is produced by a source below the product (bake mode) with a source above the product driving radiative heat transfer (broil mode). Abraham and Sparrow (2003) used CFD to model the flow-field in an electric oven on baking mode with isofood sidewalls, employing temperature boundary conditions consistent with experimental measurements. These simulations allowed insight into the relevant contributions of convective and radiative heat transfer to be gained as well as determining the accuracy of predictions made by simple elementary surface correlations. Predictions were obtained with a steady-state solver, and the solutions were then used as input into a quasi-steady model to permit the timewise temperature variation of the foodstuff to be analyzed from which excellent agreement between experimental and numerical results was achieved. Abraham and Sparrow (2003) also found that for a blackened food load, radiative heat transfer contributed to 72% of the total heat transfer whereas this was 8% when the load was reflective. Heat flux distribution on the lower and upper surfaces varied as a function of contributions from temperature difference between the oven and the product alongside that from the buoyant plume emanating from the heat source. From this, it was noted that steady-state heat transfer could be obtained at the top surface whereas the under surface was highly unsteady. More recently, Mondal and Datta (2010) developed a 2-D CFD model for crustless bread, in which they computed the heat and mass transfer during baking. It was found that a simulated model was able to predict very well the pattern of temperature and moisture profile during the bread-baking process.

12.5.3.2 Forced Convection Ovens

Verboven et al. (2000a) developed a CFD model of an empty, isofood forced-convection oven, in which numerous submodels were incorporated, including those describing the airflow through the heating coils and fan, alongside swirl in the flow regime added by the fan. Due to the complexity of the flow dynamics involved in the oven, a detailed validation exercise was performed. An important outcome of the study was the observation that the swirl model was a necessary requisite for accurate quantitative results as the swirl actually improves the airflow rate while having no direct effect on the mechanical energy balance. In a later study, they conducted a CFD investigation into the temperature and airflow distribution in the same oven during its operation cycle, loaded with product (Verboven et al. 2000b). Prior to the CFD modeling, a lumped FEM model of heat transfer within a food was designed to calculate the appropriate heat sources and boundary conditions, which were then applied in the CFD model. The prediction errors averaged to 4.6°C for the entire cavity at the end of the warm-up stage with this reducing to 3.4°C at the end of the process as the variability in the system leveled out. As expected, the products subject to the highest heat transfer were situated closest to the fan where they experienced uniform high velocity air flow. Unfortunately, the CFD predictions of the surface heat transfer coefficients were grossly under-predicted and were attributed to the inaccuracy of the wall function calculations. However, in order to circumvent these errors, Verboven et al. (2000b) used the local turbulence intensity and Reynolds number, obtained from the CFD model, to compute the heat transfer coefficients.

Recently, Smolka et al. (2010) presented an experimentally validated 3-D CFD analysis of the flow and food processes in a laboratory drying oven with a forced-air circulation from a rotating fan. Because the numerical results show very good agreement with the measured values, a number of new device modifications were simulated to improve temperature uniformity within the chamber of the device.

12.5.3.3 Commercial Ovens

For the bakery industry, it is necessary to maintain consistency in product quality throughout processing. This requires quality-conscious control strategies to be implemented, which should permit the adjustment of process parameters in response to disturbances, such as a change in the oven load. The successful implementation of such control strategies has been the long-term goal of process modeling via CFD for some time now. The tone of the first studies with this aim was set out by Therdthai et al. (2003), who solved the two-dimensional flow and temperature field in conventional "u-turn" bread-baking oven. They used the simulations to provide information for temperature control as well as for control sensor position. However, a number of simplifications limited the veracity of the model with regards to its representation of the physical process. For example, a steady-state flow regime was assumed even though the process is inherently transient. Furthermore, the contribution from radiative heat transfer was ignored despite the fact that this mode significantly influences food exchange. These assumptions were, however, addressed in a subsequent publication, in which a full three-dimensional model was built, using a moving grid technique to simulate the transient baking process (Therdthai et al. 2004a). Radiative heat transfer was also incorporated. Therdthai et al. (2002) also used mathematical models, which they had developed in a previous study, in conjunction with the CFD results to predict the quality attributes of bread. Using both sets of predictions, they provided recommendations to reduce energy use during the process while not compromising bread quality. More specifically, with reference to Figure 12.3, the propositions were that the duct temperature

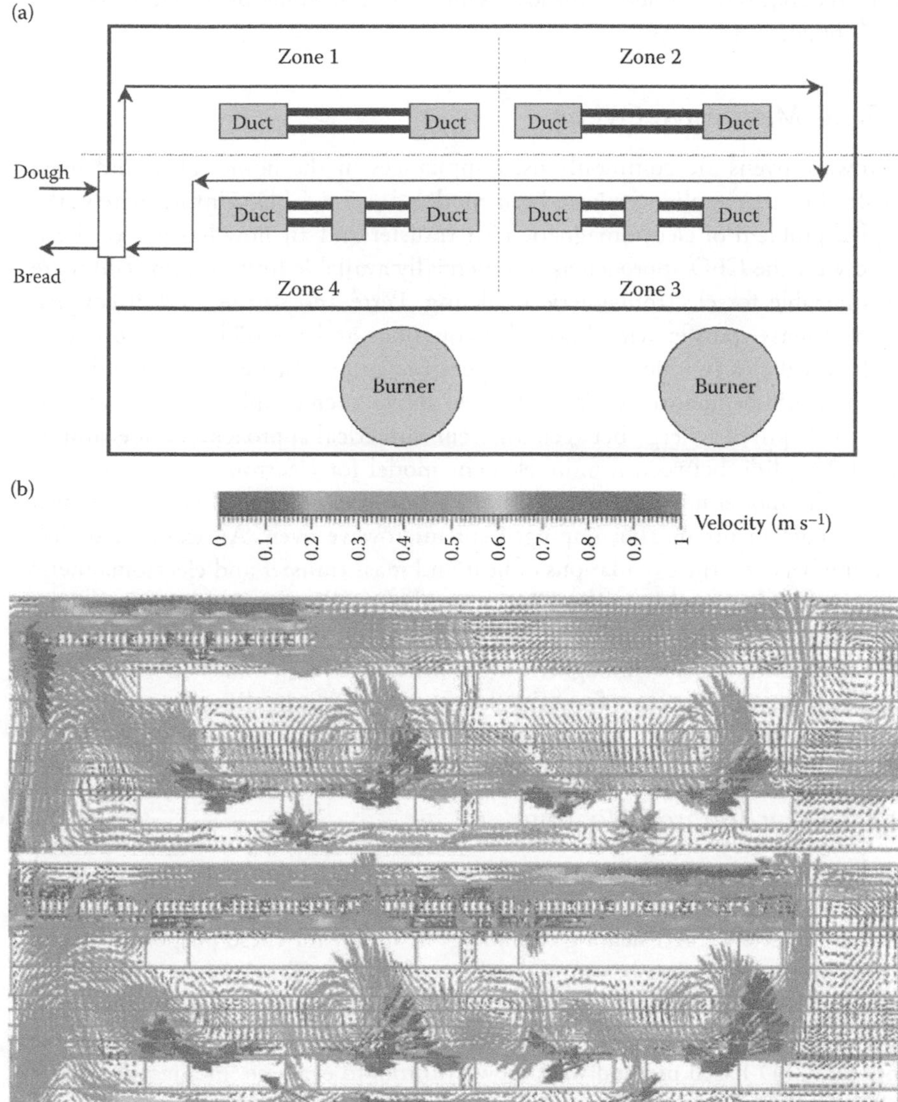

Figure 12.3 Industrial bread baking oven (a) schematic and (b) simulation results. (From Therdthai, N. et al., *Journal of Food Engineering*, 65(4), 599–608, 2004.)

in zones 1, 3, and 4 should be reduced by about 10°C so that the tin temperature in zones 1 and 3 would decrease, thereby reducing weight loss in this zone. At the same time, their observations showed that the flow rate of the convection fan in zone 3 should be increased by doubling its throughput so that the tin temperatures in zones 2 and 4 could be maintained at high levels. This optimization was predicted to produce a weight loss of 7.95% with the lightness values of the crust color on the top, bottom and side of the loaf being around *L*-values of 50.68, 55.34, and 72.34, respectively.

12.5.3.4 Microwave Ovens

Microwave ovens are commonly used appliances in the home. Heat and mass transfer in these appliances have been modeled using CFD. Unfortunately, the coupled problem of electromagnetic heat transfer and air flow has not yet been solved with the CFD approach as commercially available finite volume codes are not amenable for electromagnetic modeling. Perré and Turner (1999) coupled heat and mass transfer with Maxwell's equations and the dielectric properties, which varied as a function of both temperature and moisture so that the drying of wood could be quantified. As mentioned above, such coupling in food simulations has required synergy between different numerical approaches. For example, a weak coupling between a finite element model for electromagnetic radiation and a CFD model has been achieved by Verboven et al. (2007) via surface heat transfer coefficients in a jet impingement microwave oven. An example of full coupling between the calculations of heat and mass transfer and electromagnetic radiation has been achieved by Dincov et al. (2002), who mapped the (moisture and temperature dependent) porous media dielectric properties from the CFD mesh onto the electromagnetic finite difference mesh and then mapped the microwave heating function from the electromagnetic finite difference mesh onto the CFD mesh.

12.5.3.5 Far-Infrared Heating

Post-harvest heat treatments have become increasingly popular to control insect pests, prevent fungal spoilage, and affect the ripening of fruits and vegetables. Far infrared radiation (FIR) heating technology is useful for these purposes because it can achieve contactless heating (Tanaka et al. 2007). Tanaka et al. (2007) performed advanced 3-D food radiation calculations to assess the quality and efficiency of FIR heating for post-harvest treatments of fruits. It was shown that the proposed CFD-based method was a powerful tool to evaluate, in a fast and comprehensive way, complex heating configurations that include radiation, convection, and conduction. However, the authors also found that the FIR method was found inferior to water-bath heating in terms of efficiency and uniformity.

12.6 Modeling Emerging Food Technologies with CFD

12.6.1 High-Pressure Food Processing

High-pressure processing (HPP) is a novel food processing method, which has shown great potential in the food industry. During compression/decompression phases, the internal energy of a high-pressure processing system changes, resulting in heat transfer between the internal system and its boundaries. These food-hydraulic characteristics were studied with CFD by Hartmann (2002); the technique was deemed necessary in gaining a thorough understanding of the phenomena inherent in HPP, especially when the scale-up phenomena need to be analyzed, e.g., layout and design of high-pressure devices, packages, etc. Hartmann and Delgado (2002) used CFD and dimensional analyses to determine the time scales of convection, conduction, and bacterial inactivation, the relative values of which contribute to the efficiency and uniformity of conditions during HPP. Conductive and convective time scales were directly compared to the inactivation time scale in order to provide a picture of the food-hydraulic phenomena in the HP vessel during bacterial inactivation. The results showed that pilot scale systems exhibited a larger convection time scale than the inactivation time scale and that the intensive fluid motion and convective heat transfer resulted in homogeneous bacterial inactivation. Furthermore, the simulations of industrial scaled systems showed greater efficiency in bacterial inactivation as the compression heating subsisted for greater time periods when compared to smaller laboratory systems. Other CFD simulations showed that the food properties of the HP vessel boundaries have considerable influence on the uniformity of the process and insulated material promoted the most effective conditions (Hartmann et al. 2004). As well as this, the insulated vessel was found to increase the efficiency of the HPP by 40%. A CFD and dimensionless analysis of the convective heat transfer mechanisms in liquid food systems under pressure was also performed by Kowalczyk and Delgado (2007), who advised that HP systems with a characteristic dimension of 1 m alongside a low viscous medium should be used to avoid heterogeneous processing of the product.

12.6.2 Ohmic Heating

The basic principle of ohmic heating is that electrical energy is converted to food energy within a conductor. In food processing, foodstuffs act as the conductor. The main advantage of ohmic heating is that, because heating occurs by internal energy generation within the conductor, this processing method leads itself to even distribution of temperatures within the food, and it does not depend on heat transfer mechanisms (Jun and Sastry 2005). Jun and Sastry (2005) were the first to model the ohmic heating process with the aim of enhancing heating techniques for use by cabin crews during long-term space missions. They developed a two-dimensional transient model for chicken noodle soup (assumed single phase) and black beans under the ohmic heating process by solving the electric field via the Laplace equation. From this solution, the internal heat generation was obtained, which was added as a source to the Fourier equation and was then numerically solved by the CFD code. Electrical conductivity was allowed to vary as a function of temperature. The CFD model was able to predict regions of electric field overshoot in the food as well as the non-uniformities in the predicted food field. Moreover, they noted that as the electrode got wider, the cold zone area developed in the middle of the packaging diminished to a minimum and then appeared and grew at the corners of the packaging, clearly illustrating the existence of a threshold value for electrode size optimization. They later expanded the model so that a three-dimensional representation of pouched tomato soup could be simulated (Jun and Sastry 2007), in which they found the electric field strength near the edges of electrodes to overshoot as it got close to the maximum value as predicted by their first model. On the other hand, the food between the V-shaped electrodes experienced a weak electric field strength, which gave rise to cold zones in the food. Jun and Sastry (2007) recognized that the presence of these cold zones merited further research on pouches via modeling and package redesign.

12.7 Challenges for CFD in Food Process Modeling

12.7.1 Improving the Efficiency of the Solution Process

As can be seen above, the physical mechanisms that govern food processes generally include any combination of fluid flow, heat, mass, and scalar transport. As CFD involves the modeling of such mechanisms, which mostly occur on different time scales, the temporal accuracy and stability of a solution is usually bounded by the ability of the model to capture the mechanism that is the quickest to occur. This is done through time stepping, which has therefore to be optimized during model development. The time step must be small enough to resolve the frequencies of importance during a transient process. To do this, an appropriate characteristic length and velocity of the problem is necessary, which can be obtained from nondimensional numbers, such as the Stroudal number, from experimental data, or from experience. For example, when Szafran and Kmiec (2005) developed a CFD model of a spouted bed dryer, they found that extremely small time steps of the order of 1×10^{-4} were required to resolve the instabilities in the flow regime as well as the circulation of phases. However, this meant that a year of computation would result in a solution for only one hour of drying. Such a time frame is excessive by any length of the imagination, and overcoming this difficulty required considerable insight into the physical process. To do so, Szafran and Kmiec (2005) transformed the process from time-dependent into time-independent from which drying curves could be developed as described above.

12.7.2 Turbulence

One of the main issues faced by the food industry over the last two decades is the fact that most turbulence models have been shown to be application specific. At the present time, there are many turbulence models available; however, until a complete turbulence model capable of predicting the average field of all turbulent flows is developed, the CFD optimization of many food processes will be hampered. The reason is that in every application, many different turbulence models must be applied until the one that gives the best predictions is found. The closest to the complete turbulence model thus far is LES, which uses the instantaneous Navier-Stokes equations to model large-scale eddies with smaller scales solved with a sub-grid model. However, using this model demands large amounts of computer resources, which may not be presently achievable.

For many cases, the k–ε model and its variants have proven to be successful in applications involving swirling flow regimes once the mesh is considerably fine. This is because, as pointed out by Guo et al. (2003), when the mesh is very fine, the k–ε model performs like the sub-grid model of a LES and handles small-scale turbulence while the largest scales are solved by the transient treatment of the averaged equations. In practice, therefore, most of the important energy-carrying eddies can be solved by this means. However, k–ε models have lacked performance when predicting impinging flows or in flows with large adverse pressure gradients. In such instances, models such as the k–ω or LES should be considered more closely.

12.7.3 Boundary Conditions

In CFD simulations, the boundary conditions must be adequately matched to the physical parameters of the process with the precision of similarity being conditioned by the mechanism under study and the level of accuracy required. Even when this is done, the CFD solution still may not be a correct physical representation of the physical system. This was shown by Mistry et al. (2006), who found that an artificial pressure differential was required to predict the correct flow patterns in an oven, which was heated by natural convection. Such results suggest the importance of sensitivity analysis studies alongside experimental measurements in the early stages of model development. Sensitivity analyses are also necessary for turbulence model specification or turbulence model tuning via inlet conditions and for CFD model simplification.

12.8 Summary

CFD has played an active part in the design of food processes for more than a decade now. In recent years, simulations have reached higher levels of sophistication as application-specific models can be incorporated into the software with ease via user-defined files. Undoubtedly, with current computing power progressing unrelentingly, it is conceivable that CFD will continue to provide explanations for transport phenomena, leading to better design of food processes in the food industry.

Nomenclature

A: area (m^2)
C: water concentration (mol m^{-3})
C_μ: empirical turbulence model constant
c_p: specific heat capacity (W kg^{-1} K^{-1})
d: width of jet (m)
D: width of cylinder (m)
g: acceleration due to gravity (m s^{-2})
M: molecular weight (kg $kmol^{-1}$)
Nu: Nusselt number
p: pressure (Pa)
R: gas constant (J $kmol^{-1}$ K^{-1})
Re: Reynolds number
k: turbulent kinetic energy (m^2 s^{-2})
s_T: food sink or source (W m^{-3})
s_c: concentration sink or source (mol m^{-3})
T: temperature (K)
t: time (s)
U: velocity component (m s^{-1})
V: volume (m^3)
$\vec{v}_i$: velocity component (m s^{-1})
x: Cartesian coordinates (m)

Greek Letters

ρ: density (kg m^{-3})
μ: dynamic viscosity (kg m^{-1} s^{-1})
β: food expansion coefficient (K^{-1})
λ: food conductivity (W m^{-1} K^{-1})
α: water molar latent heat of vaporization (J mol^{-1})
ε: turbulent dissipation rate (m^2 s^{-3})
ϕ: the transported quantity
Γ: diffusion coefficient of transported variable
ϑ: the diffusivity of the mass component in the fluid (m^2 s^{-1})
ω: specific dissipation (s^{-1})
μ_t: turbulent viscosity (kg m^{-1} s^{-1})

Subscripts

i: Cartesian coordinate index
bf: bulk fluid
s: surface
V: mesh element volume
A: area of mesh element
m: mean
0: with no turbulence

References

Abdul Ghani, A.G., and Farid, M.M. (2006). Numerical simulation of solid-liquid food mixture in a high measure processing unit using computational fluid dynamics. *Journal of Food Engineering*, 80(4), 1031–1042.

Abdul Ghani, A.G., and Farid, M.M. (2007). Food sterilisation of foods using computational fluid dynamics. In *Computational Fluid Dynamics in Food Processing*, D.-W. Sun (Ed.), CRC Press, Boca Raton, FL, pp. 347–381.

Abdul Ghani, A.G., Farid, M.M., and Chen, X.D. (2002). Theoretical and experimental investigation of the food inactivation of *Bacillus stearothermophilus* in food pouches. *Journal of Food Engineering*, 51(3), 221–228.

Abdul Ghani, A.G., Farid, M.M., Chen, X.D., and Richards, P. (1999). An investigation of deactivation of bacteria in a canned liquid food during sterilization using computational fluid dynamics (CFD). *Journal of Food Engineering*, 42, 207–214.

Abdul Ghani, A.G., Farid, M.M., Chen, X.D., and Richards, P. (2001). Food sterilization of canned food in a 3-D pouch using computational fluid dynamics. *Journal of Food Engineering*, 48, 147–156.

Abraham, J.P., and Sparrow, E.M. (2003). Three-dimensional laminar and turbulent natural convection in a continuously/discretely wall-heated enclosure containing a food load. *Numerical Heat Transfer Part A: Applications*, 44(2), 105–125.

Afonso, I.M., Hes, L., Maia, J.L., and Melo, L.F. (2003). Heat transfer and rheology of stirrred yoghurt during cooling in plate exchangers. *Journal of Food Engineering*, 57, 179–187.

Anon. (2007). *A Brief History of Computational Fluid Dynamics*. Internet article, www.fluent.com, Fluent Inc. 10 Cavendish Court, Lebanon, NH 03766, USA.

Aversa, M., Curcio, S., Calabro, V., and Iorio, G. (2007). An analysis of the transport phenomena occurring during food drying process. *Journal of Food Engineering*, 78(3), 922–932.

Chhanwal, N., Anishaparvin, A., Indrani, D., Raghavarao, K.S.M.S., and Anandharamakrishnan, C. (2010). Computational fluid dynamics (CFD) modelling of an electrical heating oven for bread-baking process. *Journal of Food Engineering*, 100(3), 452–460.

Choudhury, D. (1993). Introduction to the renormalization group method and turbulence modelling. Technical Memorandum TM-107. Fluent Inc., Lebanon, NH.

Denys, S., Pieters, J.G., and Dewettinck, K. (2003). Combined CFD and experimental approach for determination of the surface heat transfer coefficient during food processing of eggs. *Journal of Food Science*, 68(3), 943–951.

Denys, S., Pieters, J.G., and Dewettinck, K. (2004). Computational fluid dynamics analysis of combined conductive and convective heat transfer in model eggs. *Journal of Food Engineering*, 63(3), 281–290.

Denys, S., Pieters, J.G., and Dewettinck, K. (2005). Computational fluid dynamics analysis for process impact assessment during food pasteurization of intact eggs. *Journal of Food Protection*, 68(2), 366–374.

Denys, S., Pieters, J., and Dewettinck, K. (2007). CFD analysis of food processing of eggs. In *Computational Fluid Dynamics in Food Processing*, D.-W. Sun (Ed.), CRC Press, Boca Raton, FL, pp. 347–381.

Dincov, D.D., Parrott, K.A., and Pericleous, K.A. (2002). Coupled 3-D finite difference time domain and finite volume methods for solving microwave heating in porous media. *Lecture Notes in Computer Science*, 2329, 813–822.

Farid, M., and Abdul Ghani, G.A. (2004). A new computational technique for the estimation of sterilization time in canned food. *Chemical Engineering and Processing*, 43(4), 523–531.

Feng H., and Tang, J. (1998). Microwave finish drying of diced apples in a spouted bed. *Journal of Food Science*, 63, 679–683.

Fernandes, C.S., Dias, R.P., Nobrega, J.M., Afonso, I.M., Melo, L.F., and Maia, J.M. (2006). Food behaviour of stirred yoghurt during cooling in plate heat exchangers. *Journal of Food Engineering*, 69, 281–290.

Fernandes, J., Simoes, P.C., Mota, J.P.B., and Saatdijan, E. (2008). Application of CFD in the study of supercritical fluid extraction with structured packing: Dry pressure drop calculations. *Journal of Super Critical Fluids*, 47(1), 17–24.

Ferziger, J.H., and Peric, M. (2002). *Computational Methods for Fluid Dynamics*, Springer-Verlag, Berlin Heidleberg, pp. 1–100.

Fletcher, D.F., Guo, B., Harvie, D.J.E., Langrish, T.A.G., Nijdam, J.J., and Williams, J. (2006). What is important in the simulation of spray dryer performance and how do current CFD models perform? *Applied Mathematical Modelling*, 30(11), 1281–1292.

Geedipalli, S.S.R., Rakesh, V., and Datta, A.K. (2007). Modelling the heating uniformity contributed by a rotating turntable in microwave ovens. *Journal of Food Engineering*, 82(3), 359–368.

Grijspeerdt, K., Kreft, J.U., and Messens, W. (2005). Individual-based modeling of growth and migration of *Salmonella Enteritidis* in hens' eggs. *International Journal of Food Microbiology*, 100(1–3), 323–333.

Guo, B., Langrish, T.A.G., and Fletcher, D.F. (2003). Simulation of gas flow instability in a spray dryer. *Part A, Chemical Engineering Research and Design*, 81, 631–638.

Hartmann, C. (2002). Numerical simulation of thermodynamic and fluid-dynamic processes during the high-pressure treatment of fluid food systems. *Innovative Food Science and Emerging Technologies*, 3(1), 11–18.

Hartmann, C., and Delgado, A. (2002). Numerical simulation of convective and diffusive transport effects on a high-pressure-induced inactivation process. *Biotechnology and Bioengineering*, 79(1), 94–104.

Hartmann, H., Derksen, J.J., Montavon C., Pearson, J., Hamill, I.S., Van den Akker, H.E.A. (2004). Assessment of large eddy and RANS stirred tank simulations by means of LDA. *Chemical Engineering and Science*, 59, 2419–2432.

Jun, S., and Puri, V.M. (2005). 3D milk-fouling model of plate heat exchangers using computational fluid dynamics. *International Journal of Dairy Technology*, 58(4), 214–224.

Jun, S., and Sastry, S. (2005). Reusable pouch development for long term space missions: A 3D ohmic model for verification of sterilization efficacy. *Journal of Food Engineering*, 80(4), 1199–1205.

Jun, S., and Sastry, S. (2007). Modeling and optimization of ohmic heating of foods inside a flexible package. *Journal of Food Processing Engineering*, 28(4), 417–436.

Kaya, A., Aydin, O., and Dincer, I. (2007). Numerical modeling of forced-convection drying of cylindrical moist objects. *Numerical Heat Transfer Part A: Applications*, 51(9), 843–854.

Kenneth, J.B. (2004). Heat exchanger design for the process industries. *Journal of Heat Transfer*, 126(6), 877–885.

Kowalczyk, W., and Delgado, A. (2007). Dimensional analysis of thermo-fluid-dynamics of high hydrostatic pressure processes with phase transition. *International Journal of Heat and Mass Transfer*, 50(15–16), 3007–3018.

Langrish, T.A.G., and Fletcher, D.F. (2003). Prospects for the modelling and design of spray dryers in the 21st century. *Drying Technology*, 21(2), 197–215.

Launder, B.E., and Spalding, D.B. (1974). The numerical computation of turbulent flows. *Computer Methods in Applied Mechanics and Engineering*, 3, 269–289.

Li, Z., Su, W., Wu, Z., Wang, R., and Mujumdar, A.S. (2010). Investigation of flow behaviors and bubble characteristics of a pulse fluidized bed via CFD modeling. *Drying Technology*, 28(1), 78–93.

Madhiyanon, T., Soponronnarit, S., and Tia, W. (2002). A mathematical model for continuous drying of grains in a spouted bed dryer. *Drying Technology*, 20, 587–614.

Maia, J.M., Nobrega, J.M., Fernandes, C.S., and Dias, R.P. (2007). CFD simulation of stirred yoghurt processing in plate heat exchange. In *Computational Fluid Dynamics in Food Processing*, D.-W. Sun (Ed.), CRC Press, Boca Raton, FL, pp. 381–402.

Markowski, M., Białobrzewski, I., and Modrzewska, A. (2010). Kinetics of spouted-bed drying of barley: Diffusivities for sphere and ellipsoid. *Journal of Food Engineering*, 96(3), 380–387.

Marra, F., Lyng, J., Romano, V., and McKenna, B. (2007). Radio-frequency heating of foodstuff: Solution and validation of a mathematical model. *Journal of Food Engineering*, 79(3), 998–1006.

Marra, F., Valeria, M., De Bonis, M., and Ruocco, G. (2010). Combined microwaves and convection heating: A conjugate approach. *Journal of Food Engineering*, 97(1), 31–39.

Marra, F., Zell, M., Lyng, J.G., Morgan, D.J., and Cronin, D.A. (2009). Analysis of heat transfer during ohmic processing of a solid food. *Journal of Food Engineering*, 91(1), 56–63.

Mathur, K.B., and Epstein, N. (1974). *Spouted Beds*, Academic Press, New York.

Matz, S. (1989). *Equipment for Bakers*, Elsevier Science, Waltham, MA.

May, B.K., and Perré, P. (2002). The importance of considering exchange surface area reduction to exhibit a constant drying flux period in foodstuffs. *Journal of Food Engineering*, 54(4), 271–282.

Menter, F.R. (1994). Two-equation eddy-viscosity turbulence models for engineering applications. *AIAA Journal*, 32(8), 1598–1604.

Mistry, H., Ganapathi-subbu Dey, S., Bishnoi, P., and Castillo, J.L. (2006). Modeling of transient natural convection heat transfer in electric ovens. *Applied Food Engineering*, 26(17–18), 2448–2456.

Modest, M. (1992). *Radiative Heat Transfer*, Academic Press, New York.

Mondal, A., and Datta, A.K. (2010). Two-dimensional CFD modeling and simulation of crustless bread baking process. *Journal of Food Engineering*, 99(2), 166–174.

Mullineux, G., and Simmons, M.G.J. (2007). Surface representation of time dependent thixotropic materials properties. *Food Manufacturing Efficiency*, 1(2).

Nema, P.K., and Datta, A.K. (2006). Improved milk fouling simulation in a helical triple tube heat exchanger. *International Journal of Heat and Mass Transfer*, 49, 3360–3370.

Nicolaï, B.M., Verboven, P., and Scheerlinck, N. (2001). Modelling and simulation of food processes. In *Food Technologies in Food Processing*, P. Richardson (Ed.), Woodhead Publishing, Cambridge, pp. 91–109.

Norton, T., and Sun, D.-W. (2007). An overview of CFD applications in the food industry. In *Computational Fluid Dynamics in Food Processing*, D.-W. Sun (Ed.), CRC Press, Boca Raton, FL, pp. 1–43.

Ozalpa, A.A., and Dincer, I. (2010). Hydrodynamic-thermal boundary layer development and mass transfer characteristics of a circular cylinder in confined flow. *International Journal of Thermal Sciences*, 49(9), 1799–1812.

Park, K., Choi, D.-H., and Lee, K.-S. (2004). Optimum design of plate heat exchanger with staggered pin arrays. *Numerical Heat Transfer Part A: Applications*, 45(4), 347–361.

Patankar, S.V. (1980). *Numerical Heat Transfer*, Hemisphere Publishing, Washington, DC.

Perré, P., and Turner, I.W. (1999). A 3-D version of TransPore: A comprehensive heat and mass transfer computational model for simulating the drying of porous media. *International Journal of Heat and Mass Transfer*, 42(24), 4501–4521.

Ressing, H., Ressing, M., and Durance, T. (2007). Modelling the mechanisms of dough puffing during vacuum microwave drying using the finite element method. *Journal of Food Engineering*, 82(4), 498–508.

Shih, T.H., Liou, W.W., Shabbir, A., and Zhu, J. (1995). A new k–ε eddy viscosity model for high Reynolds number turbulent flows: Model development and validation. *Computers and Fluids*, 24(3), 227–238.

Siegel, R., and Howell, J.R. (1992). *Food Radiation Heat Transfer*, Hemisphere Publishing, Washington, DC.

Smolka, J., Nowak, A.J., and Rybarz, D. (2010). Improved 3-D temperature uniformity in a laboratory drying oven based on experimentally validated CFD computations. *Journal of Food Engineering*, 97(3), 373–383.

Suresh, H.N., Narayana, P.A.A., and Seetharamu, K.N. (2001). Conjugate mixed convection heat and mass transfer in brick drying. *Heat and Mass Transfer*, 37, 205–213.

Szafran, R.G., and Kmiec, A. (2004). CFD modeling of heat and mass transfer in a spouted bed dryer. *Industrial and Engineering Chemistry Research*, 43(4), 1113–1124.

Szafran, R.G., and Kmiec, A. (2005). Point-by-point solution procedure for the computational fluid dynamics modeling of long-time batch drying. *Industrial and Engineering Chemistry Research*, 44(20), 7892–7898.

Tanaka, F., Verboven, P., Scheerlinck, N., Moritaa, K., Iwasakia, K., and Nicolaï, B. (2007). Investigation of far infrared radiation heating as an alternative technique for surface decontamination of strawberry. *Journal of Food Engineering*, 79(2), 445–452.

Tattiyakul, J., Rao, M.A., and Datta, A.K. (2001). Simulation of heat transfer to a canned corn starch dispersion subjected to axial rotation. *Chemical Engineering and Processing*, 40, 391–399.

Therdthai, N., Zhou, W.B., and Adamczak, T. (2002). Optimisation of the temperature profile in bread baking. *Journal of Food Engineering*, 55(1), 41–48.

Therdthai, N., Zhou, W.B., and Adamczak, T. (2003). Two-dimensional CFD modelling and simulation of an industrial continuous bread baking oven. *Journal of Food Engineering*, 60(2), 211–217.

Therdthai, N., Zhou, W.B., and Adamczak, T. (2004). Three-dimensional CFD modelling and simulation of the temperature profiles and airflow patterns during a continuous industrial baking process. *Journal of Food Engineering*, 65(4), 599–608.

Turnbull, J., and Thompson, C.P. (2005). Transient averaging to combine large eddy simulation with Reynolds averaged Navier-Stokes simulations. *Computers and Chemical Engineering*, 29, 379–392.

Verboven, P., Datta, A.K., Anh, N.T., Scheerlinck, N., and Nicolaï, B.M. (2003). Computation of airflow effects on heat and mass transfer in a microwave oven. *Journal of Food Engineering*, 59(2–3), 181–190.

Verboven, P., Datta, A.K., and Nicolaï, B.M. (2007). Computation of airflow effects in microwave and combination heating. In *Computational Fluid Dynamics in Food Processing*, D.-W. Sun (Ed.), CRC Press, Boca Raton, FL, pp. 313–331.

Verboven, P., de Baerdemaeker, J., and Nicolaï, B.M. (2004). Using computational fluid dynamics to optimise food processes. In *Improving the Food Processing of Foods*, P. Richardson (Ed.), Woodhead Publishing, Cambridge, pp. 82–102.

Verboven, P., Scheerlinck, N., De Baerdemaeker, J., and Nicolaï, B.M. (2000a). Computational fluid dynamics modelling and validation of the isofood airflow in a forced convection oven. *Journal of Food Engineering*, 43(1), 41–53.

Verboven, P., Scheerlinck, N., De Baerdemaeker, J., and Nicolaï, B.M. (2000b). Computational fluid dynamics modelling and validation of the temperature distribution in a forced convection oven. *Journal of Food Engineering*, 43(2), 61–73.

Versteeg, H.K., and Malalasekera, W. (1995). *An Introduction to Computational Fluid Dynamics*, Longman Group, Harlow, pp. 1–100.

Wang, L., and Sun, D.-W. (2003). Recent developments in numerical modelling of heating and cooling processes in the food industry: A review. *Trends in Food Science and Technology*, 14, 408–423.

Welti-Chanes, J., Vergara-Balderas, F., and Bermúdez-Aguirre, D. (2005). Transport phenomena in food engineering: Basic concepts and advances. *Journal of Food Engineering*, 67, 113–128.

Wilcox, D.C. (1993). Comparison of 2-equation turbulence models for boundary-layer flows with pressure gradient. *AIAA Journal*, 31(8), 1414–1421.

Xia, B., and Sun, D.-W. (2002). Applications of computational fluid dynamics (CFD) in the food industry: A review. *Computers and Electronics in Agriculture*, 34, 5–24.

Zheng, L., Delgado, A., and Sun, D.-W. (2007). Surface heat transfer coefficients with and without phase change. In *Food Properties Handbook* (2nd ed), S. Rahman (Ed.), CRC Press, Boca Raton, FL, pp. 717–758.

FOOD SAFETY AND QUALITY ASSESSMENT

Chapter 13

Safety and Quality Management in Food Processing

Alok Jha, Catherine W. Donnelly, Ashutosh Upadhyay, and D.S. Bunkar

Contents

13.1 Introduction

The world food situation is being rapidly redefined day by day on account of new driving forces. Food habits, consumption patterns, production practices, marketing systems, and safety measurements are changing due to changes in income growth, climate, globalization, and urbanization throughout the world. Food safety and quality management is also witnessing increased attention due to consumer awareness about safe, minimally processed, additive-free, convenient, and highly nutritional foods, which is governed by the changing lifestyle of people and expanding world trade and its impact on public health. The World Health Organization (WHO) in 2006 estimated that 20% of deaths among children under the age of five in India were caused by diarrhea, frequently related to contamination of food and drinking water.

Food quality encompasses those characteristics of food that are acceptable to consumers. This includes external factors, such as physical appearance, including size, shape, color, gloss, consistency, texture, and flavor; regulatory provisions at the federal or state level; and internal factors, such as chemical, physical, and microbial. These internal factors in the modern world are described under the term "food safety." It is a scientific discipline describing the handling, processing, and storage of food in ways that prevent foodborne illness. Safety in foods includes a number of routines to be followed to avoid potentially severe health hazards. Food can transmit disease from person to person as well as serve as a growth medium for bacteria that can cause food poisoning.

Unlike most processed foods, fresh agricultural produce and products require additional care and attention. Apart from the productivity and quality considerations at the production level, there are some major precautions that need to be taken when agricultural produce and products are stored, transported, and distributed. Absence of such cautionary measures adversely affects the quality of the food products by increasing rejections and/or decreasing the market value of the produce. Further, this holds true for both raw and processed food products. Thus, it is in the interest of the food producers as well as the exporters to ensure that certain hygienic and safety conditions are met while processing foods for the intended final products. World-over, food standards acquire greater importance, giving high concerns to food safety on the back of outbreaks of diseases, such as foot and mouth disease (FMD), bovine spongiform encephalopathy (BSE), avian influenza, salmonellosis, etc., on the one hand, and growing consumer demand for food products, which are healthy on the other. Therefore, compliance with international food standards is a prerequisite for gaining a higher share of the world trade.

13.2 Post-Harvest Management

Food quality is an important food manufacturing requirement because consumers are susceptible to any form of contamination that may occur during the manufacturing process. Table 13.1 summarizes the various sources of bacterial contamination during handling of fresh produce. Quality and safety aspects initiate from the production of produce to the distribution of the processed product. The following fundamental requirements must be maintained to assure sanitary hygiene during post-harvest processing of fresh produce:

i. *Worker hygiene:* The cleanliness of workers and their personal hygiene throughout all stages of processing, packaging, storage, and distribution is to be maintained.
ii. *Packing house:* The packing and storage rooms are to be maintained clean by adopting proper sanitary system, such as washing, rinsing, and disinfecting the same with suitable disinfectant.
iii. *Cooling and cold storage:* The cold chain should be disinfected, and storage should be fumigated at regular intervals with approved chemicals to avoid contamination generated from such places.
iv. *Transport of produce from farm to market:* Transport vehicles are to be clean and sanitary. The vehicles should also be disinfected before using the same for transporting the farm produce.
v. *Grading and sorting at producer's level:* Foreign and undesirable substances are required to be removed. The damaged, bruised, diseased, and culled produce are to be sorted out and segregated on the basis of their shape, size, color, weight, or variety. Use of artificial coloring or flavoring substances is to be avoided. Samples may be analyzed for color, moisture, density, viscosity, oil, protein, fat, lactose, °Brix, total solids, etc., depending on the produce type, prior to consumption or processing.
vi. *Packaging:* Produce should be packed in clean and contamination-free packing material. Wherever necessary, moisture-proofing bags with polythene lining should be preferred. Use of hooks to handle the bags at any stage of distribution should be avoided. Packages should bear all needed information, such as name of commodity, place of origin, name of the producer/processor,

Table 13.1 Sources of Bacterial Contamination During Handling of Food Materials

Genus	Humans				Animals				Water		Soil	Foods
	Skin	*Intestine*	*Feces*	*Others*	*Skin*	*Intestine*	*Feces*	*Others*	*Salt*	*Fresh*		
Acetobacter												+
Acinetobacter	+	+	+	+						+	+	+
Aeromonas			+	+		+	+			+	+	+
Alcaligenes			+	+		+	+			+	+	+
Altreomonas					+				+			+
Arthrobacter											+	
Bacillus			+				+		+	+	+	+
Brevibacterium			+		+				+		+	+
Brochotrix					+		+				+	
Campylobacter		+	+	+		+	+	+		+	+	+
Clostridium		+				+			+	+	+	
Corynebacterium	+		+	+	+			+				+
Desulfotomaculum								+		+	+	+
Enterobacter			+	+			+	+		+	+	+

Erwinia				+				+				+
Escherichia		+				+						
Flavobacterium				+				+	+	+	+	
Glucobacterium											+	+
Halobacterium									+			
Klebsiella		+				+					+	
Lactobacillus		+	+	+		+	+	+				+
Leuconocostoc												+
Listeria		+	+	+		+	+	+			+	+
Microbacterium											+	
Micrococcus	+				+					+	+	
Moraxella	+			+				+				+
Pediocococus									+			+
Photobacterium							+	+				
Propionibacterium	+	+	+	+		+	+	+			+	
Proteus			+				+				+	

(continued)

Table 13.1 (Continued) Sources of Bacterial Contamination During Handling of Food Materials

Genus	*Humans*				*Animals*				*Water*			
	Skin	*Intestine*	*Feces*	*Others*	*Skin*	*Intestine*	*Feces*	*Others*	*Salt*	*Fresh*	*Soil*	*Foods*
Pseudomonas				+					+	+	+	
Salmonella		+	+		+	+			+			
Sarratia										+	+	+
Shigella		+				+						
Staphylococcus	+			+	+						+	+
Streptococcus		+	+	+	+		+					+
Vibrio		+				+			+	+		
Yersinia		+	+	+		+		+				+

Source: Frazier, W.C. and Westhoff, D.C., *Food Microbiology*. 4th edn. Tata McGraw-Hill, New Delhi, 2008.

compositions/intergradient, date of harvesting and expiry, weight, variety, and grade, etc.

vii. *Customer hygiene:* Customers should adopt all possible hygienic practices to avoid health hazards. Fresh farm produce is more prone to foodborne disease. Produce-associated outbreaks can be caused by bacteria, viruses, or other parasites. The outbreaks are due to poor agricultural practices or improper handling of the produce. Effective post-harvest strategies should be adopted to prevent contamination. Once contaminated, produce cannot be sanitized completely. The complex food marketing system potentially increases the exposure of consumers to different types of microorganisms during their distribution.

viii. *Effectiveness of cleaning and disinfection:* The effectiveness of cleaning and disinfection procedures should be regularly monitored by conducting microbiological analysis at various levels, starting from production to retailing. This will enable the producer and trading partners on either side of the border to assess the effectiveness of sanitary and phyto-sanitary measures. Produce should not be harvested where the presence of potentially harmful substances may lead to an unacceptable level of such substances in the food.

13.3 Good Agricultural and Animal Husbandry Practices

Good Agricultural and Animal Husbandry Practices (GAAHP) ensure safe and sound processing of raw materials by preventing introduction of undesirable contaminants, such as pesticide residues, metallic impurities, and spoilage microorganisms during food processing. The GAAHP approach to quality management includes safe use of authorized pesticides under actual conditions for effective and reliable pest control. It also encompasses a range of pesticide applications up to the highest authorized levels, applied in a manner that leaves a residue in the smallest amount possible. In the manufacture of food products, problems often arise due to agro-climatic conditions, cultivation and rearing techniques, and non-uniformity within the inputs concerned. This puts additional burden on the system to insure consistent, safe processing of raw materials.

13.4 Good Manufacturing Practices

Good Manufacturing Practices (GMP) applied to food processing define a code of practice for controlling and optimizing the process that recognizes the need to have a HACCP system in place to produce safe and cost-effective foods. In order to meet the requirements of GMP, regulatory bodies provide well-defined guidelines for food-processing operations. The major features of GMP are listed below:

i. Employees working in a food plant are required to wash their hands with a sanitized soap prior to beginning or returning to handling of food.
ii. All persons working or visiting the production area must wear authorized head covering to avoid contact of loose hair with food products.
iii. Sanitary precautions are required to be taken by all employees when sneezing or coughing. An employee with any infectious skin eruption, communicable disease, or other infected conditions must have plant management clearance before being allowed to handle food.
iv. Employees in production areas should wear clean uniforms. The uniforms should be changed daily or sooner if soiled for any reason. Shirts are required to be buttoned and tucked into trousers.
v. Workers in the production area will not wear rings and jewelry. Watches, pens, pencils, and loose materials should be removed prior to entry to the production areas.
vi. Smoking, spitting, or chewing of tobacco is prohibited in the production as well as storage areas.
vii. Consumption of beverages or food is allowed exclusively in designated areas.
viii. Nail polish and/or perfume is not allowed in production or storage areas.
ix. Containers and equipment made of glass, including glass thermometers, should not be in the production area.
x. Good housekeeping in the production area is necessary for work efficiency and workers' safety.

13.5 Food Additives and Colorants

An additive is a substance or mixture of substances, other than a base foodstuff, which is present in food as a result of any aspect of production, processing, storage, or packing. A food additive may also be used to improve or maintain the nutritional quality of a food keeping calorie value unaltered, to maintain the palatability and wholesomeness of a food, to provide leavening and pH control for baked foods, and to enhance the flavor and desired color in foods. More than 3000 different chemical compounds as recommended by the FDA are used as food additives and have GRAS (generally recognized as safe) status. Some of the additives are antimicrobial agents, antioxidants, appearance/control agents, substances other than natural color, flavor and flavor modifiers, moisture-control agent, nutrients, pH-control agents, sequestrants, surface tension control agents, and emulsifiers.

Today about 75% of the Western diet is made up of various processed foods; each person is consuming an average of 3.5–4.5 kg of food additives per year (Miller 1985; Feingold 1973). Although most food additives are apparently harmless, the rapidly increasing number and types of chemicals added to our food has increased concern regarding their harmfulness. The Food and Drug Administration (FDA), which is responsible for regulating the use of food additives, states that there are more than 3000 food additives currently cleared for use in the food industry. Once a food additive is approved by the FDA, it is considered to be fit for human consumption. But it is important to note that they might not be entirely safe for consumption. Some foods and coloring additives have been known to induce allergic reaction, and others are suspected to cause cancer, asthma, or birth defects. Although all of the additives have been approved for human consumption, many of them still provoke many emotional responses from consumers. The effects of food additives and colorants are summarized in Table 13.2.

To meet the growing demands of consumers for safe and healthy foods, the criteria of safety and quality has to be taken into consideration. Therefore, while

Table 13.2 Effects of Food Additives and Colorants on Human Health

Additives and Colorants	*Effect*	*Reference*
Colorants		
Tartrazine	Asthma, urticaria, rhinitis and childhood hyperactivity, mental and physical disorder	Freedman (1977), Feingold (1993)
Curcumin	Thyroid damage	Miller and Millstone (1987)
Sunset yellow	Kidney and adrenal damage	Miller (1985)
Carmoisine	Carcinogenic effect	Miller (1985)
Amaranth	Cancer, birth defect, still-birth, sterility and early fetus death	Miller (1985), Miller and Millstone (1987)
Ponceau	Carcinogenic effect	Miller (1985), Miller and Millstone (1987)
Erythrosine	Childhood hyperactivity	Lafferman and Silbergeld (1979), Shekim et al. (1977)
Caramels	Level of white blood cells	Miller and Millstone (1987)
Brown	Mutagenic and carcinogenic effect	Miller (1985), Miller and Millstone (1987)

(*continued*)

Table 13.2 (Continued) Effects of Food Additives and Colorants on Human Health

Additives and Colorants	*Effect*	*Reference*
Preservatives		
Benzoates	Provoke urticaria, angioedema and asthma	Miller and Millstone (1987), Michaelsson and Juhlin (1973)
Sulphite	Provoke urticaria, angioedema and asthma	Miller (1985)
Nitrates and nitrites	Cancer	Taylor (1983)
Butylated hydroxyanisole (BHA)	Provoke urticaria, angioedema and asthma	Miller (1985)
Monosodium glutamate (MSG)	Headache and asthma	Collins-Williams (1983), Allen and Baker (1981)
Sweeteners		
Saccharin	Cancer, mutagenic and growth inhibitor	Wynn and Wynn (1981), Reuber (1978)
Aspartame	Mental and physical disorder	Prinz and Roberts (1980)
Sucrose	Hyperactivity, neuroses, panic attacks, agoraphobia and schizophrenic episodes	Tuormaa (1991)

developing a new processed food, linkage among R&D organizations, producers and processors should be considered to improve the safety and quality of the processed foods. Although food additives are essential in food processing systems to increase the food productivity in terms of increased shelf life, quality, nutritive value, convenience, and economic saving, etc., it is necessary to consider if the food additives are safe and how to protect the consumer from unsafe additives. In order to minimize or avoid food additives in the everyday diet, the following steps may be taken into consideration while consuming processed foods.

i. The eating of fresh, unprocessed foods, grown by local farmers should be promoted. Because these foods are not transported thousands of miles, they do not need to be packaged or treated with preservatives. And because they are whole and unprocessed, they won't contain coloring or artificial flavors.

ii. The amount of publicity on television to buy and eat processed foods should be balanced by providing information about the advantages of buying and consuming foods grown locally.
iii. All nonessential food additives should be banned, particularly all cosmetic agents, such as food colorants. Foods that are consumed by infants and/or young children should be regulated to assure safety.
iv. All foods should be formulated only with additives having GRAS status. Consumer awareness regarding carcinogenic, mutagenic, and teratogenic properties of class II preservatives and other additives will further help in assuring food safety.
v. The government should introduce free, nutritious school meals, preferably using organic foods, which should be available to all school-age children.
vi. Consumers should learn more about the health and safety aspects and what they are consuming while buying any food. They should read the food ingredients labeled on food products and find out what is in the foods when they buy any food product.
vii. Consumption of ready-to-eat processed foods in the everyday diet should be moderated.
viii. Organic farming and organic foods should be promoted among the consumers in order to avoid pesticide or chemical effects in the foods.
ix. Food processors should take care to minimize the use of additives while processing any food.

13.6 Inspection and Certification System

In most countries, it is not customary to apply continuous inspection as in the case of meat. Few importing countries require it, and the nature of the products themselves is such that only part-time inspection is required during processing. In any event, inspection of raw materials should be carried out at the commencement of each processing run to ensure that only sound fruit or vegetables of sufficient maturity are used for processing. Sampling checks of raw materials should be carried out as frequently as the inspector thinks necessary.

The inspector must ensure that adequate hygiene practices are followed during the processing of the product. For example, in the case of canned and frozen products, raw materials should be washed absolutely clean so that fruit and vegetables entering the processing line are free from dirt, superficial residues of agricultural chemicals, insects, and extraneous plant materials. In the case of dried product, especially when the raw material is sun-dried on drying greens or racks, care must be taken to minimize contamination by bird and animal droppings, dust, and extraneous plant materials. It is often necessary to wash the dried product to ensure cleanliness of the final product. In the case of canning and freezing, the inspector must obtain full details of the processing program for at least the following day from management, so that an adequate inspection program can be scheduled. The inspector must also be aware of the pesticides and other chemicals used in the production of the raw materials. Necessary laboratory analyses can be arranged to ensure that the residue levels in the final product do not exceed tolerances adopted by importing countries.

At the commencement of and during processing, the inspector should pay attention to the state of raw materials, the preparation of raw materials for processing, preparation and density of packing medium, the state of cans or containers to be used (cleanliness and strength), the cooking or freezing process (time-temperature relationship), can filling and closure, and can or container storage. After processing, the inspector should check the final product to ensure the drained or thawed weight, the vacuum and headspace, packing medium strength, and that can/container integrity conditions are satisfactory. Statistically based sampling plans should be adopted for the examination of the final product to ensure it meets the requirements of the export standards and regulations. The labeling applied to cans or containers should also be checked to ensure both their correctness and compliance with the export regulations and the requirements of those countries in which the product is to be marketed. Cans should also be examined to make sure that the correct embossing relating to the product, its date of production, and the registered number of the export establishment has been applied.

Each establishment registered for the export of processed fruit and vegetables or for canned or frozen foods should have its own quality laboratory sufficiently equipped and staffed to carry out physical, chemical, and microbiological examinations of the goods.

Inspectors should have access to the laboratory facilities and the establishment's quality control records as and when required. Independent laboratory examination of product should be made by the agency having responsibility for export on the basis of a statistically developed sampling plan. Prior to export, the exporter should be required to notify the export quality and inspection agency of his intention to export in accordance with the provisions of the export processed fruits and vegetables regulations and on the prescribed "Notice of Intention to Export" form. The notice should be submitted in sufficient time before the shipment date to enable the product to be inspected satisfactorily, the intensity of inspection depending on the original state of the product, the conditions under which it has been stored, and the length of storage. When the product is approved, the agency will issue the exporter an "Export Permit" authorizing customs clearance of the product.

13.7 Package Labeling

Labeling is any written, electronic, or graphic communication on the packaging materials of the products. A label is simply a piece of paper, polymer, cloth, metal, or other material affixed to a container or article, on which is printed a legend, information concerning the product, addresses, etc. A label may also be printed directly on the container or article (Saroka 2002).

The purposes of packaging and package labels are

i. *Information transmission:* Packages and labels communicate how to use the food products, transport, recycle, or dispose of the package or product. With pharmaceuticals, food, medical, and chemical products, some types of information are required by governments.
ii. *Marketing:* The packaging and labels can be used by marketers to encourage potential buyers to purchase the product. Marketing communications and graphic designs are applied to the surface of the package and the point of sale display.
iii. *Security:* Packages can be made with improved tamper-resistance to deter tampering and have tamper-evident (Lee et al. 1998) features to help indicate tampering. Packages can be engineered to help reduce the risks of package pilferage: Some package constructions are more resistant to tampering, and some have tampering indicating seals. Packages may include authentication seals and use security printing to help indicate that the package and contents are not counterfeit.
iv. *Convenience:* Packages can have features that add convenience in distribution, handling, stacking, display, sale, opening, reclosing, use, and reuse.
v. *Portion control:* Single serving or single dosage packaging has a precise amount of content to control usage. Bulk commodities (such as salt) can be divided into packages that are a more suitable size for individual households. It is also aids the control of inventory.

Labels can be attached by

- Heat-activated adhesives, for example, "in-mold labeling," can be part of blow-molding containers and employ heat-activated adhesives.
- Pressure-sensitive adhesives (PSAs) are applied with light pressure without activation or heat. PSA labels often have release liners, which protect the adhesive and assist label handling.
- Rivets used to attach information plates to industrial equipment.
- Shrink wrap for printed shrinkable labels placed over packages and then heated to shrink them.
- Sewing for clothing, tents, mattresses, industrial sacks, etc.
- Wet glue (starch, dextrin, PVA) or water moistenable gummed adhesive.
- Yarn or twine for tying on a label.

13.7.1 PSA Adhesive Types

Pressure-sensitive label adhesives are commonly made from water-based acrylic adhesives with a smaller volume made using solvent-based adhesives and hot-melt adhesives. The most common adhesive types are

- *Permanent:* Typically not designed to be removed without tearing the stock, damaging the surface, or using solvents. The adhesion strength and speed can also be varied.
- *Peelable:* The adhesive is usually strong enough to be applied again elsewhere. This type is frequently known as "removable." There are many different types of removable adhesives, some are almost permanent; some are almost ultra-peelable.
- *Ultra-peelable:* Designed principally for use on book covers and glass; when removed, these adhesives labels do not leave any residue whatsoever. Adhesion is weak and only suitable for light-duty applications.
- *Freezer:* Most permanent and peelable adhesives have a service temperature limit of –10°C whereas freezer-appropriate adhesives, known as frost-fix adhesives, have a service temperature of –40°C and are suitable for deep-freeze use.
- *High Tack:* This is a permanent type of adhesive that exhibits a high initial grab to the application surfaces and is commonly used at higher coat weights to enable labels to adhere strongly to difficult, rough, or dirty surfaces.
- *Static Cling:* This is not actually an adhesive. The material (usually PVA) has a static charge to enable its adhesion to flat, smooth surfaces, such as glass. It is not sticky as such and is commonly used for window advertising, window decorations, oil change labels, etc.

13.7.2 Stock Types

The label stock is the carrier, which is commonly coated on one side with adhesive and usually printed on the other side. Label stocks can be a wide variety of papers, films, fabric, foils, etc. The stock type will affect the types of ink that will print well on them. Corona-treating or flame-treating some plastics makes them more receptive to inks and adhesives by reducing surface tension.

Labels can be supplied separately on a roll or on a sheet. Many labels are pre-printed by the manufacturer. Others have printing applied manually or automatically at the time of application. Some labels have protective overcoats, laminates, or tape to cover them after the final print is applied. This is sometimes before application and sometimes after. Labels are often difficult to peel and apply. Most companies use a label dispenser to speed up this task. Specialized high-speed application equipment is available for certain uses.

13.7.3 Symbols Used on Packages and Labels

Many types of symbols for package labeling are nationally and internationally standardized. For consumer packaging, symbols exist for product certifications, trademarks, proof of purchase, etc. Some requirements and symbols exist to communicate aspects of consumer use and safety (Choi and Burgess 2007). Recycling directions, Resin identification code (below), and package environmental claims

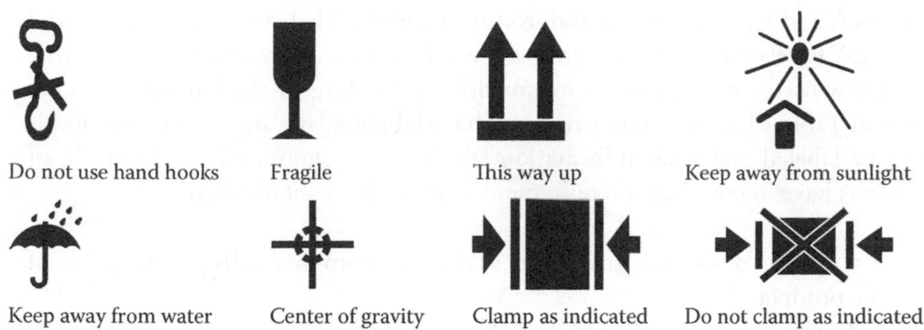

Figure 13.1 Some common symbols used in packing materials of food products.

have special codes and symbols. Barcodes (below), universal product codes, and RFID labels are common to allow automated information management.

"Print and Apply" corner wrap UCC (GS1-128) label application to a pallet load technologies related to shipping containers are identification codes, barcodes, and electronic data interchange (EDI). These three core technologies serve to enable the business functions in the process of shipping containers throughout the distribution channel. Each has an essential function: identification codes either relate product information or serve as keys to other data, barcodes allow for the automated input of identification codes and other data, and EDI moves data between trading partners within the distribution channel.

Elements of these core technologies include UPC and EAN item identification codes, the SCC-14 (UPC shipping container code), the SSCC-18 (serial shipping container codes), Interleaved 2 of 5 and UCC/EAN-128 (newly designated GS1-128) barcode symbols, and ANSI ASC X12 and UN/EDIFACT EDI standards (Severin 2007). Small parcel carriers often have their own formats. For example, United Parcel Service has a MaxiCode 2-D code for parcel tracking. RFID labels for shipping containers are also increasing in usage.

Shipments of hazardous materials or dangerous goods have special information and symbols (labels, placards, etc.) as required by UN, country, and specific carrier requirements. With transport packages, standardized symbols are also used to aid in handling (Figure 13.1).

13.8 Food Labeling

Consumers expect the labeling on food to be a true description of what they are buying. Hence, labels must declare the amounts per serving for calories, calories from fat, total fat, saturated fat, cholesterol, sodium, total carbohydrates, sugars, dietary fibers, and protein. Also the percentage daily reference values (DVR) must be shown to a 2000 cal and 2500 cal day^{-1} diets for the above nutrient as well as for vitamins A and C, and calcium and iron to make the label consumer-friendly and useful. DVR relative to various products are based on an evaluation of scientific data. DVR used for calculations for nutritional labeling in the United States are shown in Table 13.3. Most countries now have labeling laws stipulating how foods are to be labeled and what information labels must contain. Most, if not all, of those laws have in common a requirement that the label should bear

i. A statement of identity and a true, as distinct from misleading, description of the product
ii. A declaration of net contents (weight or number of pieces)
iii. The name and address of the manufacturer, packer, distributor, or consignee
iv. A list of ingredients (in descending order of volume or weight)

Table 13.3 Daily Reference Values (DRV) for Nutritional Labeling (Based on 2000 cal Diet) in the United States

Nutrients	*US Daily Values*
Fat	65 g
Saturated fat	20 g
Cholesterol	300 mg
Protein	50 g
Carbohydrate	50 g
Dietary fiber	30 g
Calcium	25 g
Iron	1000 mg
Sodium	18 mg
Potassium	2400 mg
Phosphorous	3500 mg
Iodine	1000 mg
Magnesium	150 mcg
Zinc	15 g
Copper	2 mg
Vitamin A	5000 IU[a]

(*continued*)

Table 13.3 (Continued) Daily Reference Values (DRV) for Nutritional Labeling (Based on 2000 cal Diet) in the United States

Nutrients	*US Daily Values*
Vitamin C	60 mg
Vitamin D	4000 IU[b]
Vitamin E	30 UM[c], 15 mg
Thiamin	1.7 mg
Riboflavin	20 mg
Niacin	32 g
Vitamin B_6	0.4 mg
Folate	6 mcg
Vitamin B_{12}	0.3 mg
Biotin	10 mg
Pantothenic acid	10 mg

Source: Aneja, R.P. et al., *Technology of Indian Milk Products*. A Dairy India Publication, Delhi, 2008.

[a] 1 mcg = 3.33 IU Vitamin A.

[b] 1 mcg = 40 IU Vitamin D.

[c] 1 mcg = 1 IU Vitamin E acetate or 0.65 α-tocopherol equivalent.

In addition, labels may also be required to include the country of origin, date of manufacture or packing, a use-by or expiry date, nutritional qualities or values of the food, storage directions, a quality grade, and directions for preparing the food. Table 13.4 represents the classification of the major foods in order of increasing microbial keeping quality.

13.9 Total Quality Management (TQM)

This is defined as that aspect of the overall management that determines and implements the quality policy and as such is the responsibility of top management. TQM is an organizational concern and not a domain of any specialist or specific function. TQM is not just a question of achieving standards, but one of survival and being strong all the time. Quality is a managerial responsibility in business organizations, but the scope of this responsibility is not just

Table 13.4 Classification of the Major Foods in Order of Increasing Microbial Keeping Quality

Class	*Processing, Including Heat Treatments, Compositional Modification, and Packing*	*Stability Characteristics*		*Examples*	*Predominant Microbial Flora at Retail*
		Temperature °C	*Time of Spoilage-Free Storage*		
1.	None of functional nature	< 10	10 – 40 hr	Fresh meat, milk, fish, poultry, eggs, vegetables	Psychrotrophic non-fermentative Gram positive Gram negative rods
2.	Pasteurization, followed by hermetic packing	< 10	3 days to 2 weeks	Dairy products, sliced cured meat products	Sporing rods and Lancefield group D. Streptococcci
3.	Reduction of water activity to ca. 0.95, pH reduction and addition of preservatives, in combination with hermetic packing	< 10	A few weeks	Gaffelbitter and similar semi-preserved fish products	Lactobacilli, Streptococcci, yeasts and molds
4.	Reduction of water activity to ca. 0.85, pH/water activity/lactic acid combinations of equivalent microbistatic effect, pasteurization	25	Many weeks	Condensed milk, mayonnaise, margarine, smoked sausage	Yeast and molds

5.	Reduction of water activity to ca. 0.85, sometimes in combination with pH reduction	25	Unlimited, i.e., unit chemical reaction interfere	Shelf-stable products such as salami, stock, fish, sauces	Molds
6.	Reduction of water activity to less than 0.60	35	Unlimited	Dehydrated foods	Bacilli, group D. Streptocococci, mold spores
7.	Appertization	35	Unlimited	Canned cured meat products and fruits	An occasional spore, i.e., counts of less than 10^2/g
8.	Sterilization	Any	Unlimited	Canned milk, soups, meat, vegetables, and fish	None

Source: Mossel, D.A.A., Essential and preservatives of microbial spoilage of foods. In *Food Microbiology: Advances and Prospects.* T.A. Roberts and F.A. Skinner (eds.). Academic Press, New York, 1983.

focusing on one particular aspect of the business. According to Fellows (2005), the TQM system covers (a) raw materials, purchasing, and control (including agreed specifications, supplier auditing, raw material storage, stock control, traceability, inspection, investigation of non-uniformity to specification); (b) process control (including identification, verification, and monitoring critical control points in a HACCP system, hygienic design of plant and layouts, cleaning, schedules, recording of critical production data, sampling procedures, and contingency plans to cover safety issues); (c) processing premises (including methods of construction to minimize contamination, maintenance, waste disposal); (d) quality control (including product specifications and quality standards for non-safety quality issues, monitoring, and verification of quality before distribution), (e) personnel (including training, personal hygiene, clothing, and medical screening); (f) final product (including types and levels of inspection to determine conformity with quality specifications, isolating nonconforming products, packaging checks, inspection records, complaints monitoring systems); and (g) distribution (to maintain the product integrity throughout the chain, batch traceability, and product recall systems). The benefit of a properly implemented TQM system are briefly summarized as follows:

i. Economic (more cost-effective production by getting it right the first time, reduction in wasted materials, fewer customer complaints, improved machinery efficiency, and increased manufacturing capacity)
ii. Marketing (consistently meeting customer needs, increased customer confidence, and sales)
iii. Internal (improved staff morale, increased levels of communication, better trained staff and awareness or commitment to quality, improved management control, and confidence in the operation)
iv. Legislative (demonstrating due diligence, providing evidence of commitment to quality and ability to improve)

A TQM system is broader than quality control and is a management philosophy that seeks to continually improve the effectiveness and competitiveness of the business as a whole. It is an integrated system, which ensures that all areas of the business are controlled to enable customers to consistently receive quality products that meet their needs and expectations. The approach requires collective responsibility and commitment at all levels in a business and can only be achieved through trusting working relationships and good communications (Fellows 2005). The basic elements of a TQM system are listed in Table 13.5. Guiding principles for a successful TQM are listed in Table 13.6. Performance of a TQM system may be evaluated in terms of (a) effectiveness, (b) resource utilization, (c) quality, (d) productivity, (e) quality of work, and (f) profitability.

A TQM system is necessary for establishing a culture of continuous improvement by identifying waste and eliminating it, aiming for a zero-defect

Table 13.5 Elements of a TQM System

Management commitment	It should be clear, and the management must disseminate the quality policy at all levels.
Teamwork participation	All employees must be deeply involved in improvement of quality systems.
Quality tools and techniques	All modern tools and techniques are adopted.
Continuous education and training	Documented procedures to be adopted to identify training needs, especially for skilled tasks.
Customer satisfaction	Methods to generate and evaluate customer satisfaction must be devised.

Source: Aneja, R.P. et al., *Technology of Indian Milk Products*. A Dairy India Publication, Delhi, 2008.

Table 13.6 Guiding Principles for Successful TQM

Customer orientation	An obsession for customer needs through regular feedback
Vision	Future vision of the organization
Quality culture	Dedication to continuously strive for improvement and professional excellence
Leadership	Coach the team and support the efforts
Quality strategy	Leads to strategic goals and plans
Values	Translates the organization strategies into thrust areas
Employee	Empowerment and participation
Teamwork approach	For efficiency and maximizing productivity
Training	Education for learning
Process centered	Each function to be assessed for its purpose
Communication	Essential for active involvement of whole organization and people in company's vision and values
Continuous improvement	Fellow dynamic concept of improvement, raise benchmark
Measurement and audit	Meticulously planned and periodic monitoring systems to ensure excellence

Source: Aneja, R.P. et al., *Technology of Indian Milk Products*. A Dairy India Publication, Delhi, 2008.

objective. The internal objective of redefining the business operation changes the work culture of the organization, when competitiveness is based on quality criteria, characterized by an optimization of internal and external pursuit of excellence.

13.10 Export Quality Control and Inspection Systems

With the advent and development of health consciousness among consumers, stimulated by the work of the Joint FAO/WHO Codex Alimentarius Commission through its elaboration of food standards, Code of Hygienic Practice, and Code of Ethics for International Trade in food, an increasing number of countries have adopted sophisticated food laws and regulations and established various food control departments, some with the aid of the FAO. Consequently, those countries no longer accept products on trust that they are satisfactory, but instead, demand that food imports meet the requirements of their food laws and pass inspection by their control departments. Moreover, many of them require exporting countries to certify that products comply with their national legislation and some also require additional special declarations.

As a result of these developments, the emphasis of activity of export quality control and inspection systems has changed. Although most of them still establish their own standards of quality control and adopt standards for foods for export, most of their efforts and resources are now directed at ensuring that foods for export meet the mandatory requirements of importing countries and providing the necessary associated certification. To do otherwise is to invite either the detention or, at worst, rejection of product at point of entry.

13.11 Detentions and Rejections of Foods

Food exporting countries can no longer assume that there is a good chance that products not complying with the requirements of importing countries will escape the inspection at the point of entry. Details of food imports released by the United States Food and Drug Administration (FDA) indicate that significant quantities of product are at least detained, and at worst rejected, because they fail to meet US food laws and regulations. Reasons given for the detentions include (i) noncompliance with labeling requirements, (ii) decomposition, (iii) insect and animal filth and damage, (iv) use of prohibited additives, (v) noncompliance with requirements of the US low acid canned food regulations, (vi) heavy-metal contamination, (vii) excessive levels of pesticide residues, (viii) excessive levels of mycotoxin, (ix) mold infestation, (x) microbiological contamination, and (xi) swollen and otherwise faulty cans. Table 13.7 summarizes the certain microbiological requirements of some food products.

Table 13.7 Microbiological Requirements of Food Products

Products	*Parameters*	*Limits*
Thermally processed fruits and vegetables	Total plate count	Nil
	Incubation at 37°C and 55°C for 7 days	No change in pH
Dehydrated fruits and vegetable products and soup powders	Total plate count	Not more than 40,000/g
Carbonated beverages, ready-to-serve beverages, including fruit beverages	Total plate count	No more than 50/ml
	Yeast and mold count	Not more than 2/ml
	Coliform count	Absent in 100 ml
Tomato juice and soup	Mold count	Not more than 30% of the field examined
	Yeast count	Not more than 125 per 1/50 cmm
Tomato puree and paste	Mold count	Not more than 60% of the field examined
Tomato ketchup and tomato sauce	Mold count	Not more than 40% of the field examined
	Yeast and spores	Not more than 125 per 1/60 cmm
	Bacteria	Not more than 100 million/ml
Jam/marmalade/fruit jellies/fruit chutney and sauces	Mold count	Not more than 40% of the field examined
	Yeast and spores	Not more than 125 per 1/60 cmm
Other fruit and vegetable products	Yeast and mold count	Absent in 25 g
Frozen fruit and vegetable products	Total plate count	Not more than 40,000/g
Preserves	Mold count	Nil

(*continued*)

Table 13.7 (Continued) Microbiological Requirements of Food Products

Products	*Parameters*	*Limits*
Pickles	Mold count	Nil
Fruits cereal flakes	Mold count	Nil
Candied and crystallized or glazed fruit and peel	Mold count	Nil
All fruits and vegetable products and ready-to-serve beverages, including fruit beverages and synthetic products	Flat sour organisms	Nil
	Staphylococcus aureus	Absent in 25 g or ml
	Salmonella	Absent in 25 g or ml
	Shigella	Absent in 25 g or ml
	Clostridium botulinum	Absent in 25 g or ml
	Patulin (in apple and apple products)	Absent in 25 g or ml
	E. coli	Absent in 25 g or ml
	Vibrio chlorea	Absent in 25 g or ml

Source: Manjunath, M.N., Food additives in processed fruits and vegetable. Lecture Document on *Approaches towards Value Addition to Fruits and Vegetables in Food Chain*. CFTRI, India, September 19–30, pp. 204–208, 2005.

13.12 Certification System

A Certificate of Export will generally indicate that the product is regulated by the Food and Drug Administration (FDA) and that the company is not at this time the subject of any enforcement action by the FDA. Such certificates are neither guarantees nor a certification of a product's safety and quality. They are issued at the request of a domestic (US) company.

The FDA will issue certificates assuming the product meets the requirements of 801(e) of the Federal Food, Drug, and Cosmetics Act as follows:

i. According to the specifications of the foreign purchaser
ii. Not in conflict with the laws of the country to which it is intended for export

iii. Labeled on the outside of the shipping package that it is intended for export
iv. The particular shipment is not sold or offered for sale in domestic commerce

Each request should be accompanied by

i. An original label or, if no labels have been printed, a detailed draft version of the current label
ii. Sufficient information for each product for which a certificate is requested so the reviewer can properly identify the product
iii. Adequate identification of the actual manufacturer of each product
iv. The following statement:

> "The requester hereby presents and acknowledges that the company is aware that in making this request the company is subject to the terms and provisions of Title 18, Section 1001, United States Code which makes it a criminal offense to falsify, conceal, or cover up a material fact; make any material false, fictitious, or fraudulent statement or representation; or make or use any false writing or document knowing the same to contain any materially false, fictitious, or fraudulent statement or entry."

The BIS product certification scheme is essentially voluntary in nature and is largely based on ISO Guide 28, which provides general rules for a third-party certification system of determining conformity with product standards through initial testing and assessment of a factory quality management system and its acceptance followed by surveillance that takes into account the samples from the factory and open market. All BIS certification is carried out on Indian Standards, which have been found amenable to product certification. A sizable number of Indian standards have, however, been harmonized with ISO/International Electrotechnical Commission standards.

13.13 Some Quality Management Systems

13.13.1 ISO 9000 Quality Management System

The International Organization for Standardization (ISO), a federation of national standards body, was set up in 1979. The ISO is a technical committee on quality management systems to evolve international quality standards. Figure 13.2 represents the various ISO 9000/10000 standards for quality and safety management for food processing operation, which are enforced nowadays. With the changed focus on quality issues worldwide, the ISO 9000 standards serve as a basis for ensuring consistent quality of goods and services as well as improving productivity. Their aim is to harmonize quality management practices on an international scale

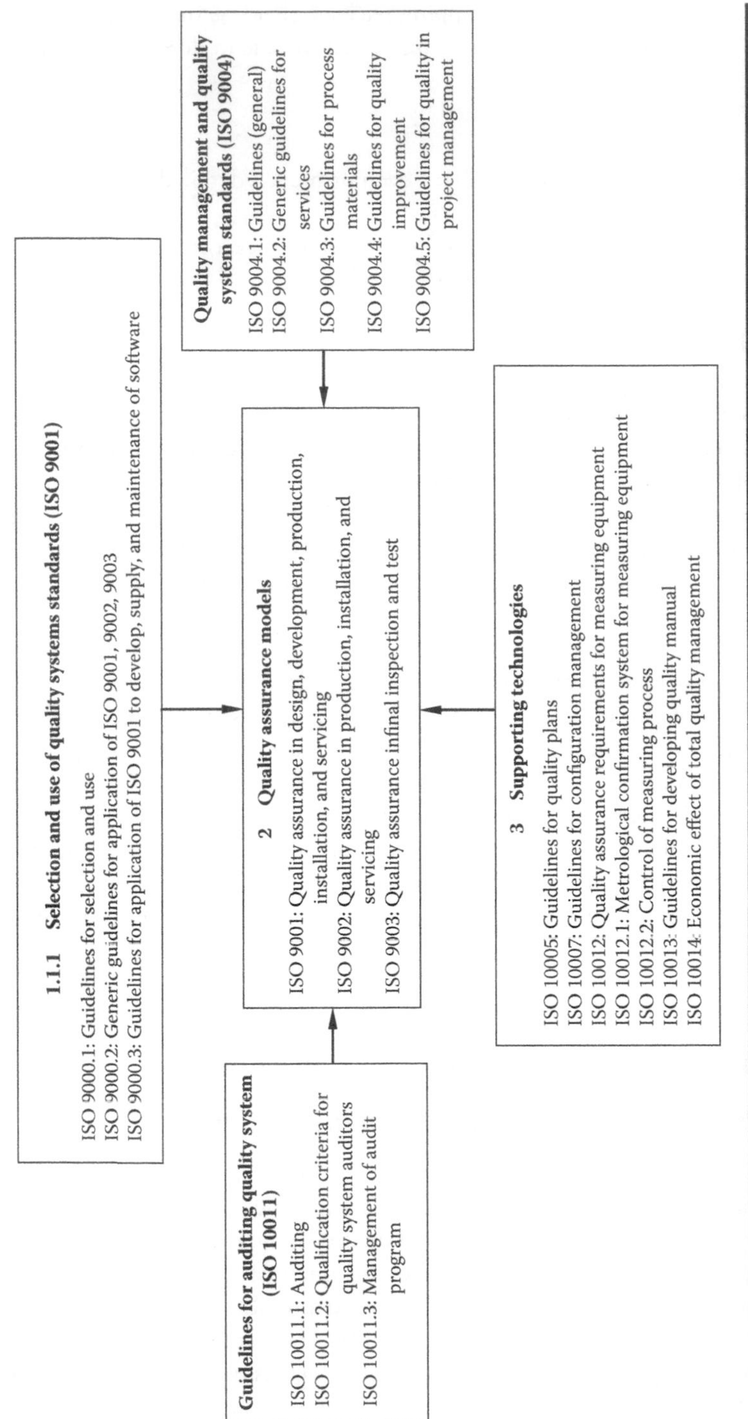

Figure 13.2 ISO 9000/10000 standards for quality and process parameters.

and establish quality as a factor in global trade. These standards necessitate organizations to reorient themselves to follow the process-centered approach to quality management systems and place the system on a continual improvement mode.

After widespread consultations and deliberations, an international consensus was reached on the ISO 9000 series of standards, which were brought out in 1987. Subsequently, standards on supporting system were brought out to make ISO 9000 models more efficiently operational. It can be thought of as a base-level common denominator of quality system requirements meant for all enterprises but not covering technology and competitive elements. It builds a baseline system for managing quality. The focus, therefore, is on designing a total quality management system, one that complies with external standards but includes the specific requirements of industry and integrates elements of competitiveness. The requirements of ISO 9000 for laying the foundation for excellence and a total quality management system are given in Figure 13.2.

In a short time, the ISO 9000 has become an internationally recognized benchmark for measuring the quality of products in markets around the globe. More than 250,000 enterprises are certified under ISO 9000 in more than 100 countries, including India. With experiences gained worldwide, the ISO committee on quality management systems has decided to launch the ISO 9000:2000 to meet the challenges of the new millennium.

How well a company is organized is reflected in the quality of its products or services. Quality in the product is impossible without quality in the process. Quality in the process is impossible without the right organization, which is built on proper leadership. Strong bottom-up commitment is the driver for all products, processes, organization, and leadership. These are steps to emulate the excellence model. The millennium standard (ISO 9000:2000) has therefore changed the focus from procedure to process. Its standards have been structured to provide a comprehensive model for bringing consistency into operation and continual improvement. The core of a quality management system is ISO 9001:2000 under which organizations could be certified. It is, depicted in Figure 13.2, built on four pillars:

- Management responsibility
- Resource management
- Product and service realization
- Measurement, analysis, and improvement

13.13.2 ISO 22000:2005 Food Safety Management System

Food safety is linked to the presence of foodborne hazards in food at the point of consumption. Because food safety hazards can occur at any stage in the food chain, it is essential that adequate controls be in place. Therefore, a combined effort of all parties through the food chain is required. ISO 22000:2005 specifies requirements for a food safety management system in which an organization in the food chain needs to demonstrate its ability to control food safety hazards in order to ensure that food is safe at the time of human consumption. It is applicable to all organizations, regardless of size, which are involved in any aspect of the food chain and want to implement systems that consistently provide safe products. The means of meeting any requirements of ISO 22000:2005 can be accomplished through the use of internal and/or external resources.

The ISO 22000 international standards specify the requirements for a food safety management system that involves the following elements:

- Interactive communication
- System management
- Prerequisite programs
- HACCP principles

Critical reviews of the above elements have been reported in the scientific literature. Communication along the food chain is essential to ensure that all relevant food safety hazards are identified and adequately controlled at each step within the food chain. This implies communication between organizations both upstream and downstream in the food chain. Communication with customers and supplies about identified hazards and control measures will assist in clarifying customer and supplier requirements.

Recognition of the organization's role and position within the food chain is essential to ensure effective interactive communication throughout the chain in order to deliver safe food products to the final consumer. The most effective food safety systems are established, operated, and updated within the framework of a structured management system and incorporated into the overall management activities of the organization. This provides maximum benefit for the organization and interested parties. ISO 22000 has been aligned with ISO 9001 in order to enhance the compatibility of the two standards.

ISO 22000 can be applied independently of other management system standards or integrated with existing management system requirements. ISO 22000 integrates the principles of the Hazard Analysis and Critical Control Point (HACCP) system and application steps developed by the Codex Alimentarius Commission. By means of auditable requirements, it combines the HACCP plan with prerequisite programs. Hazard analysis is the key to an effective food safety management system because conducting a hazard analysis assists in organizing the knowledge required to establish an effective combination of control measures. ISO 22000 requires that all hazards that may be reasonably expected to occur in the food chain, including hazards that may be associated with the type of process and facilities used, are identified and assessed. Thus, it provides the means to determine and document why certain identified hazards need to be controlled by a particular organization and why others need not. During hazard analysis, the organization determines the strategy to be used to ensure hazard control by combining the prerequisite programs and the HACCP plan.

ISO is developing additional standards that are related to ISO 22000. These standards will be known as the ISO 22000 family of standards. At the present time, the following standards will make up the ISO 22000 family of standards:

- ISO 22000: Food safety management systems – Requirements for any organization in the food chain
- ISO 22001: Guidelines on the application of ISO 9001:2000 for the food and drink industry (replaces: ISO 15161:2001)
- ISO TS 22003 – Food safety management systems for bodies providing audit and certification of food safety management systems
- ISO TS 22004: Food safety management systems – Guidance on the application of ISO 22000:2005
- ISO 22005: Traceability in the feed and food chain – General principles and basic requirements for system design and implementation
- ISO 22006: Quality management systems – Guidance on the application of ISO 9002:2000 for crop production

In comparison with ISO 9001, the standard is a more procedurally orientated guidance than a principle-based one. Apart from that, ISO 22000 is an industry-specific risk management system for any type of food processing and marketing, which can be closely incorporated with the quality management system of ISO 9001. The detailed similarities and differences of the two standards can be found elsewhere.

13.13.3 Codex Alimentarius Commission

The Codex Alimentarius brings together all the interested parties: scientists, technical experts, governments, consumers, and industry representatives to help develop standards for food manufacturing and trade. These standards, guidelines, and recommendations are recognized worldwide for their vital role in protecting the consumer and facilitating international trade. The Codex Alimentarius is recognized by the World Trade Organization (WTO) as an international reference point for the resolution of disputes concerning food safety and consumer protection (FAO 2005). As the Codex Alimentarius represents a consensus of food and trade experts from around the world, these standards are increasingly being used in international trade negotiations and also for settling of disputes by the WTO. The Codex Alimentarius, food law, or food code is a collection of food standards developed and presented in a unified, codified manner, together with associated materials, such as the code of hygienic and good manufacturing practices, recognized methods of analysis and sampling, general principles, and guidelines. The Codex Alimentarius currently comprises more than 300 standards, guidelines, and other recommendations relating to food quality, composition, and safety. The FAO and WHO jointly established the Codex Alimentarius Commission in 1962 to implement the joint FAO/WHO Food

Standard Programme. The aim of the Commission is to protect the health of consumers by ensuring observance of fair practices in the food trade. It promotes coordination of work on the formulation of food standards undertaken by international governmental and nongovernmental organizations. Adoption of HACCP standards, formulated by CAC, under the sanitary and phytosanitary (SPS) measures has made the HACCP system an instrument of food safety. It has become incumbent on signatory countries of the SPS agreement to implement these standards.

The Codex Alimentarius officially covers all foods, whether processed, semi-processed or raw, but far more attention has been given to foods that are marketed directly to consumers. In addition to standards for specific foods, the Codex Alimentarius contains general standards covering matters such as food labeling, food hygiene, food additives, and pesticide residues, and procedures for assessing the safety of foods derived from modern biotechnology. It also contains guidelines for the management of official (i.e., governmental) import and export inspection and certification systems for foods (UN 2005).

13.13.4 Codex Standards

Codex standards define the identity of the product and describe the basic composition and quality factors required for international trade. To protect consumers' health, provisions on food additives, contaminants, and hygiene requirements from a central core of standard.

The FAO, through the Codex Alimentarius Commission (CAC), plays a role in facilitating trade and improving the quality and safety of domestic and exported food products. This contributes to the harmonization, equivalency, and transparency of national food standard requirements and import/export inspection and certification schemes. These schemes are included in the WTO agreement on the application of sanitary and phytosanitary measures (SPS Agreement) and the Agreement on Technical Barriers to Trade (TBT Agreement).

13.13.5 HACCP Analysis

Hazards are biological, physical, or chemical properties that may cause food to be unsafe for human consumption. The goal of a food safety management system is to control certain factors that lead to out-of-control hazards. Foods are agricultural products and travel from farm to fork, passing through the environment. Foods can become contaminated by microbes, pathogens, toxic chemicals, toxins in the establishment or in the environment. Physical objects may also contaminate food and cause injury. Food may become naturally contaminated from the soil in which it is grown or from harvest, storage, or transportation practices. Some foods undergo further processing and, at times, despite best efforts, become contaminated. These inherent hazards, along with the other hazards that may be introduced in the system, can lead to injury or illness. Hazards are a big threat as ticking bombs in the food system. Food hazards include the following:

- *Biological agents:* Bacteria and their toxins, parasites, viruses, fungi, and molds
- *Physical objects:* Bandages, jewelry, stones, glass, bone, metal fragments, and packaging materials
- *Chemical contamination:* Natural plant and animal toxins, unlabeled allergens (allergen-causing protein), nonfood-grade lubricants, cleaning compounds, food additives, and insecticides

Lack of food-safety systems costs the food industry millions of dollars annually through waste, reprocessing, recalls, litigation, and resulting loss of sales. It is now recognized internationally that the most cost-effective approach to food safety is through the application of the HACCP technique. By adopting an effective food-safety system based on HACCP, the food industry can minimize the potential for things to go wrong and ensure the safety of food products. HACCP is entirely complementary and adds essential safety elements to existing processing systems, such as Good Manufacturing Practices and ISO 9000 standards.

Food producers as well as retailers and consumers benefit as a result of an effective HACCP system. The benefits of the HACCP system are summarized below:

i. Ensure safety of food products through final inspection and testing
ii. Capable of identifying all potential hazards
iii. Easy to introduce technological advances in equipment design and processing procedures related to food products
iv. Directs resources to the most critical part of the food processing system
v. Encourages confidence in food products by improving the relationship among regulatory bodies, food processors, and the consumer
vi. Promotes continuous improvement of the system through regular audits
vii. Focuses on safety issues in the whole chain from raw materials to consumption
viii. Complements the quality management system (e.g., ISO 9000)

Procedures for HACCP System

Figure 13.3 shows the procedure for preparation of a HACCP plan. The following steps must be considered to adopt a HACCP system in the food industry:

i. Acquire a good knowledge of the product flow sheet, from raw and pack material suppliers to consumers
ii. Assess each step and establish associated potential risk for
   A. Consumer safety (foreign materials, microbiology, aflatoxins, pesticides, heavy metals, monomers, etc.)
   B. Edible quality of the product (taste, texture, color, smell, appearance, etc.)

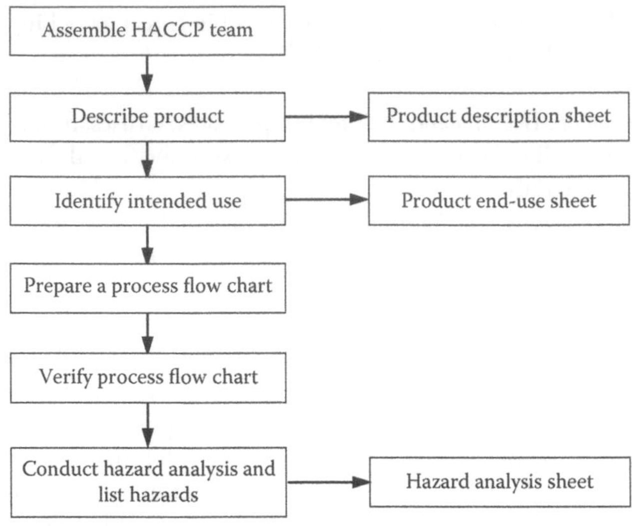

Figure 13.3 Procedure for preparation of a HACCP plan.

iii. Plot a complete product flow sheet and establish the critical control points (CCPs)
iv. Associate with each CCP an action in order to eliminate risks and prepare an action plan
v. Implement the agreed actions and keep records of results
vi. Evaluate the results of actions vs. risks and prepare an ongoing working plan
vii. Decide on a permanent monitoring of CCPs

13.13.6 Hurdle Technology

In traditionally preserved foods, such as smoked fish or meat, jams, and other preserves, there are a combination of various factors that ensure microbial safety and stability of the food and, thus, enable it to be preserved. In smoked products, this combination includes heat, reduced moisture content, and antimicrobial chemicals deposited from the smoke onto the surface of the food. Some smoked foods may also be dipped or soaked in brine or rubbed with salt before smoking to impregnate the flesh with salt and thus add further preservative mechanisms. Smoked products may also be chilled or packed in modified atmospheres to further extend the shelf life.

The concept of combining several factors or barriers (hurdles) to preserve foods has been developed, which is termed hurdle technology. In hurdle technology, each factor is a hurdle that microorganisms must overcome. The hurdles are also used to improve the quality of foods and the economic properties of the foods. To be successful, the hurdles must take into account the initial numbers and the type of microorganisms that are likely to be present in the food. The hurdles that are

Table 13.8 Examples of Some Hurdles in Foods

Type of Hurdle	*Examples*
Physical hurdles	Aseptic packaging, electromagnetic energy (microwave, radio frequency, pulsed magnetic field, high electric field), high temperature (blanching, pasteurization, sterilization, evaporation, extrusion, baking, frying), ionizing radiation, low temperature (chilling, freezing), modified atmospheres, packaging films (active packaging, edible packaging), photodynamic inactivation, ultra-high pressure, ultra-sonication, ultraviolet radiation
Physico-chemical hurdles	Carbon dioxide, ethanol, lactic acid, lactoperoxidase, low pH, low redox potential, low water activity, Maillard reaction products, organic acids, oxygen, ozone, phenols, phosphates, salt, smoking, sodium nitrite/nitrate, sodium/potassium sulfite, spices and herbs, surface treatment agents
Microbially derived hurdles	Antibiotics, bacteriocins, competitive flora, protective cultures

Source: Leistner, L. and Gorris, L.G.M., *Trends in Food Science and Technology*, 6, 41–46, 1995.

selected should be high enough so that the anticipated numbers of these microorganisms cannot overcome them. This approach involves manipulating pH, redox potential, water activity, solute type and concentration, heat treatments, chemical preservatives, effective packaging techniques and maintaining chilled conditions of storage, each treated as a hurdle. Table 13.8 summarizes some of the hurdles used to preserve foods.

The combination of hurdle technology and HACCP in process design is more successful for processing of foods. By combining hurdles, the intensity of individual preservation techniques can be maintained comparatively low to minimize loss of products quality while overall there is sufficient impact on controlling microbial growth.

13.14 Summary

Food safety is a fundamental public health concern. Food safety and quality has become an area of priority and necessity for consumers, retailers, manufacturers, and regulators. Changing global patterns of food production, international trade, technology, public expectations for health protection, and many other factors has created a huge demand for food safety and quality auditing professionals. The primary objective of food safety and management is that a food is safe and suitable for human consumption. It should ensure fair trade practices in the food trade and follow the food chain—farm to fork. It should take into account the wide diversity of food safety standards, CODEX activities, and varying degrees of risk involved in food production and lay a firm foundation for ensuring food hygiene with each specific code of hygiene practice applicable to each sector. In addition, a food safety and management program should meet performance of effectiveness (extent to which planned activities are realized and planned results achieved) and efficiency (relationship between the results achieved and the resources needed), apply proven management principles aimed at continually improving performance over the long term by focusing on customers while addressing the needs of all other stakeholders, be well established, documented, implemented, maintained, and continually improved or updated, and have products or services that actually meet their intended usage.

References

Allen, D.H. and Baker, G.J. (1981). Chinese restaurant asthma. *New England Journal of Medicine*, 305: 1154–1155.

Aneja, R.P., Mathur, B.N., Chandan, R.C. and Banerjee, A.K. (2008). *Technology of Indian Milk Products*. A Dairy India Publication, Delhi.

CAC (2005). 28th Session, FAO Headquarters, Rome, Italy.

Choi, S.-J. and Burgess, G. (2007). Practical mathematical model to predict the performance of insulating packages. *Packaging Technology and Science*, 20(6): 369–380.

Collins-Williams, C. (1983). Intolerance to additives. *Annals Allergy*, 51: 315–316.

FAO (2005). The state of food and agriculture. Food and Agriculture Organization of the United Nations, Rome, Italy.

FDA (2002). US food and drug adulterations. Center for Food Safety and Applied Nutrition Industry Activities, Staff Brochure.

Feingold, B.F. (1973). Food additives and child development. *Hospital Practice*, 21(11–12): 17–18.

Feingold, B.F. (1993). Adverse Reactions to Hyperkinesis and Learning Disabilities (H-LD) Congressional Record, S-1973, 39–42.

Fellows, J.P. (2005). *Food Processing Technology: Principles and Practice*. 2nd edn. Woodhead Publishing, Cambridge.

Frazier, W.C. and Westhoff, D.C. (2008). *Food Microbiology*. 4th edn. Tata McGraw-Hill, New Delhi.

Freedman, B.J. (1977). Asthma induced by sulphur dioxide, benzoate and tartazine contained in orange drinks. *Clinical Allergy*, 7: 407–415.

Lafferman, J.A. and Silbergeld, E.K. (1979). Erythrosin B inhibits dopamine transport in rat caudate synaptosomes. *Science*, 205: 410–412.

Lee, D.S., Hwang, Y.I. and Cho, S.H. (1998). Developing antimicrobial packaging film for curled lettuce and soybean sprouts. *Food Science and Biotechnology*, 7: 117–121.

Leistner, L. and Gorris, L.G.M. (1995). Food preservation by hurdle technology. *Trends in Food Science and Technology*, 6: 41–46.

Manjunath, M.N. (2005). Food additives in processed fruits and vegetable. Lecture Document on *Approaches towards Value Addition to Fruits and Vegetables in Food Chain.* CFTRI, India, September 19–30, pp. 204–208.

Michaelsson, G. and Juhlin, L. (1973). Urticaria induced by preservatives and dye additives in foods and drugs. *British Journal of Dermatology*, 88: 525–532.

Miller, M. (1985). *Danger: Additives at Work.* London Food Commission, London.

Miller, M. and Millstone, E. (1987). Food Additivcs Campaign Team: Report on Colour Additives. FACT, London.

Mossel, D.A.A. (1983). Essential and preservatives of microbial spoilage of foods. In *Food Microbiology: Advances and Prospects.* T.A. Roberts and F.A. Skinner (eds.). Academic Press, New York.

Prinz, R.J. and Roberts, H.E. (1980). Dietary correlates of hyperactive behavior in children. *Journal of Consulting and Clinical Psychology*, 48(6): 760–769.

Reuber, M.D. (1978). Carcinogenicity of saccharin. *Environmental Health Perspective*, 25: 173–200.

Severin, J. (2007). New methodology for whole-package microbial challenge testing for medical device trays. *Journal of Testing and Evaluation*, doi:10.1520/JTE100869.

Shekim, W.O., DeKirmeryian, H. and Chapel, J.L. (1977). Urinary catecholamine metabolites in hyperkinetic boys treated with d-amphetamine. *American Journal of Psychiatry*, 134: 1276–1279.

Soroka, W. (2002). *Fundamentals of Packaging Technology.* Institute of Packaging Professionals, Naperville, IL.

Taylor, G. (1983). Nitrates, nitrites, nitrosamines and cancer. *Nutrition and Health*, 2: 1.

Tuormaa, T.E. (1991). *An Alternative to Psychiatry.* The Book Guild Ltd., Hove, Sussex.

UN (2005). Commission adopts safety guidelines for vitamin and food supplements. UN, Geneva.

WHO (2006). Understanding the Codex Alimentarius Preface. 3rd edn. UN, Geneva.

Wynn, M. and Wynn, A. (1981). The prevention of handicap of early pregnancy origin: Some evidence for the value of good health before conception. Foundation for Education and Research in Child Bearing, London.

Chapter 14

Biosensors for Food Safety

Amit Singh, Somayyeh Poshtiban, and Stephane Evoy

Contents

14.1 Introduction

Bacteria are ubiquitous entities that coexist in a symbiotic relationship with a majority of plants and animals and therefore contribute to the ecological balance. The human gastrointestinal tract, for example, houses a large and diversified population of aerobic and anaerobic microbes that maintain a complex interaction with their host and perform several vital functions in the body, such as developing the immune system and facilitating digestion and absorption of food (Sekirov et al. 2010). Many bacteria, however, are capable of causing diseases to plants, animals, and humans, thereby having a detrimental effect on the health and safety of life forms. Such disease-causing bacteria are often termed as pathogens and are estimated to be responsible for the loss of nearly 13.3 million lives annually, 25% of the total deaths globally. The food industry is a prime source that is affected with microbial contamination, which, if not detected in a timely manner, would lead to a widespread dreadful effect of the pathogens. The Center for Disease Control and Prevention (CDC) recognizes at least 31 known and several unidentified foodborne pathogens that cause nearly 50 million cases of illnesses and disease due to food contamination. The United States Food and Drug Administration (FDA) acknowledged the prevalence, severity, and implications of the food contamination by laying down a "*Food Protection Plan*" that comprehensively illustrates an integrated strategy for containment, prevention, intervention, and response to a foodborne pathogen outbreak (FDA 2007).

Early detection of a foodborne pathogen is of utmost importance for timely intervention and quarantine of the contaminated food supply. The conventional methods for pathogen detection largely rely on microbiological culturing and subsequent biochemical analysis of the sample to give a precise identification of the contaminant. These methods, however, are extremely time consuming and labor intensive, need trained and skilled personnel for accurate pathogen identification and prediction, and most importantly, are not amenable to miniaturization and development of point-of-care diagnostics. Advancement in molecular-level detection, such as polymerase chain reaction (PCR) and enzyme-linked immunosorbent assay (ELISA), were therefore looked upon as alternative methods for a rapid and reliable detection of pathogens from food samples. PCR-based methods are used for amplification of minute quantities of genetic material from the contaminant pathogen to make multiple copies that could be detected using a suitable gene-specific probe. While PCR-based methods have enjoyed tremendous success in pathogen identification, their performance depends on the fidelity of duplication of the genetic material and can give false positive results. PCR-based systems, however, are neither useful for detecting toxins that could contaminate a food sample nor differentiate between a live or dead bacterial load in the food sample. ELISA alternatively exploits the recognition specificity of antibodies and can detect bacterial pathogens as well as toxin presence in a food sample. Both these detection methods, most importantly, are performed at very low sample volumes and require sample concentration and pre-enrichment steps to realize the detection.

Alternative technologies to overcome the limitations of conventional and biomolecular techniques have gained tremendous popularity in recent years, and advancement in nanotechnology has led to the development of biosensors as an attractive choice. A biosensor has two key elements, namely a bio-probe that provides specificity of recognition toward the analyte in question and a transduction

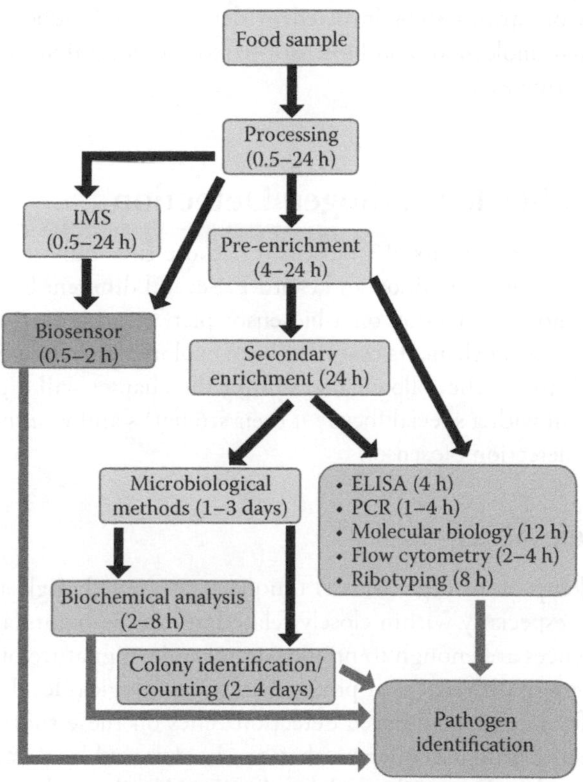

Figure 14.1 Flowchart elucidating the processing steps involved and relative time taken in detecting a pathogen in a food sample. IMS stands for immune-magnetic separation in which particles with magnetic properties are modified with target-specific antibody/antibody fragments for capture and subsequent purification using an external magnetic field. (Reprinted from Singh, A. et al., *Sensors*, 13, 1763–1786, 2013. With permission from MDPI.)

system that translates the recognition and binding event of an analyte onto a biosensor platform into a measurable signal. Biosensors offer several advantages over conventional pathogen-detection systems, such as high levels of sensitivity and selectivity toward the target, rapid real-time detection, portability for in situ detection devices, low detection limits, and minimum sample pre-processing. Selectivity toward a pathogen is an extremely important aspect of a biosensor because it governs the level of detection with minimum cross-reactivity to minimize false positives and false negatives. Several bio-inspired elements have been studied as recognition probes to impart specificity to a biosensor platform toward a pathogen. While nucleic acids and antibodies have been classically used for such applications, the recent progress in bacteriophage technology has led to several novel bio-probes, such as genetically engineered reporter phages, phage-display peptides, and, most recently, the phage receptor binding proteins (RBPs). Figure 14.1 outlines a comparative analysis of various steps involved in detection of foodborne pathogens using conventional, molecular, and biosensor-based methods along with the time involved for executing each step.

14.2 Bio-Probes for Pathogen Detection

Bio-probes are key components of a biosensor because they impart the selectivity and specificity of target recognition and capture. Several different bio-probes have been used for pathogen detection on a biosensor platform. Figure 14.2 is a schematic showing biological elements that have been explored as recognition elements for such an application. The following section of the chapter will highlight these bio-probes in detail with a special focus on their strengths and weaknesses in realizing a pathogen detection biosensor.

14.2.1 Nucleic Acids

The genetic makeup of all organisms is unique, and even though a majority of genetic material, especially within closely related groups, is redundant and similar, subtle differences are enough to provide a molecular signature of a particular organism resulting in differences at physical and physiological levels. The underlying principle for nucleic acid-based detection relies on these subtle differences in organism-specific unique sequences that can be detected by careful design of complementary probes. Deoxyribonucleic acids (DNA), ribonucleic acids (RNA), and peptide nucleic acids (PNA) have been classically used for such detections. The capability of amplifying DNA by PCR and RNA by reverse transcriptase PCR (RT-PCR) makes these probes a popular choice because even minute quantities of nucleic acids can be used to make multiple copies, which augments the signal given by the transducer in the event of positive identification. DNA- and RNA-based detection schemes are therefore simple yet versatile, powerful, and

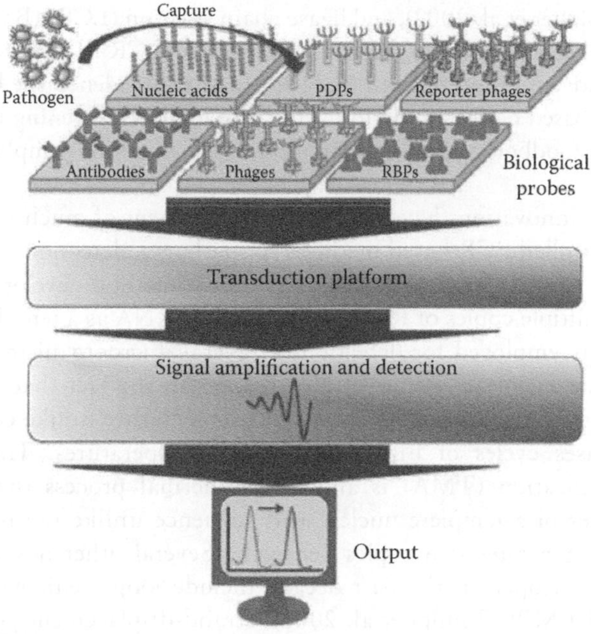

Figure 14.2 Schematic of various components of a typical biosensor highlighting the available phage-based molecular probes for pathogen detection. (Reprinted from Singh, A. et al., *Sensors*, 13, 1763–1786, 2013. With permission from MDPI.)

cost-effective for designing biosensors for pathogen detection. The recent advances in microarray technology (Yu et al. 2004) and multiplex-PCR (Taha 2000) further facilitate screening of several pathogens on a single biosensor platform for a rapid detection. Nucleic acids, especially DNA, show remarkable stability in a variety of solvents (Bonner and Klibanov 2000) and buffers, thereby expanding their utility as a recognition element to different types of food samples. These advantages have led to extensive use of nucleic acid-based platforms for designing biosensors for pathogen detection in clinical as well as food samples.

14.2.1.1 DNA/RNA

A typical DNA-based probing biosensor platform design involves immobilization of a single-stranded DNA (ssDNA) sequence complementary to a unique sequence of the intended target bacterium. RNA from target bacterium can be sometimes used instead of DNA to make multiple copies of complementary DNA using RNA polymerase as an enzyme for reverse transcription, which is subsequently detected by the ssDNA probe on the biosensor surface. PCR methods have been significantly improved since its invention, which has led to several novel methods, such as real time PCR (rtPCR) (Mothershed et al. 2004), RT-PCR (Nakaguchi et al. 2004), nested PCR (Apfalter et al. 2002), and ligase chain reaction (LCR) (Boyadzhyan et al. 2004) as well as PCR in combination with ELISA (PCR-ELISA) for pathogen detection. The advent of multiplexed PCR has further broadened the horizons for DNA- or RNA-based biosensor platforms because it allows screening for the presence of multiple foodborne pathogens simultaneously in a food sample (Benson et al. 2008).

Some recent innovations have led to the development of much more sophisticated and controlled PCR-based methods for pathogen detection. Nucleic acid sequence-based amplification (NASBA) is one such method developed in 1991, and it makes multiple copies of RNA using DNA or RNA as a template and has been successfully employed for detection of several microorganisms (Keer and Birch 2003). The major advantage of NASBA lies in the fact that it facilitates the amplification of nucleic acid at a constant temperature unlike conventional PCR, which uses cycles of higher and lower temperatures. Transcription-mediated amplification (TMA) is another isothermal process that generates 100–1000 copies of a template nucleic acid sequence unlike the regular PCR method, which generates two copies per cycle. Several other novel amplification methods developed in the past decade include loop-mediated isothermal amplification (LAMP) (Tomita et al. 2008), strand-displacement amplification (SDA) (McHugh et al. 2004), rolling-circle amplification (RCA) (Demidov 2002), cycling probe technology (Louie et al. 2000), branched DNA (Zheng et al. 1999), and hybrid capture (Van Der Pol et al. 2002). Research on DNA- or RNA-based systems initially focused primarily on detection of pathogens in clinical and pathological samples, which led to several commercialized products that have been listed in Table 14.1. Use of these methods for detection of pathogens from food samples is still in its infancy and requires extensive research and development to exploit its full potential. Table 14.2 gives a list of DNA- or RNA-based commercial products that are being routinely used for pathogen detection from food samples. Mothershed and Whitney (2006) have given a comprehensive outlook on a DNA- and RNA-based pathogen detection system, which is highly recommended to the readers.

14.2.1.2 PNA

Peptide nucleic acids are artificially synthesized DNA mimics that contain multiple units of N-(2-aminoethyl) glycines pseudo peptide backbones and were first reported by Nielsen et al. (1991). PNAs bind to the complementary DNA or RNA sequence through Watson-Crick pairing with much higher affinity and specificity than corresponding DNA or RNA strands and thus have been used for a variety of applications, such as bio-probes for sensors and in antisense technology (Wittung et al. 1994). Besides, PNA exhibits better stability to chemical, physical, biological, and environmental factors and therefore has been looked upon as an attractive substitute to DNA- or RNA-based bio-probes. Application of

PNAs is, however, limited due to the high cost involved in their synthesis. There are no commercialized products on PNA-based biosensors for pathogen detection yet, but they have been actively researched on the laboratory scale for detection of bacteria. *Staphylococcus aureus* (Oliveira et al. 2002) and *Klebsiella pneumoniae* (Sogaard et al. 2007) have been successfully detected from a blood culture, and biofilm-forming bacteria have been detected from human chronic skin wounds (Malic et al. 2009) using PNA probes by fluorescence in situ hybridization (FISH). Detection of pathogens from food samples using PNA as molecular probes, however, is yet to be realized.

14.2.2 Antibodies

Antibodies (Abs) have been an obvious choice as recognition elements due to their ease of integration into a biosensor platform and their high selectivity and specificity toward the target epitope on the pathogen surface. Abs can detect the target pathogen as well as the toxins produced by them in a sample, unlike DNA-based probes, but still fail to distinguish between a dead or alive bacterium. Polyclonal and monoclonal Abs, their fragments, genetically engineered Abs, and phage-display Abs have been successfully employed for bacterial detection. The application of Abs, however, is limited largely due to involvement of sample concentration and pre-enrichment steps, cross-reactivity in case of polyclonal antibodies, their relative instability at higher pH and temperature, and ethical issues involved with their production in animals. Despite these limitations, Abs have enjoyed tremendous popularity as probes for designing biosensors for pathogen detection.

14.2.2.1 Polyclonal Abs (pAbs)

Polyclonal Abs are generally produced by immunization of small mammals, such as a mouse, goat, or rabbit, with an antigen of choice to direct host B-cells to produce immunoglobulins against the antigen. The serum collected from the immunized animals contains a collection of IgG produced by B-cells of different lineages that recognize different epitopes on the same antigen and are therefore referred to as polyclonal Abs. Besides their direct application as probes on a biosensor surface, pAbs have also been used for immunomagnetic separation (IMS) of bacterium from a complex food matrix during the concentration or pre-enrichment steps (Gu et al. 2006). pAbs have been successfully integrated on several different biosensor platforms for pathogen detection applications. They have been a popular choice as recognition elements mainly due to their low cost of production but suffer from cross-reactivity especially within pathogens closely related to the target bacterium (Mutharia et al. 1993). Monoclonal antibodies (mAbs) are therefore preferred over pAbs because they recognize a single epitope of the target with minimal cross-reactivity.

Table 14.1 Commercially Available Nucleic Acid-Based Biosensors for Pathogen Detection with Their Mode of Detection and the Sample Source

Organism	*Company*	*Product Name*	*Detection Method*	*Sample Source*
Candida sp.	Beckton Dickinson, Inc.	BD Affirm™ APIII	DNA hybridization	Vaginal swab
Chlamydia trachomatis	Qiagen	HC2 CT-ID	Chemi-luminescence	Endocervical swab
	Gen-probe	APTIMA® CT	TMA[a]/16S RNA	Urine/urethral swab
	Gen-probe	PACE2 CT	HPA[b]	Endocervical swab
	Beckton Dickinson, Inc.	BD ProbeTec™ CT	SDA[c]	Endocervical swab
	Roche	COBASAMPLICOR CT	PCR	Endocervical/ urethral swab
Escherichia coli O157:H7	Qualicon, Inc.	BAX system	Real-time PCR	Water
Gardnerella	Beckton Dickinson, Inc.	BD Affirm™ APIII	DNA hybridization	Vaginal swab
Mycobacterium avium	Accuprobe®	Gen-probe	TMA/RNA	Culture
Mycobacterium gordonae	Accuprobe®	Gen-probe	TMA/RNA	Culture
Mycobacterium intracellulare	Accuprobe®	Gen-probe	TMA/RNA	Culture
Mycobacterium kansasii	Accuprobe®	Gen-probe	TMA/RNA	Culture

Mycobacterium tuberculosis	Accuprobe® MTD	Gen-probe	TMA	Sputum
	BD ProbeTec™ ET	Beckton Dickinson, Inc.	SDA	Respiratory and nonrespiratory
	COBAS AMPLICORMTB	Roche	PCR	Respiratory and nonrespiratory
Neisseria gonorrhoeae	Qiagen	HC2 GC-ID	Chemi-luminescence	Endocervical swab
	Gen-probe	APTIMA® GC	TMA/16S RNA	Urine/urethral swab
	Gen-probe	PACE2 GC	HPA	Endocervical swab
	Beckton Dickinson, Inc.	BD ProbeTec™ GC	SDA	Endocervical swab
	Roche	COBASAMPLICOR NG	PCR	Endocervical/ urethral swab
Streptococci Group A	Gen-probe	GASDirect®	HPA	Pharyngeal swab
Streptococci Group B	Infectio Diagnostic, Inc.	IDI-StrepB	Real-time PCR	Vaginal swab
Trichomonasvaginalis	Gen-probe	APTIMA®	TMA/16S RNA	Urine/vaginal swab
	Beckton Dickinson, Inc.	BD Affirm™ APIII	DNA hybridization	Vaginal swab

Source: Reprinted with permission from MDPI, Singh, A., Poshtiban, S. & Evoy, S., *Sensors*, 13, 1763–1786, 2013.

[a] Transcription mediated amplification (TMA); [b] hybridization probe assay (HPA); [c] strand displacement amplification (SDA).

14.2.2.2 Monoclonal Antibodies (mAbs)

Monoclonal antibodies are produced by hybridoma technology in which B-cells of the same lineage are immunized with the target epitope of interest and are immortalized by fusing them with myeloma cells. The hybrid cells are selected by screening, following which they are cloned and used for production of mAbs in high

Table 14.2 Nucleic Acid and Protein-Based Commercial Products for Foodborne Pathogen Detection with Their Method and Limit of Detection

Organism	*Product*	*Company*	*Method of Detection*	*Limit of Detection (cfu mL^{-1})*
E. coli O157:H7	BAX®	Dupont	DNA Hybridization	10^4
	Lateral Flow System	Dupont	Immunoassay	1 (per 25 g food)
	Reveal®	Neogen	Immunoassay	10^4
	GeneQuence®	Neogen	Enzyme based	1 (per 25 g food)
	VIDAS	Biomérieux	Immunoassay	–
Campylobacter	BAX®	Dupont	DNA Hybridization	10^4
	VIDAS	Biomérieux	Immunoassay	–
	ACCUPROBE	Biomérieux	DNA Hybridization	–
Listeria	BAX®	Dupont	DNA Hybridization	10^4
	Lateral Flow System	Dupont	Immunoassay	1 (per 25 g food)
	Reveal®	Neogen	Immunoassay	10^6
	ANSR™	Neogen	DNA Hybridization	10^4
	VIDAS	Biomérieux	Immunoassay	–

(continued)

Table 14.2 (Continued) Nucleic Acid and Protein-Based Commercial Products for Foodborne Pathogen Detection with Their Method and Limit of Detection

Organism	*Product*	*Company*	*Method of Detection*	*Limit of Detection (cfu mL^{-1})*
Salmonella	ANSR™	Neogen	DNA Hybridization	10^4
	GeneQuence®	Neogen	Enzyme based	1 (per 25 g food)
	Reveal®	Neogen	Immunoassay	10^6
	BAX®	Dupont	DNA Hybridization	10^4
	Lateral Flow System	Dupont	Immunoassay	1–4 (per 25 g food)
Enterobacter	BAX®	Dupont	DNA Hybridization	–
	VIDAS	Biomérieux	Immunoassay	–
Vibrio	BAX®	Dupont	DNA Hybridization	10^4

Source: Reprinted with permission from MDPI, Singh, A., Poshtiban, S. & Evoy, S., *Sensors*, 13, 1763–1786, 2013.

concentrations (Kohler and Milstein 1975). Figure 14.3 shows a schematic of events involved in the production of mAbs for a specific target epitope. mAbs are highly specific to their target epitope and do not show cross-reactivity like pAbs and therefore prove to be a great candidate as bio-probes for developing biosensors. The high specificity of mAbs has been demonstrated recently as they were able to distinguish and detect the spores of *Bacillus anthracis* from those from other *Bacillus* species (Swiecki et al. 2006). Despite such a high level of specificity of mAbs, their application as a bio-probe is restricted due to several disadvantages. mAbs are easily affected by the environmental condition and therefore demonstrate poor shelf life. They are prone to structural damage and loss of functionality due to changes in pH or temperature or due to proteolytic degradation. Therefore, a biosensor platform functionalized with mAbs as a recognition element has to be stored in a very controlled environment, which is neither cost-effective nor feasible for point-of-care analysis applications. Besides, the cost involved in the production of mAbs is significant and therefore limits their use as a bio-probe for pathogen detection.

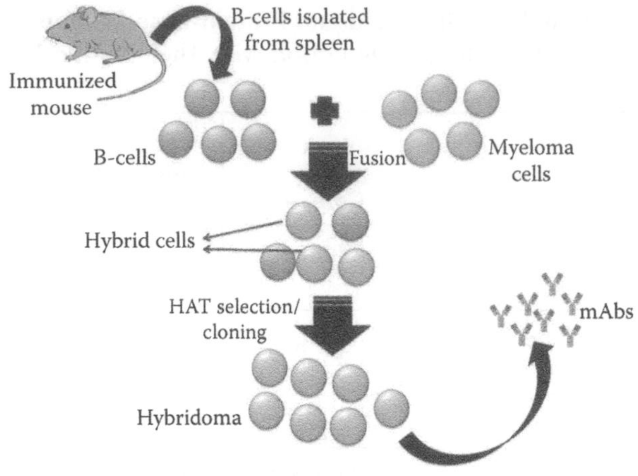

Figure 14.3 Schematic demonstrating hybridoma technology for monoclonal antibody production.

A recently published review summarizing the use of antibodies for pathogen detection and critically elaborating the advantages and disadvantages of the detection system is recommended to readers for more information (Byrne et al. 2009).

14.2.2.3 Genetically Engineered Antibodies

Our ability to tailor the properties of biomolecules using the tools for genetic manipulation has given the required boost to develop engineered antibodies that overcomes some of the limitation of pAbs and mAbs. The thermal instabilities of antibodies is a major disadvantage that limits their applicability as probes for integration on to the biosensor platform and adds significantly to the cost. A new class of antibodies developed in camels (Hamers-Casterman et al. 1993) and sharks (Greenberg et al. 1995) shows excellent stability against thermal and proteolytic degradation and has been used as a useful replacement for the conventional antibodies. These antibodies, often referred to as llama bodies, lack the light chain (V_L) and contain a single heavy chain (V_H) with a small antigen-binding domain. The hyper variable domain of the V_H region of the llama bodies has been successfully cloned to form 12–15 kDa single-domain Abs (sdAbs), and results demonstrate that they retain their functionality even at temperatures as high as 90°C, and they are resistant to proteolytic degradation. These sdAbs have been used for detection of cholera toxin and *Staphylococcus enterotoxin* B (Goldman et al. 2006); heat shock proteins of *Mycobacterium tuberculosis* (Trilling et al. 2011) and seven different serotypes of *botulinum neurotoxin* (Conway et al. 2010). These sdAbs also show minimum cross-reactivity due to their small antigen recognition and binding domain and thus are looked upon as a potential replacement to mAbs as probes.

14.2.2.4 Phage Display Antibodies (pdAbs)

Phage display is a robust and well-studied technology that enables production and display of antibodies or parts of them on the surface of a bacteriophage (Winter et al. 1994). Figure 14.4 shows a schematic of an affinity-based selection procedure for phages displaying the protein/peptide of choice. cDNA encoding the V_L and V_H regions of an antibody are cloned from the B-cells and fused in proximity with DNA sequence coding for a surface protein of a phage so that the antibody is expressed as fusion protein and displayed on the phage surface. Phage display antibodies show several advantages over conventional antibodies. The antibodies displayed on the phage surface are much smaller in size than regular antibodies and thus show better stability to environmental changes. Besides, their production does not require use of animals and therefore avoids the ethical issues associated with antibody production. Tailoring the property of the antibodies at molecular level also gives the freedom to manipulate their binding characteristics toward the target antigen and the possibility of making a library of pdAbs with variable binding affinity to the same target. From a commercialization viewpoint, pdAbs are much cheaper to produce, which, in effect, will reduce the final manufacturing cost of the biosensor. Low-yield and batch-to-batch variations are two major disadvantages that should be overcome before they can be explored as commercially viable option as molecular probes for pathogen detection. The filamentous phage M13

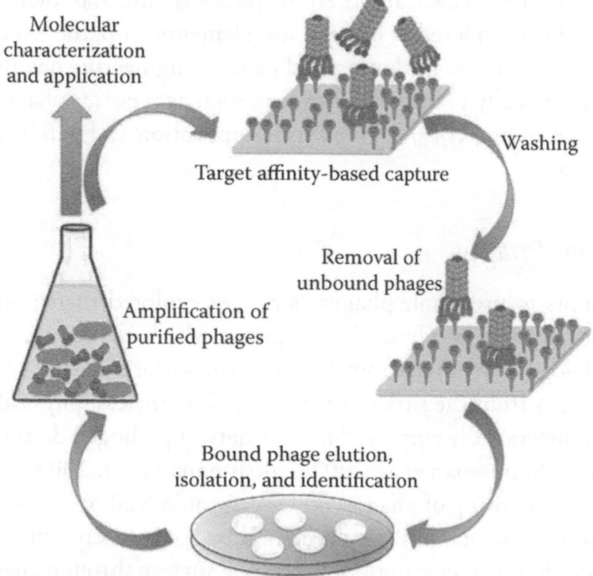

Figure 14.4 Schematic of affinity-based selection procedure adapted in phage display technology. (Reprinted from Singh, A. et al., *Sensors*, 13, 1763–1786, 2013. With permission from MDPI.)

and fd have been most extensively studied for phage-display applications. Phage displayed single-chain variable fragment (scFv) against ActA, a protein virulence factor produced by *Listeria monocytogenes*, was used for the pathogen detection in the SPREETA™ sensor (Nanduri et al. 2007a). A comprehensive review by Petrenko and Vodyanoy (2003) on the phage display antibody technology and the underlying principles and challenges and their potential application in the detection of biological threat agents is recommended to the readers.

14.2.3 Bacteriophages

Bacteriophages are obligate parasitic viruses that infect bacteria and sabotage the bacterial cellular machinery for their sustenance and multiplication and the propagation of new mature virions. Most phages are highly specific in their host recognition down to the strain level with some exceptions, such as *Listeria* phage A511, which infects within the whole genus, and some others that show inter-species recognition and binding. The phages' life cycle within their host could be either lytic in which they infect the host, inject their genomic material, propagate new virions, and lyse the bacterial cells eventually or lysogenic in which the genomic material integrates into the host genome and remains dormant until triggered for multiplication and propagation. The high specificity of the phages toward their host bacterium renders them excellent bio-probes for bacterial identification, recognition, and binding. Phages have been classically used for phage typing and identification and recently they have been explored as recognition elements for pathogen detection on a biosensor platform. Progress in cloning and genetic engineering has also led to the development of genetically manipulated probes such as reporter phages and phage receptor binding proteins (RBPs) for biosensor application and will be discussed in detail in this section.

14.2.3.1 Whole Phages

The earlier attempts to use whole phages as probes explored their immobilization on the biosensor platform by physical adsorption, which is a simple and straightforward methodology but suffers from inconsistent surface density of phages and leaking of the phages from the surface over time. Nonetheless, physically adsorbed phages have been successfully employed for a variety of pathogen detection in a biosensor setup (Balasubramanian et al. 2007; Lakshmanan et al. 2007; Nanduri et al. 2007b). Chemical anchoring of phages, however, would lead to a robust and stable phage immobilization, which has been affirmed by some recent efforts. A systematic study revealed that phages immobilized on the surface through chemical bonds result in higher phage density and an improved host bacterial capture compared to physical adsorption (Singh et al. 2009). Cysteine and cysteamine functionalized gold surfaces activated by glutaraldehyde were able to specifically capture the host *E. coli* K12 bacteria. The chemical anchoring methods were easily translated to a biosensor platform to show detection of the host bacteria with a detection limit of 700 cfu mL^{-1} (Arya et al. 2011). The chemical method applied for phage immobilization is also governed by the biosensor surface. While a metal surface such as gold or silver popularly uses thiol chemistry, a silicon or glass surface relies on silane chemistry for functionalization (Handa et al. 2008).

Genetically engineering phages to introduce a ligand that can then be exploited for anchoring them to the biosensor surface is yet another approach that has been studied. The advantage of this approach lies in the precise placement of the ligand that would aid in oriented immobilization of the phages on the surface, thus, improving the pathogen-capturing efficiency. Most tailed phages recognize and bind to their host bacterium through their tail region, and therefore, their proper orientation on the surface is of utmost importance for retention of their recognition ability. T4 phages expressing biotin in their head region were bound to a biotin-modified surface using streptavidin as a linker such that the tail region is away from the surface and free for bacterial capture (Gervais et al. 2007). In yet another report, the biotin-expressing phages were functionalized on streptavidin-modified magnetic microparticles for efficient capture, extraction, and detection of pathogens (Tolba et al. 2010). Cellulose-binding module (cbm)-expressing phages were similarly functionalized on cellulose particles to capture and analyze the host bacteria.

Purity of the phage lysate is another important factor that affects the immobilization density. Phages are grown in their host bacterial culture to obtain higher titers and thus can be easily contaminated with bacterial macromolecules, such as proteins, lipopolysaccharides, flagella, and peptidoglycan fragments that would interfere with the chemical functionalization. Naidoo et al. (2012) recently demonstrated that the phage lysate purified by several ultracentrifugation cycles still showed contamination from bacterial cell products, resulting in poor surface immobilization density. They further showed that immobilization of three different phage lysates purified by size exclusion chromatography yielded significantly higher density on the biosensor platform, which, in turn, resulted in better efficiency of bacterial capture (Naidoo et al. 2012). Such purified phage lysates not only aid better surface immobilization density but also allow a detailed and rigorous study of the surface binding kinetics, which is important for understanding and optimizing any immobilization protocol. These newly developed approaches for surface immobilization of whole phages have opened up exciting avenues to develop novel biosensors for pathogen detection.

Whole phages prove to be excellent probes for pathogen detection because they can be grown easily, are cheap to produce, are found in abundance in the environment (approximately 10^{31} phage types estimated), and show very high stability to environmental variations, such as temperature, pH, solvent, etc. They were found to survive exposure to organic solvents (Olofsson et al. 1998) as well as protease (Schwind et al. 1992). Studies also suggest that phage lysates maintained in a pH range of 3–11 were not only able to retain their host specificity but also their infection capability (Verma et al. 2009). A whole phage as a bio-probe, however, also has certain drawbacks that need to be addressed. Phages recognize and capture their host through their tail protein, and therefore their proper orientation is key to adequate functioning of the biosensor. Some advances have been made to orient the phage immobilization on the surface, but a general and universal approach to achieve it has not been devised. Whole phages are also biologically active and, being natural parasites, prefer to infect their host by various mechanisms. The P22 phage shows endorhamnosidase enzymatic activity that cleaves the target O-antigen on the surface of *Salmonella enterica*, which leads to poor bacterial capture on biosensor platform (Singh et al. 2010). Besides, whole phages tend to lyse bacteria in their lytic phase, which would result in the loss of signal from the biosensor and underestimation of pathogen load in a sample. Studies have also indicated that the drying of a whole phage derivitized surface leads to loss of phage's ability to capture bacteria.

14.2.3.2 Reporter Phages

Reporter phages are a class of genetically modified whole phages that can be viewed as cellular-level beacons for pathogen identification and typing. The unique recognition ability of the phages is coupled with a reporting gene, which is introduced into the host bacterium in the event of position recognition and binding, facilitating their identification. The reporting gene is injected into the host bacterium along with the phage genome and incorporated into the bacterial genome and usually codes for a fluorescent or substrate activated colorimetric signal. Because the phages depend on the host cellular machinery to perform their physiological functions, the reporter gene is not expressed until and unless they have successfully infected their target host. A positive expression of the reporter gene therefore confirms the presence and identity of the host bacterium in a sample. Figure 14.5 shows a schematic highlighting the underlying principle of reporter phage technology for pathogen detection.

A number of different reporter genes have been incorporated into the phages, including luciferase-expressing gene (*lux* and *luc*), *E. coli* β-galactosidase gene (lacZ), bacterial ice nucleation gene (*inaW*), and green fluorescent protein gene (*gfp*). The reporter phage technology has been successfully used for detection of foodborne pathogens from different sample food sources. The A511::luxAB recombinant reporter phage was used to successfully detect one *Listeria* bacterium per gram of sample in artificially spiked ricotta cheese, chocolate pudding, and cabbage. In a more complex food matrix, such as minced meat, up to 10 bacteria per gram of meat could be detected. However, the sample processing and pre-enrichment steps involved took nearly 20 h and a total estimated detection time of 24 h (Loessner et al. 1997). Other pathogens successfully detected by reporter phages include *Salmonella* (Kuhn et al. 2002), *E. coli* (Waddell and Poppe 2000), and *Mycobacterium* (Jacobs et al. 1993). The ability to distinguish live bacteria from

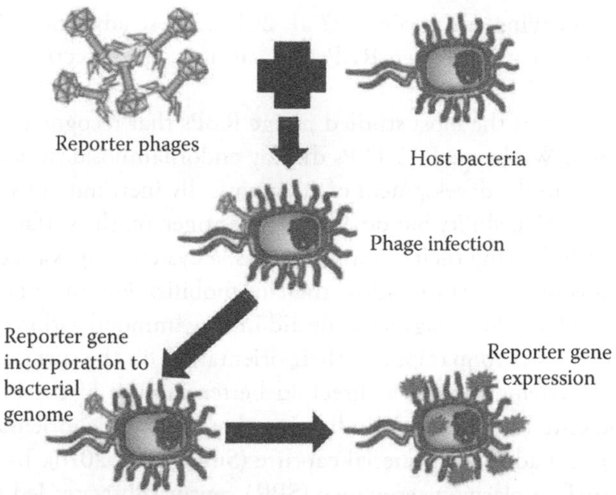

Figure 14.5 Schematic explaining the underlying principle of reporter phage-based detection of a pathogen of interest. (Reprinted from Singh, A. et al., *Sensors*, 13, 1763–1786, 2013. With permission from MDPI.)

dead is the greatest advantage of reporter phage technology because phages will not be able to infect dead bacteria for expression of the reporter gene. This technology, however, suffers from certain limitations, such as inhibition of phage multiplication due to prophage presence (Harvey et al. 1993), presence of restriction-modification system (Kim et al. 2012), specific phage inhibitory genes (Chopin et al. 2005), or antiviral bacterial immunity system (Barrangou and Horvath 2012).

14.2.3.3 Phage Receptor Binding Proteins (RBPs)

The host-specific recognition of tailed phages is attributed to the protein assembly located on the tail fiber that recognizes the surface receptor on host bacteria and effects phage binding. These protein assemblies are called RBPs because they bind to the receptor on the host surface, and their binding triggers the events leading to translocation of the phage genome into the bacterium. Due to their location at the tip of the tail or on the tail fiber, they are also referred to as tail spike proteins (TSPs). These RBPs are being considered as a new class of recognition elements for biosensor design and pathogen detection. They offer several advantages over antibodies, such as better stability to environmental factors, such as pH and temperature, as well as resistance to proteases. Genetic engineering allows tailoring of their binding affinity and specificity, incorporation of multi-valency for identification and capture of different pathogens, and the possibility of adding tags to facilitate oriented immobilization on the biosensor surface. Compared to whole phage, RBPs lack infecting capability and thus cannot induce bacterial cell lysis, and they also do not exhibit any drying effect (Singh et al. 2010). These advantages have led to some recent efforts in using phage RBPs for capture and detection of bacterial pathogens.

P22 TSPs are one of the most studied phage RBPs that recognize and bind to *Salmonella enterica*. Wild-type P22 TSPs display endorhamnosidase activity, but a recent work has led to the development of enzymatically inert mutant versions that show similar recognition ability but do not cleave antigen on the surface of bacteria (Waseh et al. 2010). During their cloning process, a cysteine tag was added to the N and C terminus of the TSPs to allow their immobilization on to gold or silver surfaces by thiol linkage. Such tags not only aid in easy immobilization of the RBPs but also provide avenues to manipulate their orientation on the surface. The TSPs with N-terminal cysteine showed a threefold better capture of the host bacteria compared to the C-terminal variant, indicating that preferential orientation of the TSP is important for adequate bacterial capture (Singh et al. 2010). Integration of these TSPs on surface plasmon resonance (SPR) sensor substrates led to sensitive and selective detection of *Salmonella enterica* with a detection limit of 10^3cfu mL^{-1}. In another study, RBP (Gp48) of phage NCTC12673 was identified, cloned, overexpressed, and purified as a glutathione-*S*-transferase-GP48 (GST-Gp48) fusion protein (Kropinski et al. 2011). The GST tags were then exploited for oriented immobilization on a glutathione self-assembled monolayer modified gold surface, which showed nearly a threefold better capture of target bacteria compared to randomly oriented RBPs using SPR-based biosensor (Singh et al. 2011). These RBPs were further immobilized on tocyl-activated Dynabeads® M-280 to specifically capture the target bacterium while the non-host control bacteria showed minimal binding. In a very recent study, these RBPs were functionalized on the surface of micro-resonators for sensitive and selective detection of host *Campylobacter jejuni* while the non-host bacterium showed negligible binding to the surface (Poshtiban et al. 2013). Although the application of RBPs for pathogen detection is still in its infancy, their superior properties make them promising candidates for their use in development of biosensors for pathogen detection. A recent review gives a detailed account of phage RBPs and is recommended to the interested readers (Singh et al. 2012).

14.3 Biosensors for Foodborne Pathogen Detection

Rapid monitoring of food products with high levels of reliability, sensitivity, and selectivity is critical for inspection of food products in industrial firms considering the short shelf time of products and low infection dose of food-borne pathogens. Biosensors have found great interest, and efforts have been focused on optimizing the biosensor transducers to improve the detection sensitivity. Bio-probes have been combined with various analytical methods to provide the specificity of recognition. We will review the biosensor transduction platforms that have been successfully employed for specific detection of food-borne pathogens. Table 14.3 summarizes various organisms that have been detected with these biosensor platforms.

14.3.1 Optical Biosensors

Optical biosensors are rapid, sensitive, and adaptable to a wide variety of assay conditions and have been widely investigated for bacterial pathogen detection. Optical techniques are divided into two main subcategories based on their working principles: labeled and label-free. The most commonly employed techniques for bacterial detection are surface plasmon resonance (SPR), fluorescence/phosphorescence spectrometry, and bio/chemi-luminescence. In the following section, we will focus on optical biosensors that are employed for detection of food-borne pathogens.

14.3.1.1 Surface Plasmon Resonance Sensors

Surface plasmon resonance is the oscillation phenomenon that occurs at the interface between two media with oppositely charged dielectric constants. SPR sensors monitor events happening at surfaces and interfaces by excitation of plasmons by light in thin metal films, which generates evanescent electromagnetic waves. These evanescent waves propagate along the interface between metal film and the ambient medium (Salamon and Tollin 2001). SPR sensors measure the refractive index near the sensor surface that changes as a result of interaction of target analyte in solution with receptors on the transducer surface. SPR has been widely used for real-time monitoring of biochemical interactions of small analytes, such as DNA hybridization, cell-ligand, protein-peptide, and protein-lipid. SPR systems have also been modified to enable the direct label-free detection of larger biomarkers, such as bacterial pathogens.

Successful detection of pathogenic bacteria has been reported on antibody-immobilized SPR sensors (Oh et al. 2005; Taylor et al. 2005). Bacteriophage-based probes have also been immobilized on SPR sensors, and successful detection of *S. aureus* (Balasubramanian et al. 2007), *E. coli* K12 (Arya et al. 2011), *E. coli* O157:H7, and methicillin-resistant *Staphylococcus aureus* (MRSA) (Tawil et al. 2012) has been reported. The limit of detection was typically in the range of 10^2–10^3 cfu mL^{-1}. Bacteriophage receptor binding proteins have also been used as biorecognition probes on SPR. For example, Singh et al. (2010, 2011) immobilized genetically engineered tailspike proteins (TSP) from P22 bacteriophage and *Campylobacter* bacteriophage NCTC 12673 onto the gold-coated SPR plates and demonstrated a selective real-time detection of *Salmonella* and *Campylobacter jejuni* bacteria with the sensitivity of 10^3 cfu mL^{-1} and 10^2 cfu mL^{-1} of bacteria, respectively.

Recently, some efforts have been focused on simplifying the bacterial detection with SPR to protein–protein interaction by combining a subtractive inhibition assay with a SPR immunosensor, and successful detection of *E. coli* O157:H7 cells

Table 14.3 Foodborne Pathogens Detected Using Various Biosensors Highlighting the Transduction Platform Used and Limit of Detection Achieved

Transducer	*Organism*	*Bioreceptor*	*Limit of Detection*	*Ref.*
SPR	*E. coli* K12	T4 Phage	7×10^2 cfu mL^{-1}	(Arya et al. 2011)
SPR	*E. coli* O157:H7	T4 Phage	10^3 cfu mL^{-1}	(Tawil et al. 2012)
SPR	MRSA	BP14 Phage	10^3 cfu mL^{-1}	(Tawil et al. 2012)
SPR	*Salmonella*	P22 Phage TSP	10^3 cfu mL^{-1}	(Singh et al. 2010)
SPR	*C. jejuni*	Phage NCTC 12673 TSP	10^2 cfu mL^{-1}	(Singh et al. 2011)
SPR	*S. aureus*	Lytic phage (phage 12600)	10^4 cfu mL^{-1}	(Balasubramanian et al. 2007)
SPR combined with subtractive inhibition assay	*E. coli* O157:H7	Goat polyclonal antibodies	3×10^4 cfu mL^{-1}	(Wang et al. 2011)
SPR combined with subtractive inhibition assay	*Listeria monocytogenes*	Antibodies	10^5 cfu mL^{-1}	(Leonard et al. 2004)
Bioluminescence	*E. coli*	*E. coli* phage	10^3 cfu mL^{-1}	(Blasco et al. 1998)
Bioluminescence	*Salmonella newport*	Felix phage or Newport phage	10^3 cfu mL^{-1}	(Blasco et al. 1998)
Bioluminescence	*Salmonella enteritidis*	Phage SJ2	10^3 cfu mL^{-1}	(Wu et al. 2001)
Bioluminescence	*E. coli* G2-2	AT20	10^3 cfu mL^{-1}	(Wu et al. 2001)

Fluorescent	*Staphylococcal enterotoxin B* (SEB)	Phage-displayed peptides	1.4 ng	(Goldman et al. 2000)
Fluorescent	*E. coli*	QD-labeled lambda phage	N/A	(Yim et al. 2009)
Fluorescent	*E. coli*	T7 phage	20 cell mL^{-1}	(Edgar et al. 2006)
QCM	*Salmonella typhimurium*	Filamentous phage	10^2 cell mL^{-1}	(Olsen et al. 2006)
Magnetoelastic sensors	*Salmonella typhimurium*	Filamentous E2 phage	5×10^2 cfu mL^{-1}	(Lakshmanan et al. 2007; Li et al. 2010)
Magnetoelastic sensors	*Bacillus anthracis* spores	Filamentous phage, clone JRB7	N/A	(Shen et al. 2009)
Amperometric combined with pre-filtration	*E. coli* K12	Phage lambda	1 cfu 100 mL^{-1}	(Neufeld et al. 2003)
Amperometric by monitoring oxygen consumption	*Salmonella typhimurium*	N/A	$1–2 \times 10^0$ cfu mL^{-1}	(Ruan et al. 2002)
Amperometric immunosensor	*E. coli*	Polypyrrole-NH2-anti-*E. coli* antibody	10 cfu mL^{-1}	(Abu-Rabeah et al. 2009)
Amperometric magnetoimmunosensor	*Staphylococcus aureus*	Anti-Prot A antibody	1 cfu mL^{-1}	(Esteban-Fernández de Ávila et al. 2012)
Impedimetric	*E. coli*	T4 Phage	10^4 cfu mL^{-1}	(Lee et al. 2006; Shabani et al. 2008)

Source: Adapted from Singh, A., Poshtiban, S. & Evoy, S., *Sensors*, 13, 1763–1786, 2013.

(Wang et al. 2011) and *Listeria monocytogenes* (Leonard et al. 2004) have been reported. In this method, the target bacteria are incubated in antibodies for a short amount of time, and then the cells bound to antibodies are separated from unbound antibodies by stepwise centrifugation. The remaining free antibodies are exposed to the surface of the antibody-immobilized SPR sensor chip, and the general response is inversely proportional to the bacterial cell concentration. The lowest limit of detection reported is 3×10^4 cfu mL^{-1}.

14.3.1.2 Bioluminescence Sensors

Bioluminescence assays quantitatively detect bacteria by measuring the level of light emission from intercellular components. This assay involves two major steps of bacterial cell lysis for release of interacellular components, followed by measuring the content level using a bioluminescent reaction with luciferase. The major drawback of this technique is the lack of specificity. Using lytic phage specific to target bacteria as a recognition probe overcomes this limitation. Blasco et al. (1998) demonstrated successful detection of *E. coli* and *Salmonella newport* combining an ATP bioluminescence assay with lytic phage as a bioprobe and lysis agent. Ten- to a hundredfold better sensitivity was reported when adenylate kinase (AK) was used as an alternative cell marker, and fewer than 10^4 cfu mL^{-1} *E. coli* could be detected in less than 1 h (Blasco et al. 1998). A similar assay for *Salmonella* was slower and took up to 2 h (Blasco et al. 1998). Wu et al. (2001) showed that various factors, such as the bacterial type, the growth stage, the phage type, and the infection time, influence the amount of AK released from bacterial cells. The use of lytic phage as a biorecognition probe provides sensitivity and eliminates the need for lengthy conventional microbiological methods and selective media.

14.3.1.3 Fluorescent Bioassay

Fluorescent bioassay is a sensitive and selective detection method that involves staining bacteria with fluorescence labels. For example, the fluorescently stained bacteriophages are used to recognize and bind to their host bacteria. The flow cytometry or epifluorescent filter techniques are then used to detect the complex of phage-bacteria with the average sensitivity of around 10^2–10^3 cfu mL^{-1} for epifluorescent microscopy and is 10^4 cfu mL^{-1} for flow cytometric detection (Hennes and Suttle 1995; Hennes et al. 1995; Lee et al. 2006). This technique is further combined with an immunomagnetic separation method and improved the detection limit to 10–10^2 cfu mL^{-1} *E. coli* O157:H7 in artificially contaminated milk after 10 h enrichment (Goodridge et al. 1999b) and 10^4 cfu mL^{-1} concentration of *E. coli* O157:H7 in broth (Goodridge et al. 1999a).

Edgar et al. (2006) and Yim et al. (2009) further improved the sensitivity of this approach by tagging bacteriophage with fluorescent quantum dots (QDs). A QD improves the sensitivity of detection platforms, such as flow cytometry and epifluorescence microscopy by boosting the intensity and stability of fluorescent signal. The streptavidin-coated QDs were bound to the head of bacteriophage modified with biotin binding peptide. This method enabled detection of as low as 20 *E. coli* cells in 1 mL water sample in 1 h (Edgar et al. 2006).

The fluorescent assays have also been used for detection of bacterial toxins. Goldman et al. (2000) applied phage display to select a 12-mer peptide that could bind to *Staphylococcal enterotoxin B* (SEB), which causes food poisoning. They could detect as low as 1.4 ng of SEB per sample well in a fluorescence-based immunoassay using a fluorescently labeled SEB-binding phage. Array biosensors were also developed based on a similar principle to simultaneously detect *Bacillus globigii*, MS2 phage, and SEB (Rowe et al. 1999).

14.3.2 Micromechanical Biosensors

14.3.2.1 Quartz Crystal Microbalance Biosensors

Quartz crystal microbalance (QCM) sensors are very sensitive with a capability for detection of nanogram changes in mass. A QCM sensor is composed of two metallic electrodes coated on two sides of a thin piezoelectric plate. Applying an electrical field across the quartz crystal excites the mechanical resonance. The fundamental wavelength (λ) and resonance wavelength ($\lambda = 2d/n$) are determined based on the plate thickness, d, and thus the corresponding resonant frequency:

$$f_n = n f_0 = \frac{n\, v}{2d} \tag{14.1}$$

where v is the sound velocity, and f_0, f_n are the fundamental and the nth overtone resonant frequency, respectively. The resonance frequency shifts to lower frequencies after adsorption of mass onto the electrode surface. The rate of frequency change is proportional to the adsorbed mass with uniform distribution according to the Sauerbrey equation:

$$\Delta f = \frac{-2 f_0^2 \Delta m}{A(\mu\rho)^{1/2}} \tag{14.2}$$

where μ is the shear modulus of quartz (2.947×10^{11} g cm^{-1} s^{-2}), A is the piezoelectrically active crystal area, and ρ is the density of the quartz (2.648 g cm^{-3}). QCM sensors are used to measure the mass of various target analytes by immobilizing specific probes on a sensor surface. Bacteriophage probes are immobilized onto QCM sensors for specific detection of bacteria. Olsen et al. (2006) developed a sensitive platform for rapid detection of *Salmonella typhimurium* by physical adsorption of ~3×10^{10} phages cm^{-2} on piezoelectric transducer surface. This phage-immobilized QCM sensor had a low detection limit of 10^2 cells mL^{-1} with a wide linear range of $10–10^7$ cells mL^{-1} and a rapid response time of less than 180 s (Olsen et al. 2006).

14.3.2.2 Phage Immobilized Magnetoelastic Sensors

Magnetoelastic sensors are mechanical oscillators that are excited with an AC magnetic field. The sensor resonates when the applied field frequency equals the natural frequency of sensors. The fundamental resonant frequency of longitudinal oscillations is given by

$$f = \sqrt{\frac{E}{\rho(1-\sigma^2)}}\frac{1}{2L} \tag{14.3}$$

where E is the Young's modulus of elasticity, ρ is the density of the sensor material, σ is the Poisson's ratio, and L is the long dimension of the sensor. The mechanical oscillation is damped when non-magnetoelastic materials are added to the sensor surface resulting in frequency shift:

$$\Delta f = -\frac{f}{2}\frac{\Delta m}{M} \tag{14.4}$$

where f is the initial resonance frequency, M is the initial mass, Δm (smaller than M) is the mass change, and Δf is the shift in the resonant frequency of the sensor. The response of the magnetoelastic sensors can be measured wirelessly, making the real-time and *in vivo* bio-detection feasible.

Filamentous bacteriophages are immobilized on magnetoelastic sensors for the detection of various bacteria, including *Salmonella typhimurium* and *Bacillus anthracis* spores in different food matrixes, such as fat-free milk and fresh tomato (Lakshmanan et al. 2007; Shen et al. 2009; Li et al. 2010). The limit of detection was typically in the range of 10^3 cfu mL^{-1}.

14.3.3 Electrochemical Biosensors

14.3.3.1 Amperometric Biosensors

Amperometric biosensors usually rely on an enzyme-based system that converts the analyte to an electrochemically active product that can be reduced or oxidized at a working electrode. They are usually composed of a pair of reference and working electrodes. They apply a bias voltage between two electrodes to generate a current through the analyte that directly depends on the rate of electron transfer, which changes with a variation in ionic concentration of analyte. Various amperometric biosensors are developed for detection of foodborne pathogens, including phage-based, microbial methabolism-based, and antibody-based biosensors.

Phage-based amperometric sensors work based on the principle that the over-exposure of phages to bacterial cells results in bacteria lysis leading to release of bacteria cell content, such as enzymes, into the surrounding medium. This enzymatic activity can be measured and quantified using a specific substrate. The product of the reaction between the substrate and enzyme is oxidized at the carbon anode at the reference electrode, producing a current. Neufeld et al. (2003) combined an amperometric technique with phage typing for specific detection of *E. coli* K12, *Mycrobacterium smegmatis*, and *Bacillus cereus* bacteria. They could achieve a limit of detection of 1 cfu mL^{-1} within 6–8 h using this technique in combination with filtration and pre-incubation before infecting the bacteria with phage.

Amperometric biosensors are also employed to detect bacteria by monitoring a specific metabolic process. Ruan et al. (2002) developed a new in situ method for monitoring of *Salmonella typhimurium* in selective media by measuring the oxygen reduction peak on a gold electrode surface using electrochemical cyclic voltammetry during proliferation of bacteria.

Antibody-based amperometric sensors immobilize antibodies specific to bacteria directly on an electrode surface. Croci et al. (2001) used 3,3′,5,5′-tetramethylbenzidine as a substrate to measure the activity of horseradish peroxidase as a label enzyme for detection of *Salmonella* in artificially contaminated pork, chicken, and beef. Abu-Rabeah et al. (2009) reported a double-layered amperometric immunosensor for detection of *E. coli*, consisting of a polypyrrole-NH_2-anti-*E. coli* antibody (PAE) inner layer followed by an alginate-polypyrrole (Alg-Pply) outer packing layer. Presence of bacterial enzyme, *p*-aminophenyl β-D-galactropyranoside (PAPG), β-D-galactosidase produces the p-aminophenol (PAP) product resulting in an amperometric signal. The low limit of detection of 10 cfu mL^{-1} is reported with this technique (Abu-Rabeah et al. 2009).

A disposable amperometric magnetoimmunosensor was developed for detection of Staphyloccocal protein A and *Staphylococcus aureus* (*S. aureus*) with a very low limit of detection (1 cfu mL^{-1} of raw milk sample). This assay involved immobilization of anti-protein A antibody onto protein A-functionalized magnetic beads, followed by a competitive immunoassay involving protein A labeled with HRP. The modified magnetic beads were captured on the surface of tetrahiafulvalene-modified Au/SPEs, and the amperometic response was obtained with respect to the silver pseudo-reference electrode of the Au/SPE after the addition of H_2O_2 (Esteban-Fernández de Avila et al. 2012).

14.3.3.2 Impedimetric Biosensors

Electrochemical impedance spectroscopy (EIS) biosensors monitor biomolecular interactions by measuring the resulting changes in impedance over a range of frequencies. For example, EIS bacterial biosensors monitor the changes in the solution–electrode interface due to the capture of microorganisms on the sensor surface. The capture of target analyte, such as bacteria on sensor increases the insulation resulting in increase in the impedance. Shabani et al. (2008) and Mejri et al. (2010) developed an EIS bacterial detection platform by immobilizing T4 phage onto the functionalized screen-printed carbon electrode and showed specific detection of *E. coli* bacteria with limit of detection of approximately 10^4 cfu mL^{-1}. Although EIS offers label-free detection of pathogens compared to amperometry technique; its application for pathogen detection is limited due to its lower detection limit compared to other techniques.

14.4 Conclusions and Future Outlook

The conventional microbiological and biochemical methods for pathogen detection are labor intensive, time consuming, and cost ineffective. Nucleic acid and antibody-based detection systems improve the time scale of accurate detection but suffer from several other drawbacks, such as susceptibility of the probe to environmental variations, cross-reactivity, cost-ineffectiveness, and requirement of trained personnel. Biosensors, on the other hand, show several advantages, including rapid, accurate, and sensitive detection; cost-effectiveness; consistency, and being amenable to miniaturization; and integration multiplexing for point-of-care detection application. Biosensors have come a long way in detecting pathogens in clinical samples, but their application in foodborne pathogen detection is yet to reach its full potential. Different types of recognition elements have been actively researched and interfaced to the biosensor platform to improve the performance and consistency of diagnosis. The chapter summarized these bio-probes critically, highlighting their strengths and weakness. Some recent efforts have led to the development of RBPs as novel recognition elements for pathogen detection. RBPs offer several advantages over other recognition elements, such as better stability and ease of oriented immobilization to different platforms and improved bacterial capture. Besides, they can be genetically manipulated to customize their binding affinity, selectivity, and sensitivity to their target. Recent efforts have resulted in the development of RBPs against the three major and most potent foodborne pathogens, namely *Salmonella, Shigella,* and *Campylobacter*. The advent of new and sensitive detection platforms and development of RBPs against economically relevant pathogens have paved the way for significantly improved biosensors with the capability of detecting a single bacterium from a food sample.

The recent developments have made a promising start, but there is lot to be done yet. RBPs are greatly preferred as bio-probes due to their better stability and ease of oriented immobilization. However, identifying, cloning, over-expressing, and purifying a phage RBP requires years of dedicated research, which impedes the progress of the field considerably. A robust generic method to identify the part of phage genome encoding the RBPs is yet to be developed. Specificity of recognition is yet another problem of phages and phage derived bio-probes. Phages are often extremely specific to their host, which limits their application to detection of one pathogen at a time. Although specificity of recognition is highly desirable for any biosensor, a very high degree of specificity implies development of an individually addressable multiplexed system to be able to screen for common and more prevalent foodborne pathogens, which has its own challenges. Besides, much work is still to be done to develop consistent surface immobilization methods for phages and their derivatives to maintain batch-to-batch reliability from the perspective of commercial biosensors. From a biosensor viewpoint, improvement of detection limit of pathogens is extremely crucial. Micro- and nano-cantilevers are known to be very sensitive and show the capability to detect minute quantities of mass but are difficult to integrate into a portable biosensor for in situ applications. Besides, reliable methods to functionalize these devices are far from reality. Although much is yet to be done, the progress made so far has been favorable, and the newly developed direction of research hold tremendous promise for the future.

Acknowledgments

S.P. thanks the University of Alberta and the National Institute for Nanotechnology for their support throughout her PhD program. She also acknowledges the financial support received from the Alberta Livestock and Meat Agency (ALMA).

References

Abu-Rabeah, K., Ashkenazi, A., Atias, D., Amir, L. and Marks R.S. (2009). Highly sensitive amperometric immunosensor for the detection of *Escherichia coli*. *Biosensors and Bioelectronics*, 24, 3461–3466.

Apfalter, P., Assadian, O., Blasi, F., Boman, J., Gaydos, C.A., Kundi, M., Makristathis, A., Nehr, M., Rotter, M.L. and Hirschl, A.M. (2002). Reliability of nested PCR for detection of Chlamydia pneumoniae DNA in atheromas: Results from a multicenter study applying standardized protocols. *Journal of Clinical Microbiology*, 40, 4428–4434.

Arya, S.K., Singh, A., Naidoo, R., Wu, P., McDermott, M.T. and Evoy, S. (2011). Chemically immobilized T4-bacteriophage for specific *Escherichia coli* detection using surface plasmon resonance. *The Analyst*, 136, 486–492.

Balasubramanian, S., Sorokulova, I.B., Vodyanoy, V.J. and Simonian, A.L. (2007). Lytic phage as a specific and selective probe for detection of *Staphylococcus aureus*: A surface plasmon resonance spectroscopic study. *Biosensors and Bioelectronics*, 22, 948–955.

Barrangou, R. and Horvath, P. (2012). CRISPR: New horizons in phage resistance and strain identification. *Annual Review of Food Science and Technology*, 3, 143–162.

Benson, R., Tondella, M.L., Bhatnagar, J., Carvalho Mda, G., Sampson, J.S., Talkington, D.F., Whitney, A.M., Mothershed, E., McGee, L., Carlone, G., McClee, V., Guarner, J., Zaki, S., Dejsiri, S., Cronin, K., Han, J. and Fields, B.S. (2008). Development and evaluation of a novel multiplex PCR technology for molecular differential detection of bacterial respiratory disease pathogens. *Journal of Clinical Microbiology*, 46, 2074–2077.

Blasco, R., Murphy, M.J., Sanders, M.F. and Squirrell, D.J. (1998). Specific assays for bacteria using phage mediated release of adenylate kinase. *Journal of Applied Microbiology*, 84, 661–666.

Bonner, G. and Klibanov, A.M. (2000). Structural stability of DNA in nonaqueous solvents. *Biotechnology and Bioengineering*, 68, 339–344.

Boyadzhyan, B., Yashina, T., Yatabe, J.H., Patnaik, M. and Hill, C.S. (2004). Comparison of the APTIMA CT and GC assays with the APTIMA combo 2 assay, the Abbott LCx assay, and direct fluorescent-antibody and culture assays for detection of *Chlamydia trachomatis* and *Neisseria gonorrhoeae*. *Journal of Clinical Microbiology*, 42, 3089–3093.

Byrne, B., Stack, E., Gilmartin, N. and O'Kennedy, R. (2009). Antibody-based sensors: Principles, problems and potential for detection of pathogens and associated toxins. *Sensors*, 9, 4407–4445.

Chopin, M.C., Chopin, A. and Bidnenko, E. (2005). Phage abortive infection in lactococci: Variations on a theme. *Current Opinion in Microbiology*, 8, 473–479.

Conway, J.O., Sherwood, L.J., Collazo, M.T., Garza, J.A. and Hayhurst, A. (2010). Llama single domain antibodies specific for the 7 botulinum neurotoxin serotypes as heptaplex immunoreagents. *PloS One*, 5, e8818.

Croci, L., Delibato, E., Volpe, G. and Palleschi, G. (2001). A rapid electrochemical ELISA for the detection of salmonella in meat sample. *Analytical Letters*, 34, 2597–2607.

Demidov, V.V. (2002). Rolling-circle amplification in DNA diagnostics: The power of simplicity. *Expert Review of Molecular Diagnostics*, 2, 542–548.

Edgar, R., McKinstry, M., Hwang, J., Oppenheim, A.B., Fekete, R.A., Giulian, G., Merril, C., Nagashima, K. and Adhya, S. (2006). High-sensitivity bacterial detection using biotin-tagged phage and quantum-dot nanocomplexes. *Proceedings of the National Academy of Sciences*, 103, 4841–4845.

Esteban-Fernández de Ávila, B., Pedrero, M., Campuzano, S., Escamilla-Gómez, V. and Pingarrón, J.M. (2012). Sensitive and rapid amperometric magnetoimmunosensor for the determination of *Staphylococcus aureus*. *Analytical and Bioanalytical Chemistry*, 403, 917–925.

Food and Drug Administration (2007). *Food Protection Plan* [Electronic Version] from http://www.fda.gov/oc/initiatives/advance/food/plan.html.

Gervais, L., Gel, M., Allain, B., Tolba, M., Brovko, L., Zourob, M., Mandeville, R., Griffiths, M. and Evoy, S. (2007). Immobilization of biotinylated bacteriophages on biosensor surfaces. *Sensors and Actuators B: Chemical*, 125, 615–621.

Goldman, E.R., Anderson, G.P., Liu, J.L., Delehanty, J.B., Sherwood, L.J., Osborn, L.E., Cummins, L.B. and Hayhurst, A. (2006). Facile generation of heat-stable antiviral and antitoxin single domain antibodies from a semisynthetic llama library. *Analytical Chemistry*, 78, 8245–8255.

Goldman, E.R., Pazirandeh, M.P., Mauro, J.M., King, K.D., Frey, J.C. and Anderson, G.P. (2000). Phage-displayed peptides as biosensor reagents. *Journal of Molecular Recognition*, 13, 382–387.

Goodridge, L., Chen, J. and Griffiths, M. (1999a). Development and characterization of a fluorescent-bacteriophage assay for detection of *Escherichia coli* O157:H. *Applied Environmental Microbiology*, 65, 1397–1404.

Goodridge, L., Chen, J. and Griffiths, M. (1999b). The use of a fluorescent bacteriophage assay for detection of *Escherichia coli* O157:H7 in inoculated ground beef and raw milk. *International Journal of Food Microbiology*, 47, 43–50.

Greenberg, A.S., Avila, D., Hughes, M., Hughes, A., McKinney, E.C. and Flajnik, M.F. (1995). A new antigen receptor gene family that undergoes rearrangement and extensive somatic diversification in sharks. *Nature*, 374, 168–173.

Gu, H., Xu, K., Xu, C. and Xu, B. (2006). Biofunctional magnetic nanoparticles for protein separation and pathogen detection. *Chemical Communications*, 941–949.

Hamers-Casterman, C., Atarhouch, T., Muyldermans, S., Robinson, G., Hamers, C., Songa, E.B., Bendahman, N. and Hamers, R. (1993). Naturally occurring antibodies devoid of light chains. *Nature*, 363, 446–448.

Handa, H., Gurczynski, S., Jackson, M.P., Auner, G. and Mao, G. (2008). Recognition of *Salmonella typhimurium* by immobilized phage P22 monolayers. *Surface Science*, 602, 1392–1400.

Harvey, D., Harrington, C., Heuzenroeder, M.W. and Murray, C. (1993). Lysogenic phage in *Salmonella enterica* serovar Heidelberg (*Salmonella heidelberg*): Implications for organism tracing. *FEMS Microbiology Letters*, 108, 291–295.

Hennes, K.P. and Suttle, C.A. (1995). Direct counts of viruses in natural waters and laboratory cultures by epifluorescence microscopy. *Limnology and Oceanography*, 1995, 1050–1055.

Hennes, K.P., Suttle, C.A. and Chan, A.M. (1995). Fluorescently labeled virus probes show that natural virus populations can control the structure of marine microbial communities. *Applied Environmental Microbiology*, 61, 3623–3627.

Jacobs, W.R., Jr., Barletta, R.G., Udani, R., Chan, J., Kalkut, G., Sosne, G., Kieser, T., Sarkis, G.J., Hatfull, G.F. and Bloom, B.R. (1993). Rapid assessment of drug susceptibilities of *Mycobacterium tuberculosis* by means of luciferase reporter phages. *Science*, 260, 819–822.

Keer, J.T. and Birch, L. (2003). Molecular methods for the assessment of bacterial viability. *Journal of Microbiological Methods*, 53, 175–183.

Kim, J.W., Dutta, V., Elhanafi, D., Lee, S., Osborne, J.A. and Kathariou, S. (2012). A novel restriction-modification system is responsible for temperature-dependent phage resistance in *Listeria monocytogenes* ECII. *Applied and Environmental Microbiology*, 78, 1995–2004.

Kohler, G. and Milstein, C. (1975). Continuous cultures of fused cells secreting antibody of predefined specificity. *Nature*, 256, 495–497.

Kropinski, A.M., Arutyunov, D., Foss, M., Cunningham, A., Ding, W., Singh, A., Pavlov, A.R., Henry, M., Evoy, S., Kelly, J. and Szymanski, C.M. (2011). Genome and proteome of *Campylobacter jejuni* bacteriophage NCTC 12673. *Applied and Environmental Microbiology*, 77, 8265–8271.

Kuhn, J., Suissa, M., Wyse, J., Cohen, I., Weiser, I., Reznick, S., Lubinsky-Mink, S., Stewart, G. and Ulitzur, S. (2002). Detection of bacteria using foreign DNA: The development of a bacteriophage reagent for Salmonella. *International Journal of Food Microbiology*, 74, 229–238.

Lakshmanan, R.S., Guntupalli, R., Hu, J., Kim, D.J., Petrenko, V.A., Barbaree, J.M. and Chin, B.A. (2007). Phage immobilized magnetoelastic sensor for the detection of *Salmonella typhimurium*. *Journal of Microbiological Methods*, 71, 55–60.

Lakshmanan, R.S., Guntupalli, R., Hu, J., Petrenko, V.A., Barbaree, J.M. and Chin, B.A. (2007). Detection of *Salmonella typhimurium* in fat free milk using a phage immobilized magnetoelastic sensor. *Sensors and Actuators B*, 126, 544–550.

Lee, S.H., Onuki, M., Satoh, H. and Mino, T. (2006). Isolation, characterization of bacteriophages specific to *Microlunatus phosphovorus* and their application for rapid host detection. *Letters in Applied Microbiology*, 42, 259–264.

Leonard, P., Hearty, S., Quinn, J., and O'Kennedy, R. (2004). A generic approach for the detection of whole *Listeria monocytogenes* cells in contaminated samples using surface plasmon resonance. *Biosensors and Bioelectronics*, 19, 1331–1335.

Li, S., Li, Y., Chen, H., Horikawa, S., Shen, W., Simonian, A. and Chin, B.A. (2010). Direct detection of *Salmonella typhimurium* on fresh produce using phage-based magnetoelastic biosensors. *Bioengineering and Bioelectronics*, 26, 1313–1319.

Loessner, M.J., Rudolf, M. and Scherer, S. (1997). Evaluation of luciferase reporter bacteriophage A511::luxAB for detection of *Listeria monocytogenes* in contaminated foods. *Applied and Environmental Microbiology*, 63, 2961–2965.

Louie, L., Matsumura, S.O., Choi, E., Louie, M. and Simor, A.E. (2000). Evaluation of three rapid methods for detection of methicillin resistance in *Staphylococcus aureus*. *Journal of Clinical Microbiology*, 38, 2170–2173.

Malic, S., Hill, K.E., Hayes, A., Percival, S.L., Thomas, D.W. and Williams, D.W. (2009). Detection and identification of specific bacteria in wound biofilms using peptide nucleic acid fluorescent in situ hybridization (PNA FISH). *Microbiology*, 155, 2603–2611.

McHugh, T.D., Pope, C.F., Ling, C.L., Patel, S., Billington, O.J., Gosling, R.D., Lipman, M.C. and Gillespie, S.H. (2004). Prospective evaluation of BDProbeTec strand displacement amplification (SDA) system for diagnosis of tuberculosis in non-respiratory and respiratory samples. *Journal of Medical Microbiology*, 53, 1215–1219.

Mejri, M.B., Baccar, H., Baldrich, E., Del Campo, F.J., Helali, S., Ktari, T., Simonian, A., Aouni, M. and Abdelghani, A. (2010). Impedance biosensing using phages for bacteria detection: Generation of dual signals as the clue for in-chip assay confirmation. *Biosensors and Bioelectronics*, 26, 1261–1267.

Mothershed, E.A., Sacchi, C.T., Whitney, A.M., Barnett, G.A., Ajello, G.W., Schmink, S., Mayer, L.W., Phelan, M., Taylor, T.H., Jr., Bernhardt, S.A., Rosenstein, N.E. and Popovic, T. (2004). Use of real-time PCR to resolve slide agglutination discrepancies in serogroup identification of *Neisseria meningitidis*. *Journal of Clinical Microbiology*, 42, 320–328.

Mothershed, E.A. and Whitney, A.M. (2006). Nucleic acid-based methods for the detection of bacterial pathogens: Present and future considerations for the clinical laboratory. *Clinica Chimica Acta; International Journal of Clinical Chemistry*, 363, 206–220.

Mutharia, L.W., Raymond, B.T., Dekievit, T.R. and Stevenson, R.M. (1993). Antibody specificities of polyclonal rabbit and rainbow trout antisera against *Vibrio ordalii* and serotype 0:2 strains of *Vibrio anguillarum*. *Canadian Journal of Microbiology*, 39, 492–499.

Naidoo, R., Singh, A., Arya, S.K., Beadle, B., Glass, N., Tanha, J., Szymanski, C.M. and Evoy, S. (2012). Surface-immobilization of chromatographically purified bacteriophages for the optimized capture of bacteria. *Bacteriophage*, 2, 15–24.

Nakaguchi, Y., Ishizuka, T., Ohnaka, S., Hayashi, T., Yasukawa, K., Ishiguro, T. and Nishibuchi, M. (2004). Rapid and specific detection of tdh, trh1, and trh2 mRNA of *Vibrio parahaemolyticus* by transcription-reverse transcription concerted reaction with an automated system. *Journal of Clinical Microbiology*, 42, 4284–4292.

Nanduri, V., Bhunia, A.K., Tu, S.I., Paoli, G.C. and Brewster, J.D. (2007a). SPR biosensor for the detection of *L. monocytogenes* using phage-displayed antibody. *Biosensors and Bioelectronics*, 23, 248–252.

Nanduri, V., Sorokulova, I.B., Samoylov, A.M., Simonian, A.L., Petrenko, V.A. and Vodyanoy, V. (2007b). Phage as a molecular recognition element in biosensors immobilized by physical adsorption. *Biosensors and Bioelectronics*, 22, 986–992.

Neufeld, T., Schwartz-Mittelmann, A., Biran, D., Ron, E.Z. and Rishpon, J. (2003). Combined phage typing and amperometric detection of released enzymatic activity for the specific identification and quantification of bacteria. *Analytical Chemistry,* 75, 580–585.

Nielsen, P.E., Egholm, M., Berg, R.H. and Buchardt, O. (1991). Sequence-selective recognition of DNA by strand displacement with a thymine-substituted polyamide. *Science,* 254, 1497–500.

Oh, B.-K., Lee, W., Chun, B.S., Bae, Y.M., Lee, W.H. and Choi, J.-W. (2005). The fabrication of protein chip based on surface plasmon resonance for detection of pathogens. *Biosensors and Bioelectronics*, 20, 1847–1850.

Oliveira, K., Procop, G.W., Wilson, D., Coull, J. and Stender, H. (2002). Rapid identification of *Staphylococcus aureus* directly from blood cultures by fluorescence in situ hybridization with peptide nucleic acid probes. *Journal of Clinical Microbiology,* 40, 247–251.

Olofsson, L., Ankarloo, J. and Nicholls, I.A. (1998). Phage viability in organic media: Insights into phage stability. *Journal of Molecular Recognition: JMR,* 11, 91–93.

Olsen, E.V., Sorokulova, I.B., Petrenko, V.A., Chen, I., Barbaree, J.M. and Vodyanoy, V.J. (2006). Affinity-selected filamentous bacteriophage as a probe for acoustic wave biodetectors of *Salmonella typhimurium. Biosensors and Bioelectronics,* 21, 1434–1442.

Petrenko, V.A. and Vodyanoy, V.J. (2003). Phage display for detection of biological threat agents. *Journal of Microbiological Methods,* 53, 253–262.

Poshtiban, S., Singh, A., Fitzpatrick, G. and Evoy, S. (2013). Bacteriophage tail-spike protein derivitized microresonator arrays for specific detection of pathogenic bacteria. *Sensors and Actuators B: Chemical,* 181, 410–416.

Rowe, C.A., Tender, L.M., Feldstein, M.J., Golden, J.P., Scruggs, S.B., MacCraith, B.D., Cras, J.J. and Ligler, F.S. (1999). Array biosensor for simultaneous identification of bacterial, viral, and protein analytes. *Analytical Chemistry,* 71, 3846–3852.

Ruan, C., Yang, L. and Li, Y. (2002). Rapid detection of viable *Salmonella typhimurium* in a selective medium by monitoring oxygen consumption with electrochemical cyclic voltammetry. *Journal of Electroanalytical Chemistry,* 519, 33–38.

Salamon, Z. and Tollin, G. (2001). Plasmon resonance spectroscopy: Probing molecular interactions at surfaces and interfaces. *Spectroscopy,* 15, 161–175.

Schwind, P., Kramer, H., Kremser, A., Ramsberger, U. and Rasched, I. (1992). Subtilisin removes the surface layer of the phage fd coat. *European Journal of Biochemistry/FEBS,* 210, 431–436.

Sekirov, I., Russell, S.L., Antunes, L.C. and Finlay, B.B. (2010). Gut microbiota in health and disease. *Physiological Reviews,* 90, 859–904.

Shabani, A., Zourob, M., Allain, B., Marquette, C.A., Lawrence, M.F. and Mandeville, R. (2008). Bacteriophage-modified microarrays for the direct impedimetric detection of bacteria. *Analytical Chemistry,* 80, 9475–9482.

Shen, W., Lakshmanan, R.S. Mathison, L.C., Petrenko, V.A. and Chin, B.A. (2009). Phage coated magnetoelastic micro-biosensors for real-time detection of *Bacillus anthracis* spores. *Sensors and Actuators B,* 137, 501–506.

Singh, A., Arutyunov, D., McDermott, M.T., Szymanski, C.M. and Evoy, S. (2011). Specific detection of *Campylobacter jejuni* using the bacteriophage NCTC 12673 receptor binding protein as a probe. *The Analyst*, 136, 4780–4786.

Singh, A., Arutyunov, D., Szymanski, C.M. and Evoy, S. (2012). Bacteriophage based probes for pathogen detection. *The Analyst,* 137, 3405–3421.

Singh, A., Arya, S.K., Glass, N., Hanifi-Moghaddam, P., Naidoo, R., Szymanski, C.M., Tanha, J. and Evoy, S. (2010). Bacteriophage tailspike proteins as molecular probes for sensitive and selective bacterial detection. *Biosensors and Bioelectronics*, 26, 131–138.

Singh, A., Glass, N., Tolba, M., Brovko, L., Griffiths, M. and Evoy, S. (2009). Immobilization of bacteriophages on gold surfaces for the specific capture of pathogens. *Biosensors and Bioelectronics*, 24, 3645–3651.

Singh, A., Poshtiban, S. and Evoy, S. (2013). Recent advances in bacteriophage based biosensors for food-borne pathogen detection. *Sensors*, 13, 1763–1786.

Sogaard, M., Hansen, D.S., Fiandaca, M.J., Stender, H. and Schonheyder, H.C. (2007). Peptide nucleic acid fluorescence *in situ* hybridization for rapid detection of *Klebsiella pneumoniae* from positive blood cultures. *Journal of Medical Microbiology*, 56, 914–917.

Swiecki, M.K., Lisanby, M.W., Shu, F., Turnbough, C.L., Jr. and Kearney, J.F. (2006). Monoclonal antibodies for *Bacillus anthracis* spore detection and functional analyses of spore germination and outgrowth. *Journal of Immunology*, 176, 6076–6084.

Taha, M.K. (2000). Simultaneous approach for nonculture PCR-based identification and serogroup prediction of *Neisseria meningitidis*. *Journal of Clinical Microbiology*, 38, 855–857.

Tawil, N., Sacher, E., Mandeville, R. and Meunier, M. (2012). Surface plasmon resonance detection of *E. coli* and methicillin-resistant *S. aureus* using bacteriophages. *Biosensors and Bioelectronics*, 37, 24–29.

Taylor, A.D., Yu, Q., Chen, S., Homola, J. and Jiang, S. (2005). Comparison of *E. coli* O157:H7 preparation methods used for detection with surface plasmon resonance sensor. *Sensors and Actuators B*, 107, 202–208.

Tolba, M., Minikh, O., Brovko, L., Evoy, S. and Griffiths, M. (2010). Oriented immobilization of bacteriophages for biosensor applications. *Applied and Environmental Microbiology*, 76, 528–535.

Tomita, N., Mori, Y., Kanda, H. and Notomi, T. (2008). Loop-mediated isothermal amplification (LAMP) of gene sequences and simple visual detection of products. *Nature Protocols*, 3, 877–882.

Trilling, A.K., De Ronde, H., Noteboom, L., Van Houwelingen, A., Roelse, M., Srivastava, S.K., Haasnoot, W., Jongsma, M.A., Kolk, A., Zuilhof, H. and Beekwilder, J. (2011). A broad set of different llama antibodies specific for a 16 kDa heat shock protein of *Mycobacterium tuberculosis*. *PloS One*, 6, e26754.

Van Der Pol, B., Williams, J.A., Smith, N.J., Batteiger, B.E., Cullen, A.P., Erdman, H., Edens, T., Davis, K., Salim-Hammad, H., Chou, V.W., Scearce, L., Blutman, J. and Payne, W.J. (2002). Evaluation of the digene hybrid capture II assay with the rapid capture system for detection of *Chlamydia trachomatis* and *Neisseria gonorrhoeae*. *Journal of Clinical Microbiology*, 40, 3558–3564.

Verma, V., Harjai, K. and Chhibber, S. (2009). Characterization of a T7-like lytic bacteriophage of *Klebsiella pneumoniae* B5055: A potential therapeutic agent. *Current Microbiology*, 59, 274–281.

Waddell, T.E. and Poppe, C. (2000). Construction of mini-Tn10luxABcam/Ptac-ATS and its use for developing a bacteriophage that transduces bioluminescence to *Escherichia coli* O157:H7. *FEMS Microbiology Letters*, 182, 285–289.

Wang, Y., Ye, Z., Si, C. and Ying, Y. (2011). Subtractive inhibition assay for the detection of *E. coli* O157:H7 using surface plasmon resonance. *Sensors*, 11, 2728–2739.

Waseh, S., Hanifi-Moghaddam, P., Coleman, R., Masotti, M., Ryan, S., Foss, M., Mackenzie, R., Henry, M., Szymanski, C.M. and Tanha, J. (2010). Orally administered P22 phage tailspike protein reduces salmonella colonization in chickens: Prospects of a novel therapy against bacterial infections. *PloS One*, 5, e13904.

Winter, G., Griffiths, A.D., Hawkins, R.E. and Hoogenboom, H.R. (1994). Making antibodies by phage display technology. *Annual Review of Immunology*, 12, 433–455.

Wittung, P., Nielsen, P.E., Buchardt, O., Egholm, M. and Norden, B. (1994). DNA-like double helix formed by peptide nucleic acid. *Nature*, 368, 561–563.

Wu, Y., Brovko, L. and Griffiths, M.W. (2001). Influence of phage population on the phage-mediated bioluminescent adenylate kinase (AK) assay for detection of bacteria. *Letters in Applied Microbiology*, 33, 311–315.

Yim, P.B., Clarke, M.L., McKinstry, M., De Paoli, L.S.H., Pease, L.F., III, Dobrovolskaia, M.A., Kang, H., Read, T.D., Sozhamannan, S. and Hwang, J. (2009). Quantitative characterization of quantum dot-labeled lambda phage for *Escherichia coli* detection. *Biotechnology and Bioengineering*, 104, 1059–1067.

Yu, X., Susa, M., Knabbe, C., Schmid, R.D. and Bachmann, T.T. (2004). Development and validation of a diagnostic DNA microarray to detect quinolone-resistant *Escherichia coli* among clinical isolates. *Journal of Clinical Microbiology*, 42, 4083–4091.

Zheng, X., Kolbert, C.P., Varga-Delmore, P., Arruda, J., Lewis, M., Kolberg, J., Cockerill, F.R. and Persing, D.H. (1999). Direct mecA detection from blood culture bottles by branched-DNA signal amplification. *Journal of Clinical Microbiology*, 37, 4192–4193.

Chapter 15

Machine Vision Systems for Food Quality Assessment

C. Karunakaran, N.S. Visen, J. Paliwal,
G. Zhang, D.S. Jayas, and N.D.G. White

Contents

15.1 Introduction

Food (including agricultural produce) products are highly perishable. Large quantities of food products are produced every year. These products need to be sorted, graded, stored, and marketed. Increasing consumer awareness for fresh and high-quality products combined with strict grading policies by the marketing agencies results in a demand for inspection of all food commodities. Mechanization of harvesting, processing, and packaging systems results in a greater percentage of immature, misshapen, and damaged products than by doing the tasks manually. This is due to the lack of high discrimination power of machines (Humburg and Reid 1990). The outcome of the mechanized system is the increased burden on the sorting and grading systems. Human inspection systems pose problems of high labor cost, fatigue, inconsistency, and variability in the sorting and grading processes (Heinemann et al. 1994; Keagy et al. 1996; Kim and Schatzki 1999). Moreover, getting trained workers during the peak harvesting seasons is also very difficult.

A machine vision system (MVS), on the other hand, is fast and gives objective results. Sorting and grading require determination of product qualities that depend on external and internal features and characteristics. MVSs can be used to determine external features of products such as color, size, and shape (Zayas et al. 1990; Vande Vooren et al. 1992; Tao et al. 1995a, b; Shatadal et al. 1995; Luo et al. 1999; Majumdar and Jayas 2000a, b; Paliwal et al. 2003a, b, 2005; Visen et al. 2004; Choudhary et al. 2008; Chen et al. 2010; Mebatsion et al. 2012). Different nondestructive techniques can determine the products' internal characteristics such as moisture content, maturity or ripeness, nutrient contents, and foreign materials

(Finney and Norris 1978; Schatzki and Wong 1989; Karunakaran et al. 2004a; Zhang et al. 2007; Manickavasagan et al. 2008, 2010). Therefore, MVS and nondestructive quality evaluation techniques are becoming increasingly common in food industries. Manual inspection and grading of food products (i.e., fruits, vegetables, grains, meat, fish, and processed and packed foods) has a potential to be replaced by MVSs combined with nondestructive quality testing techniques in the future with further research and development.

15.2 Machine Vision Systems

15.2.1 Definition

Machine vision, also known as computer vision, is the science that deals with object recognition and classification by extracting useful information of objects from its image or image set. It is a branch of artificial intelligence technology with the combination of image processing and pattern recognition techniques (Davies 2000). The major tasks performed by an MVS can be grouped as three processes, namely, image acquisition, processing or analysis, and recognition. The success of a machine vision application is the result of well-designed interdependent processes mentioned above. An MVS simulates the human vision system for identification of different objects or scenes (Figure 15.1). The sensor or camera in an MVS represents the human eye in acquiring the object image or scenes. The information perceived by the camera is processed using a computer analogous to the human brain. Various characteristics of the objects are extracted, and final decisions are made using different image processing algorithms and pattern recognition techniques, respectively.

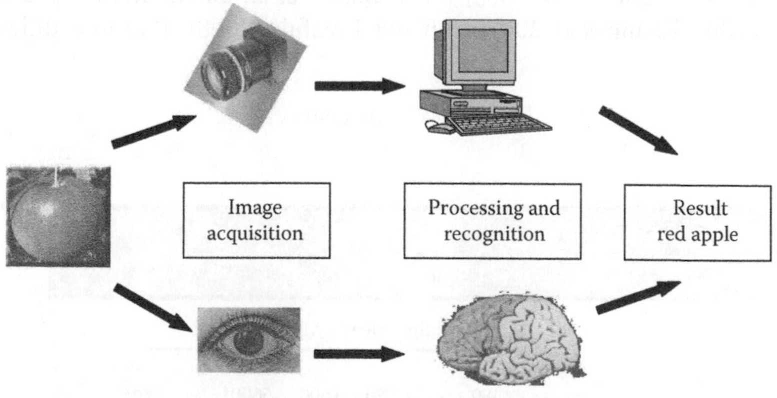

Figure 15.1 Schematic of human and machine vision systems.

15.2.2 Image Acquisition

The elements of the image acquisition system are the illumination system, sensor, frame grabber or digitizer, and computer.

15.2.2.1 Illumination System

A proper illumination system has to be designed to produce uniform and stable illumination without any glare or shadow of the object in the resulting image. This reduces the image preprocessing time and helps in better feature identification and extraction from the objects. Various types of illumination systems are used in different applications (Gunasekaran 2001). Incandescent, fluorescent, halogen, and polarized lamps are used in different applications as front-lighting or backlighting sources. Front-lighting and backlighting are used in obtaining object images for the evaluation of surface (color, texture, cracks, scars, scales, etc.) and internal characteristics of objects (stress cracks, bruises, water core, etc.), respectively (Leemans et al. 2002; Xing et al. 2007; Chen et al. 2010; Lak et al. 2010). The suitability of a particular illumination system should be evaluated for uniform lighting using objects of known color. Different objects such as a flat metal disc, yellow ball, and Kodak standard color cards can be used to calibrate the illumination systems.

15.2.2.2 Sensor

Human eyes can perceive only the visible light band in the electromagnetic spectrum (Figure 15.2). But determination of internal characteristics and compositions of products necessitates the use of infrared (IR) and near infrared (NIR), ultraviolet (UV), and x-ray imaging techniques (Schatzki and Wong 1989; Miller and Delwiche 1991; Singh and Delwiche 1994; Tollner 1993; Schatzki et al. 1997; Tao et al. 2001; Varith et al. 2003; Karunakaran et al. 2004a; Manickavasagan and Jayas 2007; Zhang et al. 2007; Haff and Toyofuku 2008). Therefore, different

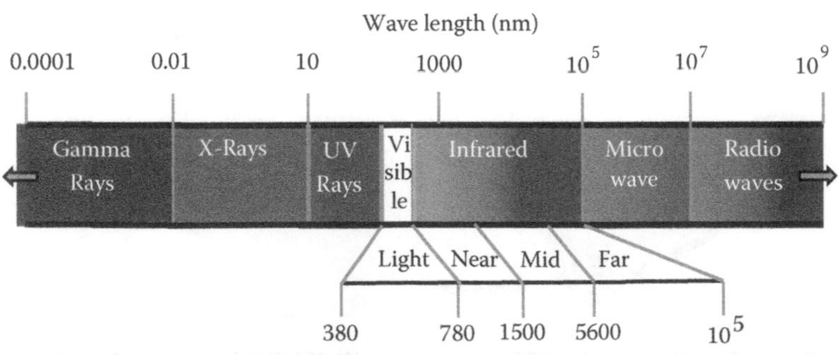

Figure. 15.2 Electromagnetic spectrum.

types of sensors that are sensitive to visible, NIR, IR, UV, and x-ray bands in the electromagnetic spectrum are used in the MVS.

Charge-coupled device (CCD) cameras that are sensitive to the visible light bands are available in monochrome (black and white) and color modes and have been used extensively in imaging applications. A CCD camera consists of arrays of sensor elements called photosites. Thc intensity of incident light on the photosites is converted into its equivalent output voltage. Monochrome cameras are adequate in applications that determine the size, shape, and surface damage in objects (Rigney et al. 1992; Ghate et al. 1993; Singh and Delwiche 1994; Heinemann et al. 1994; Pearson and Slaughter 1996). Color cameras are used for color inspection of objects (Miller and Delwiche 1989; Shearer and Payne 1990; Tao et al. 1995b; Luo et al. 1999; White et al. 2006; Chen et al. 2010). Monochrome images are less effective in identification of surface defects such as greening in white potatoes and damage in apples (Rehkugler and Throop 1986; McClure and Morrow 1987). Area and line scan CCD cameras are used in different applications. Area scan cameras have photosites or sensors in the horizontal and vertical directions and are useful for recording images of stationary objects. Line scan cameras have a row of photosites and produce two-dimensional images by relative motion of the object with respect to the camera. Recent developments in the introduction of progressive scan and complementary metal oxide semiconductor (CMOS) cameras offer promise for cost effectiveness and precision for high speed in machine vision applications.

Different types of filters that attenuate a particular range of wavelengths can be used with the sensors for better performance. Better discrimination between good and defected asparagus surfaces and enhancement of the wrinkles in peanuts is achieved by the use of special filters attached to the conventional CCD cameras (Rigney et al. 1992). Back-illuminated CCD (BCCD) cameras are more sensitive than conventional CCD cameras and are used in fluorescence and NIR imaging. NIR images are obtained using a CCD camera with a band-pass filter centered at 750 nm (Miller and Delwiche 1991; Singh and Delwiche 1994). X-ray images are recorded with the combination of x-ray detection screens (fluoroscopes that convert the incident x-rays into IR or visible images) and CCD cameras (Shahin and Tollner 1997; Karunakaran et al. 2000; Haff and Toyofuku 2008).

15.2.2.3 Frame Grabber

The analog video signal output from the camera needs to be converted into digital form suitable for further analysis using a computer. Digital signal processing boards called frame grabbers or digitizers sample analog video signals at specific intervals and store them in digital form (two-dimensional array of picture elements). The criteria for selecting a frame grabber for a particular application depend on the output type from the camera, spatial and gray level resolutions of the images required, and signal handling capabilities of frame grabbers (Gunasekaran 2001). The frame grabbers have memory space available for temporary storage of images and implementation of image analysis operations that make them ideal for fast real-time imaging applications. The use of frame grabbers can be replaced by the use of digital cameras in the MVS.

15.3 Image Processing

15.3.1 Digital Monochrome and Color Images

A digital image is a two-dimensional representation of the light intensity of an object or scene. The elements of the digital images are called "pixels" or "picture elements." The digital images are represented in the form of two-dimensional arrays that are suitable for further analysis using the image processing algorithms (Figure 15.3a and b). The row and column numbers of the arrays indicate the position of the pixel in the image, and its corresponding value (pixel value) is the intensity of the object at that point. The term "gray level" represents the scalar measurement of the intensity of the radiant energy from the object.

Monochrome (black and white) images commonly used are 8-bit (per pixel) images and have 256 (2^8) gray levels. The gray value 0 represents pure black, and 255 represents pure white (Figure 15.4a). The intermediate values represent the shades of gray between black and white.

Color images are important as human eyes are sensitive to thousands of color shades and help in easy automated object identification (Gonzalez and Woods 1998). Color images became very popular after 1980 because of the availability of color cameras that use three frames for the three primary colors, namely, red (R), green (G), and blue (B). Color images are usually 24-bit (24 bits per pixel; 8 bits for each red, green, and blue colors) images and have red, green, and blue color intensity values at each pixel (Figure 15.4b). Red, blue, and green colors are called the "primary colors," and all other colors seen are variable combinations of the primary

$$\begin{pmatrix} 10 & 0 & 15 & \cdots\cdots & 200 \\ 12 & 20 & 10 & \cdots\cdots & 100 \\ \vdots & & & & \\ 110 & 100 & \cdots\cdots & & 70 \end{pmatrix}$$

(a)

$$\begin{pmatrix} (0,85,100) & (255,0,100) & \cdots & (200,25,10) \\ \vdots & & & \\ (10,63,100) & (25,0,100) & \cdots & (0,25,10) \end{pmatrix}$$

(b)

Figure 15.3 Two-dimensional digital representation of monochrome (a) and color (b) images.

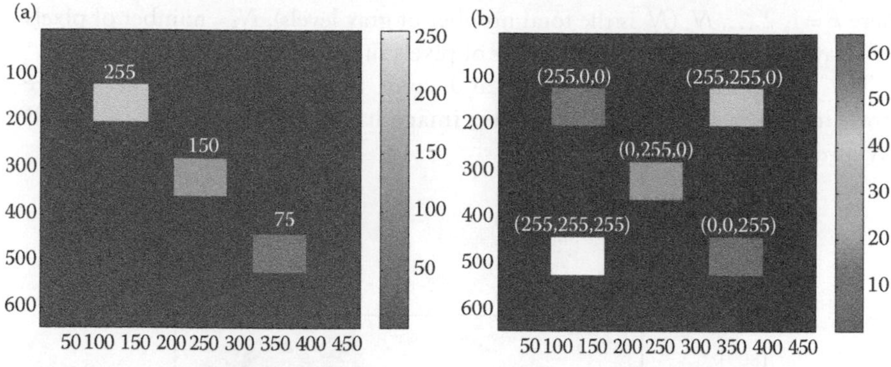

Figure 15.4 Two-dimensional graphical representation of monochrome (a) and color (b) images.

colors and are called "secondary colors." The size of monochrome and color images depends on the capability of the cameras.

15.3.2 Preprocessing

Various operations can be performed initially on the digital images to increase the identification accuracy of the objects in the later stages of image analysis. Preprocessing operations are applied to images to remove unwanted noise, artifacts, and details in an image. Noise in the digital images is introduced during image acquisition. Different shapes of objects can produce shades during imaging. These undesirable characteristics must be removed or reduced for better end results. Average and median filters are the common filters used for noise reduction in images (Gonzalez and Woods 1998). An average filter replaces the pixel value in an image with the average gray value of a small region ($n \times n$ pixels, where $n = 3, 5, 7,\ldots$) surrounding the pixel under consideration. A median filter sorts the pixel intensities in the $n \times n$ region and replaces the central pixel with the median value. Object shades can be removed or reduced by deleting a set of boundary layer pixels from the image.

15.3.3 Segmentation

15.3.3.1 Histogram

The histogram of an image describes the occurrence of a particular gray level and represents the brightness distribution of an object in an image. The histogram can also be defined as the probability density function and is given by

$$p(k) = \frac{N_k}{N} \tag{15.1}$$

where $k = 1, 2, \ldots, N_g$ (N_g is the total number of gray levels), N_k = number of pixels with gray level k, and N = total number of pixels in the image.

The histograms of monochrome and color images of a wheat kernel are shown in Figure 15.5a and b. The color image has histograms for R, G, and B gray levels.

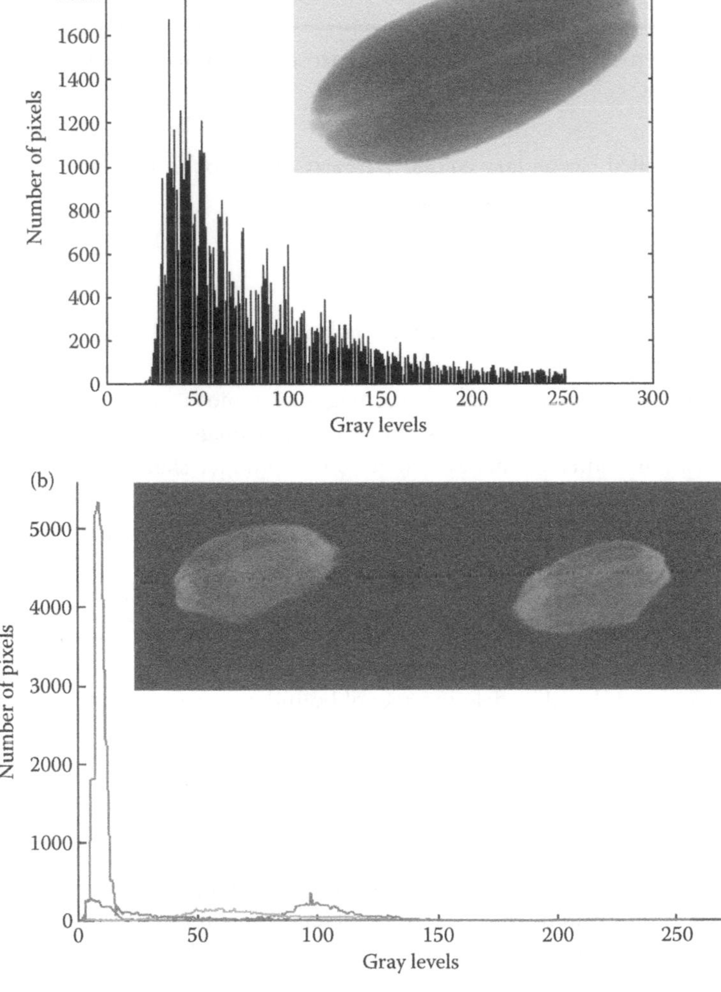

Figure 15.5 Histograms of (a) monochrome and (b) color images of wheat kernels.

15.3.3.2 Thresholding

Thresholding is the process of converting the monochrome or color images into binary images to distinguish the objects from the background, that is, assigning a gray value of 255 (object) to the pixels with gray levels greater than the threshold value *T*, and 0 (background) to other pixels in an image for images with black backgrounds and vice versa for images with white backgrounds. Thresholding, otherwise called histogram-based segmentation, can be achieved by manual or adoptive thresholding methods (Gonzalez and Woods 1998; Parker 1994). Manual or simple thresholding is done by a trial-and-error method by selecting a threshold value *T* using the histogram of the original image. The thresholded image is then compared with the original image, and the process is repeated until the entire object is differentiated from the background. In the adoptive thresholding method, the threshold value *T* is determined by an iterative process. The two histogram peaks (object and background) are identified from the original image. The mean of the two peaks is used as the first threshold value T_1. The second threshold value T_2 is determined as the average value of the mean object and background gray levels of the thresholded image. The iteration process is repeated until the threshold values in two consecutive iterations are the same or stopped after a predetermined number of iterations.

Thresholding of a color image is achieved by converting the color image into a monochrome or gray scale image by assigning different weights to the R, G, and B gray levels. For example, the gray level pixel values of the Pistachio nut color image are determined by the sum of 0.299R, 0.587G, and 0.114B gray values of each pixel (Pearson and Slaughter 1996).

15.3.3.3 Object Identification

If more than one object is present in an image, they can be assigned a unique label after thresholding using a region labeling algorithm (Haralick and Shapiro 1992). The region labeling algorithm scans the binary image from the top-left to the bottom-right. The first object pixel (i, j) encountered in the image is assigned a unique label (i.e., a value of 100). The same label value is propagated, and the region is grown to those pixels that possess the same pixel value as that of the (i, j) pixel using the eight-neighbor connectivity method (Section 15.4.1.2). In the next step, the eight neighbors of those previously labeled pixels are examined, and those that have the same pixel value are labeled. The process is repeated continuously until no more neighboring pixels have the same pixel value as that of (i, j). The scanning of the binary image is continued until all the objects in the image are uniquely labeled.

After region labeling, objects with regions less than a predetermined number of pixels depending on the size of the objects imaged should be removed, as there will be some isolated pixels in the images that are not the objects of interest. The original gray values of the labeled objects can then be superimposed on the thresholded images so that each object has its original gray value with the background value set to zero.

15.4 Feature Extraction (Image Analysis)

15.4.1 Morphological Features

Morphological features are the physical features that describe the appearance of an object. The commonly used morphological features are area, perimeter, length, and width (major and minor axis lengths) of the objects.

15.4.1.1 Area

The area of an object is defined as the number of pixels contained within its region. Area is determined by counting the number of object pixels:

$$\text{Area} = \sum_i \sum_j X(i,j) \tag{15.2}$$

where $X(i, j) = 1$ in a binary image.

Area cannot be used as the only feature for size grading, as misshapen objects can have the same area as flawless ones (Figure 15.6).

15.4.1.2 Perimeter

The perimeter of an object is defined as its boundary length. Boundary length is the sum of distance between successive boundary pairs of pixels of an object. Boundary pixels can be identified using the four-neighbor or eight-neighbor connectivity methods. In the four-neighbor connectivity method, the gray level of each pixel relative to its four neighbors is examined. A pixel $X(i, j)$ is considered a boundary pixel if $X(i, j + 1)$ or $X(i, j - 1)$ and $X(i + 1, j)$ or $X(i - 1, j)$ is a background pixel (gray level 0) (Figure 15.7a). In the eight-neighbor connectivity method, in addition to the four neighbors, the four corner pixels are considered (Figure 15.7b). The perimeter length of objects is determined using the Euclidean distance principle. If the adjacent boundary pixels occur in the horizontal or vertical positions (0, 4 or 2, 6 in Figure 15.7b), a perimeter length of 1 is added. The perimeter lengths of 1.414 (1, 5 or 3, 7) and 1.207 (2, 5 or 2, 7 or

Figure 15.6 Area (6400 pixels).

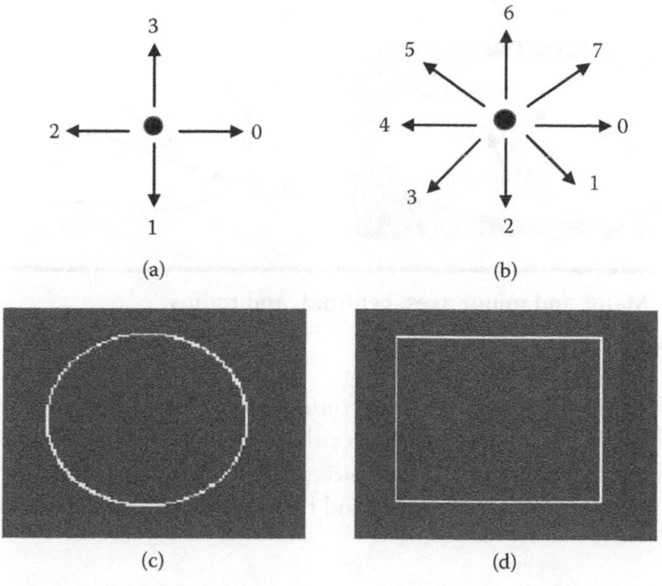

Figure 15.7 Perimeter (316 pixels).

3, 6 or 3, 0) are added if the neighboring pixels occur in the diagonal or nondiagonal positions, respectively. Two different-shaped objects, for example, a circle and square, can have the same number of perimeter pixels (Figure 15.7c and d). However, the area can be used to distinguish these two shapes.

15.4.1.3 Major and Minor Axes

Major axis is the longest length of an object that runs through its centroid or center of mass. Minor axis is the longest length of the object through the centroid that is perpendicular to the major axis (Figure 15.8a). The centroid of an object cannot be used as a feature in object recognition but is required for determining major and minor axes and other features of an object. The centroid (x_c, y_c) of an object is determined as

$$x_c = \frac{1}{N}\sum_{i=1}^{N} x_i \tag{15.3a}$$

$$y_c = \frac{1}{N}\sum_{i=1}^{N} y_i \tag{15.3b}$$

where N = total number of object pixels and x_i, y_i = x, y coordinates of the ith pixel.

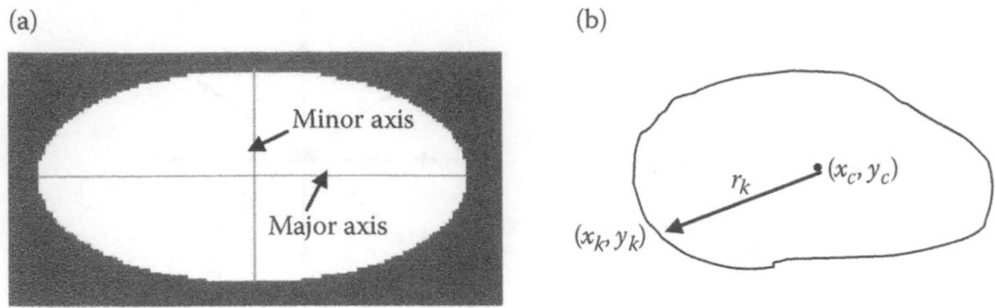

Figure 15.8 Major and minor axes, centroid, and radius.

Tao et al. (1995a) reported that determining the centroid of an object using Green's theorem is computationally faster than considering the whole object as described above. The boundary of the objects is extracted using the four-neighbor or eight-neighbor connectivity method, and the centroid can be determined as

$$x_c \approx \frac{\sum_{k=1}^{n}\left[y_k\left(x_k^2 - x_{k-1}^2\right) - x_k^2(y_k - y_{k-1})\right]}{2\sum_{k=1}^{n}\left[y_k(x_k^2 - x_{k-1}^2) - x_k(y_k - y_{k-1})\right]} \tag{15.4a}$$

$$y_c \approx \frac{\sum_{k=1}^{n}\left[y_k^2(x_k - x_{k-1}) - x_k\left(y_k^2 - y_{k-1}^2\right)\right]}{2\sum_{k=1}^{n}\left[y_k(x_k - x_{k-1}) - x_k(y_k - y_{k-1})\right]} \tag{15.4b}$$

where n = total number of boundary pixels and x_k, y_k = x, y coordinates of the kth pixel.

Another definition of major axis is determining it as the longest distance through the object or the length of the bounding rectangle that surrounds the object (Majumdar and Jayas 2000a). The radius (r) of the object is determined as the distance of the boundary pixels from the centroid of the object (Figure 15.8b):

$$r(k) = [(x_k - x_c)^2 + (y_k - y_c)^2]^{1/2} \tag{15.5}$$

The mean, maximum, and minimum radii of the objects can be used as morphological features for size grading and sorting.

15.4.1.4 Derived Features

The following morphological features can be determined from the primary features mentioned above and are called derived features (Majumdar and Jayas 2000a):

$$\text{Thinness ratio} = \text{Perimeter}^2/\text{Area} \tag{15.6}$$

$$\text{Aspect ratio} = \text{Length of major axis/Length of minor axis} \tag{15.7}$$

$$\text{Radius ratio} = \text{Maximum radius/Minimum radius} \tag{15.8}$$

$$\text{Haralick ratio} = \text{Mean radii/Standard deviation of radii} \tag{15.9}$$

The primary and derived morphological features are extensively used in the MVS for grading of fruits, vegetables, and cereal grains (e.g., McClure and Morrow 1987; Singh and Delwiche 1994; Zayas et al. 1990; Rigney et al. 1992; Majumdar and Jayas 2000a).

15.4.1.5 Moments

Moments are the statistical representation of objects. They can be determined for binary and gray scale representation of objects. Moments of binary objects describe their shape, and moments of gray level images describe the gray level distribution of objects. The general moments (m_{pq}) of different orders are determined as

$$m_{pq} = \sum_i \sum_j i^p j^q X(i,j) \tag{15.10}$$

where p, q = 0, 1, 2,… is the order of the moments and $X(i, j)$ = gray level of the object at coordinate (i, j).

In binary images, the gray level of the object $X(i, j)$ is 1 for all pixels. For the monochrome and color images, the gray scale or R, G, B gray values are substituted for the value of $X(i, j)$.

The zero-order spatial moment m_{00} of a binary object is the sum of all pixels, which is the area of the object. The first-order moments m_{10} and m_{01} are the sum of x and y coordinates, and m_{10}/m_{00} and m_{01}/m_{00} represents the centroid (x_c, y_c) of the object. The second-order moments m_{20} and m_{02} represent the moment of inertia.

The moments m_{pq} are dependent on the position of the object in the image. For comparison and identification of objects, the moments have to be independent of position and orientation in the image and size of the objects.

The central moments μ_{pq} that are invariant to translation (position of the object in a given image) and normalized central moments η_{pq} that are invariant to translation and size of the object are given by

$$\mu_{pq} = \sum_i \sum_j (i - x_c)^p (j - y_c)^q \tag{15.11}$$

$$\eta_{pq} = \frac{\mu_{pq}}{\mu_{00}^{\gamma}} \tag{15.12}$$

where $\gamma = \frac{1}{2}(p+q)+1$.

The moment invariants ϕ are independent of translation, size, and orientation of the object in an image and are determined as (Gonzalez and Woods 1998)

$$\begin{aligned}
\phi_1 &= \eta_{20} + \eta_{02} \\
\phi_2 &= (\eta_{20} - \eta_{02})^2 + 4\eta_{11}^2 \\
\phi_3 &= (\eta_{30} - 3\eta_{12})^2 + (3\eta_{21} - \eta_{03})^2 \\
\phi_4 &= (\eta_{30} + \eta_{12})^2 + (\eta_{21} + \eta_{03})^2
\end{aligned} \tag{15.13}$$

Shape moments are successfully used to detect misshapen fruits and vegetables and to identify different fish species (e.g., Singh and Delwiche 1994; Heinemann et al. 1994; White et al. 2006).

15.4.1.6 Fourier Descriptors

Fourier descriptors (FDs) are used to differentiate objects with distinct boundary shapes. Boundary shapes can be represented on a quantitative basis using FDs (Table 15.1). Slow variations or smooth boundaries are represented by the low harmonic components, and complex variations along a boundary are represented by the high harmonic components of the FD (Tao et al. 1995a).

FDs can be determined by considering the coordinates of the boundary pixels or their lengths from the centroid of the objects. The discrete Fourier transform $F(u)$ of an object is given by

$$F(u) = \frac{1}{n}\sum_{k=0}^{n-1} p(k) e^{-j2\pi uk/n} \tag{15.14}$$

Table 15.1 Shape Representation by Fourier Descriptors

Fourier Descriptor	*Shape Features*
F(0)	Average radius
F(1)	Bendingness
F(2)	Elongation
F(3)	Triangle
F(4)	Square

Source: Tao, Y. et al., *Transactions of the ASAE* 38(3):949–957, 1995a. Reprinted with permission.

where $p(k) = x(k) + iy(k)$; $x(k)$, $y(k) = x$, y coordinates of the boundary pixel (k); $k = 1, 2, \ldots, n$ (n is the total number of boundary pixels); and $u = 0, 1, 2, \ldots, (n - 1)$ is the order of the FD.

The magnitude of $F(u)$ is the square root of the sum of squares of its real and imaginary values. Fourier transform determined from the radius of the boundary pixels is given by

$$F(u) = \frac{1}{n}\sum_{k=0}^{n-1} r(k)e^{-j2\pi uk/n} \tag{15.15}$$

where r_k = radius of the boundary pixel, that is, the distance of boundary pixel k from the centroid of the object.

Tao et al. (1995a) used the first 10 harmonic functions of the FD to combine into a shape factor, and different threshold values of the shape factors were used to classify potatoes and apples into different classes. Over 90% of potatoes and apples were correctly identified into different shapes, and raisins were graded into different classes using FD (Tao et al 1995a; Okamura et al. 1993).

15.4.2 Gray Scale and Color Features

The mean and standard deviation of the gray levels can be used to compare the brightness and brightness distribution in different objects. The gray level intensity across the longitudinal profiles has been used to detect the early splits in pistachio nuts, viel opening in mushrooms, and insect infestations in grain kernels (Heinemann et al. 1994; Pearson and Slaughter 1996; Haff and Slaughter 1999; Karunakaran et al. 2003a, b, 2004b). The mean and standard deviation of the gray levels of the defective surface have been used to determine the size and types of defects such as cut, scar, bruise, and wormholes in fruits (Singh and Delwiche 1994).

The gray values of monochrome images and color images can be used as features in object recognition. The histogram groups of monochrome images have been used to detect infestations in pistachio nuts (Casasent and Sipe 1996).

The ratios of primary colors (R, G, B) are used as features for color grading of peaches (Miller and Delwiche 1989).

$$R = \frac{R}{R+G+B} \tag{15.16}$$

$$G = \frac{G}{R+G+B} \tag{15.17}$$

$$B = \frac{B}{R+G+B} \tag{15.18}$$

Human beings perceive color as three independent attributes, namely, hue, saturation, and intensity (HSI), and the use of primary colors is computationally inefficient (Thomas and Connoly 1986; Morrisey 1988). Hue represents the dominant wavelength, that is, pure color; saturation refers to the amount of white light mixed with the hue or the pure color; and intensity is the brightness of the achromatic light. Tao et al. (1995b), for color grading of potatoes and apples, converted the RGB into the HSI system in two steps. RGB values are first converted into IUV values using the relation given by

$$\begin{bmatrix} I \\ V \\ U \end{bmatrix} = \begin{bmatrix} 1/3 & 1/3 & 1/3 \\ 1 & -1/2 & -1/2 \\ 0 & \sqrt{3}/2 & -\sqrt{3}/2 \end{bmatrix} \begin{bmatrix} R \\ G \\ B \end{bmatrix} \tag{15.19}$$

The IUV coordinate system is then converted into the HSI system. I values are the same in IUV and HSI systems, and the H and S values are determined as

$$H = 90° + \tan^{-1}\left\{(2R - G - B)/(\sqrt{3}(G - B))\right\} \times 255/360 \tag{15.20}$$

$$H = H + 180° \text{ if } G < B$$

$$S = \left(1 - \frac{\min(R, G, B)}{I}\right) \times 255/360 \tag{15.21}$$

Table 15.2 Classification of Hue Classes from RGB Color Components

R > I	*G > I*	*(R – I) > (G – I)*	*Hue (Color)*
1	1	1	Orange
1	1	0	Yellow
0	1	0	Green
0	1	1	Cyan
0	0	1	Blue
0	0	0	Violet
1	0	0	Magenta
1	0	1	Red

Source: Shearer, S.A. and F.A. Payne, *Transactions of the ASAE* 33(6):2045–2050, 1990. Reprinted with permission.

The hue histogram values were then used as features to detect green surfaces in potatoes and maturity in apples. The classification accuracy of more than 90% was achieved in classifying potatoes and apples into different groups (Tao et al. 1995b).

Shearer and Payne (1990) obtained eight groups of hue values for sorting bell peppers into different classes from the RGB images. The intensity value I was the average of R, G, and B gray values of each pixel. The I value of each pixel was then compared with its red and green gray levels, and the pixels were grouped into eight categories (Table 15.2).

15.4.3 Textural Features

Textural properties of an object can be defined as the spatial distribution of gray level intensities. They can be used to quantify properties such as smoothness, coarseness, fineness, and granulation.

The statistical approach of measuring texture is the simplest method that takes into consideration the moments of the gray level histograms. The average intensity (m) of an object is given by

$$m = \sum_{i=1}^{L} z_i p(z_i) \tag{15.22}$$

where z_i = gray levels of the image (i = 1, 2,…, g) and $p(z_i)$ = probability density function.

The *n*th-order moment of gray level histogram is given by

$$\mu_n(z) = \sum_{i=1}^{L} (z_i - m)^n p(z_i) \tag{15.23}$$

The zero- and first-order moments (μ_0 and μ_1) are 1 and 0, respectively, for all objects. The second moment (μ_2), also called as variance, describes the relative smoothness or the distribution pattern of gray level intensities. The third-order moment (μ_3), also called skewness, describes the symmetry of the histogram curve about the mean gray level. The fourth-order moment (μ_4), the kurtosis, represents the degree of peakness or flatness of the histogram of the object. Histogram moments have been used to identify mature and immature peanuts and freshness of mushrooms (Rigney et al. 1992; Heinemann et al. 1994). More than 80% of the mature peanuts are correctly identified using the histogram moments.

The statistical approach of textural measurement cannot differentiate different objects at times. Two objects can have the same morphological and color features. They can be differentiated using spatial gray level co-occurrence and run length matrix methods (Figure 15.9).

15.4.3.1 Gray Level Co-Occurrence Matrix

The gray level co-occurrence matrix (CM) provides information about the distribution of gray level intensities with respect to the relative position of the pixels with equal intensities. The matrix elements of CM(i, j) are the number of occurrence of pixels with gray level i encircled by pixels with gray level j at a distance d in 0°, 45°, 90°, and 135° directions (Figure 15.10a). Considering a 4 × 4 image with gray levels in the range 0 to 3, the co-occurrence matrices in the 0°, 45°, 90°, and 135° directions are shown in Figure 15.10b. The co-occurrence matrices in the four directions are combined, and each element is divided using a normalizing constant k given by

$$k = 2N_x(N_y - 1) + 2N_y(N_x - 1) + 4(N_x - 1)(N_y - 1) \tag{15.24}$$

where N_x = number of object pixels in the horizontal direction and N_y = number of object pixels in the vertical direction.

Figure 15.9 Textural features of two different objects.

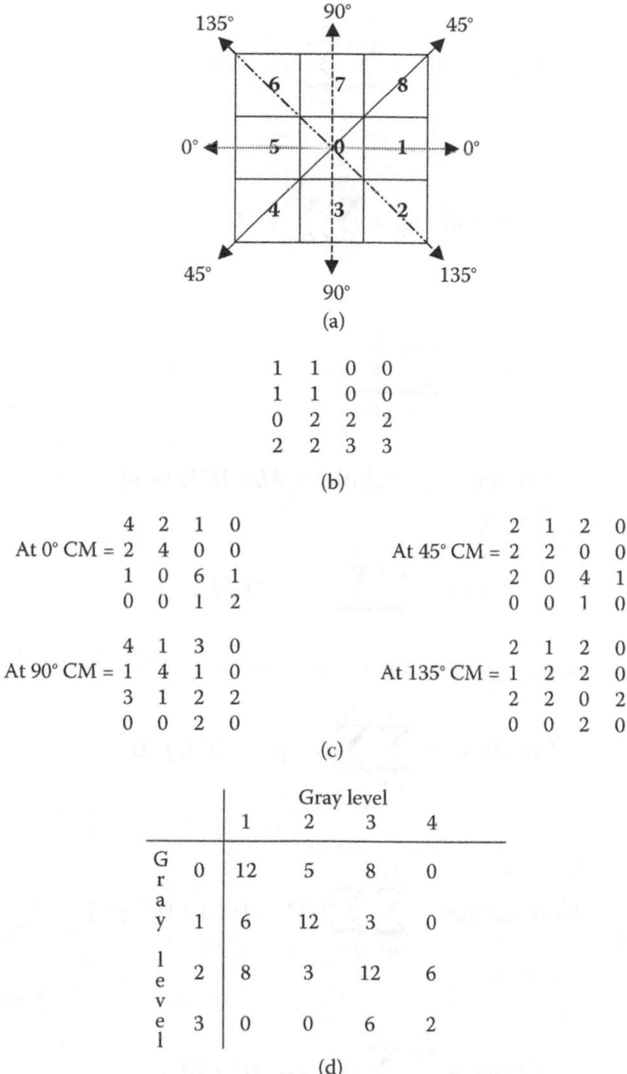

Figure 15.10 Co-occurrence matrix for textural features.

The following features, which are invariant to orientation of objects, can be determined from the normalized CM (Haralick 1979; Unser 1986; Majumdar and Jayas 2000c):

$$\text{Mean } (\mu) = \sum_{i=1}^{N_g} \sum_{j=1}^{N_g} iCM(i, j) \tag{15.25}$$

$$\text{Vaiance }(\sigma^2) = \sum_{i=1}^{N_g}\sum_{j=1}^{N_g}(i-\mu)^2 CM(i,j) \tag{15.26}$$

$$\text{Uniformity} = \sum_{i=1}^{N_g}\sum_{j=1}^{N_g} CM(i,j)^2 \tag{15.27}$$

$$\text{Entropy} = -\sum_{i=1}^{N_g}\sum_{j=1}^{N_g} CM(i,j)\log\{CM(i,j)\} \tag{15.28}$$

$$\text{Maximum probability} = Max\,\{CM\,(i,j)\} \tag{15.29}$$

$$\text{Inertia} = \sum_{i=1}^{N_g}\sum_{j=1}^{N_g}(i-j)^2 CM(i,j) \tag{15.30}$$

$$\text{Correlation} = \sum_{i-1}^{N_g}\sum_{j=1}^{N_g}(i-\mu)^2 CM(i,j)/\sigma^2 \tag{15.31}$$

$$\text{Homogeneity} = \sum_{i=1}^{N_g}\sum_{j=1}^{N_g} CM(i,j)/[(1+(i-j)^2] \tag{15.32}$$

$$\text{Cluster} = \sum_{i=1}^{N_g}\sum_{j=1}^{N_g}(i+j-2\mu)^3 CM(i,j) \tag{15.33}$$

where N_g = maximum gray level of the image.

15.4.3.2 Gray Level Run Length Matrix

Gray level run length matrix (GLRM) is a representation of the occurrence of collinear and consecutive pixels of the same or similar gray levels in an object. The matrix elements of GLRM (i, j) are the number of occurrences of a gray level (i)

$$\begin{matrix} 1 & 1 & 0 & 0 \\ 1 & 1 & 0 & 0 \\ 0 & 2 & 2 & 2 \\ 2 & 2 & 3 & 3 \end{matrix}$$

(a)

$$\text{At } 0^\circ \text{ RM} = \begin{matrix} 1 & 2 & 0 & 0 \\ 0 & 2 & 0 & 0 \\ 0 & 1 & 2 & 0 \\ 0 & 1 & 0 & 0 \end{matrix} \qquad \text{At } 45^\circ \text{ RM} = \begin{matrix} 3 & 1 & 0 & 0 \\ 2 & 1 & 0 & 0 \\ 1 & 2 & 0 & 0 \\ 2 & 0 & 0 & 0 \end{matrix}$$

$$\text{At } 90^\circ \text{ RM} = \begin{matrix} 1 & 2 & 0 & 0 \\ 0 & 2 & 0 & 0 \\ 3 & 1 & 0 & 0 \\ 2 & 0 & 0 & 0 \end{matrix} \qquad \text{At } 135^\circ \text{ RM} = \begin{matrix} 3 & 1 & 0 & 0 \\ 2 & 1 & 0 & 0 \\ 5 & 0 & 0 & 0 \\ 2 & 0 & 0 & 0 \end{matrix}$$

(b)

Gray level	Run length 1	2	3	4
0	8	6	0	0
1	4	6	0	0
2	9	4	1	0
3	6	1	0	0

(c)

Figure 15.11 Run length matrix for textural features.

with a run length (j) in a given direction. The row index of the GLRM is the gray level, and the column index is the run length (Figure 15.11). The following features can be extracted from the GLRM (Galloway 1975; Majumdar and Jayas 2000c):

$$\text{Run length (R)} = \sum_{i=1}^{N_g} \sum_{j=1}^{N_r} \text{RM}(i, j) \tag{15.34}$$

$$\text{Short run} = \sum_{i=1}^{N_g} \sum_{j=1}^{N_r} \{\text{RM}(i, j)/j^2\}\text{R} \tag{15.35}$$

$$\text{Long run} = \sum_{i=1}^{N_g} \sum_{j=1}^{N_r} \{j^2\text{RM}(i, j)\}/\text{R} \tag{15.36}$$

$$\text{Gray level nonuniformity} = \sum_{i=1}^{N_g} \left(\sum_{j=1}^{N_r} \text{RM}\,(i, j) \right)^2 /\text{R} \qquad (15.37)$$

$$\text{Run percent} = \text{R} / \sum_{i=1}^{N_g} \sum_{j=1}^{N_r} j\text{RM}(i, j) \qquad (15.38)$$

$$\text{Gray level entropy} = \sum_{i=1}^{N_g} \sum_{j=1}^{N_r} \text{RM}(i, j) \log\left\{\text{RM}(i, j)\right\} /\text{R} \qquad (15.39)$$

$$\text{Run length nonuniformity} = \sum_{j=1}^{N_r} \left(\sum_{i=1}^{N_g} \text{RM}(i, j) \right)^2 /\text{R} \qquad (15.40)$$

where N_r = maximum number of run lengths.

15.5 Object Recognition and Classification

The statistical classifiers, artificial neural networks (ANNs), and fuzzy logic are the common decision methods used for object recognition using the extracted features in MVS. The statistical classifiers use the discrimination (DISCRIM) procedure to classify a set of patterns containing one or more quantitative variables and a classification variable into one or more distinct groups (SAS 2000). The discriminant function or classification criterion is derived from the pattern set called "training set" and applied to other data sets called "testing sets." The DISCRIM procedure uses Bayes' theorem to determine the probability of an observation belonging to a particular group by assuming the prior probabilities of its group membership and the group-specific densities. However, the probability density functions and the posterior probabilities are unknown in the practical applications and are estimated using parametric or nonparametric methods. The parametric method is used to estimate the probability parameters, if the within-class distributions of the variables are approximately multivariate normal. If the within-class distributions are not multivariate normal or no assumptions can be made about the distributions of observations, the nonparametric method is used to generate discriminant functions.

ANN is an extension of the statistical technique. This is widely used for pattern recognition problems for food products (Jayas et al. 2000). An ANN is a parallel computing network, made up of processing elements called nodes or neurons whose functions are analogous to the neurons of the human brain (Bishop 1997). The network can be trained using a training set, and the processing ability of the network is stored in the interunit connection strengths or weights. The trained network can be saved and applied on independent testing sets. There are different types of networks depending on the number of nodes and the type of connections between them. Based on architecture, they can be classified as full connectivity layer, non-full connectivity layer, layered, or nonlayered networks (Gurney 1997). Back-propagation, probabilistic, Kohonen, and general regression networks are different types of networks available commercially (Paliwal et al. 2001). Back-propagation or multilayer feed-forward networks are widely used for quality inspection and classification of agricultural products. Jayas et al. (2000) reviewed the use of multilayer networks in different MVS applications for agricultural products and compared their advantages with the statistical classifiers.

15.6 Application of Machine Vision for Food Products

A wide range of research work has been reported in the literature that determines the feasibility of using MVS to sort and grade fruits, vegetables, cereal grains, meat, fish, and processed and packed foods. Promising results have been obtained in different studies where MVSs are equivalent to or better than humans in performing the tasks (Tao et al. 1995b; Blasco et al. 2003; Brosnan and Sun 2004; Sun 2008).

Applications of MVS using images obtained in the visible light spectrum to sort, grade, and detect defects in apples (Taylor et al. 1984; Tao and Wen 1999), peaches (Miller and Delwiche 1989, 1991; Han et al. 1992; Singh and Delwiche 1994; Varith et al. 2003; Xing et al. 2007; Kavdir and Buyer 2008), citrus fruits (Blasco et al. 2007), potatoes (Tao et al. 1995a, b), bell peppers (Shearer and Payne 1990; Wolfe et al. 1992), carrots (Howarth et al. 1992), mushrooms (Heinemann et al. 1994), tomatoes (Laykin et al. 2002), pistachio nuts (Pearson and Slaughter 1996; Ghazanfari and Irudayaraj 1996; Pearson and Schatzki 1998), almonds (Pearson and Young 2002), peanuts (Ghate et al. 1993), raisins (Okamura et al. 1993), cereal grains (Zayas et al. 1990; Majumdar and Jayas 2000d; Wigger et al. 1988), meat (Tao et al. 2001; Chen et al. 2010), and fish (White et al. 2006) are reported in the literature.

NIR (700–1100 nm) images are used to detect defects in fruits (Miller and Delwiche 1991; Singh and Delwiche 1994; Lu and Ariana 2002), seed damage (Delwiche 2003; Zhang et al. 2007), insect infestation (Paliwal et al. 2004; Singh et al. 2009a, 2010a), and fungal infection (Singh et al. 2009b, 2010b, 2012) in seeds. The NIR images show better discrimination between good and defective peach tissues than the color images obtained with visible light (Miller and Delwiche 1991; Blasco et al. 2007). The NIR spectrum with wavelengths 700–2500 nm is used in spectroscopic techniques to quantify chemical constituents in agricultural products. Hence, NIR imaging has high potential to be effectively used to quantify and determine the distribution of chemicals or nutrients in the products in the future.

Soft x-ray images are used to determine maturity of tomato, peach, and lettuce by measuring the density changes at different maturity levels (Brecht et al. 1991; Barcelon et al. 1999; Lenker and Adrian 1971). The mechanical harvester with the x-ray system resulted in harvesting only 4% soft heads compared to 13% soft heads picked by human experts (Han et al. 1992). Nearly 100% of hollow hearts in potatoes and over 80% of the infested almonds are correctly detected with x-ray images (Finney and Norris 1978; Kim and Schatzki 1999). The use of x-ray technique to detect bruises, water core, and stem rot in apples is reported in different studies (Diener et al. 1970; Tollner 1993; Schatzki et al. 1997). In seeds, soft x-rays have been used to detect insect infestations, vitreousness, and cracks (Karunakaran et al. 2003b; Neethirajan et al. 2006). X-rays are used to detect contaminants and foreign materials in packaged foods and in meat products (Schatzki and Wong 1989; Tao et al. 2001).

15.6.1 MVS for Cereal Grain Classification

The image acquisition system used for acquiring the grain images is shown in Figure 15.12. The diffuse illumination system consisted of an illumination chamber with a flat sample platform. A semispherical steel bowl (≈0.39 m diameter) with its inner side painted white and smoked with magnesium oxide was used as the diffuser. A circular fluorescent lamp was used, and its voltage was maintained constant (120 ± 0.1 V) by a variac. A three-chip CCD color camera (DXC 3000A, Sony, Japan) with a 10–120 mm focal length zoom lens was used to capture the images. The illumination chamber had an opening at the top (0.125 m diameter), and the camera was mounted over the opening on a stand. The output from the camera had three parallel RGB analog

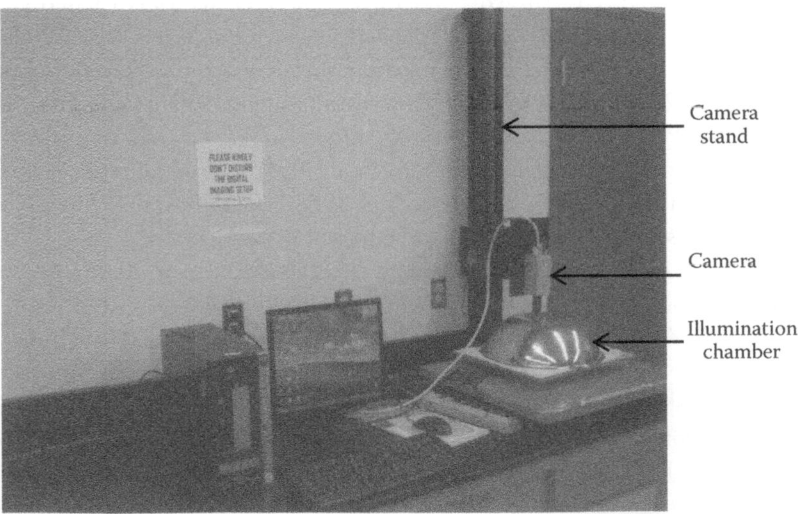

Figure 15.12 An example of an area scan imaging system.

National Television System Committee (NTSC) video signals. The frame grabber (DT2871 and DT2858, DATA Translation, Marlboro, MA) digitized the RGB analog signals into three 8-bit 640 × 480 digital images at a rate of 30 frames per second. A Kodak white card (E 152-7795, Eastman Kodak Co., Rochester, NY) with 90% reflection was used as the reference for the calibration of the illumination system.

Grain classification using MVS was studied using images of individual and bulk grain kernels. Bulk (a Petri dish filled with grain kernels) and individual grain kernels were imaged by placing them manually on the sample platform. The RGB images at a resolution of 60 μm/pixel and size of 640 × 480 were saved as tiff images (Figure 15.13a and b). Algorithms were developed using C++ language to extract 230 morphological (area, perimeter, major and minor axis lengths, shape moments, first 20 FDs using radius lengths and boundary pixel coordinates), color (mean and variances of R, G, and B colors; mean and variances of HSI values; and color moments), and textural features (using gray level co-occurrence and run length matrices). The image was first scanned by the program, and the pixel values were stored in an array (size 3 × 640 × 480). The program was designed to get the input from the user to determine whether bulk or individual grain kernel images were

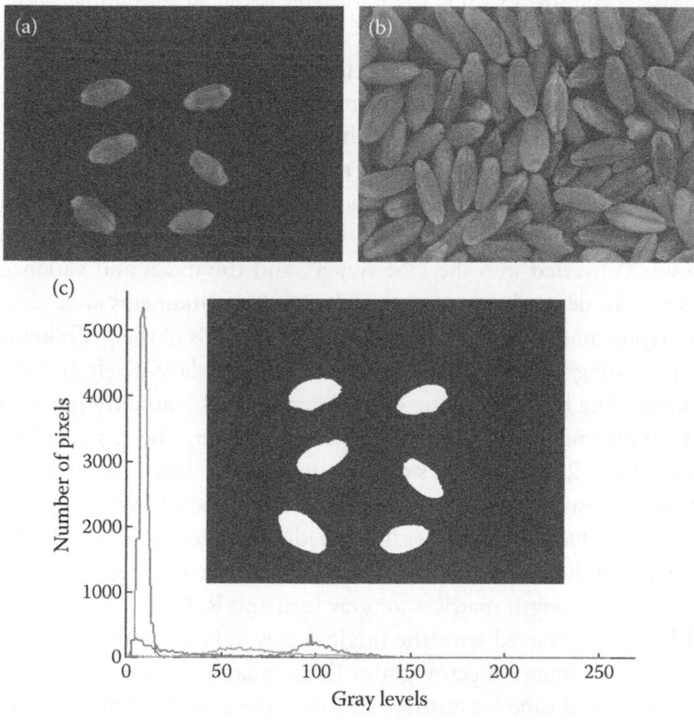

Figure 15.13 Individual kernel (a) and bulk grain (b) images of wheat. (c) Histogram and thresholded image of individual wheat kernels.

used for analysis. For bulk images, the entire image was treated as one object, and only the color and textural features were extracted.

The adoptive thresholding method was used to extract the individual grain kernels from the black background. As the red pixels were dominant in representing the grain kernels, weights of 0.80, 0.16, and 0.04 were used for the R, G, and B colors (Figure 15.13c). The mean gray value of the image was obtained by dividing the total gray value the number of pixels in the image. The iteration process started with the mean gray value as the first threshold value T_1. The mean gray value of the pixels whose values were less than and greater than T_1 was calculated as T_{11} and T_{12}, respectively, and their mean value was used as the second threshold value T_2. The process was repeated until two consecutive threshold values were the same or at the end of 40 iterations, whichever occurred first. The final threshold value was 20 levels less than the value obtained by the iteration process to extract the entire grain kernels from the background. The background pixels were assigned a value of 0, and the object pixels were assigned a value of 1. The thresholded image is shown in Figure 15.13c. The region labeling algorithm was applied to the thresholded image to mark the different kernels with their unique labels. All the pixels that belong to individual kernels were given the unique values (100, 200,..., 600, respectively, for the six kernels) (Figure 15.13c). The perimeter pixels of individual kernels were determined using the four-neighbor connectivity method, and they were marked with "1" as the last digit (101 for perimeter pixels for objects with labels 100) of the object number. The Euclidean distance principle was used to determine the perimeter lengths. The mean, maximum, and minimum radii and the major and minor axis lengths of each grain kernel were determined.

The R, G, and B gray values of the grain kernels were grouped into 32 histogram groups, and the mean and variances of the gray values were determined. The RGB color system was converted into the HSI system, and the mean and variances of H, S, and I values were determined. A total of 16 invariant moments were determined for the binary representation and R, G, B colors of the pixels of the grain kernels. FDs were determined using the coordinate values of the boundary pixels and their radii from the centroid. The FD values were determined for 128 boundary pixels that were equispaced at an incremental angle of $\pi/64$ from each other. The x, y coordinates and radius of the selected 128 pixels were stored in an array and were used to calculate the FD for the first and last 11 harmonic functions. Normalized FD values were obtained by dividing the FD values by zero-order (mean radius) FD for radius and by first-order FD for coordinates of the boundary pixels. Textural features were determined using the co-occurrence and run length matrices for gray level and R, G, B colors of the images. The textural features extracted with the original gray values (256) of the pixels gave poor classification percentage of cereal grains (Majumdar and Jayas 2000c). Therefore, to reduce computational time for feature extraction, the co-occurrence and run length matrices were determined for a maximum of eight gray level and R, G, B pixel values (the original gray level and RGB values of all pixels were divided by 32). The run lengths greater than eight were counted as having run lengths of eight.

15.6.1.1 Algorithm Testing

The accuracy of the features extracted by the algorithm was tested using images of different shapes and colors in different orientations and sizes created using the Paint Shop Pro (1998) and MATLAB (1997) software. Objects of defined shapes such

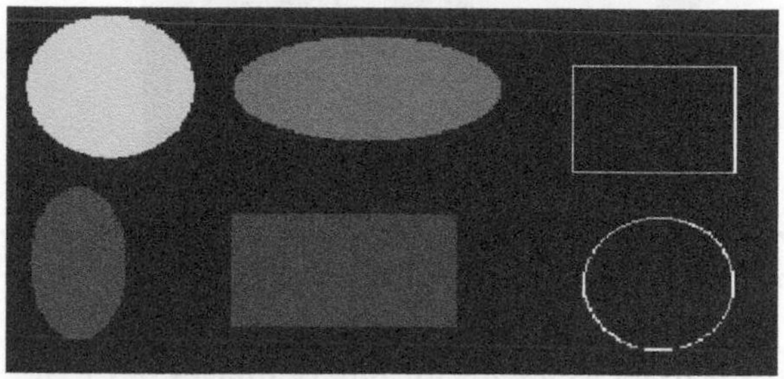

Figure 15.14 Image for testing the morphological features.

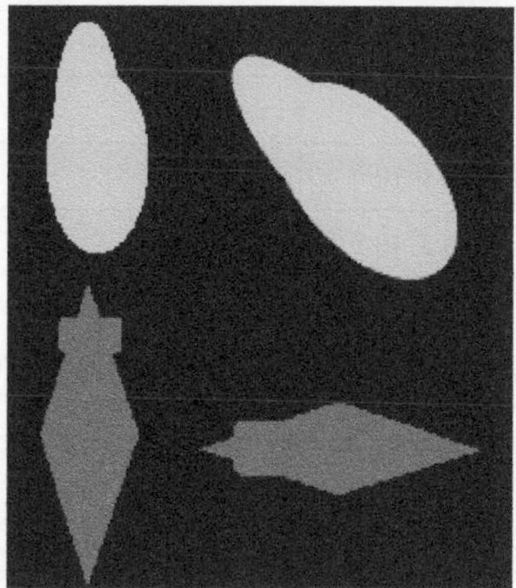

Figure 15.15 Image for testing major and minor axes.

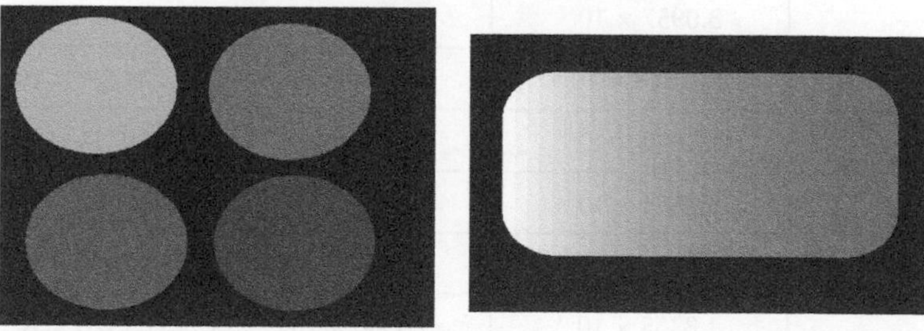

Figure 15.16 Images for testing the color features.

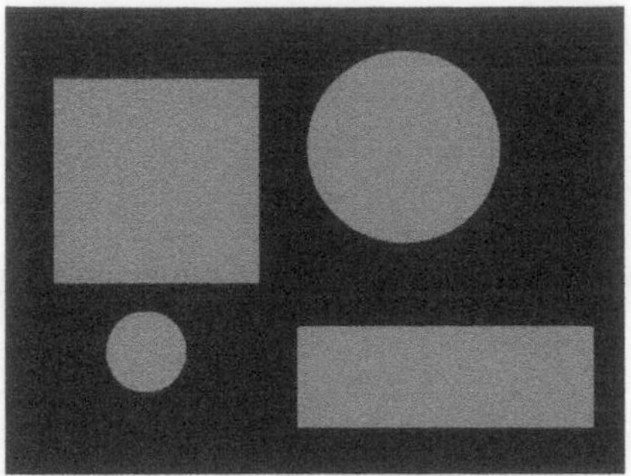

Figure 15.17 Image for testing the shape moments.

Table 15.3 Invariant Moments for Binary Images of Circle, Square, and Rectangle

Moments	*Circle*	*Square*	*Rectangle*
ϕ_1	1.5916×10^{-1}	1.6667×10^{-1}	3.0894×10^{-1}
ϕ_2	0	0	6.7783×10^{-2}
ϕ_3	5.3821×10^{-8}	0	0
ϕ_4	1.3003×10^{-7}	0	0

Table 15.4 Normalized Fourier Descriptor Values Determined from Radius of Boundary Pixels of Different Object Shapes

Normalized FD	*Circle*	*Square*	*Rectangle*
FD (1)	8.7876×10^{-2}	3.5263×10^{-2}	3.9591×10^{-1}
FD (2)	4.1430×10^{-4}	2.3306×10^{-2}	2.8652×10^{1}
FD (3)	2.6166×10^{-2}	6.8457×10^{-2}	1.1709×10^{-1}
FD (4)	3.0957×10^{-2}	6.9909	6.7966
FD (5)	3.97841×10^{-2}	4.0159×10^{-2}	5.4254×10^{-3}
FD (6)	7.6022×10^{-3}	6.7267×10^{-3}	3.0713×10^{-1}
FD (7)	1.3681×10^{-2}	9.9319×10^{-3}	9.2452×10^{-2}
FD (8)	3.5779×10^{-2}	2.1873	3.2864×10^{-1}
FD (9)	1.8365×10^{-3}	3.3338×10^{-2}	1.1528×10^{-1}
FD (10)	9.3619×10^{-4}	5.1582×10^{-3}	2.0135

Table 15.5 Normalized Fourier Descriptor Values Determined from Coordinates of Boundary Pixels of Different Object Shapes

Normalized FD	*Circle*	*Square*	*Rectangle*
FD (1)	6.2517×10^{-1}	1.3259×10^{-1}	2.8569×10^{-1}
FD (2)	2.5994×10^{-2}	1.0084×10^{-1}	3.2409×10^{1}
FD (3)	8.2717×10^{-2}	3.2191×10^{-2}	3.6724×10^{-1}
FD (4)	3.5093×10^{-1}	6.9233	1.2739×10^{1}
FD (5)	2.9277×10^{-1}	1.1068×10^{-1}	4.1084×10^{-2}
FD (6)	7.8689×10^{-2}	4.4566×10^{-2}	3.8073×10^{-1}
FD (7)	3.9899×10^{-2}	8.0952×10^{-2}	1.8633×10^{-2}
FD (8)	8.2338×10^{-3}	2.1762	8.4410×10^{-1}
FD (9)	1.5603×10^{-1}	3.5419×10^{-2}	3.2279×10^{-1}
FD (10)	1.8422×10^{-1}	7.5384×10^{-2}	2.6833

as a circle, ellipse, square, and rectangle were created using MATLAB programs. The objects were created with a known number of area, perimeter, radius pixels, and position of the centroids (Figure 15.14). The area, radius, perimeter, and minor and major axis lengths and the centroid coordinates were verified with the values extracted by the program. "ImageTool" (2001) software was also used to verify the features extracted by the program. The program was then tested on irregular shaped objects created using Paint Shop Pro and verified with the results obtained using the "ImageTool" software (Figure 15.15). The mean RGB and HSI values were verified by creating images of known RGB and HSI values created using the Paint Shop Pro software (Figure 15.16). The range of RGB values was verified using images filled with linear shades of different colors (Figure 15.16).

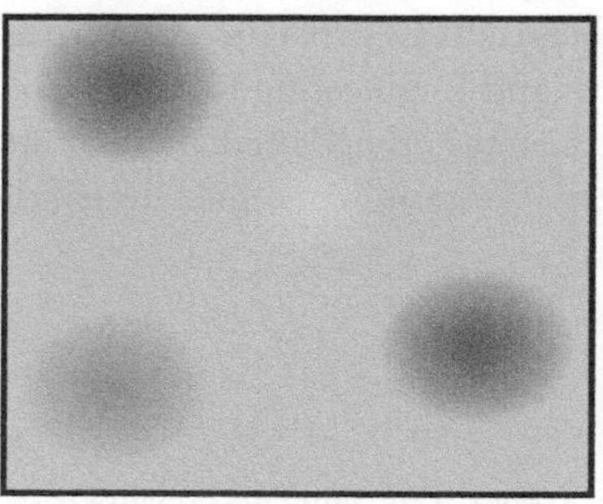
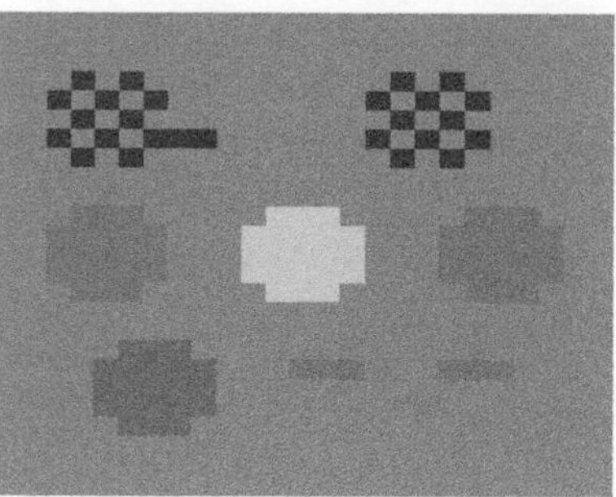

Figure 15.18 Images for testing the textural features.

The invariant moment values were checked for their accuracy with the calculated values for binary images of circle, square, and rectangular shapes of different sizes and orientations (Figure 15.17). The first four invariant moments for different shapes of binary images are listed in Table 15.3.

The FDs of different orders were extracted using the same images used for determining invariant moments. The zero-order FD derived from the radius of an object gives the mean radius. Hence, this value of FD of zero-order was compared with the mean radius of the object. The correct value of an FD of zero-order implies that the algorithm was correctly implemented in the program. The normalized FD values of different orders determined from the radius and coordinates of the boundary pixels are listed in Tables 15.4 and 15.5. The co-occurrence and run length matrix values were verified by using images of defined objects with known color and pixel lengths (Figure 15.18).

15.7 Summary

The demand for high quality products by consumers has increased the need for automated inspection of large quantities of products. Machine vision systems are more consistent and faster than human inspectors in quality evaluation of products. Machine vision systems and common image processing algorithms used for the quality evaluation of food products and classification of cereal grains are described using different examples. Algorithms for extracting size, shape, and colour features; shape and colour moments; Fourier descriptors; and textural features using co-occurrence and run length matrices for the colour and gray scale images of objects and testing procedures for the image processing algorithms are discussed in detail using digital images and examples.

Acknowledgments

We thank the Natural Sciences and Engineering Research Council of Canada for partial funding of this study and Li Wang for her technical assistance. The Canadian Light Source is funded by the Natural Sciences and Engineering Research Council of Canada, the National Research Council Canada, the Canadian Institutes of Health Research, the Province of Saskatchewan, Western Economic Diversification Canada, and the University of Saskatchewan.

References

Barcelon, E.G., S. Tojo, and K. Watanabe. 1999. X-ray CT imaging and quality detection of peach at different physiological maturity. *Transactions of the ASAE* 42(2):435–441.

Bishop, C.M. 1997. *Neural Networks for Pattern Recognition*. New York: Oxford University Press.

Blasco, J., N. Aleixos, J. Gomez, and E. Molto. 2007. Citrus sorting by identification of the most common defects using multispectral computer vision. *Journal of Food Engineering* 83:384–393.

Blasco, J., N. Aleixos, and E. Molto. 2003. Machine vision system for automatic quality grading of fruit. *Biosystems Engineering* 85(4):415–423.

Brecht, J.K., R.L. Schewfelt, J.C. Garner, and E.W. Tollner. 1991. Using X-ray computed tomography to nondestructively determine maturity of green tomatoes. *HortScience* 26:45–47.

Brosnan, T. and D. Sun. 2004. Improving quality inspection of food products by computer vision–A review. *Journal of Food Engineering* 61:3–16.

Casasent, D. and M. Sipe. 1996. Neural net classification of X-ray pistachio nut data. In *Proceedings of SPIE—The International Society for Optical Engineering*, 2907:217–227. Bellingham, WA: The Society of Photo-Optical Instrumentation Engineers.

Chen, K., X. Sun, Ch. Qin, and X. Tang. 2010. Color grading of beef fat by using computer vision and support vector machine. *Computers and Electronics in Agriculture* 70:27–32.

Choudhary, R., J. Paliwal, and D.S. Jayas. 2008. Classification of cereal grains using wavelet, morphological, colour and textural features of non-touching kernel images. *Biosystems Engineering* 99(3):330–337.

Davies, E.R. 2000. *Image Processing for the Food Industry*. New York: World Scientific Publishing.

Delwiche, S.R. 2003. Classification of scab- and other mold damaged wheat kernels by near-infrared spectroscopy. *Transactions of the ASAE* 46(3):731–738.

Diener, R.G., J.P. Mitchell, and M.L. Rhoten. 1970. Using X-ray image scan to sort bruised apples. *Agricultural Engineering* 51(6):356–357, 361.

Finney, E.E. and K.H. Norris. 1978. X-ray scans for detecting hollow heart in potatoes. *American Potato Journal* 55:95–105.

Galloway, M.M. 1975. Textural analysis using gray level run lengths. *Computer Vision Graphics and Image Processing* 4:172–179.

Ghate, S.R., M.D. Evans, C.K. Kvien, and K.S. Rucker. 1993. Maturity detection in peanuts (*Arachis Hypogaea L.*) using machine vision. *Transactions of the ASAE* 36(6):1941–1947.

Ghazanfari, A. and J. Irudayaraj. 1996. Classification of pistachio nuts using a string matching technique. *Transactions of the ASAE* 39(3):1197–1202.

Gonzalez, R.C. and R.E. Woods. 1998. *Digital Image Processing*, 2nd edn. New York: Addison-Wesley.

Gunasekaran, S. 2001. *Nondestructive Food Evaluation—Techniques to Analyze Properties and Quality*, 1st ed. New York: Marcel Dekker.

Gurney, K. 1997. *An Introduction to Neural Networks*, 1st edn. London: UCL Press.

Haff, R.F. and D.C. Slaughter. 1999. X-ray inspection of wheat for granary weevils. ASAE Paper No. 99-3060. St. Joseph, MI: ASAE.

Haff, R.P. and N. Toyofuku. 2008. X-ray detection of defects and contaminants in the food industry. *Sensing and Instrumentation for Food Quality and Safety* 2:262–273.

Han, Y.J., S.V. Bowers III, and R.B. Dodd. 1992. Nondestructive detection of split-pit peaches. *Transactions of the ASAE* 35(6):2063–2067.

Haralick, R.M. 1979. Statistical and structural approaches to texture. *Proceedings of the IEEE* 67:786–804.

Haralick, R.M. and L. Shapiro. 1992. *Computer and Robot Vision*. Reading, MA: Addison-Wesley.

Heinemann, P.H., R. Hughes, C.T. Morrow, H.J. Sommer III, R.B. Beelman, and P.J. Wuest. 1994. Grading of mushrooms using a machine vision system. *Transactions of the ASAE* 37(5):1671–1677.

Howarth, M.S., J.R. Brandon, S.W. Searcy, and N. Kehtarnavaz. 1992. Estimation of tip shape for carrot classification by machine vision. *Journal of Agricultural Engineering Research* 53:123–139.

Humburg, D.S. and J.F. Reid. 1990. Field performance of machine vision for the selective harvest of asparagus. ASAE Paper No: 90-7523. St. Joseph, MI: ASAE.

ImageTool. 2001. Developed by the University of Texas Health Science Center at San Antonio, Texas. Available at http://ddsdx.uthscsa.edu/dig/itdesc.html.

Jayas, D.S., J. Paliwal, and N.S. Visen. 2000. Multi-layer neural networks for image analysis of agricultural products. *Journal of Agricultural Engineering Research* 77(2):119–128.

Karunakaran, C., D.S. Jayas, and N.D.G. White. 2000. Detection of insect infestations in wheat kernels using soft X-rays. CSAE/SCGR Paper No. AFL122. Mansonville, QC: CSAE/SCGR.

Karunakaran, C., D.S. Jayas, and N.D.G. White. 2003a. X-ray image analysis to detect infestations caused by insects in grain. *Cereal Chemistry* 80(5):553–557.

Karunakaran, C., D.S. Jayas, and N.D.G. White. 2003b. Soft X-ray inspection of wheat kernels infested by *Sitophilus oryzae*. *Transactions of the ASAE* 46(3):739–745.

Karunakaran, C., D.S. Jayas, and N.D.G. White. 2004a. An on-line X-ray system for grain inspection—A future perspective. In *Proceedings of 2004 CIGR International Conference*, Paper No. 30-211A, Beijing, China.

Karunakaran, C., D.S. Jayas, and N.D.G. White. 2004b. Detection of internal wheat seed infestation by *Rhyzopertha dominica* using X-ray imaging. *Journal of Stored Products Research* 40:507–516.

Kavdir, I. and D.E. Buyer. 2008. Evaluation of different pattern recognition techniques for apple sorting. *Biosystems Engineering* 99:211–219.

Keagy, P.M., T.F. Schatzki, L. Le, D. Casasent, and D. Weber. 1996. Expanded image database of pistachio X-ray images and classification by conventional methods. In *Proceedings of SPIE—The International Society for Optical Engineering*, 2907:196–204. Bellingham, WA: The Society of Photo-Optical Instrumentation Engineers.

Kim, S. and T.F. Schatzki. 1999. Detection of insect damages in almonds. In *Proceedings of SPIE—The International Society for Optical Engineering*, 3543:101–110. Bellingham, WA: The Society of Photo-Optical Instrumentation Engineers.

Lak, M.B., S. Minaei, J. Amiriparian, and B. Beheshti. 2010. Apple fruits recognition under natural luminance using machine vision. *Advance Journal of Food Science and Technology* 2(6):325–327.

Laykin, S., V. Alchanatis, E. Fallik, and Y. Edan. 2002. Image-processing algorithms for tomato classification. *Transactions of the ASAE* 45(3):851–858.

Leemans, V., H. Magein, and M.F. Destain. 2002. On-line fruit grading according to their external quality using machine vision. *Biosystems Engineering* 83(4):397–404.

Lenker, D.H. and P.A. Adrian. 1971. Use of X-rays for selecting mature lettuce heads. *Transactions of the ASABE* 14(5):894–898.

Lu, R. and D. Ariana. 2002. A near-infrared sensing technique for measuring internal quality of apple fruit. *Applied Engineering in Agriculture* 18(5):585–590.

Luo, X.Y., D.S. Jayas, T.G. Crowe, and N.R. Bulley. 1997. Evaluation of light sources for machine vision. *Canadian Agricultural Engineering* 39(4):309–315.

Luo, X., D.S. Jayas, and S.J. Symons. 1999. Identification of damaged kernels in wheat using a color machine vision system. *Journal of Cereal Science* 30(1):49–59.

Majumdar, S. and D.S. Jayas. 2000a. Classification of cereal grains using machine vision: I. Morphology models. *Transactions of the ASAE* 43(6):1669–1675.

Majumdar, S. and D.S. Jayas. 2000b. Classification of cereal grains using machine vision: II. Color models. *Transactions of the ASAE* 43(6):1677–1680.

Majumdar, S. and D.S. Jayas. 2000c. Classification of cereal grains using machine vision: III. Texture models. *Transactions of the ASAE* 43(6):1681–1687.

Majumdar, S. and D.S. Jayas. 2000d. Classification of cereal grains using machine vision: IV. Morphology, color, and texture models. *Transactions of the ASAE* 43(6): 1689–1694.

Manickavasagan, A. and D.S. Jayas. 2007. Infrared thermal imaging for agricultural and food applications. *Stewart Postharvest Review* 3(5):1–8.

Manickavasagan, A., D.S. Jayas, and N.D.G. White. 2008. Thermal imaging to detect infestation by *Cryptolestes ferrugineus* inside wheat kernels. *Journal of Stored Products Research* 44(2):186–192.

Manickavasagan, A., D.S. Jayas, N.D.G. White, and J. Paliwal. 2010. Wheat class identification using thermal imaging. *Food and Bioprocess Technology* 3:450–460.

McClure, J.E. and C.T. Morrow. 1987. Computer vision sorting of potatoes. ASAE Paper No: 87-6501. St. Joseph, MI:ASAE.

Mebatsion, H.K., J. Paliwal, and D.S. Jayas. 2012. A novel, invariant elliptic Fourier coefficient based classification of cereal grains. *Biosystems Engineering* 111:422–428.

Miller, B.K. and M.J. Delwiche. 1989. A color vision system for peach grading. *Transactions of the ASAE* 32(4):1484–1490.

Miller, B.K. and M.J. Delwiche. 1991. Peach defect detection with machine vision. *Transactions of the ASAE* 34(6):2588–2597.

Morrisey, M.M. 1988. (HSI) Color processing on personal computers. *Applications of Digital Image Processing,* 974:173–176. Bellingham, WA: SPIE—The International Society for Optical Engineering.

Neethirajan, S., C. Karunakaran, S. Symons, and D.S. Jayas. 2006. Classification of vitreousness in durum wheat using X-rays and transmitted light images. *Computers and Electronics in Agriculture* 53(1):71–78.

Okamura, N.K., M.J. Delwiche, and J.F. Thompson. 1993. Raisin grading by machine vision. *Transactions of the ASAE* 36(2):485–492.

Paliwal, J., N.S. Visen, and D.S. Jayas. 2001. Evaluation of neural network architectures for cereal grain classification using morphological features. *Journal of Agricultural Engineering Research* 79(4):361–370.

Paliwal, J., N.S. Visen, D.S. Jayas, and N.D.G. White. 2003a. Comparison of a neural network and a non-parametric classifier for grain kernel identification. *Biosystems Engineering* 85(4):405–413.

Paliwal, J., N.S. Visen, D.S. Jayas, and N.D.G. White. 2003b. Cereal grain and dockage identification using machine vision. *Biosystems Engineering* 85(1):51–57.

Paliwal, J., D.S. Jayas, N.S. Visen, and N.D.G. White. 2005. Quantification of variations in machine-vision-computed features of cereal grains. *Canadian Biosystems Engineering* 47:7.1–7.6.

Paliwal, J., W. Wang, S.J. Symons and C. Karunakaran. 2004. Insect species and infestation level determination in stored wheat using near-infrared spectroscopy. *Canadian Biosystems Engineering* 46:7.17–7.24.

Parker, J.R. 1994. *Practical Computer Vision using C.* New York: John Wiley & Sons.

Pearson, T.C. and D.C. Slaughter. 1996. Machine detection of early split pistachio nuts. *Transactions of the ASAE* 39(3):1203–1207.

Pearson, T.C. and T.F. Schatzki. 1998. Machine vision system for automated detection of aflatoxin-contaminated pistachios. *Journal of Agricultural and Food Chemistry* 46:2248–2252.

Pearson, T. and R. Young. 2002. Automated sorting of almonds with embedded shell by laser transmittance imaging. *Applied Engineering in Agriculture* 18(5):637–641.

Rehkugler, G.E. and J.A. Throop. 1986. Apple sorting with machine vision. *Transactions of the ASAE* 35(6):1388–1397.

Rigney, M.P., G.H. Brusewitz, and G.A. Kranzler. 1992. Asparagus defect inspection with machine vision. *Transactions of the ASAE* 35(6):1873–1878.

SAS. 2000. *SAS User's Guide: Statistics.* Cary, NC: SAS Institute Inc.

Schatzki, T.F. and R.Y. Wong. 1989. Detection of submilligram inclusions of heavy metals in processed foods. *Food Technology* 43(11):72–76.

Schatzki, T.F., R.P. Haff, R. Young, I. Can, L.C. Le, and N. Toyofuku. 1997. Defect detection in apples by means of X-ray imaging. *Transactions of the ASAE* 40(5):1407–1415.

Shahin, M.A. and E.W. Tollner. 1997. Apple classification based on water core features using fuzzy logic. ASAE Paper No. 97-3077. St. Joseph, MI: ASAE.

Shatadal, P., D.S. Jayas, J.L. Hehn, and N.R. Bulley. 1995. Seed classification using machine vision. *Canadian Agricultural Engineering* 37(3):163–167.

Shearer, S.A. and F.A. Payne. 1990. Color and defect sorting of bell peppers using machine vision. *Transactions of the ASAE* 33(6):2045–2050.

Singh, C.B., D.S. Jayas, J. Paliwal, and N.D.G. White. 2009a. Detection of insect-damaged wheat kernels using near-infrared hyperspectral imaging. *Journal of Stored Products Research* 45:151–158.

Singh, C.B., D.S. Jayas, J. Paliwal, and N.D.G. White. 2009b. Detection of sprouted and midge-damaged wheat kernels using near-infrared hyperspectral imaging. *Cereal Chemistry* 86(3):256–260.

Singh, C.B., D.S. Jayas, J. Paliwal, and N.D.G. White. 2010a. Identification of insectdamaged wheat kernels using short-wave near-infrared hyperspectral and digital color imaging. *Computers and Electronics in Agriculture* 73:118–125.

Singh, C.B., D.S. Jayas, J. Paliwal, and N.D.G. White. 2010b. Detection of midge-damaged wheat kernels using short-wave near-infrared hyperspectral and digital color imaging. *Biosystems Engineering* 105:380–387.

Singh, C.B., D.S. Jayas, J. Paliwal, and N.D.G. White. 2012. Fungal damage detection in wheat using short-wave near-infrared hyperspectral and digital color imaging. *International Journal of Food Properties* 15:11–24.

Singh, N. and M.J. Delwiche. 1994. Machine vision methods for defect sorting stonefruit. *Transactions of the ASAE* 37(6):1989–1997.

Sun, D. 2008. *Computer Vision Technology for Food Quality Evaluation.* San Diego, CA: Academic Press.

Tao, Y., Z. Chen, H. Jing, and J. Walker. 2001. Internal inspection of deboned poultry using X-ray imaging and adaptive thresholding. *Transactions of the ASAE* 44(4):1005–1009.

Tao, Y., P.H. Heinemann, Z. Varghese, C.T. Morrow, and H.J. Sommer III. 1995b. Machine vision for color inspection of potatoes and apples. *Transactions of the ASAE* 38(5):1555–1561.

Tao, Y., C.T. Morrow, P.H. Heinemann, and H.J. Sommer III. 1995a. Fourier-based separation technique for shape grading of potatoes using machine vision. *Transactions of the ASAE* 38(3):949–957.

Tao, Y. and Z. Wen. 1999. An adoptive spherical image transform for high-speed fruit detection. *Transactions of the ASAE* 42(1):241–246.

Taylor, R.W., G.E. Rehkugler, and J.A. Throop. 1984. Apple bruise detection using digital line scan camera system. In *Agricultural Electronics—1983 and Beyond. Volume II*. St. Joseph, MI: ASAE.

Thomas, W.V. and C. Connoly. 1986. Applications of color processing in optical inspection. *Applications of Digital Image Processing*, 654:116–122. Bellingham, WA: SPIE—The International Society for Optical Engineering.

Tollner, E.W. 1993. X-ray technology for detecting physical quality attributes in agricultural produce. *Postharvest News and Information* 4(6):149N–155N.

Unser, M. 1986. Sum and difference histograms for texture classification. *IEEE Transactions on Pattern Analysis and Machine Intelligence* PAMI-8:118–125.

Vande Vooren, J.G., G. Polder, and G.W.A.M. Vander Heijden. 1992. Identification of mushroom cultivars using image analysis. *Transactions of the ASAE* 35(1):347–350.

Varith, J., G.M. Hyde, A.L. Baritelle, J.K. Fellman, and T. Sattabongkot. 2003. Non-contact bruise detection in apples by thermal imaging. *Innovative Food Science and Emerging Technologies* 4:211–218.

Visen, N.S., D.S. Jayas, J. Paliwal, and N.D.G. White. 2004. Comparison of two neural network architectures for classification of singulated cereal grains. *Canadian Biosystems Engineering* 46:3.7–3.14.

White, D.J., C. Svellingen, and N.J.C. Strachan. 2006. Automated measurement of species and length of fish. *Fisheries Research* 80:203–210.

Wigger, W.D., M.R. Paulsen, J.B. Litchfield, and J.B. Sinclair. 1988. Classification of fungal-damaged soyabeans using colour image processing. ASAE Paper No. 88-3053. St. Joseph, MI:ASAE.

Wolfe, R.R., V.N. Ruzhitsky, and C.A. Hoernlein. 1992. Measurement of surface areas of interest such as bell pepper coloration using multiple orthogonal images. *Transactions of the ASAE* 35(5):1723–1727.

Xing, J., W. Saeys, and J. Baerdemaeker. 2007. Combination of chemometric tools and image processing for bruise detection on apples. *Computers and Electronics in Agriculture* 56:1–13.

Zayas, I., H. Converse, and J. Steele. 1990. Discrimination of whole from broken corn kernels with image analysis. *Transactions of the ASAE* 33:1642–1646.

Zhang, H., J. Paliwal, D.S. Jayas, and N.D.G. White. 2007. Classification of fungal infected wheat kernels using near-infrared reflectance hyperspectral imaging and support vector machines. *Transactions of the ASABE* 50(5):1779–1785.

Chapter 16

Vibrational Spectroscopy for Food Processing

Wenbo Wang and J. Paliwal

Contents

16.1 Introduction

Vibrational spectroscopy, which probes energy transitions due to molecular vibrations, is a valuable tool for quality control and screening in food processing. The molecular vibrations give rise to absorption bands in the infrared region of the electromagnetic spectrum. With incident photons of the right frequency, the vibrational energy of molecules could be promoted to the next allowed energy level. These energy transitions are studied by popular absorption spectroscopy techniques, such as mid-infrared (MIR) and near-infrared (NIR) spectroscopy. Besides absorption techniques, Raman spectroscopy, which is based on an inelastic scattering effect, has also proven to be a rather useful technique in food studies. Raman and infrared (IR) techniques are complementary techniques because their selection rules for molecular vibration are different. However, Raman shift and MIR absorption share the same energy range. One fundamental difference between absorption and Raman scattering is probability, with the Raman effect being far less likely to happen, which makes Raman a weak phenomenon (McCreery 2000). This lack of sensitivity can be remediated with special techniques, such as resonance Raman or surface-enhanced Raman scattering (SERS).

Both the frequency and intensity of molecular vibrations induced by external radiation are subject to the chemistry and environment around the individual atoms. Thus, vibrational spectroscopy can be used to study chemical compositions of various food materials and products (Li-Chan 1996). There exist a number of organic compound and functional groups that are active to Raman and IR techniques. Each functional group can be identified by its unique pattern of either absorption or Raman scattering. The intensity of these spectral features can be used to calculate the relative concentration in the sample (Wetzel and LeVine 1999). Therefore, both qualitative and quantitative information of the target sample are obtainable using vibrational spectroscopy. The spectroscopic methods hold their distinct advantages over chemical analysis as a routine characterization tool. One significant advantage is that vibrational spectroscopy takes samples with little preparation. Types of samples classically measured by vibrational spectroscopy include gases, liquids, and bulk and powdered solids. Nowadays, samples, such as interfacial species, microsamples, and trace analyte, can also be analyzed on modern spectrometers. As routine analytical methods, different branches of vibrational spectroscopy have achieved amazingly short measurement time varying from minutes to fractions of seconds (Griffiths and de Haseth 2007).

This chapter starts with a brief description of the fundamental theory and principles behind three vibrational techniques, i.e., MIR spectrometry, NIR spectroscopy, and the Raman technique. From the practical aspects, to measure a sample, an incident beam is projected onto the surface of a sample and gets transmitted, absorbed, reflected, and scattered and may also cause emission. The basic components of a spectroscopic instrument comprise a light source, wavelength separating device, detector, and sampling accessories. The spectrometer designs have so many variations that an entire book could be dedicated to the topic. In most literature, those prevailing designs are primarily categorized into dispersive and nondispersive instruments. Therefore, it is the purpose of the authors to describe the basic components instead of specific designs in this chapter. The steps to interpret spectral data are as important as having an instrument capable of acquiring quality data. In most cases, the true spectra are often mixed with background signal and noises. There is possible interference due to the physical properties of a sample, such as multiplicative scattering. Sometimes, the spectra might not have enough spectral resolution or signal-to-noise ratio. All these unfavorable aspects could be improved by pretreating spectral data before subjecting them to further analysis. As mentioned earlier, vibrational spectroscopy is capable of qualitative and quantitative analysis. A number of chemometric tools favorite to the spectroscopists are discussed later in the chapter. To demonstrate the utility and potential of vibrational spectroscopy, a non-exhaustive review of recent publications on the application of vibrational spectroscopy in the agriculture and food industries is provided.

16.2 Theory and Principles

For transmission measurement by IR spectroscopy, the sample has to be prepared as a thin film with a thickness of about 10 μm. An IR spectrum of such thin film could yield absorption bands that are neither saturated nor too weak. The utilization of attenuated total reflection (ATR) for sampling IR spectrum is the most widely practiced technique in IR spectroscopy. Although not foolproof, the technique does not require much sample preparation while consistent results could be easily obtained. On the practical side, the internal reflection element (IRE) is susceptible to contamination by chemical agents, such as hand lotion, silicon grease, or even greasy fingerprints, etc. Therefore, the possibility of sampling real-world samples using ATR FT-IR seems to be restrained by practical conditions imposed on the ATR devices. In cases like that, NIR spectra are sampled instead because absorption of NIR light is weak, which allows the light to penetrate deeper into the sample. The reason that NIR absorption is weak is that it measures overtones, combinations, and difference bands instead of fundamentals. As a matter of fact, overtones, combinations, and difference bands are weak spectral features compared to fundamentals. For example, the first overtones and combinations are usually 10 to 100 times weaker than the fundamentals. Higher-order overtones and combinations are even weaker. Although the NIR region is sometimes called the overtone region, many of the overtone and combination bands absorb in the IR region. Only the overtones of O-H, N-H, C-H, and S-H stretching modes along with the combination bands originating from the stretching-bending modes of these bonds are found in the region above 4000 cm^{-1}, i.e., the NIR range.

The IR spectroscopy studies only one transition, which is the direct transition between two energy levels through the process of absorption. The Raman scattering process involves two transitions and three different vibrational states. To ensure that the Raman effect occurs, an intermediate level called the virtual state with a higher level of energy than the initial level and a final level of transition must exist. Otherwise, the transition defined in the Raman process is forbidden. In Figure 16.1, the absorption and scattering processes are illustrated. In Raman spectroscopy, the sample is illuminated with a monochromatic beam of intense radiation, which is usually a laser source. The frequency of the excitation radiation is much higher than the vibrational frequencies encountered in IR spectroscopy but is usually lower than the electronic frequencies. All incident photons that interact with the sample molecules increase the vibrational and rotational energy of the molecules to the virtual state by the amount of energy equal to $h\nu_0$, where ν_0 is the frequency of incident photons. Immediately after the transition to the virtual state, most molecules involved in the interaction return to the ground state through the process of emission of photons at the same frequency of the incident photons. This process is called Rayleigh scattering and is an elastic collision process between the incident photon and the molecule. The energy of the molecules that undergo Rayleigh scattering is unchanged, and the scattered photon has identical frequency with the incident photon. Rayleigh scattered photons account for the largest portion of all scattered photons. The Raman effect is an inelastic collision process between the incident photons and the sample

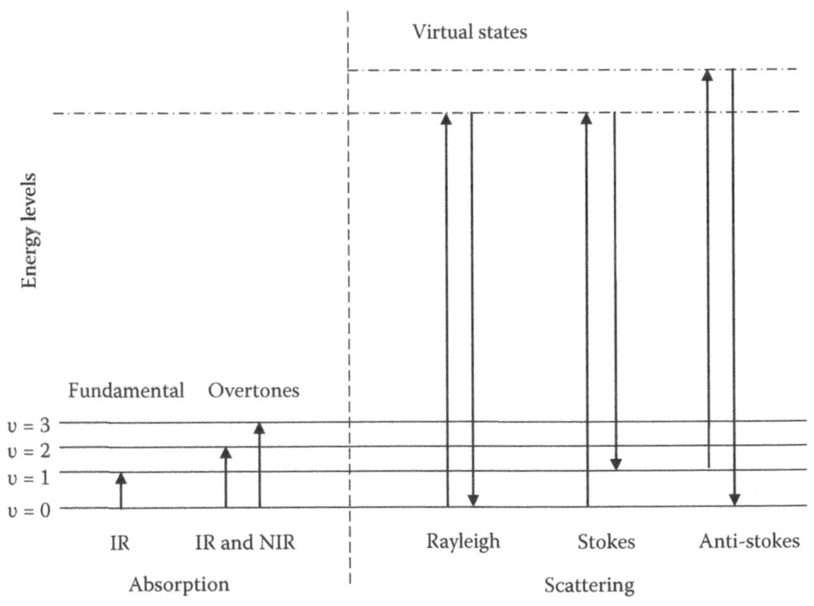

Figure 16.1 Absorption and scattering processes during vibration energy transitions.

molecule. During the inelastic scattering, the energy of the incident photons and the vibrational-rotational energy of the molecule are altered. A small fraction of the incident photons loses energy as a result of a small proportion of molecules falling to the first excited energy level instead of the ground energy level. Correspondingly, the scattered photons with less energy appear in the Raman spectrum as a source of emitted radiation with increased wavelength. This process is known as Stokes Raman scattering, and the spectral lines are called Stokes lines. Similarly, it is also possible for some of the molecules already in an excited vibrational state to be exposed to the laser radiation and promoted to a virtual level with unstable high energy and then returned to the ground state. This type of inelastic collision is called anti-Stokes Raman scattering. In anti-Stokes Raman scattering, the molecules lose energy, which gives rise to spectral lines of increased frequency compared to excitation radiation in the Raman spectrum. According to Boltzmann distribution, most molecules exist in the ground state instead of excited states at ordinary temperatures. Therefore, Stokes lines have greater intensities than anti-Stokes lines because the excited energy level has a lower population compared to the ground state.

16.2.1 Molecular Vibration

Energy transfer among the molecular vibration states contributes to the absorption lines and bands throughout most of the infrared region of the spectrum. The infrared region is further divided into the near-infrared, mid-infrared, and far-infrared regions according to different wavelength ranges. The corresponding frequencies in Hz, which is usually large, are commonly replaced by wavenumber (cm^{-1}). Near-infrared covers the region of 780–2500 nm (12,800–4000 cm^{-1}). Mid-infrared, which is commonly referred to as IR, covers the region of 2500–50,000 nm (4000–200 cm^{-1}). Far-infrared covers the region of 0.05–1 mm (200–10 cm^{-1}). The vibrational frequencies of molecules occupy the IR region. Such frequency features relating to the molecular structures could be studied using IR absorption spectroscopy and Raman spectroscopy.

Suppose there are N atomic nuclei in a molecule; then there exists $3N$ degrees of freedom of motion in total for all the nuclear masses in that molecule. Taking out the total independent degrees of freedom from the total of $3N$ degrees of freedom for a molecule, we obtain $3N - 6$ internal degrees of freedom for a nonlinear molecule and $3N - 5$ internal degrees of freedom for a linear molecule. It was shown that the number of internal degrees of freedom corresponds to the number of independent normal modes of vibration for the molecule. Considering the simple model for a diatomic molecule by connecting two masses m_1 and m_2 using a massless spring, the classical vibrational frequency formula could be derived. It is a simple and straightforward way to look at the molecular vibration using such a mechanical model that describes the diatomic molecule as two masses attached to ends of a spring with a force constant (k). Any pair of bound atoms in a molecule exhibits a natural oscillation and vibrates at a specific frequency given by

$$v = \frac{1}{2\pi}\sqrt{\frac{k}{\mu}} \tag{16.1}$$

where k is the force constant (N m^{-1}) and μ (kg) is the reduced mass given by

$$\mu = \frac{m_1 m_2}{m_1 + m_2} \tag{16.2}$$

where m_1 and m_2 are two masses in a diatomic molecule (kg).

According to the mechanical model, the vibrating system only absorbs energy from externally applied forces that have an oscillating frequency matching that of the natural frequency of the ball and spring system. External forces at non-resonant frequencies do not increase the energy of the system. Similar to the external forces applied to the ball and spring system, the infrared radiation absorbed by the molecule must match specific frequencies to increase the molecular vibration energy. Therefore, when irradiating samples with incident light having a continuous spectrum in the infrared, only radiation that matches the vibrational frequencies of molecules in the samples is absorbed. When a spectrometric system is used to receive the transmitted infrared photons, a spectrum with absorption features at specific frequencies is recorded. The absorption intensities at the natural vibrational frequencies reflect the effectiveness of energy transfer from external IR radiation to the molecular vibration system.

16.2.2 Anharmonicity, Overtones, Combination, and Difference

A simplified diatom vibration model treats the molecular vibration as harmonic. The diatom system could only be used to explain fundamental vibrations of molecules. In the real molecular system, the repulsive and attractive forces between atoms are not linearly proportional to the nuclear displacement coordinate. This introduces mechanical anharmonicity. Similarly, if the dipole moment changes in a nonlinear way with the change in nuclear displacement coordinates, electrical anharmonicity is introduced into the system. Either mechanical or electrical anharmonicity will cause the dipole moment to change periodically with time. The periodic wave could be decomposed into simple sine terms with frequencies equal to integral multiples of the fundamental vibrational frequencies of the molecule. Therefore, the dipole moment oscillates at the fundamental frequency and its integral multiples for anharmonic vibration of the molecule. The integral multiples are referred to as the first overtone, second overtone, etc.

16.2.3 Vibrational Potential Energy, Quantum Effect

According to the harmonic vibration model of a diatomic molecule, the potential energy is a function of the internuclear distance. The origin of the internuclear potential energy could be explained by the existence of electronic and nuclear repulsion energy. When two atoms move closer to each other, the internuclear distance becomes smaller, resulting in internuclear repulsion. Vice versa, when two atoms move away from each other, the resulting larger internuclear distance leads to two atomic components attracting each other. At an equilibrium distance r_e, vibrational potential energy is at its minimum. The potential energy changes tend to level off when the internuclear distances become large. When the distance grows large enough, the force breaks the chemical bond and dissociates the molecule. For internuclear distances around the equilibrium distance r_e for a diatomic model, the harmonic oscillator of potential function provides a good first approximation. The energy could be expressed by

$$E = \frac{1}{2}k(r - r_e)^2 \tag{16.3}$$

where E is the potential energy, k is a constant that depends on the strength of the molecular force field holding two atoms, and r is the internuclear distance.

Further away from the equilibrium position, the Coulomb repulsion between the two nuclei causes the potential energy to rise rather rapidly than the prediction from the harmonic equation. On the other hand, the potential energy levels off when the internuclear distance approaches dissociation spacing. An empirical function, due to Morse, offers a better approximation:

$$E = E_D\left(1 - e^{-\beta(r - r_e)}\right)^2 \tag{16.4}$$

where E is the potential energy, E_D is the dissociation energy, and β is a constant.

Using a Taylor series to expand the equation, the following equation is derived:

$$E = E_e + \left(\frac{\partial E}{\partial S}\right)_e S + \frac{1}{2!}\left(\frac{\partial^2 E}{\partial S^2}\right)_e S^2 + \frac{1}{3!}\left(\frac{\partial^3 E}{\partial S^3}\right)_e S^3 + \ldots \tag{16.5}$$

where S is the internal coordinate, i.e., $r - r_e$. If starting from the term S^3, the higher-order terms in the expansion are neglected, and the term $(\partial^2 E/\partial S^2)_e$ is replaced with a force constant k, the harmonic approximation equation in Equation 16.5 is obtained.

From quantum mechanics theory, several conclusions are introduced to explain the harmonic molecular oscillation. Using Schrödinger's generalized wave equation and assuming constant potential energy, it could be derived that molecular energy only has discrete values, and the quantum number could only allow integral numbers. That means the molecular energy is not continuous but quantized instead, which is a consequence of the wave-particle duality of matter. The same restriction, i.e., energy levels are quantized, applies to the molecular rotation as well. Another conclusion is that the momentum can never be zero for the oscillating particle in the model because wavelength cannot be infinite. Therefore, as part of molecular energy, vibrational energy can only have discrete values, and the lowest possible value cannot be zero. The discrete energy levels are described in the following equation:

$$E = \left(\upsilon + \frac{1}{1} \right) h\nu \tag{16.6}$$

where E is the discrete energy level value; υ is the quantum number, which is only allowed to have integer values (i.e., 0, 1, 2, …); h is the Planck's constant (6.63×10^{-34} J s); and ν is the classical vibration frequency. When the quantum number equals zero, the oscillator possesses the lowest vibrational energy of $h\nu/2$.

Making the necessary adaptation to the anharmonic oscillator equations according to the quantum mechanics theory yields the following equation:

$$E = \left(\upsilon + \frac{1}{2} \right) h\nu - \left(\upsilon + \frac{1}{2} \right)^2 xh\nu + \ldots \tag{16.7}$$

where x is called the anharmonicity constant, which is dimensionless and has a value between 0.001 and 0.02. One direct result of the anharmonic oscillator model is that energy levels are not equally spaced, and frequency is not completely independent of vibration amplitudes. Therefore, overtone transitions, e.g., from $\upsilon = 0$ to $\upsilon = 2$ or 3, which are allowed in the anharmonic model, do not appear at exactly two or three times the fundamental frequency.

16.2.4 Polarizability

A classical theoretical treatment of Raman and Rayleigh scattering involves the fact that scattered photons are the results of an oscillating dipole moment of the molecule induced by the electric field of the incident light wave. When a molecule is subject to the electric field of incident radiation, its protons and electrons receive electric forces in the opposite directions. Such oppositely directed forces cause a change in the charge spacing, which induces a dipole moment for the polarized molecule. Polarizability is used to describe the relationship between the electric field of the incident light wave and the induced molecular dipole moment. According to Equation 16.8, polarizability is defined as the induced dipole moment divided by the electric field strength that leads to the induced dipole moment.

$$\alpha = \frac{\mu}{E} \tag{16.8}$$

where α is the polarizability, μ is the induced dipole moment, and E is the strength of the electric field for the incident radiation. The molecular polarizability is not a constant, and its value depends on the displacements the electrons make relative to the protons during vibrations and rotations. The dependence of polarizability on normal coordinates Q could be expressed in an expanded form using the Taylor series.

$$\alpha = \alpha_0 + \left(\frac{\partial\alpha}{\partial Q}\right)Q + \frac{1}{2!}\left(\frac{\partial^2\alpha}{\partial Q^2}\right)Q^2 + \frac{1}{3!}\left(\frac{\partial^3\alpha}{\partial Q^3}\right)Q^3 + \ldots \tag{16.9}$$

where α_0 is the polarizability near equilibrium, Q is the normal coordinate, and $\partial\alpha/\partial Q$ is the molecular polarizability changing rate with respect to Q taken under the equilibrium condition. For small amplitude vibrations near the equilibrium, higher order terms in Equation 16.9 are neglected, and linear dependence of polarizability on Q is assumed.

$$\alpha = \alpha_0 + \left(\frac{\partial\alpha}{\partial Q}\right)Q \tag{16.10}$$

The normal vibrations are treated as harmonic and the normal coordinate Q varies periodically according to

$$Q = Q_0 \cos 2\pi v_\upsilon t \tag{16.11}$$

where Q_0 is the maximum for Q, ν_υ is the frequency for the normal coordinate vibration, and t is time. Substituting Q in Equation 16.10 with that in Equation 16.11, the polarizability is expressed as

$$\alpha = \alpha_0 + \left(\frac{\partial\alpha}{\partial Q}\right)Q_0 \cos 2\pi v_\upsilon t \tag{16.12}$$

Now consider the time domain. The induced dipole moment caused by the electric field of the incident radiation varies with time. The strength of electric field around the molecule varies with time and is given by

$$E = E_0 \cos 2\pi v t \tag{16.13}$$

where E_0 is the maximum value of the electric field, ν is the frequency of the radiation, and t is time. Under the influence of the electric field of the incident radiation, the time dependence of induced dipole moment is expressed as Equation 16.14 by inserting Equation 16.13 into Equation 16.12:

$$\mu = \alpha_0 E_0 \cos 2\pi v t + \frac{\partial \alpha}{\partial Q} Q_0 E_0 (\cos 2\pi v t)(\cos 2\pi v_\upsilon t) \tag{16.14}$$

Using trigonometric identity on the second term containing two cosine functions in Equation 16.14, the following is obtained:

$$\mu = \alpha_0 E_0 \cos 2\pi v t + \frac{1}{2}\frac{\partial \alpha}{\partial Q} Q_0 E_0 [\cos 2\pi (v - v_\upsilon)t + \cos 2\pi (v + v_\upsilon)t] \tag{16.15}$$

From Equation 16.15, there are three cosine functions with three different component frequencies, i.e., ν, $(\nu + \nu_\upsilon)$, and $(\nu - \nu_\upsilon)$. In the classical treatment, the polarized induced dipole radiates light at the frequency of their oscillation. Therefore, Equation 16.15 predicts three distinct frequencies for the simultaneous scattering process. The first term describes Rayleigh scattering, which is observed at the frequency of incident radiation ν. The second and the third terms give rise to anti-Stokes and Stokes Raman scattering. It is noted that ν_υ is the frequency that could be detected using IR absorption spectroscopy for the quantum mechanical transition when $\Delta\upsilon = \pm 1$ if allowed by the molecular symmetry. The dependence of the second and third terms on $\partial\alpha/\partial Q$ could explain the selection rules for molecular vibrations to be Raman active. To yield Raman scattering at the Stokes and anti-Stokes frequencies, the molecular vibrations must change the polarizability, i.e., $\partial\alpha/\partial Q \neq 0$. Otherwise, the terms predicting Raman component frequencies in Equation 16.15 would have zero amplitude, which indicates no Raman scattering can be generated.

16.3 Instrumentation

Despite the popularity of NIRS today, spectroscopic study in the mid-infrared region first gained wide acceptance. For the first few decades in the early 20th century, mid-infrared spectroscopy continued to receive constant research and instrumentation development. On the other hand, study of the light absorption in the NIR region was largely neglected by researchers in optical spectroscopy. Until the 1950s, only a few NIR spectroscopic systems were investigated, and the application of NIRS did not begin until the computer was invented (Davies 1998). In year 1983, Professor David Wetzel suggested that the great potential of NIRS had long been neglected, describing NIRS as the sleeper among spectroscopic techniques (Wetzel 1983). In the last quarter of the 20th century, owing to the fast development in electronics, photonics, and computer technologies, inexpensive and robust NIRS systems could be made readily available to many industries. Because Raman scattering is a weak phenomenon and competing fluorescence effect, Raman spectroscopy was deemed to be more theoretical than being a useful tool for the spectroscopists before the mid-1980s. It was the renaissance of NIR absorption spectroscopy in about year 1985 that somehow gave analytical Raman spectroscopy a chance of rebirth. A surge of interest in the analytical applications of Raman spectroscopy during the period from 1986 to 1990 was attributed to the introduction of FT-Raman systems, which are modified FTIRs. The technological developments, including novel NIR lasers, Fourier transform techniques, CCD array detectors, and multichannel detectors, largely overcome the two fundamental problems associated with the Raman effect. These renovations actually accelerated an explosion in Raman instrumentation, which resulted in a wide array of configurations suited for various applications. A general overview of major components to the many available instrumentation designs is included below. The choice of components and instrument configuration is often decided by its application and the type of sample to be measured.

16.3.1 Light Source

For the needs of IR radiation to conduct transmission and reflection measurement, the most popular source currently used is a resistively heated silicon carbide rod, which is known as a Globar. The typical operating temperature for a Globar is ~1300 K. The actual spectral energy density of an IR source depends on its operating temperature and the blackbody spectral energy density. Therefore, both the operating temperature and spectral energy density of any incandescent IR source must be as high as possible to maximize the radiation output in the IR spectral region and thus achieve the best SNR of the spectrum. Other factors that also need to be considered are the lifetime and stability of the source, etc. In recent years, some new sources have been proposed, such as the synchrotron and free electron laser (Williams 2002). The tunable IR diode laser is also a possibility as an amazing radiation source in the future. Combined with high-sensitivity detectors and grating spectrometer, the tunable semiconductor lasers can provide real-time spectra with superior spectral resolution and SNR. However, the tunable IR diode laser is still under development at present. The major technical difficulties are insufficient output power and inability of the device to operate under room temperature or thermoelectrically cooled conditions (Mantz 2002).

The choice for radiation source in the NIR range is application dependent. Manley et al. (2008) put the radiation sources into two categories, i.e., thermal or non-thermal. The thermal source can be as simple as an inexpensive tungsten filament light bulb. Non-thermal sources include light emitting diodes (LEDs), emitting diode arrays (EDAs), laser diodes, lasers, and discharge lamps, etc. (McClure 2001).

Thermal radiation sources for today's NIR instruments are mainly quartz-tungsten-halogen (QTH) lamps. The NIR radiation sources run at temperatures much higher than that of IR sources. The color temperature that a QTH lamp operates at is between 2900 and 3400 K. The QTH lamp produces continuous spectra from 300 to 3500 nm, which covers the entire NIR region, making it a suitable radiation choice for FT-NIR system and grating based instruments (DeThomas and Brimmer 2002). Compared to ordinary tungsten-filament light bulbs, QTH lamps are more intense. On the other hand, such intensity could result in problems, such as response nonlinearity.

Light emitting diodes (LEDs) and EDAs consume much less power than the thermal radiation sources. The reason that LEDs and EDAs are more efficient is that most of the energy is converted into radiation emission within a rather limited range of wavelength. Their low power consumption coupled with their compact sizes makes them a perfect radiation source choice for miniaturized non-scanning instruments. As electronic devices, LEDs can be easily turned on and off electronically to switch among wavelengths without having to shift any hardware components, e.g., filters or gratings. The typical spectral bandwidth of LEDs is 50 nm. Therefore, appropriate band-pass filters need to be installed to narrow down the bandwidth to around 12 nm to make it useful for quantitative analysis of samples, such as ground grains.

Lasers are mainly used as light sources for Raman spectroscopy. Martin (2002) listed some of the latest applications of NIR laser absorption spectroscopy. The described applications focused especially on solving environmental problems while the author also pointed out its possible application in other areas, such as chemistry, physics, material science, and medical monitoring. According to McClure and Tsuchikawa (2007), diode lasers are still in the research stage for NIR spectroscopy due to its nontrivial cost. Sato et al. (2003) agreed that NIR lasers, especially tunable NIR lasers, have not been applied in NIR absorption spectroscopy due their operational complexity and cost. For Raman spectroscopy, there exist a wide array of lasers with varying laser wavelengths and operating principles. When considering a laser for Raman analytical application, things such as wavelength, frequency stability, output power, and output line width should be taken into account. A shorter excitation wavelength could mean a stronger Raman signal and better sensitivity. However, it more likely excites fluorescence. Therefore, the laser wavelength should be as long as the required sensitivity permits. Another limiting factor is the detector noise, i.e., using a laser with a longer wavelength, e.g., 1064 nm, could mean a noisier detector compared to a CCD detector using a 785 nm excitation laser. Therefore, lasers of wavelengths less than 900 nm are commonly used for dispersive systems, and lasers with wavelengths of 1064 nm and above are mostly used as light sources for FT-Raman systems. The accepted lasers could be classified as gas lasers, solids-state lasers, and diode lasers (Coates 1998). Gas lasers include helium-neon (HeNe) lasers and Ar+/Kr+ ion lasers. The HeNe laser is less powerful than the Ar+/Kr+, which, however, requires water cooling. A few strong emission lines for the Ar+/Kr+ laser are at 488, 514.5, and 647.1 nm, and the HeNe laser has a well-known output at 632.8 nm. A neodymium-doped yttrium aluminium garnet (Nd:YAG) laser pumped by a diode laser with a 1064 nm fundamental lasing line is widely used for FT-Raman systems. The frequency-doubled Nd:YAG laser of 532 nm is a most popular rare-earth laser used with dispersive Raman systems. Diode lasers usually have relatively long wavelengths in the range from 800 to 1500 nm. They are perfect choices for integrated Raman systems due to their modest power consumption and low cost. However, frequency stability due to temperature effects and mode hopping could restrain the suitability of diode lasers for Raman applications.

16.3.2 Detectors

Infrared detectors are divided into two basic types: thermal detectors and quantum detectors (Lerner 1996). Thermal detectors measure the change in temperature of material that absorbed the total energy from the IR beam. Although they are inexpensive and need little cooling to operate, with a response time of several milliseconds, the thermal detectors are slow for today's FT-IR spectrometers (Theocharous and Birch 2002). There are several types of thermal detectors, i.e., thermocouples, bolometers, pyroelectric detectors, and pneumatic detectors. Quantum detectors rely on the interaction of individual photons with the electrons in the substrate of the detector. Only when the wavelength of the incident radiation is less than a critical value, the energy of the photon could excite the electrons from a valence band to a conduction band. Phototubes and photomultipliers require high energy to excite electrons through photoemission, and they are suitable only for ultraviolet, visible, and short-NIR regions up to 1000 nm. The IR quantum detectors include lead sulfide (PbS), lead selenide (PbSe), indium antiminide (InSb), and mercury cadmium telluride (MCT). Low energy of the photons in the IR region makes the detectors susceptible to random thermal noise and most of these detectors require cooling, either thermoelectrically cooled or using liquid nitrogen. There also exist many detectors for NIR spectroscopy, such as indium gallium arsenide (InGaAs), silicon (Si), indium arsenide (InAs), germanium (Ge), and MCT detectors. For all the detector material mentioned here, MCT is the most commonly used quantum detector for IR spectrometry, and InGaAs is widely used for NIR measurement because it can operate at room temperature with quite good sensitivity.

As mentioned earlier in the chapter, instrumentation for modern Raman spectroscopy was inspired by the rapid development of NIR techniques during the 1980s. Before 1985, photomultiplier tubes (PMTs) had been the detector of choice and the only significant one for Raman spectroscopy. Currently used FT-Raman detectors are Ge and InGaAs detectors with InGaAs detectors being the most popular. Cooling reduces inherent detector noise at the cost of reduced spectral range. Therefore, cooling of these detectors is an option, but a choice upon the trades between spectral range versus sensitivity needs to be made.

Compared to single-channel detectors, detector arrays offer advantages, such as simultaneous measurement across the entire spectral range and no moving parts in spectrometer designs. Silicone-based photodiode arrays (PDAs) dominated the use for spectrometric detection in the UV, visible, and short-wavelength NIR regions for some time. For measuring spectral signal in the longer wavelength range, an InGaAs array detector is available. Array detectors also come in different formats, i.e., linear or two-dimensional with a variety of pixel numbers. Generally speaking, as the number of pixels increases, the cost of the array detector device quickly adds up. The Raman renaissance during the post-1986 period was heralded by technological advancement, such as the adoption of multichannel detectors. Early multichannel detectors for Raman were intensified photodiode arrays (IPDA) and charge transfer devices (CTDs). The charge coupled device, one type of CTDs, quickly surpassed its counterpart, i.e., the charge injection device (CID), and the IPDA in performance and dictates the detector configurations for dispersive Raman spectrometers (Bilhorn et al. 1987). Some of the major parameters for a CCD detector include number and size of pixels, gain, dark current, readout noise, bias, and dynamic range. To further increase the quantum efficiency of a CCD detector, the photoactive area could be back thinned through ion etching to produce back-illumninated CCDs. However, the improved sensitivity means significantly increased cost compared to front-illuminated devices. With a two-dimensional CCD detector, functions, such as binning and optical multiplexing, could be achieved. Binning refers to the process of combining signals belonging to the same column, which corresponds to the common wavelength separated by the dispersion device, e.g., diffraction gratings, to increase the SNR. Binning could be either performed on-chip or in software. When multiple optical fiber probes are used as input devices coupled to the spectrograph in front of the CCD array, spectra of more than one sample can be produced at the same time. Likewise, more than one spectral image of a single sample could be collected if sample homogeneity is an issue. It is also possible that multichannel detectors, such as CCDs, could be used for FT spectrometer configuration, which is called the multichannel Fourier transform (MCFT) technique (Archibald et al. 1988; Zhao and McCreery 1996).

16.3.3 Wavelength Separating Devices and Mechanism

From the generic spectrometer design perspective, there are two fundamentally different light analyzing mechanisms, i.e., dispersive or non-dispersive (Wang and Paliwal 2007). Dispersive spectrometers achieve separating of wavelengths by spatially spreading light at different frequencies onto a single detector or across the photosensitive area of an array detector. The major components for dispersive spectrometers are diffraction gratings or prisms. It is worth noting that nearly all dispersive spectrometer designs are based on diffraction gratings. A diffraction grating is a collection of finely spaced grooves deposited on a glass substrate (Palmer 2002). Classical grating spectrometer designs have the Czerny-Turner geometry or Littrow configuration in a scanning monochromator. A diffraction grating can also be coupled with a focal plane array detector in a spectrograph configuration (Mayes and

Callis 1989). Prisms are not used as a single light-separating device in contemporary instrumentation, e.g., it has been used in conjunction with a diffraction grating as a prism-grating-prism (PGP) spectrograph for hyperspectral imaging. For non-dispersive spectrometer designs, there exist two most popular light analyzing mechanisms, i.e., use of band-pass filters and the Fourier transform interferometer. There are two kinds of filters used to isolate light of different wavelengths, i.e., optical interference filters and electronically tunable filters. Fabry-Perot (FP) dielectric filters work on the principle of light interference to cut down the transmission of light outside the pass band (O'Shea 1985). Typical interference filters possess a transmission efficiency up to 70% and a pass band of 10 nm (Wetzel 2001). Commercial filter instruments employ a filter wheel on which a number of interference filters are installed to produce light at multiple predefined wavelengths (Watson et al. 1976). Interference filters are considered the simplest and least expensive way to separate light, and NIRS instrumentation using interference filters compete well for online and in-field applications. However, optical interference filters are rarely used for Raman spectroscopy except for special applications, such as dedicated devices that monitor only a few Raman shifts. Currently, there are two prevailing technologies, i.e., acousto-optical tunable filters (AOTF) and liquid crystal tunable filters (LCTF), as electronically tunable filters (ETF) (Gat 2000). An AOTF utilizes radio frequency (RF) to modulate the refractive index of a bi-refringent crystal of tellurium dioxide (TeO_2), which can be considered as a longitudinal transmission diffraction grating (Wetzel et al. 2002). The LCTF is composed of a stack of polarizers and retardation crystal plates. Both AOTF and LCTF have relatively large optical apertures, good spectral resolution, and a wide spectral working range. In addition to sequential scanning of multiple wavelengths, an ETF allows instant access (in microseconds) to random wavelength data. An ETF device is compact, digitally controlled, and has no moving parts. All these technical strengths make it an appealing choice for both NIRS and Raman instrumentation for online-processing monitoring, hyperspectral imaging systems, Raman microscopy. A FT system is based on a Michaelson interferometer to acquire an interferogram and uses Fourier transform to deconvolve frequency information encoded in it. Spectrometers based on FT technique are perhaps the most universal instrument for different branches of vibrational spectroscopy. Particularly, FT systems dominate the IR spectroscopy measurement as nowadays spectroscopists refer to the technique as FT-IR rather than simply as IR spectroscopy. The FT-Raman instruments are considered modified FT-IR systems. Before 1986, FT-Raman was not widely accepted because of the increase in shot noise while the instrument operates in shot noise limit. With the introduction of a NIR laser at 1064 nm, FT-Raman received a considerable amount of development during the period 1986 to 1990. Today, the technology is still going strong. The FT-NIR technology has only become available within the past 20 years or so. The performance of early FT-NIR instrumentation was not comparable with their grating-based instruments initially. The FT-NIR instruments only gained more and more importance during the last decade as technology recently improves. The FT instrument has multiplex and throughput advantages over the aforementioned dispersive systems. Excellent wavelength resolution and accuracy are additional advantages for a FT system. Traditional FT systems have a relatively high capital cost, and their performance is subject to mechanical stability and dust (Coates 1998). Technological advancements have been able to gradually move FT systems out of the laboratory into industrial settings.

16.3.4 Chemometrics

Workman et al. (1996) gave a concept of chemometrics stating that chemometrics can generally be described as the application of mathematical and statistical methods to 1) improve chemical measurement processes and 2) extract more useful chemical information from chemical and physical measurement data. Handling spectral data can be divided into two steps: spectral data preprocessing and exploratory or quantitative analysis of transformed data.

16.4 Spectra Preprocessing Techniques

The main purposes of preprocessing spectral data are to reduce noise levels, smooth spectra, remove baseline, linearize spectral data, and correct for effects due to light scattering (Ozaki et al. 2007). Common denoising methods include taking multiple measurements to calculate averages (accumulated averages) and apply Fourier transform (FT) (McClure and Davies 1988). A smoothing operation removes components in the high-frequency domain regardless of its amplitude. Smoothing can usually be done by using a running average, i.e., boxcar smooth, or a Savitzky-Golay smooth with bionomial filtering and FT as alternative methods (Savitzky and Golay 1964; Hruschka 2001). The first and second derivatives of spectra are used to remove constant or sloping baselines with the added effect in resolution enhancement and SNR deterioration (Giese and French 1955; Holler et al. 1989). The linearization of spectra is necessary because measured signals are often not proportional to sample properties of interest and concentrations of chemical compounds. Transmittance (T) is converted to absorbance (A), log(1/T); reflectance (R) is converted to apparent absorbance (A), log(1/R); and the Kubelka-Munk (KM) equation is used for diffuse reflectance conversion (Isaksson and Næs 1988; Osborne et al. 1993; Law et al. 1996). Wavelet transform (WT) is a relatively new mathematical technique, and its basis functions are dually localized in both the time and frequency domains. As a chemometric tool, WT has been used to denoise, smooth, and compress spectra (Barclay and Bonner 1997). Other utilities of WT include baseline removal, resolution enhancement, improving pattern recognition, and a combination of these for quantitative analysis of NIR spectra (Walczak et al. 1996; Depczynski et al. 1999; Chau et al. 2004). Research has shown the importance of spectral pretreatment to classification and regression; however, the choice of the best preprocessing techniques relies on statistical testing and the chemometrician's experience (Ootake and Kokot 1998; Delwiche and Reeves 2004).

For the IR spectrum, some data preprocessing may be required. It is common to find an IR spectrum that rides on a constant or sloping baseline. The discrete data, absorption data, are only available at discrete wave numbers and have limited resolution that may cause inaccurate data representation at sampling points and abrupt change of slope. For the same reason, it is also difficult to pick out the actual center for absorption peaks at the peak wave numbers. It is also sometimes demanded that overlapping spectral features, e.g., due to overtone and combination bands or limited spectral resolution of the spectrometer, be resolved to conduct accurate analysis. These operations include baseline removal, spectral smoothing, taking derivatives, interpolation, peak searching, and Fourier self-deconvolution (FSD) of spectra, etc. The constant baseline is removed using subtraction of the offset, and the linear baseline is often dealt with using a two-point baseline correction. It is also common to observe a nonlinear baseline, generally removed by fitting higher-order polynomials or other functions to the baseline and then subtract it from the spectra. To further enhance the spectral resolution of the data, procedures, such as zero padding, derivatives, and FSD, could be applied. Besides some of the data preprocessing techniques mentioned for IR spectra, a significant amount of literature has been devoted to the correction of multiplicative scattering for NIR spectra. The reason is that most of the sampling devices for NIR measurement of solid samples choose to use the reflectance mode for its ease in application. Because a significant amount of variance due to scatter noise exists in NIR reflectance spectra, it is important to correct for the light-scattering effect to decrease multivariate complexity (Geladi et al. 1985). Commonly used light-scattering correction techniques are multiplicative scatter correction (MSC) (Helland et al. 1995) and standard normal variate transform with detrending (Barnes et al. 1989). Rinnan et al. (2009) provided an updated overview of the prevailing preprocessing techniques for handling NIR data. Raman measurement faces similar problems to the IR. The well-known problems associated with Raman spectroscopy are poor SNR, undesirable background signals, and line broadening due to instrument response. Because Raman is a weak phenomenon, frequency shift and instrument response should be calibrated in order to make meaningful comparison of spectra obtained on different Raman systems.

16.4.1 Qualitative and Quantitative Analysis Techniques

Multivariate analysis techniques applied to optical spectroscopy can be categorized into qualitative methods and quantitative methods. The qualitative methods consist mainly of three groups of techniques: exploratory data analysis, unsupervised pattern recognition, and supervised classification (Brereton 2003). Principal component analysis (PCA) and factor analysis (FA) are the tools for exploratory analysis. More frequently, PCA is used for reducing data dimensionality and feature extracting for subsequent classification or regression analysis (Wold et al. 1987). The main technique for unsupervised classification is cluster analysis, which detects similarities between spectral data (Lee et al. 2005). The commonly used techniques in supervised classification include linear discriminant analysis (LDA), quadratic discriminant analysis (QDA), categorical regression, K-nearest neighbor analysis (KNN), and artificial neural networks (ANNs) (Downey 1996; Indahl et al. 1999). There are also high-dimensional classifiers currently used in chemometrics that circumvent the need for feature extraction, using techniques such as PCA, FT, and WT. High-dimensional classification techniques include soft independent modeling of class analogy (SIMCA), regularized discriminant analysis (RDA), and penalized discriminant analysis (PDA) (Mallet et al. 1996). The application of vibrational spectroscopy in the food industry has been a historical driving force for development of multivariate calibration techniques (Brereton 2003). Currently, the principal methods applied in multivariate calibration are multiple linear regression (MLR), principal components regression (PCR), and partial least squares regression (PLSR) (Næs and Martens 1984). Martens and Næs (1984) explain some key concepts in multivariate calibration, e.g., direct vs. indirect calibration and controlled vs. natural calibration. Geladi and Kowalski (1986) provide a tutorial on PLSR and compared MLR, PCR, and PLSR. In practice, spectral components for components in a mixture invariably overlap or are altered from spectral features of pure components, whether it is IR, Raman, or NIR. However, the principle of linear superposition works for Raman spectroscopy because Raman cross-sections are small. Therefore, methods, such as classical least squares (CLS), sometimes called the K-matrix method, could be appropriate when spectra for the pure components are known. For NIRS applications, PCR and PLSR solve the co-linearity problem in MLR. Compared to the two-step PCR, i.e., PCA and regression, PLSR also considered the inner relationship between sample properties, e.g., compound concentrations, and spectral data during matrix decomposition. As a physicochemical method, measurement of NIR spectra is subject to deviations from the Beer-Lambert law because of physical effects, such as sample particle size, homogeneity, and temperature. These effects cause nonlinearity in collected data (Borggaard 2001). Fuller et al. (1988) provided a comparative overview of various methods for multicomponent quantitative analysis of IR data, i.e., K-matrix, multivariate least squares, PCR, and PLS. The advantages and disadvantages of each method were highlighted. Other than pretreating data to linearize spectra response, nonlinear modeling methods, such as ANNs, can be used for classification and calibration in vibrational spectroscopy. The commonly used ANN architectures are back-propagation (BP) networks, radial basis function (RBF) networks, and Kohonen self-organizing map networks (Massart et al. 1988; Dou et al. 2006; He et al. 2007).

Some advanced topics in chemometrics have been treated in previous research. Design of calibration and selection of samples for training models could improve calibration robustness with a minimum number of measurements. The general principle is to cover the possible variations in prediction population (Næs and Isaksson 1989). Cluster analysis could be a useful tool in selecting calibration samples (Isaksson and Næs 1990). Standardization of NIR instruments is crucial to the applicability of calibration models developed on a master instrument. Fearn (2001) reviewed the fundamentals of standardization and calibration transfer for NIRS instruments. The author found that pretreatment of spectral data, building calibrations with pooling data from several instruments, and using methods such as piecewise direct standardization are three effective approaches. For industrial applications, using a calibration model trained with a few wavelength variables had a significant advantage over those built over hundreds or even thousands of wavelengths. Lestander et al. (2003) studied the selection of NIR wavelength variables using genetic algorithms (GA) in order to build filter spectrometers for moisture measurement in seeds.

16.4.2 Recent Advances in Chemometrics

Martens et al. (2003) proposed a preprocessing algorithm, extended multiplicative signal correction (EMSC), capable of separating light-scattering effects from chemical absorbance effects in powdered samples. Pedersen et al. (2002) reported application of extended inverted signal correction (EISC) to NIR transmittance spectra of single weed seeds. Their results showed that EISC is more promising than the two-step second derivative followed by MSC and classical spectra pretreatment methods, such as MSC. Fearn and Davies (2003) examined the data compression performance of FT and WT on NIR spectra. They found that although WT is more popular than FT in NIRS research, WT is no more efficient than FT in compressing NIR spectra. Burger and Geladi (2007) lately studied the effect of different spectra pretreatment methods, e.g., KM, SNV, first and second derivatives, and variants of MSC, on eliminating scattering effects in hyperspectral imaging data.

The support vector machine (SVM) is a relatively new classification technique that has only recently been introduced to the chemometrics society to solve both classification and calibration problems (Xu et al. 2006). The SVM is a boundary method based on statistical learning theory with the aim of determining optimal separating boundaries between classes (Vapnik 1995). The advantages of SVM methods are their optimal generalization performance in ill-posed situations with a limited amount of training data. With an appropriate choice of kernel functions, SVM is capable of modeling highly nonlinear classification boundaries (Belousov et al. 2002). Although initially designed for binary classification problems, SVMs can be used for multi-class by employing strategies, such as one against the rest or one against one (Pal and Mather 2005). Least squares support vector machines (LS-SVM) have been recently introduced as a multivariate calibration technique (Cogdill and Dardenne 2004; Thissen et al. 2004). Chauchard et al. (2004) compared LS-SVM, MLR, and PLSR in the development of a portable NIR sensor that captured spectral data with nonlinearity. A combination of LS-SVM, SNV preprocessing, and selected variables produced the most accurate and robust regression models (Chauchard et al. 2004). However, for linear problems that are well understood in analytical chemistry, SVMs are unnecessarily complicated and conventional approaches, such as LDA, PCA, and PLSR are more effective (Xu et al. 2006).

16.5 Applications

It is interesting to note that all three techniques, i.e., NIR, Raman, and IR spectroscopy, seem to be appropriate tools for food processing. Nevertheless, technically speaking, they possess certain advantages and disadvantages, which depend on specific applications. Theoretically, the three techniques tend to complement each other rather than competing against each other in revealing chemical information about the sample. Therefore, it is of interest to the readers to appreciate their utilization for certain applications and see their differences. In this section, some comparative studies among the three are given, followed by literature reviews in specific fields related to food processing.

The similarity in the chemical information provided by the three vibrational spectroscopy techniques has led to comparative study in various fields. Schulz et al. (2003) explored the potential of vibrational spectroscopy techniques to identify and quantify substances in 11 different basil chemotypes. The NIR data of crushed samples were taken using rectangular cups, and isolated basil oils were measured under transflection mode using quartz cuvettes. An ATR sample cell was used to carry the sample extract for FT-IR measurement. The ATR/FT-IR spectrometer operated in a spectral range of 650–3500 cm^{-1} with a spectral resolution of 2 cm^{-1}. Raman spectra were collected using a NIR-FT Raman spectrometer with a Nd:YAG laser excitation emitting light at 1064 nm. Because different basil chemotypes have special chemical signatures of main volatile components, all three vibrational spectroscopy methods were able to distinguish the 11 basil chemotypes. Due to the characteristic bands of individual volatiles present in IR and Raman spectra, discrimination of basil chemotypes was possible without subjecting IR and Raman data to chemometric analysis. On the other hand, NIR data of various basil chemotypes could only be interpreted by coupling it with chemometric algorithms. It was also demonstrated that NIR calibrations built using a PLS algorithm were capable of quantitatively predicting the main valuable volatiles in the air-dried basil leaves. The authors finally recommended that NIR and Raman calibrations developed from essential oils could be used for online monitoring of the distillation process.

Baranska et al. (2006) compared FT-Raman, ATR-IR, and NIR spectroscopy in determining the lycopene and β-carotene content in tomato fruits and products. Mashed tomato samples were presented to the sample stage for FT-Raman spectrometers, ATR cells of the IR instruments, and a spinning sample cup under the NIR fiber optic probe separately for spectra acquisition. The acquired spectral data were mean centered, and PLS models were subsequently calibrated. For qualitative analysis purposes, the results showed that FT-Raman spectroscopy could identify carotenoids in plant tissues and food products with no preliminary sample preparation. Water was a strong interferent for ATR-FT-IR and NIR measurements. In addition, no absorption bands characteristic for carotenoids are found in IR and NIR spectra. Interestingly, IR spectroscopy outperformed FT-Raman in quantifying the lycopene and β-carotene contents in tomato samples despite its lack of relevant spectral features. The quantitative prediction results using NIR spectroscopy was the worst in their study. Paradkar et al. (2002) examined the utility of FTIR, FT-Raman, and NIR spectroscopy in detecting adulteration of maple syrup. A total of 54 samples of maple syrup adulterated with different quantities of corn syrup were prepared. The addition was expressed in percentage by the weight of maple syrup ranging from 0% to 27% in steps of 0.5%. The IR and Raman data were collected in the spectral range from 400 cm^{-1} to 4000 cm^{-1}. The NIR measurement was performed in the 600–1700 nm region. It was reported that all three techniques were able to detect adulteration of corn syrup in the maple syrup. From a quantitative analysis perspective, IR and Raman techniques were superior to NIR in predicting adulteration percentage.

Qiao and van Kempen (2008) compared three techniques, i.e., FT-IR, NIR, and Raman, for analyzing amino acid content in animal meals. Raw samples were presented to NIR and Raman instruments directly for scanning while finely ground samples were prepared for collecting FT-IR spectra. The authors found a high level of noise in both Raman and IR data and only classified the Raman technique as unsuitable. It was concluded that FT-IR and NIR has comparable performance in predicting amino acid content when both techniques used full spectral range. If the FT-IR wave number variables exhibiting high noise level were excluded from calibration, FT-IR offered improved calibration compared to NIR spectroscopy. The authors also noted that preparing a sample for FT-IR was more time consuming. Yang et al. (2005) conducted a qualitative analysis of edible oils and fats using FT-IR, FT-NIR, and FT-Raman spectroscopy. Little sample preparation was needed for all three techniques. The authors employed linear discriminant analysis (LDA) and canonical variate analysis (CVA) to segregate different oils based on their spectral features. It was found that FT-IR combined with chemometric analysis offered the highest successful classification rate (98%) followed by FT-Raman (94%) and FT-NIR (93%). The paper concluded that all three techniques are capable of rapidly discriminating edible oils and fats while FT-IR and Raman offer more exquisite structural information of the functional groups of oils than FT-NIR spectroscopy.

Chung and Ku (2000) compared NIR, IR, and Raman for their utilization in analyzing American petroleum institute (API) gravity of atmospheric residue (AP) in crude oil. The authors found that Raman spectroscopy was unsuitable because of the strong fluorescence caused by asphaltenes. Although it used a FT-Raman system with a Nd:YAG excitation laser operating at 1064 nm and a low radiation power of 5 mW, the fluorescence was overwhelming the Raman signal. Infrared spectra were acquired on a FT-IR spectrometer with an attenuated total reflection (ATR) probe.

The poor reproducibility and high noise level made IR spectroscopy impractical for their quantitative application despite the rich spectral features it provided. The authors claimed that NIR spectroscopy was most successful for determining the AR content in crude oil due to the high reproducibility of spectral data.

Gresham et al. (1999) provided a comparison of NIR absorption spectroscopy versus Raman spectroscopy for industrial process monitoring purposes. In their experiment, two individual solid-state instruments, i.e., one for NIR absorption and the other for Raman scattering spectroscopy, were used to simultaneously quantify components in xylene isomer mixtures. Their results demonstrated the utility of both the NIR and Raman technique for quantitative analysis of ortho-, meta-, and para-xylene with superior absolute error at ±0.05, ±0.12, ±0.09 *w/v* using NIR and ±0.08, ±0.04, ±0.07 *w/v* using Raman. The authors suggested the use of solid-state NIR and Raman spectrometers for online raw material monitoring. The choice between two techniques was mainly determined by the desired investment cost and actual monitoring environment.

Ku and Chung (1999) compared the performance of NIR and FT-Raman in determining the chemical and physical properties of naphtha. The authors selected six different chemical compositions, i.e., total paraffin, total naphthene (cyclo-alkane), total aromatic, C6 paraffin, benzene, and cyclopentane, and a physical parameter, i.e., specific gravity, to compare NIR and FT-Raman methods. It was found that NIR calibration outperformed that of FT-Raman because of the high SNR and reproducibility of NIR data. However, the paper pointed out that Raman possesses great potential as an online quantitative analytical tool because of the richer chemical information that it provides and the convenience of interfacing using fiber optics for remote sensing.

Afseth et al. (2005) tested the capacity of NIR and Raman spectroscopy for determining fatty acid and main constituents in a complex food model system. The food model consisted of 70 different mixtures of protein, water, and oil blends to imitate fish and meat samples. For fatty acid determination, both NIR and Raman spectroscopy were able to predict content of saturated, monounsaturated, and poly-unsaturated fatty acids in the samples with validation errors ranging from 2.4% to 6.1%. Raman spectroscopy provided the best results, predicting the iodine values with a validation error of 2.8%. For predicting the main constituents, the NIR technique demonstrated a substantial advantage over Raman spectroscopy.

16.5.1 Cereal and Cereal Products

Although it is feasible and there is great potential for using IR and Raman spectroscopy for cereal grain study, NIR spectroscopy currently dominates research fields associated with quality evaluation and processing of grains and their products. The difficulties in popularizing IR and Raman in grain research are technological and high capital cost.

Büchmann et al. (2001) applied European ANN calibrations of moisture and protein in cereals to the Danish NIR transmission network. The authors reported that European ANN had more accurate prediction for moisture in wheat and barley than PLS models. Prediction of protein in wheat and barley was also more accurate with European ANN calibrations than PLS models, but the difference is less pronounced. Miralbés (2003) applied NIR transmittance spectroscopy to a determination of moisture, protein, wet gluten, dry gluten, and alveograph parameters of whole wheat. A modified PLSR method was used. The PLSR models, according to the author, had sufficient prediction accuracy for screening wheat according to its quality parameters at the receiving stage at mills or elevators. Maertens et al. (2004) tested the feasibility of using NIRS for predicting moisture and protein content in wheat during harvesting. A NIRS sensor was installed in a bypass unit of the clean grain elevator in a combine harvester. The online NIR measurement of whole wheat quality offered prediction accuracies comparable with those achieved with a similar stationary unit.

Armstrong et al. (2006) compared dispersive and FT-NIR instruments for evaluating grain and flour attributes. The authors developed PLSR models on protein, moisture, and hardness index of whole wheat and protein, ash, and amylase of wheat flour. Comparable prediction performance was achieved for both dispersive and FT-NIR systems. Maghirang et al. (2003) reported an automated NIR reflectance sensor for detection of single wheat kernels containing insects. Their findings suggested that calibration models built with wheat samples having dead internal insects can be used to detect live insects inside wheat kernels without sacrificing accuracy. Rittiron et al. (2004) provided useful information on developing a high-speed single brown rice kernel sorting NIRS system. A scanning range from 1100 nm to 1800 nm in the transmittance mode is recommended for protein and moisture measurement of brown rice. Wang et al. (2004) demonstrated the ability of NIRS to classify healthy and fungal-damaged soybean seeds. In the study, two classes of categorical PLSR models and five classes of ANN models can accurately detect fungal-damaged seeds and discriminate the type of fungal damage. Cogdill et al. (2004) applied a NIR hyperspectral imaging technique for determining the moisture and oil content in single kernel of maize. The experimental results showed that NIR hyperspectral imaging may be useful for moisture measurement but unable to predict oil content due to poor reference method.

Ma and Philips (2002) provided a review paper on the application of FT-Raman spectroscopy in cereal chemistry. The paper briefly went through the fundamentals of Raman spectroscopy. It was pointed out in the article that Raman technique, compared to IR spectroscopy, is more suitable as a routine analytical tool to study biological systems and requires little sample preparation. The authors also listed the limitations of NIR spectroscopy and stated the availability of information from Raman could alleviate the situation. Some applications of FT-Raman were presented, and the authors focused on two important fields, which were a measurement of the chemical changes of chemically modified starches and investigation of conformational and structural changes induced in oat globulin. Barron and Rouau (2008), however, performed a rather interesting experiment on the study of wheat grain peripheral tissues using IR and Raman spectroscopy. The authors dissected pure samples of the aleurone layer, hyaline layer, outer pericarp, and testa + inner pericarp composite layer. On these layer samples, IR and Raman spectra were acquired. The spectral features showed in the data could be related to the known biochemical composition of each tissue. Supervised classification methods, i.e., LDA and partial least square discriminant analysis (PLS-DA) were applied on IR data and layers, such as pericarp, aleurone, hyaline, and testa, could be clearly identified for four different wheat cultivars. Piot et al. (2002) conducted research on the molecular composition of cereal grain using a confocal Raman microspectroscopy system. The study provided new insight into how the grain is formed. The significance of their research lies in that the hardness of grain could be determined from molecular assessment of its microstructure. Based on their findings, the authors suggested that the α-helical secondary structure of protein may be related to grain hardness.

16.5.2 Fruit and Vegetables

Moros et al. (2005) studied the nutritional parameters, e.g., carbohydrate content and energetic value, of fruit juice and milk using ATR-FTIR spectroscopy. In order to give readers a better overview of the research status, the authors reviewed some of the traditional methods for fruit juice analysis. Also included in the article was a thorough list of previous research on using FT-IR spectroscopy for examining the carbohydrate content in a wide variety of fruit juice, e.g., apple juice, mango juice, sugar cane juice, etc. Pointing out that most of the previous research concentrated on determining the main sugar content instead of the total content of carbohydrates, the authors decided to evaluate ATR-FTIR spectroscopy that allowed them to determine carbohydrates and energetic value in a fast and accurate manner. It was demonstrated that the proposed method offered a favorable performance compared with that obtained in previous work that employed much-restricted samples. The authors even stated that their method complied with statuary values set out by the food and drug administration (FDA) of the United States. Bureau et al. (2009) reported the use of FT-IR spectroscopy for simultaneous determination of sugar and organic acid in apricot slurries. In addition, the complementary quality traits, i.e., fruit firmness, skin color, ethylene production, soluble solids, and titratable acidity were also predicted and evaluated against standard methods. Partial least squares models for individual sugar and organic acids showed prediction errors of 12% for glucose, malic acid, and citric acid; 16% for sucrose; and 18% for fructose. The authors also found good correlation among IR spectra and the complementary quality traits although the prediction results varied. The paper concluded with findings that suggested the possibility of using ATR-FTIR spectroscopy as a fast and economic alternative to the current enzymatic assay methods for sugars and organic acids.

Di Egidio et al. (2009) conducted research on the shelf life of fresh-cut pineapple using FT-NIR and FT-IR spectroscopy. The authors selected three batches of fresh-cut pineapples and analyzed the samples during storage at three different temperatures (5.3°C, 8.6°C, and 15.8°C). At the same time, traditional analytical methods were used to monitor the microbiological and chemical changes. To remove baseline, spectral data were preprocessed with second derivative transformation. Principal component analysis was then applied to the transformed data set. From the score plots for the first two principal components, two groups of samples, i.e., fresh and old, could be distinguished for each sample batch under the three storage temperatures. The initial freshness decay time revealed by IR and NIR techniques was confirmed by microbial decay, which resulted in moisture loss and composition modification. Research works on other kinds of fruit or fruit products were also reported. Defernez and Wilson (1995) used diffuse reflectance infrared and ATR infrared spectroscopy to detect adulteration of strawberry jam with cheaper ingredients. Manrique and Lajolo (2002) determined the methyl esterification degree (MED) of pectins in papaya fruit. Irudayaraj and Tewari (2003) used ATR-FTIR to simultaneously monitor organic acids and sugars in fresh and processed apple juice. Duarte et al. (2002) quantified the sugar content in mango juice using FTIR spectroscopy.

Muik et al. (2004) attempted to sort the quality of olive fruits using Raman spectroscopy and pattern recognition techniques. Sample sets consisted of sound olives, olives with frostbite, olives collected from the ground, fermented olives, and diseased olives. Olive samples were milled with a hammer mill before being presented to the FT-Raman sample cell. Raman spectra were subjected to hierarchical cluster analysis (HCA), PCA, KNN, and SIMCA. Under unsupervised cluster methods, i.e., HCA and PCA, sound olives, frostbit ones, olive samples collected from ground, and fermented olives formed clusters. These clusters partly overlapped, but the diseased samples did not form a group. For supervised cluster analysis, SIMCA outperformed KNN and offered a lower false negative error for all five classes. The authors believed that their work demonstrated the potential of Raman spectroscopy as a quality screening tool for olives. In an experiment performed by Veraverbeke et al. (2005), Raman spectroscopy was studied alongside FT-IR and NIR spectroscopy to evaluate the surface quality of an apple. Three different cultivars of apples were harvested at three different picking dates and stored under three different controlled atmosphere durations. First derivative was applied to the NIR and IR spectra and canonical discriminant analysis was carried out. It was found that differences caused by storage times and cultivars were significant. Raman, however, did not provide satisfactory results.

Sirisomboon et al. (2007) measured the pectin constituents of Japanese pears using NIRS. The NIR spectra of intact pears and juice were collected using an interactance fiber optic probe. The developed MLR equations could accurately predict the pectin constituents. Zude et al. (2006) employed miniaturized vis/NIR spectrometers to predict fruit firmness and solid soluble content of apples on the tree and in storage. The results suggested that the predicted fruit maturity stage allowed determination of optimal harvesting date and fruit quality in shelf life. Lu and Peng (2005) studied peach firmness using a multi-spectral scattering technique. The authors utilized a push broom multispectral imaging system working in the 400 to 1000 nm wavelength range to determine the scattering profiles of soft and firm red haven peaches. Experiments showed that soft fruits possessed a broader scattering profile than firm ones, and the difference was most pronounced at 680 nm. In another two-part paper, Peng and Lu (2006) assembled a LCTF–based multispectral imaging system to acquire scattering profiles of apple fruit for wavelengths between 650 and 1000 nm. The experimental results demonstrated that this Vis/NIR multispectral scattering technique is superior to conventional NIRS in predicting fruit firmness. El Masry et al. (2007) tested a visible/NIR (400–1000 nm) hyperspectral imaging system to determine the moisture content, total soluble solids, and acidity in strawberries. The PLSR models showed good accuracy in prediction of these strawberry quality parameters. Qin and Lu (2005) experimented on detecting pits in tart cherries using a hyperspectral transmission technique. The trained ANN classifiers had good prediction performance for tart cherries of the same size, color, or defect. Hyperspectral transmission images were more effective than single spectra in cherry pit detection. Nicolaï et al. (2006) developed a hyperspectral NIR imaging system for bitter pit lesion detection on apples. With a discriminant PLS model, the hyperspectral NIR imaging technique was able to identify bitter pit lesions invisible to the naked eyes. Clark et al. (2004) utilized visible/NIR spectroscopy to predict the storage potential of kiwifruit during a 24-week cold storage period. By using a canonical discriminant classifier on relative reflectance intensities, the incident of storage disorder was greatly reduced. The authors suggested using NIRS to identify the least mature kiwifruit at harvest to reduce the storage disorder incident rate.

Liu et al. (2005) developed an algorithm for detecting chilling injuries in cucumbers using Visible/NIR hyperspectral imaging technique. It was found that a large spectral difference exists between good, smooth skins and chilling-injured skins in the 700–850 nm region. Both a simple band ratio method and PCA were successful in detecting chilling-injured skins. Kavdir et al. (2007) studied the measurement of the firmness, skin and flesh color, and dry matter content of pickling cucumbers using visible/NIR spectroscopy. It was found that PLSR calibrations trained with spectra in the NIR region (800–1650 nm) gave a better prediction of firmness and dry matter than that of the visible/NIR region (550–1100 nm). Skin and flesh color were highly correlated with measurements in the visible/NIR region. Xie et al. (2007) attempted to discriminate transgenic tomatoes from nontransgenic tomatoes using visible/NIR spectroscopy. Three classifiers, i.e., PCA, SIMCA, and discriminant PLSR based on PCA scores, were tested on spectral data of two groups of tomatoes with different genes. The discriminant PLSR method trained on second derivative spectra achieved a 100% classification rate of transgenic tomatoes and conventional ones. Hartmann and Büning-Pfaue (1998) determined the chemical constituents of potatoes using NIR reflectance spectroscopy. Diffuse reflectance spectra of homogenized peeled potatoes were collected to analyze the dry matter, starch, crude protein, and sugar content using modified PLSR models. Their results indicated the possibility of measuring potato quality using a transportable NIRS system. However, the transferability of calibration models is limited by the potato cultivar.

16.5.3 Milk and Dairy Products

The use of NIR spectroscopy for quality monitoring of dairy products is widely accepted. Rodriguez-Otero et al. (1997) provided an overview on the application of NIR spectroscopy for dairy products. The authors tried to demonstrate the capability of the technique by citing literatures published on utilizing NIR techniques for powder dairy products, liquid milk, cheese, butter, and fermented milk. The paper concluded that NIR spectroscopy is an adequate technique for analyzing the major components of dairy products except the mineral contents. The main components in raw milk that could be quantified using NIR spectroscopy are fat and protein (Laporte and Paquin 1999). Chen et al. (2002) studied the fat content in raw milk using the NIR technique. Díaz-Carrillo et al. (1993) experimented to build NIR calibrations for total protein, total casein, casein fractions (αs-, β-, and κ-caseins), fat, and lactose in goats' milk. Satisfactorily low prediction errors and high coefficients of determination were obtained for the models for individual components. The authors suggested the NIR technique as a suitable tool for analysis of dairy products because of its low cost and speed. Some other researchers used the NIR technique to determine total bacteria count (Saranwong and Kawano 2008) and identify possible disease and pathogen (Tsenkova et al. 2006). From a microsystems perspective, Brennan et al. (2002) discussed the potential application of a NIR microsystems optical device for online process control of milk product. Initial results showed satisfactory sensitivity to the fat content. Karoui and De Baerdemaeker (2007) wrote a review article on methods for dairy product application. Spectroscopic techniques, including NIR and IR spectroscopy, are briefly described. It is specially mentioned in the article that IR spectroscopy is the most prevailing technique for milk and dairy product analysis.

Iňón et al. (2004) carried out a study on the nutritional parameters of various types of milk, e.g., whole, semi, skim, etc., using FT-IR spectroscopy and chemometrics. Total fat, total protein, total carbohydrates (CH), calories, and calcium were determined using PLS regression. The authors conclude that one single PLS model was sufficient to simultaneously predict all attributes. In 1985, Kennedy and White published a paper using an ATR IR quantitative analyzer to determine the fat, protein, and lactose content in samples prepared with skimmed milk powder in distilled water. The method was considered a cheap and simple way for routine analysis of dairy products. Luinge et al. (1993) demonstrated in-line application of FT-IR spectroscopy for determination of fat, protein, and lactose content in milk. Raman spectroscopy is a great tool for determining the secondary structure of milk proteins. The conformation of major milk proteins, i.e., casein, was studied using IR and Raman spectroscopy (Byler and Susi 1988). Identification and quantitative determination of concentration of conjugated linoleic acid (CLA) of milk fat was carried out by Meurens et al. (2005). The authors were able to accurately determine the CLA content and claimed their study on CLA using Raman spectroscopy was the first of its kind. The determination of nutritional parameters, including energetic value, carbohydrate, protein, and fat contents of powdered milk, was evaluated using FT-Raman spectroscopy coupled with PLS regression. Quality and sensory attributes of cheese are investigated and reported. Downey et al. (2005) used NIR spectroscopy to predict the maturity and sensory attributes of cheddar cheese. The moisture, fat, and protein content of Greek feta cheese were determined using NIR spectroscopy (Adamopoulos et al. 2001). The authors suggested that continuous monitoring of quality parameters at critical control points was possible. Čurda and Kukačková (2004) conducted a similar study to determine dry matter (DM) fat (F), crude protein (CP), pH, and rheological properties of processed cheeses. Karoui et al. (2006) compared and jointly used NIR and IR spectroscopy in determining fat, dry matter (DM), pH, total nitrogen (TN), and water-soluble nitrogen (WSN) contents in soft cheese. Results showed that a combination of NIR and IR data only slightly improved the prediction for pH values. The authors also mentioned that further research was required because this was only a feasibility study. Fontecha et al. (1993) characterized the secondary structure of casein in various types of cheeses made from ewes' milk using IR and Raman spectroscopy.

16.5.4 Meat and Meat Products

Raman spectroscopy is a valuable tool that offers detailed information concerning the secondary structure of proteins. In addition, Raman spectroscopy provides structural information on water and lipids of meat products during deterioration (Herrero 2008). The technique could also be used for sensory attribute and authenticity evaluation (Ellis et al. 2005). Herrero et al. (2009) also conducted an experiment on the modifications of the protein secondary structure in smoked salmon subject to electron-beam radiation. Beattie et al. (2004) predicted the sensory quality of beef silverside using Raman spectroscopy. The degree of tenderness, degree of juiciness, and overall acceptability of cooked beef samples were evaluated using Raman spectra. The coefficients of determination (R^2) were slightly better than 0.6, and the work was considered successful in prediction texture and tenderness of beef using the Raman method. Pedersen et al. (2003) predicted the water-holding capacity in porcine meat using Raman and FT-IR spectroscopy. The authors advised that spectral regions with high absorption, e.g., water absorption regions, should be avoided in IR spectroscopy application. It was also doubtful that FT-IR could be applied in a rough environment at a slaughterhouse, where temperature and humidity change constantly. Water barely interferes with the Raman process; however, Raman spectra suffer from poor SNR and fluorescence. Ripoche and Guillard (2001) evaluated both FT-NIR and ATR-FTIR spectroscopy for quantitatively determining the fatty acid content in fat extract and fat slices. With good prediction results obtained from validation samples of PLS regression models, the paper concluded that both techniques were capable of predicting fatty acids in pork fat. Infrared spectroscopy is a faster, easier, and cheaper method compared to other techniques. Downey et al. (2000) managed to identify homogenized meat samples of five species, i.e., chicken, turkey, pork, beef, and lamb, using IR, NIR, and visible reflectance spectra. Four supervised classification methods, i.e., factorial discriminant analysis (FDA), soft independent modeling of class analogy (SIMCA), K-nearest neighbor (KNN) analysis, and discriminant partial least squares (PLS) regression were tried. Correct classification rates of more than 90% were achieved in identifying different species.

Prieto et al. (2009) reviewed the use of NIR spectroscopy for determination of quality attributes of meat and meat products over the past three decades. The potential of NIR spectroscopy for assessing the chemical composition, technological parameters, and sensory attributes of meat and meat products were demonstrated. Venel et al. (2001) studied the application of NIRS for determination of a number of quality attributes of beef. The best results were achieved with predicting the tenderness of beef. Park et al. (2001) applied PCR on NIR reflectance spectra of beef to predict its tenderness. It was found that tough meat had an overall higher absorption than tender meat at all wavelengths. An obvious absorption difference exists between tough and tender meats at the protein, fat, and water absorption bands. Qiao et al. (2007) studied the quality parameters of pork using a visible/NIR hyperspectral imaging system. They used PCA scores for different number of PCs as input to an ANN classifier to discriminate reddish, firm, and non-exudative (RFN) and reddish, soft, and exudative (RSE) samples. A successful classification rate of 85% was achieved based on scores of 10 PCs. Marbling scores were also automatically determined. Savenije et al. (2006) predicted drip-loss, pH, intramuscular fat, and color values of pork with a visible/NIR reflectance spectrometer working in the 400–2500 nm range. The modified PLSR models could predict pH, intramuscular fat, and color of pork quantitatively. Viljoen et al. (2005) attempted to determine ash, dry matter, crude protein, and fat content in freeze-dried ostrich meat. The PLSR models were accurate in predicting the crude protein and fat content but not for ash and dry matter. Shackelford et al. (2004) tried to set up optimal protocol for using visible/NIR spectroscopy in meat quality evaluation. The authors declared that the use of high-intensity probe, averaging 20 spectrum per observation, and obtaining spectra at a standardized bloom time were critical factors for online NIRS measurement of meat.

16.6 Summary

Vibrational spectroscopy is a great tool for quality control and process monitoring in food and agricultural industries. Infrared (IR), near-infrared (NIR), and Raman spectroscopy are three branches of vibration spectroscopy, and they are popular techniques in food analysis. Today's technology improvements have enabled many of the unthinkable tasks to become a reality. It is the purpose of this chapter to introduce the reader to the fantastic world of vibrational spectroscopy. Fundamental concepts and theoretical background are covered to bring to readers a basic understanding of the underlying mechanisms of light matter interaction. Instruments are the tools to collect the spectral data, and it is very important to have the right tool for the right task. Instead of focusing on instrumentation designs, the topic was broken down to sections describing the main components of a spectrometer. Therefore, scientists and engineers in food processing could know exactly what is most appropriate for their application and leave the rest of the work to instrument manufacturers. Chemometrics is an important aspect of interpreting and understanding spectral information in spectroscopy study. An overview of data preprocessing and analysis, both qualitative and quantitative techniques, is provided. To give an idea of the width and depth as to how vibrational spectroscopy has penetrated the fields of food processing, a section on its application is included in the chapter.

References

Adamopoulos, K.G., Goula, A.M. and Petropakis, H.J. (2001). Quality control during processing of feta cheese – NIR application. *Journal of Food Composition and Analysis*, 14, 431–440.

Afseth, N.K., Segtnan, V.H., Marquardt, B.J. and Wold, J.P. (2005). Raman and near-infrared spectroscopy for quantification of fat composition in a complex food model system. *Appl. Spectrosc.*, 59, 1324–1332.

Archibald, D.D., Lin, L.T. and Honigs, D.E. (1988). Raman spectroscopy over optical fibers with the use of a near-IR FT spectrometer. *Appl. Spectrosc.*, 42(8), 1558–1563.

Armstrong, P.R., Maghirang, E.B., Xie, F. and Dowell, F.E. (2006). Comparison of dispersive and Fourier-Transform NIR instruments for measuring grain and flour attributes. *Appl. Eng. Agric*, 22(3), 453–457.

Baranska, M., Schütze, W. and Schulz, H. (2006). Determination of lycopene and β-carotene content in tomato fruits and related products: Comparison of FT-Raman, ATR-IR, and NIR spectroscopy. *Anal. Chem.*, 78, 8459–8461.

Barclay, V.J. and Bonner, R.F. (1997). Application of wavelet transforms to experimental spectra: Smoothing, denoising, and data set compression. *Anal. Chem.*, 69(1), 78–90.

Barnes, R.J., Dhanoa, M.S. and Lister, S.J. (1989). Standard normal variate transformation and de-trending of near-infrared diffuse reflectance spectra. *Appl. Spectrosc.*, 43(5), 772–777.

Barron, C. and Rouau, X. (2008). FTIR and Raman signatures of wheat grain peripheral tissues. *Cereal Chem.*, 85(5), 619–625.

Beattie, R.J., Bell, S.J., Farmer, L.J., Moss, B.W. and Patterson, D. (2004). Preliminary investigation of the application of raman spectroscopy to the prediction of the sensory quality of beef silverside. *Meat Science*, 66(4), 903–913.

Belousov, A.I., Verzakov, S.A., and von Frese, J. (2002). A flexible classification approach with optimal generalisation performance: Support vector machines. *Chemom. Intell. Lab. Syst.*, 64(1), 15–25.

Bilhorn, R.B., Sweedler, J.V., Epperson, P.M. and Denton, M.B. (1987). Charge transfer device detectors for analytical optical spectroscopy: Operation and characteristics. *Appl. Spectrosc.*, 41(7), 1114–1125.

Borggaard, C. (2001). Neural networks in near-infrared spectroscopy. In *Near-infrared Technology in the Agricultural and Food Industries*, eds. P.C. Williams and K.H. Norris, 101–108, St. Paul, MN: AACC.

Brennan, D., Alderman, J., Sattler, L., O'Connor, B. and O'Mathuna, C. (2002). Issues in development of NIR micro spectrometer system for on-line process monitoring of milk product. *Measurement*, 33(1), 67–74.

Brereton, R.B. (2003). *Chemometrics: Data Analysis for the Laboratory and Chemical Plant.* Hoboken, NJ: John Wiley & Sons.

Büchmann, N.B., Josefsson, H. and Cowe, I.A. (2001). Performance of european artificial neural network (ANN) calibrations for moisture and protein in cereals using the danish near-infrared transmission (NIT) network. *Cereal Chem.*, 78(5), 572–577.

Bureau, A.S., Ruiz, D., Reich, M., Gouble, B., Bertrand, D., Audergon, J. and Renard, C.M.G.C. (2009). Application of ATR-FTIR for a rapid and simultaneous determination of sugars and organic acids in apricot fruit. *Food Chem.*, 115, 1133–1140.

Burger, J. and Geladi, P. (2007). Spectral pre-treatments of hyperspectral near infrared images: Analysis of diffuse reflectance scattering. *J. Near Infrared Spectrosc.*, 15(1), 29–38.

Byler, D.M. and Susi, H. (1988). Application of computerized infrared and Raman spectroscopy to conformation studies of casein and other food proteins. *Journal of Industrial Microbiology*, 3, 73–88.

Chau, F., Liang, Y., Gao, J. and Shao, X. (2004). *Chemometrics: From Basics to Wavelet Transform.* Hoboken, NJ: John Wiley & Sons.

Chauchard, F.R. Cogdill, S., Roussel, J.M.R. and Bellon-Maurel, V. (2004). Application of LS-SVM to non-linear phenomena in NIR spectroscopy: Development of a robust and portable sensor for acidity prediction in grapes. *Chemom. Intell. Lab. Syst.*, 71(2), 141–150.

Chen, J.Y., Iyo, C., Terada, F. and Kawano, S. (2002). Effect of multiplicative scatter correction on wavelength selection for near infrared calibration to determine fat content in raw milk. *J. Near Infrared Spectrosc.*, 10(4), 301–307.

Chung, H. and Ku, M. (2000). Comparison of near-infrared, infrared, and Raman spectroscopy for the analysis of heavy petroleum products. *Appl. Spectrosc.*, 54(2), 239–245.

Clark, C.J., McGlone, V.A., De Silva, H.N., Manning, M.A., Burdon, J. and Mowat, A.D. (2004). Prediction of storage disorders of kiwifruit (*Actinidia chinensis*) based on visible-NIR spectral characteristics at harvest. *Postharvest Biol. Technol.*, 32(2), 147–158.

Coates, J. (1998). Vibrational spectroscopy: Instrumentation for infrared and Raman spectroscopy. *Appl. Spectrosc. Rev.*, 33(4), 267–425.

Cogdill, R.P. and Dardenne, P. (2004). Least-squares support vector machines for chemometrics: An introduction and evaluation. *J. Near Infrared Spectrosc.*, 12(2), 93–100.

Cogdill, R.P., Hurburgh, C.R. Jr., Rippke, G.R., Bajic, S.J., Jones, R.W., McClelland, J.F., Jensen, T.C. and Liu, J. (2004). Single-kernel maize analysis by near-infrared hyperspectral imaging. *Trans. ASAE*, 47(1), 311–320.

Čurda, L. and Kukačková, O. (2004). NIR spectroscopy: A useful tool for rapid monitoring of processed cheeses manufacture. *J. Food Eng.*, 61, 557–560.

Davies, T. (1998). The history of near infrared spectroscopic analysis: Past, present and future—from sleeping technique to the morning star of spectroscopy. *Analysis Magazine,* 26(4), M17–M19.

Defernez, M. and Wilson, R.H. (1995). Mid-infrared spectroscopy and chemometrics for determining the type of fruit used in jams. *J. Sci. Food Agric.,* 67, 461–467.

Delwiche, S. and Reeves, J.B. III. (2004). The effect of spectral pre-treatments on the partial least squares modelling of agricultural products. *J. Near Infrared Spectrosc.,* 12(3), 177–182.

Depczynski, U., Jetter, K., Molt, K. and Niemöller, A. (1999). Quantitative analysis of near infrared spectra by wavelet coefficient regression using a genetic algorithm. *Chemom. Intell. Lab. Syst.,* 47(2), 179–187.

DeThomas, F.A. and Brimmer, P.J. (2002). Monochromators for near-infrared spectroscopy. In *Handbook of Vibrational Spectroscopy,* eds. J.M. Chalmers and P.R. Griffiths, 304–315, Chichester: Wiley.

Díaz-Carrillo, E., Muñoz-Serrano, A., Alonso-Moraga, A. and Serradilla-Manrique, J.M. (1993). Near infrared calibrations for goat's milk components: Protein, total casein, αs-, β- and κ-caseins, fat and lactose. *J. Near Infared Spectrosc.,* 1(3), 141–146.

Di Egidio, V., Sinelli, N., Limbo, S., Torri, L., Franzetti, L. and Casiraghi, E. (2009). Evaluation of shelf-life of fresh-cut pineapple using FT-NIR and FT-IR spectroscopy. *Postharvest Biol. Technol.,* 54, 87–92.

Dou, Y., Mi, H., Zhao, L., Ren, Y. and Ren, Y. (2006). Radial basis function neural networks in non-destructive determination of compound aspirin tablets on NIR spectroscopy. *Spectrochim. Acta Part A,* 65(1), 79–83.

Downey, G. (1996). Review: Authentication of food and food ingredients by near infrared spectroscopy. *J. Near Infrared Spectrosc.,* 4, 47–61.

Downey, G., McElhinney, J. and Fearn, T. (2000). Species identification in selected raw homogenized meats by reflectance spectroscopy in the mid-infrared, near-infrared, and visible ranges. *Appl. Spectrosc.,* 54(6), 894–899.

Downey, G., Sheehan, E., Delahunty, C., O'Callaghan, D., Guinee, T., and Howard, V. (2005). Prediction of maturity and sensory attributes of cheddar cheese using near-infrared spectroscopy. *International Dairy Journal,* 15, 701–709.

Duarte, I.F., Barros, A., Delgadillo, I., Almeida, C. and Gil, A.M. (2002). Application of FTIR spectroscopy for the quantification of sugars in mango juice as a function of ripening. *J. Agric. Food Chem.,* 50(11), 3104–3111.

Ellis, D.I., Broadhurst, D. and Clarke, S.J. (2005). Rapid identification of closely related muscle foods by vibrational spectroscopy and machine learning. *Analyst,* 130, 1648–1654.

El Masry, G., Wang, N., El Sayed, A. and Ngadi, M. (2007). Hyperspectral imaging for nondestructive determination of some quality attributes for strawberry. *J. Food Eng.,* 81(1), 98–107.

Fearn, T. (2001). Review: Standardisation and calibration transfer for near infrared instruments: A review. *J. Near Infrared Spectrosc.,* 9(4), 229–244.

Fearn, T. and Davies, A.M.C. (2003). A comparison of fourier and wavelet transforms in the processing of near infrared spectroscopic data: Part 1. Data compression. *J. Near Infrared Spectrosc.,* 11(1), 3–15.

Fontecha, J., Bellanato, J. and Juarez, M. (1993). Infrared and Raman spectroscopic study of casein in cheese: Effect of freezing and frozen storage. *Journal of Dairy Science,* 76(11), 3303–3309.

Fuller, M.P., Ritter, G.L. and Draper, C.S. (1988). Partial least-squares quantitative analysis of infrared spectroscopic data. Part I: Algorithm implementation. *Appl. Spectrosc.*, 42(2), 199–369.

Gat, N. (2000). Imaging spectroscopy using tunable filters: A review. In *Wavelet Applications VII*, SPIE Proceedings 4056, eds. H.H. Szu, M. Vetterli, W.J. Campbell and J.R. Buss, 50–64. Bellingham, WA: SPIE.

Geladi, P. and Kowalski, B. (1986). Partial least-squares regression: A tutorial. *Analytica Chimica Acta*, 185, 1–17.

Geladi, P., MacDougall, D. and Martens, H. (1985). Linearization and scatter-correction for near-infrared reflectance spectra of meat. *Appl. Spectrosc.*, 39(3), 491–500.

Giese, A.T. and French, C.S. (1955). The analysis of overlapping spectral absorption bands by derivative spectrophotometry. *Appl. Spectrosc.*, 9(2), 78–96.

Gresham, C.A., Gilmore, D.A. and Denton, M.B. (1999). Direct comparison of near-infrared absorbance spectroscopy with Raman scattering spectroscopy for the quantitative analysis of xylene isomer mixtures. *Appl. Spectrosc.*, 53(10), 1177–1182.

Griffiths, P.R. and de Haseth, J.A. (2007). *Fourier Transform Infrared Spectroscopy*. New York: John Wiley & Sons.

Hartmann, R. and Büning-Pfaue, H. (1998). NIR determination of potato constituents. *Potato Res.*, 41, 327–334.

He, Y., Li. X. and Deng, X. (2007). Discrimination of varieties of tea using near infrared spectroscopy by principal component analysis and BP model. *J. Food Eng.*, 79(4), 1238–1242.

Helland, I.S., Næs, T. and Isaksson, T. (1995). Related versions of the multiplicative scatter correction method for preprocessing spectroscopic data. *Chemom. Intell. Lab. Syst.*, 29(2), 233–241.

Herrero, A.M. (2008). Raman spectroscopy a promising technique for quality assessment of meat and fish: A review. *Food Chem.*,107, 1642–1651.

Herrero, A.M., Carmona, P., Ordóñez, J.A., de la Hoz, L. and Cambero, M.I. (2009). Raman spectroscopic study of electron-beam irradiated cold-smoked salmon. *Food Research International*, 42(1), 216–220.

Holler, F., Burns, D.H. and Callis, J.B. (1989). Direct use of second derivatives in curve-fitting procedures. *Appl. Spectrosc.*, 43(5), 877–882.

Hruschka, W.R. (2001). Data analysis: Wavelength selection methods. In *Near-infrared Technology in the Agricultural and Food Industries*, eds. P.C. Williams and K.H. Norris, 2nd edn, 39–58, St. Paul, MN: AACC.

Indahl, U.G., Sahni, N.S., Kirkhus, B. and Naes, T. (1999). Multivariate strategies for classification based on NIR-spectra with application to mayonnaise. *Chemom. Intell. Lab. Syst.*, 49, 19–31.

Iñón, F.A., Garrigues, S. and de la Guardia, M. (2004). Nutritional parameters of commercially available milk samples by FTIR and chemometric techniques. *Analytica Chimica Acta*, 513, 401–412.

Irudayaraj, J. and Tewari, J. (2003). Simultaneous monitoring of organic acids and sugars in fresh and processed apple juice by fourier transform infrared-attenuated total reflection spectroscopy. *Appl. Spectrosc.*, 57(12), 1599–1604.

Isaksson, T. and Næs, T. (1988). The effect of multiplicative scatter correction (MSC) and linearity improvement in NIR spectroscopy. *Appl. Spectrosc.*, 42(7), 1273–1284.

Isaksson, T. and Næs, T. (1990). Selection of samples for calibration in near-infrared spectroscopy. Part II: Selection based on spectral measurements. *Appl. Spectrosc.*, 44(7), 1152–1158.

Karoui, R. and De Baerdemaeker, J. (2007). Issues in development of NIR micro spectrometer system for on-line process monitoring of milk product. *Food Chem.*, 102, 621–640.

Karoui, R., Mouazen, A.M., Dufour, É., Schoonheydt, R. and De Baerdemaeker, J. (2006). A comparison and joint use of VIS-NIR and MIR spectroscopic methods for the determination of some chemical parameters in soft cheeses at external and central zones: A preliminary study. *Eur. Food Res. Technol.*, 223, 363–371.

Kavdir, I., Lu, R., Ariana, D. and Ngouajio, M. (2007). Visible and near-infrared spectroscopy for nondestructive quality assessment of pickling cucumbers. *Postharvest Biol. Technol.*, 44(2), 165–174.

Kennedy, J.F. and White, C.A. (1985). Application of infrared reflectance spectroscopy to the analysis of milk and dairy products. *Food Chem.*, 16, 115–131.

Ku, M. and Chung, H. (1999). Comparison of near-infrared and Raman spectroscopy for the determination of chemical and physical properties of naphtha. *Appl. Spectrosc.*, 53(5), 557–564.

Laporte, M. and Paquin, P. (1999). Near-infrared analysis of fat, protein, and casein in cow's milk. *J. Agric. Food Chem.*, 47(7), 2600–2605.

Law, D.P., Blakeney, A.B. and Tkachuk, R. (1996). The Kubelka–Munk equation: Some practical considerations. *J. Near Infrared Spectrosc.*, 4, 189–193.

Lee, K., Herrman, T.J., Lingenfelser, J. and Jackson, D.S. (2005). Classification and prediction of maize hardness-associated properties using multivariate statistical analyses. *J. Cereal Sci.*, 41(1), 85–93.

Lerner, E.J. (1996). Infrared detectors offer high sensitivity. *Laser Focus World*, 32(6), 155–164.

Lestander, T.A., Leardi, R. and Geladi, P. (2003). Selection of near infrared wavelengths using genetic algorithms for the determination of seed moisture content. *J. Near Infrared Spectrosc.*, 11(6), 433–446.

Li-Chan, E.C.Y. (1996). The applications of Raman spectroscopy in food science. *Trends in Food Science and Technology*, 7, 361–370.

Liu, Y., Chen, Y., Wang, C.Y., Chan, D.E. and Kim, M.S. (2005). Development of a simple algorithm for the detection of chilling injury in cucumbers from visible/near-infrared hyperspectral imaging. *Appl. Spectrosc.*, 59(1), 78–85.

Lu, R. and Peng, Y. (2005). Assessing peach firmness by multi-spectral scattering. *J. Near Infrared Spectrosc.*, 13(1), 27–36.

Luinge, H.J., Hop, E., Lutz, E.T.G., van Hemert, J.A. and De Jong, E.A.M. (1993). Determination of the fat, protein and lactose content of milk using fourier transform infrared spectrometry. *Analytica Chimica Acta*, 284, 419–433.

Ma, C.Y. and Phillips, D.L. (2002). FT-Raman spectroscopy and its applications in cereal science. *Cereal Chem.*, 79(2), 171–177.

Maertens, K., Reyns, P. and De Baerdemaeker, J. (2004). On-line measurement of grain quality with NIR technology. *Trans. ASAE*, 47(4), 1135–1140.

Maghirang, E.B., Dowell, F.E., Baker, J.E. and Throne, J.E. (2003). Automated detection of single wheat kernels containing live or dead insects using near-infrared reflectance spectroscopy. *Trans. ASAE*, 46(4), 1277–1282.

Mallet, Y., Coomans, D. and de Vel, O. (1996). Recent developments in discriminant analysis on high dimensional spectral data. *Chemom. Intell. Lab. Syst.*, 35(2), 157–173.

Manley, M., Downey, G. and Baeten, V. (2008). Spectroscopic technique: Near-infrared (NIR) spectroscopy. In *Modern Techniques for Food Authentication* ed. D.-W. Sun, 65–116, Amsterdam: Elsevier/Academic Press.

Manrique, G.D. and Lajolo, F.M. (2002). FT-IR spectroscopy as a tool for measuring degree of methyl esterification in pectins isolated from ripening papaya fruit. *Postharvest Biol. Technol.*, 25, 99–107.

Mantz, A.W. (2002). Diode laser spectrometers for mid-infrared spectroscopy. In *Handbook of Vibrational Spectroscopy*, eds. J.M. Chalmers and P.R. Griffiths, 304–315, Chichester: Wiley.

Martens, H. and Næs, T. (1984). Multivariate calibration. I. Concepts and distinctions. *Trends Anal. Chem.*, 3(8), 204–210.

Martens, H., Nielsen, J.P. and Engelsen, S.B. (2003). Light scattering and light absorbance separated by extended multiplicative signal correction. Application to near-infrared transmission analysis of powder mixtures. *Anal. Chem.*, 75, 394–404.

Martin, P.A. (2002). Near-infrared diode laser spectroscopy in chemical process and environmental air monitoring. *Chemical Society Reviews*, 31, 201–210.

Massart, D.L., Vandeginste, B.G.M., Deming, S.N., Michotte, Y. and Kaufman, L. (1988). *Chemometrics: A Textbook*. Amsterdam: Elsevier.

Mayes, D.M. and Callis, J.B. (1989). A photodiode-array-based near-infrared spectrophotometer for the 600–1100 nm wavelength region. *Appl. Spectrosc.*, 43(1), 27–32.

McClure, W.F. (2001). Near-infrared instrumentation. In *Near-infrared Technology in the Agricultural and Food Industries*, eds. P. Williams and K. Norris, 109–117, St Paul, MN: American Association of Cereal Chemists.

McClure, W.F. and Davies, A.M.C. (1988). Fast fourier transforms in the analysis of near-infrared spectra. In *Analytical Applications of Spectroscopy*, eds. C.S. Creaser and A.M.C. Davies, 414–436, London: Royal Society of Chemistry.

McClure, W.F. and Tsuchikawa, S. (2007). Instruments. In *Near-infrared Spectroscopy in Food Science and Technology*, eds. Y. Ozaki, W.F. McClure and A.A. Christy, 75–108, Hoboken, NJ: Wiley-Interscience.

McCreery, R.L. (2000). *Raman Spectroscopy for Chemical Analysis*. New York: John Wiley & Sons.

Meurens, M., Baeten, V., Yan, S.H., Mignolet, E. and Larondelle, Y. (2005). Determination of the conjugated linoleic acids in cow's milk fat by fourier transform Raman spectroscopy. *J. Agric. Food Chem.*, 53(15), 5831–5835.

Miralbés, C. (2003). Prediction chemical composition and alveograph parameters on wheat by near-infrared transmittance spectroscopy. *J. Agric. Food Chem.*, 51(21), 6335–6339.

Moros, J., Iñón, F.A., Garrigues, S. and de la Guardia, M. (2005). Determination of the energetic value of fruit and milk-based beverages through partial-least-squares attenuated total reflectance-fourier transform infrared spectrometry. *Analytica Chimica Acta*, 538, 181–193.

Muik, B., Lendl, B., Molina-Díaz, A., Ortega-Calderón, D. and Ayora-Cañada, M.J. (2004). Discrimination of olives according to fruit quality using fourier transform Raman spectroscopy and pattern recognition techniques. *J. Agric. Food Chem.*, 52, 6055–6060.

Næs, T. and Isaksson, T. (1989). Selection of samples for calibration in near-infrared spectroscopy. Part I: General principles illustrated by example. *Appl. Spectrosc.*, 43(2), 328–335.

Næs, T. and Martens, H. (1984). Multivariate Calibration. II. Chemometric Methods. *Trends Anal. Chem.*, 3(10), 266–271.

Nicolaï, B.M., Lötze, E., Peirs, A., Scheerlinck, N. and Theron, K.I. (2006). Non-destructive measurement of bitter pit in apple fruit using NIR hyperspectral imaging. *Postharvest Biol. Technol.*, 40(1), 1–6.

Ootake, Y. and Kokot, S. (1998). Discrimination between glutinous and non-glutinous rice by vibrational spectroscopy. II: Effects of spectral pretreatment on the classification of the two types of rice. *J. Near Infrared Spectrosc.*, 6(1), 251–258.

Osborne, B.G., Fearn, T. and Hindle, P.H. (1993). *Practical NIR Spectroscopy: With Applications in Food and Beverage Analysis*, 2nd edn. Harlow: Longman Scientific & Technical.

O'Shea, D.C. (1985). *Elements of Modern Optical Design*. New York: John Wiley & Sons.

Ozaki, Y., Morita, S. and Du, Y. (2007). Spectral analysis. In *Near-Infrared Spectroscopy in Food Science and Technology*, eds. Y. Ozaki, W. F. McClure and A.A. Christy, 43–72, Hoboken, NJ: Wiley-Interscience.

Pal, M. and Mather, P.M. (2005). Support vector machines for classification in remote sensing. *Int. J. Remote Sensing*, 26(5), 1007–1011.

Palmer, C. (2002). *Diffraction Grating Handbook*, 5th edn. Rochester, NY: Richardson Grating Laboratory.

Paradkar, M.M., Sakhamuri, S. and Irudayaraj, J. (2002). Comparison of FTIR, FT-Raman, and NIR spectroscopy in a maple syrup adulteration study. *J. Food Sci.*, 67(6), 2009–2015.

Park, B., Chen, Y.R., Hruschka, W.R., Shackelford, S.D. and Koohmaraie, M. (2001). Principal component regression of near-infrared reflectance spectra for beef tenderness prediction. *Trans. ASAE*, 44(3), 609–615.

Pedersen, D.K., Martens, H., Nielsen, J.P. and Engelsen, S.B. (2002). Near-infrared absorption and scattering separated by extended inverted signal correction (EISC): Analysis of near-infrared transmittance spectra of single wheat seeds. *Appl. Spectrosc.*, 56(9), 1206–1214.

Pedersen, D.K., Morel, S., Andersen, H.J. and Engelsen, S.B. (2003). Early prediction of water-holding capacity in meat by multivariate vibrational spectroscopy. *Meat Science*, 65, 581–592.

Peng, Y. and Lu, R. (2006). An LCTF-based multispectral imaging system for estimation of apple fruit firmness: Part I. Acquisition and characterization of scattering images. *Trans. ASABE*, 49(1), 259–267.

Piot, O., Autran, J.C. and Manfait, M. (2002). Assessment of cereal quality by micro-Raman analysis of the grain molecular composition. *Appl. Spectrosc.*, 56(9), 1132–1138.

Prieto, N., Roehe, R., Lavín, P., Batten, G. and Andrés, S. (2009). Application of near infrared reflectance spectroscopy to predict meat and meat products quality: A review. *Meat Science*, 83(2), 175–186.

Qiao, J., Ngadi, M.O., Wang, N., Gariépy, C. and Prasher, S.O. (2007). Pork quality and marbling level assessment using a hyperspectral imaging system. *J. Food Eng.*, 83(1), 10–16.

Qiao, Y. and van Kempen, T.A.T.G. (2008). Technical note: Comparison of Raman, mid, and near infrared spectroscopy for predicting the amino acid content in animal meals. *J Anim. Sci.*, 82, 2596–2600.

Qin, J. and Lu, R. (2005). Detection of pits in tart cherries by hyperspectral transmission imaging. *Trans. ASAE*, 48(5), 1963–1970.

Rinnan, A., van den Berg, F. and Engelsen, S.B. (2009). Review of the most common pre-processing techniques for near-infrared spectra. *Trends in Analytical Chemistry*, 28(10), 1201–1222.

Ripoche, A. and Guillard, A.S. (2001). Determination of fatty acid composition of pork fat by fourier transform infrared spectroscopy. *Meat Science*, 58(3), 299–304.

Rittiron, R., Saranwong, S. and Kawano, S. (2004). Useful tips for constructing a near infrared-based quality sorting system for single brown-rice kernels. *J. Near Infrared Spectrosc.,* 12(2), 133–139.

Rodriguez-Otero, J.L., Hermida, M. and Centeno, J. (1997). Analysis of dairy products by near-infrared spectroscopy: A review. *J. Agric. Food Chem.,* 45(8), 2815–2819.

Saranwong, S. and Kawano, S. (2008). Interpretation of near infrared calibration structure for determining the total aerobic bacteria count in raw milk: Interaction between bacterial metabolites and water absorptions. *J. Near Infrared Spectrosc.,* 16(6), 497–504.

Sato, H., Saito, N., Akagawa, K., Wada, S. and Tashiro, H. (2003). Electronically tunable-laser light sources for near infrared spectroscopy. *J. Near Infrared Spectrosc.,* 11(4), 295–308.

Savenije, B., Geesink, G.H., van der Palen, J.G.P. and Hemke, G. (2006). Prediction of pork quality using visible/near-infrared reflectance spectroscopy. *Meat Science,* 73(1), 181–184.

Savitzky, A. and Golay, M.J.E. (1964). Smoothing and differentiation of data by simplified least squares procedures. *Anal. Chem.,* 36(8), 1627–1639.

Schulz, H., Schrader, B., Quilitzsch, R., Pfeffer, S. and Krüger, H. (2003). Rapid classification of basil chemotypes by various vibrational spectroscopy methods. *J. Agric. Food Chem.,* 51, 2475–2481.

Shackelford, S.D., Wheeler, T.L. and Koohmaraie, M. (2004). Development of optimal protocol for visible and near-infrared reflectance spectroscopic evaluation of meat quality. *Meat Science,* 68(3), 371–381.

Sirisomboon, P., Tanaka, M., Fujita, S. and Kojima, T. (2007). Evaluation of pectin constituents of Japanese pear by near infrared spectroscopy. *J. Food Eng.,* 78(2), 701–707.

Theocharous, E. and Birch, J.R. (2002). Detectors for mid and far infrared spectroscopy: Selection and use. In *Handbook of Vibrational Spectroscopy*, eds. J.M. Chalmers and P.R. Griffiths, 349–367, Chichester: Wiley.

Thissen, U., Üstün, B., Melssen, W.J. and Buydens, L.M.C. (2004). Multivariate calibration with least-squares support vector machines. *Anal. Chem.,* 76(11), 3099–3105.

Tsenkova, R., Atanassova, S., Morita, H., Ikuta, K., Toyoda, K., Iordanova, I.K. and Hakogic, E. (2006). Near infrared spectra of cow's milk for milk quality evaluation: Disease diagnosis and pathogen identification. *J. Near Infrared Spectrosc.,* 14(6), 363–370.

Vapnik, V.N. (1995). *The Nature of Statistical Learning Theory*. New York: Springer-Verlag.

Venel, C., Mullen, A.M.. Downey, G. and Troy, D.J. (2001). Prediction of tenderness and other quality attributes of beef by near infrared reflectance spectroscopy between 750 and 1100 nm: Further studies. *J. Near Infrared Spectrosc.,* 9(3), 185–198.

Veraverbeke, E.A., Lammertyn, J., Nicolaï, B.M. and Irudayaraj, J. (2005). Spectroscopic evaluation of the surface quality of apple. *J. Agric. Food Chem.,* 53(4), 1046–1051.

Viljoen, M., Hoffman, L.C. and Brand, T.S. (2005). Prediction of the chemical composition of freeze dried ostrich meat with near infrared reflectance spectroscopy. *Meat Science,* 69(2), 255–261.

Walczak, B., van den Bogaert, B. and Massart, D.L. (1996). Application of wavelet packet transform in pattern recognition of near-IR data. *Anal. Chem.,* 68(10), 1742–1747.

Wang, D., Dowell, F.E., Ram, M.S. and Schapaugh, W.T. (2004) Classification of fungal damage soybean seeds using near-infrared spectroscopy. *Int. J. Food Properties,* 7(1), 75–82.

Wang, W. and Paliwal, J. (2007). Near-infrared spectroscopy and imaging in food quality and safety. *Sensing and Instrumentation for Food Quality and Safety,* 1(4), 193–207.

Watson, C.A., Carville, D., Dikeman, E., Daigger, G. and Booth, G.D. (1976). Evaluation of two infrared instruments for determining protein content of hard red winter wheat. *Cereal Chem.*, 53(2), 214–222.

Wetzel, D.L. (1983). Near-infrared reflectance analysis – sleeper among spectroscopic techniques. *Anal. Chem.*, 55, 1165A–1171A.

Wetzel, D.L. (2001). Contemporary near-infrared instrumentation. In *Near-infrared Technology in the Agricultural and Food Industries*, eds. P.C. Williams and K.H. Norris, 2nd ed., 129–144, St. Paul, MN: AACC.

Wetzel, D.L., Eilert, A.J. and Sweat, J.A. (2002). Tunable filter and discrete filter near-IR spectrometers. In *Handbook of Vibrational Spectroscopy*, eds. J.M. Chalmers and P.R. Griffiths, 436–452, Chichester: Wiley.

Wetzel, D.L. and LeVine, S.M. (1999). Microspectroscopy: Imaging molecular chemistry with infrared microscopy. *Science*, 285, 1224–1225.

Williams, G.P. (2002). Synchrotron and free electron laser sources of infrared radiation. In *Handbook of Vibrational Spectroscopy*, eds. J.M. Chalmers and P.R. Griffiths, 341, Chichester: Wiley.

Wold, S., Esbensen, K. and Geladi, P. (1987). Principal component analysis. *Chemom. Int. Lab. Syst.*, 2(1–3), 37–52.

Workman, J.J., Mobley, P.R., Kowalski, B.R. and Bro, R. (1996). Review of chemometrics applied to spectroscopy: 1985–1995 (Part 1). *Appl. Spectrosc. Rev.*, 31, 73–124.

Xie, L., Ying, Y. and Ying, T. (2007). Combination and comparison of chemometrics methods for identification of transgenic tomatoes using visible and near-infrared diffuse transmittance technique. *J. Food Eng.*, 82(3), 395–401.

Xu, Y., Zomer, S. and Brereton, R.G. (2006). Support vector machines: A recent method for classification in chemometrics. *Crit. Rev. Anal. Chem.*, 36, 177–188.

Yang, H., Irudayaraj, J. and Paradkar, M.M. (2005). Discriminant analysis of edible oils and fats by FTIR, FT-NIR and FT-Raman spectroscopy. *Food Chem.*, 93, 25–32.

Zhao, J. and McCreery, R.L. (1996). Multichannel fourier transform Raman spectroscopy: Combining the advantages of CCDs with interferometry. *Appl. Spectrosc.*, 50(9), 1209–1214.

Zude, M., Herold, B., Roger, J., Bellon-Maurel, V. and Landahl, S. (2006). Non-destructive tests on the prediction of apple fruit flesh firmness and soluble solids content on tree and in shelf life. *J. Food Eng.*, 77(2), 254–260.

Chapter 17

Chemosensor (Electronic Nose) for Food Quality Evaluation

P. Kumar Mallikarjunan

Contents

17.1 Introduction

One of the most critical food-quality attributes is the smell or aroma of the food as it determines the acceptability of the food by the consumers even before they place the food in their mouth. Several food products have specific aroma attributes associated with them, and any alteration to the perceived aroma characteristics will result in the rejection of the food product and, subsequently, the success of the food product in the marketplace. Traditionally, the food industry relies on trained and untrained consumer panels for aroma validation of food products for compliance, for aiming to produce acceptable aroma profiles, and for quality assurance. The aroma attribute has many descriptive terms: smell, aroma, odor, and flavor, and they are used interchangeably. Of these terms, "flavor" refers to the combined volatiles released during smelling and during mastication. The smell is the result of the reaction between volatile chemicals and the nose. The volatiles are often released from food due to both physical and chemical changes occurring in the food material. Primarily, the volatile compounds defining a particular smell are organic in nature and are comprised of aldehydes, ketones, and esters. In addition, other volatile compounds that are not organic in nature also contribute to the smell or odor. Inorganic chemical compounds that are associated with smell include sulfur-based compounds and ammonia, hydrogen sulfide, etc.

Humans can distinguish more than 10,000 different smells with the specific odor recognition through personal experiences in their life. Due to the complex nature of odor recognition in biological systems (including humans), it always has been a challenge to correlate the biological experience with analytical methods. The most common analytical method used in odor classification includes the use of hyphenated gas chromatography techniques (GC, GC-MS, GC-O). These methods require time-consuming, labor-intensive, and complex steps for analysis and depend on the effective separation of the volatiles using an appropriate chromatographic column. Further interpreting the chromatogram and taking the decision back to the sampling site to act on a decision requires sophistication and time. Complex samples have to be separated into their individual chemical components before a decision can be made. Due to the interactions of several volatiles in forming the sensory experience, even with the use of hyphenated mass spectrographic methods, the correlation of sensory smell with instrumental data has been not very successful. Aroma of a particular sample is so complex that a large number of statistical calculations or multiple sniff ports could not yield the exact smell of the sample.

Electronic nose technology has greater potential in mimicking animal olfactory sensing and can be used to identify, characterize, and measure aroma compounds more effectively than traditional volatile analysis techniques. Electronic nose technology comprises gas multi-sensor arrays that are able to measure aroma compounds in a manner that is closer to human olfactory analysis. In general, an electronic nose system has several sensors that may be set to achieve various levels of sensitivity and selectivity for a set of aroma compounds. The adsorption of volatiles on the sensor surface causes a physical or chemical change and produces an electric signal (voltage or current) and a specific reading to be obtained for a given sample (Bartlett et al. 1997) and generates a unique pattern (Haugen and Kvaal 1998). Although an electronic nose is not a substitute for human sensory panels, which are most reliable and sensitive in measuring aroma, it can be used as a rapid, automated, and objective alternative to detect, identify, and monitor aroma.

In recent decades, there have been major advances in sensor technologies for odor analyses. Electronic noses have been around for approximately 40 years, and the last 10 years have seen dynamic advancements in both sensor technology and information processing systems. Sensor arrays using the conducting polymer technology were initially developed for chemosensory applications in the early 1980s (Payne 1998). Initial work on an electronic nose with conducting polymers stemmed from polymer development advances achieved by the US Air Force. The Air Force was attempting to develop airplanes that could evade radar by using polymers that could conduct electricity. Researchers at Britain's Warwick University in the mid-1980s used the findings from the military research to develop the first electronic nose system based on conducting polymer sensors (Pope 1995). Since that time, other new sensor technologies have been developed that have properties more suitable for particular applications.

Metal oxide semiconductor (MOS) gas sensors were first used in the 1960s in Japan in home gas alarms. The conducting ceramic or oxide sensors were invented by Taguchi and produced by the company Figaro (Schaller and Bosset 1998). Electronic nose instruments have been tested successfully for use as a complementary tool in the discrimination of many consumer products. Current developments focus on developing electronic nose technologies based on the advances in microfluidics and nano-technologies.

17.2 Electronic Nose System

Electronic noses are comprised of (i) chemical sensors that are used to measure aromas, (ii) electronic system controls, and (iii) information processing systems and pattern-recognition systems that are capable of recognizing complex aromas. Although there are various sensor technologies used among the current manufactured instruments, most systems work using the same series of steps. They analyze compounds in a complex aroma and produce a simple output.

Usually, the steps in using an electronic nose system include (i) generating an odor from a sample, (ii) exposing the sensor array to the aroma, (iii) measuring changes in an array of sensors when they are exposed to the odor, (iv) establishing a recognition pattern for the sample from the responses of all or a number of sensors in the system, and (v) using this information in statistical analyses or pattern recognition neural networks to compare to a database of other chemosensory measurements.

Each of the sensors in the electronic nose are made with a unique material and, when exposed to a particular vapor mixture, each sensor reacts in a different but reproducible manner producing a "smell print" (a combination of responses from all sensors) for each volatile mixture. A database of smell prints or the digital images of a chemical vapor mixture is created by training the electronic nose system. Then using multivariate statistical techniques (e.g., principal component analysis, canonical discriminant analysis, etc.), a prediction model can be developed. When the sensor array is exposed to a new unknown vapor mixture to be identified, the electronic nose system digitizes the vapor mixture and compares this digital image with the previously established database (model) of smell prints in its memory. The unique feature of the electronic nose system is that its response takes into account all the characteristic features (chemical and physical properties) of a sample but does not provide information about the composition of the complex mixture. Thus, this system can be used when the decision about a chemical vapor of a sample is more important than its contents, such as a spoiled vs. non-spoiled food sample, age of a fruit, type of cheese, and adulteration in the product, etc. The electronic nose technologies cannot be used for a quantitative analysis of concentration of certain chemicals due to this inherent nature of statistical algorithms employed in developing the prediction models. Researchers have attempted to develop such quantitative analysis using statistical methods such as the partial least square method but have not been successful in implementation of electronic nose technologies in such applications.

The electronic nose has both advantages and disadvantages over the use of human sensory panels as well as GC/MS analyses. Therefore, it finds use as a complementary instrument to monitor odors.

The human olfaction system, which is the basis of sensory panels, is still the most sensitive device available for aroma measurement. It is also the odor measurement method used by consumers when assessing the odor of consumable products. Therefore, it is important that any odor-monitoring methods used in quality control or quality assurance is capable of detecting odors that may be found to be offensive by the human olfactory system. This fact is also the reason that human sensory panels are still the basis of aroma measurements in the food industry.

Although electronic noses cannot compete with the sensitivity, selectivity, and final correlation of sensory panels and replace them, they are objective instruments and involve primarily a small capital investment compared to having a standing sensory panel. They can also be used on the production floor and even can be implemented for in-line measurements. Work performed by Strassburger (1998) demonstrated that an MOS sensor-based instrument showed great potential in aiding in flavor analysis going from the research and development phases to the production floor as it produced results that were directly correlated to sensory and analytical results.

Sensory panels are inherently subjective, and the physical condition of panelists may vary from one day to another. This brings inherent error into any scientific quantification of experimental results. Human panels require sustained training for each type of product or sample, and standardization between different panels at different sites is extremely difficult. Sensory panels have high costs associated with training, maintaining, and testing, and they experience fatigue. Therefore, they are not run continuously for extended periods of time. A trained electronic nose provides a complementary objective tool available for 24 h complex aroma analysis (Payne 1998). Newman et al. (1999) used a conducting polymer sensor-based electronic nose as a complementary tool to sensory analysis in the odor analysis of raw tuna quality. Electronic nose measurements were successfully correlated with sensory scores with correct classification rates of 88%, 82%, and 90% for raw tuna stored at three temperatures.

The electronic nose is an instrument that can be portable or be connected to an auto-sampler to reduce the need for human involvement in multiple sample testing. It is also an instrument available to potentially test odors that a human sensory panel would not be willing to test although this facet is not particularly pertinent to the food industry. An aroma may be taken at ambient conditions to mimic what the human nose would experience under normal circumstances, or the sample may be heated to intensify aroma concentrations. Aroma exposure to the sensor array is generally accomplished by one of two methods: static headspace analysis or flow injection analysis. Static headspace analysis involves direct exposure to a saturated vapor taken from the headspace above a sample. Flow injection analysis involves injecting the aroma sample into a control gas that is constantly pumped through the sensor chamber (Payne 1998).

Past objective odor monitoring analyses options involved the use of analytical GC/MS techniques. These techniques offer identification and quantification of compounds comprising an aroma. However, GC/MS techniques often find difficulties in identifying which of the constituent compounds contribute to the recognized odor and to what extent, particularly if they are complex odors. Electronic noses have a unique advantage over GC and MS techniques because they use an analytical technique that samples an entire aroma rather than identifying it by its constituent components. It is also a faster method of aroma analysis (Payne 1998). A portable electronic nose unit could also be used to directly sample headspace aromas from bulk raw materials or food containers in which sampling for GC/MS analysis becomes difficult (Hodgins and Conover 1995).

Electronic nose analysis is also a technique that may be nondestructive and incurs low operational costs. Overall, it fills a number of gaps in odor analysis not achieved by use of sensory panels and GC/MS techniques in conjunction. While the electronic nose has a number of weak points that inhibit its ability to be used exclusively, it is a powerful tool that enhances aroma monitoring when used as a complementary tool to sensory panels and GC/MS techniques.

Currently, three major technologies are being used in commercially available electronic nose systems: MOS, semiconducting polymers, and quartz crystal microbalance sensors. A select number of companies also produce systems using MOS field effect transistor (MOSFET) technology. In addition, technologies using surface acoustic wave sensors provide additional opportunities to convert traditional GC methods to develop fast-GCs that are similar to electronic nose systems. The determination of which of the aforementioned types of electronic nose systems, if any, that would be most suitable as a discriminatory analysis tool to be used in quality control is of interest in the food industry. The other major issue with the electronic nose technology is it is perceived as market-pushed and not demand-driven. Many companies have sprung up but failed to sustain a market presence due to the push versus pull of the technology. The electronic nose manufacturers are trying to find applications to employ their systems and trying to convince the user base, especially the food industry, to adopt their systems. As a result, one can see rapid changes in the marketplace. A few examples include Cyrano Sciences merging with Smith Detection Systems and is now a new company, Intelligent Optical Systems; AromaScan changed to Osmetech; Nordic Sensors merged into Applied Sensors (and moved to other areas); Perkin Elmer (1999) dropped the support for their system with HKR GmbH, letting that company go out of business; and Agilent 4440A from Hewlett Packard is no longer available. The 2012 Institute of Food Technologists Annual Meeting and Food Exposition had only one exhibitor promoting the electronic nose technology (Alpha MOS) when previous years had many companies on the expo floor promoting the technology.

Current commercial electronic nose system manufacturers that are most involved in the market include Airsense (Germany), Alpha MOS (France), Applied Sensor (US) merged from Nordic Sensor Technologies (Sweden) and Motech GmbH (Germany), AromaScan (UK), Bloodhound Sensors (UK), HKR Sensor Systems (Germany), Lennartz Electronic (Germany), Neotronics (USA, UK), RST Rostock (Germany), and OligoSense (Belgium) (Wilson and Baietto 2009). A list of current manufactures of electronic nose systems with different technologies is shown in Table 17.1.

Table 17.1 Commercially Available Electronic Nose Systems

Manufacturer	*Enose Technology*	*Current Models Offered*
Airsense Analytics GmbH, Schwerin, Germany	MOS	Pen 3
Alpha MOS, Toulouse, France	MOS	FOX 3000, GEMINI
	Soft Ionized Mass Spec	AIRSENSE
Applied Sensor GmbH, Reutlingen, Germany	MOS	Air Quality Module
	QCM	VocChek (not available now)
CSIRO, Australia	Receptor-based array	CyberNose (under development)
Intelligent Optical Systems, California, USA	Conducting Polymer	Cyranose 320
Technology Translators, West Yorkshire, UK	Conducting Polymer	BloodHound
Electronic Sensor Technology, Pennsylvania, USA	SAW	ZNose 4300, 7100

17.3 Types of Electronic Nose

The major types of chemosensory-based electronic nose technology include MOS sensors, conducting polymer (CP) sensors, quartz microbalance (QMB) sensors, and MOSFETs. Certain manufacturers in recent years have also been developing hybrid or modular chemosensory systems that use multiple sensor types. The MOS and MOSFET sensors are considered to be "hot" sensors, and the remaining sensor technologies, CP and QMB sensors, are considered to be "cold" sensors due to their operating temperatures (Schaller and Bosset 1998). Recently, there has been an increase in the development of nanoscale sensors (primarily using metal oxide bases) with an aim to miniaturizing the sensing device.

MOS sensors and CP sensors are the two technologies that have been used the longest in commercial electronic nose systems. Conducting polymer sensors are easily fabricated and are fabricated with a high degree of reproducibility. They also have the greatest range of selectivity and sensitivity. However, the MOS-based systems are less susceptible to water vapor variations, are more robust, have a longer useful life, and are cheaper to replace.

17.3.1 Metal Oxide Sensors

MOS sensors consist of a ceramic substrate heated by wire and coated by a metal oxide semiconducting film. The metal oxide coatings used are often n-type oxides that include zinc oxide, tin dioxide, titanium dioxide, or iron (III) oxide. P-type oxides, such as nickel oxide or cobalt oxide, are also used.

The main difference between sensors using the two types of oxide coatings are the types of compounds with which they react. The sensors using n type (n = negative electron) coatings respond to oxidizing compounds because the excitation of these sensors results in an excess amount of electrons in its conduction band. The p type (p = positive hole) sensors develop an electron deficiency when excited and therefore are more prone to react with reducing compounds (Schaller and Bosset 1998).

MOS sensors have a low sensitivity to moisture and are robust. They typically operate at temperatures ranging from 400°C to 600°C to avoid moisture effects. These sensors are not typically sensitive to nitrogen- or sulfur-based odors, but they are sensitive to alcohols and other combustibles (Bartlett et al. 1997).

17.3.2 Conducting Polymer Sensors

Conducting polymer sensors are composed of a conducting polymer, a counter ion, and a solvent that are grown from a solution onto an electrode bridging a 10-μm gap to produce a resistor. Measurements are made by measuring changes in resistance. Altering one or more of the three constituent materials produces different sensors. The single-stage fabrication technique allows the reproducibility from the production of one sensor to the next to be consistent.

Conducting polymer sensors are formed electrochemically onto a silicon or carbon substrate. This results in a polymer in an oxidized form that has cationic sites and anions from the electrolyte. Sensors made from polymers based on aromatic or hetero-aromatic compounds, such as polypyrrole, polythiophene, and polyaniline, are sensitive to many volatile compounds and experience a reversible change in conductions (Persaud et al. 1999). Conduction is achieved in the electrically conductive polymer by electron transport, not ion transport. The charge carriers are associated with the cation sites (Hodgins and Simmonds 1995).

Although the conducting polymer sensors have the greatest range and balance between selectivity and sensitivity, they are more sensitive to water vapor and are more expensive to produce and replace. They can be used at room temperatures and temperatures moderately higher. This allows for future development of handheld electronic nose instruments and avoids problems associated with the breakdown of volatiles at the sensor surface of systems using increased heating (Persaud et al. 1999).

17.3.3 Quartz Crystal Microbalance

Quartz microbalance (QMB) sensors or quartz crystal microbalance (QCM) sensors are sensors that evolved from a larger group of piezoelectric crystal sensors. These sensors use crystals that can be made to vibrate in a surface acoustic mode (SAW) or bulk acoustic mode (BAW). The sensors are made from thin discs composed of quartz, lithium niobate ($LiNbO_3$), or lithium tantalite ($LiTaO_3$) and then coated. The coating materials are usually gas chromatographic stationary phases but may be any nonvolatile compounds that are chemically and thermally stable (Schaller and Bosset 1998).

The quartz microbalance sensors respond to an aroma through a change in mass. When an alternating voltage is applied at a constant temperature, the quartz crystal vibrates at a very stable and measurable frequency. This is dependent upon the assumption that viscoelastic effects are negligible (Bartlett et al. 1997). Upon exposure to volatile compounds in an aroma, the volatiles adsorb onto the GC phase coating of the sensor, which causes a change in the mass of the sensor. The change in mass results in a measurable change of the oscillating frequency of the sensor. QMB sensors have developed as a useful electronic nose technology because they produce stable responses and are formed through a simple fabrication process.

In reporting on trends and developments in quartz microbalance chemosensory systems, Nakamoto and Morizumi (1999) performed work examining QMB sensor responses with different aroma injection systems as well as model development for response prediction. QMB sensor technology continues to improve and Applied Sensor has released a handheld unit this year that is currently the least expensive electronic nose system on the market.

17.3.4 Metal Oxide Semiconductors Field Effect Transistors

MOSFET sensors respond to aroma volatiles with a measurable change in electrostatic potential. Each sensor in a MOSFET system consists of three layers, including a silicon semiconductor, a silicon oxide insulator, and a catalytic metal. The catalytic metal component is also called the gate and is usually palladium, platinum, iridium, or rhodium (Schaller and Bosset 1998). The standard transistor is an example of an "active" circuit component, a device that can amplify, producing an output signal with more power than the input signal.

The field-effect transistor (FET) controls the current between two points but does so differently than the bipolar transistor. The FET operates by the effects of an electric field on the flow of electrons through a single type of semiconductor material. Current flows within the FET in a channel from the source terminal to the drain terminal. A gate terminal generates an electric field that controls the current.

Placing an insulating layer between the gate and the channel allows for a wider range of control (gate) voltages and further decreases the gate current (and thus increases the device input resistance). The insulator is typically made of an oxide (such as silicon dioxide, SiO_2). This device is the MOSFET. MOSFET sensors are similar to MOS sensors in that they are also robust and have a low sensitivity to water.

17.4 Electronic Nose Technology Implementation Issues

Major problems that exist with the use of the electronic nose systems are sensor drift, poor repeatability and reproducibility due to system sensitivity to changes in operational conditions, and poor gas selectivity and sensitivity (Roussel et al. 1998). In order to overcome these difficulties, it is necessary to develop a successful and efficient testing methodology at optimum parameter settings and periodic calibration, or retraining the nose is warranted.

Issues such as sensor drift and the nature of the instruments' discriminatory abilities are major concerns as electronic nose technology is selected and developed for particular applications. To achieve the necessary repeatability, it is necessary that the sensors in the electronic nose systems react reversibly with the compounds in a sample aroma. Sensor drift occurs when the sensors experience small additive changes over time and usage. The aging of sensors, or sensor drift, has been a major issue of concern throughout the history of the development of electronic nose systems. However, some of the most recent advancements in electronic nose technology have been developed to deal with this issue. Advances in the design and manufacture of sensors have to increase the useful life of sensors, and calibration standards and artificial neural networks have been better developed to increase the reliability and longevity of measurements to which unknowns are compared.

In addition, optimization of systems and experimental parameters can establish more stable conditions and combat sensor drift. Mielle and Marquie (1998) performed work examining several parameters or dimensions of electronic nose analysis, including sensor temperature, number of sensors, and sample incubation time, in order to stabilize system response and lengthen the useful life of library patterns in the system database.

Roussel et al. (1999) examined the influence of various experimental parameters on the multi-sensor array measurements using an electronic nose with SnO_2 sensors and attempted to quantify them. Volatile concentration in the headspace increases as the sample temperature is increased. In screening factorial experimental designs, it is necessary that the response of the experimental parameters be monotonic within a studied range. Alternatively, response surface designs must be generated in order to develop a model of the multi-sensor response.

The discriminatory power of any electronic nose chemosensory system is based upon its ability to respond measurably and repeatedly to components of aromas and to respond differently to aromas with varying components. The chemical nature and concentration of the volatiles in an aroma and reaction kinetics and dynamics of those volatiles as well as system parameters and sample preparation affect the fundamental response of each sensor. Schaak et al. (1999) examined the effects of the system parameters, injection volume, incubation time, and incubation temperature, and their effect on sensor response and discriminatory power for the MOS sensor-based Alpha MOS FOX 3000. Optimizing the system response of sensors in an electronic nose system through controlling the system and experimental parameters is a key to it being a useful analytical tool in most applications. Nakamoto and Morrizumi (1999) reported that the QMB sensor responses could be predicted using computational chemistry. This allows for the ability of optimal sensor selection for target odors. Hansen and Wiedemann (1999) performed optimization work in using the Alpha MOS (Toulouse, France) FOX 4000. The experimental and system parameters were investigated to optimize the response range of the sensors and enhance their discriminatory power. This work was performed using a full factorial design and examined four experimental parameters: incubation time, incubation temperature, sample mass, and sample agitation rate. It was found in this work that only the oven temperature had a major influence on volatile generation in the sample headspace. Bazzo et al. (1999) performed optimization work for the MOS sensor-based FOX 4000 system in analyzing high-density polyethylene (HDPE) packaging. The optimization work allowed the selection of discriminating sensors as well as appropriate sample throughput conditions.

It is necessary to optimize electronic nose instrumentation to ensure sensitivity at the lowest detection thresholds. The threshold detection levels of 30 food aroma compounds with varying chemical structures for a MOS sensor-based electronic nose system were found to be similar to reported ortho-nasal human detection thresholds (Harper and Kleinhenz 1999). Harper and Kleinhenz also found that the matrix solution used strongly influences electronic nose threshold levels, and the use of a 4% ethanol matrix solution resulted in the sensor response resistance changes above their useable range. Subsequently, it is necessary to find a workable range of sensitivity for the sensors in a chemosensory array for particular samples in order to achieve an appropriate sensor response. It must also be acknowledged that although electronic nose technology continues to improve, it still responds very differently to many compounds than does the human nose. For example, the human nose is not sensitive to water vapor as well as several other compounds. However, such compounds affect most electronic nose systems, particularly those operating at lower temperatures. Consequently, electronic nose systems may be blinded by such compounds or not suited to discriminating others to which the human nose is sensitive.

The other major issue with the adaptation of the commercially available systems is the limitations from the software and user interface. Many systems allow only a limited number of samples (around 10) to develop the smell print for that particular aroma. However, when dealing with biological products having wide variations, this becomes very limited for practical use. In addition, the presence of moisture in the biological materials also creates unique problems with respect to identifying the aroma.

17.5 Response Analysis

17.5.1 Response Surface Analysis

Problems exist with the use of electronic nose sensors, such as sensor drift and poor repeatability and reproducibility due to system sensitivity to changes in operational conditions or poor gas selectivity and sensitivity (Roussel et al. 1998). In order to overcome these difficulties, it is necessary to develop a successful and efficient testing methodology at optimum parameter settings. Roussel et al. (1999) examined the influence of various experimental parameters on the multi-sensor array measurements using an electronic nose with SnO_2 sensors and attempted to quantify them. Volatile concentration in the headspace increases as the sample temperature is increased. In screening factorial experimental designs, it is necessary that the response of the experimental parameters be monotonic within a studied range. Alternatively, response surface designs must be generated in order to develop a model of the multi-sensor response.

Response surface analysis involves the investigation of linear and quadratic effects of two or more factors. The fundamental principle of response surface methodology is to develop a simple mathematical expression, usually first- or second-order polynomials, that approximate the relationship between response and the examined factors (Devineni et al. 1997). An experimental design is selected that allows a minimal number of experiments to be used to examine a full range of values for a particular factor. Popular Box-Behnken designs are fractions of 3^N designs used to estimate a full quadratic model in N factors. They consist of all 2^k possible combinations of high and low levels for different subsets of the factors of size k with all other factors at their central levels; the subsets are chosen according to a balanced incomplete block design for N treatments in blocks of size k. A number of center points with all factors at their central levels may also be added (Box and Draper 1987; SAS System Help 1988). The response surface analysis procedure uses the method of least squares to fit quadratic response surface regression models. The models focus on characteristics of the fit response function and, in particular, when optimum estimated response values occur.

17.5.2 Multivariate Factor Analyses

Statistical analysis is a key to understanding the sensor responses in an electronic nose instrument and realizing their discriminatory power. Discrimination and identification of sample recognition patterns requires the use of multivariate factor analysis. Factor analysis is a type of multivariate analysis that is concerned with the internal relationships of a set of variables (Lawley and Maxwell 1971). There are several multivariate statistical methods used among electronic nose systems.

Multivariate discriminant analyses and PCA are factor analysis methods that are most common to electronic nose data analysis software and will be the primary discussion topics. Other types of factor analysis, such as cluster analysis, partial least squares, soft independent modeling of class analogy, and artificial neural networks, will also be discussed briefly. Great length will be given to discussion of the descriptive statistics quantifying the amount of separation between sample classes and identification of unknowns, particularly the Mahalanobis distance.

17.5.3 Principal Components Analysis

Principal components analysis (PCA) allows data exploration and was initially developed and proposed by Hotelling (1933). It is the extraction of principal factors through the use of a component model. This analysis process does so by assessing the similarities between samples and the relationships between variables. It is a linear technique and uses the assumption that response vectors are well described in Euclidean space (Bartlett et al. 1997). The object is to determine if samples are similar or dissimilar and can be separated in homogeneous groups and to determine which variables are linked and the degree to which they correlate. PCA summarizes information contained in a database in subspaces with the object of reducing the number of variables and eliminating redundancy (Gorsuch 1983; Jolliffe 1986).

In PCA, the principal factor method is applied to a correlation matrix with unit values as the diagonal elements. The factors then give the most suited least squares fit to full correlation matrix with each factor ranked based upon the amount of the total correlation matrix that it accounts for. The principal components of the analysis are linear combinations of the original variables, and the discerned information from the analysis are presented in two- or three-dimensional spaces relative to the chosen components, which are classified based upon the level of information that they produce. The smaller factors are generally dropped from the model because they carry a trivial portion of the total variance and do not provide significant information (Gorsuch 1983; AlphaSoft Manual 2000). PCA is a form of dimension reduction factor analysis in which the relationships of a set of quantitative variables are examined and transformed into factors based on the amount of contributed variability to the system. Although PCA does not ignore co-variances and correlations, it concentrates on variances. The principal components are selected and ranked based on the amount of total variation, not the variation that most discriminates among classes of observations. This method of analysis is commonly used to reduce the number of variables used prior to performing discriminant analysis in order to make the calculations in the latter more manageable (Jolliffe 1986).

17.5.4 Discriminant Analyses

Multivariate discriminant analysis, also known as discriminant function analysis, discriminant factorial analysis (DFA), Gaussian discriminant function (GDF), and canonical discriminant analysis (CDA), originated with the work of Spearman

(1904) and is the most common analysis method used by electronic nose systems to separate classes of observations in a database (Lawley and Maxwell 1971).

CDA is a dimension-reduction technique that creates new canonical variables by taking special linear combinations of the original response variables. The canonical variables of the CDA, in some sense, are similar to principal components of the PCA. The principal advantage of CDA is its ability to allow the researcher to visualize the observations, which are classified into the different categories, in 2-D or 3-D space. Another advantage of CDA is that the output from a PCA can be used as an input for the CDA, thus the data visualized. If possible, one can attempt to interpret the canonical variables (Johnson 1998).

Canonical correlation analysis (CCA) is generally performed when there is a need to compare groups of variables. It helps in reducing the dimensionality of the data. CCA can be used to summarize the underlying relationship between groups of variables by creating new variables from the existing groups of variables. These new variables are called canonical functions. While performing the CCA, the optimum number of canonical functions can be known only after performing a preliminary CCA. Generally the option NCAN = 2 is used to limit the number of canonical functions generated to two. Interpretation of canonical functions is generally considered to be difficult (Johnson 1998).

CDA may be used to determine descriptive variables that predict the divisions between groupings when information regarding sample groupings is known ahead of time. An algorithm is used to determine linear combinations of new descriptive variables that separate the predetermined groups as much as possible. A set of data N_x is partitioned into m subsets $\left\{N_x^1, N_x^2, \ldots, N_x^k, \ldots, N_x^{m-1}, N_x^m\right\}$ that represent different quality descriptors. CDA proposes to then develop an algorithm with new variables $\{F^1, F^2, \ldots, F^j, \ldots, F^s\}$ that correspond to the directions that separate the subsets. This method allows the classification prediction of an unknown as one of these groups through the computation of the distance to the centroid of each of the groups. The unknown is classified with the closest associated group (Lawley and Maxwell 1971; Harman 1976; AlphaSoft Manual 2000).

CDA is a dimension-reduction type of factor analysis related to principal component analysis and canonical correlation. The manner in which the canonical coefficients are derived parallels that of a one-way MANOVA. In a CDA, linear combinations of the quantitative variables are found that provide maximal separation between the classes or groups. Given a classification variable and several quantitative variables, this procedure derives canonical factors, linear combinations of the quantitative variables that summarize between-class variation in much the same way that principal components summarize total variation. The discriminant function procedure in the SAS Software (Cary, North Carolina), PROC DISCRIM, develops a discriminant criterion to classify each observation into one of the groups for a set of observations containing one or more quantitative variables and a classification variable defining groups of observations. The derived discriminant criterion from a training or calibration data set can be applied to an unknown data set (Harman 1976; SAS System Help 1988).

In CDA, the classification of an unknown observation involves plotting the unknown observation and determining if the point falls near the mean point of one of the groups. If the unknown observation is close enough, the sample can be classified as being the same material. If the point is far away from all groups in a database, the sample may be a different material or have a different concentration from the sample observations used as the training set data. The approach is relatively straightforward except for the concept of how being near a group is actually defined. Visual inspection of a CDA projection plot provides useful initial information. However, it is not a viable method for real world discriminant analysis applications. Quantification with a mathematical equation is needed to measure the closeness of the unknown point to the mean point of the groups in a database.

The Euclidean distance is one such measurement technique. The unknown response can be used in a formula to calculate the distance of the unknown point to the group mean point. This would be an acceptable method except for two facts. The Euclidean distance does not give any statistical measurement of how well the unknown matches the training set, and the Euclidean distance only measures a relative distance from the mean point in the group. The method does not take into account the distribution of the points in the group as the variation along one axis is often greater than the variation along another axis. The training set group points tend to form an elliptical shape. However, the Euclidean distance describes a circular boundary around the mean point. The Euclidean distance method is not an optimum discriminant analysis algorithm as it does not take into account the variability of the values in all dimensions (Jolliffe 1986; Marcus 2001).

Mahalanobis distance (*D*), however, does take the sample variability into account. It weights the differences by the range of variability in the direction of the sample point. Therefore, the Mahalanobis distance constructs a space that weights the variation in the sample along the axis of elongation less than in the shorter axis of the group ellipse. In terms of Mahalanobis measurements, a sample point that has a greater Euclidean distance to a group than another sample point may have a significantly smaller distance to the mean if it lies along the axis of the group that has the largest variability (Jolliffe 1986; Marcus 2001).

Mahalanobis distances examine not only variance between the samples within a group, but also the co-variance among groups. Another advantage of using the Mahalanobis measurement for discrimination is that the distances are calculated in units of standard deviation from the group mean. Therefore, the calculated circumscribing ellipse formed around the cluster of a class of observations actually defines a preset number of standard deviations as the boundary of that group (Jolliffe 1986). The user can then assign a statistical probability to that measurement. For relatively large samples and normality assumptions, $D/2$ behaves like a normal multivariate z with standard deviation 1. In theory, $D/2$ can be examined to obtain an indication of the separation between samples and their estimated populations and the probability of incorrect assignment. A D value of 5 would correspond to about five standard deviation separations, which cover approximately 99% of a population given a multivariate normal distribution. Separation of groups quantified with a Mahalanobis distance greater than 5 would indicate very little overlap. In practice, the determination of the cutoff value depends on the application and the type of samples (Marcus 2001).

The Mahalanobis distance, expressed as D^2 or D, is consequently the statistic most often used in multivariate analyses to identify unknown samples and to quantify the probability that they belong to the identified class. Mahalanobis distance is the most appropriate measure of multivariate relationships when data are normally distributed, homoskedastic, and have equality among co-variance matrices. Most software gives a classification matrix of the Mahalanobis distances to each group centroid and identifies the sample as belonging to the group with the smallest distance (Jolliffe 1986).

The Mahalanobis metric in a minimum-distance classifier is generally used as follows. Let m_1, m_2, ..., m_c be the means for the c classes, and let C_1, C_2, ..., C_c be the corresponding co-variance matrices. An unknown vector x is classified by measuring the Mahalanobis distance from x to each of the means and assigning x to the class for which the Mahalanobis distance is minimum (Knapp 1998).

Articles are often not specific about reporting whether D or D^2 is being used, and usually it is only discerned in context. Mahalanobis D or D^2 is a descriptive measure of similarity or adjusting for pooled values within variance and covariance. D^2 is calculated first and often preferred as variance to standard deviation because of its additive and has known distribution. Yet only the standard deviation is in original measure units. Also, D is used as the ruler in canonical variate space or canonical projection plots and so is in the more useful form when examining the data graphically (Marcus 2001).

D^2 is approximately chi-square distributed with p degrees of freedom. Therefore, an unknown is still assigned to the population with the smallest D^2. Furthermore, using this idea, one can decide not to assign an unknown if all D^2 are larger than some cutoff based on the chi-square distribution with p degrees of freedom. It is also a useful statistic for finding multivariate outliers in a sample. If the data does follow a multivariate normal distribution, then the D^2 values will be approximately chi-squared distributed with p degrees of freedom. For standardized principal components, D^2 is the sum of squares (Jolliffe 1986).

One problem with the Mahalanobis distance is that because it is a summation of coefficient products the number of observations and independent variables used in the calculation affects it. As with many multivariate quantitative methods, the Mahalanobis distance solves for multiple dimensions simultaneously. However, the Mahalanobis model tends to become overfit very quickly as more independent variables are added. This is similar to an increased R^2 value for models when an increased number of independent variables are used and only logical when the method of calculating the Mahalanobis matrix is considered. D^2 cannot be 0 as it is always a quantity greater than zero. Therefore, D^2 is a biased estimate whether the null hypothesis is true or not. The size of the bias can be substantial when the sample sizes are small relative to the number of variables measured. An unbiased Mahalanobis distance $\left(D_u^2\right)$ is given by Equation 17.1 (Marcus 2001):

$$D_{u(1|2)}^2 = \frac{(n_1 + n_2 - p - 3)D^2}{(n_1 + n_2 - 2)} - \frac{(n_1 + n_2)p}{n_1 n_2} \tag{17.1}$$

where D^2 = Mahalanobis distance between classes 1 and 2, dimensionless; n_1 and n_2 = number of observations in class 1 and 2; and p = number of independent variables.

The answer to combining these apparently opposing necessities into one method for sample discrimination lies in first reducing the sensor data in electronic nose systems into its component variations with principal component analysis. A commercially available electronic nose, the Cyranose 320, uses this method in order to avoid overfitting and instability in calculations. The PCA method indirectly performs a sensor selection and reduces the number of variables used in building a canonical model. It is recommended for that instrument that a breakpoint of 5 for the Mahalanobis distance, D, indicates well-separated groups (Cyranose 320 User's Manual 2000).

For cases in which the number of observations or independent variables used in the discriminant analysis differ, it is not fair to compare Mahalanobis distances, and so a standardized value must be compared. An average or unbiased Mahalanobis distance calculated by using a proportionality constant accounting for the number of independent variables and observations may be used or the F value for the Mahalanobis distances may also be used for comparison as well. Hotelling's T square, T^2, equals this distance except for an included proportionality constant. A problem with both D^2 and T^2 is that they are based on the inverse of the variance–co-variance matrix, $S = X'X$, with the assumption that X has been centered and scaled. This inverse can be calculated only when the number of variables, p, is small compared to the number of training set samples, N.

Mahalanobis D^2 is also part of the formula for finding the two-sample extension of a student's t test, testing that the centroids of two multivariate populations have the same mean. This is Hotelling's T^2 test and is a maximum likelihood test. D^2 can be converted to T^2 and then to an F statistic, which has p, and $n_1 - n_2 - p - 1$ degrees of freedom by multiplying by appropriate constants based on the number of observations in each class and the number of variables used. This statistic is known to be fairly robust to violations of normality assumptions but is more sensitive to equality of variance assumptions, particularly for disparate sample sizes (SAS System Help 1988).

Because the F value incorporates the Mahalanobis distance and also takes into account the number of observations used, the degrees of freedom, and the number of independent variables used, it is a useful term for comparing discriminant analyses performed by different systems. The Wilks' lambda value is calculated from the inverse of the product of each of the eigenvalues incremented by one. Because lambda is a type of inverse measure, values of lambda that are close to zero denote high discrimination between groups. The F value for the Wilks' lambda provides a quantitative value for the overall discrimination of all the classes involved in the discriminant analysis. While it is a useful number for quickly quantifying the amount of separation between classes, it denotes total discrimination and does not indicate if the total amount of separation is due to a balanced separation of all the classes or a very large separation of some classes while having little separation between other classes. Consequently, the most useful value in comparing Mahalanobis distances from different systems are the F values for the Mahalanobis distances as they give a standardized value of the separation between each of the three classes analyzed in the discriminant analysis.

The percentage correct during cross-validation also provides additional information regarding the degree of separation. After the discriminant model is developed, the most common method of validation is to use what is commonly referred to as the "leave one out method" or cross-validation. In this cross-validation, each data point is removed and tested as an unknown to the model developed with the remaining data points. A value of 100% indicates complete separation of all classes. A value of 90% is usually considered sufficient validation for a database model. The user sets the actual required percentage of recognition for a training set validation based on the application requirements. This is also often called the "leave one out" procedure as each observation is left out in turn in the analysis and then identified using all of the remaining data (SAS System Help 1988).

Equations 17.2 through 17.6 used to calculate the discussed terms are given as follows (Jolliffe 1986):

$$D^2_{(1|2)} = \left(\bar{x}_1 - \bar{x}_2\right)' \text{COV}^{-1}\left(\bar{x}_1 - \bar{x}_2\right) \tag{17.2}$$

$$F_{\text{Mahanalobis}(1|2)} = \frac{(n_1-1)+(n_2-1)+(n_3-1)-p+1}{(n_1-1)+(n_2-1)+(n_3-1)p} \frac{n_1 n_2}{n_1+n_2} D^2 \tag{17.3}$$

$$\Lambda = \frac{1}{1+\lambda_1} \frac{1}{1+\lambda_2} \tag{17.4}$$

$$F_\Lambda = \frac{1-\Lambda^{1/t}}{\Lambda^{1/t}} \frac{[N-1-0.5(p+k)]t - 0.5[p(k-1)-2]}{p(k-1)_1} \tag{17.5}$$

$$t = \sqrt{\frac{p^2(k-1)^2-4}{p^2+(k-1)^2-5}} \tag{17.6}$$

where $D^2_{(1|2)}$ = Mahalanobis distance between classes 1 and 2, dimensionless; $F_{\text{Mahalanobis}(1|2)}$ = F value for Mahalanobis distance between classes 1 and 2, dimensionless; $\overline{x}_1$ and $\overline{x}_2$ = the geographic means of classes 1 and 2, dimensionless; n_1, n_2, and n_3 = number of observations in each class; p = number of independent variables; Λ = Wilks' lambda, dimensionless; λ_1 and λ_2 = first and second eigenvalues derived from the discriminant analysis, dimensionless; F_Λ = F value for Wilks' lambda, dimensionless; COV = pooled variance matrix, dimensionless; k = number of classes; and N = total number of observations in all classes.

Discriminant analysis is used primarily to answer three basic questions: (i) is the number of sensors and the sensor data obtained from the training set useful for building a model to classify the apples into its maturity level or stage? (ii) Can the model classify correctly the unknown apples of varying maturity levels? (iii) If not, what is the miscalculated percentage? Discriminant analysis, also known as classification analysis, is a multivariate method for classifying observations into appropriate categories (apples into appropriate maturity levels) (Johnson 1998).

The concept of discriminant analysis is analogous to regression analysis as the goal of the latter is to predict the value of the dependent variable while that of the former is to predict the category of the individual observation (Johnson 1998). The main difference is that the multivariate (discriminant analysis) approach is used when the variables are not independent.

According to Johnson (1998), there are four nearly equivalent ways to develop a discriminant rule to classify observations into categories (likelihood rule, linear discriminant function rule, Mahalanobis distance rule and posterior probability rule). There are three different methods that can be used to verify or estimate the probability of the correct classification of the observations and are described in detail below (Johnson 1998).

17.5.5 Resubstitution Method

The resusbstitution method employs a discriminant rule to the same data that were used to develop the rule and check how many observations were correctly classified by the rule into the correct categories. This method presumes that if a rule cannot classify properly on the original data used to build the rule, then there is a poor chance of it doing well with a new data set. The major drawback with this method is its overestimation of the probabilities when it classifies correctly. In SAS, this method can be invoked using the DATA = option (lists).

17.5.6 Holdout Method

This method uses a holdout set or a test data set, in which we know which observation belongs to which particular category, and the holdout data set is not used to develop the discriminant rule. The major drawback for this method is that one has to sacrifice the holdout data in order to build the discriminant rule and thus is not able to develop the best possible discriminant rule. In SAS, this method can be invoked using the DATA = option (test data).

17.5.7 Cross-Validation Method

Lachenbruch (1968) first proposed the cross-validation method, also known as jack-knifing. This is the preferred method when compared to the above two discriminant rules. The first observation vector is held out, and the remaining data is used to construct the discriminant rule; then the rule is used to classify the first observation, and then it checks whether the observation is correctly classified into the particular category. In the next step, the second observation vector is removed, but the first observation is replaced back into the original data, and then the discriminant rule is constructed. The rule thus developed is used to classify the second observation and thus check whether the observation is classified correctly. Thus, the same process is continued for the entire data set and also noting down the category into which it is being classified. It is claimed that this method is almost unbiased. In SAS, this method can be invoked using the DATA = option (cross lists).

17.6 Variable Selection Procedure

Because the number of variables involved in this study is high (32), a variable selection procedure is used to reduce the number of variables that are really necessary for effective discrimination of the data. The three types of variable selection procedures are the forward selection procedure, backward elimination procedure, and stepwise selection procedure. Johnson (1998) recommends the stepwise selection procedure when the number of variables exceeds 15.

Other statistical analyses used include partial least squares, soft independent modeling of class analogy, cluster analysis, and the use of artificial neural networks. Discriminant analysis should not be confused with cluster analysis and principal component analysis as discriminant analysis requires prior knowledge of the classes. The data used in cluster analysis do not include information on class membership. Its purpose is to construct the classification (Guertin and Bailey 1970; SAS System Help 1988).

Partial least squares is a statistical method that may be used to extract quantitative information. It is an algorithm based on linear regression and can be used to extract concentration sensory score predictions. Partial least squares (PLS) is a statistical method that may be used to extract quantitative information, such as concentration or sensory scores. The PLS algorithm, based on linear regression techniques, attempts to correlate a matrix containing quantitative measurements to a predictive matrix using a matrix of sensor measurements from the electronic nose instrument. After building the model, the predictive matrix is used to predict quantitative information contained in an unknown sample (Gorsuch 1983; AlphaSoft Manual 2000).

Soft independent modeling of class analogy (SIMCA) is a factor analysis method that is similar to PCA and CDA. This method classifies unknown samples using a comparison to a database composed of one group only. PCA is first performed on the data with the objective to find the subspace that most precisely contains samples. Each sample is explained in terms of its projection on the subspace and its projection on the orthogonal subspace. This matrix composed of a set of sensor observations induces two new matrices. The threshold identification criteria are set with theoretical values for the norm of the residual part of the predictive matrix and the Mahalanobis distance of the quantitative scores matrix to the centroid of the values projected in the subspace. SIMCA modeling works with as few as five observations from each population with no restriction on the number of independent variables (Jolliffe 1986; AlphaSoft Manual 2000).

Cluster analysis deals with data sets that are to be divided into classes when very little is known beforehand about the groupings. It provides an entry into factor analysis by establishing groupings within a data set. Within cluster analysis, principal components are calculated and used to provide an ordination or graphical representation of the data or to construct distance measures. The majority of cluster analysis techniques require the computation of similarity or dissimilarity among each pair of observations with the objective of clearly identifying group structures. The PCA graphical representation is often useful in verifying a cluster structure. This method of analysis is also often used in conjunction with artificial neural networks to perform the classifications (Guertin and Bailey 1970; Jolliffe 1986).

17.6.1 Partial Least Square Method

The partial least square method (PLS) is gaining broader use in electronic nose analysis and is seen as a good alternative to multivariate analysis, such as principal component analysis, due to its robustness in developing prediction models (Geladi and Kowalski 1986). The parameters of the prediction models developed by PLS do not vary with the addition of new calibration samples from the total population. The PLS algorithm uses the score matrices to represent the data matrix. A simplified PLS model consists of a regression between the scores for the X and Y block. The number of components to be used is a very important aspect of a PLS model. It is possible to calculate as many PLS components as the rank of X block, but not all of them are useful in prediction model development as the measured data are noisy. Because smaller components will describe only the noise in the data set, it is common to leave out small components. In order to decide the number of PLS components and stop the analysis, several methods can be employed, such as smaller error term, or the reduction in error from a previous iteration is very small or a combination of different criteria. In addition, the F test can also be used to validate the model.

17.6.2 Artificial Neural Networks

In many applications, there may be many references or combinations of sensor data to which an unknown needs to be compared. In these cases, an artificial neural network (ANN) is often used to analyze data from the sensor array. ANNs are particularly useful in analyzing data from hybrid electronic nose instruments when combined data must be analyzed. They are also particularly useful when the data to be analyzed exhibits a non-Gaussian distribution. The artificial neurons carry out a summation or other simple equation using predetermined weighted factors. The weighted factors are determined during the training of the neural network and are set arbitrarily before it is trained (Hodgins and Simmonds 1995). The training process for any ANN is a defining factor for its success.

The training of an ANN is accomplished by inputting data from the sensor array to the artificial neurons defined by a set answer for that data. The neural network calculates the values at all the neurons in the hidden and output layers. A back-propagation technique is then used to adjust the weighted factors until the correct output is achieved. This is repeated for all sensor data for all samples in a training set. A common breakpoint value for determining if the ANN is sufficiently trained for an application is if the weighting factors vary by no more than 10% during a training run (Hodgins and Simmonds 1995).

A trained ANN can then be used to identify an unknown sample by comparing it to all of the references in the training set. In practice, ANNs do not always identify unknowns that are one of its references with 100% confidence. However, ANNs do provide a means for performing numerous comparison calculations quickly to provide identification.

17.7 Applications in the Food Industry

As mentioned earlier, this technology is more a push from the instrument makers than a market demand driven from the users, and thus the use of this technology in the food industry is not widespread. Many food industry professionals are skeptical about the claims and capabilities of the technology, and the need for developing training sets for each application is also slowing down the adaptation of this technology on a wider scale.

The technology has excellent potential to be used in quality assurance and quality control applications with compliance of ingredients from suppliers. A summary of electronic nose applications developed for food industry adaptation is shown in Table 17.2. Baby et al. (2005b) evaluated the discrimination ability of the modular sensor system (MOSES) to classify medicinal plants and found that the discrimination between *Valeriana officinalis* and *Valeriana wallichii* types was achieved very successfully. Classification of milk samples from one dairy product and based on fat content within a particular dairy product was developed using a support vector machines (SVM) approach using metal oxide sensors (Brudzewski et al. 2004).

Table 17.2 Application of Electronic Nose Technology in Food Processing

Application	*Technology*	*Detection Target*	*Reference*
Grains	MOSFET	Off-odors in oats and barley resulting from mold infestation	Johnson et al. 1997
Beef	MOS	Spoilage in beef	Blixt and Borch 1999
Medicinal plants	MOSES II	*Valeriana* classification	Baby et al. 2005a
Milk	MOS	Fat content of milk	Brudzewski et al. 2004
Milk	MOS	Milk and yogurt samples	Collier et al. 2003
Olive oil	Conducting Polymer	Rancidity in olive oil	Aparicio et al. 2000
Milk	MOS Sol-Gel	Rancidity in milk	Capone et al. 2001
Yogurt	MOSFET	Fermenter monitoring	Cimander et al. 2002
Composting biosolids	MOSES II	Thermophillic period in the reactor	Baby et al. 2005b
Emmental cheese	QCM	Ripening state	Bargon et al. 2003
Crescenza cheese	MOS	Shelf life of cheese	Benedetti et al. 2005
Apples	CP	Maturity and harvest date	Pathange et al. 2006
Tomato	QCM	Shelf life	Berna et al. 2005
Pink lady apples	MOS	Shelf life	Brezmes et al. 2001
Peaches and pears	MOS	Fruit ripeness	Brezmes et al. 2000
Oranges and apples	TSMR	Post-harvest defects	Di Natale et al. 2001c
Peaches and nectarines	TSMR	Sensory quality	Di Natale et al. 2001b
Tomato paste	QCM	Sensory quality	Di Natale et al. 1998

(continued)

Table 17.2 (Continued) Application of Electronic Nose Technology in Food Processing

Application	*Technology*	*Detection Target*	*Reference*
Codfish fillets	TSMR	Freshness	Di Natale et al. 2001a
Red wine	TSMR	Wine analysis	Di Natale et al. 2004
Cheddar cheese	QCM	Aging	Drake et al. 2003
Pinapple slices	MOS	Shelf life	Torri et al. 2010
Red wine	Mass Spec	Yeast spoilage	Cynkar et al. 2007
Mandarin Oranges	MOS	Fruit maturity	Gomez et al. 2006
Long Jing Green Tea	MOS	Tea quality grade	Yu and Wang 2007
Peach	MOS	Fruit quality	Zhanga et al. 2008
Peach	MOS	Fruit maturity	Benedetti et al. 2008
Wine grapes	CP and SAW	Viticulture practices on wine quality	Zoecklein et al. 2012; Devarajan et al. 2011; Gardner et al. 2011
Oyster	CP and QCM	Quality and spoilage	Hu et al. 2008
Blue crab	CP	Quality	Sarnoski et al. 2008
Red wine	QCM	Quality	Martin et al. 2008
Wine grapes	CP	Maturity	Athamneh et al. 2008
Frying oil	QCM	Frying oil quality	Innawong et al. 2003
Food packaging	CP, QCM, and MOS	Volatiles from printing ink on plastic films	Van Deventer and Mallikarjunan 2002a, 2002b

Similarly, Collier et al. (2003) attempted using metal oxide sensors to classify various dairy products and compared that to screen-printed electrochemical arrays. In addition, the technology can also be used to monitor quality changes, especially oxidative changes in foods having adequate lipid content, and off-flavor development in food products due to spoilage microorganisms. Aparicio et al. (2000) studied the rancidity in olive oils using conduction polymer-based sensors and found that they could detect the rancidity at very low levels. Similarly, oxidative rancidity in milk could be detected using a MOS thin film-based sensor (Capone et al. 2001).

Another promising area in which this technology is finding widespread adaptation is evaluating fruit maturity. Tin oxide-based sensors were used by Simon et al. (1996) to monitor blueberry flavor. Benady et al. (1992) related the data derived from electronic senses to various ripeness indices, such as slip pressure, and classical volatile measurements in melons. Data from sensory panels were correlated to the electronic nose data that registered gases from the degradation reactions in tomatoes (Simon et al. 1996). Young et al. (1999) demonstrated that electronic nose technology using metal oxide sensors could be used as a potential maturity indicator to predict the harvest date for Royal Gala apples. According to Young et al. (1999), the EN analysis was approximately 40 times more sensitive than the headspace/gas chromatography.

With higher correlation to human sensory panels, this technology also has great potential in product development activities. Recent developments in real-time sensing with the gas sensors and online implementation for process control is very promising. Cimander et al. (2002) developed a system using MOSFET-based sensors integrated with a near infrared sensing system for online monitoring of yogurt fermentation. Results showed proposed online sensor fusion improves monitoring and quality control of yogurt fermentation with implications for other fermentation processes.

To determine the ripening stage in Emmental cheese, a quartz microbalance-based sensing system was developed by Bargon et al. (2003), which monitors the ripening process continuously. Similarly, the shelf life of Crescenza cheese stored at different temperatures was measured by a metal oxide-based sensing system (Benedetti et al. 2005). The technology also has the potential to be used for detecting pathogen contamination in selected foods with a sufficiently larger population but within the risk level for human illnesses.

In addition to applications in the food and bioprocess industries, use of electronic nose technology has been explored in the medical field as well. An electronic nose can examine odors from the body and identify possible health-related problems. Odors in the breath can be indicative of gastrointestinal, sinus, infections, diabetes, and liver problems.

Infected wounds and tissues emit distinctive odors that can be detected by an electronic nose. Odors coming from body fluids, such as blood and urine, can indicate liver and bladder problems (Blood Gas Analyzer). There is extensive literature available in this area. As the scope of this book is limited to applications of electronic nose technology in the food industry, information related to medical applications are not included here.

17.8 Case Studies

The researchers at Virginia Tech have used the electronic nose technology for various food-related applications, including but not limited to detection of plasticizers in packaging material (Van Deventer and Mallikarjunan 2002); discrimination of oil quality (Innawong et al. 2003); determination of fruit maturity (Pathange et al. 2006; Athamneh et al. 2008); detection of oxidation-related quality changes in meat, peanuts, and milk (Ballard et al. 2005; Mallikarjunan et al. 2006); and spoilage detection in seafood products (Hu et al. 2008; Sarnoski et al. 2008).

17.8.1 Detection of Retained Solvent Levels in Printed Packaging Material

The packaging suppliers use plasticizers to make the printing stay on the packaging material. Some of the plasticizer could transfer into the food product and alter the taste and flavor. In addition, at higher levels, these plasticizers pose health risks, and the industry wants to limit the level of plasticizer in printed packaging materials. Currently, the industry uses a human sensory panel-based sniff test and found that the chromatographic techniques did not correlate with sensory results. It was decided to explore the feasibility of using an electronic nose for this application, and three different types of electronic nose systems were tested for their ability to discriminate the packaging based on the contamination level, ease of use for training and prediction, and repeatability (Figure 17.1). First and foremost, each system was optimized for its performance to obtain maximum sensor response. The results of the optimization are described by Van Deventer and Mallikarjunan (2002).

Performance analyses of these systems, which use three leading sensor technologies, showed that the conducting polymer sensor technology demonstrated the most discriminatory power. All three technologies proved able to discriminate among different levels of retained solvents. Each complete electronic nose system was also able to discriminate between assorted packaging having either conforming or non-conforming levels of retained solvents. Each system correctly identified 100% of unknown samples. Sensor technology had a greater effect on performance than the number of sensors used. Based on discriminatory power and practical features, the FOX 3000 and the Cyranose 320 were superior (Van Deventer and Mallikarjunan 2002).

17.8.2 Discrimination of Frying Oil Quality Based on Usage Level

Various criteria are being used to judge when frying oils needs to be discarded. In restaurants and food services, changes in physical attributes of frying oils, such as oil color, odor, and foam level have been used as an indicator of oil quality. In the food industry, not only physical tests, but chemical tests are also used to measure oil quality, including acidity, polymer content, and/or total polar content. Many of the methods do not correlate with oil quality effectively, and many times, they are time-consuming and expensive. Previous monitoring methods used to analyze the volatile compounds and aromas in food needed either a highly trained sensory

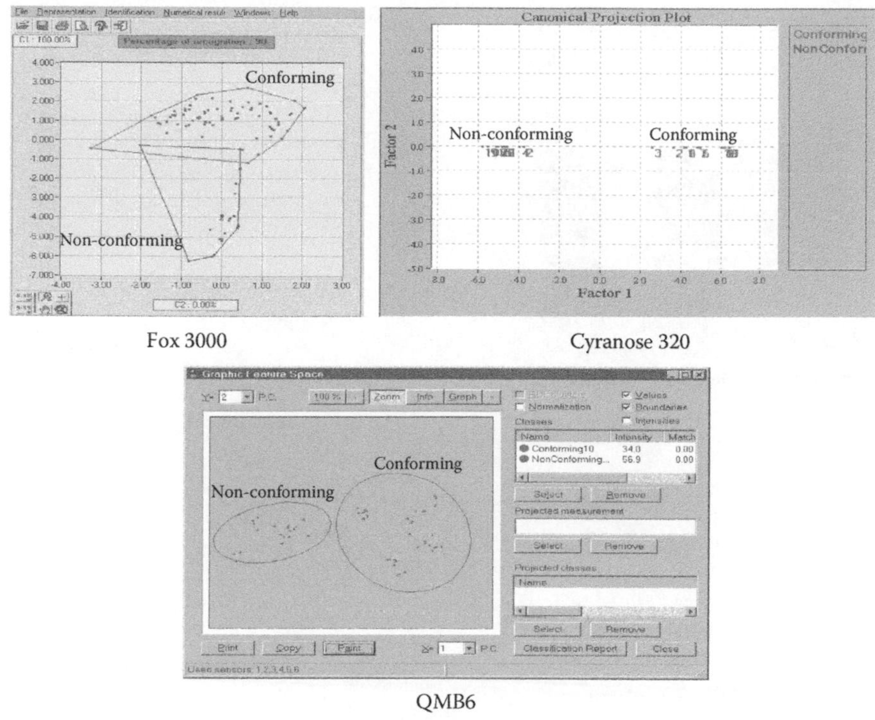

Figure 17.1 Comparison of three types of electronic nose systems in discriminating packaging material based on the level of plasticizers.

panel or GC/MS techniques. Thus, there has been a genuine need for a quick, simple, and powerful objective test for indicating deterioration of oil.

This study was conducted to determine the possibility of using a chemosensory system to differentiate among varying intensities of oil rancidity and investigate discrimination between good, marginal, and unacceptable frying oils (Figure 17.2). Fresh, one day, two days used, and discarded frying oils were obtained from a fast food restaurant in each frying cycle for four weeks. The oil samples were analyzed using a quartz microbalance-based chemosensory system. The discrimination between good, marginal, and unacceptable frying oils with regard to rancidity was examined, and the results were compared to their physicochemical properties, such as dielectric constant, peroxide value, and free fatty acid content. The different qualities of frying oils were successfully evaluated and discriminated using the chemosensory system. Good correlations (0.87 to 0.96) were found between changes in physico-chemical properties of oil and the sensor signals (Innawong et al. 2003). Based on the results, oils from two different restaurants were obtained with different usage levels to discriminate between the usage levels (Bengtson et al. 2005).

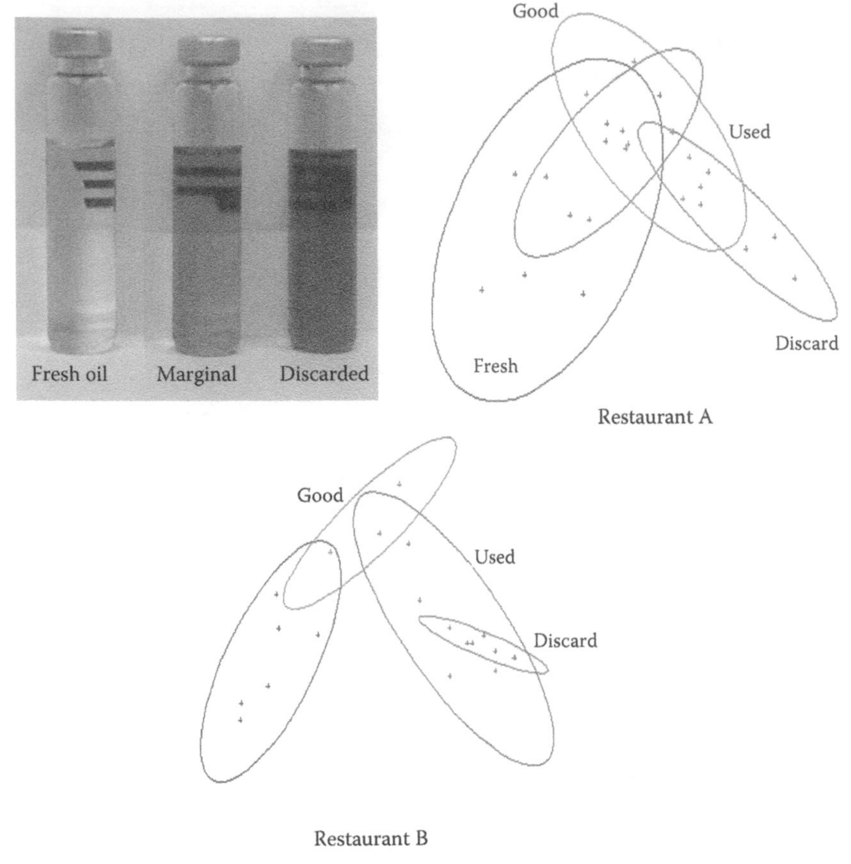

Figure 17.2 Discrimination of frying oil using a chemosensory system.

17.8.3 Evaluating Apple Maturity

Sometimes harvested apples are a mixture of mature, immature, and overly mature fruits, and the quality of an apple fruit depends primarily on its level of maturity at the time of harvest. Even though the external appearance of an immature fruit may look perfect to harvest, store, and sell due to its pre-climacteric physiological condition, these apples do not ripen normally, and thus their taste is strongly impaired due to lack of full-flavor compounds. On the other hand, overly mature fruits have a shorter storage life; soften rapidly; develop storage disorders, such as off-flavor and lack of firm texture; and are unattractive in appearance.

Currently, random and destructive sampling techniques are used to evaluate apple quality. Thus, there exists a need to develop a nondestructive technique to assess apple quality. Gala and York apples were harvested at different times to obtain different maturity groups (immature, mature, and ripe). Headspace evaluation was

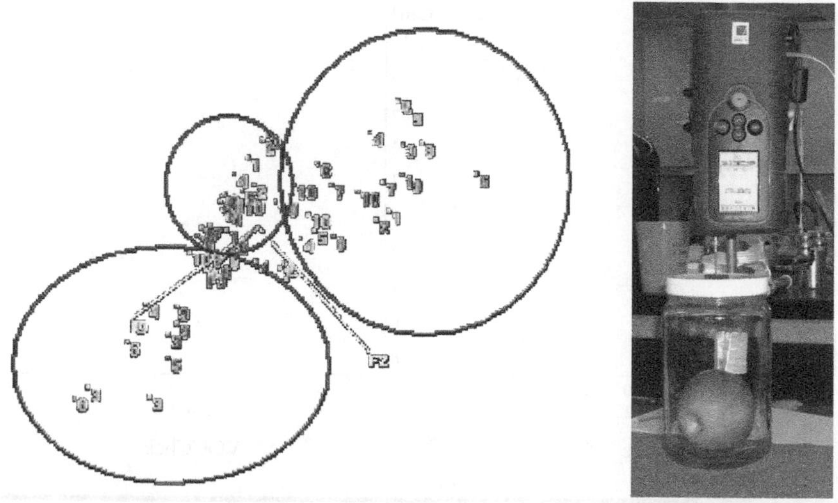

Figure 17.3 Evaluation of apple maturity using a conducting polymer-based sensing system.

performed first, and maturity indices were measured within 24 h after harvest. Individual apples were placed in a 1.5-liter glass bottle, and the headspace gas from the glass bottle was exposed to the electronic nose. A conducting polymer-based sensing system was used for apple maturity evaluation (Figure 17.3). Maturity indices, such as starch index, puncture strength, total soluble solids, and titratable acidity, were used to categorize apples into three maturity groups referred to as immature, mature, and overly mature fruits.

Multivariate analysis of variance (MANOVA) of the electronic nose sensor data indicated that there were different maturity groups (Wilks lambda $F = 3.7$, $P < 0.0001$). From the discriminant analysis (DA), the electronic nose could effectively categorize Gala apples into the three maturity groups with the correct classification percentage of 83% (Pathange et al. 2006).

17.8.4 Detection of Spoilage and Discrimination of Raw Oyster Quality

The effectiveness of two hand-held electronic nose systems to assess the quality of raw oysters was studied on live oysters stored at 4°C and 7°C for 14 days. Electronic nose data were correlated with a trained sensory panel evaluation by quantitive description analysis (QDA) and with microbial enumeration. Oysters stored at both temperatures exhibited varying degrees of microbial spoilage with bacterial load reaching 10^7 CFU/g at day seven for 7°C storage (Figure 17.4).

The Cyranose 320 electronic nose system was capable of generating characterized smell prints to differentiate oyster qualities of varying age (100% separation).

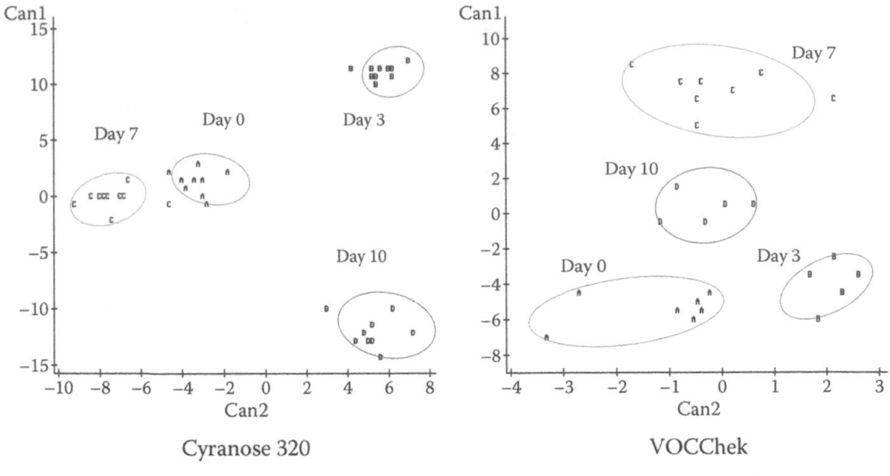

Figure 17.4 Discrimination of raw oyster quality by two types of hand-held electronic nose systems.

The validation results showed that the Cyranose 320 can identify the quality of oysters in terms of storage time with 93% accuracy. Comparatively, the correct classification rate for the VOCChek electronic nose was only 22%. Correlation of electronic nose data with microbial counts suggested the Cyranose 320 was able to predict the microbial quality of oysters. Correlation of sensory panel scores with electronic nose data revealed that the electronic nose has demonstrated potential as a quality assessment tool by mapping varying degrees of oyster quality.

17.9 Summary

Electronic nose technology is an emerging analytical tool that has great potential to be implemented in the food industry for quality control, quality assurance, product safety evaluation, new product development, and process control. Many systems are available in the marketplace using a multitude of sensing technologies, software capabilities, and hardware configurations. The cost of the system also ranges from \$5000 to \$120,000 with varying capabilities and sensitivities. The technology has not been recognized by the market mainly due to the lack of confidence in the technology and limited available applications for immediate adaptation in the industry. In addition, the technology is perceived as a technology push from the instrument manufacturers without clear implementation strategies in the food industry. Successful development and implementation of this technology to a wide range of applications and subsequent research publications in the near future will provide enough support from the food industry.

References

Alpha M.O.S. (2000). Alpha M.O.S. Introductory Manual. Alpha M.O.S., Toulouse, France.

Aparicio, R., Rocha, S.M., Delgadillo, I. and Morales, M.T. (2000). Detection of rancid defect in virgin olive oil by the electronic nose. *Journal of Agriculture and Food Chemistry*, 48(3), 853–860.

Athamneh, A.I., Zoecklein, B.W. and Mallikarjunan, P.K. (2008). Electronic nose evaluation of Cabernet Sauvignon fruit maturity. *Journal of Wine Research*, 19, 69–80.

Baby, R.E., Cabezas, M.D., Labud, V., Marqui, F.J. and Walsoe de Reca, N.E. (2005a). Evolution of thermophilic period in biosolids composting analyzed with an electronic nose. *Sensors and Actuators B: Chemical*, 106(1), 44–51.

Baby, R.E., Cabezas, M.D., Castro, E., Filip, R. and Walsoe de Reca, N.E. (2005b). Quality control of medicinal plants with an electronic nose. *Sensors and Actuators B: Chemical*, 106(1), 24–28.

Ballard, T.S., Mallikarjunan, P., O'Keefe, S. and Wang, H. (2005). Optimization of electronic nose response for analyzing volatiles arising from lipid oxidation in cooked meat. Presented at the Institute of Food Technologists Annual Meeting, July 16–20, New Orleans, LA.

Bargon, J., Brascho, S., Florke, J., Herrmann, U., Klein, L., Loergen, J.W., Lopez, M., Maric, S., Parham, A.H., Piacenza, P., Schaefgen, H., Schalley, C.A., Silva, G., Schlupp, M., Schwierz, H., Vogtle, F. and Windschief, G. (2003). Determination of the ripening state of Emmental cheese via quartz microbalances. *Sensors and Actuators B: Chemical*, 95(1–3), 6–19.

Bartlett, P.N., Elliot, J.M. and Gardner, J.W. (1997). Electronic noses and their application in the food industry. *Food Technology*, 51(12), 44–48.

Bazzo, S., Loubet, F., Labreche, S. and Tan, T.T. (1999). Optimization of Fox sensor array system for QC packaging in factory environment. Robustness, sample throughput and transferability. In Hurst, J. (ed.). *Electronic noses and sensor array based systems: Design and applications*, Lancaster, PA: Technomic, pp. 225–234.

Benady, M., Simon, J.E., Charles, D.J. and Miles, G.E. (1992). Determining melon ripeness by analyzing headspace gas emissions. American Society of Agricultural Engineers. Paper no. 92-6055.

Benedetti, S., Buratti, S., Spinardi, A., Mannino, S. and Mignani, I. (2008). Electronic nose as a non-destructive tool to characterize peach cultivars and to monitor their ripening stage during shelf-life. *Postharvest Biology and Technology*, 47(2), 181–188.

Benedetti, S., Sinelli, N., Buratti, S. and Riva, M. (2005). Shelf-life of crescenza cheese as measured by electronic nose. *Journal of Dairy Science*, 88(9), 3044–3051.

Bengtson, R., Mallikarjunan, P., Moreira, R., Muthukumarappan, K., Wilson, L. and Weisenborn, D. (2005). Measurement of frying oil quality by various objective methods. Presented at the Annual Meeting of American Society of Agricultural Engineers, Paper No. 05-6168, July 17–20, Tampa, FL.

Berna, A.Z., Buysens, S., Di Natale, C., Grun, I.U., Lammertyn, J. and Nicolai, B.M. (2005). Relating sensory analysis with electronic nose and headspace fingerprint MS for tomato aroma profiling. *Postharvest Biology and Technology*, 36(2), 143–155.

Blixt, Y. and Borch, E. (1999). Using an electronic nose for determining the spoilage of vacuum-packaged beef. *International Journal of Food Microbiology*, 46(2), 123–134.

Box, G. and Draper, N.R. (1987). *Empirical model-building and response surfaces*. New York: John Wiley and Sons.

Brezmes, J., Llobet, E., Vilanova, X., Orts, J., Saiz, G. and Correig, X. (2001). Correlation between electronic nose signals and fruit quality indicators on shelf-life measurements with Pink Lady apples. *Sensors and Actuators B: Chemical*, 80(1), 41–50.

Brezmes, J., Llobet, E., Vilanova, X., Saiz, G. and Correig, X. (2000). Fruit ripeness monitoring using an electronic nose. *Sensors and Actuators B: Chemical*, 69(3), 223–229.

Brudzewski, K., Osowski, S. and Markiewicz, T. (2004). Classification of milk by means of an electronic nose and SVM neural network. *Sensors and Actuators B: Chemical*, 98(2–3), 291–298.

Capone, S., Epifani, M., Quaranta, F., Siciliano, P., Taurino, A. and Vasanelli, L. (2001). Monitoring of rancidity of milk by means of an electronic nose and a dynamic PCA analysis. *Sensors and Actuators B: Chemical*, 78(1–3), 174–179.

Cimander, C., Carlsson, M. and Mandenius, C.O. (2002). Sensor fusion for on-line monitoring of yoghurt fermentation. *Journal of Biotechnology*, 99(3), 237–248.

Collier, W.A., Baird, D.B., Park-Ng, Z.A., More, N. and Hart, A.L. (2003). Discrimination among milks and cultured dairy products using screen-printed electrochemical arrays and an electronic nose. *Sensors and Actuators B: Chemical*, 92(1–2), 232–239.

Cynkar, W., Cozzolino, D., Dambergs, B., Janik, L. and Gishen, M. (2007). Feasibility study on the use of a head space mass spectrometry electronic nose (MS e_nose) to monitor red wine spoilage induced by Brettanomyces yeast. *Sensors and Actuators B: Chemical*, 124(1), 167–171.

Cyrano Sciences. (2000). Cyranose 320 User's Manual. Cyrano Sciences, Pasadena, CA.

Devarajan, Y.S., Zoecklein, B.W., Mallikarjunan, K. and Gardner, D.M. (2011). Electronic nose evaluation of the effects of canopy side on Cabernet Franc (*Vitis vinifera* L.) grape and wine volatiles. *American Journal of Enology and Viticulture*, 62, 73–80.

Devineni, N., Mallikarjunan, P., Chinnan, M.S. and Phillips, R.D. (1997). Supercritical fluid extraction of lipids from deep-fried food products. *Journal of American Oil and Chemical Society*, 74(12), 1517–1523.

Di Natale, C., Macagnano, A., Martinelli, E., Proietti, E., Paolesse, R., Castellari, L., Campani, S. and D'Amico, A. (2001b). Electronic nose based investigation of the sensorial properties of peaches and nectarines. *Sensors and Actuators B: Chemical*, 77, (1–2), 561–566.

Di Natale, C., Macagnano, A., Martinelli, E., Proietti, E., D'Amico, A. and Paolesse, R. (2001c). The evaluation of quality of post-harvest oranges and apples by means of an electronic nose. *Sensors and Actuators B: Chemical*, 78(1–3), 26–31.

Di Natale, C., Macagnano, A., Paolesse, R., Mantini, A., Tarizzo, E., D'Amico, A., Sinesio, F., Bucarelli, F.M., Moneta, E. and Quaglia, G.B. (1998). Electronic nose and sensorial analysis: Comparison of performances in selected cases. *Sensors and Actuators B: Chemical*, 50(3), 246–252.

Di Natale, C., Olafsdottir, G., Einarsson, S., Martinelli, E., Paolesse, R. and D'Amico, A. (2001a). Comparison and integration of different electronic noses for freshness evaluation of cod-fish fillets. *Sensors and Actuators B: Chemical*, 77(1–2), 572–578.

Di Natale, C., Paolesse, R., Massimiliano, B., Martinelli, E., Pennazza, G. and D'Amico, A. (2004). Application of metalloporphyrins-based gas and liquid sensor arrays to the analysis of red wine. *Analytica Chimica Acta*, 513(1), 49–56.

Drake, M.A., Gerard, P.D., Kleinhenz, J.P. and Harper, W.J. (2003). Application of an electronic nose to correlate with descriptive sensory analysis of aged Cheddar cheese. *LWT – Food Science and Technology*, 36(1), 13–20.

Gardner, D.M., Zoecklein, B.W. and Mallikarjunan, K. (2011). Electronic nose analysis of Cabernet Sauvignon (*Vitis vinifera* L.) grape and wine volatile differences during cold soak and post-fermentation. *American Journal of Enology and Viticulture*, 62, 81–90.

Geladi, P. and Kowalski, B.R. (1986). Partial least-squares regression: A tutorial. *Analytica Chimica Acta*, 185, 1–17.

Gómez, A.H., Wang, J., Hu, G. and Pereira, A.G. (2006). Electronic nose technique potential monitoring mandarin maturity. *Sensors and Actuators B: Chemical*, 113(1), 347–353.

Gorsuch, R.L. (1983). *Factor analysis*, 2nd edn. Hillsdale, NJ: Lawrence Erlbaum.

Guertin, W. and Bailey, J.P. (1970). *Introduction to modern factor analysis.* Ann Arbor, MI: Edwards Brothers, Inc., pp. 72.

Hansen, W.G. and Wiedemann, S.C. (1999). Evaluation and optimization of an electronic nose. In Hurst, J. (ed.). *Electronic noses and sensor array based systems: design and applications.* Lancaster, PA: Technomic, pp. 131–144.

Harman, H.H. (1976). *Modern factor analysis.* Chicago, IL: University of Chicago Press.

Harper, W.J. and Kleinhenz, J.P. (1999). Factors affecting sensory and electronic nose threshold values for food aroma compounds. In Hurst, J. (ed.). *Electronic noses and sensor array based systems: Design and applications.* Lancaster, PA: Technomic, pp. 308–317.

Haugen, J.E. and Kvaal, K. (1998). Electronic nose and artificial neural network. *Meat Science*, 49, 273–286.

Hodgins, D. and Conover, D. (1995). Evaluating the electronic NOSE. *Perfumer and Flavorist*, 20(6), 1–8.

Hodgins, D. and Simmonds, D. (1995). Sensory technology for flavor analysis. *Cereal Foods World*, 40(4), 186–191.

Hotelling, H. (1933). Analysis of a complex of statistical variables into principal components. *Journal of Educational Psychology*, 24, 417–441, 498–520.

Hu, X., Mallikarjunan, P. and Vaughan, D. (2008). Development of non-destructive methods to evaluate oyster quality by electronic nose technology, sensing instrument. *Food Quality and Safety*, 2(1), 51–57.

Innawong, B., Mallikarjunan, P. and Marcy, J.E. (2003). The determination of frying oil quality using a chemosensory system. *Lebensmittel-Wissenschaft und-Technologie*, 37(1), 35–41.

Johnson, A., Winquist, F., Schnurer, J., Sundgren, H. and Lundstrom, I. (1997). Electronic nose for microbial quality classification of grains. *International Journal of Food Microbiology*, 35(2), 187–193.

Johnson, D.E. (1998). *Applied multivariate methods for data analysis.* Pacific Grove, CA: Brooks/Cole Publishing Company, pp. 217–511.

Jolliffe, I.T. (1986). *Principal component analysis.* New York: Springer-Verlag.

Knapp, R.B. (1998). Mahalanobis metric. http://www.engr.sjsu.edu/~knapp/HCIRODPR/PR_Mahal/M_metric.htm. San Jose State University. Posted July 16, 1998.

Lawley, D.N. and Maxwell, A.E. (1971). *Factor analysis as a statistical method.* New York: American Elsevier Publication Co.

Lachenbruch, A.H. (1968). Preliminary geothermal model of the Sierra Nevada. *Journal of Geophysical Research*, 73(22), 6977–6989.

Mallikarjunan, S., Mallikarjunan, P. and Duncan, S.E. (2006). Evaluating levels of photo oxidation in milk using an electronic nose. Abstract No. 78E-05, Presented at the Annual Meeting of Institute of Food Technologists, Orlando, FL, June 24–28.

Marcus, L. (2001). Mahalanobis distance. http://www.qc.edu/Biology/fac_stf/marcus/multisyl/fourth.htm. Queens College. Posted August 3, 2001.

Martin, A., Mallikarjunan, K. and Zoecklein, B.W. (2008). Discrimination of wines produced from Cabernet Sauvignon grapes treated with aqueous ethanol post-bloom using an electronic nose. *International Journal of Food Engineering*, 4(2), 14.

Mielle, P. and Marquie, F. (1998). Electronic nose: Improvement of the reliability of the product database using new dimensions. *Seminar in Food Analysis*, 3, 93–105.

Nakamoto, T. and Morizumi, T. (1999). Developments and trends in QCM odor sensing systems. In Hurst, J. (ed.). *Electronic noses and sensor array based systems: Design and applications*. Lancaster, PA: Technomic, pp. 123–130.

Newman, D.J., Luzuriaga, D.A. and Balaban, M.O. (1999). Odor and microbiological evaluation of raw tuna: Correlation of sensory and electronic nose data. In Hurst, J. (ed.). *Electronic noses and sensor array based systems: Design and applications*. Lancaster, PA: Technomic, pp. 170–184.

Pathange, L., Mallikarjunan, P., Marini, R., O'Keefe, S. and Vaughan, D. (2006). Nondestructive evaluation of apple maturity using an electronic nose system—A statistical approach. *Journal of Food Engineering*, 77, 1018–1023.

Payne, J.S. (1998). Electronic nose technology: An overview of current technology and commercial availability. *Food Science and Technology Today*, 12(4), 196–200.

Perkin-Elmer Corporation. (1999). Operators manual for the chemosensory-system QMB6/HS40XL. Norwalk, CT: HKR Sensorsysteme GmbH.

Persaud, K.C., Bailey, R.A., Pisanelli, A.M., Byun, H.G., Lee, D.H. and Payne, J.S. (1999). Conducting polymer sensor arrays. In Hurst, J. (ed.). *Electronic noses and sensor array based systems: Design and applications*. Lancaster, PA: Technomic, pp. 318–328.

Pope, K. (1995). Technology improves on the nose as scientists try to mimic smell. *Wall Street J*, March 1, B1.

Roussel, S., Forsberg, G., Grenier, P. and Bellon-Maurel, V. (1999). Optimization of electronic nose measurements. Part II: Influence of experimental parameters. *Journal of Food Engineering*, 39, 9–15.

Roussel, S., Forsberg, G., Steinmetz, V., Grenier, P. and Bellon-Maurel, V. (1998). Optimization of electronic nose measurements. Part I: Methodology of output feature selection. *Journal of Food Engineering*, 37, 207–222.

Sarnoski, P., Jahncke, M., O'Keefe, S., Mallikarjunan, P. and Flick, G. (2008). Determination of quality attributes of blue crab (*Callinectes sapidus*) meat by electronic nose and draeger-tube analyses. *Journal of Aquatic Food Product Technology*, 17(3), 234–252.

SAS System Help. (1988). SAS Software Version 7. SAS Institute Inc., Cary, NC.

Schaak, R.E., Dahlberg, D.B. and Miller, K.B. (1999). The electronic nose: Studies on the fundamental response and discriminative power of metal oxide sensors. In Hurst, J. (ed.). *Electronic noses and sensor array based systems: Design and applications*. Lancaster, PA: Technomic, pp. 14–26.

Schaller, E. and Bosset, J.O. (1998). Electronic noses and their application to food: A review. *Seminar in Food Analysis*, 3, 119–124.

Simon, J.E., Hetzroni, A., Bordelon, B., Miles, G.E and Charles, D.J. (1996). Electronic sensing of aromatic volatiles for quality sorting of blueberries. *Journal of Food Science*, 61(5), 967–970.

Spearman, C. (1904). General intelligence, objectively determined and measured. *American Journal of Psychology*, 15, 201–293.

Strassburger, K.J. (1998). Electronic nose technology in the flavor industry: moving from R&D to the production floor. *Seminars in Food Analysis*, 3(1), 5–13.

Torri, L., Sinelli, N. and Limbo, S. (2010). Shelf-life evaluation of fresh-cut pineapple by using an electronic nose. *Postharvest Biology and Technology*, 56(3), 239–245.

Van Deventer, D. and Mallikarjunan, P. (2002a). Comparative performance analysis of three electronic nose systems using different sensor technologies in odor analysis of retained solvents on printed packaging. *Journal of Food Science*, 67(8), 3170–3183.

Van Deventer, D. and Mallikarjunan, P. (2002b). Optimizing an electronic nose for analysis of volatiles from printing inks on assorted plastic films. *Innovative Food Science and Emerging Technologies*, 3(1), 93–99.

Wilson, A.D. and Baietto, M. (2009). Applications and advances in electronic-nose technologies. *Sensors*, 9, 5099–5148.

Young, H., Rossiter, K., Wang, M. and Miller, M. (1999). Characterization of Royal Gala apple aroma using electronic nose technology: Potential maturity indicator. *Journal of Agricultural and Food Chemistry*, 47, 5173–5177.

Yu, H. and Wang, J. (2007). Discrimination of Long Jing green-tea grade by electronic nose. *Sensors and Actuators B: Chemical*, 122(1), 134–140.

Zhang, H., Chang, M., Wang, J. and Ye, S. (2008). Evaluation of peach quality indices using an electronic nose by MLR, QPST and BP network. *Sensors and Actuators B: Chemical*, 134(1), 332–338.

Zoecklein, B.W., Devarajan, Y.S., Mallikarjunan, K. and Gardner, D.M. (2011). Monitoring effects of ethanol spray on cabernet franc and merlot grapes and wine volatiles using electronic nose systems. *American Journal of Enology and Viticulture*, 62, 351–358.

WASTE AND BYPRODUCT MANAGEMENT AND ENERGY CONSERVATION

Chapter 18

Waste Management in Food Processing

Lijun Wang

Contents

18.1 Introduction

Large quantities of liquid and solid wastes are produced by the food processing industry. Food wastes comprise approximately 40% of the total municipal solid wastes (Mata-Alvarez et al. 2000). Food processing wastes contain principally biodegradable organic matter, and disposal of them may create serious environmental problems (Hang 2004). Management of food processing wastes may involve (1) reduction of wastes from production, (2) conversion of wastes into food byproducts, (3) utilization of wastes as a fuel source or feedstock to produce energy products, and (4) treatment of wastes as sewage (Zall 2004).

Various methods, including landfill, incineration, and complete decomposition, have been used for the treatment of food wastes (Park et al. 2002). Waste characteristics, governmental regulations, and disposal costs are the primary considerations for the selection of waste treatment methods in the food manufacturing industry. Traditionally, large amounts of solid food processing wastes are buried in landfills while liquid food processing wastes are released untreated into rivers, lakes, and oceans and disposed of in public sewer systems. Public concern and government legislation on environmental protection has reduced the number of wastes that are considered safe for disposal. Use of incineration is restricted because it is expensive to operate. The increasing energy prices and waste disposal costs have dramatically increased food production costs in recent years. A survey of the United States Department of Agriculture showed that 50 million US$ annually could be saved alone in solid waste disposal costs for landfills if 5% of the total amount of $43.54*10^9$ kg food loss from processing, retail, food service, and consumer foods in 1995 were recovered (Kantor et al. 1997).

Technology choice for the treatment of food processing wastes depends on capital investment, operating costs, and governmental regulations. Much of the wastes generated in food processing facilities can become a resource for producing bioenergy and biochemicals (Matteson and Jenkins 2007). Conversion of food processing wastes into useful energy products, such as bioethanol, biodiesel, bio-oil, biogas, syngas, steam, and electricity in a food processing facility could result in significant savings for the food manufacturing industry in terms of reducing the amount of energy purchased and waste disposal costs. The energy utilizations of food processing wastes are dependent upon a basic understanding of (i) operating conditions of a food processing facility; (ii) waste types, availability, and energy potential; and (iii) process and equipment for handling and conversion of the waste. Comprehensive technical and economic analyses are essential to determine the technical feasibility and economics of an energy conversion process for utilization of food process wastes (Wang 2008).

The types of raw food materials and the operations of a food processing facility greatly influence the kinds and amounts of wastes produced. Because the marginal profit in the food industry is usually small, the costs versus the return on investment may not be favorable for some alternative waste treatment technologies. However, the selection of a food waste treatment technology is affected not only by the net return on the investment but also by the emphasis on sustainable development and governmental regulations. In this chapter, wastes from different food processing sectors are reviewed in terms of their availability, quality, and current utilizations. The procedure of auditing wastes in a food processing facility is briefly introduced. Technologies for value-added processing of different food wastes are then reviewed. Finally, the environmental, economical, and sustainable impacts of managing food processing wastes are discussed.

18.2 Processing Wastes in the Food Industry

18.2.1 Availability and Quality of Processing Wastes in Different Food Processing Sectors

18.2.1.1 Fruit and Vegetable Processing Wastes

The fruit and vegetable processing facilities generate large amounts of liquid and solid wastes. During processing of fruit and vegetables, solid residues are produced from operations, such as peelings, trimmings, cores, stems, pits, culls of undesirable fruits or vegetables, nut shells, kernel fragments, and grain hulls. Liquid wastes are produced from operations, such as hydro-handling, product cleaning, and blanching. Water is used in washers, blanchers, French pumps, graders, peeler flumes, and expressates during processing of vegetables and fruits (Sargent and Steff 1986). For processing each 1000 kg of raw apples, 20,861 liters of waste water are generated, which contain 2.5 kg of suspended solids and 300 kg of solid residues. For processing each 1000 kg of raw citrus, 12,517 liters of waste water are produced, which contain 2.5 kg of suspended solids and 220 kg of solid residues (Woodruff and Luh 1975). On average, the processing waste is as high as 72.4% of the original mass for sweet corn. Approximately half of the potato, carrot, and beet enter their waste streams during processing. Snap beans and green peas generate less waste, which is 28.2% and 22% of their original weights, respectively (Blodin et al. 1983). Table 18.1 gives the waste quantities in selected countries (Laufenberg et al. 2003).

Solid processing wastes from fruits and vegetables have high moisture and carbohydrate contents. The moisture content of fresh fruits and vegetables is as high as 95% on a wet basis. Pomace or press cakes from juice production may have moisture content below 50% depending on the method and efficiency of the press. Most of the carbohydrates in fruit and vegetable wastes are composed of soluble sugars and other easily hydrolysable polysaccharides (Cooper 1976). A major portion of the carbohydrates is dissolved or suspended in the processing wastewaters. Appropriate disposal of these wastewaters are a costly burden to the fruit and vegetable processing industry (Blondin et al. 1983). Excessively dilute wastewaters (e.g., less than 0.2% sugar) would be too costly to recover the dissolved sugars.

18.2.1.2 Oilseed Processing Wastes

Vegetable oils and fats are important constituents of human foods and animal feeds. Vegetable oils and animal fats are also used to produce industrial products, such as biodiesel. Approximately 125 million tons of oils and fats were produced worldwide in 2003. Soybean, palm, rapeseed, and sunflower are main oil sources as shown in Table 18.2. In the United States, the average production capacity of vegetable oils was 10.7 million tons year^{-1} between 1995 and 2000. Soybean was the dominant oil source in the United States, which was 8.3 million tons year^{-1} or about 78% of the total oil production in the United States and 27% of the total soybean production in the world (Canakci 2007).

The oil contents of oil seeds are related to their varieties, genotype, and environmental influences. Sunflower seeds typically have 40%–50% of oil by weight (Gercel 2002). Rapeseed has 40% oil (Onay et al. 2001). Soybean has only 15%–25% oil, but it is a rich source of protein for foods and feeds. In oilseed milling facilities, oil cakes are solid residue generated after oil extraction. There are two types of oil cakes: edible and nonedible (Ramachandran et al. 2007). Oil cakes are rich in protein and fiber. The composition of oil cakes depends on their variety, growing condition, and extraction methods. The amino acid profiles significantly vary from one variety to another. The chemical compositions of selected oil cakes are given in Table 18.3.

18.2.1.3 Grain Processing Wastes

Grains are an abundant renewable resource of starch for production of foods, feeds, and fuels. Corn, wheat, and rice account for 87% of all grain production worldwide.

Table 18.1 Fruit and Vegetable Processing Wastes in Selected Countries

Country	*Year*	*Quantity and Waste Type*
Germany	1997	380,000 t/a organic waste from potato, vegetable, and fruit processing
		1,954,000 t/a spent malt and hops in breweries
		1,800,000 t/a grape pomace in viniculture
		3,000,000 t/a crude fiber residues in sugar production
		100,000 t of wet apple pomace (~25,000 t dry apple pomace) if 400,000 t apples are processed into apple juice
Belgium	1992	105,000 t/a biowaste, including vegetable, garden, and fruit waste
		280,000 t/a estimations due to legislation of separate household collection
Thailand	1993	386,930 t/a empty fruit bunches
		165,830 t/a palm press fiber
		110,550 t/a palm kernel shells
		1,000,000 t/a cassava pulp
Spain	1997	>250,000 t/a olive pomace
EEC	1996	14,000,000 t/a dry sugar beet pulp
Portugal	1994	14,000,000 t/a tomato pomace
Jordan	1999	36,000 t/a olive pomace
Malaysia	1996	2,520,000 t/a palm mesocarp fiber
		1,440,000 t/a oil palm shells
		4,140,000 t/a empty fruit bunches
Australia	1995	400,000 t/a pineapple peel
USA	–	300,000 t/a grape pomace in California only in 1994
		9525 t/a cranberry pomace in 1998
		200,000 t/a almond shells in 1997
		3,300,000 t/a orange peel in Florida in 1994

Source: Laufenberg, G. et al., *Bioresource Technology*, 87, 167–198, 2003.

Table 18.2 World Production of Oils and Fats in 2003

Oil and Fat	*Amount (million tons)*
Soybean	31.3
Palm oil	27.8
Rapeseed oil	12.5
Sunflower oil	8.9
Palm kernel oil/coconut oil	6.5
Other plant oils	14.9
Tallow	8.0
Butter	6.3
Other animal fats	8.2
Total	125

Source: Hill, K.: Industrial development and application of biobased oleochemicals. Volume 2. In *Biorefineries – Industrial Processes and Products,* eds. B. Kamm, P. R. Gruber, and M. Kamm. Pages 291–314. 2006. Copyright Wiley-VCH Verlag GmbH & Co. KGaA. Reproduced with permission.

The compositions of main grains are given in Table 18.4. Grain processing facilities extract carbohydrates, proteins, fats, and other materials from grains and convert them into multiple products including foods, fuels, and high-value chemicals and materials. Grains, such as corn and wheat are processed through either a wet milling or dry milling process (Johnson 2006). Wet milling has been used for many years in the starch industry. It has also been adapted and modified for ethanol production. In the corn wet milling process, corn is first soaked in water containing sulfur dioxide for 24–48 h. It is then ground, and the components are physically separated using a series of centrifuges, screens, and washes. Some of the separated fractions then undergo additional refining steps. The corn wet milling process converts corn into starch, oil, gluten meal, gluten feed, and germ meal (Kraus 2006). For wheat wet milling, the bran and germ are generally removed by dry processing in a flour mill before steeping in water. The dry milling process has been widely used in the ethanol industry. The dry milling process usually does not fractionate the different components of the grains. However, because the germ contains high contents of proteins and lipids, the germ is removed by sieving and aspiration and/or by gravity methods for foods and feeds (Koseolu et al. 1991). A dry milling process involves grinding of the grain, followed by the addition of water and treatment with heat. Enzymes are added to the slurry, and the sugar, which results from starch

Table 18.3 Composition of Oil Cakes

Oil Cake	*Dry Matter (%)*	*Crude Protein (%)*	*Crude Fiber (%)*	*Ash (%)*	*Calcium (%)*	*Phosphorous (%)*
Canola	90	33.9	9.7	6.2	0.79	1.06
Coconut	88.8	25.2	10.8	6.0	0.08	0.67
Cottonseed	94.3	40.3	15.7	6.8	0.31	0.11
Ground nut	95.6	49.5	5.3	4.5	0.11	0.74
Mustard	89.8	38.5	3.5	9.9	0.05	1.11
Olive	85.2	6.3	40.0	4.2	–	–
Palm kernel	90.8	18.6	37	4.5	0.31	0.85
Sesame	83.2	35.6	7.6	11.8	2.45	1.11
Soybean	84.8	47.5	5.1	6.4	0.13	0.69
Sunflower	91	34.1	13.2	6.6	0.30	1.30

Source: Ramachandran, S. et al., *Bioresource Technology*, 98, 2000–2009, 2007.

Table 18.4 Composition of Selected Grains

Components (%)	*Corn*	*Wheat*	*Rice*	*Potato*	*Tapioca*
Moisture	16	13	14	75	70
Starch	62	60	77	19	24
Protein	8.2	13	7	2	1.5
Fat	4.2	3	0.4	0.1	0.5
Fiber	2.2	1.3	0.3	1.6	0.7
Minerals/ash	1.2	1.7	0.5	1.2	2
Sugars	2.2	8	0.3	1.1	0.5

Source: Grull, D. R., F. Jetzinger, M. Kozich, and M. M. Wastyn: Industrial starch platform-status quo of production, modification and application. In *Biorefineries – Industrial Processes and Products*, eds. B. Kamm, P. R. Gruber, and M. Kamm. Pages 61–95. 2006. Copyright Wiley-VCH Verlag GmbH & Co. KGaA. Reproduced with permission.

hydrolysis, is fermented to ethanol by the addition of yeast. Fuel ethanol is recovered by distillation and evaporation. The processing wastes after ethanol recovery are distiller grains and solubles and traditionally used as animal feeds.

A large amount of corn is used to produce ethanol in the United States, and wheat is considered the primary raw material for ethanol production in Europe and Australia. The type of co-products that are produced in the grain processing industry depends on a number of factors, including conversion technology, feedstock, and milling processing (Turhollow and Heady 1986). Grain processing facilities may generate four main co-products besides starch:

- Gluten with a high protein content
- Bran with a high fiber content
- Germ with high protein and lipid contents
- Distillers grains and solubles in ethanol production

Protein is seen as the major co-product in ethanol production. Corn has about 8% of protein by weight and a concentrate containing 90% protein has been isolated from corn gluten meal (Satterlee 1981). Wheat has an approximate protein content of 13% by weight, 80%–90% of which is made up of gluten (Weegels et al. 1992). Cereal bran has a high fiber content (Dexter et al. 1994). Fiber is composed mainly of non-starch polysaccharides and lignin. Different grains have different proportions of different types of fibers. The insoluble dietary fiber includes cellulose, lignin, and hemicelluloses while the soluble fiber consists of pectins, beta-glucans, gums, and mucilages. Germ is particularly attractive because it contains high concentrations of protein and minerals and a number of high-value lipid compounds. The drying milling process generates distiller grains and solubles as byproducts, which contain high concentrations of dietary fiber, protein, and fats (Rasco et al. 1987). In a corn dry mill, one third of the original corn mass becomes corn distiller dried grains and solubles (Wang et al. 2005).

Starch processing plants produce a large amount of diluted wastewater, which may cause environmental problems. The need for fresh water in a corn wet milling plant is as high as 1.5 m^3 ton^{-1} of corn (Kollacks and Rekers 1988). Typically, wheat milling results in 78%–80% flour, 19% bran, and 1% wheat germ. The flour is washed with water to produce starch, and the starch processing residue is a commercially accepted feedstock for ethanol production (Nguyen 2003). The starch processing wastewater contains some solids, which can be recovered as a potential renewable source (Nguyen 2003; Verma et al. 2007). Nguyen (2003) reported that the distillery effluent from a starch-based ethanol plant had 3.3% total solids.

18.2.1.4 Meat Processing Wastes

The consumption of pork, beef, and poultry in developed countries has been on the increase. Figure 18.1 shows the per capita consumption of pork, beef, and poultry

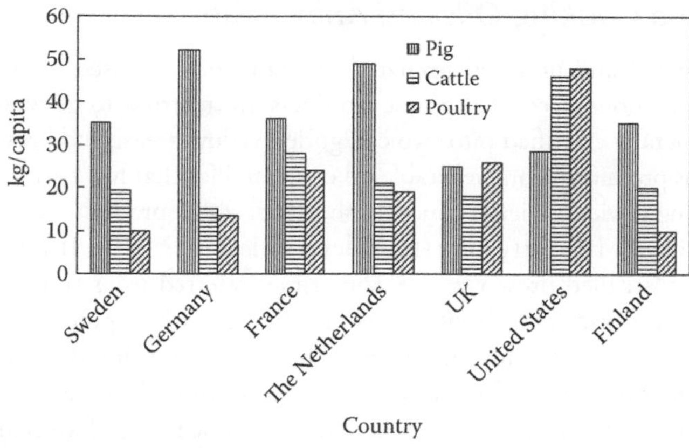

Figure 18.1 Per capita consumption of pork, beef, and poultry in several developed countries. (Adapted from Salminen, E., and J. Rintala, *Bioresource Technology*, 83, 13–26, 2002.)

in several developed countries in 1997 (Salminen and Rintala 2002). In the meat industry, slaughterhouses are producing increasing amounts of organic solid byproducts and wastes. The organic solid wastes generated in a poultry slaughterhouse include blood, feather, offal, feet, head, trimmings, and bones. A broiler is about 1.8–1.9 kg before it is slaughtered. During slaughtering and processing, about 40% of the broiler mass enters into the waste stream. Table 18.5 gives the quantities and characteristics of organic solid wastes produced in poultry slaughterhouses (Salminen and Rintala 2002).

Table 18.5 Quantities and Characteristics of Organic Solid Wastes Produced in Poultry Slaughterhouses

	Live Weight (%)	*TS (%)*	*VS (%)*	*Protein (% of TS)*	*Fat (% of TS)*
Blood	2	22	91	48	2
Feather	10	24.3	96.7	91	1–10
Offal, feet, and heart	21	39	95	32	54
Trimmings and bone	7	22.4	68	51	22

Source: Salminen, E., and J. Rintala, *Bioresource Technology*, 83, 13–26, 2002.

Note: TS: total solids; VS: volatile solids.

18.2.1.5 Used Cooking Oils and Animal Fats

Both hydrogenated and non-hydrogenated vegetable oils are used in commercial food frying operations. Recycled grease products are referred to as waste grease. Greases are generally classified into two categories: yellow grease and brown grease. Yellow grease is produced from vegetable oil or animal fat that has been heated and used for cooking a wide variety of meat, fish, or vegetable products. Yellow grease is required to have a free fatty acid (FFA) level of less than 15%. If the FFA level exceeds 15%, it is called brown grease, sometimes referred to as trap grease, and it may be sold at a discount. Brown grease is often cited as a potential feedstock for biodiesel production because it currently has very low value (Canakci 2007). The main sources of animal fats are primarily meat animal processing facilities. Another source of animal fats is the collection and processing of animal mortalities by rendering companies. In general, all greases and oils are classified as lipids. Oils are generally considered to be liquids at room temperature while greases are solid at room temperature. Many animal fats and hydrogenated vegetable oils tend to be solid at room temperature. Chemically, greases, oils, and fats are classified as triacylglycerides (Canakci 2007).

The food processing industry generates a large amount of waste cooking oils and animal fats. Approximately 1.135 million tons of waste restaurant fats are collected annually from restaurants and fast food establishments in the United States (Haumann 1990). An annual average of 4.1 kg person^{-1} of yellow grease and 5.9 kg person^{-1} of brown grease were produced in metropolitan areas in the United States in 1998 (Wiltsee 1998). The production of inexpensive nonedible feedstocks, including grease and animal fats, was about 5.284 million tons year^{-1}, which represents one third of the US total oils and fats production as shown in Table 18.6 (Canakci 2007). In France, fatty residues from both plants and animals represent an overall production of 0.55 million tons year^{-1}. The food industry generates 29% of the total fatty residues. The catering industry, wastewater treatment plants and autonomous

Table 18.6 Average Production Capacity of Nonedible Oils and Fats in the United States between 1995 and 2000

Animal Fats and Grease	*Amount (Million tons)*
Tallow	2.489
Yellow grease	1.195
Lard and grease	0.593
Poultry fat	1.006
Total	5.284

Source: Canakci, M., *Bioresource Technology*, 98, 183–190, 2007.

sanitation have shares of about 32%, 23%, and 16%, respectively (Mouneimne et al. 2003).

Animal fats and greases contain a large amount of saturated fatty acids, which are almost 50% of the total fatty acids. Due to the higher content of saturated fatty acids, the animal fats and greases have the unique properties of high melting point and high viscosity. They are solid at room temperature. Vegetable oils usually contain higher levels of unsaturated fatty acids. They are liquids at room temperature. However, it is more prone to oxidation due to the unsaturated bonds. Polyunsaturated fatty acids were very susceptible to polymerization and gum formation caused by oxidation during storage or by complex oxidative and thermal polymerization at the higher temperature and pressure of combustion. The gum did not combust completely, resulting in carbon deposits and lubricating oil thickening (Ma and Hanna 1999).

The contents of free fatty acids and moisture of feedstocks are two main parameters used to evaluate the quality of the feedstock. Natural vegetable oils and animal fats are extracted or pressed to obtain crude oil or fat. These usually contain free fatty acids, phospholipids, sterols, water, odorants, and other impurities. Even refined oils and fats contain small amounts of free fatty acids and water. Research conducted by Canakci (2007) showed that the content of free fatty acids in nonedible oils and fats varied from 0.7% to 41.8% and moisture content from 0.01% to 55.38%.

During high-temperature cooking, various chemical reactions, such as hydrolysis, polymerization, and oxidation, can occur in vegetable oils. The physical and chemical properties of the oil change during cooking. The percentage of FFAs has been found to increase due to the hydrolysis of triglycerides in the presence of food moisture and oxidation. As an example, the FFA level of fresh soybean oil changed from 0.04% to 1.51% after 70 h of frying at 190°C (Tyagi and Vasishtha 1996). Increases in viscosity were also reported due to polymerization, which resulted in the formation of higher molecular weight compounds. Other observations were that the acid value, specific gravity, and saponification value of the frying oil increased, but the iodine value decreased. The peroxide value increased to a maximum, and then started to decrease (Mittelbach et al. 1992).

18.2.2 Waste Auditing and Monitoring in Food Processing Facilities

18.2.2.1 Significance of Waste Auditing and Monitoring

Food processing wastes are commodities like other raw materials that should be measured regularly in facility operation. The audit of food processing waste not only quantifies the loss of food materials but also serves as a measure to determine the process-control effectiveness. Monitoring waste continuously and keeping records of the waste flows are also important to determine whether the effluent from a facility meets local, regional, and national pollution abatement regulations (Zall 2004).

Monitoring waste is a management tool. Waste monitoring can be done to answer the questions: (i) What is the waste source? (ii) How much waste is produced? (iii) What is its composition or concentration? (iv) Where does it come from? The amount and quality of the waste are the important information to identify opportunities for recovering the unused resources into value-added products. The conversion of raw materials into finished food products involves a series of unit operations. Material loss can occur at each unit operation. Some food processing facilities first collect wastes from individual unit operations to a sump within the facility and then transport it to a central waste treatment area. To quantify loss at only the central discharge of a facility neglects the ability to determine the loss to a specific unit operation. Therefore, it is necessary to establish waste sampling stations to gather and test representative samples of waste materials sent to the central discharge in a food processing facility (Zall 2004).

18.2.2.2 Auditing the Amount of Food Processing Wastes

To manage food processing waste, the first step is to find out how much waste the facility produces. Most food processing facilities fail to measure their wastes as a separate identifiable byproduct (Zall 2004). For liquid food processing wastes, flow measurement is used to determine the quantity of the wastes. A simple technique for the flow measurement is to record the time that a pump of known capacity is in operation to deliver the wastes. A number of waste pumps in a large facility can be timed, and their total output capacity is then calculated. The quantity of the liquid wastes is the flow rate of pump(s) times the operating time. A weir, which has a dam-like structure, is a common system to measure flow (Zall 2004). During measurement, a flow is restricted through a weir, and the height of flow above a discharge outlet is a measure of volume passing over or through the weir. Solid wastes in a food processing facility may be combined. The solid waste may contain food fragments, damaged packaged goods, and a mixture of raw materials, paper, plastics, and others. The mass of solid wastes can be weighted directly.

18.2.2.3 Auditing the Quality of Food Processing Wastes

For liquid food processing wastes, the common quality parameters are biological oxygen demand (BOD), chemical oxygen demand (COD), pH value, and temperature. The BOD to COD ratio determines the biodegradability of wastes. Other common analyses of liquid wastes include total solids, total suspended solids, volatile suspended solids, fixed suspended solids, or ash and dissolved solids.

Temperature of wastes should be measured because heat may be recovered from high-temperature liquid wastes. Hot water can be generated from high-temperature wastes. The hot water can be used as the feed water of a boiler and plant wash water. The record of the temperature of an effluent from a facility can also help to isolate causes of fuel waste in the facility if an unusual temperature rise occurs (Zall 2004).

Waste concentration or composition can be determined through sampling. For a small amount of liquid wastes, the best sampling method would be to contain waste in a single vessel, agitate the mass, and then collect a well-mixed representative portion of the waste. For a large amount of liquid waste, periodic sampling may be used. Periodic sampling is time and flow dependent. Proper sampling equipment is a unit in which both sampling time and sample size can be varied (Zall 2004). The composition of solid waste can be determined by grinding the waste and sample homogeneous material. The food processing wastes contain carbohydrate, fat, proteins, and minerals besides waste. The different components have variable values depending on the ingredient sources (Wang 2008).

18.2.3 Current Utilizations and Challenges in Food Waste Management

18.2.3.1 Animal Feeds

Most fruit and vegetable processing residuals are presently used for animal feeds. Cannery wastes, such as potato and corn byproducts, and pomace could be used as a fiber substitute in animal feeds. The fruit and vegetable wastes can be fed to animal directly in the same form as they are produced. The wastes can also be used to make prepared animal feeds after drying and treatment with nutrient additives, which have longer storage periods. However, more research should be conducted to investigate any negative effects of the feeds made from fruit and vegetable wastes on the health of animals. It may need some physical and/or chemical pretreatment to increase the digestibility of solid food processing wastes as animal feeds (Lahmar et al. 1994).

Edible oil cakes, such as soy bean cakes, are usually rich in protein. Edible oil cakes have been used for feed in the poultry, fish, and swine industries. Some oil cakes can also be considered for food supplementation (Ramachandran et al. 2007). Co-products from grain processing facilities are another protein-rich source. Traditionally, the animal feed industry has provided a market for co-products from ethanol production (Rasco et al. 1987).

Because slaughterhouse wastes are rich sources of protein and vitamins, some low-risk wastes are preserved with formic acid and used as animal feed for fur animals and pets. Waste vegetable oils and fats are generally low in cost and are currently collected from large food processing and service facilities. They are then rendered to separate fat from meat by heating the meat and used almost exclusively in animal feed in developed countries (Canakci 2007). The Commission of the European Communities (1990) required that rendering at 133°C and 300 kPa for a minimum of 20 min is needed for high-risk materials, such as dead and stillborn animals intended for animal feed or as an intermediate product for the manufacture of organic fertilizer and other products. The fat may be used for animal feeds or produce chemicals and fuel products. However, caution should be taken to reduce the risk of disease transmission via the feeds (Salminen and Rintala 2002).

18.2.3.2 Organic Fertilizer

Nonedible cakes are used as organic nitrogenous fertilizers due to their high N, P, and K contents (Ramachandran et al. 2007). Incineration and composting have been used to convert slaughterhouse wastes into fertilizers (Ritter and Chinside 1995). Incineration is a thermal degradation technology for effectively destroying potential infectious microorganisms in the wastes.

Composting, which is an aerobic biological process, has been used to reduce pathogens and decompose organic slaughterhouse and other food processing wastes. During composting, the food processing wastes are treated aerobically and converted into a hygienic, stable, and odor-free product of fertilizer (Schaub and Leonard 1996; Mohaibes and Heinonen-Tanski 2004). Ideal composting conditions are the carbon to nitrogen ratio of a composting material between 20 and 40, the moisture content between 50% and 60%, adequate oxygen supply, small particle sizes, and enough void space for air to flow through (Chang et al. 2006). Biofilters are usually used to treat the air during composting for odor control. The composted materials can be used as soil conditioner or fertilizer (Tritt and Schuchardt 1992). Currently, nearly 300 composting facilities in the US accept food waste. The large majority (~80%) of these facilities process <5000 Mg of food waste per year. About half of the food waste-composting facilities in the United States are commercial or municipal facilities. Most of the remainder are colleges and universities as well as farms. The tipping fees at the composting facilities in the United States ranged from about $20 to $50 per ton of food wastes (Levis et al. 2010).

18.2.3.3 High-Value Co-Products

It has been estimated that co-product revenue can account for up to 40% of the income of a grain-based ethanol plant (Chang et al. 1995). Wheat gluten with a high protein content is an important additive in bakery and breakfast foods, fish and meat products, and dairy products (Gras and Simmonds 1980; Satterlee 1981). Wheat gluten has also been used in the non-food industries for production of adhesives and films (Krull and Inglett 1971). Corn protein has not established a market to be used in foods. Zein is the only corn protein that has been developed for industrial production of packaging films, linoleum tiles, coatings, ink, and textile fibers. Corn germ in wet milling is extracted to produce corn oil, which is the most valuable co-product of the corn wet milling process. Corn germ makes up 10%–20% of the total product generated when corn is dry milled (Blessin et al. 1973).

18.2.3.4 Challenges in Food Waste Management

The maintenance of food hygiene and product quality presents a challenge to the minimization of raw material wastage in a food processing facility because the former requires the rejection of substandard raw materials. However, the reduction of food processing wastes through the improvement of resource efficiency in a food processing facility can improve its environmental performance, profitability, and competitiveness. Food manufacturers must give greater priority to reducing raw material wastage as far as practically possible and to finding good alternative uses or composting solutions for wasted raw material (Henningsson et al. 2004).

Without proper treatment, food processing wastes create odor and hygiene concerns and cause negative environmental impacts. Large amounts of solid food processing wastes are disposed of as wastes in landfills, on the land, or burned onsite. The costs of disposal of solid wastes on land vary considerably and are dependent on the moisture content of the waste materials, the distance that the wastes are transported, and the methods of delivery and application. One of the most promising approaches to minimizing the disposal costs and environmental problems is to remove excess moisture from the waste materials using dewatering methods, such as filtration and centrifugation (Hang 2004). It was reported that 60% of solid fatty residues generated in France were stored in landfill sites (Mouneimne et al. 2003). In most developing countries, a large proportion of the waste oils and fats are disposed of inappropriately (Al-Widyan and Al-Shyoukh 2002). Fatty wastes are a major source of pollution, and landfilling of fatty wastes is not acceptable in some developed countries. Food processing waste, like other municipal solid wastes, produces methane when it decomposes anaerobically in a landfill. As methane has a global warming potential (GWP) 25 times greater than CO_2, landfills have become the major contributor to GWP from the waste management sector (Themelis 2003).

There is significant variation in the landfill fees in the United States. Landfill tipping fees in some parts of the Northeastern United States approach $110 per ton, which includes hauling waste up to 1000 km to Virginia and central Pennsylvania. The landfill tipping fees in Ohio are closer to $28 per ton (Levis et al. 2010). High transport and landfill costs have led to alternative utilization of food processing wastes (Laufenberg et al. 2003). Compost facilities sited in the New York, New Jersey, and Connecticut areas may be more economical than current landfill alternatives. In some regions, as there are more landfills than composting facilities, it is more difficult for composting to compete with landfills (Levis et al. 2010).

Solid organic wastes from slaughterhouses may contain several species of microorganisms, including potential pathogens (Salminen and Rintala 2002). In addition, animals may accumulate various metals, drugs, and other chemicals added in their feed for nutritional and pharmaceutical purposes (Haapapuro et al. 1997). The disposal and processing of animal wastes should destroy potential pathogens present in the wastes for prevention of animal and public health problems. Burial of slaughterhouse wastes should also be strictly controlled to avoid groundwater contamination. Incineration and composting have been used to treat slaughterhouse wastes. Caution should be taken to control the air emissions, incineration conditions, and disposal of solid and liquid residues after incineration. The emission to air, water, and soil associated with composting may cause a pollution problem and reduction of nitrogen in the compost fertilizer (Tritt and Schuchardt 1992).

Most of the liquid wastes from food processing facilities are released into rivers, lakes, and oceans and disposed of in public sewer systems without any treatment. A small amount of liquid wastes are utilized in onsite treatment or irrigation. If the liquid wastes contain a high amount of suspended solids, some treatments are required before being released into the environment. If the liquid wastes contain toxic constituents, such as salt brines and lye, they must be detoxified prior to disposal (Sargent and Steff 1986). The wastewater from the grain processing facilities is usually disposed of by spray irrigation. The problems caused by the spray irrigation include strong odor, insect invasion, increase of soil acidity, salt leaching, buildup of sulfates, and putrescence (Nguyen 2003).

18.3 Value Added Processing of Food Wastes

18.3.1 Extraction and Purification of Value-Added Products from Food Processing Wastes

Plants produce a large amount of primary metabolites of lipids, proteins, and carbohydrates. Plants also produce a broad range of high-value bioactive compounds of secondary metabolites, which can be classified into three main groups of phenolics, terpenoids, and alkaloids. Extraction techniques have been widely investigated to obtain such valuable natural compounds from plants to be used in the food, pharmaceutical, cosmetics, herbicide, and pesticide industries (Wang and Weller 2006; Wang 2010). Many of those phytochemicals, such as flavonoids and carotenoids, have been determined to be beneficial to the human body in preventing or treating one or more diseases or improving physiological performance (Wildman 2000).

High-value phytochemicals can be identified and extracted from the waste residuals in the food processing facilities before they are used for energy production for the increase of economic profitability (Wang et al. 2005). Aqueous methanol was used to extract citrus junos fruit waste after juice extraction. It was found that aqueous methanol extracts of the citrus junos peels inhibited the growth of several weed species. Therefore, citrus junos waste may possess allelopathic potential, and the extracts from the waste may be potentially useful for weed management (Kato-Noguchi and Tanaka 2004).

Grain kernels, such as sorghum, corn germ, and wheat germ, are particularly attractive for use in food products because they contain a high concentration of proteins and minerals and a number of high-value lipid components. One of the promising areas for co-products from grain processing facilities is that nutraceuticals, such as wheat fiber, phytosterols, policosanols, and free fatty acids extracted from grains or their co-products, have been proved to have either health or medical advantages (Wrick 1993; DeFelice 1995; Zbasnik et al. 2009). More than 1.3 million metric tons of grain sorghum is used annually to produce ethanol in the United States, and the number is expected to increase in the future. Approximately 8.2 kg of dry residual in the form of distiller dried grains with solubles (DDGS) remains from each 25 kg of grain sorghum used to produce ethanol. The lipids in sorghum DDGS contain considerable amounts of long-chain fatty acids, fatty aldehydes, fatty alcohols (policosonols), triacylglycerols, and other valuable components, such as phytosterols, tocols, and diacylglycerols. Interest is increasing in the lipid compounds, such as phytosterols, tocopherol, policosanols, and unsaturated fatty acids, which could play a preventive role in many diseases (Zbasnik et al. 2009). Hexane and supercritical CO_2 have been used to extract these high-value lipid compounds from sorghum DDGS (Wang et al. 2005, 2007, 2008a).

18.3.2 Biological Conversion of Food Processing Wastes into High-Value Chemicals

Food processing wastes, which are rich in sugars, vitamins, and minerals, are easily assimilated for microorganisms. Food processing wastes are very suitable as raw materials for the production of high-value chemicals or metabolites by microorganisms. Food processing wastes have been used to produce organic acids, enzymes, antibiotics, antioxidants, vitamins, and mushrooms.

Lactic acid is one of the useful compounds utilized in food, pharmaceutical, and chemical industries. Food wastes from a cafeteria were used to produce lactic acid by *Lactobacillus manihotivorans* LMG18011. At the optimum initial pH of 5.0–5.5 and the fermentation pH of 5.0, 19.5 g L(+)-lactic acid was produced from 200 g food wastes. It was also found that the addition of manganese stimulated the direct fermentation significantly and enabled complete bioconversion within 100 h (Ohkouchi and Inoue 2006).

Laccases are copper-containing oxidase enzymes that are found in many plants, fungi, and microorganisms. Kiwifruit wastes were used to produce laccase by white-rot fungus *Trametes hirsuta* under solid-state fermentation. The highest laccase value (around 90,000 nkat L^{-1}) was obtained operating at an initial ammonium concentration of 0.150 g L^{-1} and with 2.5 g of pretreated peelings of kiwifruit (Rosales et al. 2005). α-Amylase and glucoamylase are two important enzymes to hydrolyze starch into glucose in the bioenergy industry. Food wastes collected from a university cafeteria was used as substrate for the glucoamylase production by *Aspergillus niger* UV-60 under submerged fermentation. At the optimum concentration of 2.50% (dry basis), smashed food waste produced glucoamylase of 126 U ml^{-1} after 96 h fermentation whereas 137 U mL^{-1} of glucoamylase could be attained within the same time from raw food waste of 3.75% (Wang et al. 2008b). Brewery wastewater and meat wastewater supplemented with different starch concentrations were used to synthesize amylase and protease by *Aspergillus niger* strain UO-1. The highest amylase yields of 70.29 and 60.12 EU mL^{-1} were obtained using the brewery wastewater and meat processing wastewater supplemented with 40 g of starch L^{-1} of medium after 88 h of fermentation. At the same conditions, the highest protease yields were 6.11 and 6.03 EU mL^{-1}, respectively. The initial chemical oxygen demand (COD) in both wastes was reduced by more than 92% (Hernandez et al. 2006). Oil cakes are also an ideal source as a support matrix for various biotechnological processes for production of enzymes, antibiotics, vitamins, antioxidants, and mushrooms (Ramachandran et al. 2007).

18.3.3 Conversion of Food Processing Wastes into Energy Products

18.3.3.1 Production of Ethanol from Carbohydrate-Rich Processing Wastes

Because fruit and vegetable processing wastes are rich in carbohydrates, they can be used as a source for the production of fermentable sugars. Ethanol can be produced directly from fruit pomace with a high sugar content (Hang et al. 1986; Ngadi and Correia 1992; Nigam 2000). Starchy and cellulosic waste materials from fruit and vegetable processes, however, must be hydrolyzed first to fermentable sugars before a yeast culture can ferment them to ethanol. Fischer and Bipp (2005) used acidic hydrolysis of organic acids and monosaccharides from several food processing residues, including sugar beet molasses, whey powder, wine yeast, potato peel sludge, spent hops, malt dust, and apple marc. The yields of ethanol depend on the initial carbohydrate content in the wastes.

Grain residuals are rich in cellulose and hemicellulose that can be a renewable source for enzymatic production of soluble sugars as feedstocks for ethanol fermentation (Hang 2004). Wheat bran consists of three main components of residual starch, cellulose, and hemicellulose. Choteborska et al. (2004) produced sugar solution from wheat bran for ethanol fermentation. They first treated the wheat bran with starch-degrading enzymes to remove the starch from the bran. The maximum yield of sugars was 52.1 g $100g^{-1}$ of starch-free wheat bran obtained using 1% of sulfuric acid at 130°C for 40 min. The furfural and 5-hydroxy-methyl-2-furaldehyd, which cause inhibition of fermentation, was as low as 0.28 g L^{-1} and 0.05 g L^{-1}, respectively. A dry mill generates a large amount of distiller grains. Tucker et al. (2004) used a dilute acid pretreatment process to hydrolyze sugars from the residual starch and fiber in the distiller grains for additional ethanol production. The pretreatment of distiller grain at 140°C with 3.27% H_2SO_4 for 20 min could hydrolyze 77% of available carbohydrate in the distillers grains. The yeast of *Saccharomyces cerevisiae* D_5A was further used to ferment sugars. The ethanol yield was 73% of the theoretical value from available glucans.

Fruit, vegetable, and grain starch processing facilities generate a large amount of processing wastewaters. Because a major portion of carbohydrates is dissolved or suspended in the processing wastewaters during processing of fruits and vegetables, fruit and vegetable processing wastewater is also a potential feedstock for alcohol fermentation (Blondin et al. 1983). Reverse osmosis (RO) technology has been used to recover and concentrate soluble sugars from dilute effluent and juice wastewaters. The recovered sugar concentrates are sterile, easily storable or transportable, and excellent feedstocks for bioconversion processes. Using the reverse osmosis technology in the United States, approximately 1.42 million tons of fermentable sugars could be recovered as a 20% sugar concentrate, which is suitable for bioconversion to useful liquid fuels. The annual fuel alcohol potential of these sugars is between 750 and 900 million liters. Economic analysis indicates that overall alcohol production costs, on average, should be only 40% of current prices for bulk fuel alcohol. The coincident major reduction in wastewater biological oxygen demand (BOD) and associated disposal costs, and the recovery of more than 430 billion liters of reusable RO permeate water affords added incentives for industrial participation in the production of sugar concentrates from fruit and vegetable byproduct wastewaters (Blondin et al. 1983). As wastewater from starch processing facilities contains some residual starch, it can be converted to ethanol by fermentation. The solids in the wastewater can economically be recovered by ultrafiltration (Nguyen 2003). Moreover, the recovery of sugars can remove most of the BOD-contributing solids from effluents in fruit, vegetable, and grain starch processing facilities and significantly reduce wastewater disposal costs.

18.3.3.2 Anaerobic Digestion of Food Processing Wastes into Biogas

Food wastes, such as fruit and vegetable processing wastes, meat processing wastes, and slaughterhouse wastes, can be treated anaerobically to reduce pollutant and pathogen risk and recover energy from the wastes. Anaerobic digestion is a biological process in which an organic matter is degraded to a gaseous mixture of biogas in the absence of oxygen. Biogas, which mainly consists of methane and carbon dioxide, can be used as an energy source to replace fossil natural gas. The methane in the biogas could be as high as 50%–60% by volume (Gebauer 2004). If biogas produced by anaerobic digestion of biomass is used for electricity generation, the overall conversion efficiency from biomass to electricity is about 10%–16% (McKendry 2002).

One great advantage of anaerobic digestion is that it can be used to treat very wet and pasty organic wastes or liquid wastes (Shih 1993; Braber 1995). Anaerobic digestion is a commercially proven technology and is widely used for treating organic wastes with a high moisture content (i.e., >80%–90% moisture) (McKendry 2002). For the treatment of food processing wastes, anaerobic digestion can not only produce methane for energy, but it also destroys pathogenic bacteria presenting in the wastes and reduces pollutant emission from the wastes. Recent advances in anaerobic digestion technologies have made it possible to compete well with other methods for the treatment of a diversity of food processing wastes.

Anaerobic digestion has been used to treat solid fish wastes (Gebauer 2004), slaughterhouse wastes (Salminen and Rintala 2002), fruit and vegetable processing wastes (Kalia et al. 2000), and food processing wastewater (Di Berardino et al. 2000). Co-digestion of wastes with different characteristics is an approach to dilute toxicants and supply required nutrients (Mata-Alvarez et al. 2000). Fruit and vegetable wastes and animal manure are a promising combination to be used as co-digestion feedstock. Callaghan et al. (2002) investigated the co-digestion of fruit and vegetable wastes and chicken manure in a completely stirred tank reactor at 35°C, HRT of 21 days, and loading rate in the range of 3.19–5.01 kg VS m^{-3} day. They found that with increasing the fruit and vegetable wastes from 20% to 50%, the methane yield increased from 0.23 to 0.45 m^3 kg^{-1} VS added. The increase of fruit and vegetable wastes in the feedstock can decrease the ammonia inhibition from chicken manure. Nonedible oil cakes have been used to adjust the carbon to nitrogen ratio of feedstocks to anaerobic digestion processes for maximum microbiological activity and fuel output (Lingaiah and Rajasekaran 1986).

Anaerobic digestion effluents are generally not suitable to directly be disposed of on the land because they are too wet, contain some phytotoxic volatile fatty acids, and are not hydrogenated if the digestion does not occur at a thermophilic temperature. Therefore, aerobic post-treatment or composting after anaerobic digestion is needed (Mata-Alvarez et al. 2000). Compared to direct aerobic composting, anaerobic digestion technology is complex and requires a large investment, but it can recover an amount of energy and reduce pollutant emissions from the wastes.

18.3.3.3 Thermochemical Conversion of Food Processing Wastes into Biofuels

Thermochemical conversion provides a competitive way to produce chemical and energy products from low-value and highly distributed biomass resources with large variations in properties (Wang et al. 2011). Combustion, pyrolysis, gasification, and thermochemical liquefaction are four main thermochemical conversion methods (Wang 2008). Combustion is to convert the chemical energy stored in an organic matter into heat, generating carbon dioxide and water as final products. Combustion usually produces hot gas at temperatures around 800°C–1000°C. Pyrolysis is the conversion of biomass to liquid, solid, and gaseous fractions by heating the biomass in the absence of air or oxygen to around 500°C. The gasification process falls between complete combustion and pyrolysis. Gasification is the partial oxidation of organic matter at a high temperature to convert the organic matter into a combustible gas mixture called syngas, which mainly consists of carbon monoxide, hydrogen, methane, and carbon dioxide. Gasifiers are operated at approximately 800°C–900°C, although a non-catalytic entrained flow gasifier could be operated at a temperature as high as 1300°C. Solvents, such as water and alcohols, at an elevated temperature and pressure can convert solid biomass into liquid fuels via a thermochemical liquefaction process. Thermochemical conversion technologies have been used for the reduction of the environmental impact of food processing wastes and the recovery of energy from the wastes (Shinogi and Kanri 2003). It is possible to convert any type of organic food wastes in a thermochemical process. However, it could be practical and economic to thermochemically convert a feedstock with a low moisture content (e.g., <50%–60%) if the biomass is not dried before the conversion (Wang et al. 2009).

Crop residues, such as bran, husk, and bagasse, can be used as energy feedstocks for production of liquid and gaseous fuels and supply of heat and power in the processing facilities. Solid food processing wastes can either directly be combusted or first converted into gaseous and liquid fuels for further combustion. Rice husk was used as a fuel in a 30 kWh bubbling fluidized bed combustor. Combustion efficiency itself was higher than 97% (Armesto et al. 2003). Liquid processing wastes, such as waste vegetable oils and fats, can be converted to alcohol ester by transesterification to be used as a combustion fuel (Tashtoush et al. 2003).

Food wastes with a low moisture content are also suitable for gasification. Mansaray et al. (1999) used a fluidized bed gasifier for air gasification of rice husk at a fluidization velocity from 0.22 to 0.33 m s^{-1}, air equivalence ratio from 0.25 to 0.35, and temperature from 665°C to 830°C. The gas yield and carbon conversion were in the range of 1.30 to 1.98 Nm^2 kg^{-1} and 55% to 81%, respectively. The higher heating value of the syngas was 3.09–5.03 MJ Nm^{-3}. A sugar factory produces nearly 30% of bagasse out of its total crushing. Many research efforts have been attempted to use the bagasse as a renewable feedstock for power generation and for the production of bio-based materials. A cyclone gasifier was used to gasify bagasse powder at 39–53 kg h^{-1} and 820°C–850°C. The heating values of the syngas at an oxygen equivalence ratio from 0.18 to 0.25 were in the range of 3.5–4.5 MJ Nm^{-3} dry gas (Gabra et al. 2001a, b). Thermal energy consumption is significant in the grain milling facilities. It is a promising way to supply heat and power from the grain processing residues (Eggeman and Verser 2006).

Pyrolysis has been used to convert oil seed cakes into a liquid fuel of bio-oil (Ozbay et al. 2001; Gercel 2002; Ozcimen and Karaosmanoglu 2004). Gercel (2002) used a fixed-bed tubular reactor to pyrolyze sunflower oil cake. The maximum bio-oil yield was 48.89% by weight obtained at a pyrolysis temperature of 550°C and a heating rate of 5°C s^{-1}. The calorific value of bio-oil produced from sunflower cake was measured at 32.15 MJ kg^{-1}, which is very close to those of petroleum fractions. Direct solvent liquefaction provides a method for production of liquid fuels, chemical feedstocks, and carbon materials from biomass. Kucuk (2001) used supercritical methanol, ethanol, and acetone to liquefy verbascum and sunflower stalks. The liquid yields varied from 40% to 60.5% of the feedstock mass at temperatures of 260°C and 300°C and with a catalyst of 10% NaOH or without catalyst.

18.3.3.4 Conversion of Used Cooking Oil and Animal Fats into Biodiesel

Biodiesel is currently produced from food-grade vegetable oils, such as soybean oil in the United States and rapeseed in Europe. Because food-grade vegetable oils are expensive, biodiesel produced from food-grade vegetable oil is not economically feasible. Animal fats, waste cooking oils, and restaurant grease are potential feedstocks for biodiesel production. Waste vegetable oils from restaurants and rendered animal fats are inexpensive compared with food-grade vegetable oils. The price of yellow grease varied widely from \$0.09 to \$0.20 lb^{-1} in 2000, compared to \$0.35 lb^{-1} for soybean oil. Brown grease is usually discounted \$0.01–\$0.03 lb^{-1} below the price of yellow grease. One pound of most fats and oils can be converted to a pound of biodiesel. If all of the 5.284 million ton $year^{-1}$ of grease and animal fats in the United States were converted to biodiesel, it would replace about 1.5 billion gallons of diesel fuel (Canakci 2007).

The free fatty acid and water contents have significant effects on the trans-esterification of glycerides with alcohols using alkaline in a traditional biodiesel production process. They also interfere with the separation of fatty acid esters and glycerol. Free fatty acids and moisture reduce the efficiency of trans-esterification reaction to convert a feedstock into biodiesel using traditional alkaline catalysts. Therefore, an efficient process for converting waste grease and animal fats must tolerate a wide range of feedstock properties (Wang 2008).

18.4 Environmental, Economical and Sustainable Impacts of Managing Food Processing Wastes

18.4.1 Environmental Impacts

There are numerous environmental factors to be considered in evaluating alternatives for organic waste management. Without proper treatment, food wastes create odor, hygiene concerns, and cause negative environmental impacts. Organic food wastes produce methane when they decompose anaerobically in a landfill. As methane has a global warming potential (GWP) 25 times greater than CO_2, landfills have become the major contributor to GWP from the waste management sector. Research shows that every ton of municipal solid wastes processed can avoid one ton of CO_2 equivalent (Themelis 2003). The methane captured at landfills or anaerobic digestion facilities can be used to generate electricity or to provide steam for district heating (Levis et al. 2010).

Life cycle assessment (LCA) is an emerging tool to measure and compare the environmental impacts of human activities. LCA consists of two procedures: the selection of impact indicators and the analysis of inventory data for emissions. The impact indicators used in the LCA may include global warming potential, acidification, eutrophication, photochemical oxidation, and energy use (Lundie and Peters 2005). Kim and Kim (2010) evaluated different food waste disposal technologies from the perspective of global warming and resource recovery using LCA. Their results showed that 200 kg of CO_2 equivalent could be produced from the dry animal feeding process, 61 kg of CO_2 equivalent from the wet animal feeding process, 123 kg of CO_2 equivalent from the composting process, and 1010 kg of CO_2 equivalent from landfilling for processing 1 ton of food wastes. LCA was used to identify and quantify the potential environmental impacts of differential food waste management methods, including incineration, anaerobic digestion, and composting (Khoo et al. 2010). The LCA shows that anaerobic digestion of food wastes could significantly reduce the global warming impacts of food wastes, compared to incineration and composting.

Case Study 1: Life Cycle Analysis of Combined Anaerobic Digestion and Composting of Food Processing Wastes

Anaerobic digestion combined with composting has been widely used to treat food wastes. In the combined anaerobic digestion and composting process used in Singapore, food wastes are pretreated and then fed into an anaerobic digester. The biogas from the anaerobic digester is transferred into gas engines to generate energy. The solid residues from the anaerobic digester are converted into bio-compost (Khoo et al. 2010). The whole system produces two products: electrical energy and bio-compost. The electricity generated will be used to replace fossil-based electricity on the existing grid. The bio-compost material can be used as a replacement of mineral fertilizers. The nutrient contents of the bio-compost are 0.0076 kg N kg^{-1} bio-compost and 0.0011 kg P kg^{-1} bio-compost. Greenhouse gas savings in terms of CO_2 equivalence are 5.3 kg kg^{-1} N mineral fertilizer and 0.52 kg kg^{-1} P mineral fertilizer (Khoo et al. 2010). Below is the process information (Khoo et al. 2010):

- Treatment capacity: 109,500 ton $year^{-1}$
- Amount of compost produced: 66,000 ton $year^{-1}$
- Electrical energy consumption for pretreatment: 25 kWh ton^{-1}
- Electrical energy consumption for AD and composting: 32 kWh ton^{-1}
- Electrical energy output: 260.82 kWh ton^{-1}

LCA is used to investigate the environmental performance of this combined anaerobic digestion and compositing process for the treatment of food processing wastes. The LCA considers the positive environmental impacts of this process, including (i) useful energy of electricity from the process and (ii) carbon dioxide mitigation from the bio-compost, and negative environmental impacts, including (i) air emissions of the process and (ii) energy consumption by the process. The electricity grid emissions and the emissions to air for combined anaerobic digestion and composting are given in Table 18.7. Main assumptions for the LCA include

- Transportation emissions are not included.
- Construction materials for the waste treatment plant and chemical used are not included.
- Both CO_2 and CH_4 contribute to global warming, and the global warming potential of CH_4 is 25 times that of CO_2.

Table 18.7 Emissions to Air by an Electricity Grid and the Combined Anaerobic Digestion and Composting Process for the Treatment of Food Wastes

Pollutant	*CO*	*CO_2*	*CH_4*	*N_2O*	*NO_x*	*SO_x*	*PM*	*NMVOC*
Electricity grid emission (kg kWh^{-1})	$1.78*10^{-4}$	$5.00*10^{-1}$	$8.15*10^{-6}$	$3.03*10^{-6}$	$1.19*10^{-4}$	$7.34*10^{-4}$	$1.16*10^{-5}$	$2.17*10^{-5}$
Waste conversion (kg kg^{-1})	$1.13*10^{-6}$	$6.00*10^{-3}$	$1.64*10^{-5}$	$8.15*10^{-5}$	$4.50*10^{-5}$	$3.19*10^{-5}$	$1.65*10^{-7}$	$2.70*10^{-5}$

Source: Khoo, H. H. et al., *Science of the Total Environment*, 408, 1367–1373, 2010.

The life cycle assessment has two steps. The first step is to select the impact indicators. In this case study, global warming potential is selected as an impact indicator. The second step is to analyze the inventory data. The LCA of the inventory data for the global warming potential is given below.

(i) *Annual saving of CO_2 equivalence by electricity generated*:
$109{,}500$ tons $a^{-1} \times 260.82$ kWh $ton^{-1} \times 5.00 \times 10^{-1}$ kg CO_2 $kWh^{-1} = 1.43 \times 10^7$ kg CO_2 a^{-1} (CO_2 emission)
$109{,}500$ tons $a^{-1} \times 260.82$ kWh $ton^{-1} \times 8.15 \times 10^{-6}$ kg CH_4 $kWh^{-1} \times 25$ kg $CO_2/CH_4 = 5.82 \times 10^3$ kg CO_2 a^{-1} (CH_4 emission)

(ii) *Annual saving of CO_2 equivalence by bio-composting generated*:
N: $66{,}000$ tons $a^{-1} \times 1000$ kg $ton^{-1} \times 0.0076$ kg N $kg^{-1} \times 5.3$ kg CO_2 kg^{-1} N $= 2.66 \times 10^6$ kg CO_2 a^{-1}
P: $66{,}000$ tons $a^{-1} \times 1000$ kg $ton^{-1} \times 0.0011$ kg P $kg^{-1} \times 0.52$ kg CO_2 kg^{-1} P $= 3.78 \times 10^4$ kg CO_2 a^{-1}

(iii) *Annual cost of CO_2 equivalence for the electricity consumption*:
$109{,}500$ tons $a^{-1} \times (25 + 32)$ kWh $ton^{-1} \times 5.00 \times 10^{-1}$ kg CO_2 $kWh^{-1} = 3.12 \times 10^6$ kg CO_2 a^{-1} (CO_2 emission)
$109{,}500$ tons $a^{-1} \times (25 + 32)$ kWh $ton^{-1} \times 8.15 \times 10^{-6}$ kg CH_4 $kWh^{-1} \times 25$ kg $CO_2/CH_4 = 1.27 \times 10^3$ kg CO_2 a^{-1} (CH_4 emission)

(iv) *Annual cost of CO_2 equivalence for the combined AD and composting of food wastes*:
$109{,}500$ tons $a^{-1} \times 1000$ kg $ton^{-1} \times 6.00 \times 10^{-3}$ kg CO_2 kg^{-1} waste $= 6.57 \times 10^5$ kg CO_2 a^{-1} (CO_2 emission)
$109{,}500$ tons $a^{-1} \times 1000$ kg $ton^{-1} \times 1.64 \times 10^{-5}$ kg CH_4 kg^{-1} waste $\times 25$ kg $CO_2/CH_4 = 4.49 \times 10^4$ kg CO_2 a^{-1} (CH_4 emission)

(v) *The net saving of global warming due to the combined AD and composting conversion*:
$(1.43 \times 10^7 + 5.82 \times 10^3 + 2.66 \times 10^6 + 3.78 \times 10^4) - (3.12 \times 10^6 + 1.27 \times 10^3 + 6.57 \times 10^5 + 4.49 \times 10^4)$ kg CO_2 $a^{-1} = 1.32 \times 10^7$ kg CO_2 a^{-1}

Therefore, the combined anaerobic digestion and composting process with a loading capacity to treat 109,500 tons of food wastes per year will reduce the CO_2 emission by $1.32*10^7$ kg CO_2 each year.

18.4.2 Economical Impacts

The investments can be broadly categorized into two classifications of capital costs and operating costs. Capital investments for land, building, and equipment are generally more strategic and have long-term effects. The capital investment is usually a function of the capacity or the scale of the project. Operating costs include labor, power, repairs, taxes, and other expenses. The costs versus return on investment may not be favorable for some alternative waste treatment technologies. However, the selection of a food waste treatment technology is affected not only by the net return on the investment but also by the emphasis a sustainable development and governmental regulations. As the costs of landfill, which is widely used to treat solid food wastes, are rising higher, alternative methods, such as pyrolysis and gasification, to treat food processing wastes become more favorable. Many governments have focused on diverting waste away from landfill through regulation, taxation, and public awareness (Mena et al. 2011). The economics of the treatment technologies for recovering energy from food processing wastes also depends on energy costs.

There are a number of methods for evaluating economic performance. These methods include the simple payback period, life cycle cost method, net benefit or net savings method, benefit/cost ratio method, internal rate of return method, overall rate of return method, and payback method. Usually, several methods are used to provide better understanding of an investment's worth. The simple payback period is commonly used by businesses. However, the primary criterion mandated for assessing the effectiveness of waste management technologies may be the minimization of life cycle costs (Wang 2008). A life cycle cost analysis is to quantify costs over the entire life cycle of the project investment. The life cycle cost method sums the costs of acquisition, maintenance, repair, replacement, energy, and other costs, such as salvage value, that are affected by the investment decision. All amounts are usually measured either in a present value or annual value. The time value of money must be taken into account for all amounts over the relevant period. Two important parameters for a life cycle cost analysis are the lifetimes of the equipment and the interest rate. The life cycle cost method is particularly useful for decisions that are made primarily on the basis of cost-effectiveness to determine whether an investment will lower total cost. However, it cannot be used to find the best investment in general. Numerous alternatives may be compared. The alternative with the lowest life cycle cost that meets the investor's objectives and constraints is the preferred investment (Wang 2008).

A formula for calculating the life cycle costs at the present value is given by Wang (2008):

$$LCC_p = I_p + CF_p \tag{18.1}$$

where I_p = present value of investment costs, and CF_p = present value of expected total cash flow. The expected annual cash flow is given by Gebrezgabher et al. (2010).

$$CF_a = \sum_{i=1}^{n} p_i C_i + FC - \sum_{j=1}^{m} p_j R_j \tag{18.2}$$

where p_i = the unit cost of the ith item; C_i = the total input of the ith item; P_j = the unit price of the jth item; R_j = the total output of the jth item; i = index of input items, such as feedstock, labor, and maintenance; j = index of output items, such as electricity and heat from a CHP unit; and FC = fixed cost, such as labor cost and maintenance cost. The present value of expected total cash flow is calculated by Wang (2008):

$$CF_p = CF_a \frac{(1+r)^t - 1}{r(1+r)^t} \tag{18.3}$$

where r = the internal rate of return, and t = the life of the investment. The capital investment cost consists of equipment cost and installation cost. The equipment cost is a function of its size, which is calculated by Wang (2008):

$$\frac{\text{Cost}}{\text{Cost}_{\text{ref}}} = \left(\frac{\text{Size}}{\text{Size}_{\text{ref}}}\right)^n \tag{18.4}$$

where Cost_{ref} is reference cost at a reference unit size of Size_{ref}, and n is cost scaling factor. The cost scaling factors were usually between 0.6 and 0.8 (Wang 2008).

Case Study 2: Economic Analysis of Combined Anaerobic Digestion and Composting of Food Processing Wastes

The life cycle cost method is used to analyze the economics of the combined anaerobic digestion and composting process described in case study 1.

(i) *Total capital investment*:

The capital investment for an anaerobic digestion–based power plant at a loading capacity of 70,000 tons per year is about US$ 9 million given in literature (Gebrezgabher et al. 2010). Using a cost-scaling factor of 0.6, the capital investment for a similar unit at a loading capacity of 109,500 tons year^{-1} is

$$I_p = 9 \times \left(\frac{109{,}500}{70{,}000}\right)^{0.6} = \$\ 11.8 \text{ million}$$

(ii) *Expected annual cash flow*:

The total revenue of this investment consists of the revenues of generated electricity, compost fertilizer, and the saving of disposal costs for the food wastes.

(a) Electricity: The unit price of electricity is assumed to be \$0.1 kWh^{-1}. The annual revenue of the electricity generated is 109,500 tons year^{-1} × 260.82 kWh ton^{-1} × \$0.1 kWh^{-1} = \$2.86 million year^{-1}.

(b) Compost: The price of compost as fertilizer is assumed to be \$10 ton^{-1}, excluding transportation costs. The annual revenue of the compost is 66,000 tons year^{-1} × \$10 ton^{-1} = \$0.66 million year^{-1}.

(c) Saving of landfill cost: The cost for the disposal of food processing wastes is assumed to be \$25 ton^{-1}. The total revenue for the saving of disposal costs is 109,500 tons year^{-1} × \$25 ton^{-1} = \$2.74 million year^{-1}.

The operating cost includes the costs for feedstock and energy, personnel, and facility maintenance. There is no cost for the feedstock of food processing wastes because it requires paying \$25 ton^{-1} to dispose of the wastes in landfills. The saving of landfill cost using the anaerobic digestion method is considered as revenue.

(a) Cost for electricity consumption: The annual cost of the electricity consumed is 109,500 tons year^{-1} × (25 + 32) kWh ton^{-1} × \$0.1 kWh^{-1} = \$0.62 million year^{-1}

(b) Labor cost: It is assumed that six people are hired to work for three shifts, and the average cost for salary and benefits is \$100,000 person^{-1} year^{-1}. The total annual labor cost is \$0.6 million year^{-1}

(c) Maintenance cost: The annual maintenance cost is calculated at 1.5% of the total capital investment, which is \$11.8 million × 1.5% year^{-1} = \$0.18 million year^{-1}.

The expected net annual cash flow is given by

$$CF_a = \sum_{i=1}^{n} p_i C_i + FC - \sum_{j=1}^{m} p_j R_j = (0.62 + 0.60 + 0.18)$$
$$- (2.86 + 0.66 + 2.74) = \$ - 4.86 \text{ million year}^{-1}$$

(iii) *The present value of expected total cash flow*:

It is assumed that the life of the plant, *t*, is 20 years, and the internal rate of return is 10%. The present value of expected total cash flow is calculated by

$$CF_p = CF_a \frac{(1+r)^t - 1}{r(1+r)^t} = -4.86 \times \frac{(1+0.1)^{20} - 1}{0.1(1+0.1)^{20}} = \$ - 41.38 \text{ million}$$

Therefore, the present value of the life cycle cost of the project is

$$LCC_p = I_p + CF_p = 11.8 - 41.38 = \$ - 29.58 \text{ million}$$

That is, the net present value of the project is \$29.58 million over 20 years.

18.4.3 Sustainable Development in Food Processing Facilities

The food processing industry depends heavily on raw materials and energy. Most of the energy sources used in food processing facilities have a limited availability. The food industry is under increasing pressure from governments and environmental groups to improve the sustainability of its process and create sustainable business practices. Gerbens-Leenes et al. (2003) used three indicators that address global environmental sustainability of a food production system: (i) the total energy from both fossil and renewable sources, (ii) the water, and (iii) land requirements per kilogram of available foods. This method can be used to compare the trends of production sustainability over time.

Energy resources, energy consumption, and the utilization of food processing wastes are intimately related to the sustainable development of the food industry. Many research projects have been conducted around the world with the general goals of (i) improving the thermodynamic efficiency of energy systems, and (ii) developing alternatives to fossil sources used in the food industry. One of the major factors undermining the sustainability of a food production process is the depletion of resources it uses. The difference between renewable and nonrenewable resources is that renewable resources are created at least as fast as they are consumed while nonrenewable resources are consumed faster than they are created (de Swaan Arons et al. 2004).

Three parameters can be used to characterize the sustainability of a process (de Swaan Arons et al. 2004). These three parameters are

- The thermodynamic efficiency of the process
- The extent of the use of the renewable resources
- The extent of a process cycle to be closed

Energy conservation is vital for sustainable development. Reduced energy consumption through conservation can benefit not only energy consumers by reducing their energy costs, but also a society. The most direct benefits that a society would have from the improvement of the energy efficiency is the reduction in the use of energy resources and the emission of many air pollutants, such as CO_2. To develop a sustainable society, much effort must be devoted not only to discovering renewable energy resources, but also to increasing the energy efficiency of devices and processes utilizing these resources (Wang 2008).

All unit operations associated with food processing facilities are ultimately limited by our ability to supply useful energy to the processes. It is not enough to operate processes efficiently. Improvement of thermodynamic efficiency is not the only contribution to sustainability. The improvement of thermodynamic efficiency may lower the consumption rate of nonrenewable natural resources. In terms of the sustainability of the process, the net energy entering the process to make it proceed has to come from renewable resources. This means the technology itself should be sustainable. The food processing facilities driven by nonrenewable resources should be transformed into one based on renewable resources. The food industry can adapt the zero-waste concept in its processing facilities. The main idea of the zero-waste concept is that the waste generated by one industry could become a raw material for other industries (Pauli 1998). Each food processing facility produces large masses of waste streams, which can be reused or recycled (Wang 2008).

18.5 Summary

The food processing facilities generate large amounts of organic wastes in a solid or liquid form. Traditionally, part of those wastes is processed as animal feeds. Large amounts of solid food processing wastes are buried in landfills at a cost while liquid food processing wastes are released untreated into rivers, lakes, and oceans and disposed of in public sewer systems. Due to the problems associated with the greenhouse gas emission, leachate, and limited availability of landfill sites, it becomes important to develop alternative methods for treating food processing wastes. Energy utilizations of food processing wastes can reduce both fossil fuel costs and waste disposal costs in a food processing facility. Integration of energy conversion processes into the food processing facilities will achieve not only economic profitability but also environmental benefits. Conversion of food processing wastes into energy is growing in the food manufacturing industry. Food processing wastes can be converted to heat and power, liquid and gaseous fuels, using a biological, thermochemical, or chemical conversion process, depending on the characteristics of the wastes, the quantity of wastes available, the desired form of energy and chemical products, the efficiency of the conversion process, product demands in the market, and economical feasibility. Wet food processing wastes are usually better suited for biological processes, such as anaerobic digestion, and dry wastes are better for thermochemical conversion processes, such as combustion, gasification, and pyrolysis. Thermochemical liquefaction can also be used to convert wet food processing wastes at high pressure and moderate temperature into a bio-oil of partly oxygenated hydrocarbons. Fermentation, trans-esterification, pyrolysis, and liquefaction produce liquid fuels for use as transportation fuels. Combustion, gasification, and anaerobic digestion produce gaseous energy products, which are suitable to be used at the production location. High-value phytochemicals can be identified and extracted from the waste residuals in the food processing facilities before they are used for energy production for the increase of economic profitability. Food processing wastes, which are rich in sugars, vitamins, and minerals, are very suitable as raw materials for the production of high-value chemicals or metabolites, such as organic acids, enzymes, antibiotics, antioxidants, vitamins, and mushrooms by microorganisms.

References

Al-Widyan, M.I., and A.O. Al-Shyoukh. 2002. Experimental evaluation of the transesterification of waste palm oil into biodiesel. *Bioresource Technology* 85: 253–256.

Armesto, L., A. Bahillo, K. Veijonen, A. Cabanillas, and J. Otero. 2003. Combustion behavior of rice husk in a bubbling fluidized bed. *Biomass and Bioenergy* 23: 171–179.

Berardino, S., S. Costa, and A. Converti. 2000. Semi-continuous anaerobic digestion of a food industry wastewater in an anaerobic filter. *Bioresource Technology* 71: 261–266.

Blessin, C.W., W.J. Garcia, W.L. Deatherage, J.F. Cavins, and G.E. Inglett. 1973. Composition of three food products containing defatted corn germ flour. *Journal of Food Science* 38: 602–606.

Blondin, G.A., S.J. Comiskey, and J.M. Harkin. 1983. Recovery of fermentable sugars from process vegetable wastewaters. *Energy in Agriculture* 2: 21–36.

Braber, K. 1995. Anaerobic digestion of municipal solid waste: A modern waste disposal option on the verge of breakthrough. *Biomass and Bioenergy* 9: 365–376.

Callaghan, F.J., D.A.J. Wase, K. Thayanithy, and C.F. Forster. 2002. Continuous co-digestion of cattle slurry with fruit and vegetable wastes and chicken manure. *Biomass and Bioenergy* 27: 71–77.

Canakci, M. 2007. The potential of restaurant waste lipids as biodiesel feedstocks. *Bioresource Technology* 98: 183–190.

Chang, D., M.P. Hojilla-Evangelista, L.A. Johnson, and D.J. Myers. 1995. Economic-engineering assessment of sequential extraction processing of corn. *Transactions of the ASAE* 38: 1129–1138.

Chang, J.I., J.J. Tsai, and K.H. Wu. 2006. Thermophilic composting of food waste. *Bioresource Technology* 97: 116–122.

Choteborska, P., B. Palmarola-Adrados, M. Galbe, G. Zacchi, K. Melzoch, and M. Rychtera. 2004. Processing of wheat bran to sugar solution. *Journal of Food Engineering* 61: 561–565.

Commission of the European Communities. 1990. Council Directive 90/667/EEC. Official Journal, No. L 363: 51.

Cooper, J.L. 1976. The potential of food processing solid wastes as a source of cellulose for enzymatic conversion. *Biotechnology and Bioengineering* 6: 251–271.

de Swaan Arons, J., H. van der Kooi, and K. Sankaranarayanan. 2004. *Efficiency and Sustainability in the Energy and Chemical Industries.* New York: Marcel Dekker.

DeFelice, S.L. 1995. The time has come for nutraceutical cereals. *Cereal Foods World* 40: 51–52.

Dexter, J.E., D.G. Martin, G.T. Sadaranganey, J. Michaelides, N. Mathieson, J.J. Tkac, and B.A. Marchylo. 1994. Preprocessing: Effect on durum wheat milling and spaghetti-making quality. *Cereal Chemistry* 71: 10–16.

Eggeman, T., and D. Verser. 2006. The importance of utility systems in today's biorefineries and a vision for tomorrow. *Applied Biochemistry and Biotechnology* 129–132: 361–381.

Fischer, K., and H.P. Bipp. 2005. Generation of organic acids and monosaccharides by hydrolytic and oxidative transformation of food processing residues. *Bioresource Technology* 96: 831–842.

Gabra, M., E. Pettersson, R. Backman, and B. Kjellstrom. 2001a. Evaluation of cyclone gasifier performance for gasification of sugar cane residue – Part 1: Gasification of bagasse. *Biomass and Bioenergy* 21: 351–369.

Gabra, M., E. Pettersson, R. Backman, and B. Kjellstrom. 2001b. Evaluation of cyclone gasifier performance for gasification of sugar cane residue – Part 2: Gasification of cane trash. *Biomass and Bioenergy* 21: 371–380.

Gebauer, R. 2004. Mesophilic anaerobic treatment of sludge from saline fish farm effluents with biogas production. *Bioresource Technology* 93: 155–167.

Gebrezgabher, S.A., M.P.M. Meuwissen, B.A.M. Prins, and G.J.M.O. Lansink. 2010. Economic analysis of anaerobic digestion – a case of green power biogas plant in the Netherlands. *NJAS-Wageningen Journal of Life Science* 57: 109–115.

Gerbens-Leenes, P.W., H.C. Moll, and A.J.M. Schoot Uiterkamp. 2003. Design and development of a measuring method for environmental sustainability in food production systems. *Ecological Economics* 46: 231–248.

Gercel, H.F. 2002. The production and evaluation of bio-oils from the pyrolysis of sunflower-oil cake. *Biomass and Bioenergy* 23: 307–314.

Gras, P.W., and D.H. Simmonds. 1980. The utilization of protein-rich products from wheat carbohydrate separation processes. *Food Technology Australia* 32: 470–472.

Grull, D.R., F. Jetzinger, M. Kozich, and M.M. Wastyn. 2006. Industrial starch platform-status quo of production, modification and application. In *Biorefineries-Industrial Processes and Products*, eds. B. Kamm, P.R. Gruber, and M. Kamm, 61–95. Weinheim: Wiley-VCH Verlag.

Haapapuro, E.R., N.D. Barnard, and M. Simon. 1997. Review: Animal waste used as livestock feed: Danger to human health. *Preventive Medicine* 26: 599–602.

Hang, Y.D. 2004. Management and utilization of food processing wastes. *Journal of Food Science* 69: 104–107.

Hang, Y.D., C.Y. Lee, and E.E. Woodams. 1986. Solid state fermentation of grape pomace for ethanol production. *Biotechnology Letters* 8: 53–56.

Haumann, B.F. 1990. Renderers give new life to waste restaurant fats. *Inform* 1: 722–725.

Henningsson, S., K. Hyde, A. Smith, and M. Campbell. 2004. The value of resource efficiency in the food industry: A waste minimisation project in East Anglia, UK. *Journal of Cleaner Production* 12: 505–512.

Hernandez, M.S., M.R. Rodriguez, N.P. Guerra, and R.P. Roses. 2006. Amylase production by *Aspergillus niger* in submerged cultivation on two wastes from food industries. *Journal of Food Engineering* 73: 93–100.

Hill, K. 2006. Industrial development and application of biobased oleochemicals. In *Biorefineries – Industrial Processes and Products*, eds. B. Kamm, P. R. Gruber, and M. Kamm, 291–314. Weinheim: Wiley-VCH Verlag.

Johnson, D.L. 2006. The corn wet milling and corn dry milling industry-a base for biorefinery technology developments. In *Biorefineries – Industrial Processes and Products*, eds. B. Kamm, P.R. Gruber, and M. Kamm, 344–352. Weinheim: Wiley-VCH Verlag.

Kalia, V.C., V. Sonakya, and N. Raizada. 2000. Anaerobic digestion of banana stem waste. *Bioresource Technology* 73: 191–193.

Kantor, L.S., K. Lipton, A. Manchester, and V. Oliveira. 1997. Estimating and addressing America's food losses. Economic Research Service, United States Department of Agriculture. Available at http://151.121.66.126:80/whatsnew/feature/ARCHIVES/JULAUG97/INDEX.HTM.

Kato-Noguchi, H., and Y. Tanaka. 2004. Allelopathic potential of *Citrus junos* fruit waste from food processing industry. *Bioresource Technology* 94: 211–214.

Khoo, H.H., T.Z. Lim, and R.B.H. Tan. 2010. Food waste conversion options in Singapore: Environmental impacts based on an LCA perspective. *Science of the Total Environment* 408: 1367–1373.

Kim, M.H., and J.W. Kim. 2010. Comparison through a LCA evaluation analysis of food waste disposal options from the perspective of global warming and resource recovery. *Science of the Total Environment* 408: 3998–4006.

Kollacks, W.A., and C.J.N. Rekers. 1988. Five years of experience with the application of reverse osmosis on light middlings in a corn wet milling plant. *Starch* 40: 88–94.

Koseolu, S.S., K.C. Rhee, and E.W. Lusas. 1991. Membrane separations and applications in cereal processing. *Cereal Foods World* 36: 376–383.

Kraus, G.A. 2006. Phytochemicals, dyes, and pigments in the biorefinery context. In *Biorefineries – Industrial Processes and Products*, eds. B. Kamm, P.R. Gruber, and M. Kamm, 315–323. Weinheim: Wiley-VCH Verlag.

Krull, L.H., and G.E. Inglett. 1971. Industrial uses of gluten. *Cereal Science Today* 16: 232–236, 261.

Kucuk, M.M. 2001. Liquefaction of biomass by supercritical gas extraction. *Energy Sources* 23: 363–368.

Lahmar, M., V. Fellner, R.L. Belyea, and J.E. Williams. 1994. Increasing the solubility and degradability of food processing biosolids. *Bioresource Technology* 50: 221–226.

Laufenberg, G., B. Kunz, and M. Nystroem. 2003. Transformation of vegetable waste into value added products: (A) the upgrading concept; (B) practical implementations. *Bioresource Technology* 87: 167–198.

Levis, J.W., M.A. Barlaz, N.J. Themelis, and P. Ulloa. 2010. Assessment of the state of food waste treatment in the United States and Canada. *Waste Management* 30: 1486–1494.

Lingaiah, V., and P. Rajasekaran. 1986. Biodigestion of cow dung and organic wastes mixed with oil cake in relation to energy. *Agricultural Wastes* 17: 161–173.

Lundie, S., and G.M. Peters. 2005. Life cycle assessment of food waste management options. *Journal of Cleaner Production* 13: 275–286.

Ma, F., and M.A. Hanna. 1999. Biodiesel production: A review. *Bioresource Technology* 70: 1–15.

Mansaray, K.G., A.E. Ghaly, A.M. Al-Taweel, F. Hamdullahpur, and V.I. Ugursal. 1999. Air gasification of rice husk in a dual distributor type fluidized bed gasifier. *Biomass and Bioenergy* 17: 315–332.

Mata-Alvarez, J., S. Mace, and P. Llabres. 2000. Anaerobic digestion of organic solid wastes. An overview of research achievements and perspectives. *Bioresource Technology* 74: 3–16.

Matteson, G.C., and B.M. Jenkins. 2007. Food and processing residues in California: Resource assessment and potential for power generation. *Bioresource Technology* 98: 3098–3105.

McKendry, P. 2002. Energy production from biomass (part 2): Conversion technologies. *Bioresource Technology* 83: 47–54.

Mena, C., B. Adenso-Diaz, and O. Yurt. 2011. The causes of food waste in the supplier–retailer interface: Evidences from the UK and Spain. *Resources, Conservation and Recycling* 55: 648–658.

Mittelbach, M., B. Pokits, and A. Silberholz. 1992. Production and fuel properties of fatty acid methyl esters from used frying oil. In: *Liquid Fuels from Renewable Resources. Proceedings of an Alternative Energy Conference,* 74–78. Nashville: ASAE Publication.

Mohaibes, M., and H. Heinonen-Tanski. 2004. Aerobic thermophilic treatment of farm slurry and food wastes. *Bioresource Technology* 95: 245–254.

Mouneimne, A.H., H. Carrere, N. Bernet, and J.P. Delgenes. 2003. Effect of saponification on the anaerobic digestion of solid fatty residues. *Bioresource Technology* 90: 89–94.

Ngadi, M.O., and L.R. Correia. 1992. Kinetics of solid-state ethanol fermentation from apple pomace. *Journal of Food Engineering* 17: 97–116.

Nguyen, M.H. 2003. Alternatives to spray irrigation of starch waste based distillery effluent. *Journal of Food Engineering* 60: 367–374.

Nigam, J.N. 2000. Continuous ethanol production from pineapple cannery waste using immobilized yeast cells. *Journal of Biotechnology* 80: 189–193.

Ohkouchi, Y., and Y. Inoue. 2006. Direct production of L(+)-lactic acid from starch and food wastes using *Lactobacillus manihotivorans* LMG18011. *Bioresource Technology* 97: 1554–1562.

Onay, O., S.H. Beis, and O.M. Kochar. 2001. Fast pyrolysis of rape seed in a well-swept fixed-bed reactor. *Journal of Analytical and Applied Pyrolysis* 58–59: 995–1007.

Ozbay, N., A.E. Putun, and E. Putun. 2001. Biocrude from biomass: Pyrolysis and steam pyrolysis of cottonseed cake. *Journal of Analytical and Applied Pyrolysis* 60: 89–101.

Ozcimen, D., and F. Karaosmanoglu. 2004. Production and characterization of bio-oil and biochar from rapeseed cake. *Renewable Energy* 29: 779–787.

Park J.I., Y.S. Yun, and J.M. Park. 2002. Long-term operation of slurry bioreactor for decomposition of food wastes. *Bioresource Technology* 84: 101–104.

Pauli, G. 1998. Upsizing: The road to zero emissions. *More jobs, More Income and No Pollution.* London: Greenleaf Publishing.

Ramachandran, S., S.K. Singh, C. Larroche, C.R. Soccol, and A. Pandey. 2007. Oil cakes and their biotechnological applications – a review. *Bioresource Technology* 98: 2000–2009.

Rasco, B.A., F.M. Dong, A.E. Hashisaka, S.S. Gazzaz, S.E. Downey, and M.L. San Buenaventura. 1987. Chemical composition of distillers' dried grains with solubles (DDGS) from soft white wheat, hard red wheat and corn. *Journal of Food Science* 52: 235–237.

Ritter, W.F., and A.E.M. Chinside. 1995. Impact of dead bird disposal pits on groundwater quality on the Delmarva Peninsula. *Bioresource Technology* 53: 105–111.

Rosales, E., S.R. Couto, and M.A. Sanroman. 2005. Reutilisation of food processing wastes for production of relevant metabolites: Application to laccase production by Trametes hirsute. *Journal of Food Engineering* 66: 419–423.

Salminen, E., and J. Rintala. 2002. Anaerobic digestion of organic solid poultry slaughterhouse waste – a review. *Bioresource Technology* 83: 13–26.

Sargent, S.A., and J.F. Steffe. 1986. Energy generation from direct combustion of solid food processing wastes. In *Energy in Food Processing*, ed. R.P. Singh, 247–266. New York: Elsevier Science.

Satterlee, L.D. 1981. Proteins for use in foods. *Food Technology* 35: 53–70.

Schaub, S.M., and J.J. Leonard. 1996. Composting: An alternative waste management option for food processing industries. *Trends in Food Science & Technology* 7: 263–268.

Shih, J.C.H. 1993. Recent development in poultry waste digestion and feather utilization – a review. *Poultry Science* 72: 1617–1620.

Shinogi, Y., and Y. Kanri. 2003. Pyrolysis of plant, animal and human waste: Physical and chemical characterization of the pyrolytic products. *Bioresource Technology* 90: 241–247.

Tashtoush, G., M.I. Al-Widyan, and A.O. Al-Shyoukh. 2003. Combustion performance and emissions of ethyl ester of a waste vegetable oil in a water-cooler furnace. *Applied Thermal Engineering* 23: 285–293.

Themelis, N.J. 2003. An overview of the global waste-to-energy industry. *Waste Management World* 7–8: 40–47.

Tritt, W.P., and F. Schuchardt. 1992. Materials flow and possibilities of treating liquid and solid wastes from slaughterhouses in Germany: A review. *Bioresource Technology* 41: 235–245.

Tucker, M.P., N.J. Nagle, E.W. Jennings, K.N. Ibsen, A. Aden, Q.A. Nguyen, K.H. Kim, and S.L. Noll. 2004. Conversion of distiller's grain into fuel alcohol and a higher-value animal feed by dilute-acid pretreatment. *Applied Biochemistry and Biotechnology* 113–116: 1139–1159.

Turhollow, A.F., and E.O. Heady. 1986. Large-scale ethanol production from corn and grain sorghum improving conversion technology. *Energy in Agriculture* 5: 309–316.

Tyagi, V.K., and A.K. Vasishtha. 1996. Changes in the characteristics and composition of oils during deep-fat frying. *Journal of the American Oil Chemists' Society* 73: 499–506.

Verma, M., S.K. Brar, R.D. Tyagi, R.Y. Surampalli, and J.R. Valero. 2007. Starch industry wastewater as a substrate for antagonist *Trichoderma viride* production. *Bioresource Technology* 98: 2154–2162.

Wang, L.J. 2008. *Energy Efficiency and Management in Food Processing Facilities*. Boca Raton, FL: CRC Press.

Wang, L.J. 2010. Advances in extraction of plant products in nutraceutical processing, In *Handbook of Nutraceuticals: Volume II: Scale up, Processing and Automation*, ed. Y. Pathak. Boca Raton, FL: CRC Press.

Wang, L.J., M.A. Hanna, C.L. Weller, and D.D. Jones. 2009. Technical and economical analyses of combined heat and power generation from distillers grains and corn stover in ethanol plants. *Energy Conversion and Management* 50: 1704–1713.

Wang, L.J., A. Shahbazi, and M.A. Hanna. 2011. Characterization of corn stover, distiller grains and cattle manure for thermochemical conversion. *Biomass and Bioenergy* 35: 171–178.

Wang, L.J., and C.L. Weller. 2006. Recent advances in extraction of natural products from plants. *Trends in Food Science and Technology* 17: 300–312.

Wang, L.J., C.L. Weller, and K.T. Hwang. 2005. Extraction of lipids from grain sorghum DDG. *Transactions of the ASABE* 48: 1883–1888.

Wang, L.J., C.L. Weller, V.L. Schlegel, T.P. Carr, S.L. Cuppett, and K.T. Hwang. 2007. Comparison of supercritical CO_2 and hexane extraction of lipids from sorghum distillers grains. *European Journal of Lipid Science and Technology* 109: 567–574.

Wang, L.J., C.L. Weller, V.L. Schlegel, T.P. Carr, and S.L. Cuppett. 2008a. Supercritical carbon dioxide extraction of grain sorghum DDGS lipids. *Bioresource Technology* 99: 1373–1382.

Wang, Q., X. Wang, X. Wang, and H. Ma. 2008b. Glucoamylase production from food waste by *Aspergillus niger* under submerged fermentation. *Process Biochemistry* 43: 280–286.

Weegels, P.L., J.P. Marseille, and R.J. Hamer. 1992. Enzymes as a processing aid in the separation of wheat flour into starch and gluten. *Starch* 44: 44–48.

Wildman, R.E.C. 2000. *Handbook of Nutraceuticals and Functional Foods*. Boca Raton, FL: CRC Press.

Wiltsee, G. 1998. Urban waste grease resource treatment. Final Report to the National Renewable Energy Laboratory, NREL/SR-50-26141.

Woodruff, J.G., and B.S. Luh. 1975. *Commercial Fruit Processing*. Westport, CT: AVI Publishing.

Wrick, K.L. 1993. Functional foods: Cereal products at the food-drug interface. *Cereal Foods World* 38: 205–214.

Zall, R.R. 2004. *Managing Food Industry Waste*. Ames, IA: Blackwell Publishing Professional.

Zbasnik, R., T. Carr, C. Weller, K.T. Hwang, L.J. Wang, S. Cuppett, and V. Schlegel. 2009. Antiproliferation properties of grain sorghum dry distiller's grain lipids in caco-2 cells. *Journal of Agricultural and Food Chemistry* 57: 10435–10441.

Chapter 19

Waste Minimization and Utilization in the Food Industry: Valorization of Food Industry Wastes and Byproducts

Nóra Pap, Eva Pongrácz, Liisa Myllykoski, and Riitta L. Keiski

Contents

19.1 Introduction

The food and drinks industry is one of the most important industry sectors throughout the European Union. In Finland, the food industry ranks fourth, after the metal, forest, and chemical industries, in terms of the value of its output. The food supply chain is also a major employer: The entire chain provides work for some 300,000 wage and salary earners, 40,000 of whom are employed by the food industry (Elintarviketeollisuusliitto 2010).

The study on food insecurity by the Food and Agriculture Organization highlighted that approximately 870 million people in the world suffered from under-nutrition during 2010–2012. The vast majority of the hungry, 852 million people, live in developing countries, around 15% of their population, while 16 million people are undernourished in developed countries (FAO, WFP and IFAD 2012). At the same time, Gustavsson et al. (2011) reported that close to one third of the food produced globally will virtually become waste, totalling 1.3 billion tons per year. This is clearly unsustainable because the wasting of food will have serious social, environmental, and economic impacts. Food wastage also entails the waste of resources used to produce the food, such as water, energy, land for agricultural production, and other inputs. In addition, the environmental impacts of the food chain would have been meaningless if the produced goods became waste.

Compared with other manufacturing industries, the food sector is a minor contributor to environmental loads as its outputs are generally not hazardous. However, the food industry's negative impacts are significant in terms of water consumption and energy use. Throughout the supply chain, the food industry uses significant amounts of water and, correspondingly, generates large amounts of wastewater. Food industry wastewaters contain biological materials and dissolved organic and inorganic solids that constitute a significant environmental load. Concerning waste generation, the largest producers are from the dairy, cocoa, chocolate or sugar confections, brewing, distillation, and meat processing sectors (Niranjan and Shilton 1994).

Waste prevention and pollution prevention, energy-efficient process technologies, and water saving are always a priority. However, waste minimization and utilization are also desirable strategies as the waste streams from the food industry are high in organic content and usually rich in valuable compounds, such as oils, sugars, antioxidants, etc.

19.2 Wastes in the Food Sector

A distinction is to be made whether the loss of resources happens in the early stages of the food supply chains (FSC) or the resource was wasted by the action of the retail sector or consumers. In the first case, we speak about food losses while in the latter case about food waste.

Food loss should mean the decrease in *edible food mass* throughout the food chain. Food losses take place in the production, post-harvest, and processing stages in the FSC (Parfitt et al. 2010). Based on this definition, food losses do not include the parts of the goods not intended for human consumption, such as the peels or seeds of fruits, bones of animal-origin products, etc. Food losses can be avoided by a correct action, e.g., by maintaining the cold supply chain or ensuring correct storage conditions for products. Based on this definition, food loss also occurs if the product that was originally intended for human consumption is recovered in the form of feed, fertilizer, or energy.

On the other hand, the term *food waste* is more comprehensive, and includes all resources that are lost in the different sectors of the food supply chain and will include also those parts that were originally not intended for human consumption.

The share of different sectors in food waste production is illustrated in Figure 19.1. Food is wasted throughout the FSC from initial agricultural production down to final household consumption.

Food losses in industrialized countries are as high as in developing countries, but in developing countries, more than 40% of the food losses occur at post-harvest

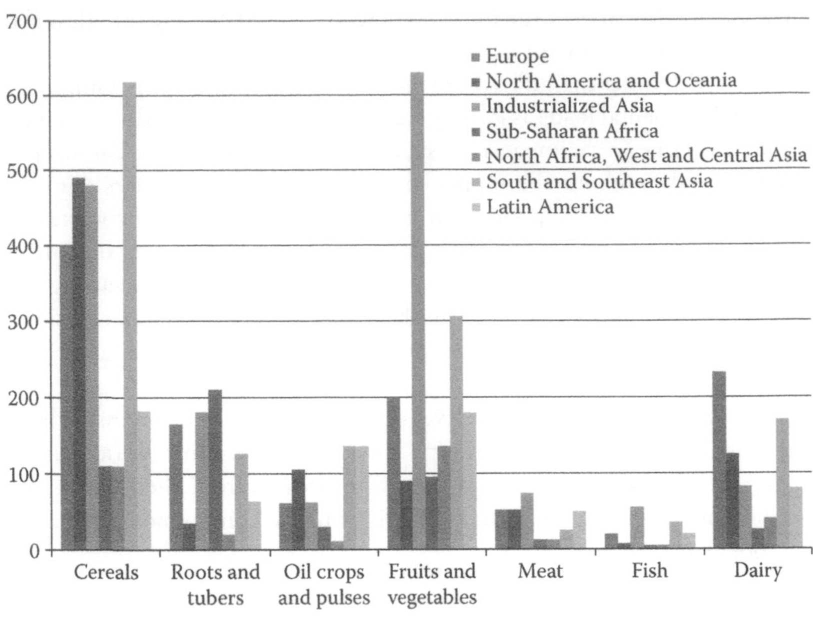

Figure 19.1 Production volumes (million tons) of each commodity group, per region in 2007. (From Gustavsson, J. et al., Global food losses and food waste. Extent, causes and prevention. Food and Agriculture Organization of the United Nations, Rome, Italy. Available at: http://www.fao.org/docrep/014/mb060e/mb060e00.pdf, 2011.)

and processing levels, and in industrialized countries, more than 40% of the food losses occur at retail and consumer levels. Food waste at the consumer level in industrialized countries (222 million tons) is almost as high as the total net food production in sub-Saharan Africa (230 million tons) (Gustavsson et al. 2011). Table 19.1 highlights wastage in different food sectors with the stages of the highest wastage and the reasons for wastage.

Figure 19.2 illustrates the part of it that is wasted. The total per capita production of edible parts of food for human consumption in Europe and North America is about 900 kg per year, of which the per capita food loss is 280–300 kg per year (31%–33%). Of this amount of wastage, consumers are responsible for 34%–38%, some 95–115 kg per capita waste a year.

In sub-Saharan Africa and South or Southeast Asia, the per capita food production is 460 kg per year, of which food loss accounts to 120–170 kg per year (26%–37%). However, food wastage by consumers is only 5%–6.5%, some 6–11 kg per year per capita (Gustavsson et al. 2011).

Table 19.1 Food Loss in Different Sectors: The Stage of Food Chain with the Highest Loss and the Reason for Wastage

Food Sector	*Part of Initial Produce Lost*	*Stage of Highest Food Loss and Percentage of Loss at that Stage*		*Reason for Wastage*
Cereals	20%–35%	Consumption	2%–25%	Wasted by consumers
Roots and tubers	32%–60%	Agricultural production	5%–20%	Post-harvest crop grading due to quality standards set by retailers
Oily crops and pulses	18%–30%	Agricultural production	6%–12%	Lost during harvest
Fruits and vegetables	35%–55%	Consumption	2%–15%	Wasted by consumers
		Agricultural production	10%–20%	Post-harvest fruit and vegetable grading
Meat and meat products	20%–28%	Consumption	2%–10%	Wasted by consumers
		Animal production	2%–15%	Animal mortality during breeding and transportation
Fish and seafood	30%–50%	Fisheries	5%–15%	Discard of marine catches
		Consumption	2%–25%	Wasted by consumers
Dairy products	10%–25%	Consumption	2%–15%	Wasted by consumers

Source: Gustavsson, J. et al., Global food losses and food waste. Extent, causes and prevention. Food and Agriculture Organization of the United Nations, Rome, Italy. Available at: http://www.fao.org/docrep/014/mb060e/mb060e00.pdf, 2011.

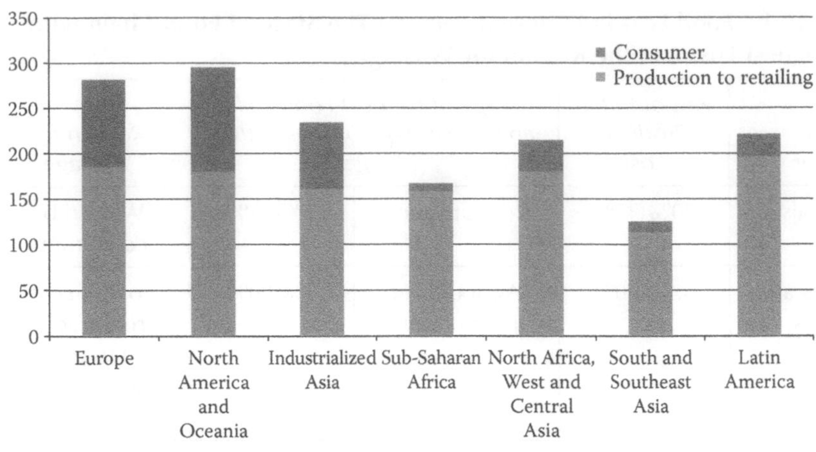

Figure 19.2 Per capita food losses and waste at consumption and pre-consumption stages in different regions. (From Gustavsson, J. et al., Global food losses and food waste. Extent, causes and prevention. Food and Agriculture Organization of the United Nations, Rome, Italy. Available at: http://www.fao.org/docrep/014/mb060e/mb060e00.pdf, 2011.)

19.3 Waste Minimization in European Waste Legislation

Environmental legislation has significantly contributed to the introduction of sustainable waste management practices throughout the European Union. The primary aim of waste legislation is the prevention of waste generation. The Waste Framework Directive 2008/98/EC defines waste prevention as measures taken before a substance, material, or product has become waste that reduce

- The quantity of waste, including reuse or extension of life span
- The adverse impacts of the generated waste
- The content of harmful substances in materials and products

Once waste is formed, it should be recycled or recovered for better environmental and economic performance. Figure 19.3 illustrates the position of waste minimization amid different waste management strategies.

As Figure 19.3 indicates, both waste prevention and recycling activities will contribute to waste minimization. In terms of food waste management, the key pressure is provided by the Landfill Directive 99/31/EC, which forbids disposal of untreated organic waste in landfills starting from 1.5.2005. The ban of organic waste from landfills is enforced in order to reduce greenhouse gas emissions associated with the landfill of biodegradable wastes. By July 2016, biodegradable municipal waste disposal

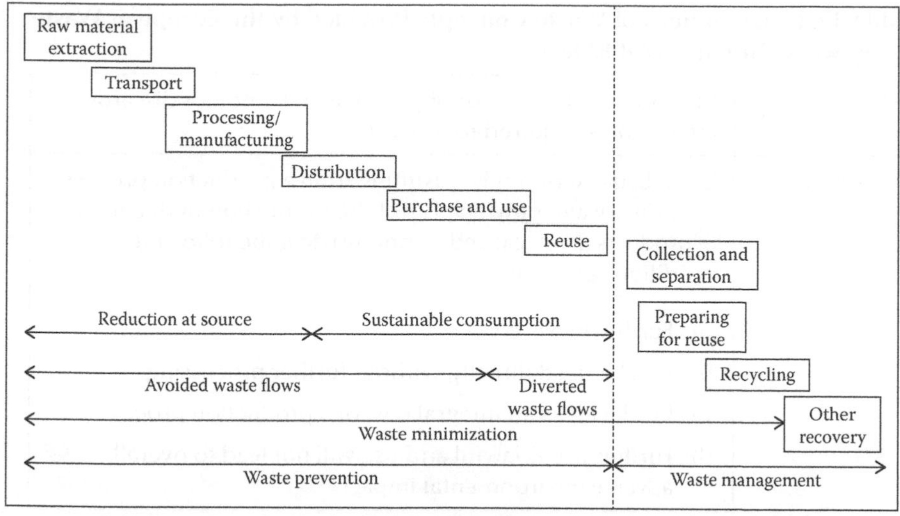

Figure 19.3 Waste management strategies. (Based on European Commission 2012, quoting ADEME, the French Environment Agency.)

going to landfill has to be reduced by 35% (European Council). In addition, Council decision 2003/33/EC on waste acceptance criteria (WAC) prescribes that waste with total organic carbon content of more than 6% is no longer accepted for landfill. In Finland, the Finnish bio-waste strategy (Ympäristöministeriö 2004) prescribes that, in 2016, a maximum of 25% of all biodegradable waste generated can go to landfill. Considering the challenges in the food industry, efforts will have to be made to optimize processing technologies to minimize the amount of waste and, in particular, to emphasize efforts on recycling and recovery of bio-waste.

19.4 Defining Waste Concepts in the European Waste Legislation

In Europe, the Waste Directive has been codified in 2006. Codification is a process by which legal texts that have been revised several times are codified into one new text that replaces all the previous versions. The codified Waste Framework Directive (WFD) (2006/12/EC) has been revised in order to modernize and streamline its provisions in 2008. The revised WFD 2008/98/EC sets the basic concepts and definitions related to waste management and lays down waste management principles such as the waste management hierarchy (WMH). The WMH is the order of preference of waste management options. The currently defined WMH is (i) waste prevention, (ii) preparing for reuse, (iii) recycling, (iv) disposal, and (v) other recovery (e.g., energy recovery). Table 19.2 summarizes the definitions provided by the WFD.

Table 19.2 Definitions of Waste Concepts Provided by the European Waste Framework Directive (2008/98/EC)

Waste	Means any substance or object, which the holder discards or intends or is required to discard
Byproduct	Is a substance or object resulting from a production process, the primary aim of which is not the production of that item. Byproducts are regarded as non-waste if the following conditions are met: a) Further use is certain b) Can be used directly, without further processing c) Produced as an integral part of a production process d) Further use is lawful and use will not lead to overall adverse environmental impacts
Prevention	Means measures taken before a substance, material, or product has become waste that reduce a) The quantity of waste, including reuse or extension of life span b) The adverse impacts of the generated waste c) The content of harmful substances in materials and products
Reuse	Means any operation by which products or components that are not waste are used again for the same purpose for which they were conceived
Treatment	Means recovery or disposal operations, including preparation prior to recovery or disposal
Preparing for reuse	Means checking, cleaning, or repairing operations, by which products or components of products have become waste are prepared so that they will be reused without any other preprocessing
Recycling	Means any recovery operation by which waste materials are reprocessed into products, materials, or substances whether for the original or other purposes. It includes the reprocessing of organic material but does not include energy recovery and the reprocessing into materials that are to be used as fuels or for backfilling operations

(*continued*)

Table 19.2 (Continued) Definitions of Waste Concepts Provided by the European Waste Framework Directive (2008/98/EC)

Recovery	Means any operation the principal result of which is waste serving a useful purpose by replacing other materials that would otherwise have been used to fulfill a particular function or waste being prepared to fulfill that function in the plant or in the wider economy
Disposal	Means any operation, which is not recovery, even when the operation has as a secondary consequence the reclamation of substances or energy

19.5 Bio-Waste Management in European Waste Legislation

The food industry is perhaps one of the most regulated sectors in Europe. Hygiene and safety are the key objectives, which take priority to reducing environmental impacts. In terms of wastes, the European Waste Framework Directive (WFD) defines bio-waste as follows: biodegradable garden and park waste; food and kitchen waste from households, restaurants, caterers, and retail premises; and comparable waste from food processing plants. In its Article 22, the WFD prescribes that Member States shall take measures to encourage

- The separate collection of bio-waste with a view to the composting and digestion of bio-waste
- The treatment of bio-waste in a way that fulfills a high level of environmental protection
- The use of environmentally safe materials produced from bio-waste

In the food sector, recycling and recovery shall mean the following operations (2008/98/EC):

- Composting
- Reclamation of organic substances
- Oil re-refining and reuse
- Fuel or energy generation
- Spreading on land resulting in benefit to agriculture or ecological improvement

The European Commission still aims at examining the requirements for bio-waste management, such as quality criteria for compost and digestate from bio-waste.

19.6 Waste Management in the Food Industry

Waste management strategies aim at reducing wastage, recovering resources, and treating waste before final disposal (Niranjan and Shilton 1994). The high volumes of solid wastes generated in the agro-food industry are not only a potential environmental problem but also an economic burden for companies in terms of their management. Therefore, the advantages of proper waste management go beyond environmental benefits and include cost savings and resource efficiency. To minimize costs, the food industry will have to concentrate on waste avoidance. Utilization of byproducts and wastes as raw materials are the next most preferable methods. Value recovery operations can include additional processing steps, such as separation, concentration, biological, or chemical conversion, that all add costs and require energy and other inputs.

19.6.1 Waste Prevention

The focus should always be on prevention as waste always costs money, and most companies are unaware of the true costs of waste (WRAP 2013). While the principle of waste prevention is universally accepted, the practice has lagged far behind. Table 19.3 illustrates the main causes of food loss in different stages of the food chain and their prevention.

Within the food processing stage, good housekeeping practices, including equipment maintenance, process optimization, and design for efficiency are the best options. It has been demonstrated that a combination of equipment maintenance and design changes can reduce up to 20%–30% of costs and efficient process control can reduce costs by an additional 5%. Conversely, inefficient packaging systems can cause a 4% loss (WRAP 2013).

19.6.2 Solid Waste Treatment

Considering the landfill ban on organic wastes, the strategy should be to utilize solid wastes by recovering valuable components, nutrients, or energy. Composting is currently widely used in Europe as a bio-waste treatment method. It is foreseen that energy use of bio-waste will be preferred in the future. However, priority should be given to valorization to recover the maximum value embedded in wastes. The following sections will focus on valorization technologies and products that can be recovered from food processing byproduct and wastes, followed by an outline of waste-to-energy technologies. Figure 19.4 summarizes the technologies covered.

19.6.3 Waste and Byproduct Valorization

Valorization is a term with origins in economics, meaning generating value. This expression is used in connotation to recovering valuable components from waste

Table 19.3 Causes of Food Losses in Different Stages of the Food Supply Chain (FSC) and Their Prevention

Stage of FSC	*Cause of Food Loss or Wastage*	*Prevention Options*
Agriculture	Production exceeds demand	Communication and cooperation between farmers
	Premature harvesting	Organizing small farmers and diversifying and up-scaling their production and marketing
Post-harvest	Poor storage facilities and lack of infrastructure	Investment in infrastructure and transportation
Distribution	High "appearance quality standards" from supermarkets for fresh products	Consumer surveys by supermarkets to ascertain that consumers will not buy food that has the "wrong" weight, size, or appearance
		Selling farm crops closer to consumers without having to pass the strict quality standards set up by supermarkets
	Unsafe food is not fit for human consumption	Develop knowledge and capacity of food chain operators to apply safe food-handling practices
	Disposing is cheaper than "using or reusing" attitude in industrialized countries	Develop markets for "substandard" products that are still safe and of good taste and nutritional value
	Large quantities on display and a wide range of products/brands in supply	Marketing cooperatives and improved market facilities
Consumption	Abundance leads to wasteful consumer behavior in industrialized countries	Public awareness to change people's attitudes

Source: Gustavsson, J. et al., Global food losses and food waste. Extent, causes and prevention. Food and Agriculture Organization of the United Nations, Rome, Italy. Available at: http://www.fao.org/docrep/014/mb060e/mb060e00.pdf, 2011.

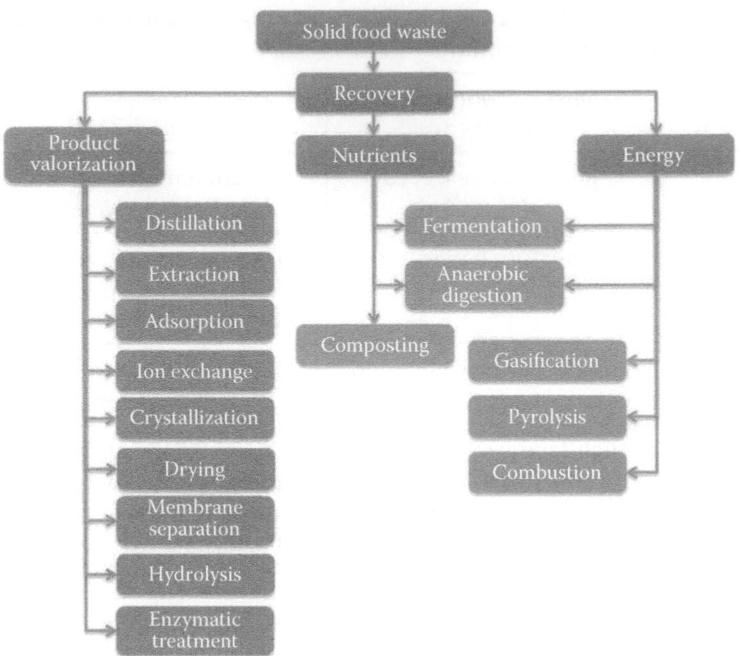

Figure 19.4 Recovery technologies for solid food waste.

biomass. The agro-food wastes minimization and recycling network (AWARENET) defined valorization as *"Increase of technical and/or economic value of by-products and wastes that are generated in different agro-food industries."* Valorization techniques include mechanical and diffusional separation technologies as well chemical and biochemical modifications.

19.6.4 Mechanical Separation Methods

From the mechanical separations, mechanical extraction or pressing is used often for the solid-liquid phase separation method. Fat and oil in different oil seeds are separated by pressing, and it can be used for fruit juice extraction.

19.6.5 Diffusional Separation Methods

Diffusional separation methods, such as distillation, different forms of extraction, adsorption, ion exchange, evaporation, crystallization, drying, and membrane separation, are widely used in different areas of the food industry for byproduct valorization. Table 19.4 summarizes the application areas of these technologies.

Table 19.4 Main Technologies and Applications for Diffusional Separation Methods

Technology	*Application*
Distillation	Separation of solvent mixtures for recovery and reuse
	Removal of volatile compounds from aqueous feed streams
	Production of essential oils
	Alcohol production from wine pomace and starch-rich solid byproducts
Extraction	Solid-liquid: isolation of flavors, fragrances, pharmaceuticals
	Microwave-assisted extraction: anthocyanins, caffeine extraction, flavonoids
	Ultrasound-assisted extraction: isoflavone derivatives, phenolic compounds, anthocyanins
	Pressurized liquid extraction: isoflavones, flavonoids, phenolic compounds
	Enzyme-assisted extraction: extraction of oils, phenolic compounds
	Supercritical fluid extraction: spice oils and oleoresins, essential oils, herbal medicines, natural pesticides, vitamin E (tocopherols), nicotine/tar-free tobacco, decaffeinated coffee and tea, cholesterol-free food products, bitter from hops
	Water extraction: collagen and gelatin
Adsorption	Removal of organic components from drinking water
	Removal of color-promoting components from sugar solutions
Ion exchange	Demineralization of whey for whey powder and lactose production
Evaporation	Dewatering of salt streams
	Concentration of highly contaminated wastewaters
	Concentration of saline effluents (e.g., wastewater from fish and meat industry)

(*continued*)

Table 19.4 (Continued) Main Technologies and Applications for Diffusional Separation Methods

Technology	*Application*
Crystallization	Lactose production
	Production of natural sweeteners from pomace
Drying	Lyophilization: preservation and drying of food products (meat, vegetables, fish, fruits, instant coffee products)
	Spray-drying: blood meals, whey protein and powders, soluble and refined fibers
	Flash drying: fast and suitable for heat sensitive or easily oxidized substances, e.g., fibers from potato pulp
Membrane separation	Whey demineralization
	Water purification
	Juice clarification, sterilization, concentration

19.6.6 Chemical Separation Methods

From chemical modifications, hydrolysis, i.e., the breaking up of a chemical compound influenced by water, is one of the most often used chemical separation methods in the food industry. The addition of strong acids, bases, or steam will often be applied if ordinary water has no effect. Some examples of the use of hydrolysis include the conversion of starch into sugars in the presence of a strong acid catalyst; the conversion of animal fats or vegetable oils to glycerol and fatty acids by reaction with steam; or just the conversion of proteins, fats, oils, or carbohydrates by enzymes. Another application of chemical modifications can be calcination, which is used for production of construction materials from mollusk shells.

19.6.7 Biochemical Methods

Biochemical modification for food byproduct valorization includes pasteurization, fermentation, biogas production, and enzymatic treatment. Pasteurization, such as HTST (high temperature short time) and LTLT (low temperature long time) pasteurization, is usually applied in the dairy industry for utilization of byproducts. The product of fermentation processes varies greatly on the particular microorganisms involved, e.g., yeast and fungi results in ethanol production from glucose.

Enzymatic treatment is the other application field of biochemical methods. Enzymes can be used for the degradation of food waste consisting of proteins, lipids, and carbohydrates. Enzymes are also used to degrade plant cell walls and help, e.g., the extraction of oils or other valuable compounds from seeds, skins, and peels.

19.7 Extraction of Bioactive Compounds from Plant Material

Various extraction processes can be used for the recovery of valuable compounds from plant sources. Beside the conventional methods, such as the Soxhlet extraction or conventional solid-liquid extraction, many efforts have been made during the last 50 years to develop more environmentally friendly and efficient extraction techniques. Bioactive compounds were extracted from plant sources using advanced nonconventional extraction processes; ultrasound for the extraction of bioactive compounds from herbs and grape peel (Vinatoru et al. 1997; Ghafoor et al. 2011), pulsed electric field (Toepfl et al. 2006), enzyme-assisted extraction of edible oils (Gaur et al. 2007), extrusions for oil seeds (Lusas and Watkins 1988), microwave-assisted extraction and pressurized solvent extraction (Kaufmann and Christen 2002), ohmic heating for oil extraction (Lakkakula et al. 2004), supercritical fluid extraction (Marr and Gamse 2000; Lang and Wai 2001), and superheated water (Smith 2002).

Microwaves are nonionizing electromagnetic waves with a frequency between 300 MHz and 300 GHz that heat up the molecules in the presence of polar solvent by the combination of two actions: ionic conduction and dipole rotation. Ionic conduction is the migration of dissolved ions under the influence of an electromagnetic field and is influenced by the ion concentration, ion mobility, and solution temperature. On the other hand, dipole rotation is the realignment of dipoles of the molecules with the changing electric field—as the electric field of the microwaves increases, the molecules become polarized, and when the field decreases, thermally induced disorder is restored (Lebovka et al. 2011).

Heating efficiency of the microwaves is determined on the dielectric properties of the materials, which is described by the dielectric constant and the dielectric loss. The dielectric constant (ε) describes the ability of the molecules to be polarized by the electric fields. The dielectric loss (ε') describes the efficiency to turn the energy of the electromagnetic radiation into heat. The dissipation factor, $\tan\delta$, is the ratio of the dielectric loss to the dielectric constant of the material, i.e., $\tan\delta = \varepsilon'/\varepsilon$: therefore, the higher the dissipation factor of the sample is, the less microwave energy will penetrate into it.

The efficiency of the microwave-assisted extraction will depend on a number of process parameters, such as the applied power, the treatment time, solvent, and size distribution of the sample. From all the above, a crucial role has the applied solvent because if the solvent is not able to absorb the microwave energy, no heating will occur, and therefore the extraction of the target compounds will fail. The temperature of the extraction plays also an important role because it will have an effect on the degradation of the thermo-sensitive components and on the solubilization of the substances. To this end, microwave-assisted extraction devices should have a capability for the controlling of the process temperature. The effect of the microwave power on treatment time will be opposite: the more power is applied the less time will be required to reach the desirable process yield.

Microwave devices can be classified into two categories; in multimode systems, the microwave radiation is dispersed randomly, and in single-mode systems, the microwaves are restricted to the zone in which the sample is held. Multimode systems are usually applied in closed-vessel type devices in which the extraction is performed at high pressure, and the single-mode systems are applied in open-vessel type applications, usually at atmospheric pressure. This operation condition will also result in some advantages of the closed vessels across the open-vessel type because, due to the construction of the system, the loss of material due to evaporation is completely avoided, and therefore, there is no need to replace the evaporated solvent and, therefore, the total amount of solvent used is decreased. In addition, higher temperatures can be achieved during the treatments due to the higher pressure applied. On the other hand, the open-vessel type system is more safe due to the low pressure applied, and if needed, the excess of solvent can be removed and the sample dried. Considering the process economics, it also requires lower capital costs when compared to the closed-vessel system.

One recent contribution of the authors was to exploit the applicability of microwave-assisted extraction in fruit byproduct processing for the recovery of anthocyanins from blackcurrant marc (Pap et al. 2012). The MAE was optimized using response surface methodology, and its efficiency was compared to a conventional solvent extraction system.

19.8 Products Recovered from Food Industry Wastes and Byproducts

While the industrial processing of foodstuff is not the one with the highest percentage of wastage, there are some sectors in which wastage is significant in processing as shown in Table 19.5. In the case of these wastes and byproducts, waste prevention options are either limited or not possible. This will necessitate a different approach toward valorization of this waste fraction.

19.9 Fruit Juice Processing

The amount of the byproduct during fruit processing is usually 30%–50%, depending on the fruit, and we can distinguish two groups of byproducts: preprocessing byproducts that include stems, stalks, and rotten fruits from sorting processes and byproducts that occur during processing, such as seeds, pulp, pomace, and peels (Pap et al. 2004).

While the preprocessing byproducts are usually applied for composting, biogas and fertilizer production by anaerobic treatment, animal feed production, and as

Table 19.5 Highest Waste and Byproduct Formation in Food Industrial Processes

Process	*Waste & Byproduct (%)*	*Waste/byproduct*
Fruit and vegetable juice production	30%–50%	Stem, stalks, rotten fruit, peels, seeds, pomace
Cheese production	85%–90%	Whey
Fish canning	30%–65%	Rejected fish, heads, offal, tails, skins, bones
Beef slaughtering	40%–52%	Head, tail, udder, testicles, hooves, hides, intestines, red and white offal
Sugar production from sugar beet	86%	Beet pulp, carbonation lime, molasses

Source: AWARENET, Handbook for the prevention and minimization of waste and valorization of by-products in the European agro-food industries. Summary of the final report of the AWARENET Thematic network for 2002–2004.

a final option for landfilling, the processing byproducts have a wider application area, and different foodstuffs and pharmaceuticals can be produced from them.

These products are the following:

Pectin: Pectin is usually produced from citrus peel or apple pomace that comes from juice processing. The processing includes the extraction of pectin with hot acidified water, followed by filtrations and centrifugations. Finally, pectin is precipitated with alcohol.

Natural sweeteners: Sweeteners are produced from byproducts of chicory processing by extraction of sugar alcohols from pomace. The resulting liquid is evaporated, and the sugars are crystallized and dried.

Antioxidants: Fruits are usually rich in antioxidants that are used for health prevention and in prevention of food deterioration. They are extracted from fruit pomace using different organic solvents, such as ethanol, hexane, acetone, etc. After the extraction, the extract is fractionated and purified.

Essential oils: Many essential oils have medicinal properties, for example, antiseptic properties, and are widely applied for health purposes. They are extracted from pomace or berry seeds by different techniques, such as cold pressing, supercritical fluid extraction, or distillation.

Fibers: Fibers, both soluble and insoluble, are important byproducts from fruit processing. They can be obtained from fruit pomace by grinding and centrifugation or by mechanical dewatering until 1% dry matter.

19.10 Case Study: Recovery of Anthocyanins from Black Currant Marc by MAE

Black currants are very popular berries among consumers due to their high vitamin C content, but they also contain other vitamins, such as P, B1, and B2 vitamins. From different types of currants, black currant was shown to have the highest anthocyanin aglycons and flavonol glycosides. The most relevant anthocyanins present in black currant are delphinidin-3-glucoside (17%), delphinidin-3-rutinoside (33%), cyaniding-3-glucoside (11%), and cyaniding-3-rutinoside (39%) (Määttä et al. 2001). The main flavonols in these berries are myricetin or quercetin, followed by kaempferol, depending on the varieties (Hermann 1976; Häkkinen et al. 1999; Määttä-Riihinen et al. 2004). Black currant anthocyanins may play a beneficial role in maintaining brain health in mice (Vepsäläinen et al. 2012), improve the blood circulation and the peripheral muscle work during typing work (Matsumoto et al. 2005) and modulates exercise-induced oxidative stress and lipopolysaccharide-stimulated inflammatory responses (Lyall et al. 2009). Flavonols, such as quercetin, kaempferol, and myricetin were proven to reduce the risk of some chronic diseases (Knekt et al. 2002).

Also, the oil from black currant seeds may have a beneficial effect in reducing blood pressure and blood cholesterol. The oil was shown to reduce the blood pressure in hypertensive rats (Engler 1993). The consumption of 6 g per day of black currant seed oil was shown to reduce cardiovascular reactivity in mildly hypertensive male (Deferne and Leeds 1996). Fa-lin et al. (2010) described that the consumption of 1.8 g per day black currant seed oil for six weeks reduced the total cholesterol and triacylglycerol level if patients had a slight hyperlipidemia and low BMI.

Despite the many health-promoting effects associated with black currant consumption, the problem that arises is the availability of health-promoting compounds in the processed berries. Black currants are rarely consumed in their fresh form; rather they are processed into juices, jams, and jellies. Usually, these operations include heat or other treatments that may have a negative effect on the anthocyanin and flavonol concentration; heat treatment was reported to decrease the anthocyanin content in different juices (Bakker and Bridle 1992; Garzon and Wrolstad 2002), and the decrease of anthocyanin content of strawberry jam was observed during storage (Patras et al. 2009). Meyer and Bagger-Jorgenssen (2002) reported a 21% of total phenol and 19%–29% of anthocyanin decrease in the black currant juice clarification process when gelatin-silica sol treatment was followed by vacuum filtration.

Because the valuable compounds of black currants are affected to a great extent during processing, it is relevant to design processes that are using mild conditions to avoid the loss of nutrients and, in this way, to contribute to waste minimization. During juice pressing, part of the anthocyanins and flavonols will be recovered in the juice. However, the black currant press residue is a still a potential source for phenolic extraction (Kapasakalidis et al. 2006). In small-scale industrial processing, the black currant juice yield was approximately 75%, and the rest stays behind as the press cake. Even though the press cake is utilized in some form, e.g., fertilizer or feed production, the bioactive compounds that may provide nutritive and health-promoting effects for humans virtually vanish. On the other hand, the benefits associated with black currant oil will be only available if the seeds are processed and extracted. Based on these considerations, currently available extraction and concentration methods should form a part of industrial practices to recover anthocyanins, flavonols, and oils from black currant berries.

Presently, the common utilization of marc is alcoholic fermentation, fodder production, or composting. During these processes, the bioactive components, antioxidants, and vitamins virtually vanish. The extraction of valuable components can be beneficial from economic and environmental aspects.

Microwave-assisted extraction (MAE) has become one of the most popular extraction methods for phenols, pharmaceuticals, and natural products. In addition, conventional solvent extraction systems can also be used for the recovery of phenolic compounds. However, these technologies require unusually high operation temperatures, large organic solvent volumes, and long processing times, which may, in contrast, impact adversely on the quality of the end product.

The results from the MAE process indicated that the highest yield of anthocyanins of 22 mg g^{-1} marc was reached when 700 W power was used, and the treatment was applied for 10 min using an aqueous solvent with a pH of 2 as illustrated in Figure 19.5.

On the other hand, in the conventional extraction process, the concentration of the monomeric anthocyanin pigments increased during the 5 h of extraction until it reached the maximum value, but extraction rate was higher in the first 3 h as illustrated in Figure 19.6.

The reason behind this might be the continuous decrease in driving force, i.e., the concentration difference between the soluble and solid phases. On the other

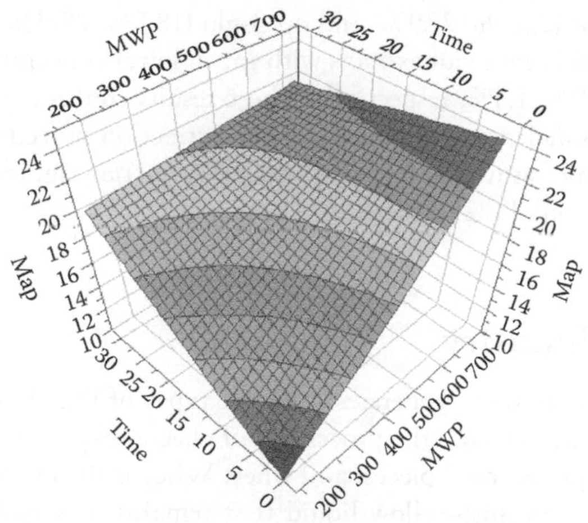

Figure 19.5 Response surface plot of anthocyanin yield by microwave-assisted extraction from black currant marc. (Copyright Food and Bioprocess Technology.)

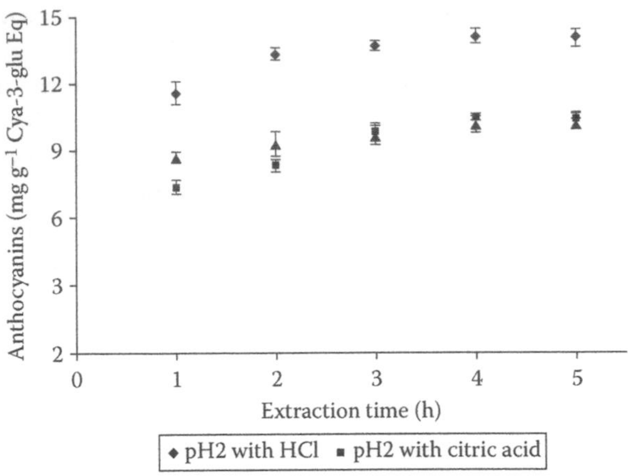

Figure 19.6 Monomeric anthocyanin pigment content of the extracts during conventional extraction using different acidic solvents. (Copyright Food and Bioprocess Technology.)

hand, the prolonged exposure time to the high temperature might have caused thermal degradation of heat-resistant anthocyanins. In the extraction of anthocyanins from blackcurrant marc, MAE was suitable to reduce extraction time, minimize the volume of solvent used, and increase the final concentration of anthocyanins in the extracts compared to the conventional solvent extraction process.

The results of the analysis showed that the composition of the extracts were similar to that reported in the literature. The most abundant anthocyanin compound was dp-3-rut with a relative distribution 37.5%–47.1% in the different extracts, followed by cy-3-rut (23.7%–27.9%) and dp-3-glu (18.5%–28.3%) while cy-3-glu and mal-3-rut was present in all extracts with much lower concentrations between 5.8%–9.7% and 0.5%–1.4%, respectively. These results confirmed that MAE has only increased the efficiency and the rate of the extraction procedure, but neither the application of microwave nor the conventional extraction significantly compromised the composition of the extracts.

19.11 Dairy Industry

Production of dairy products generates different types of liquid and solid wastes and byproducts. If we consider the processing of cheese, byproducts we get are in the form of cheese pieces, curd pieces, and whey. Whey is the byproduct of cheese manufacture. It is a greenish-yellow liquid that remains in solution after casein precipitation during cheese processing. It is rich in lactose and proteins (mainly β-lactoglobulin and β-lactalbumin) and is formed in large amounts during processing because about 85%–90% of the milk becomes whey during cheese manufacture.

Therefore, its utilization is of great economical and environmental importance in the dairy industry. Recently, the development of whey processing technologies focused on the biological treatment without valorization (aerobic digestion) or with a valorization process, physicochemical treatment (protein fractionation), or direct application on field (Prazeres et al. 2012).

Different products are obtained during the whey valorization process.

Whey powder is obtained mainly with spray drying and has a protein content of 11%–12%. It is used as animal feed or for ethanol production.

Lactic ferments: Lactic bacteria are used as starters to transform lactose to lactic acid. The culture medium can be skimmed milk or whey. Lactic ferments are used in fermentation processes.

Demineralized whey powder: Ion exchange or electro-dialysis is applied to the demineralization of whey in order to make it more suitable for human consumption. Foodstuffs for lactose intolerant people are produced with demineralized whey.

Lactose: Lactose is obtained by crystallization of concentrated whey, from which the protein was previously removed or from whole whey and used for syrups.

Whey protein concentrate: Whey proteins are used as food and pharmaceutical ingredients.

Whey cheeses: Whey is condensed to produce cheese. Examples are Ricotta cheese or Messoer (Swedish soft whey cheese).

Pressure-driven membrane processes can be used in the whey valorization process as illustrated in Figure 19.7. In the first step, fat and caseins are removed from

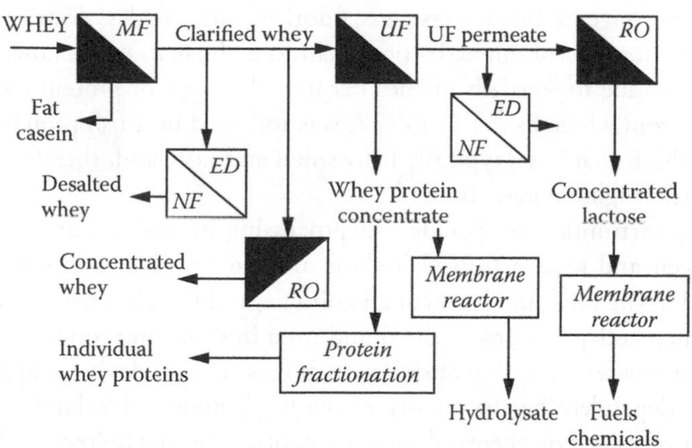

Figure 19.7 Whey processing by membrane technologies. (From Cheyran, M., *Ultrafiltration and Microfiltration Handbook*, 2nd Edn. CRC Press, Boca Raton, FL, 1998.)

the waste stream by microfiltration; after which whey can simply be concentrated by reverse osmosis. Purification of the whey is carried out in the hybrid process of electro-dialysis and nano-filtration. Ultrafiltration is used mostly for the production of whey protein concentrates (WPC) with a protein content of 35%–85%.

In this process, whey proteins are separated from non-protein nitrogen, lactose, and minerals based on the difference in their molecular weight. The lactose is in a diluted form; therefore, it is concentrated by reverse osmosis. Although lactose is often used in the food industry as an additive, concentrated lactose can also be used for fuel and chemical production. The WPC can be partially digested in a membrane bioreactor and thus be more readily absorbed in humans.

19.12 Fish Industry

The solid byproducts generated during fish processing consist of rejected fish, scraps, offal, heads, skin, tails, and bones. The amount varies between 30%–65% in the case of fish canning and 50%–75% in the case of fish filleting, salting, and smoking.

Most of the byproducts are used for fish meal and fish oil production. Fish oil is known to have beneficial effects to human health by preventing coronary heart disease due to its unsaturated fatty acid content. The raw materials are the rejected fish and other fish parts, and the first step is cooking to coagulate the proteins and broke liberating oil and physico-chemically bound water. In the pressing, the oil and water will be released from the solid mass and is further separated by centrifugal separation. The oil, in order to be utilized, has to be refined. The byproduct from the oil pressing—the press cake—is dried to obtain fish meal.

Fish protein concentrates can be produced from almost any kind of fish. The first fish protein concentrate was a tasteless, odorless, and colorless powder (FPC A), which was produced by solvent extraction technique. Its marketing, however, faces some challenges due to legislative issues because this type of protein concentrate was prepared from whole fish. This FPC A was followed by a fish protein concentrate called FPC B that has a typically fishy smell and taste and, therefore, its food application area might be more limited.

Further opportunities to upgrade fish processing byproducts are the production of collagen and gelatin from fish skins and bones. Collagen is a structural protein found in the skins and bones of all animals, and it is obtained by cold-water extraction. Then collagen is freeze-dried and used in different products. Gelatin is obtained by hot-water extraction of skins and bones, and air-drying is applied. The application is dependent on the quality of the final product. Norlandprod (2013) classified gelatins based on their utilization potential into the following classes:

- *Edible gelatin:* clarifier, stabilizer, preservative and texturing agent, formulation of medications and dietary supplements
- *Industrial gelatin:* micro-encapsulation of dyes

- *Photographic:* light-sensitive coating in the electronics area
- *Glue:* used for adhesive applications

The production of fish gelatin includes the pretreatment of the skin, i.e., washing in water and soaking in dilute alkaline solution (NaOH) first, followed by washing and soaking in dilute weak (sulfurous) acid. Once the skin is rinsed with water again, it is ready for extraction at about 70°C–80°C for two consecutive 30 min periods. Although fish gelatin does not have good foaming properties as gelatin derived from other domestic animal byproducts, it can form rather strong gels that make its application in diverse food products possible.

19.13 Meat Industry

Depending on the type of animal slaughtered, up to 40%–50% of the animal's live weight remains as byproducts that are sometimes undervalued as they are not utilized in food due to their physical or chemical characteristics. These byproducts can be divided into edible and nonedible categories, in which, for example, hides and bones belong to the nonedible byproduct and blood to the edible category.

Although blood is a very valuable byproduct due to its high-value protein content that corresponds up to 10% of total protein content of the animal, it is usually underutilized by the food industry. The lack of its utilization is due to ethical reasons and the non-acceptance of the consumers rather than in the shortage of economical ways to recover blood proteins. Blood consist of a plasma fraction and red cell fractions that represent about 60% and 40% of the total volume, respectively. Blood plasma has a protein content of 6%–8% and consists of albumins (50%), globulins (23%–27%), and fibrinogen (17%–23%) as major proteins. If the fibrinogen is removed from the plasma by the salting out method (Pares et al. 2012), the remaining protein mixture is called serum, which is further processed by ultra- or diafiltration in order to reduce its salt content. Ultrafiltration can also be successfully applied in the concentration of animal plasma (Pihlanto et al. 2012); using spiral-wound polyethersulfone and regenerated cellulose membranes, the initial volume of the plasma was reduced to one third. Both plasma and serum have good functional properties, such as good gelling, emulsifying, or foaming properties, that call for their applications in food products. They can be used as emulsifiers to replace egg albumen or soy protein or as fat replacer in the production of light products. Serum was used successfully to replace caseinates in phosphate-free frankfurter production. The use of hemoglobin is somewhat more challenging due to its undesirable dark color and typical taste. Therefore, their major application area should be in products that are of dark color origin, such as black puddings, blood cookies, or black sausages.

Bones and hide from the animal processing can be used to obtain collagen and gelatin as gelatin is produced by the controlled hydrolysis of the collagen. For this aim, three different technologies can be used: treatment in the alkaline process (Type B gelatin), the acid process (Type A gelatin), and the high pressure steam extraction (Ockerman and Hansen 2000). The application of the derived gelatin is broad; it can be used in the food industry as a texture modifier (ice cream) or to bind water in meat products. Beside, it is also used in the pharmaceutical industry to manufacture capsules. Other fields of application include the cosmetics industry or the manufacturing of glue.

19.14 Beet Sugar Processing Byproducts

Due to their high energy content, the tendency in sugar processing byproduct valorization is use as feedstock in biofuel production. Molasses has been the principal raw material used in industrial-scale production of biobutanol (Qureshi et al. 2001), and sugar beet pulp and other beet sugar byproducts are increasingly considered as feedstock for ethanol production (Grahovac et al. 2012; Vučurović and Razmovski 2012). Due to the enormous potential and interest in using food wastes in energy generation, the next section will focus on waste-to-energy technologies for the valorization of energy content.

19.14.1 Waste-to-Energy Technologies

In the last decade, tertiary biomass has become a prime source of interest in bio-energy generation, and efforts are extended to utilizing the energy content of organic waste and effluents. Biomass-based waste materials from the agro-food sector can be converted to energy by thermo-chemical, biochemical, mechanical, chemical, or electro-chemical process routes. From these, thermo-chemical and biochemical conversion technologies are suited to a wide range of waste biomass while the others have more limitations on the feedstock. Thermochemical conversion processes take place at high temperatures and occur in an environment characterized by very different concentrations of oxygen. Selected technologies that fall into these categories are combustion, pyrolysis, and gasification. Thermo-chemical conversion methods are suited to relatively dry woody and herbaceous biomass. In biochemical conversion, microorganisms convert biomass into biofuels. Biochemical technologies, such as anaerobic digestion and alcohol fermentation can also handle biomass with high moisture content. These wet processing techniques are more economical and efficient than thermochemical conversion processes for high moisture materials (McKendry 2002).

19.14.2 Fermentation

Fermentation is a process in which sugar-containing biomass is converted to alcohol, e.g., ethanol, by the metabolism of microorganisms. The fermentation process is usually anaerobic, but also aerobic conditions can be feasible. Fermentation processes can be batch, fed-batch, or continuous processes (Nag 2007). In the fermentation process, microbes (usually yeast or bacteria) or, less frequently, fungi and split organic matter typically produce alcohol as a final product (Lampinen and Jokinen 2006). This process is particularly suitable for crops and plants with a high content of sugar or starch (e.g., sugar beet, corn, potatoes, etc.).

The conventional fermentation process consists of hydrolysis, fermentation, separation, and purification steps. If the process uses lignocellulosic raw materials, milling and acid or enzymatic hydrolysis is required. After pretreatment, sugars go through the fermentation process. In the fermentation process, the process conditions, such as temperature and water content, have to be optimal for microbial growth and action (Scragg 2006). Lack of nutrients can cause inefficient ethanol yields. Microorganisms need several nutrients and trace elements, such as carbon, hydrogen, phosphorus, sulfur, vitamins, potassium, and calcium. An adequate pH level is also vital for fermentative microorganisms. The accurate pH level can be controlled by adding ammonia to the input, for example (Scragg 2006).

The fermentation process must be free from oxygen. Otherwise, the presence of oxygen restricts the production of ethanol considerably. Stirring is usually needed to improve mass and heat transfer in a bio-reactor, especially in continuous reactors (Scragg 2006).

The produced alcohol is removed from the process at concentrations of around 6% because 15% ethanol concentration already starts to be toxic for the microbes. In the end, ethanol is enriched to 99% bioethanol (Nag 2007). Ethanol is the most common fuel produced through commercial fermentation thanks to its versatility as transportation fuel and a fuel additive. Beside ethanol, butanol is being developed as a fuel substitute. The most conventional process to separation of water and ethanol is distillation. Distillation is an energy-intensive solution. Current research efforts concentrate on low-energy separation processes, such as membrane processes, in particular pervaporation (García et al. 2011).

19.14.3 Anaerobic Digestion

Anaerobic digestion (AD) is a biochemical process in which biogas is produced from organic matter by microorganisms in the absence of oxygen. Biogas consists mainly of methane (CH_4) and carbon dioxide (CO_2). Biogas can be produced from post-consumer food waste, food industry wastewaters, and organic byproducts from industry and agriculture (Tavitsainen 2006). Anaerobic digestion occurs in a bioreactor, which can be classified as a wet or a dry reactor. Organic food waste and vegetable waste are used in dry reactors, and wet reactors are more commonly used for manure and sludge waste. The operating temperatures are also divided to mesophilic (35°C) and thermophilic (55°C) temperatures. The advantage of thermophilic reactors is shorter retention time, but maintaining a higher temperature requires higher energy input (EUBIA 2011).

Food industry wastewaters may require concentration prior to AD treatment to increase the overall capacity of digestion with the higher organic matter content of AD feed. Moreover, the specific biogas yield and the rate of the AD process can be improved by pretreatment, which could enhance the biogas production and the methane content of the produced biogas (Beszédes et al. 2010). Low-risk raw materials from the food industry, such as raw meat that has passed meat inspection; waste from food manufacturers and food retailers; certain other animal-based byproducts, which do not have any disease, is sterilized before further anaerobic digestion to reach adequate hygiene standards. Sterilization kills pathogenic bacteria, and, conventionally, this is done usually at a temperature of 70°C for 60 min (Erjava 2009; Evira 2013). Moreover, the digestate from the reactor is post-treated to fulfill the regulations of the fertilizer legislation (Tavitsainen 2006). For higher risk materials, such as animal byproducts presenting a risk of contamination with other animal diseases, residues of animal medicines, thermophilic anaerobic digestion (when the temperature is above 50°C and retention time is at least 20 days), plus composting is an appropriate process combination.

In case of animal input, attention should be paid to potential inhibitors in the feed. Antibiotics appear to inhibit bacterial activity resulting in a delay and overall decline in CH_4 production (Shi et al. 2011). The anaerobic digestion process occurs mainly in four steps: hydrolysis, acidogenesis, acetogenesis, and methanogenesis. In hydrolysis, insoluble organic matter is converted to a soluble form. The main idea of acidogenesis is to produce acetate, volatile fatty acids, carbon dioxide, and hydrogen. Volatile acids are degraded to acetate and hydrogen by acetogenesis. In the final step, acetate and hydrogen are converted to methane and carbon dioxide by methanogenesis (EUBIA 2011).

The final product (biogas) can be used for combined heat and power production. The biogas can be also purified to methane and used as a fuel for vehicles. The residues from the anaerobic digestion can provide further benefits as a fertilizer.

19.14.4 Pyrolysis

Pyrolysis is a process in which organic material is heated at high temperatures in an oxygen-free environment. The products are gases, oils, and char (Ahmed and Gupta 2009). In pyrolysis, large hydrocarbon molecules (cellulose, hemicelluloses, and part of the lignin) break down into smaller and lighter molecules. Unlike combustion and gasification, pyrolysis occurs in a total absence of oxygen (Basu 2010). Gases are usually utilized for drying and pyrolysis reactions; oils are utilized for heating or, if refined, as secondary fuels. Pyrolysis can be divided into slow and fast pyrolysis. In slow pyrolysis, biomass is heated slowly to pyrolysis temperatures (400°C–800°C) with a long residence time. Slow pyrolysis produces more tar and charcoal and less gases. The purpose of fast pyrolysis is to maximize the yield of liquid or gases. In fast pyrolysis, biomass is heated rapidly to the adequate temperature (up to 650°C) and held there only for a few seconds or less than a second. Also flash and ultra-rapid pyrolysis has been researched. Known reactor types are fixed bed, moving bed, bubbling fluidized bed, and circulating fluidized bed reactors (Basu 2010; EUBIA 2011). Raw materials containing large amounts of potassium as well as other alkali metals and chlorine are not beneficial to a pyrolysis reactor due to their corrosive effects. These compounds can corrode the reactor walls, boilers, and other process equipment, causing malfunctions, leaks, and structural problems. In addition, too high moisture content (up to 30%) can inhibit the pyrolysis process and lead to higher consumption of thermal energy. The specific hydrogen-to-carbon ratio of the raw material affects also the product yield (Basu 2010).

19.14.5 Gasification

In gasification, biomass is converted by partial oxidation at high temperatures into a gas mixture called product gas or syngas. The most suitable feedstock is lignocellulosic materials, but agricultural wastes and crop residues are also suitable raw materials (Basu 2010).

A conventional gasification process consists of biomass drying, pyrolysis, oxidation, and reduction steps. In the pyrolysis chamber, large hydrocarbon molecules of biomass break up into smaller molecules in the absence of oxygen. Therefore, relatively volatile compounds of the biomass are separated from the char. Temperature in the pyrolysis chamber varies between 400°C and 650°C (Basu 2010).

Endothermic pyrolysis and gasification reactions occur in the oxidation chamber at temperatures between 900°C and 1200°C. Syngas is formed in the reduction chamber through several reactions. Gasification processes can be divided into updraft and downdraft gasifiers. Furthermore, reactors can be roughly classified to a moving bed reactor, a fluidized bed reactor, and an entrained flow reactor (Basu 2010).

The gasification process is strongly dependent on a number of factors, such as the feedstock particle size range, moisture content, gas-solid contacting mode, pressure, heating rate, temperature, and residence time (Austerman and Whiting 2007).

19.14.6 Combustion

Combustion is the oldest and still the most used way to convert biomass to energy. Combustion of biomass is a process in which oxygen reacts with carbon in the fuel and produces carbon dioxide, water, and heat (van Sjaak and Koppejan 2008). Combustion can be utilized to produce heat or electricity. Ash from the process can be utilized as fertilizer. Combustion, however, is a rather inefficient power-generation method when compared with other methods, such as, for example, gasification (Lampinen and Jokinen 2006).

19.15 Selection of Waste-to-Energy Technologies

Waste-to-energy (W2E) solutions have the potential to answer the need for energy resources while systematically reducing human impact on the environment. The technologies suitable strongly depend on feedstock characteristics, such as moisture content, calorific value, ash, residue content, etc. (Kelleher et al. 2002; McKendry 2002). Table 19.6 summarizes the characteristics of the W2E technologies.

Table 19.6 Characteristics of Waste-to-Energy Technologies

	Fermentation	*Anaerobic Digestion*	*Gasification*	*Pyrolysis*	*Combustion*
Preferable input materials	Food crops, food industry byproducts, bio-waste	Bio-waste and food industry byproducts and wastewaters	Agricultural byproducts and dry industrial wastes	Mill waste, agricultural wastes, food waste	Agricultural byproducts and dry industrial wastes
Limiting factors	Homogenous input, nutrients, pH, moisture	Total solids 4%–40%	Moisture < 45% Ash < 15%	Moisture < 45% Ash < 25%	Moisture < 50%
Operating temperature	15°C–60°C	Optimum 35°C or 55°C	650°C–1200°C	400°C–800°C	> 800°C
Oxygen requirements	Depends on microbes	Absence of oxygen	Partial oxidation	Absence of oxygen	Excess of oxygen
Scale	Ethanol yield 102–106 m^3 annually	Reactor size 50–10.000 m^3	1 kWe–150 MWe, depending on the technology used	Pilot plant of 200 kg h^{-1}, with 66% energy yield	Small to large scale
Product	Alcohol	Biogas	Syngas	Pyrolysis oil	Heat

Byproducts	Reject, gases, water	Digestate, water	Char	Gases, char	Ash
Post-treatment	Water removal	Gas purification, digestate stabilization	Particulates and tars removal	Oxygen removal	None
Product application and use	Transportation fuel, heat, and power; reject as fertilizer or animal feed	Transportation, fuel, heat, and power; digestate as fertilizer or soil conditioner	Synthetic fuel, heat, and power	Fuel, heat, and power	Electricity and heat production

Source: Austerman, S. et al., Anaerobic digestion technology for biomass projects. Commercial Assessment. Report produced by Juniper Consultancy Services Ltd for Renewables East, 2007; Austerman, S., and Whiting, K.J., Advanced conversion technology (gasification) for biomass projects. Commercial Assessment. Report produced by Juniper Consultancy Services Ltd for Renewables East, 2007; Kauriinoja, A., *Small-scale biomass-to-energy solutions for Northern Periphery areas.* Master's thesis. University of Oulu, Department of Process and Environmental Engineering, 2010; Kelleher, B.P. et al., *Bioresource Technology*, 83, 27–36, 2002; McKendry, P., *Bioresource Technology*, 83, 55–63, 2002; Mikkonen, L., and Pongrácz, E., Installation, safety and troubleshooting of biomass and waste-to-energy technologies. MicrE project report. University of Oulu, NorTech Oulu. URL: http://nortech.oulu.fi/MicrE_files/MicrE_W2E_IST.pdf, 2011; Reprinted with permission from Soltes, E.J., (Chapter 1). Of Biomass, Pyrolysis and Liquids Therefrom. In Soltes, E.J. and Milne, T.A. (Eds.) *Pyrolysis oils from biomass: producing, analyzing and upgrading*, 353. Copyright 1988 American Chemical Society; Uslu, A. et al., *Energy*, 33, 1206–1223, 2008; Ward, A.J. et al., *Bioresource Technology*, 99, 7928–7940, 2008; Wisbiorefine, Wisconsin Biorefining Development Initiative™. Fermentation of 6-carbon sugars and starches. Available at http://www.wisbiorefine.org/proc/fermentss.pdf, 2004.

Both environmental and political pressures require increased use of renewables in the energy mix. There are many ongoing attempts for measuring the sustainability aspects of biomass-based energy. Most of them focus on controlling land use impacts and greenhouse gas (GHG) emissions. This is due to the potential long-term adverse impacts of direct and indirect land-use changes on GHG emissions and, in particular, the danger of using productive croplands for biofuel feedstock production. This highlights the value of using wastes as feedstock for biofuel production. The challenge of using waste materials is in their heterogeneous nature. Research needs to intensify to increase the efficiency of W2E technologies and to address the challenges provided by the feedstock.

19.16 Summary

The food industry generates significant amounts of solid wastes and byproducts. In medium- and high-income countries, food is to a great extent wasted, meaning that it is thrown away even if it is still suitable for human consumption. Significant food losses and wastage occur also early in the food supply chain. In low-income countries, food is mainly lost during the early and middle stages of the food supply chain; much less food is wasted at the consumer level.

More and more stringent environmental regulations are calling for actions to reduce the environmental load of the food industry. In this chapter, technologies for waste prevention and minimization have been summarized as well as valorization methods for the recovery of products, nutrients, and energy from the food industry's wastes and byproducts were presented.

Fruit and vegetable processing, cheese production, and meat and fish processing as well as beet sugar production are the largest producers of wastes. However, many valuable compounds can be recovered from their wastes and byproducts. Hydrolysis, distillation, different forms of extraction, adsorption, ion exchange, evaporation, crystallization, drying, membrane separation, and enzymatic treatment are widely used in different areas of the food industry for byproduct valorization.

For example, in fruit processing, the major source of solid waste generation is the pressing process, in which peels, seeds, and pulps are separated from the fruit juice. There is a large unused potential in juice processing wastes as they contain a sizeable amount of healthy substances, such as flavonoids, colors, and pectins. When applying proper extraction technology, such as super-critical fluid or microwave-assisted extraction, these healthy compounds can be recovered and applied either by the food industry or the cosmetic or pharmaceutical industries. As illustrated with the case study of black currant marc processing, microwave-assisted extraction (MAE) can have a great potential to be used in the food industry to obtain anthocyanin extracts. Compared to conventional solvent extraction, the optimized MAE process can have 20% higher yields within a fraction of the extraction time (300 min to 10 min).

In our current political climate, there are considerable pressures to increase the use of renewables in the energy mix. Food industry wastes are considered a sustainable option for biofuel production. Commercial waste-to-energy conversion technologies include anaerobic digestion, alcohol fermentation and combustion. Also, pyrolysis and gasification are applicable but more demanding in feedstock requirements.

While economically and technologically feasible, some valorization products face challenges in terms of legislative constraints and consumer acceptance. It is concluded that research needs to intensify to increase the efficiency of food waste valorization technologies and to address the challenges provided by the feedstock as well as legislative pressures and consumer awareness.

References

Ahmed, I., and Gupta, A.K. (2009). Syngas yield during pyrolysis and steam gasification of paper. *Applied Energy*. 86 (9), 1813–1821.

Austerman, S., and Whiting, K.J. (2007). Advanced conversion technology (gasification) for biomass projects. Commercial Assessment. Report produced by Juniper Consultancy Services Ltd. for Renewables East.

Austerman, S., Archer, E., and Whiting, K.J. (2007). Anaerobic digestion technology for biomass projects. Commercial Assessment. Report produced by Juniper Consultancy Services Ltd. for Renewables East.

AWARENET (2004). Handbook for the prevention and minimization of waste and valorization of by-products in the European agro-food industries. Summary of the final report of the AWARENET Thematic network for 2002–2004.

Bakker, J., and Bridle, P. (1992). Strawberry juice colour: The effect of sulphur dioxide and EDTA on the stability of anthocyanins. *Journal of the Science of Food and Agriculture*. 60, 477–481.

Basu, P. (2010). *Biomass Gasification and Pyrolysis*. Academic Press, Burlington, MA.

Beszédes, S., Pap, N., Pongracz, E., Hodúr, C., and Keiski, R.L. (2010). Optimization of reverse osmosis process for the purification of meat processing wastewater. Proceedings of the conference PERMEA 2010, Tatranské Matliare, Slovakia, September 4–8.

Cheyran, M. (1998). *Ultrafiltration and Microfiltration Handbook*, 2nd Edn. CRC Press, Boca Raton, FL.

Deferne, J.L., and Leeds, A.R. (1996). Resting blood pressure and cardiovascular reactivity to mental arithmetic in mild hypertensive males supplemented with blackcurrant seed oil. *Journal of Human Hypertension*. 10, 531–537.

Elintarviketeollisuusliitto (2010). *Henkilökunta teollisuusaloittain*. Finnish Food and Drink Industries' Federation, Helsinki.

Engler, M.M. (1993). Comparative study of diets enriched with evening primrose, black currant, borage or fungal oils on blood pressure and pressor responses in spontaneously hypertensive rats. *Prostaglandins Leukot Essent Fatty Acids*. 49, 809–814.

European Biomass Industry Association (EUBIA) (2011). URL: http://www.eubia.org/108.0.html.

Erjava, A. (2009). Biokaasulaitoksen perustaminen kasvihuonetilalla. Bioenergiakeskuksen julkaisusarja (BDC publications) Nro 46. p. 83.

European Commission (2012). *Waste Prevention-Handbook*: Guidelines on waste prevention programmes, URL: http://ec.europa.eu/environment/waste/prevention/pdf/Waste%20 prevention%20guidelines.pdf.

Evira (2013). Finnish food safety authority Evira. Internet pages. URL: http://www.evira.fi.

Fa-lin, Z., Zhen-yu, W., Yan, H., Tao, Z., and Kang, L. (2010). Efficacy of blackcurrant oil soft capsule, a Chinese herbal drug, in hyperlipidemia treatment. *Phytotherapy Research*. 24 (2), 209–213.

FAO, WFP and IFAD. (2012). *The State of Food Insecurity in the World 2012. Economic growth is necessary but not sufficient to accelerate reduction of hunger and malnutrition.* Rome, FAO ISBN 978-92-5-107316-2.

García, V., Päkkilä, J., Ojamo, H., Muurinen, E., and Keiski, R.L. (2011). Challenges in biobutanol production: How to improve the efficiency? *Renewable and Sustainable Energy Reviews*. 15 (2), 964–980.

Garzon, G.A., and Wrolstad, R.E. (2002). Comparison of the stability of pelargonidin-based anthocyanins in strawberry juice and concentrate. *Journal of Food Science*. 67 (4), 1288–1299.

Gaur, R., Sharma, A., Khare, S.K., and Gupta, M.N. (2007). A novel process for the extraction of edible oils: Enzyme assisted three phase partitioning (EATPP). *Bioresource Technology*. 98 (3), 696–699.

Ghafoor, K., Hui, T., and Choi, Y.H. (2011). Optimization of ultrasound-assisted extraction of total anthocyanins from grape peel. *Journal of Food Biochemistry*. 35, 735–746.

Grahovac, J.A., Dodić, J.M., Dodić, S.N., Popov, S.D., Vučurović, D.G., and Jokić, A.I. (2012). Future trends of bioethanol co-production in Serbian sugar plants. *Renewable and Sustainable Energy Reviews*. 16 (5), 3270–3274.

Gustavsson, J., Cederberg, C., Sonesson, U., von Otterdijk, R., and Meybeck, A. (2011). Global food losses and food waste. Extent, causes and prevention. Food and Agriculture Organization of the United Nations, Rome, Italy. Available at: http://www.fao.org/docrep/014/mb060e/mb060e00.pdf.

Häkkinen, S., Heinonen, M., Kärenlampi, S., Mykkänen, J., Ruuskanen, J., and Törrönen, R. (1999). Screening of selected flavonoids and phenolic acids in 19 berries. *Food Research International*. 32 (5), 345–353.

Hermann, K. (1976). Flavonoids and flavones in food plants: A review. *Journal of Food Technology*. 11, 433–448.

Kapasakalidis, P.G., Rastall, R.A., and Gordon, M.H. (2006). Extraction of polyphenolds from processed black currant (*Ribes nigrum* L) residues. *Journal of Agricultural and Food Chemistry*. 54, 4016–4021.

Kaufmann, B., and Christen, P. (2002). Recent extraction techniques for natural products: Microwave-assisted extraction and pressurized solvent extraction. *Phytochemical Analysis*. 13 (2), 105–113.

Kauriinoja, A. (2010). *Small-scale biomass-to-energy solutions for Northern Periphery areas*. Master's thesis. University of Oulu, Department of Process and Environmental Engineering.

Kelleher, B.P., Leahy, J.J., Henihan, A.M., O'Dwyer, T.F., Sutton, D., and Leahy, M.J. (2002). Advances in poultry disposal technology – a review. *Bioresource Technology*. 83, 27–36.

Knekt, P., Kumpulainen, J., Järvinen, R., Rissanen, J., Heliövaara, M., and Reunanen, A. (2002). Flavonoid intake and risk of chronic diseases. *The American Journal of Clinical Nutrition*. 76 (3), 560–568.

Lakkakula, N.R., Lima, M., and Walker, T. (2004). Rice bran stabilization and rice bran oil extraction using ohmic heating. *Bioresource Technology*. 92 (2), 157–161.

Lampinen, A., and Jokinen, E. (2006). Suomen maatolojen energiatuotantopotentiaali (Energy potential of Finnish farms). University of Jyväskylä, Research reports in bio and environmental sciences, nro 84. URL: http://energiahamppu.turkuamk.fi/tutkimukset/energiapotentiaali.pdf.

Lang, Q., and Wai, C.M. (2001). Supercritical fluid extraction in herbal and natural product studies – a practical review. *Talanta*. 53 (4), 771–782.

Lebovka, F., Vorobiev, N., and Chemat, E. (2011). *Enhancing Extraction Processes in the Food Industry*. CRC Press, Boca Raton, FL.

Lusas, E.W., and Watkins, L.R. (1988). Oilseeds: Extrusion for solvent extraction. *Journal of the American Oil Chemists' Society*. 65 (7), 1109–1114.

Lyall, K.A., Hurst, S.M., Cooney, J., Jensen, D., Lo, K., Hurst, R.D., and Stevenson, L.M. (2009). Short-term blackcurrant extract consumption modulates exercise-induced oxidative stress and lipopolysaccharide-stimulated inflammatory responses. *American Journal of Physiology. Regulatory, Integrative and Comparative Physiology*. 297, 70–81.

Määttä, K., Kamal-Eldin, A., and Törrönen, R. (2001). Phenolic compounds in berries of black, red, green and white currants (Ribes sp.). *Antioxidant and Redox Signaling*. 3, 981–993.

Määttä-Riihinen, K.R., Kamal-Eldin, A., Mattila, P.H., Gonzáles-Paramás, A.M., and Törrönen, R. (2004). Distribution and contents of phenolica compounds in eighteen Scandinavian berry species. *Journal of Agricultural and Food Chemistry*. 52, 4477–4486.

Marr, R., and Gamse, T. (2000). Use of supercritical fluids for different processes including new developments – a review. *Chemical Engineering and Processing*. 39 (1), 19–28.

Matsumoto, H., Takenami, E., Iwasaki-Kurashige, K., Osada, T., Katsumura, T., and Hamaoka, T. (2005). Effects of blackcurrant anthocyanin intake on peripheral muscle circulation during typing work in humans. *European Journal of Applied Physiology*. 94, 36–45.

McKendry, P. (2002). Energy production from biomass (Part 3): Gasification technologies. *Bioresource Technology*. 83, 55–63.

Meyer, A.S., and Bagger-Jørgensen, R. (2002). Retainment of phenolic phytochemicals by new technological approaches in berry juice processing. Dias report Horticulture no. 29. Proceedings of Health promoting compounds in vegetables and fruits. 90–98.

Mikkonen, L., and Pongrácz, E. (2011). Installation, safety and troubleshooting of biomass and waste-to-energy technologies. MicrE project report. University of Oulu, NorTech Oulu. URL: http://nortech.oulu.fi/MicrE_files/MicrE_W2E_IST.pdf.

Nag, A. (2007). *Biofuels Refining and Performance*. McGraw-Hill, Columbus, OH.

Niranjan, K., and Shilton, N.C. (1994). Food processing wastes–their characteristics and assessment of processing options. In: Environmentally Responsible Food Processing, ed. E.L. Gaden. New York: American Institute of Chemical Engineers.

Norlandprod (2013). Fish gelatin. Available at http://www.norlandprod.com.

Ockerman, H.W., and Hansen, C.L. (2000). *Animal By-product Processing and Utilization*. CRC Press, Boca Raton, FL.

Pap, N., Beszédes, S., Pongrácz, E., Myllykoski, L., Gábor, M., Gyimes, E., Hodúr, C., and Keiski, R.L. (2012). Microwave-assisted extraction of anthocyanins from blackcurrant marc: Use of statistical experimental design. *Food and Bioprocess Technology*. DOI: 10.1007/s11947-012-0964-9.

Pap, N., Pongrácz, E., Myllykoski, L., and Keiski, R. (2004). Waste minimization and utilization in the food industry: Processing of arctic berries and extraction of valuable compounds from juice-processing by-products. In Pongrácz E. (ed.) Proceedings of the Waste Minimization and Resources Use Optimization Conference, June 10, University of Oulu, Finland. Oulu University Press, Oulu, pp. 159–168.

Parés, D., Saguer, E., Pap, N., Toldrà, M., and Carretero, C. (2012). Low-salt porcine serum concentrate as functional ingredient in frankfurters. *Meat Science*. 92 (2), 151–156.

Parfitt, J., Barthel, M., and Macnaughton, S. (2010). Food waste within food supply chains: Quantification and potential for change to 2050. *Phil. Trans. R. Soc.* 365, 3065–3081.

Patras, A., Brunton, N.P., Tiwari, B.K., and Butler, F. (2009). Stability and degradation kinetics of bioactive compounds and colour in strawberry jam during storage. *Food and Bioprocess Technology*. 4, 1245–1252.

Pihlanto, A., Pap, N., Silvenius, F., Kymäläinen, M., and Niemistö, M. (2012). Teurastamoista saatavien sivujakeiden uudet prosessointimenetelmät ja hyötykäyttökohteet. (New processing methods and utilization of slaughterhouse wastes). MTT Raportti 62. Available at http://jukuri.mtt.fi/bitstream/handle/10024/438267/mttraportti62.pdf?sequence = 1.

Prazeres, A.R., Carvalho, F., and Rivas, J. (2012). Cheese whey management: A review. *Journal of Environmental Management*. 110, 48–68.

Qureshi, N., Lolas, A., and Blaschek, H.P. (2001). Soy molasses as fermentation substrate for production of butanol using *Clostridium beijerinckii* BA101. *Journal of Industrial Microbiology and Biotechnology*. 26, 290–295.

Scragg, A. (2006). *Environmental Biotechnology*, 2nd edn. Oxford University Press, New York.

Shi, J.C., Liao, X.D., Wu, Y.B., and Liang, J.B. (2011). Effect of antibiotics on methane arising from anaerobic digestion of pig manure. *Animal Feed Science and Technology*. 166–167, 457–463.

Smith, R.M. (2002). Extractions with superheated water. *Journal of Chromatography A*. 975 (1), 31–46.

Soltes, E.J. (1988). Of biomass, pyrolysis and liquids therefrom. In Soltes, E.J. and Milne, T.A. (Eds.) *Pyrolysis Oils from Biomass: Producing, Analyzing and Upgrading*. American Chemical Society, Washington, DC.

Taavitsainen, T. (2006). Maatalouden biokaasulaitoksen perustaminen ja turvallisuustarkastelu. Savonia ammattikorkeakoulu (Malla2). Available at: http://portal.savonia.fi/img/amk/sisalto/teknologia_ja_ymparisto/ymparistotekniikka/Malla2Loppuraportti%281%29.pdf.

Toepfl, S., Mathys, A., Heinz, V., and Knorr, D. (2006). Review: Potential of high hydrostatic pressure and pulsed electric fields for energy efficiency and environmentally friendly food processing. *Food Review International*. 22 (4), 504–423.

Uslu, A., Faaij, A.P.C., and Bergman, P.C.A. (2008). Pre-treatment technologies, and their effect on international bioenergy supply chain logistics. Techno-economic evaluation of torrefaction, fast pyrolysis and pelletisation. *Energy*. 33, 1206–1223.

van Sjaak, L., and Koppejan, J. (2008). *The Handbook of Biomass Combustion and Co-firing*. Earthscan, London.

Vepsäläinen, S., Koivisto, H., Pekkarinen, E., Makinen, P., Dobson, G., McDougall, G.J., Steward, D., Haapasalo, A., Karjalainen, R.O., Tanila, H., and Hultunen, M. (2012). Anthocyanin-enriched bilberry and blackcurrant extracts modulate amyloid precursor protein processing and alleviate behavioural abnormalities in the APP/PS1 mouse model of Alzheimer's disease. *Journal of Nutritional Biochemistry*. 24, 360–370.

Vinatoru, M., Toma, M., Radu, O., Filip, P.I., Lazurca, D., and Mason, T.J. (1997). The use of ultrasound for the extraction of bioactive principles from plant materials. *Ultrasonics Sonochemistry*. 4 (2), 135–140.

Vučurović, V.M., and Razmovski, R.N. (2012). Sugar beet pulp as support for *Saccharomyces cerivisiae* immobilization in bioethanol production. *Industrial Crops and Products*. 39, 128–134.

Ward, A.J., Hobbs, P.J., Holliman, P.J., and Jones, D.L. (2008). Optimisation of the anaerobic digestion of agricultural resources: Review. *Bioresource Technology.* 99, 7928–7940.

Wisbiorefine. (2004). Wisconsin Biorefining Development Initiative™. Fermentation of 6-carbon sugars and starches. Available at: http://www.wisbiorefine.org/proc/fermentss.pdf.

WRAP (2013). Waste Mapping: Your Route to More Profit. March 2013. URL: http://www.wrap.org.uk/sites/files/wrap/WRAP_Waste_Mapping_Guide.pdf.

Ympäristöministeriö (Finnish Ministry of Environment) (2004). Kansallinen strategia biohajoavan jätteen kaatopaikkakäsittelyn vähentämisestä (National strategy to reduce landill of biodegradale wastes). URL: http://www.ymparisto.fi/download.asp?contentid=27161&lan=fi.

Chapter 20

Energy-Efficient Food Processing: Principles and Practices

M.K. Hazarika

Contents

20.1 Energy Needs of the Food Processing Industries

20.1.1 Food Processing Industry

The agri-food industrial sector involves a complex chain of agricultural production practices and subsequent handling and processing of the products for improving quality and value. As a whole, the agri-food industrial sector is considered to be one of the largest sectors with a significant contribution to the economic advancement of nations and with major social impact. This sector strives to ascertain adequate food production, effective food chain management, safe valued-added products, healthy choices for the consumer, and compliance with regulatory directives on matters of quality and environmental impact. Sustainable growth of the agri-food industrial sector depends heavily on the effective implementation of these actions. Food processing industries, being an important constituent of this large sector, address the issues related to target actions on ensuring safe and quality products, optionally with added values, and giving a healthy choice for the consumer within the framework of the regulatory directives on matters of quality and thereby bear significant importance in economic development.

As a business establishment, food processing industries are evaluated for economic contribution, environmental impact, social sustainability, resource security, and industrial competitiveness. For this purpose, related statistical data are collected, analyzed, and maintained in a classified manner along with similar data from various other business establishments. One such classification system is the North American Industry Classification System (NAICS). As per NAICS classification, industrial food manufacturing practices fall into one of the nine manufacturing sectors with NAICS Code 311. The nine manufacturing sectors with NAICS Code 311 (NAICS 2012) are the following:

a. 3111 Animal food manufacturing
b. 3112 Grain and oil seed milling
c. 3113 Sugar and confectionary products manufacturing
d. 3114 Fruit and vegetable preserving and specialty food manufacturing
e. 3115 Dairy product manufacturing
f. 3116 Animal slaughtering and processing
g. 3117 Seafood product preparation and packaging
h. 3118 Bakeries and tortilla manufacturing
i. 3119 Other food manufacturing

Another important classification system is NACE, which is followed by the European Union, which categorizes the food processing business establishments into two major categories C10 (food products with 09 subcategories) and C11 (beverages with 01 subcategories) in the revised system (NACE 2008). In the context of the subject matter of this chapter, it is worth mentioning that data available from the database maintained based on the above classifications and similar systems, such as ISIC (ISIC 2008), NIC (NIC 2008), etc., followed by other nations, have been used by various earlier workers for evaluating the energy consumption pattern and efficiency (Wang 2009).

20.1.2 Processing Sector Specific Energy Consumption

In a food manufacturing industry, a major fraction of the energy demand is for carrying out the unit operations, which convert and/or render the food raw materials safe and better preserved. Recorded data on cost of energy and other components of manufacturing practices along with the value of shipment, maintained by the US Department of Commerce or similar agencies across various nations reveal the current trend of energy consumption in the food processing sector. Such data indicate the food manufacturing sector is one of the top energy-intensive manufacturing industries.

Perhaps it was not intended to be so, but the general observation is that foods that have undergone energy-intensive processing have become increasingly popular in global markets. For example, shipment value to consumer-preferred high value-added foods, such as prepared foods, nonalcoholic beverages, table spreads, confectionery products, and cereal and bakery products, has been added through energy-intensive manufacturing practices. The extent of industrialization and degree of urbanization influences the preference for processed foods, consequently, the share out of the total national energy consumption in the food processing sector. For example, a developed country such as the United States finds the food sector to be among the top eight energy-intensive industries, bearing a share of 9.57% (in 2006) of the energy consumption. The corresponding value for the European Union, combining the food and tobacco sectors, was 8% in 2001. Out of the total commercial energy, the share is as high as 30% in Thailand and on the order of more than 25% in Nepal, Burma, and Haiti, and a relatively lower share, on the order of 11% in nations such as India and Brazil. This is attributed to the fact that a lesser degree of urbanization is affecting a lower consumption of processed food (Wang 2009).

For making a comparison between data groups on energy consumption, such as between countries or across different sectors or between processes or among different kinds of materials, the normalized energy unit is used. For the purpose of energy auditing, a variety of energy measures are used. In the case of primary energy, a term used to describe the energy content of raw fuels, the measure is in joules or in the form of a quantity of the fuel. Official statistics on supply and demand of energy are mostly expressed in terms of primary energy. However, electrical energy input, which is expressed as kWh, is basically measured at the application site and is not considered to be a primary energy. For making any comparison of electrical energy with primary energy, fuels used to produce the electricity, the conversion efficiency of power stations and the transmission and distribution losses also need to be taken into account.

For a meaningful comparison, a normalized energy unit is used; normalization is done by dividing the energy quantum or energy cost by a common denominator. The ratio of energy consumption to process output, such as MJ per tons of finished product, is used to quantify the trend in energy consumption in specific sectors. For example, Ramirez et al. (2006b) have made a comparison of the percentage rise in energy consumption in the meat industries of France, Germany, The Netherlands, and the UK. In another instance, Ramirez et al. (2006a) have used the energy cost per unit price of shipment value as the energy indicator to compare the energy consumption in the food processing sector in respect to other manufacturing sectors in the United States for the period 2002–2006.

Based on the energy cost per dollar of shipment value, oil seed milling (NAIC-3112) and animal slaughtering and processing (NAIC-3116) were reported as the two largest energy-consuming sectors among food industries in the United States in 2006. Data obtained from the US Census Bureau, a similar evaluation for the year 2010 (Table 20.1) reveals that they are still the two largest energy-consuming food processing sectors with shares of 20.4% and 25.7%.

The focus of this chapter is to explore the possibilities for lowering the energy indicator (energy cost/price of shipment value) of food processing industries. The energy cost over the total energy consumed during the food manufacturing process includes energy consumed in a series of unit operations and energy required to facilitate carrying out the conversion operations. Therefore, before attempting to quantitatively estimate the possible extent of savings in energy, the nature of the food processing operations and kind of energy inputs required are discussed in two subsequent sections.

20.1.3 Energy-Consuming Food Processing Operations

For transforming the edible raw material into value-added products or products with better preservation quality and convenience for packaging and storage and to ensure them to be safe for consumption, the food processing industries need to carry out a range of unit operations. These unit operations destroy the microorganisms and enzymes with the potential of causing diseases and spoilage in food and change the characteristics of foods in a way that make them resistant against microbial decomposition or natural/environment-induced degradation. They create a barrier between food ingredients and spoiling agents; change the size, shape, and other physical characteristics to render them convenient to handle and transport, and change structural and micro-structural characteristics for the convenience of the consumer. Out of necessity, a few other operations are carried out in the food processing industry to facilitate processing foods in a hygienic manner, conveying materials as well as processing media between points, facilitating temporary storage of raw materials or storage of packaged products, etc.

Table 20.1 Sector-Wise Energy Use and Indicators in Food Manufacturing Sectors in the United States in 2010 and 2006

Manufacturing Sector and NAICS Code	*Year*	*Shipment Value (million US$)*	*Total Energy Cost (million US$)*	*Energy Indicator (% energy cost/$ shipment value)*	*Electricity Cost Among Total Energy Cost (%)*	*Total Energy Cost in the Whole Energy (%)*
3111 Animal food manufacturing	2010	46,884	580	1.24	55	5.3
	2006	33,988	522	1.54	47.1	5.3
3112 Grain and oil seed milling	2010	82,999	2229	2.69	45.9	20.4
	2006	57,667	2198	3.81	37	22.3
3113 Sugar, confectionary products manufacturing	2010	28,963	589	2.03	40.9	5.4
	2006	28,225	577	2.04	36.6	5.8
3114 Fruits and vegetables preserving and specialty food manufacturing	2010	64,331	1379	2.14	50.4	12.6
	2006	56,279	1302	2.31	44.3	13.1

3115 Dairy products manufacturing	2010	94,102	1245	1.32	56.9	11.4
	2006	75,425	1184	1.57	52.4	11.9
3116 Animal slaughtering and processing	2010	175,589	2807	1.6	58	25.7
	2006	149,577	2055	1.37	53.4	20.7
3117 Seafood product preparation and packaging	2010	10,681	209	1.96	39.6	1.9
	2006	10,849	246	2.27	45.5	2.5
3118 Bakeries and tortilla manufacturing	2010	58,888	935	1.59	62.8	8.6
	2006	54,173	911	1.68	54.7	9.2
3119 Miscellaneous food manufacturing	2010	84,015	942	1.12	54.2	8.6
	2006	71,602	926	1.29	52.9	9.3

While there are more than 150 unit operations related to food processing, based on the effects produced, these operations can be grouped into a few categories as given below:

a. Operations for mechanical transport of process materials, which include pumping of fluids, hydraulic conveying, pneumatic conveying, mechanical conveying of materials, etc.
b. Mechanical processing operations carried out on raw material to modify its size/shape/structure, such as mechanical peeling, decorticating, polishing, whitening, deseeding, coring, cutting, sizing, slicing, grinding, homogenizing, extrusion, forming, mixing, agglomeration, emulsification, etc.
c. Mechanical separation processes, such as screening, cleaning, washing, sorting, grading, filtration, centrifugation, mechanical pressing, etc.
d. Heat transfer operations to heat or cool the process material include operations, such as heating, thawing, blanching, cooking, frying, roasting, baking, pasteurization, sterilization, evaporation, cooling, chilling, freezing, etc.
e. Mass transfer operations to effect chemical activity-based separation, such as drying, extraction, distillation, absorption, adsorption, crystallization, etc.
f. Non-thermal processing operations, which include pulsed electric fields, high pressure processing, application of radiation, etc.
g. Membrane processes, such as reverse osmosis, ultra-filtration, etc.
h. Operations relating to packaging and storage of foods, such as filling, forming, sealing and seaming, aseptic packaging, refrigerated storage, frozen storage, etc.
i. Cleaning and sanitizing operations, including cleaning in place, etc.

In any of the food processing sectors, for the manufacturing of a processed product, a sequence of unit operations are carried out, depending on the process technology. These unit operations carried out with the help of conversion devices and the related facilities maintained to facilitate operating the conversion devices are the major consumers of energy in the food industry.

20.1.4 Processing Equipment as Energy-Consuming Units

During processing of foods in an industry, energy is an absolute necessity. The end users of energy in the food industry are process heating, process cooling and refrigeration, machine drive, and miscellaneous users. About half of all energy input is used to process raw materials into products. Fuels are mainly used for process heat and space heating while electricity is used for refrigeration, motor drives, and automation. For carrying out the unit operations along the processing line, energy is delivered in various forms, such as mechanical drive to run pumps, compressors, and other mechanical devices; heat sources, such as steam or heated fluid; direct heat; cooling sources, such cool air, refrigerating effects, or cryogenic fluids; etc. Mechanical operations require basically mechanical power, mostly derived through electrical motor drives (EM). Heating processes include cooking, frying, blanching, baking, roasting, pasteurization, thawing, etc. While baking and roasting require direct heating (DH), processes such as cooking, blanching, or frying require a heating medium and high-temperature heat in the form of heated water or steam (ST). Processes such as cooling, chilling, or freezing require refrigeration (HV), and they consumes a significant amount of electrical energy. Drying basically depends on fossil fuels through direct heating (DH) of air. Table 20.2 lists the groups and forms of energy required by some of the unit operations.

Graphical elaboration of the group of processing operations leading to separation is shown in Figure 20.1.

Table 20.2 Group of Unit Operations and End Use Energy Form

Group of Operations	*Unit Operation*	*Form of Energy*
Mechanical transport	Pumping of fluids,	EM
	Mechanical conveying	EM
	Hydraulic conveying	EM, PW
	Pneumatic conveying	EM, CA
Processing operations leading to separation	Mechanical peeling/decorticating	EM
	Deseeding/coring	EM
	Sorting/grading	EM
	Polishing/whitening	EM
	Mechanical pressing	EM
	Filtration/ultra-filtration	EM
	Reverse osmosis	EM
	Washing	EM, PW
	Centrifugation	EM, CA/PW
	Cleaning	EM, CA
	Screening	EM, CA (optional)
Processing operations leading to size reduction	Cutting/sizing/slicing	EM
	Size reduction (grinding)	EM
	Homogenization	EM

(*continued*)

Table 20.2 (Continued) Group of Unit Operations and End Use Energy Form

Group of Operations	*Unit Operation*	*Form of Energy*
Processing operations leading to mixing	Mixing/kneading	EM
	Emulsification	EM
	Agglomeration	EM
	Extrusion/forming	EM, ST
Operations requiring heating	Heating, thawing	ST, DH
	Blanching	ST
	Cooking/frying	ST, DH
	Roasting/baking	DH
	Sterilization, pasteurization	ST, DH
	Evaporation	ST
	Drying	ST, DH,CA
Operations requiring cooling	Cooling	HV
	Chilling	HV
	Freezing	HV
Mass transfer operations	Extraction	ST, EM, HV
	Distillation	ST, EM, HV
	Absorption	ST, EM, HV
	Adsorption	ST, EM, HV
	Crystallization (from solution)	ST, EM, HV
Non-thermal preservation operations	Irradiation	RW
	Pulsed electric fields	RW
	High pressure processing	EM
Cleaning and sanitation operations	Cleaning and sanitization	ST, CA, PW
	CIP system	ST, PW
Packaging and storage operations	Filling and forming	EM
	Sealing and seaming	EM, ST
	Aseptic packaging	ST, DH, EM, CA
	Refrigerated storage	HV
	Frozen storage	HV

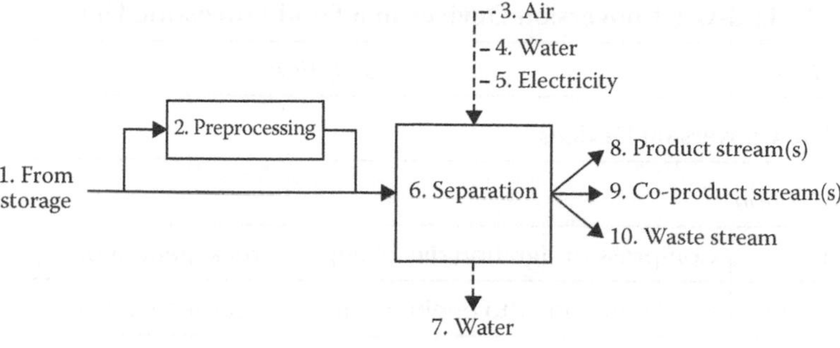

Figure 20.1 Separation process in food processing, reproduced from *Energy-Related Best Practices: A Sourcebook for the Food Industry*, Iowa State University Extension Program (ISU 2005).

20.2 Thermodynamic Evaluation of Energy Conservation Opportunities

20.2.1 Energy Chain from Fuel to Food Machineries

Energy data presented in Table 20.1 is basically consumption of primary energy in transforming the food material to a processed product, part of which has been used in converting the raw materials, part are lost in thermodynamic sense, and some are lost due to operational reasons. To explore the conservation opportunities, there is a need to study the energy chain as applicable in food processing.

From an engineering perspective, and for evaluating the energy conservation opportunities, the energy chain from fuel to service is viewed to be comprised of two major transformations: (a) conversion of energy sources into refined energy, i.e., electricity or refined fuels, and (b) conversion of this refined energy into final services. Examples of the first transformation includes burning fossil fuel to produce electricity or refining crude oil into gasoline or diesel, etc.; principles of this transformation are well understood and are dealt with by engineers specialized for the task. The second transformation involves initially converting the refined energy by some end-use device into a useful form (mainly heat or motion) to drive the activity of an engineering system (a grinder or cold room) and is called end-use conversion. The useful energy is then transported to a service site and delivered as a service (grinding effect in powdered spice or chilling effect in chilled beer, etc.).

The last engineering system is termed a *passive system*, and it refers to a system into which useful energy (in the form of heat, motion, light, cooling, or sound) is delivered and whereupon the useful energy is "lost" as low-grade heat. Whereas in *conversion devices*, input energy is converted into another useful form. Examples

Table 20.3 End-Use Conversion Devices in a Food Processing Unit

Device Name	*Description*
A. End-Use Conversion Devices	
A.1. For Motion	
Diesel engine	Compression-ignition diesel engine: truck, generator
Petrol engine	Spark-ignition Otto engine: generator, garden machinery
Other engine	Steam or natural gas-powered engine
Electric motor	AC/DC induction motor (excludes refrigeration)
A.2. For Heat	
Oil burner	Oil combustion device: boiler, biochemical reactor
Biomass burner	Wood/biomass combustion device: open fire, stove, boiler
Gas burner	Gas combustion device: open fire, stove, boiler, chemical reactor
Coal burner	Coal combustion device: open fire, stove, boiler, chemical reactor
Electric heater	Electric resistance heater, electric arc furnace
Heat exchanger	Direct heat application: district heat, heat from CHP
A.3. Other	
Cooler	Refrigeration, air conditioning, for storage/cooling application
Light device	Lighting: tungsten, fluorescent, halogen
Electronic	Computers, portable devices
B. In-Plant Energy Distribution Systems	
B.1. For Motion	
Diesel engine	Compression-ignition diesel engine: truck, generator
Petrol engine	Spark-ignition Otto engine: generator, machinery
C. Passive Energy Systems	
C.1. For Motion	
Truck/trolley	

(continued)

Table 20.3 (Continued) End-Use Conversion Devices in a Food Processing Unit

Device Name	*Description*
C.2. Factory	
Driven system	Refrigerator, air compressor, conveyor, pump
Steam system	Medium-temperature application: reaction vessel, cleaning facility
Furnace	High-temperature application: furnace, oven
Hot water system	Fuel and electric immersion boilers
Electric heater	Electric resistance heater, electric arc furnace
Heat exchanger	Direct heat application: district heat, heat from CHP
C.3. Building	
Heated/cooled space	Milk receiving tank, cold storage, CA storage
Appliance	Refrigerator, cooker, washer, dryer, dishwasher, electronic devices
Illuminated space	Commercial indoor space, processing site, outdoor space

Source: Cullen, J.M. and Allwood, J.M., *Energy Policy*, 38, 1, 75–81, 2010.

of passive systems include a cold room (excluding the refrigeration unit), which delivers cold effect to the stored food and is a service to the consumer or food preserving technician. In a household refrigerator, the rotational energy from the electric motor is used to produce cooling, and the refrigeration unit is defined as the conversion device and the insulated cold-box as the passive system.

In line with the above divisions, along the chain of flow of energy from fuel to service, five different efficiency terms are defined: (a) the extractive efficiency (E1) of conversion of fuel in the ground or extraction of power from wind or from sun in the atmosphere, etc., into the primary energy to be fed into the initial conversion device; (b) the conversion efficiency (E2) of primary energy into secondary energy; (c) the distribution efficiency (E3) of delivering the secondary energy from the point of conversion to the point of end use; (d) the end-use efficiency (E4) of converting the delivered secondary energy into desired energy services, such as a baked effect to a cake or chilling effect to a can of beer, etc.; and finally, (e) the hedonic efficiency (E5), i.e., efficiency of transformation of delivered energy services into human welfare (Cullen and Allwood 2010a, b).

Although it is theoretically possible to distinguish five divisions, in attributing the energy losses in plant operation losses occurring during the end-conversion processes, during the transportation of useful energy, and losses from passive systems are represented as embedded, and evaluated efficiency will indicate the product of three efficiency terms, i.e., (E2 × E3 × E4). For the purpose of identifying the potential opportunities for improving energy efficiency, connections between these technical categories should be mapped to enable thermodynamic analysis at these three divisions. The technical components under these three divisions are given in the Table 20.3 (Cullen and Allwood 2010a, b).

20.2.2 Examples: Conversion Stages in Thermal Processing

For implementation of the knowhow of taking an energy chain from fuel to service to explore the energy conservation opportunities, an example of a food processing operation—the batch milk pasteurization process in a steam-jacketed vat using steam from a biomass-fired boiler—is presented. The five conversions are denoted as A, B, C, D, and E.

A. In the energy chain, the first conversion involves production of biomass-based palettes from biomass.
B. The second conversion is the combustion of palettes in the boiler furnace to produce vacuum steam.
C. For making the vacuum steam usable, it has to be carried to the batch pasteurizer and circulated through the heating section of the pasteurizer.
D. A jacketed vessel with appropriate insulation and cover is required to provide the space for the process stream, i.e., milk, to be kept for heating.
E. A certain amount of energy gets transferred to the milk, making it a value-added product for use by the consumer.

From the perspective of food engineering, there are energy conservation opportunities to be explored in each of above five tasks. However, because most of time, the first operation (operation A) is carried out outside of a food processing plant, a food engineer has hardly an opportunity to intervene. Among the rest of the operations, the following considerations are to be considered.

During operation B,

i. Reducing energy loss occurs during internal heat exchange during combustion. The internal heat exchange between product molecules leaving the reaction site (carrying energy) and neighboring un-reacted molecules leads to unrecoverable loss. Such a loss can be avoided if the reactant and product streams are separated.

ii. Reducing energy loss occurring due to heat transfer from the furnace to the environmental reference state can be minimized using insulation, preventing leaks of hot gas and liquids, and ensuring reactants and products leave the system at the surrounding temperature.
iii. Reducing heat loss occurring due to escaping of water vapor from combustion site. Extracting heat from the water vapor (condensing boilers) and completely oxidizing fuel can prevent some loss.

During operation C,

i. Reduce other losses, such as friction: Friction (sliding and fluid flow), inelastic deformation, and unrestrained compression/expansion lead to nonrecoverable energy losses (e.g., in pump and pipe). Losses are reduced by reducing fluid flow velocities and resisting expansion of gases.

During operation D,

i. Reduce heat loss in the condensate produced, which is still at a temperature higher than the surrounding temperature. Throwing away the condensate would cause loss, and loss can be minimized by reusing it as feed water for the boiler.
ii. Reducing energy loss occurring due to heat transfer from a jacketed vessel to the environmental reference state can be minimized using insulation.

During operation E,

i. The supply of heat to the raw material is necessitated by the consumer's specification of sensory quality and food safety. Carrying out processing operations just enough to meet the requirement or evolving an energy-efficient alternative to give a similar satisfaction level are opportunities to save energy.

Besides this quantitative loss in energy, there are qualitative losses in energy resulting from thermodynamic irreversibility. They include the following:

During operation B,

i. Chemical interactions during combustion, i.e., the reaction of oxygen and fuel, producing irreversible changes of energy.
ii. Spontaneous mixing of reactants in the pre-combustion stage and products in the post-combustion stage cannot be reversed without additional energy input.
iii. During the process of heating, heat transfer occurs through a finite temperature (e.g., from combustion gases to steam) difference producing irreversibility. Minimizing the temperature difference reduces losses but increases the heat exchanger costs.

During operation D,

i. During the process of heating milk by heat transfer from steam, heat transfer occurs through a finite temperature, leading to irreversible loss of energy. Minimizing the temperature difference reduces losses but increases the heat exchanger costs.

20.2.2.1 Other Issues

During operation A,

i. The total energy bill, in the form of purchase of refined fuel, can be reduced by the use of energy from wastes.

During operation D,

i. Innovative processes intervention for replacing batch pasteurization

Based on this discussion, the following options can be identified:

i. Supplementing purchased nonrenewable energy with renewable energies
ii. Improving the efficiency of delivery at the service site, i.e., improvement in (E2 × E3 × E4)
iii. Loss minimization at transportation of energy and trapping waste heat, i.e., improvement in (E3 × E4)
iv. Improving (altering) the pathway for given service requirement, i.e., improvement in E4
v. Improving (altering) the service requirement without affecting hedonic value, i.e., improvement in E5

While the possible extent of raising the end-use efficiency (E4) is frequently estimated on benchmark energy-consumption data based on globally accepted norms, the opportunities for improving E2 and E3, individually or in combination, are amenable to be evaluated in thermodynamic sense. The conservation limit for conversion systems can be theoretically computed based on exergy efficiency. Exergy shows how far each device is operating from its thermodynamic ideal, allowing all energy conversion devices to be compared on an equivalent basis. Thus, whenever exergy models exist for individual conversion devices the target will be to bring exergy efficiency as close as possible to one. The following subsection introduces the principle of estimating energy efficiency for various food processing operations.

20.2.3 Thermodynamic Measure of Energy Quality: Exergy

When energy is described based on the second law of thermodynamics, quality measure of the energy becomes relevant. Based on the second law, 1 kJ of energy drawn from a thermal energy reservoir at 900 K should be attributed with more value than 1 kJ of energy drawn from a thermal energy reservoir at 600 K because while delivering the heat to a low-temperature thermal energy reservoir at 300 K, the former can give more work output. Thus, energy at a higher temperature is more valuable than energy at a low temperature.

Further, work is considered to be a higher quality form of energy than heat because its 100% conversion is possible whereas, as per the second law of thermodynamics, all heat cannot be converted to work. Work equals heat at infinite temperature. Similarly, electricity has been given a similarly high value, which, for practical purposes, is interchangeable with work.

Creation of lower-quality energy from higher-quality energy is a spontaneous process. However, for creating higher-quality energy from lower-quality energy for use in industrial processes, significant loss is incurred. For example, to produce 1 kWh of electrical energy, approximately 10.8 MJ of fuel energy has to be spent. Thus, while certain applications require a supply of energy for process heating, an energy form considered to be of lower quality, higher-quality energy, such as electrical heating should not be used for direct heating. Quantitatively, for example, to supply 3.6 MJ of process heat, if it consumes 1 kWh of electrical energy, in fact, 10.8 MJ of fuel energy has been spent. A better alternative is to supply the process heat directly from fuel without converting it to higher-quality energy. The other alternative is, to use the electrical energy to run a heat pump, which will deliver the process heat.

In terms of quality of energy, consumption of energy during the food processing operation can be considered as conversion of high-quality energy to a lower-quality energy form, meaning less work is available for any subsequent process. The degree of degradation is expressed as an increase in entropy, which means, when a process leads to a higher value of entropy change, the energy form is undergoing degradation, and its ability to produce useful work is diminishing. In other words, energy quality conservation will be indicated by minimization of entropy generation.

Entropy is defined as a measure of the disorder or randomness of a system and is an extensive state variable (proportional to the size of the system) that is definable for any material substance or any system and measured in Joules per Kelvin. Entropy is useful for defining the minimum theoretical energy requirement for a process, significance of which is described in the next section.

Example 20.1

For the requirement of process heat of 1 kW at 277 K for heating water to 300 K from available thermal energy from saturated steam at 600 K and saturated steam at 310 K, which one would you use? Will an electrical heater be more useful?

Solution

1 kW of heat from steam at 600 K, when allowed to deliver heat to a sink at 277 K can produce a work output of

$$W_1 = 1\left(1 - \frac{277}{600}\right) = 0.538\ \text{kW}$$

1 kW of heat from steam at 310 K, when allowed to deliver heat to a sink at 277 K can produce a work output of

$$W_2 = 1\left(1 - \frac{277}{310}\right) = 0.106\ \text{kW}$$

Because heat from steam at 600 K can deliver more work output than heat from steam at 310 K, in other words, steam at 600 K has a higher quality than steam at 310 K. Because the requirement is process heat, a lower quality energy, therefore, steam at 310 K should be used.

Similarly, to produce 1 kW of electrical energy, approximately 3 kW of low-quality energy was spent. To directly use 1 kW of electrical energy to generate 1 kW low-quality process heat would mean that, effectively, 3 kW of energy is spent.

In this example, electrical energy has the highest quality (1 kW can produce 1 kW work), followed by heat from steam at 600 K (1 kW can produce 0.538 kW work) and then followed by heat from steam at 310 K (1 kW can produce 0.108 kW work output).

20.2.3.1 Exergy Efficiency

Discussions on the quality of energy in Section 20.2.3 highlighted that higher-quality energy can be completely converted to useful work while lower-quality energy can only be partially converted to useful work. Based on this realization, a generalized statement for all energy forms can be made that the work equivalent of an energy form has two fractions: One fraction is available for conversion as useful work, termed as available energy, and the other fraction is not available for conversion as useful work, termed as non-available energy. The useful part of energy or the potential work that can be extracted from a system by reversible processes as the system equilibrates with its surroundings is termed as system exergy. The remaining fraction of energy is denoted as anergy. Thus mathematically,

$$\text{Energy} = \text{exergy} + \text{anergy} \tag{20.1}$$

Higher-quality energy forms such as electrical energy or potential energy are considered to be consisting of exergy only. Heat drawn from a thermal energy reservoir at T K, which is at a temperature higher than the surrounding T K, will have both exergy and anergy. In this case, the exergy fraction can be computed from Carnot's theorem as (Rotstein 1986)

$$W/_Q = \left(1 - \frac{T_0}{T}\right) \tag{20.2}$$

Heat drawn from a thermal energy reservoir that is at a temperature equal to its surroundings will have anergy only. When energy is consumed by a food processing operation, exergy of the energy form decreases whereas anergy increases.

Exergy is more useful than entropy as it measures both resource quantity and quality, and it is useful for aggregating heterogeneous energy sources and materials. Exergy uses mechanical work, the highest quality and lowest entropy form of energy, as its measure with joules as its unit.

Exergy has four components: kinetic, potential, physical (pressure or temperature), and chemical. For most energy conversion processes, only the chemical component of exergy is significant. Chemical exergy measures the available work, normally referenced relative to the standard condition of temperature 298 K and pressure 101.325 kPa. Accordingly, the tabulated values of standard chemical exergy per mole (B) as defined in reference to this standard state can be prepared. Physical exergy is the work obtainable by taking the substance through a reversible process from its initial state to the reference state. The physical exergy may have one thermal (temperature) and one mechanical (pressure) component.

20.2.3.2 Application of Exergy Analysis

Thermodynamics permits describing the efficiency of conversion of energy from one form to another. When such a performance is measured based on the first law of thermodynamics, it takes account of the quantity of energy streams, however, it will not consider how ideally the system has accomplished the conversion. Based on the first law, efficiency of a natural gas power plant can have 40% efficiency, an electric motor can be 96% efficient, and an air conditioner can have a coefficient of performance (COP) of 2.0. However, this measure of efficiency without the consideration of the quality of energy may give a wrong representation of possible improvements while, for example, in space heating applications, an electric heating system is 100% efficient. It still has scope for improvement by the use of a heat pump, which can have a COP of 3 (equivalent to an efficiency of 300%). In contrast, exergy efficiency, which uses mechanical work rather than energy as the basis for comparing devices, provides a more equitable measure of conversion efficiency.

Exergy efficiency is defined for a device as (Tagardh 1986)

$$\text{Exergetic efficiency} = \frac{\text{actual exergy exchange rate}}{\text{maxium possible exergy exchange rate}} \tag{20.3}$$

System exergy is defined as

$$B_{sys} = U + p_o V - T_o S \tag{20.4}$$

And stream exergy is defined as

$$B_q = H_q - T_o S_q \tag{20.5}$$

For evaluating the steam exergy or system exergy, it is essential to know enthalpy values and entropy values. For various utilities in food processing and for well-defined chemical species, tabulated property values are available. Thus, for a utility, such as steam, for which tabulated data are available, exergy can be computed easily. In certain cases, ideal gas exergies, calculated based on following equation are used.

$$\Delta B = \int_{T_1}^{T_2} c_p \, dT - T_0 \int_{T_1}^{T_2} \frac{c_p}{T} dT - R \ln\left(\frac{p_2}{p_1}\right) \tag{20.6}$$

For the computation of the exergy of solutions in the liquid phase, the exergy change of mixing is conveniently used. For real solutions,

$$B_{sol} = \sum x_i B_i^o - RT \sum x_i \left[\frac{\partial \ln a_i}{\partial \ln T}\right]_{p,x} - RT_o \sum x_i \ln a_i + \sum x_i \left[\frac{\partial \ln a_i}{\partial \ln T}\right]_{p,x} \tag{20.7}$$

For ideal solutions, the equation simplifies to

$$B_{sol} = \sum x_i B_i^o + RT_o \sum x_i \ln x_i \tag{20.8}$$

Most simplified expressions of the estimation of exergy, occur for the case of constant specific heat and negligible residual exergies over the temperature range under consideration. In this case,

$$\Delta B = C_p \left[(T - T_\infty) - T_\infty \ln \frac{T}{T_\infty} \right] \tag{20.9}$$

For using such an expression for foods, specific heat is approximated based on the composition of foods. In the case of moist air, the exergy change is computed from

$$\Delta B = \Delta H - T_o \Delta S \tag{20.10}$$

where the basis is the unit weight of moist air.

20.3 Energy Conservation in Food Plant Operation

The energy saving opportunities in a food processing operation include three aspects: (i) improvement of energy efficiency in existing units, (ii) replacement of energy-intensive units with novel units, and (iii) use of renewable energy sources, particularly food processing wastes (Wang 2009). The thermodynamic analyses based on exergy as a diagnostic tool addresses the aspects of improving energy efficiency within practical limits through optimized operating conditions and retrofitting. In practice, besides the thermodynamic considerations, user behavioral aspects, housekeeping practices, operations management considerations, and energy management practices are combined to evolve energy conservation programs at the food processing industry.

Wang (2009) has summarized the estimates from published studies and states that around 57% of industry's primary energy inputs are lost or diverted before reaching the intended process activities. Discussions presented in Section 20.2.2 highlight the inefficiency terms associated with this loss in the energy chain from fuel to food machineries. A strategic approach to energy management can result in significant energy savings in the energy chain as well as minimizing waste or misuse of energy at the service site, thereby reducing the energy requirement. Reported studies indicate that (i) through procedural and behavioral changes only, and without capital investment, a savings of around 20% to 30% is achievable, and (ii) through a well-structured energy management program, which combines technology, operations management practices, and energy management systems, 10% to 20% reduction in energy consumption is achievable.

Implementing energy conservation practices in a food processing facility with the thrust to savings of energy in the energy chain as well as reducing the waste or misuse of energy at the application site requires coordination among various levels within the facility. The "Energy Star" program developed jointly by the US Department of Energy and the Environmental Protection Agency (EPA) offers an energy star guide for energy and plant managers for various food products, such as fruits and vegetable products, bakery products, dairy products, corn refining processes, etc. These energy guides recommend multilevel coordinated efforts for energy-efficient plant operation. At the component and equipment level, efforts can be made to improve energy efficiency through regular preventative maintenance, proper loading and operation, and replacement of older components and equipment with efficient alternatives whenever feasible. At the process level, production operations can be ensured to run at maximum efficiency through process control and optimization. At the facility level, the efficiency of space lighting, cooling, and heating can be improved while total facility energy inputs can be minimized through process integration and combined heat and power systems wherever feasible. Above all, at the organization level, energy management systems can be implemented with a strong corporate framework for energy monitoring, target setting, employee involvement, and pursuit for continuous improvement.

Forms of energy required by the end-user devices of a food industry are highlighted in Section 20.1.4. They are a steam system (ST) for process heating; a machine drive, motor system (EM), and pump for process stream/medium circulation; a refrigeration system (HV) for processes like cooling, chilling, or freezing; and compressed air (CA) for cleaning, separation, or transportation processes. The building part of the food processing facility housing the commercial indoor spaces, storage spaces, processing sites, and outdoor spaces acts as a passive energy user. Based on the EPA guidelines, a brief discussion on the energy-efficiency improvement practices and cost-saving opportunities in these four systems is presented in this section.

20.3.1 Energy-Efficiency Measures for Steam System

Steam is required in many important unit operations in a food processing facility, such as heating applications, cleaning and sanitation, mass transfer operations, etc., as indicated in Table 20.2. A steam system comprises primarily a steam generation system, i.e., a boiler, and a distribution system, which carries the process heat. For improving boiler energy efficiency, the focus is primarily on reducing the heat loss and enhancing heat recovery and operating it with improved process control. It is recommended, that from the stage of installation of new boiler system, energy-efficient practices can be implemented by using a custom-designed configuration to meet the needs of a particular plant. A predesigned boiler cannot be fine-tuned to meet the steam generation and distribution system requirements unique to a given plant in the most efficient manner.

The second component of the steam system, i.e., the process heat distribution system may involve steam and/or water to carry the heat from the boiler to the application site. It is potentially a major contributor to energy losses within a food processing facility, which can be improved by reducing heat losses throughout the system and recovering useful heat from the system wherever feasible. The measures tabulated in Table 20.4 are some of the most significant opportunities for saving energy in industrial steam distribution systems (Masanet et al. 2008, 2012).

Because food processing facilities have multiple heating and cooling applications, the use of process integration techniques, such as recovering heat rejected in a cooling process for use in process heating applications (Das 2000) may significantly improve facility energy efficiency. Linking hot and cold process streams in a thermodynamically optimal manner for improving energy efficiency by the process integration refers to the exploitation of potential synergies that might exist in systems that consist of multiple components working simultaneously.

Table 20.4 Summary of Energy-Efficiency Measures for Steam System

(a) Efficiency Measures for Boiler	*(b) Efficiency Measures for Distribution System*
1. Boiler process control	1. Improved distribution system insulation
2. Reduction of excess air	2. Insulation maintenance
3. Blow down steam recovery	3. Steam trap improvement
4. Improved boiler insulation	4. Steam trap maintenance
5. Reduction of flue gas quantities	5. Flash steam recovery
6. Properly sized boiler systems	6. Steam trap monitoring
7. Flue gas heat recovery	7. Leak repair
8. Direct contact water heating	
9. Condensate return	
10. Boiler maintenance	
11. Boiler replacement	

Source: Masanet, E. et al., *Energy Efficiency Improvement and Cost Saving Opportunities for the Pharmaceutical Industry: An ENERGY STAR Guide for Energy and Plant Managers,* Lawrence Berkeley National Laboratory, Berkeley, CA, Report LBNL-59289-Revision, 2008.

20.3.2 Energy-Efficiency Measures for Motor Systems and Pumps

Motors are used throughout food processing facilities to drive process equipment (e.g., for mixing, peeling, cutting, pulping, filling, and packaging), conveyors, ventilation fans, compressors, and pumps as indicated in Table 20.3. The magnitude of savings in electricity consumption through improving the efficiency of motor-driven systems has been estimated to be around 5% to 15% for the manufacturing industries (US DOE 2006).

Pumps, one of the pieces of motor-driven equipment used in food processing plants, are used extensively to pressurize and transport water in cleaning, water fluming, and wastewater handling operations, for transporting liquid food streams between processes, and for circulating liquid food streams within the processes themselves. Proper management of the pumping equipment and/or pump control systems can lead to an energy savings of as much as 20% (US DOE 2002a).

A systems approach may be followed for improving a facility's motor systems' energy efficiency wherein the energy efficiency of entire motor systems (i.e., motors; drives; driven equipment, such as pumps, fans, and compressors; and controls) is optimized. This approach analyzes both the energy supply and energy demand sides of motor systems as well as how these sides interact to optimize total system

Table 20.5 Summary of Energy-Efficiency Measures for Motor Systems and Pumps

(a) Efficiency Measures for Motor Systems	*(b) Efficiency Measures for Pumps*
1. Motor management plan	1. Pump system maintenance
2. Maintenance	2. Pump system monitoring
3. Adjustable-speed drives	3. Pump demand reduction
4. Minimizing voltage unbalances	4. Controls
5. Strategic motor selection	5. Properly sized pumps
6. Properly sized motors	6. High-efficiency pumps
7. Power factor correction	7. Multiple pumps for variable loads
	8. Impeller trimming
	9. Avoiding throttling valves
	10. Replacement of belt drives
	11. Proper pipe sizing
	12. Adjustable-speed drives

Source: Masanet, E. et al., *Energy Efficiency Improvement and Cost Saving Opportunities for the Pharmaceutical Industry: An ENERGY STAR Guide for Energy and Plant Managers*, Lawrence Berkeley National Laboratory, Berkeley, CA, Report LBNL-59289-Revision, 2008.

performance, which includes not only energy use but also system uptime and productivity.

The measures tabulated in Table 20.5 highlight that the appropriate selection of motor systems and pumps match the requirements of an efficient design as well as adequate monitoring and control of the systems, which constitute the energy management practices for motor systems and pumps of a food processing facility.

20.3.3 Energy-Efficiency Measures for Refrigeration Systems

In food processing facilities, the necessity of refrigeration systems is justified for the generation of chilled water for various process cooling applications (e.g., the cooling stage in pasteurization processes) in generation of cold air for food pre-cooling and food-freezing applications and space cooling for low-temperature storage applications.

Vapor compression refrigeration systems are most widely used in food processing faculties. Ammonia is the commonly used refrigerant as it possesses some favorable properties, such as its high latent heat of vaporization, its classification as a non-ozone–depleting substance, the fact that it is noncorrosive to iron and steel, and because ammonia leaks can often be easily detected by smell (Singh and Heldman 2009).

Four primary components of the vapor compression refrigeration system are the compressor, condenser, expansion valve, and the evaporator, through which the refrigerant is circulated in a closed loop. Energy for circulating the refrigerant is provided through the compressor whereupon the cooling effect is produced by the evaporator. The condenser is supposed to give up thermal energy, which is the sum of the compressor work input and the evaporator heat input.

Some of the most significant energy-efficiency measures available for industrial refrigeration systems can be grouped under the following four major categories, based on their applicability: (1) refrigeration system management, (2) cooling load reduction, (3) compressors, and (4) condensers and evaporators.

Good housekeeping practices attempt to reduce the heat load added by electrical appliances as well as unhindered distribution of cooling effects. On the other hand, in a large retrofit or new system installation, an efficient piping design would minimize friction and pressure drops, thereby reducing energy losses in the system (Pearson 2003). As a general practice, checking the refrigeration system for refrigerant contamination or monitoring the system for causes of deterioration in system performance and timely repair constitute refrigeration system management for energy conservation.

Pipes conveying cold refrigerant (i.e., pipes between the expansion valve and evaporator) should be properly insulated to minimize heat infiltration. Galitsky et al. (2008) estimated the typical energy savings attributable to improved piping insulation at 3% with a payback period of less than two years. Reducing heat infiltration into the cold storage areas, switching off the heat-generating equipment when not needed, adequately ventilated compressors, removal of surface water before freezing, free cooling, hydro-cooling, geothermal cooling, etc., are some of the practices that would minimize the cooling load of the refrigeration system.

The compressor is the workhorse of the refrigeration system, and proper management and maintenance of the compressor often leads to a sound strategy for energy efficiency. Adjustable-speed drives used in conjunction with control systems to better match compressor loads to system cooling requirements improves the better energy efficiency. Practices such as raising system suction pressure, indirect lubricant cooling, and compressor heat recovery can be considered.

The two heat-exchanging units, i.e., the condensers and evaporators affect the thermodynamic performance of the refrigeration system. In general, a 1°C increase in condensing temperature, which may result from faulty operation, will increase operating costs by 2%–4% (EEBPP 2000). Keeping the condenser's surface clean from dirt, ice buildup, a fouling layer, scaling, etc., are considered to be effective management practices. Use of adjustable-speed drives (ASDs) on condenser fans, evaporator fans, use of axial condenser fans in air-cooled or evaporative condensers, and reducing condenser fan use, defrosting based on necessity and preferably with water instead of air are some of the practices to lower energy requirement.

20.3.4 Energy-Efficiency Measures for Compressed Air Systems

It is recommended that if compressed air is used, it should be of minimum quantity for the shortest possible time because of its low efficiency, which is only around 10% (US DOE and CAC 2003). Not only should it be constantly monitored for energy efficiency, but the quest for a better alternative should always be kept in mind.

Energy savings from compressed air system improvements can range from 20% to 50% of total system electricity consumption (US DOE 2002b). Common energy-efficiency measures for industrial compressed air systems include system improvement, improved load management, pressure drop minimization, turning off unnecessary compressed air, properly sized pipe diameters, replacement of compressed air by other sources, leak reduction, heat recovery, monitoring inlet air temperature reduction, controls, modification of system in lieu of increased operational pressure, etc.

For improving energy efficiency, natural gas engine-driven compressors can be considered as an alternative to replace electric compressors with some advantages and disadvantages: They are more expensive and can have higher maintenance costs but may have lower overall operating costs, depending on the relative costs of electricity and gas.

20.3.5 Building Energy-Efficiency Measures

Lighting systems and HVAC systems are significant consumers of electricity at many food processing facilities, together accounting for anywhere from 10% to 25% of total electricity use. Additionally, HVAC systems are expected to consume around 5% of total facility natural gas use. The energy-efficiency measures to be considered are valid for most workspaces within a typical processing facility, including manufacturing areas, offices, laboratory spaces, and warehouses (Table 20.6).

20.4 Process-Level Energy-Efficiency Measures

At the process level, production operations are amenable to thermal engineering analysis and operations research formulation for selecting optimized operating conditions, and through process control, it can be ensured that it runs at maximum efficiency. Based on engineering considerations and management decisions, the EPA recommends process-level energy-efficiency measures for various food processing operations and has listed for quite a number of unit operations. In this section, energy-efficiency measures for four commonly used unit operations, i.e., blanching, drying, evaporation, and thermal processing are presented.

20.4.1 Energy-Efficiency Measures for Blanching

On the implementation front, blanching involves heating the food material to a specified product temperature, which is accomplished by exposing the material to

Table 20.6 Summary of Building Energy-Efficiency Measures

(a) Efficiency Measures for HVAC Systems	*(b) Lighting Energy-Efficiency Measures*
1. Energy-efficient system design	1. Turning off lights in unoccupied areas
2. Fan modification	2. Replacement of mercury lights
3. Re-commissioning	3. Lighting controls
4. Efficient exhaust fans	4. High-intensity discharge voltage reduction
5. Energy monitoring and control systems	5. Exit signs
6. Use of ventilation fans	6. High-intensity fluorescent lights
7. Nonproduction hours set-back temperatures	7. Electronic ballasts
8. Cooling water recovery	8. Daylighting
8. Duct leakage repair	9. Replacement of T-12 tubes with T-8 tubes
10. Solar air heating	
11. Variable-air-volume systems	
12. Building reflection	
13. Adjustable-speed drives	
14. Low-emittance windows	
15. Heat recovery systems	

Source: Masanet, E. et al., *Energy Efficiency Improvement and Cost Saving Opportunities for the Pharmaceutical Industry: An ENERGY STAR Guide for Energy and Plant Managers*, Lawrence Berkeley National Laboratory, Berkeley, CA, Report LBNL-59289-Revision, 2008.

a heating medium. Based on considerations of improved alternative, reduced exposure time, and heat recovery, certain measures as given below can improve the energy efficiency.

20.4.1.1 Upgrading of Steam Blanchers

Modern steam blanchers are designed for better retention of heat, efficient distribution of heat in the product stream, and reduced steam losses. These blanchers offer better energy efficiency by (i) minimizing steam leakage at the entrance and exit of the blancher by the use of steam seals; (ii) minimizing heat losses through adequate insulation of the steam chamber; (iii) reducing the blanching time through forced convection of steam; (iv) optimizing the flow of steam by properly designed process controls; and (v) recovering the condensate (Rumsey 1986a; FMCITT 1997; FIRE 2005).

20.4.1.2 Intermittent Application of Heat (Heat-and-Hold Technique)

In conventional blanching operation, the supply of a heating medium is continued until the core of the product attains the desired temperature. To facilitate transfer of heat from the surface to the core, essentially there exists a temperature gradient toward the core from the surface, implying the need of a higher surface temperature. Steam requirement to achieve the target core temperature reduces if the same is accomplished with a lesser gradient between the surface and the core. It is possible by using the heat-and-hold technique, where, after raising the product surface temperature to a desired level in the heating section, the material is maintained with no supply of additional heating medium, known as adiabatic holding. During this process, with the spontaneous attempt of the system of attaining a thermal equilibrium, heat from the surface diffuses toward the core and thereby raises the inner temperature of the product even without the use of additional steam. Blanchers employing the heat-and-hold technique are reported to reduce blanching energy intensity by up to 50% and blanching time by up to 60% (Rumsey 1986a; FIRE 2005).

20.4.1.3 Heat Recovery from Blanching Water or Condensate

Steam blanchers produce condensate during the process of blanching. In case internal recirculation of condensate is not practiced, then heat from the exiting hot condensate may be recovered with the help of a heat exchanger for using in preheating of boiler water or preheating of equipment cleaning water. Similarly, for hot water blanchers, heat exchanger may be employed to recover heat from the discharge water. In both cases, it is necessary to control the fouling of the heat exchanger employed in extracting heat from the exiting stream for application in preheating of feed water or cleaning water (Lund 1986).

20.4.1.4 Steam Recirculation

In case of steam blanchers, some of the steam may not condense on the product at the first pass. This uncondensed part of steam may be recirculated to affect multiple pass of steam, thereby reducing the steam inputs into the blanching chamber (Masanet et al. 2008).

20.4.2 Energy-Efficiency Measures for Drying and Dehydration

20.4.2.1 Maintenance

Improper maintenance of drying and dehydrating equipment may result in raised energy consumption, which can be as high as 10% (ISU 2005). Regular checking and remedial follow-ups for various components of the drying and dehydration system, such as burner and combustion chamber, heat exchangers, air filter, belt and fan, temperature and humidity sensors, utility supply lines, etc., help in avoiding the raised energy consumption. For lowering the energy consumption, it is important to implement inspection and repair to avoid leakage of air through checks, to repair doors and seals to minimize leakage, to check for fouling, to monitor heat transfer efficiency, and to check and repair the insulation around the dryer, air ducts, the heat exchanger, and the burner.

20.4.2.2 Insulation

When hot surfaces of a dryer such as ducts, pipes, walls, roofs, heat exchangers, burners, etc. are exposed to air, heat is lost to the surrounding, thereby increasing the energy demand. Such exposed surfaces need to be insulated effectively to minimize the heat losses. The insulation layers may get damaged or start decaying with use, thereby reducing its effectiveness. Routine check should include a schedule of inspection of insulation, and ineffective ones may be repaired or replaced for saving energy.

20.4.2.3 Mechanical Dewatering

As a rule of thumb, if the moisture content of the feeding material to a dryer is reduced by 1%, then the energy requirement to dry the material reduces by up to 4% (BEE 2004). Hence, prior to feeding the material for drying, it may be subjected to mechanical dewatering methods, such as mechanical compression, use of centrifugal force, gravity, filtration, and high velocity air, to reduce the moisture removal load of the dryer and save energy requirement of (ISU 2005). The overall benefit in terms of energy is obtained from the fact that to remove equal mass of water, mechanical dewatering methods require lesser energy compared to drying.

20.4.2.4 Direct-Fired Dryers

It is estimated that if direct-fired dryers are used, the primary fuel requirement can be reduced by 35%–45% as compared to conventional dryers employing indirect heating methods (BEE 2004; ISU 2005), because they remove one of the inefficient intermediate processes of heating of air. However, care must be taken not to compromise the hedonic value of the dried product resulting from contact with combustion products.

20.4.2.5 Exhaust Air Heat Recovery

Thermal energy content of air exiting the dryer can be utilized to preheat the incoming drying air, which saves energy of air heating. The primary requirement for implementation of this measure is to provide additional ducts for guiding the exhaust air to enable it to exchange heat with the stream of inlet air. Thus, as retrofit application, the success of this measure may be restricted if limited space is available for additional ducts around the dryer (ISU 2005). For affecting the recovery of heat from exhaust air, it is possible to apply either of the direct or indirect heat transfer methods. In the direct method, the exhaust air is directly injected to the inlet air stream, and in the indirect method, a heat exchanger is used to recuperate the heat of exhaust air to heat the inlet air stream (EEBPP 1996).

20.4.2.6 Using Dry Air

In case of dryers requiring small volumes of air for drying application, it is possible to reduce the water vapor in air by using desiccants or dehumidifying techniques before dryer application (Traub 1999b). The use of dry air reduces energy requirement for heating as well as enhances drying capacity of air.

20.4.2.7 Heat Recovery from the Product

Similar to heat recovery from exhaust air, it is possible to recover heat from the dried product. When air is used to cool the product, heat from the resulting warm air can be recovered either by directly using it in the dryer or by recovering the heat through a heat exchanger to preheat inlet air stream. Applicability of this measure is limited to system products that are required to be cooled using forced air after drying (EEBPP 1996).

20.4.2.8 Process Controls

Precise control of the energy inputs to match the process requirement of the material being dried helps to minimize dryer energy consumption. Intermittent application of drying air with preset tempering time or preset flow reversal time for uniform drying would help reduce energy consumption during drying (Masanet et al. 2008).

20.4.3 Evaporation

20.4.3.1 Maintenance

Convective and radiation heat losses, leakage of air and of water, fouling of heat exchanging surfaces, poor separation efficiency, poor vacuum system performance, and excessive venting are the common sources of heat loss and of inefficient performance of evaporators (Rumsey 1986b). Therefore, maintenance program for an evaporator aiming at avoiding these energy losses would include activities like maintaining cleaner surfaces to ensure efficient heat transfer, inspecting and implementing measures to prevent leaking of air into the evaporator as well as to prevent leakage of water into the system that otherwise may dilute the products, replacing damaged and wet insulation in combination with regular inspection, cleaning of vapor separation vessels, operating the evaporator as per manufacturer-specified pressure, etc.

20.4.3.2 Multiple-Effect Evaporators

To interpret as a thumb rule, when 1 kg steam condenses in evaporator operation, it produces approximately 1 kg of vapor. Use of produced vapor as the heating medium would produce an effect almost similar to another kilogram of steam. When this produced vapor is used in another effect, termed as multiple-effect operation of evaporators, it leads to significant saving in energy as compared to single-effect operation of evaporators. Steam economy, defined as the ratio of mass of water vaporized to the mass of steam used, increases proportionally with the number of effects. For example, for a double-effect evaporator steam, economy is 1.8, whereas for triple-effect evaporator, it rises to 2.6. While the energy saving comes at the cost of the added capital costs of each of the added effects, from economic considerations, up to five effects are considered feasible to use in food processing (Maroulis and Saravacos 2003).

20.4.3.3 Vapor Recompression

For the purpose of energy conservation, vapors produced during evaporation process are used as the heating medium. Instead of using vapors, another evaporator is put in sequence, as is done in multiple-effect evaporators, which is an alternate arrangement to using vapor in the same effect, and it yields better energy efficiency as compared to the multiple-effect evaporators. However, to ensure a thermal gradient from the reintroduced vapor into the evaporator to the evaporating liquid, the vapors exiting the evaporator need to be compressed to raise the vapor temperature. Compression of vapor can be carried out effectively by two methods: mechanical vapor recompression (MVR) and thermal vapor recompression (TVR).

In the MVR systems, centrifugal compressors or turbo fans are used to compress the vapors mechanically before reintroducing them into the evaporator as the heating medium. Addition of a small amount of heating steam is required to make up the operational requirement of thermal energy (Maroulis and Saravacos 2003). To compensate for the addition of the heating steam, a part of the vapors obtained from the evaporator is removed before compressing the steam. Similarly, in TVR systems, a steam ejector using high-pressure steam is used to compress the vapors. In this system also, the thermal energy requirement is made up by adding a small amount of heating steam before reintroduction of compressed steam as the heating medium.

MVR systems yield a very high range of steam economy, i.e., in the range of 10 to 30, while TVR systems yield a lesser value of steam economy, i.e., in the range of 4 to 8. Either system of vapor recompression has the similar limitation in applicability; they are applicable where the product boils at a pressure close to atmospheric pressure or under moderate vacuum condition. Hence, these systems are not suitable for highly concentrated products, which may require boiling under vacuum thereby raising the cost (Blanchard 1992). When electricity is available at low cost, MVR systems can be run quite economically, and when high-pressure steam is available at low cost, TVR systems prove to be an economical choice over MVR systems (Maroulis and Saravacos 2003).

20.4.3.4 Preconcentrating Products Prior to Evaporation

Preconcentrating products prior to evaporation either by the use of membranes or freeze concentration systems reduce the amount of water to be removed by evaporation and thus reduce the energy requirement of the evaporation process. Overall energy saving is obtained because these preconcentration methods consume significantly less amount of energy to remove unit mass of water as compared to evaporation.

Reverse osmosis and ultrafiltration based on membrane separation techniques have commonly been used for preconcentration. Since these methods can affect removal of water without involving a phase change in water, they are more energy efficient than evaporation (Martin et al. 2000). The freeze concentration involves freezing of the liquid foods to produce a slurry of liquids and ice crystals, which is then separated from the liquid phase by a mechanical separation device such as a filter press or a centrifuge. Since energy requirement of crystallizing 1 kg of water is only about one-eighth the energy required to vaporize the same amount (SCE 2005).

20.4.4 Pasteurization and Sterilization

20.4.4.1 Sterilizer Insulation

When hot surfaces of a sterilizer are exposed to surrounding air, heat is lost to the surrounding, thereby increasing the energy demand. The exposed surfaces need to be effectively insulated to reduce heat losses. Insulation may get damaged or may decay with time, and thus they should be checked regularly for possible repairing or replacement. When the temperature of the exposed surfaces of sterilizers is greater than 75°C, the typical payback period of adding insulation is two to three years (UNIDO 1995).

20.4.4.2 Heat Recovery from Pasteurization

Since the pasteurization process involves both heating and cooling, almost all modern pasteurizers facilitate internal heat regeneration, permitting use of heat of heated stream to preheat the cold stream. Further, a suitable heat exchanger or heat pumps can be used to recover the heat contained in the rejected water in order to use it to preheat air or water in other applications (Masanet et al. 2008).

20.4.4.3 Compact Immersion Tube Heat Exchangers

Immersion tube compact heat exchangers consist of a coiled heat exchanger tube kept immersed in a water reservoir and a natural gas fired combustion chamber. Exhaust from the combustion chamber is circulated through the coiled heat exchanger, which facilitates transfer of heat from the gas to the water of the reservoir. Heated water in the reservoir is then a heating medium in another heat exchanger that, in turn, is used in heating foods to accomplish pasteurization or sterilization. This compact heat exchanger facilitates better utilization of heat from combustion; a reduction of up to 35% of energy requirement of centralized water heating systems has been reported (CADDET 1992).

20.4.4.4 Helical Heat Exchangers

Due to their hydrodynamic characteristics, helical heat exchangers prove to be better as compared to traditional straight tube heat exchangers including shell and tube heat exchangers. Helical heat exchangers offer an enhanced heat transfer rate, reduced fouling of exchanger surface, and reduced maintenance costs. These heat exchangers are an energy-efficient option for using in continuous operation of thermal processes like pasteurization and sterilization (Stehlik and Wadekar 2002).

20.4.4.5 Induction Heating of Liquids

The instantaneous dissipation of energy, generated during the short-circuiting of the secondary winding of a transformer, to the liquid circulating in a coil around the transformer core, affects heating of the circulating liquid. Using this principle, liquid food can be heated for carrying out thermal processing such as continuous liquid pasteurization and sterilization processes. As compared to boiler-based methods of liquid heating, the energy saving by this method is reported to be up to 17% (CADDET 1997).

20.5 Emerging Energy-Efficient Technologies

20.5.1 Advanced Rotary Burners

A new rotary burner design has been reported to be developed with a gas expansion technique to more effectively mix air and fuel for combustion (U.S. DOE 2002a). As compared to existing low emission burners in which electrical air distribution systems are employed to aid combustion, the innovative component of this burner, i.e., the gas expansion technique, is claimed to be economical in fuel consumption and have reduced emissions. In food processing, such a burner may find application in processing equipment that employs heat from combustion such as dryers, boilers, fryers, etc. According to the U.S. DOE (2002a), the benefits of the burner include

- Increased fuel efficiency up to 4% in comparison to conventional rotary and stationary burners
- Efficient transfer of heat through radiation and convection
- Near perfect mixing of gas and air, resulting in lower nitrous oxide emissions
- Suitability for limited space applications

20.5.2 Geothermal Heat Pumps for HVAC

Geothermal heat pumps have been used for space heating and cooling, taking advantage of the cool, constant temperature of the earth. While the application of this method has been remaining limited to residential and commercial sectors, it can prove to be a better replacement for the conventional HVAC system applicable to fresh produce storage systems that require cooling of warehouse spaces. In winter such a system works to extract heat from the earth to supply heat to the conditioned space, and in summer it extracts heat from the conditioned space to cool it, and delivers it to the earth. This system requires air duct within the building to be conditioned, ducts within the ground, fans and ground loop pumps. In the winter, water solution circulated through the underground piping absorbs heat and carries to the building structure where upon a heat pump system transfers this heat to air circulated in building's ductwork to warm the interior space. In the summer, extracted heat from the air of the building is transferred through the heat pump to the water in the buried piping, through which heat is transferred to the earth. Energy consumption in the process is in the form of electricity to operate fans and ground loop pumps. It is claimed that the technology can reduce space heating and cooling energy consumption by 25% to 50% compared to traditional HVAC systems (GHPC 2005).

20.5.3 Carbon Dioxide as a Refrigerant

Liquid CO_2 under high pressure when sprayed through nozzles to freezing tunnels produces a mixture of dry ice and CO_2 gas. Application of this technique of spraying pressurized liquid CO_2 over surfaces of food products yields a faster rate of cooling and is a viable option for food industries in applications such as chilling and refrigeration, surface freezing and quick freezing, etc. Liquid CO_2–based freezing equipment occupies less space as compared to mechanical refrigeration systems as it can be run without compressor systems.

20.5.4 Copper Rotor Motors

The electrical conductivity of copper is up to 60% higher than aluminum, and by replacing the aluminum rotor in the rotor "squirrel cage" structure of the motor with a copper rotor, it will produce a more energy-efficient induction motor. In addition, copper reacts with much more stability to changing loads, and especially at low speeds and frequencies, it operates cooler. Copper rotors can also require fewer repairs and rewindings, which can lead to increased motor life and decreased maintenance costs. Copper rotor motors have been shown to reduce total motor losses by 10% to 15% and energy use by 1% to 3% compared to aluminum rotor motors. Additionally, the operating costs for copper rotor motors are expected to be less than conventional aluminum motors, while the life expectancy is predicted to be 50% greater (CDA 2004; Worrell et al. 2004).

20.5.5 Magnetically Coupled Adjustable-Speed Drives

In this new type of adjustable-speed drives (ASDs), the motor and driven load is coupled magnetically, eliminating the need of the physical connection between them; instead the connection is through an air gap. For control purposes, causing a variation in the air gap distance between the rotating plates results in corresponding variation in the amount of torques transferred through the assembly. An advantage of this new kind of ASD is greater energy efficiency as compared to conventional mechanically coupled ASDs. Further, Worrell et al. (2004) list several other advantages, viz., (i) greater tolerance to motor misalignment, (ii) little impact of power quality, (iii) expected extended motor lives, and (iv) expected lower long-term maintenance costs. Application of one commercially available model, which is installed in pump, fan, and blower installations in the pulp and paper, mining, food processing, and raw materials processing industries power generation, water treatment, and HVAC systems is reported to provide energy savings of 25% to 66% (Worrell et al. 2004).

20.5.6 Advanced Motor Lubricants

Lubricants add to energy savings by reducing friction and at the same time add to the life of motor-driven equipment by reducing component wear. Synthetic and engineered lubricants are optimized for their application; thus, using them in place of conventional petroleum-based oils and greases will reduce energy consumption and equipment wear, with an extended life of the lubricant. Reported energy savings with the use of synthetic lubricants is 2% to 30%, when using in pumps, compressors, motors, and gear boxes (Martin et al. 2000).

20.5.7 Heat Pumps

Heat pumps are capable of converting low-temperature heat to better quality heat at higher temperature. Hence, they are utilized as heat recovery systems for recovering waste heat to produce better quality heat at elevated and more useful temperature for process heating applications. For accomplishing the task of recovering heat, a heat pump is supplied with high-grade energy such as electricity or fuel combustion. Yet, on the comparison of the overall energy requirement to supply the required process heat, the heat pump system is less energy intensive compared to the supply of the process heat from some high-grade energy sources (Perera and Rahman 1997).

20.5.8 Self-Generation and Combined Heat and Power

In food processing plants, on-site generation of electricity for internal use in the process plant is practiced in a very limited way. It is reported by the U.S. Census bureau that only 5% of the industry's electricity was generated at individual fruits and vegetable processing facilities (U.S. Census Bureau 2004a, b, c, d). Yet, the use of renewable energy systems or cogeneration systems is an option requiring favorable consideration from energy conservation perspectives.

The use of combined heat and power (CHP) systems is an attractive option for food processing operations, which have requirements for process heat, steam, as well as electricity. Examples include processes from the canning subsector such as blanching, pasteurization, sterilization, and evaporation, which require extensive use of steam. Benefits of CHP systems are as follows: they are less polluting and they are more energy efficient than conventional units of power generation because they can utilize the waste heat. Since CHP systems can be located near the application site, this minimizes the transmission loss of electricity. A few plant-specific and locational factors decide the overall economic benefits of using a CHP system, such as power and heat demand of the plant; power tariff for buying; and selling prices, natural gas prices, costs of interconnection, and utility charges for backup power.

On implementation front, for larger processing plants, a CHP system based on combined cycles, which combines a gas turbine and a steam turbine, can be used to produce both power and steam. For such larger sites, the combined cycle-based CHP is an attractive option depending on prices of natural gas and electricity. While these have a potential for using in smaller sites as well as for supplying energy with reduced running cost, the high capital cost of the steam turbine is a limiting factor.

Steam-injected gas turbines (STIGs) are adapted to absorb excess steam, which occurs as a result of seasonal variation in demand of heating steam. The excess steam is injected to the gas turbine to boost the power generation. A STIG makes use of the exhaust heat from the gas turbine system to generate steam from water, which eventually is fed back to mix with the combustion gas to be fed to the turbine. The advantages of such a system include increased power output, faster peak up to reaching full output, and simplified construction as compared to a combined cycle gas turbine. STIG units are available with capacities as low as 5 MW (Willis and Scott 2000).

20.6 Emerging Energy-Efficient Food Processing Technologies

Progressive developments in specific field with the aim of competitive advantage lead to technical innovations. Embodiment of the latest technical innovations in technologies for enhanced efficiency and productivity at commercial scale leads to emerging technologies. Such technologies are tested for performance and reliability under laboratory conditions and through field demonstrations. They are typically installed first in new or recently upgraded plants. These technologies may cease to be considered as emerging after having been successfully commercialized for an extended period.

Quite a number of emerging technologies for food processing are being developed that may be considered as promising ones in terms of energy efficiency in industrial applications. Even some of these emerging technologies hold promises not only for energy efficiency but also for better product quality, improved productivity, increased reliability, and savings in water consumption. However, quantified benefits with respect to existing technologies are not available in published literature. Accordingly, estimates presented in existing literatures in regards to energy savings as well as other benefits need to be considered as a representative trend only. Actual performance of an emerging technology depends on the facility, effective application of technology, and the existing processing equipment with which the new technology is integrated.

20.6.1 Heat Pump Drying

In conventional air dryers, hot air is used as the source of heat for heating the material to be dried. Heat supplied to fresh air to use it as an energy source constitutes the major share of energy consumed by such dryers. However, this energy requirement can be significantly reduced in a heat pump dehumidifying dryer. In heat pump dehumidifying dryers, dehumidified air is circulated around the product after heating it with a heat pump system. Energy for heating the air is obtained by employing the heat pump principle wherein the heat input is from the latent heat of condensation of water vapor released by the drying air during its dehumidification process. During operation of the dryer prior to heating, air undergoes the process of dehumidification, which is accomplished by passing the air over the evaporator coil of the heat pump system to cause condensation of water vapor. Hence, such a drying system essentially consists of components of a regular heat pump system, viz., evaporator, condenser, compressor, and expansion device, along with an air circulation system comprising guided air passages and fan. The product to be dried and the components of the heat pump drying system are placed in an enclosed chamber. Heated dehumidified air when circulated over the product picks up moisture and gets humidified, thereby drying the product. When the humidified air passes over the evaporator of the heat pump, dehumidification occurs, as the vapor in the air condenses giving up the latent heat of vaporization to the vaporizing refrigerant. The condenser of the heat pump system releases this heat to the flowing dehumidified cool air. Energy saving is obtained because of the enhanced rate of moisture removal obtained with the use of dehumidified air and also due to the reuse of latent heat of condensation, which reduces the requirement of external heat supply. Other advantages of this system include products with better quality as drying can be carried out with relatively lower temperature and better control over drying air conditions (Perera and Rahman 1997; Chua et al. 2002).

20.6.2 Ohmic Heating

In this method, heating of the food product being processed is achieved by allowing an alternating electrical current to pass through the material, which serves as an electrical resistor, and wherein the electrical energy is dissipated into heat, resulting in internal generation of heat. In principle, preservative action of ohmic heating is similar to the conventional thermal processing systems; however, due to internal heat generation, the method is considered to induce a uniform inside-out heating pattern as opposed to outside-in heating in conventional thermal processing methods. Also, the heating rate of the foods is rapid and uniform because of the enhanced rate of propagation of heating front. Knirsch et al. (2010) have recognized potential applications of this method as an improved heating method in thermal food processing operations such as pasteurization, dehydration, blanching, evaporation, fermentation, extraction, etc. While performance of ohmic heating–based thermal process technology will be dependent on ohmic heating parameters such as electrical and thermal properties of the food, the applied voltage, alternating current frequency, target temperature, etc., based on one of the experimental ohmic heating applications, Lima et al. (2002) have reported an increase in the freeze-drying rate of up to 25%. This indicates savings in energy use as well as in processing time.

20.6.3 Infrared Drying

In the conventional method of drying of food products, air is used as the heating medium. For this, at first, hot air is produced by heating; heated air is then circulated around the material to be dried wherein it acts as the source of heat for heating the product. Thus, this involves heating of the intermediate medium, air. With the use of an infrared heating system, the material to be dried is heated by direct application of infrared radiation to the material eliminating the necessity of the intermediate step of heating of circulating air. This, in turn, saves energy by eliminating the possibility of energy loss through the exit air. A comparison of the relative saving in energy by this method with conventional air-drying methods, while applying in drying of apple slices with equivalent processing parameters, is reported by Nowak and Lewicki (2004). The energy cost is reported to be lowered, and the drying time is reported to be lowered by up to 50%.

20.6.4 Pulsed Fluid-Bed Drying

The technique of drying of granular materials while keeping them in suspended state by continuous flow of air has been conventionally used widely in drying fruits and vegetables. In the pulsed fluid-bed drying technique, the conventional fluid bed drying system is modified so that pulsation of the gas flow takes place thereby causing high-frequency vibrations within the bed of product particles. This technique is reported to be effective in overcoming defluidization, particularly in the case of cohesive particles. Further, pulsation is expected to reduce channeling of particles and to affect easier fluidization of an irregular-shaped particle. The requirement of saving energy is achieved by saving 30% to 50% of the air requirement for fluidization, because air for fluidization in drying application consumes energy not only for its circulation but also for supplying it at elevated temperature. Additionally, dryers operating with pulsed fluid-bed technique require a smaller size. With the pulsed fluid-bed drying, energy savings is achievable without affecting the production yield due to the lower drying air requirement (Nitz and Taranto 2007).

20.6.5 Pulsed Electric Field Pasteurization

As an emerging food processing technique, pulsed electric field is showing the promise of providing a superior taste and freshness compared to juices treated by existing pasteurization techniques. In this emerging technique of pasteurization of liquid foods, high voltage pulses of electricity are applied to the liquids, which inactivates harmful microorganisms and quality-degrading enzymes, thereby rendering the food safe and preventing quality deterioration of processed fruit juices during storage. This nonthermal technique achieves the pasteurization target by operating at relatively lower temperatures than the conventional thermal method of pasteurization resulting in a lesser cooling requirement. This contributes as an energy savings option in this emerging nonthermal pasteurization process (Lung et al. 2006).

20.7 Summary

Energy applied to food materials during processing enhances the hedonic value of foods and adds to the shipment value. Within two preset extremes of the energy chain, i.e., a desired hedonic value of produced food decided by consumer preference or the market demand on one hand and available forms of primary energy for industrial use on the other, engineering analysis for design and selection may be carried out for improving the efficiencies of (a) converting primary energy into secondary energy, (b) delivering the secondary energy from the point of conversion at the point of end use, and (c) converting the delivered secondary energy into desired energy services. At the plant level, besides the engineering interventions, an organization-wide implementation of a strategic energy management program is a cost-effective way to bring about energy-efficiency improvements. The innovative energy conservation practices as applied in various manufacturing industries provide opportunities of improving energy efficiency in food industries. Innovative food processing technologies, although primarily targeted at improving hedonic values, are amenable to be modified for better energy efficiency.

References

Blanchard, P.H. (1992). *Technology of Corn Wet Milling and Associated Processes*. Elsevier, Amsterdam.

Brush, A., Masanet, E., and Worrell, E. (2011). *Energy Efficiency Improvement and Cost Saving Opportunities for Dairy Processing Industry: An ENERGY STAR Guide for Energy and Plant Managers*. Lawrence Berkeley National Laboratory, Berkeley, CA. Report LBNL- 6261E.

Bureau of Energy Efficiency (BEE) India (2004). Best Practice Manual: Dryers. New Delhi, India. http://www.energymanagertraining.com/bee_draft_codes/best_practices_manual-DRYERS.pdf.

Centre for Analysis and Dissemination of Demonstrated Energy Technologies (CADDET) (1992). *Fruit Juices Pasteurization with Natural Gas Compact Immersed Heat Exchanger.* Case Study CA-1992-015.

Centre for the Analysis and Dissemination of Demonstrated Energy Technologies (CADDET) (1997). *Compact UHT (Ultra High Temperature) Induction Steriliser.* Case Study CA-1996-513.

Chua, K.J., Chou, S.K., Ho, J.C., and Hawlader, N.A. (2002). Heat pump drying: recent developments and future trends. *Drying Technology*, 20 (8), pp. 1579–1610.

Copper Development Association (CDA) (2004). Available at http://www.copper-motor-rotor.org/.

Cullen, J.M., and Allwood, J.M. (2010a). The efficient use of energy: Tracing the global flow of energy from fuel to service. *Energy Policy*, 38 (1), pp. 75–81.

Cullen, J.M., and Allwood, J.M. (2010b). Theoretical efficiency limits for energy conversion devices. *Energy*, 35 (5), pp. 2059–2069.

Das, F. (2000). Integrated heating and cooling in the food and beverage industry. *Centre for Analysis and Dissemination of Demonstrated Energy Technologies (CADDET) Newsletter*, Number 2.

Energy Efficiency Best Practice Programme (EEBPP) (1996). Rotary Drying in the Food and Drink Industry. Carbon Trust, London. Good Practice Guide 149.

Energy Efficiency Best Practice Programme (EEBPP) (2000). Running Refrigeration Plant Efficiently—A Cost Saving Guide for Owners. Carbon Trust, London. Good Practice Guide 279.

European Communities (2008). NACE Rev. 2 (NACE) – Statistical classification of economic activities in the European Community, Eurostat Publications, Luxembourg. Available at http://ec.europa.eu/eurostat.

Fellows, P. (2000). *Food Processing Technology: Principles and Practice.* 2nd Edition. CRC Press, Boca Raton, FL.

Food Manufacturing Coalition for Innovation and Technology Transfer (FMCITT) (1997). *State-of the-Art Report: Food Blanching Process Improvement.* The Food Processing Center, University of Nebraska, Lincoln, Nebraska, June.

Food Processing Industry Resource Efficiency (FIRE) Project (2005). *Sustainable Farmers Find Food Processing Requires Large Volumes of Energy, Water Resources. Case Study of Stahlbush Farms.* Northwest Food Processors Association, Portland, Oregon.

Galitsky, C., Chang, S.C., Worrell, E., and Masanet, E. (2008). *Energy Efficiency Improvement and Cost Saving Opportunities for the Pharmaceutical Industry: An ENERGY STAR Guide for Energy and Plant Managers.* Lawrence Berkeley National Laboratory, Berkeley, CA. Report LBNL- 57260-Revision.

Galitsky, C., Worrell, E., and Ruth, M. (2003). *Energy Efficiency Improvement and Cost Saving Opportunities for Corn Wet Milling Industry: An ENERGY STAR Guide for Energy and Plant Managers.* Lawrence Berkeley National Laboratory. Berkeley, CA. Report LBNL-57307.

Geothermal Heat Pump Consortium (GHPC) (2005). *Commonly Asked Questions.* Available at http://www.geoexchange.org/about/questions.htm.

International Standard Industrial Classification of All Economic Activities, Rev. 4 (ISIC) 2008. United Nations Classification Systems, Classifications Registry. Available at http://unstats.un.org/unsd/cr/registry/regcst.asp?Cl=27&Lg=1.

Iowa State University (ISU) (2005). *Energy-Related Best Practices: A Sourcebook for the Food Industry.* Iowa State University Extension Program, Ames, IA.

Knirsch, M.C., dos Santos, C.A., Martins, A.A., Vicente, O.S., and Vessoni Pena, T.C. (2010). Ohmic heating – a review. *Trends in Food Science & Technology*, 21, pp. 436–441.

Lima, M., Zhong, T., and Rao Lakkakula, N. (2002). Ohmic heating: a value-added food processing tool. *Louisiana Agriculture Magazine*, Fall.

Luh, B.S., Feinberg, B., Chung, J.I., and Woodroof, J.G. (1986). Freezing of fruits. In: Woodroof, J.G., and B.S. Luh (Eds.) *Commercial Fruit Processing*, 2nd Edn. Van Nostrand Reinhold, New York.

Lund, D.B. (1986). Low-temperature waste-heat recovery in the food industry. In: Singh, R.P. (Ed.) *Energy in Food Processing*. Elsevier, Amsterdam.

Lung, R.B., Masanet, E., and McKane, A. (2006). The role of emerging technologies in improving energy efficiency: examples from the food processing industry. Proceedings of the industrial energy technologies conference, New Orleans, LA.

Maroulis, Z.B., and Saravacos, G.D. (2003). *Food Process Design*. Marcel Dekker, New York.

Martin, N.E., Worrell, M., Ruth, L., Price, R.N., Elliott, A.M., Shipley, and Thorne, J. (2000). *Emerging energy-efficient industrial technologies*. Lawrence Berkeley National Laboratory, Berkeley, CA. Report LBNL-46990.

Masanet, E., Therkelsen, P., and Worrell, E. (2012). *Energy Efficiency Improvement and Cost Saving Opportunities for the Baking Industry: An ENERGY STAR Guide for Energy and Plant Managers*. Lawrence Berkeley National Laboratory, Berkeley, CA. Report LBNL-6112E.

Masanet, E., Worrell, E., Graus, W., and Galitsky, C. (2008). *Energy Efficiency Improvement and Cost Saving Opportunities for the Pharmaceutical Industry: An ENERGY STAR Guide for Energy and Plant Managers*. Lawrence Berkeley National Laboratory, Berkeley, CA. Report LBNL-59289-Revision.

National Industrial Classification All Economic Activities (2008). (NIC) Central Statistical Organisation, Ministry of Statistics and Programme Implementation, Government of India. Available at: http://mospi.nic.in/Mospi_New/upload/nic_2008_17apr09.pdf.

Nitz, M., and Taranto, O. (2007). Drying of beans in a pulsed fluid bed dryer: drying kinetics, fluid-dynamic study and comparisons with conventional fluidization. *Journal of Food Engineering*, 80 (1), pp. 249–256.

North American Industry Classification System (NAICS) Association. (2012). NAICS Definitions. Available at http://www.census.gov/eos/www/naics/2012NAICS/2012_Definition_File.pdf.

Nowak, D., and Lewicki, P.P. (2004). Infrared drying of apple slices. *Innovative Food Science and Emerging Technologies*, 5 (3), pp. 353–360.

Ozyurt, O., Comakli, O., Yilmaz, M., and Karsli, S. (2004). Heat pump use in milk pasteurization: An energy analysis. *International Journal of Energy Research*, 28, pp. 833–846.

Pearson, S.F. (2003). How to Improve Energy Efficiency in Refrigerating Equipment. International Institute of Refrigeration, Paris, France. 17th Informatory Note on Refrigeration Technologies, November.

Perera, C., and Rahman, M.S. (1997). Heat pump dehumidifier drying of food. *Trends in Food Science and Technology*, 8, pp. 75–79.

Ramirez, C.A., Blok, K., Neelis, M., and Patel, M. (2006a). Adding apples and oranges: The monitoring of energy efficiency in the Dutch food industry. *Energy Policy*, 34, pp. 1720–1735.

Ramirez, C.A., Patel, M., and Blok, K. (2006b). How much energy to process one pound of meat? A comparison of energy use and specific energy consumption in the meat industry of four European countries. *Energy*, 31, pp. 2047–2063.

Rotstein, E. (1986). Exergy analysis: a diagnostic and heat integration tool. In: Singh, R.P. (Ed.) *Energy in Food Processing*. Elsevier, Amsterdam.

Rumsey, T.R. (1986a). Energy use in food blanching. In: Singh, R.P. (Ed.). *Energy in Food Processing*. Elsevier, Amsterdam.

Rumsey, T.R. (1986b). Energy use in evaporation of liquid foods. In: Singh, R.P. (Ed.). *Energy in Food Processing*. Elsevier, Amsterdam.

Singh, R.P., and Heldman, D.R. (2009). *Introduction to Food Engineering*, 4th Edn. Academic Press, San Diego, CA.

Southern California Edison (SCE) (2005). Business Tips–Food Processing. Rosemead, California. Available at http://www.sce.com/_Tips/MediumLargeBusiness/Food Processing.htm.

Stehlik, P., and Wadekar, V.V. (2002). Different strategies to improve industrial heat exchanger. *Heat Transfer Engineering*, 23 (6), pp. 36–48.

Tragardh, C. (1986). Exergy analysis for energy conservation in the food industry. In: Singh, R.P. (Ed.) *Energy in Food Processing*. Elsevier, Amsterdam.

Traub, D.A. (1999a). *Improving your dryer performance: Part 1. Process heating*. October. 1. http://www.process-heating.com/.

Traub, D.A. (1999b). *Improving your dryer performance: Part 2. Process heating*. November 1. http://www.process-heating.com/.

United Nations Industrial Development Organization (UNIDO) (1995). Handy manual for food processing: Output of a seminar on energy conservation in food processing. https://www.unido.org/fileadmin/import/userfiles/puffk/food.pdf.

United States Census Bureau (2004a). *2002 Economic Census, Manufacturing, Industry Series, Frozen Fruit, Juice, and Vegetable Manufacturing: 2002*. United States Department of Commerce, Washington, D.C. Report EC02-311-311411.

United States Census Bureau (2004b). *2002 Economic Census, Manufacturing, Industry Series, Fruit and Vegetable Canning: 2002*. United States Department of Commerce, Washington, D.C. Report EC02-311-311421.

United States Census Bureau (2004c). *2002 Economic Census, Manufacturing, Industry Series, Specialty Canning: 2002*. United States Department of Commerce, Washington, D.C. Report EC02- 311-311422.

United States Census Bureau (2004d). *2002 Economic Census, Manufacturing, Industry Series, Dried and Dehydrated Food Manufacturing: 2002*. United States Department of Commerce, Washington, D.C. Report EC02-311-311423.

United States Department of Energy (DOE) (2002a). *New Rotary Burner Design More Effectively Mixes Air and Fuel in Industrial Combustion Processes to Increase Fuel Efficiency and Reduce Emissions*. Office of Industrial Technologies, Washington, D.C. Report I-XCO-750.

United States Department of Energy (DOE) (2002c). *Energy-Efficient Food-Blanching System*. Office of Energy Efficiency and Renewable Energy, Industrial Technologies Program, Washington, D.C. Report NICE3 OT-5: March.

United States Department of Energy (DOE) (2006). *Save Energy Now in Your Motor-Driven Systems*. Office of Energy Efficiency and Renewable Energy, Industrial Technologies Program,Washington, D.C. Report DOE/GO-102006-2276.

United States Department of Energy (DOE) (2002b). Energy Efficiency and Renewable Energy, Industrial Technologies Program, "Compressed Air's Role in Productivity," *Energy Matters*, Fall. Available at http://www.nrel.gov/docs/fy03osti/32767.pdf.

United States Department of Energy and Compressed Air Challenge (DOE and CAC) (2003). Improving compressed air system performance—a sourcebook for industry. Office of Industrial Technologies, Washington, DC.

Wang, L. (2009). *Energy Efficiency and Management in Food Processing Facilities*. CRC Press, Boca Raton, FL.

Willis, H.L., and Scott, W.G. (2000). *Distributed Power Generation*. Marcel Dekker, New York.

Worrell, E., Price, L., and Galitsky, C. (2004). *Emerging Energy-Efficient Technologies in Industry: Case Studies of Selected Technologies*. Lawrence Berkeley National Laboratory, Berkeley, California. LBNL-54828.

Index

Page numbers followed by f and t indicate figures and tables, respectively.

C

E

F

G

M

N

O

P

Q

R

S

T

U

X

Y

Z